Discrete Random Signals and Statistical Signal Processing

Discrete Random Signals and Statistical Signal Processing

Charles W. Therrien

Naval Postgraduate School
Monterey, CA

PRENTICE HALL SIGNAL PROCESSING SERIES
Alan V. Oppenheim, Series Editor

PRENTICE HALL, Englewood Cliffs, NJ 07632

Library of Congress Cataloging-in-Publication Data

Therrien, Charles W.
Discrete random signals and statistical signal processing / Charles W. Therrien.
p. cm.
Includes bibliographical references and index.
ISBN 0-13-852112-3
1. Signal processing--Digital techniques. 2. Signal processing--Statistical methods.
3. Stochastic processes. I. Title TK5102.5.T48 1992
621.382'2--dc20
91-31057
CIP

Acquisitions editor: ***Peter Janzow***
Editorial/production supervision
and interior design: ***Richard DeLorenzo***
Cover design: ***Ben Santora***
Prepress buyer: ***Linda Behrens***
Manufacturing buyer: ***David Dickey***
Supplements editor: ***Alice Dworkin***
Editorial assistant: ***Phyllis Morgan***

Drawing of Norbert Wiener by Kristen J. Therrien courtesy of The MIT Museum. Reproduced by permission.

Printed in the United States of America

10 9 8 7 6 5 4 3 2

ISBN 0-13-852112-3

Prentice-Hall International (UK) Limited, *London*
Prentice-Hall of Australia Pty. Limited, *Sydney*
Prentice-Hall Canada Inc., *Toronto*
Prentice-Hall Hispanoamericana, S.A., *Mexico*
Prentice-Hall of India Private Limited, *New Delhi*
Prentice-Hall of Japan, Inc., *Tokyo*
Simon & Schuster Asia Pte. Ltd., *Singapore*
Editora Prentice-Hall do Brasil, Ltda., *Rio de Janeiro*

To Judy, Kristen,
and Lara

Norbert Wiener

Contents

Preface

Signal processing has changed! No longer are we in an era where information in the form of electrical signals is processed through traditional analog devices. We are solidly, and for the foreseeable future irrevocably, in the realm of digital (sampled, or discrete-time) signal processing.

This book had its origins in the course *Discrete-Time Random Processes* introduced to the Naval Postgraduate School in 1987. Offered as an alternative to the traditional continuous-time random process course which was then required of all electrical engineering graduate students, the new course had the purpose of introducing random processes to students from a modern, new and *purely discrete* point of view and introducing some of the newer methods of statistical signal processing at an early graduate level. At the same time, several colleages throughout the country were developing new courses dealing with various forms of statistical signal processing, but no text was yet available to cover the combination of topics that was both introductory and up-to-date. Since the introduction of *Discrete-Time Random Processes* in 1987, several texts have become available that would have suited the original purpose. In the meantime, however, this course has evolved into a sequence of two courses, since several of the topics deserved more in-depth treatment than could be provided in a single quarter. The first of these deals primarily with random processes and classical estimation theory while the second, which is closely integrated with the first, deals mainly with current topics in statistical signal processing. Both precede several advanced courses in signal processing that deal with application areas such as speech, image processing, adaptive filtering, and spectral estimation. The material presented in the first two courses, and in this book, provide a very solid and comprehensive background for the later studies and is part of a strong program in signal processing enthusiastically endorsed by our students. I expect that such a plan of study will have similar benefits at many other universities.

PHILOSOPHY AND APPROACH

Although signal processing has come a long way since the early 1970s with a host of new data-oriented and advanced mathematical ideas, and many excellent graduate courses are now available to present these ideas, a student's *first* introduction to the study of random signals has typically not kept pace with these developments. Many excellent modern texts on stochastic processes for engineers are currently available (e.g., [1]–[6]). Although some now weave in the discrete-time case along with the continuous, their basic approach was inspired by a study of communications systems as they existed in another era. These formulations of random processes for engineers, which were so well developed in the classic texts of Davenport and Root [7], Middleton [8], and the three volumes of Van Trees [9]–[11], have not changed significantly to the present day.

As mentioned above, this book arose from the desire to teach an introductory course in random signals for engineers *from a purely discrete-time point of view.* The discrete-time treatment of random signals is not only practical from a modern design and analysis viewpoint, but is also considerably easier for students to grasp since it circumvents many subtle mathematical[1] problems that exist in the modeling of continuous-time random signals. The discrete random process topics are followed by an introduction to estimation and optimal filtering and a treatment of current methods of statistical signal processing. Although the text is introductory, it attempts to be thorough. For example, it deals with complex random signals right from the beginning, explaining the assumptions underlying the complex notation, and citing specifically where results differ in the real and complex cases. In addition, since discrete signals frequently originate from sampling continuous signals, it is important to make the connection between the two. Accordingly, attention is given to developing corresponding concepts for continuous random processes early in the book, including a treatment of the Wiener process and continuous white noise, and to describing the relations of these processes to their discrete (sampled) counterparts.

SPECIAL FEATURES

In addition to its deliberate orientation to discrete random processes and its comprehensive treatment of the most important areas in statistical signal processing, this book has a number of features that enhance its status both as a conventional textbook and also as a useful reference volume. These features are described below.

[1]Note that the word is *mathematical,* not "real-world."

Deliberate Tutorial Style. Each chapter begins with a brief discussion of its intent and ends with an extensive chapter summary. Results are motivated and developed as thoroughly as possible throughout. Proofs are provided for all facts and ideas that are not obvious, either directly in the text, or as problems to be tackled by the student.

Examples. One of the features that makes the book a valuable learning device is the relatively large number of examples that are provided. These examples supply many connected ideas and show how general concepts apply to specific cases. Where computer algorithms need to be demonstrated, MATLAB is used as the computer language since it is becoming (if it is not already) the *de facto* standard in many areas of electrical engineering. In addition, many algorithms are presented in a vector or matrix format that is especially convenient for implementation (especially in a language such as MATLAB that supports vector and matrix operations directly).

Treatment of Complex Signals. A strong point of the book is the consistent treatment of both the real and complex cases. For example, considerable care is taken to develop methods for minimization and maximization of quantities depending on complex parameters in the appendix so that these problems can be addressed with complete generality. In many cases real results are easily derived from complex results by simply leaving off the complex conjugate. In some cases however, where the real and complex results have distinctly different forms, this fact is pointed out rather than left out; i.e., both forms are given. This feature of the book also makes it a valuable reference for later use in research or design. Since treating the complex case sometimes requires concepts that are less familiar (such as use of the complex gradient instead of the partial derivative) some instructors may want to restrict attention to the real case in lectures and leave corresponding treatment of the complex case as assigned reading.

Written Problems. Another important feature of the book is the ample number of original problems. These problems range from basic excercises to proof of results, guided development of new concepts, and extensions of ideas presented in the text. In some cases, a proof for the complex signal case may be considerably more difficult than its proof for the case of real signals. The instructor may, therefore, want to examine these problems carefully before assigning them and in some cases decide that it is sufficient for students to restrict their consideration to the real case. I also use appropriate problems from other texts in certain instances when teaching from this book and in at least a couple of places I have made reference to those problems without directly including them in the book. New problems are generated each time the course is offered either as exam questions or as new questions that occur in the presentation. I would be happy to provide copies of these additional problems and solutions to anyone who requests them.

Computer Assignments. A special feature of the book is the set of "Computer Assignments" that are included at the end of each chapter. Most of these use data on the disk that is included with the book and most demonstrate important ideas and provide "hands on" experience that it is just not possible to obtain from any written problems. It should be emphasized that these are not ordinary problems that happen to require a computer to solve, but are rather excercises in developing algorithms, interpreting data, and summarizing results. Our students like to call them "computer projects"; they are the modern counterparts of laboratory experiments. While some of the computer assignments require a fairly large investment in time, students have found that they are invaluable in thoroughly understanding the concepts and developing their intuition. Instructors may want to assign the problems as printed, or to assign only portions, or to expand some of the computer assignments in various ways. We have done all of these things. The manual available to instructors from Prentice Hall not only provides solutions to most of the computer assignments, but also provides a brief section that details the intention of the problems, the primary ideas to be learned, and possible modifications or extensions. An interactive workspace oriented language such as MATLAB or APL is infinitely superior to a compiled language such as FORTRAN or C in these assignments and is the only feasible approach in some cases. Since algorithms form a large part of modern signal processing it is important that students be able to develop their own algorithms instead of using "canned" programs. Therefore in the use of MATLAB for example, I do *not* recommend the use of functions in the *signal processing toolbox* to replace students' development of their own tools to solve the computer assignments.

Although the book contains a large number of topics, not every facet of signal processing is covered. For example, it does not treat adaptive filtering (not even LMS), the general vector form of the Kalman filter (although the scalar form *is* covered), or any aspects of array processing. This is in keeping with the intent to develop only *basic* concepts in this book and to leave others for more advanced graduate courses. The instuctor who wants to include some of these topics in the course however will find that the book is easy to supplement and that the background provided here makes these supplementary topics easy to introduce and develop.

PREREQUISITES

While the prerequisites for courses based on this book are not numerous, they are firm. It is assumed that the student has completed a course in discrete-time signals and systems based on a text like [12] or [13]. Further it is important that the reader has a working knowledge of linear algebra and is familiar with the statistical description of random variables through distribution functions, density functions, and moments. It is *not* assumed however that the student has been introduced to *random* signals in any form.

ORGANIZATION

Although it is not explicitly indicated in the table of contents, this book (exclusive of the introduction) can be logically divided into three parts. This organization is illustrated in the following diagram. The first part, which consists of Chapters 2 through 5, contains basic material dealing with random vectors and random processes. This introduces the reader to fundamental methods of characterizing random signals and their processing and provides the material for a basic course on stochastic processes. The second part consists of Chapters 6 and 7, which deal with classical estimation theory and optimal filtering. Although this

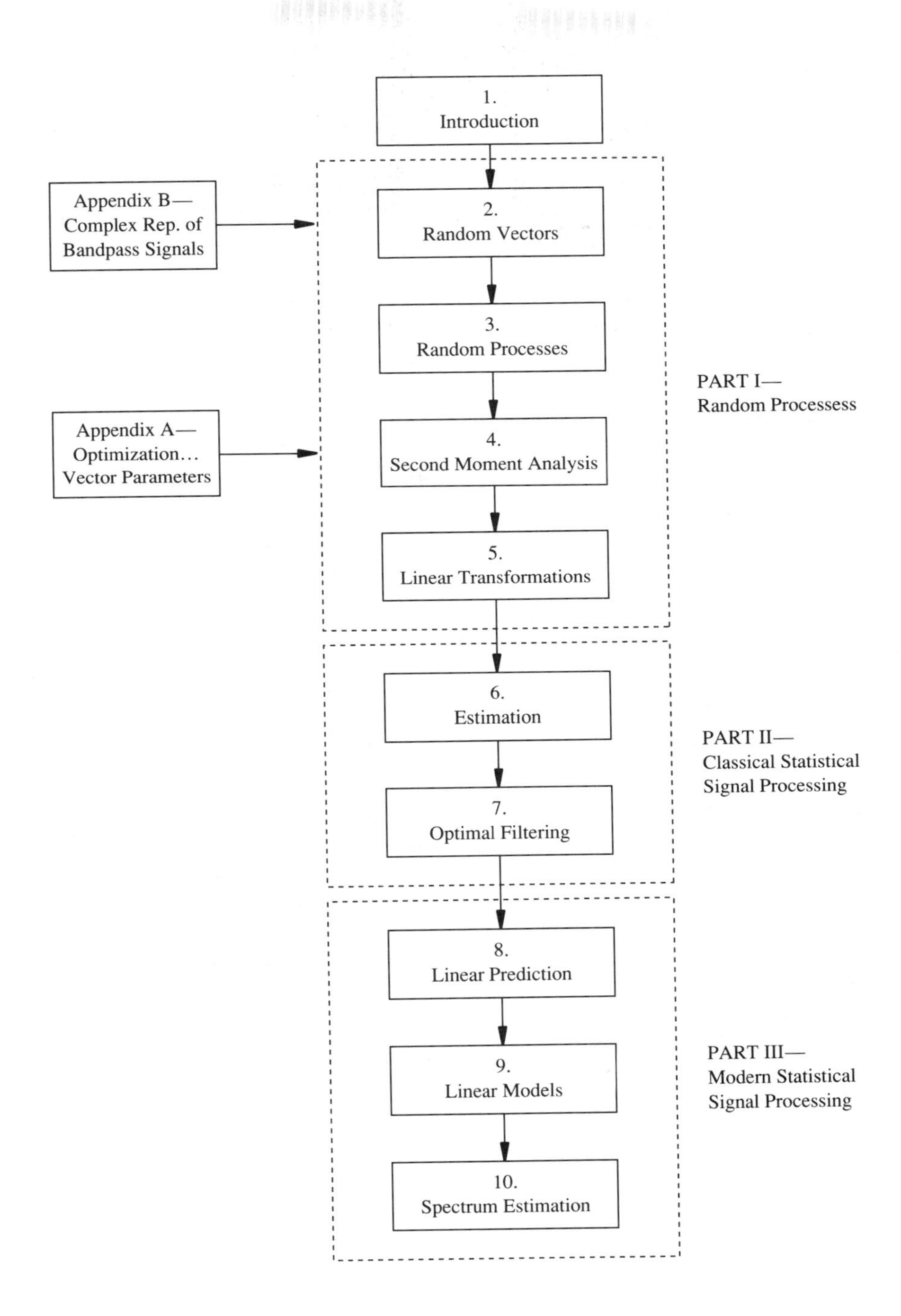
1.
Introduction
Appendix B—
Complex Rep. of
Bandpass Signals
2.
Random Vectors
3.
Random Processes
PART I—
Random Processess
Appendix A—
Optimization…
Vector Parameters
4.
Second Moment Analysis
5.
Linear Transformations
6.
Estimation
PART II—
Classical Statistical
Signal Processing
7.
Optimal Filtering
8.
Linear Prediction
9.
Linear Models
PART III—
Modern Statistical
Signal Processing
10.
Spectrum Estimation

material is basic, these chapters can be thought of as "bridge" chapters that provide the link between the study of basic stochastic processes and modern statistical methods of signal processing. This part could be called "classical statistical signal processing." The last part, which consists of Chapters 8, 9, and 10, deals clearly with current topics in modern statistical signal processing. This covers the areas of linear prediction, signal modeling, and spectrum estimation (including subspace methods). Appendices A and B support all portions of the text. Appendix A provides basic mathematical tools for dealing with optimization of expressions involving complex vectors. Since this material may differ from its treatment in other books, it should be read by both the student and instructor at or prior to its first use in Chapter 4. Appendix B provides the motivation and rationale for the treatment of *complex* random signals and processes. A reasonable place to begin covering this material in class or to assign reading of this appendix is either in Chapter 2 or in Chapter 4. The original manuscript also contained a third appendix consisting of a review of complex variable calculus. This was left out at the advice of reviewers since it is assumed that most students would have encountered this material in earlier mathematics and/or signal processing courses. I have retained this appendix as a separate set of printed notes which are handed out to students in both courses and I would be happy to send it without charge to anyone who wishes to contact me. The review is very similar to that contained in [14, Appendix A].

LEVEL AND USE

The material in this book is meant to be taught at the first-year graduate level. The book can be used in a number of different ways however depending on the specific intent of the graduate program. Three of the major options are described below.

For a Two-Course Sequence. Most university schedules are such that it would not be possible to provide in-depth coverage of all of the topics in a single course. Therefore, one alternative is is to cover the complete set of topics in Chapters 2 through 10 leaving out many of the more detailed discussions and some of the more specialized subtopics. In most cases, however, it is probably best to cover the topics in full depth in a series of two courses. (In universities on a quarter rather than a semester system, even then some of the more specialized topics may have to be left out.) The first course, on stochastic processes, would typically cover most of the first two parts of the book (i.e., up to Chapters 6 or 7) while the follow-on, more advanced course on statistical signal processing would cover the remaining chapters. Here there an option to include the material on *classical* spectrum estimation from Chapter 10 in the introductory course (say at the end of Chapter 6).

For a Single Course on Stochastic Processes. In this use the instructor would focus primarily on the topics in the first two parts of the book (Chapters 1–7) with the possible inclusion of classical spectrum estimation from Chapter 10. This provides basic knowledge of random processes and estimation to students who may not require the more advanced statistical signal processing material. At the Naval Postgraduate School this course is the first in a two-course sequence, and many students who are not in the signal processing option take only the first course (*Discrete-Time Random Processes*) and are well served by it.

For a Single Course in Statistical Signal Processing. Many universities may want to use the book in this manner since statistical signal processing has become an important topic in the graduate curriculum and students will have already taken a traditional stochastic process course based on texts such as [1]–[6]. In this case the instructor will probably want to cover the latter part of Chapter 2 on random vectors and quickly review some aspects of Chapters 4 and 5 just to establish a common frame of reference. In so doing the instructor should be certain to cover the "reversal notation" in Section 2.5 since it is unique to this book. The course can then begin with Chapter 7 or Chapter 8.

ACKNOWLEDGMENTS

This work would not be complete without the acknowledgment of *several* people who have contributed in various ways to the success of the project. Thanks is first due to Bob Strum, and department chairs John Powers and the late Harriett Rigas who supported introduction of the courses on which the material is based and thus indirectly provided the original motivation for the book. For the book itself I am indebted to Al Oppenheim, who enthusiastically supported the idea for a volume on this topic in the Prentice Hall Signal Processing series when the work was only the embryo of what it has become. In addition, I want to express appreciation to colleagues in the Electrical and Computer Engineering Department at the Naval Postgraduate School for their review and thoughtful comments on the manuscript at various stages. These persons include Professors Jeff Burl, Roberto Cristi, Ralph Hippenstiel, Bob Strum, and Murali Tummala. I also want to thank Professors Bill Gragg, Dick Hamming, Allan Kraus, and Bruno Shubert for their consultation on some problems in mathematics. Professor Shubert especially helped me see my way clear through some mathematically difficult areas in statistics, and his insight on several occasions was invaluable. I also want to acknowledge Professors Georgios Giannakis of University of Virginia, and Jerry Mendel and Max Nikias of USC for their help on matters related to cumulants, and to acknowledge Drs. Hanoch Lev Ari[2], Björn Ottersten[3], and Richard Roy of Stanford University for their help in matters related to their original work. My grateful appreciation is also extended to reviewers Dan Dudgeon of MIT Lincoln Laboratory, Jim McClellan of Georgia Tech, Murali Tummala of Naval Postgraduate School (who also used the notes for teaching), and other anonymous reviewers for their careful reading of the manuscript and extremely valuable suggestions. Thanks is also due to numerous students at NPS who endured the draft at various stages and pointed out typographical and logical errors, and especially to students Emanuel Coelho, Art Conklin, and Gary May who helped in the solution of computer assignments, and Atalla Hashad who reviewed the solutions to all of the written problems. Finally I want to thank my immediate family, my wife Judy and daughters Kristen and Lara to whom the book is dedicated, for their endurance and

[2]Currently of Northeastern University.

[3]Currently of Linköping University, Sweden.

encouragement over the last four years. The dedication of this book to them is made with the full realization that this dedication cannot compensate for the days and years of their deprivation of my time and my own deprivation of some of the joys associated with these very special times in their lives.

CHARLES W. THERRIEN
Monterey, California

REFERENCES

1. Leo Breiman. *Probability and Stochastic Processes, With a View Toward Applications.* The Scientific Press, Palo Alto, California, second edition, 1986.
2. Carl W. Helstrom. *Probability and Stochastic Processes for Engineers.* MacMillan, New York, second edition, 1991.
3. William A. Gardner. *Introduction to Random Processes with Applications to Signals and Systems.* MacMillan, New York, 1986.
4. Athanasios Papoulis. *Probability, Random Variables, and Stochastic Processes.* McGraw-Hill, New York, third edition, 1991.
5. Peyton Z. Peebles, Jr. *Probability, Random Variables, and Random Signal Principles.* McGraw-Hill, New York, second edition, 1980.
6. Henry Stark and John W. Woods. *Probability, Random Processes, and Estimation Theory for Engineers.* Prentice Hall, Inc., Englewood Cliffs, New Jersey, 1986.
7. Wilbur B. Davenport, Jr. and William L. Root. *An Introduction to the Theory of Random Signals and Noise.* McGraw-Hill, New York, 1958.
8. David Middleton. *An Introduction to Statistical Communication Theory.* McGraw-Hill, New York, 1960.
9. Harry L. Van Trees. *Detection, Estimation, and Modulation Theory, Part I.* John Wiley & Sons, New York, 1968.
10. Harry L. Van Trees. *Detection, Estimation, and Modulation Theory, Part II.* John Wiley & Sons, New York, 1971.
11. Harry L. Van Trees. *Detection, Estimation, and Modulation Theory, Part III.* John Wiley & Sons, New York, 1971.
12. Alan V. Oppenheim and Ronald W. Schafer. *Discrete-Time Signal Processing.* Prentice Hall, Inc., Englewood Cliffs, New Jersey, 1989.
13. Robert D. Strum and Donald E. Kirk. *First Principles of Discrete Systems and Digital Signal Processing.* Addison-Wesley, Reading, Massachusetts, 1988.
14. Simon Haykin. *Adaptive Filter Theory.* Prentice Hall, Inc., Englewood Cliffs, New Jersey, second edition, 1991.

Discrete Random Signals and Statistical Signal Processing

1

Introduction

Digital signal processing has become almost a "household buzzword," not only in engineering education, but even in the high-tech consumer market for high-performance electronic equipment. While digital signal processing was formerly a graduate-level course to be taken only after completion of substantial coursework in continuous signals and systems, more and more colleges and universities are now offering courses in digital (or discrete-time) signal processing at an introductory (typically an undergraduate) level. While these courses serve to introduce a student to the basic principles of processing discrete signals, the real application of modern methods of signal processing is for signals that cannot be described by the simple models to which a student is introduced in his first course.

This book is devoted to the study of discrete random signals and their subsequent analysis through statistical signal processing. This introductory chapter sets the stage for these studies by discussing the origin of discrete random signals, the need for statistical methods of processing, and several examples of applications that are based on these methods. A brief orientation to the remainder of the book is also provided in the last section.

1.1 DISCRETE RANDOM SIGNALS

Electrical engineers spend a large part of their time learning about signals and systems. Originally, the study was motivated by problems in communication and classical control and dealt almost exclusively with signals that are continuous functions of time. Little attention was given to the analysis of discrete signals and systems.[1] In the 1960s a new interest arose in control theory that was based on the state space formulation. This approach brought important ideas from linear algebra to the analysis of systems and dealt equally well with continuous or discrete formulations. Although it considerably extended the ability to analyze and design many important engineering systems, it departed significantly from the more classical methods of signal and system analysis of the previous two decades. During the late 1960s and early 1970s work was proceeding in image processing, speech, and other areas using a purely discrete approach to the analysis of signals and systems. The publication of the book *Digital Signal Processing* by Oppenheim and Schafer [1] in 1975 and the companion volume with a similar title by Rabiner and Gold [2] heralded the start of a new era in the analysis of linear signal processing systems. Although there were earlier books (e.g., [3, 4]), none were so widely accepted or have had as much influence in the teaching and development of modern signal processing. Today, young engineers are not so likely to think of a system function as the Laplace transform of a continuous impulse response but rather as the z-transform of a discrete impulse response *sequence*. They are more likely to think of stability in terms of poles inside the unit circle than in terms of poles in the left-half plane. The phrase *signal processing* has now become virtually synonymous with *digital signal processing* or *discrete-time signal processing*.

In keeping with this tendency, this book concentrates on the analysis of *random* signals from a discrete[-time] point of view. The text and the courses upon which it is based were motivated by the perception that relatively little had changed in the teaching of random processes to engineers since the introduction of the classic volume by Davenport and Root [5]. Certainly, applications and implementation of the theory have changed, but most courses on stochastic processes for engineers focus on the description of continuous signals, with discrete signals playing at most a secondary role. The analysis of discrete random signals is considerably easier than the analysis of their continuous counterparts since it is not necessary to consider many of the subtleties that deal with differentiation and integration of continuous stochastic processes. It is also a very practical approach since the trend has been to implement more and more continuous signal processing operations in discrete time. With current developments in VLSI and the cost/benefit ratios of digital implementations, this trend is unlikely to change.

The treatment of discrete random signals does not eliminate the need to *understand* continuous signals; it only simplifies their processing. Since many discrete sequences come from sampling signals that are functions of a continuous parameter (e.g., time), the connection to continuous signals is made as the need arises. Therefore, from a practical standpoint,

[1]By a discrete signal we refer to one whose domain or support is the set of integers and whose parameter (e.g., time) is correspondingly integer-valued. The values that the signal takes on may be continuous *or* discrete.

the topics covered in this book serve well in providing basic knowledge about *any* sort of random process.

1.2 STATISTICAL SIGNAL PROCESSING

Although the study of signal processing usually begins with deterministic signals, few real-world signals are actually of that form. By *deterministic* it is meant that the signal or sequence can be expressed as some explicit mathematical formula. Signals that are deterministic in this sense are usually encountered as the output of a power supply (in the case of direct current), an oscillator, or some other type of signal generator. These signals do not well represent the common signals of speech, music, images, noise, data, target motion, and so on, that are at the center of focus in many signal processing systems. For lack of a better term the latter signals are referred to here as *information-carrying signals.*

We hesitate to say that all information-carrying signals are "random" signals since one can argue endlessly and pointlessly about whether such signals are truly random. Nevertheless, it is safe to say that the design of systems to process information-carrying signals must somehow be based on the gross or "average" characteristics of a *class* of these signals. It should not be based upon the detailed characteristics of any one *particular* signal that would arise from analyzing it in a deterministic sense (if that is even possible).

Modeling an information-carrying signal as a random process and analyzing it statistically is one way to achieve the goal of effective system design. Even as these words are written, however, there is a difference in point of view about whether a signal should *truly* be considered to be a realization of a random process or simply to be observed data that is treated using statistical methods. In a sense it does not really matter if the effect of the processing is to produce the same results, and there are instances where both points of view are appropriate. The approach here begins with the random process viewpoint because it provides a fundamental framework in which to develop many needed statistical results. In later chapters some topics lend themselves well to a data-oriented least squares treatment; so that point of view is taken where it is appropriate.

It would seem negligent not to observe that statistical methods may be only one of many possible alternatives for the analysis of information-carrying signals. The newly emerging field of chaos theory [6] offers to bring other methods to signal processing that are neither deterministic (at least in the sense defined earlier) nor statistical. It is not unreasonable to suspect that even the newer methods described in the later chapters of this book will someday be regarded as "classical" and belonging to another era; but in our contemporary world statistical methods provide a very powerful set of tools. Norbert Wiener (1894–1964), whose pioneering work laid the foundation for many of the ideas studied here, predicted that engineers would one day come to learn "a new approach to communication engineering that is primarily statistical" [7] and which would prevail in their design work. His prediction was correct, and our desire is to present the theory in a way that is up-to-date with current methods of its implementation. Needless to say, these are vastly different from those that existed in Wiener's day.

1.3 APPLICATIONS OF STATISTICAL SIGNAL PROCESSING

To provide motivation for the material in the rest of the book, a brief overview of a number of topics that involve the application of statistical signal processing is given here.[2] The list of topics is by no means complete, but it is representative of the variety of applications for which the subject matter of this book is applicable. While many applications involve real-valued signals, some require the simultaneous processing of two signals separated in phase with respect to their carrier by 90°. When these signals satisfy certain statistical symmetry properties, they are treated most conveniently as complex random processes. Complex processes may arise from deliberate quadrature modulation of a communication signal or from physical processes such as reflections that affect both amplitude and phase of a traveling wave. Equal treatment is given to both real and complex sequences in this book so that when you are later faced with practical problems involving complex sequences, they will be as familiar as those involving real sequences. Some of the applications also involve statistical decision theory, which is an extensive topic in itself and is not treated specifically in this book. Although some of the examples involve principles of decision theory, they are limited in scope and self-contained. Various treatments of this theory ranging from introductory to advanced can be found in [8–14].

1.3.1 Detection of Signals in Noise

Problems in communication, radar, and sonar depend upon *detecting* the presence of a signal in noise and possibly estimating its point of arrival. Consider an observed sequence of the form[3]

$$x[n] = s[n] + w[n] \; ; \qquad 0 \leq n \leq N-1 \tag{1.1}$$

where $s[n]$ is a *deterministic* sequence representing a transmitted signal and $w[n]$ represents added noise. If the transmitted signal is not present, then the observed sequence has the form

$$x[n] = w[n] \; ; \qquad 0 \leq n \leq N-1 \tag{1.2}$$

(i.e., it consists of noise alone). If the noise consists of samples that are uncorrelated random variables (i.e., *white* noise), then it can be shown from several points of view that the optimum way to detect the presence of the signal is as illustrated in Fig. 1.1. The observed sequence is multiplied by a sequence $s_r[n]$ which is an exact replica of the transmitted sequence $s[n]$. The result is summed and at the end of the interval ($n = N-1$) the sum is compared to a threshold to make a decision. If the sum is above the threshold, it is decided that the transmitted signal was present; otherwise, it is decided that no signal was present.

[2]Readers who are already sufficiently motivated and convinced of the need for statistical signal processing may skip this section and go on to Section 1.4.

[3]Following Oppenheim and Schafer [15], discrete signals or *sequences* will be written with square brackets [ ] to emphasize the discrete integer-valued nature of their arguments.

The operation of multiplying and summing the two sequences

$$\sum_{k=0}^{N-1} s_r[k]x[k]$$

is known as cross-correlation; correspondingly the system of Fig. 1.1 is called a *correlation detector*.

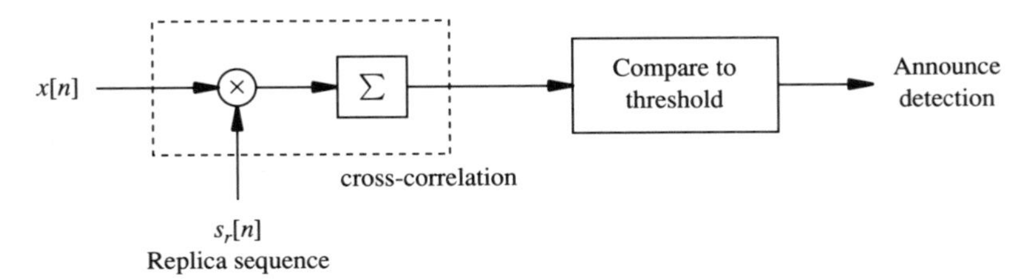

Figure 1.1 Detection of a deterministic signal in white noise.

An alternative way to implement the detector, shown in Fig. 1.2, is to build a linear shift-invariant system whose impulse response is the reversed replica signal. That is,

$$h[n] = s_r[N-1-n]\ ; \qquad 0 \leq n \leq N-1 \tag{1.3}$$

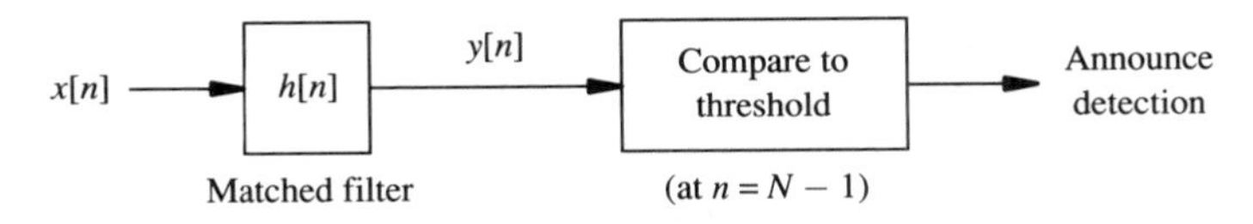

Figure 1.2 Matched filter detector for a signal in white noise.

This linear system is called a *matched filter* and its output is given by

$$y[n] = h[n] * x[n] = \sum_{k=0}^{N-1} h[n-k]x[k] \tag{1.4}$$

The output at sample $N-1$ can be seen to be identical to the cross-correlation of the input with the replica sequence:

$$y[N-1] = \sum_{k=0}^{N-1} h[N-1-k]x[k] = \sum_{k=0}^{N-1} s_r[k]x[k] \tag{1.5}$$

The matched filter realization is particularly convenient since it can be used to estimate the point of arrival of a signal when it occurs within some long observation interval. In radar and sonar, for example, a signal is transmitted and reflected from a target. The form of the returned signal is known, but its arrival time is unknown since the distance to the target is not known *a priori*. The matched filter portion of the system of Fig. 1.2 can process data continuously for any arbitrary length of time. At each point n for $n \geq N - 1$ the output $y[n]$ of the filter represents the cross-correlation of the last N points of the input sequence. It can be shown that when the filter output takes on its peak value (see Fig. 1.3), the corresponding point is an optimal estimate for the arrival of the signal.

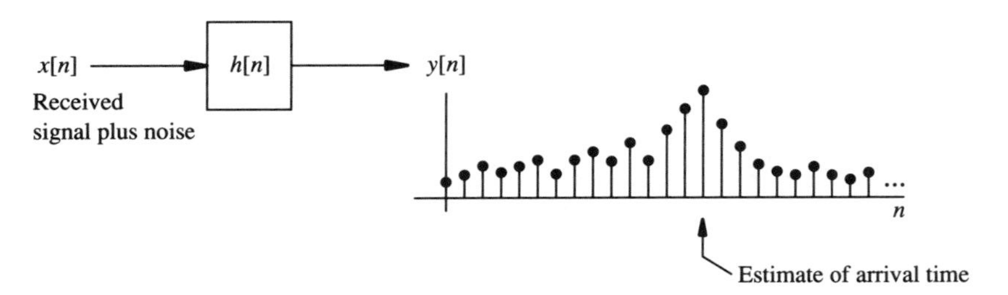

Figure 1.3 Output of a matched filter, showing estimated arrival of signal.

Statistical analysis can further be used to show that the optimal detector for a *random* signal in noise also involves a cross-correlation operation. The system, shown in Fig. 1.4 for a signal whose average value is zero, is known as an estimator-correlator detector [16]. The observed sequence $x[n]$ is passed through a noncausal filter whose output is an optimal estimate for the random sequence $s[n]$. This output is then cross-correlated with the observed sequence and compared to a threshold to perform the detection. Van Trees [13] gives a thorough discussion of estimator-correlator detectors and related structures in continuous systems. Scharf [10] provides a treatment of a number of other detectors for random signals from a contemporary discrete signal processing point of view.

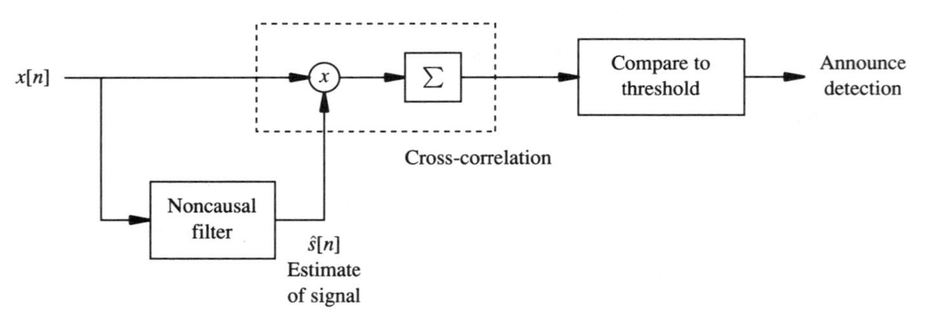

Figure 1.4 Estimator-correlator detector for a random signal in white noise.

1.3.2 Target Tracking

A high-performance method of tracking a moving target by a radar or sonar involves a kind of statistical signal processing known as *Kalman* filtering [17, 18]. Figure 1.5 shows

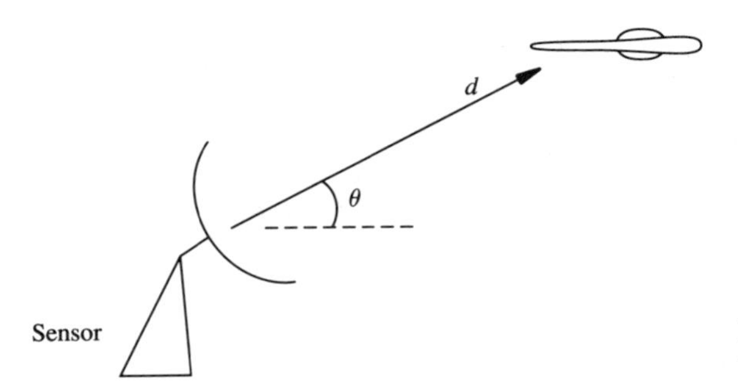

Figure 1.5 Sensor tracking a moving target.

a target being tracked in range and angle. The equations of motion of the target are known from classical physics but the exact motion is not known. In this formulation of the tracking problem the target is characterized by a known dynamic model with a (at least partially) random input. The true position variables (d, θ) are measured only with some uncertainty due to measurement errors inherent in the sensor. Thus the observed position variables are assumed to be corrupted by additive noise. The overall model for the system is shown in Fig. 1.6(a). The optimal recursive filter, which minimizes the average squared error in the estimate of the position variables, is shown in Fig. 1.6(b). The structure consists of a dynamic model identical to the one describing the target motion, which produces estimates

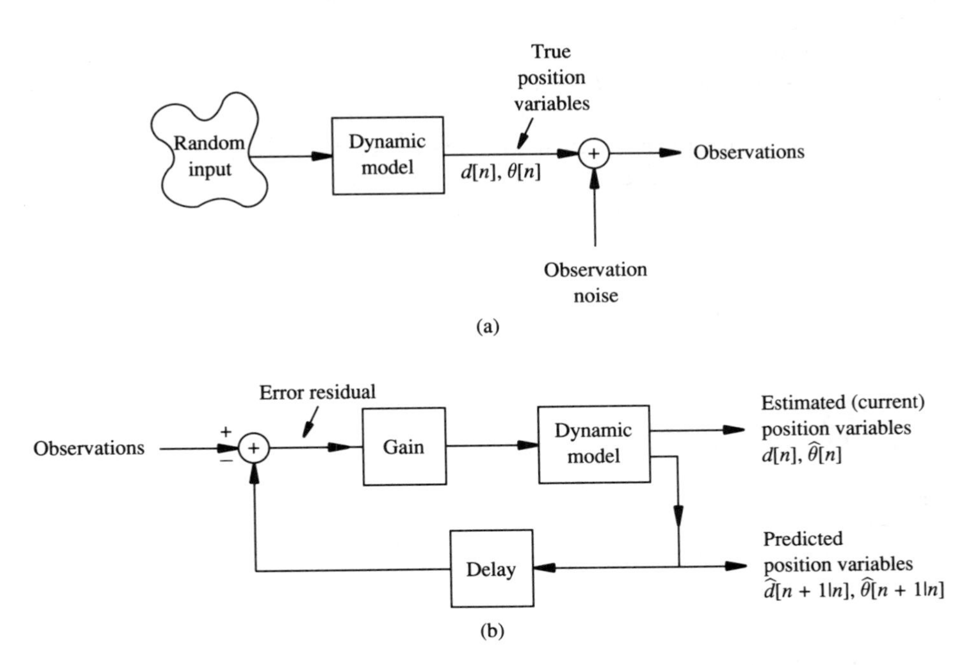

Figure 1.6 Systems involved in the Kalman filtering problem. (a) Dynamic model for target position observations. (b) Recursive tracking filter.

of the current target position and predictions of target position one step ahead. Predicted position estimates are delayed, fed back, and subtracted from the current input to form an error residual. The error residual is multiplied by a gain and drives the dynamic model in such a manner that it follows the true position variables as accurately as possible. The overall system is *not* shift-invariant but recursively produces the statistically best estimate of the target's position given the complete set of past observations.

1.3.3 Speech Modeling

Another application of statistical signal processing is in the analysis and synthesis of speech. In 1978 Texas Instruments astonished the engineering world by using this method to produce speech in a children's toy called "Speak and Spell" costing about $50 in the United States. In the analysis of speech, the speech waveform is sampled at a rate of about 8 to 20 kHz and broken up into short segments whose duration is typically 10 to 20 ms. This results in consecutive segments containing about 80 to 400 time samples. Each such segment of the waveform is first analyzed to determine if it consists of primarily *voiced* or *unvoiced* speech. Voiced speech is characteristic of vowels, while unvoiced speech is characteristic of consonants at the beginning of syllables, fricatives (/f/, /s/ sounds), and combinations of these. In the case of voiced speech a pitch period is determined by either correlation of the input sequence with itself (autocorrelation) or analysis of the spectrum. Finally, the speech segment is analyzed statistically for parameters to be used in an appropriate signal model for the segment.

In speech synthesis the reverse operations take place. A typical speech synthesis system is shown in Fig. 1.7. Each segment of the speech waveform is produced by an appropriate all-pole linear filter driven by one of two input sources. If the segment is voiced, the input is a uniform pulse train with spacing equal to the pitch period. If the segment is unvoiced, the input is a white noise source. The information thus needed to produce a typical sentence (or even a single word) is a voiced/unvoiced selection indicator, the pitch

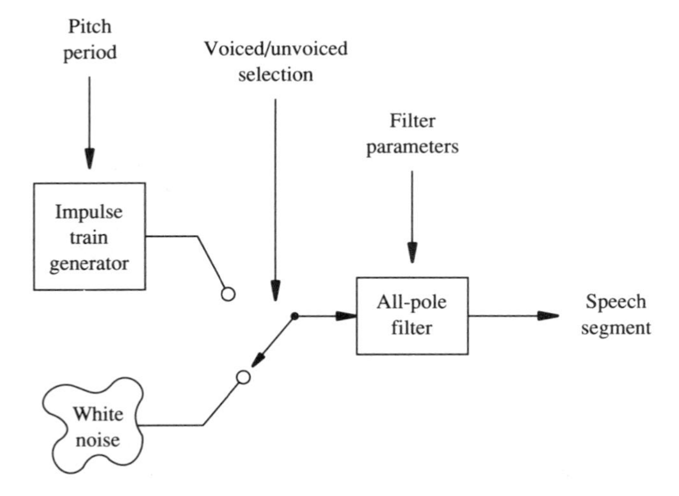

Figure 1.7 Speech synthesis model.

period, and the coefficients and gain of the all-pole filter for each segment. Needless to say, actually producing the speech requires some very fast hardware!

1.3.4 Linear Predictive Coding

Linear predictive coding of waveforms, otherwise known as differential pulse code modulation (DPCM), is another application of statistical signal processing. The components for linear predictive coding as applied in a communication system are shown in Fig. 1.8. The method could equally well be applied to the storage of coded waveform data such as speech or images on digital storage media. The basic idea of linear predictive coding is to generate a prediction $\hat{x}[n]$ of the input data $x[n]$ from P previous data samples

$$\hat{x}[n] = c_1 x[n-1] + c_2 x[n-2] + \cdots + c_P x[n-P] \tag{1.6}$$

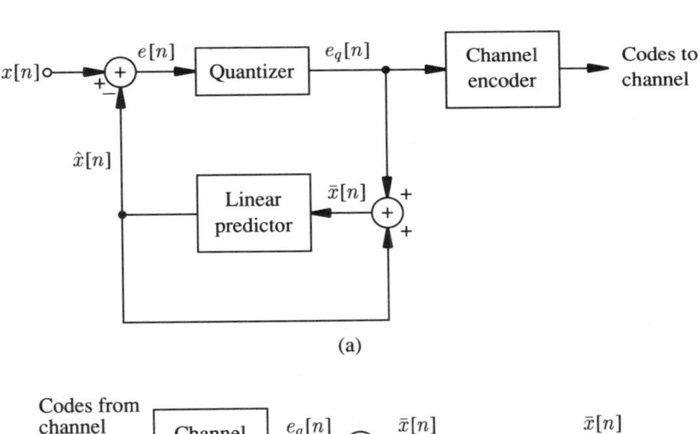

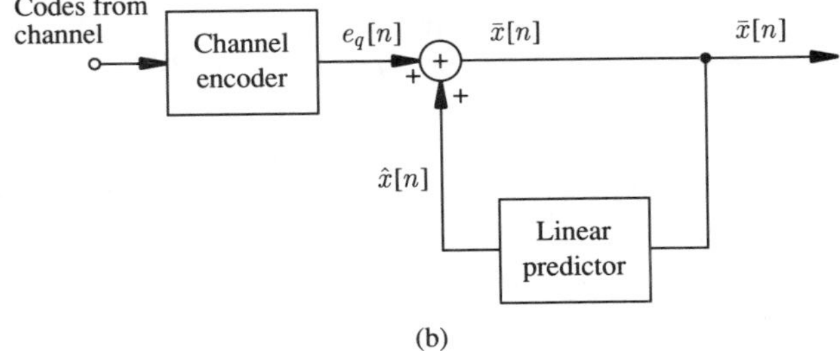

Figure 1.8 Linear predictive coding. (a) Transmitter. (b) Receiver.

where the c_i are appropriately chosen prediction coefficients, and to define an error sequence as

$$e[n] = x[n] - \hat{x}[n] \tag{1.7}$$

The error sequence is then quantized and transmitted. Since the magnitude of the error sequence is usually quite small, it is possible to get by with a much smaller average number of bits in coding the error than if the waveform were to be coded directly. If the receiver has an identical linear predictor, then (ignoring the quantization) the original waveform can be recovered from the relation

$$x[n] = e[n] + \hat{x}[n] = e[n] + c_1 x[n-1] + c_2 x[n-2] + \cdots + c_P x[n-P]$$

In truth, the receiver must work with the quantized error $e_q[n]$. Thus the signal that it constructs is actually an approximation,[4]

$$\bar{x}[n] = e_q[n] + c_1 \bar{x}[n-1] + c_2 \bar{x}[n-2] + \cdots + c_P \bar{x}[n-P] \tag{1.8}$$

To maintain consistency with the receiver, the transmitter also uses the approximation to the waveform $\bar{x}[n]$ in generating its prediction of the input sequence. Thus the predicted value $\hat{x}[n]$ is actually formed by (1.6) with the sequence $\bar{x}[n]$ substituted for the sequence $x[n]$. The entire system is thus as shown in Fig. 1.8.

Statistical signal processing is required in computing the coefficients $\{c_i\}$ of the linear predictor (and also in the analysis of the system performance). The linear predictor may have fixed coefficients that do not change and are derived from an analysis of "training data" representative of the waveforms to be coded. Alternatively, the predictor may have variable coefficients that are updated periodically and transmitted as *side information*, or coefficients that change continuously according to an adaptive filtering algorithm [19, 20]. In any of these cases the coefficients in the linear predictive filter are determined from a statistical analysis of the data.

1.3.5 Spectrum Estimation

Spectrum estimation would at first glance seem to be more a methodology that is used to support other applications than an application in itself. Although this notion is true to a large extent, there are also instances in which computation of the spectrum is the sole purpose of the processing. Examples of this are the analysis of electromagnetic or acoustically generated signals for purposes of identification or location of the source. This kind of processing is especially important in military and/or intelligence operations.

The estimation and analysis of spectra is fundamentally a statistical procedure, and several such methods are introduced in this book. Classical spectrum estimation methods are based on the DFT and frequently on the squared magnitude or *periodogram* of the data. Appropriate use of the available data to generate reliable spectral estimates is essential, and statistical methodology provides the necessary guidelines. Other more "modern" methods of spectrum estimation are based on signal models for the data, on statistical linear algebraic concepts, or on other statistically motivated ideas [21, 22].

Since many signals change their characteristics over long periods of time, a useful kind of display for these signals is the spectral content computed over a short time interval but displayed over a longer period as a function of time. While the results can be obtained by using a sliding window to compute the spectrum [23], methods are available that compute time-frequency representations directly [24, 25]. Figure 1.9 shows such a time-frequency representation for a frequency-shift keyed signal in noise. The distinct shifts in the spectral energy corresponding to the frequency transitions are quite evident in this plot.

[4]The bar over the x here denotes the approximation *not* expectation or complex conjugation.

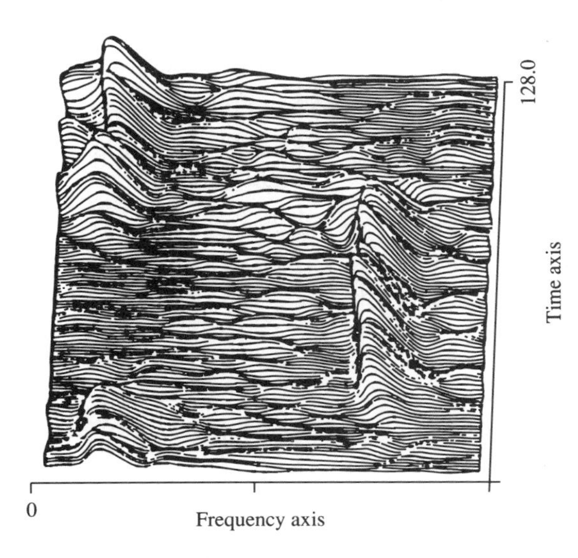

Figure 1.9 Time-frequency representation of frequency-shift keying signal in noise. (From P. M. D. M. de Oliveira, Instantaneous Power Spectrum, M.S. thesis, Naval Postgraduate School, March 1989.)

1.3.6 Beamforming

A final application to be discussed is the signal processing method used in analyzing data from an array of sensors which is known as *beamforming* [26–28]. Although the term "beamforming" seems to imply a radiation of energy, it actually refers to reception here. The basic idea is that when plane waves in the far field of a radiating source arrive at an array of sensors, the received data can be processed in a manner that is more sensitive (or less sensitive) to waves arriving from some particular direction. If multiple time samples are observed at each sensor, then the processing can also be made frequency-selective. Thus beamforming is a kind of spatial-temporal filtering.

A practical beamformer implementation that is relatively easy to understand is the frequency domain or FFT beamformer shown in Fig. 1.10(a). To explain its operation in a simple case, assume that the sensors are all arranged in a straight line with equal spacing d between sensors. This arrangement is known as a *uniform linear array*. Assume further that a sinusoidal signal of frequency f_o is propagating at an angle θ_o with respect to the centerline of the array as shown in the figure. Then from the geometry it follows that a given wavefront arrives at adjacent sensors with a relative time delay of

$$\tau = \frac{d \sin \theta_o}{c} \tag{1.9}$$

where c is the wave propagation speed in the particular medium. The phase difference between signals received at adjacent sensors is thus

$$\varphi = 2\pi f_o \tau = \frac{2\pi f_o d \sin \theta_o}{c} \tag{1.10}$$

Since the plane wave has frequency f_o, let us temporarily ignore the DFT boxes in Fig. 1.10(a) and concentrate on what happens at this particular frequency. Suppose that the complex

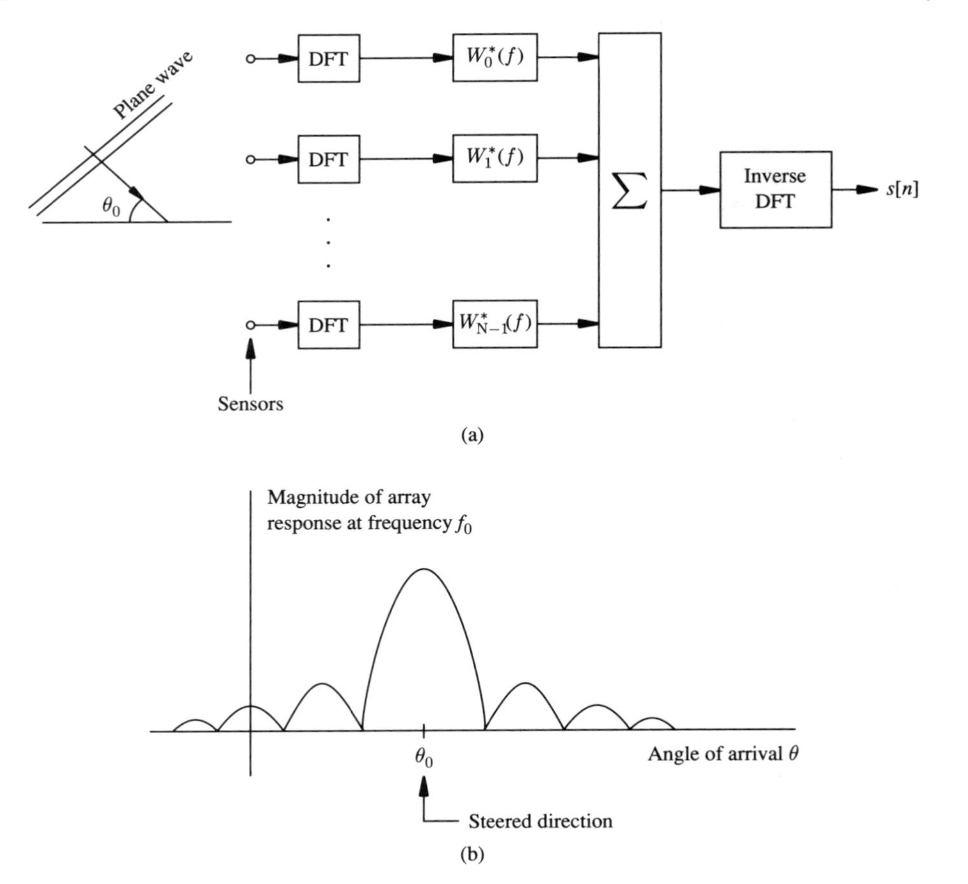

Figure 1.10 Frequency-domain beamforming. (a) Beamformer. (b) Array gain versus angle.

weights w_i^* are chosen to have a phase difference exactly equal to φ as given above. Then waves with a direction of arrival θ_o will be added *coherently* (i.e., in phase) while waves from other directions will *not* be added coherently and thus suffer some relative attenuation in the beamformer. The net effect is that the beamformer output versus arrival angle for a fixed set of weights looks something like that shown in Fig. 1.10(b). Since the direction of greatest sensitivity can be changed by changing the weights, this procedure is called "steering the beam." As long as the phases of the weights are chosen as appropriate multiples of (1.10), the largest response occurs in the steered direction θ_o. The overall *shape* of the pattern is determined by the magnitudes of the weights and the width of the main lobe can be traded off for the level of the sidelobes by varying the weights. By appropriate selection of the weights it is also possible to place nulls in the response at particular angles corresponding to an interfering source.

When the propagating wave has a wideband character or the array must process narrowband signals from sources at many different frequencies, the frequency-domain beamformer functions by performing operations similar to those described above at the center frequency of each DFT bin. The results of processing in separate DFT bins may be combined to obtain a signal domain (i.e., time domain) output as shown, or presented directly in the frequency domain. The net result is that the beamformer "filters" the input with respect to both frequency and direction of arrival. Because the beamformer has to deal both with signals of unknown amplitude and phase and with noise, the processing is necessarily statistical. Some further problems involve adapting the weights to minimize the received power from interfering sources; these also involve statistical algorithms.

A signal-domain beamformer implementation is shown in Fig. 1.11. Sampling the input wave spatially by the sensors and weighting provides the direction-of-arrival filtering, while sampling each sensor response in time and weighting provides the filtering in frequency. Beamforming for a linear array can also be viewed as a kind of two-dimensional spectral analysis, while beamforming for arrays arranged with sensors in two- and three-dimensional configurations can be related to spectral analysis in correspondingly higher dimensions (see [28]). Because of this connection, the various methods of spectrum estimation studied in Chapter 10 have counterparts in array processing, and vice versa. Some of the newer methods do not lend themselves easily to the classical beamformer structures of Figs. 1.10 and 1.11, but the ultimate purpose of the processing is the same.

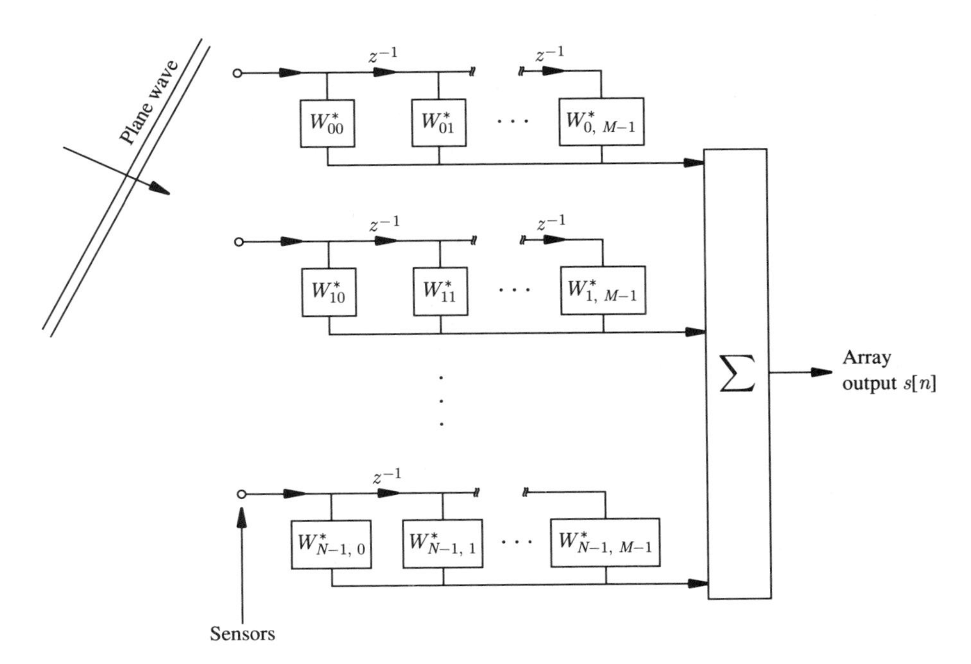

Figure 1.11 Signal-domain beamformer capable of discriminating incoming waves with respect to both frequency and direction of arrival.

1.4 WHERE TO FROM HERE?

The applications described in this chapter are meant to motivate and illustrate the importance of the topics addressed in the remainder of the book. Although there may be occasion to refer to some of the applications again in later chapters, the general purpose of the remaining chapters is to develop the tools and techniques necessary to pursue the applications. Since digital signal processing fundamentally requires *computation*, both written problems and computer assignments are included at the end of each chapter. You should *not* ignore the computer assignments. The time invested in dealing with the data and completing these assignments pays off in a lot of valuable insight that just cannot be gained from reading text and working written problems.

In the early chapters the topics covered are fairly basic, beginning with statistical properties of random vectors in Chapter 2 and proceeding to various examples of random processes in Chapter 3. Chapters 4 and 5 deal primarily with second moment analysis and processing of random signals by linear systems. The tools developed here are the main workhorses for most statistical methods of signal analysis, both past and present. In the study of estimation and optimal filtering (Chapters 6 and 7) it becomes clear that statistical methods have a lot to offer in terms of solving signal processing problems that could not have even been formulated with more elementary tools. In the last three chapters several key ideas are developed that form the mainstay for modern methods of signal processing, and we study the specific techniques built around them. These techniques include linear prediction, signal modeling, subspace decomposition, and various associated high-resolution methods of spectrum estimation.

Besides probability theory, ideas from linear algebra play a major role in the most of the newer signal processing methods. Although a fair amount of basic linear algebra is assumed, some special topics, such as singular value decomposition, are developed in the text. These ideas are brought in as needed and carried on throughout the course of study.

The study of random signals and systems has never been easy and the going may be difficult at times. However, you have a unique opportunity to learn the topic from a point of view that until recently was not even possible—that of digital signal processing. You may find, as others have, that the topic is extremely interesting and challenging and that new discoveries in both device technology and basic computational mathematics seem to meet in this area and provide new insights and surprises that constantly stimulate our creativity.

REFERENCES

1. Alan V. Oppenheim and Ronald W. Schafer. *Digital Signal Processing*. Prentice Hall, Inc., Englewood Cliffs, New Jersey, 1975.
2. Lawrence R. Rabiner and Bernard Gold. *Theory and Application of Digital Signal Processing*. Prentice Hall, Inc., Englewood Cliffs, New Jersey, 1975.
3. Bernard Gold and Charles M. Rader. *Digital Processing of Signals*. McGraw-Hill, New York, 1969.

4. Kenneth Steiglitz. *An Introduction to Discrete Systems.* John Wiley & Sons, New York, 1974.

5. Wilbur B. Davenport, Jr. and William L. Root. *An Introduction to the Theory of Random Signals and Noise.* McGraw-Hill, New York, 1958.

6. J. M. T. Thompson and H. B. Stewart. *Nonlinear Dynamics and Chaos.* John Wiley & Sons, New York, 1986.

7. Norbert Wiener. *I Am a Mathematician.* MIT Press, Cambridge, Massachusetts, 1956.

8. Anthony D. Whalen. *Detection of Signals in Noise.* Academic Press, New York, 1971.

9. Charles W. Therrien. *Decision, Estimation, and Classification: An Introduction to Pattern Recognition and Related Topics.* John Wiley & Sons, New York, 1989.

10. Louis L. Scharf. *Statistical Signal Processing: Detection, Estimation, and Time Series Analysis.* Addison-Wesley, Reading, Massachusetts, 1991.

11. Harry L. Van Trees. *Detection, Estimation, and Modulation Theory*, Part I. John Wiley & Sons, New York, 1968.

12. Harry L. Van Trees. *Detection, Estimation, and Modulation Theory*, Part II. John Wiley & Sons, New York, 1971.

13. Harry L. Van Trees. *Detection, Estimation, and Modulation Theory*, Part III. John Wiley & Sons, New York, 1971.

14. E. L. Lehmann. *Testing Statistical Hypotheses*, 2nd ed. John Wiley & Sons, New York, 1986.

15. Alan V. Oppenheim and Ronald W. Schafer. *Discrete-Time Signal Processing.* Prentice Hall, Inc., Englewood Cliffs, New Jersey, 1989.

16. R. Price. Optimum detection of random signals in noise, with application to scatter-multipath communication, I. *IRE Transactions on Information Theory*, PGIT-6:125–135, December 1956.

17. R. E. Kalman. A new approach to linear filtering and prediction problems. *Transactions of the ASME, Series D, Journal of Basic Engineering*, 82:34–45, 1960.

18. R. E. Kalman and R. S. Bucy. New results in linear filtering and prediction theory. *Transactions of the ASME, Series D, Journal of Basic Engineering*, 83:95–107, 1961.

19. Michael L. Honig and David G. Messerschmitt. *Adaptive Filters: Structures, Algorithms, and Applications.* Kluwer, Boston, 1984.

20. Simon Haykin. *Adaptive Filter Theory*, 2nd ed. Prentice Hall, Inc., Englewood Cliffs, New Jersey, 1991.

21. S. Lawrence Marple, Jr. *Digital Spectral Analysis with Applications.* Prentice Hall, Inc., Englewood Cliffs, New Jersey, 1987.

22. Steven M. Kay. *Modern Spectral Estimation: Theory and Application.* Prentice Hall, Inc., Englewood Cliffs, New Jersey, 1988.

23. S. Hamid Nawab and Thomas F. Quatieri. Short-time Fourier transform. In Jae S. Lim and Alan V. Oppenheim, editors, *Advanced Topics in Signal Processing*, pages 289–337. Prentice Hall, Inc., Englewood Cliffs, New Jersey, 1988.

24. Leon Cohen. Time-frequency distributions—a review. *Proceedings of the IEEE*, 77(7):941–981, July 1989.

25. Boualem Boashash. Time-frequency signal analysis. In Simon Haykin, editor, *Advances in Spectrum Analysis and Array Processing*, pages 418–517. Prentice Hall, Inc., Englewood Cliffs, New Jersey, 1991.

26. Barry D. Van Veen and Kevin M. Buckley. Beamforming: a versatile approach to spatial filtering. *IEEE ASSP Magazine*, 5(2), April 1988.

27. Simon Haykin, editor. *Array Signal Processing*. Prentice Hall, Inc., Englewood Cliffs, New Jersey, 1985.

28. Dan E. Dudgeon and Russell M. Mersereau. *Multidimensional Digital Signal Processing*. Prentice Hall, Inc., Englewood Cliffs, New Jersey, 1984.

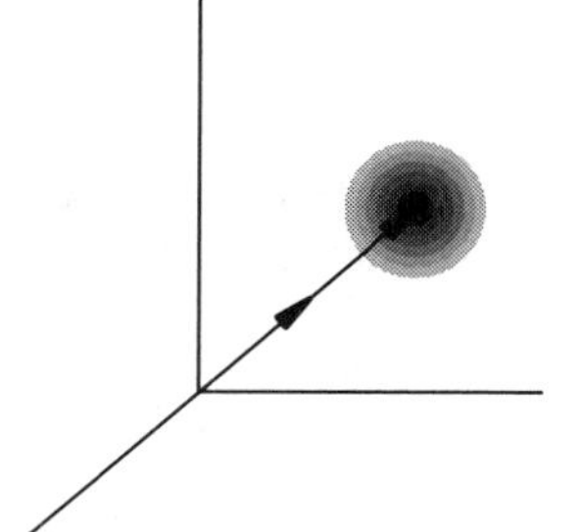

2

Random Vectors

In the analysis of random signals and sequences it is frequently convenient to represent finite-length sequences as vectors. For example, a sequence $x[n]$ defined on the interval $0 \leq n \leq N-1$ can be represented as a vector $\boldsymbol{x}$ with components $x[n]$ as shown in Fig. 2.1. Because the values of the sequence are random variables[1] and are possibly complex-valued, the vector components are also complex-valued random variables. Thus $\boldsymbol{x}$ is referred to

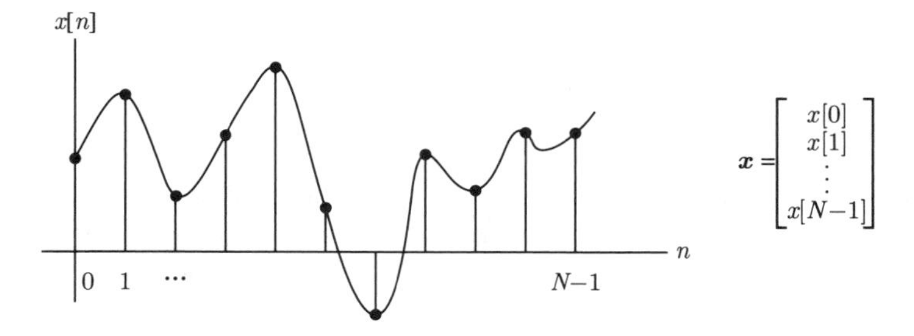

Figure 2.1 Representation of a random sequence as a random vector.

[1]A precise definition of a random sequence is given in Chapter 3.

as a *random vector*. Since random vectors are a convenient form for representing random sequences, it is important to understand how to characterize them statistically before going any further. The purpose of this chapter is to provide that understanding.

Since it is generally not necessary in this chapter to refer to the sequence from which a random vector may have been derived, the discussion here is a bit mathematical and may seem somewhat removed from our ultimate purpose of analyzing signals and systems. Rest assured, however, that the departure from dealing with signals and systems is only temporary and that the mathematical concepts developed here are essential and meant to provide a firm foundatation for the rest of the chapters to build upon.

The topics discussed here assume knowledge of probability and random variables on the level of [1, part 1] or [2, Chaps. 1–3] and basic concepts from linear algebra at least on the level of [3]. An introductory text on linear algebra that provides excellent background for most topics is [4]. Some additional more advanced references on eigenvalue problems, triangular decomposition and the like include [5–8]. General treatments of random vectors in a statistical framework can be found in [9, 10]. Finally, a wealth of information on complex random vectors and complex random processes can be found in [11].

2.1 RANDOM VECTORS AND THEIR CHARACTERIZATION

Random vectors are fundamentaly characterized by their distribution and density functions. These concepts are defined first for real random vectors and then generalized to complex random vectors.

2.1.1 Distribution and Density Functions for a Real Random Vector

A real random vector $\boldsymbol{v}$ will be denoted by the column matrix

$$\boldsymbol{v} = \begin{bmatrix} v_1 \\ v_2 \\ \vdots \\ v_N \end{bmatrix} \tag{2.1}$$

where the *components* $v_1, v_2, \ldots, v_N$ of $\boldsymbol{v}$ are real-valued random variables. Although the random vector may represent samples of a sequence beginining at $n = 0$ or some other point, it is most convenient here to index the components of the random vector from 1 to N. The vector magnitude is represented by the Euclidean norm

$$\|\boldsymbol{v}\| \stackrel{\text{def}}{=} \left(\sum_{k=1}^{N} v_k^2 \right)^{1/2} = (\boldsymbol{v}^T \boldsymbol{v})^{1/2} \tag{2.2}$$

There will be several occasions to use this in the subsequent discussion. Now let $\mathbf{v}^o$ denote a specific value for the random vector[2]

$$\mathbf{v}^o = \begin{bmatrix} \mathrm{v}_1^o \\ \mathrm{v}_2^o \\ \vdots \\ \mathrm{v}_N^o \end{bmatrix} \tag{2.3}$$

In other words, $\mathrm{v}_1^o, \mathrm{v}_2^o, \ldots, \mathrm{v}_N^o$ are fixed real numbers. The probability of the event $\boldsymbol{v} \leq \mathbf{v}^o$, which is *defined* as

$$\boldsymbol{v} \leq \mathbf{v}^o : \quad v_1 \leq \mathrm{v}_1^o, v_2 \leq \mathrm{v}_2^o, \ldots, v_N \leq \mathrm{v}_N^o \tag{2.4}$$

is clearly a function of $\mathbf{v}^o$, and that function is known as the *distribution function* for the random vector $\boldsymbol{v}$. The distribution function is denoted by $F_{\boldsymbol{v}}$ and is defined by

$$F_{\boldsymbol{v}}(\mathbf{v}^o) \stackrel{\text{def}}{=} F_{v_1 v_2 \ldots v_N}\left(\mathrm{v}_1^o, \mathrm{v}_2^o, \ldots, \mathrm{v}_N^o\right) \stackrel{\text{def}}{=} \Pr[\boldsymbol{v} \leq \mathbf{v}^o] \tag{2.5}$$

Note that the distribution function for a random vector is a joint distribution function for its components.

Figure 2.2(a) shows a typical distribution function for a two-dimensional random vector $\boldsymbol{v}$. Note that because of the definition (2.5) the distribution function has the properties

$$\begin{aligned} F_{\boldsymbol{v}}(-\infty) &= 0 \; (a) \\ F_{\boldsymbol{v}}(+\infty) &= 1 \; (b) \end{aligned} \tag{2.6}$$

While the distribution function for a random vector is a meaningful starting point, the probability *density function* is a quantity that tends to be used more frequently. The density function corresponding to the distribution function of Fig. 2.2(a) is shown in Fig. 2.2(b). The probability density function is defined as the derivative of the distribution function with respect to all of the vector components

$$f_{\boldsymbol{v}}(\mathbf{v}^o) = \frac{\partial}{\partial \mathrm{v}_1} \frac{\partial}{\partial \mathrm{v}_2} \cdots \frac{\partial}{\partial \mathrm{v}_N} F_{\boldsymbol{v}}(\mathbf{v}) \bigg|_{\mathbf{v}=\mathbf{v}^o} \tag{2.7}$$

Thus the distribution function can be obtained from the integral

$$F_{\boldsymbol{v}}(\mathbf{v}^o) = \int_{-\infty}^{\mathbf{v}^o} f_{\boldsymbol{v}}(\mathbf{v}) d\mathbf{v} \stackrel{\text{def}}{=} \int_{-\infty}^{\mathrm{v}_1^o} \int_{-\infty}^{\mathrm{v}_2^o} \cdots \int_{-\infty}^{\mathrm{v}_N^o} f_{\boldsymbol{v}}(\mathbf{v}) d\mathrm{v}_N \ldots d\mathrm{v}_2 \, d\mathrm{v}_1 \tag{2.8}$$

Note that the *definition* of the vector integral is the set of scalar integrals over all of the components of the random vector. This, like some of the previous notation, is adapted for convenience. Because of (2.6b) and (2.8) it follows that

$$\int_{-\infty}^{\infty} f_{\boldsymbol{v}}(\mathbf{v}) d\mathbf{v} = 1 \tag{2.9}$$

[2]A math italic font to represent random variables and random vectors and a roman font is used to represent a realization of the random variable or vector. When it is desired to further focus on a *specific* value for the realization, a superscript such as "o" is attached to the symbol.

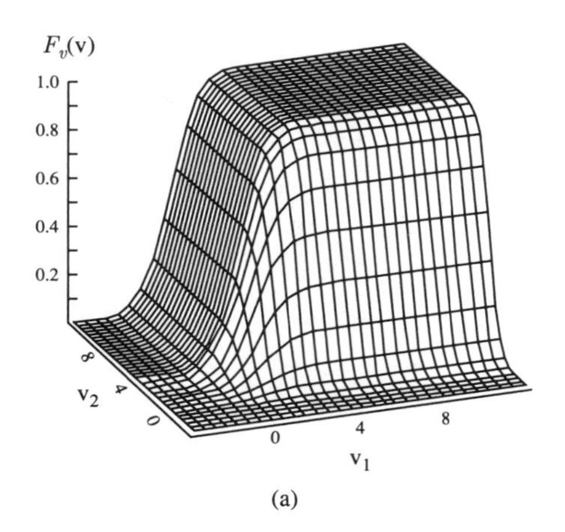

(a)

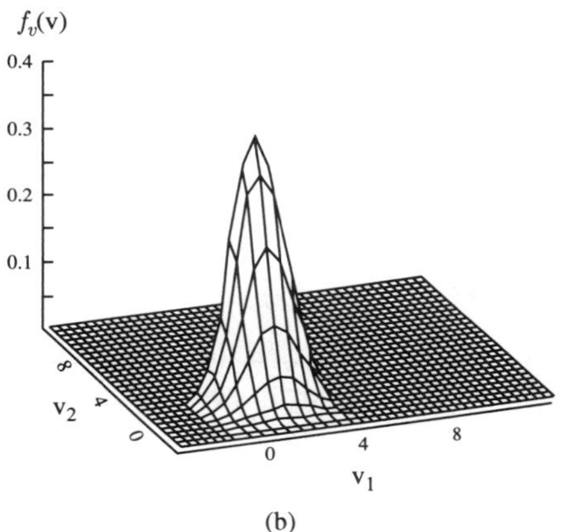

(b)

Figure 2.2 Distribution and density functions for a two-dimensional real random vector. (a) Distribution function. (b) Density function. (From Charles W. Therrien, *Decision Estimation and Classification*, copyright © John Wiley & Sons, New York, 1989. Reproduced by permission.)

To illustrate the properties of the distribution and density functions, let us consider an example. The example illustrates that it is important to be careful about regions of definition when working with some distribution and density functions.

Example 2.1

A two-dimensional real random vector $\boldsymbol{v}$ has the probability density function shown in Fig. EX2.1. Note first that the density integrates to 1 since

$$\int_{-\infty}^{\infty} f_v(\mathbf{v})d\mathbf{v} = \int_0^1 \int_0^1 \tfrac{1}{2}(\mathrm{v}_1 + 3\mathrm{v}_2)d\mathrm{v}_1 d\mathrm{v}_2$$

$$= \int_0^1 \tfrac{1}{2}(\tfrac{1}{2} + 3\mathrm{v}_2)d\mathrm{v}_2 = \tfrac{1}{2}(\tfrac{1}{2} + \tfrac{3}{2}) = 1$$

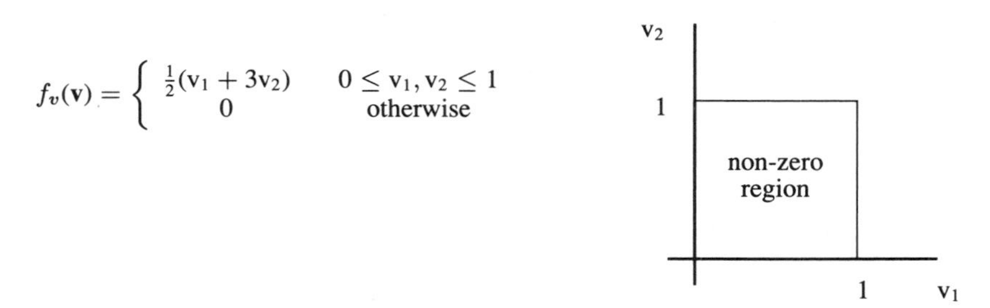

Figure EX2.1

The distribution function is obtained by integrating over both v_1 and v_2, observing the limits of the regions where the density is nonzero. First note that for $v_1 \le 0$ or $v_2 \le 0$, the density is zero; consequently, $F_v(\mathbf{v}^o)$ is zero for $v_1^o \le 0$ or $v_2^o \le 0$. For $0 < v_1^o \le 1$ and $0 < v_2^o \le 1$ the distribution function is

$$\begin{aligned} F_v(\mathbf{v}^o) &= \int_{-\infty}^{\mathbf{v}^o} f_v(\mathbf{v}) d\mathbf{v} \\ &= \int_0^{v_1^o} \int_0^{v_2^o} \tfrac{1}{2}(v_1 + 3v_2)\, dv_2 dv_1 = \tfrac{1}{4} v_1^o v_2^o (v_1^o + 3v_2^o) \end{aligned}$$

Finally, for $v_1^o > 1$ or $v_2^o > 1$ the upper limit of integration for the corresponding variable becomes equal to 1. The algebraic expression for the distribution function is thus obtained by substituting $v_1^o = 1$ or $v_2^o = 1$ in the foregoing expression. The results can be summarized as

$$F_v(\mathbf{v}) = \begin{cases} 0 & v_1 \le 0 \text{ or } v_2 \le 0 \\ \frac{1}{4} v_1 v_2 (v_1 + 3v_2) & 0 < v_1, v_2 \le 1 \\ \frac{1}{4} v_1 (v_1 + 3) & 0 < v_1 \le 1, \quad 1 < v_2 \\ \frac{1}{4} v_2 (1 + 3v_2) & 1 < v_1, \quad 0 < v_2 \le 1 \\ 1 & 1 < v_1, v_2 < \infty \end{cases}$$

where the superscript "o" on the variables has now been dropped since the equation applies for *any* value of $\mathbf{v}$.

The interpretation of a probability density function in terms of probability of events is very important in dealing with random vectors. The distribution function was *defined* by the probability of the event $\boldsymbol{v} \le \mathbf{v}^o$. To interpret the density function, consider the event $\mathbf{v}^o < \boldsymbol{v} \le \mathbf{v}^o + \Delta\mathbf{v}$ defined by

$$\mathbf{v}^o < \boldsymbol{v} \le \mathbf{v}^o + \Delta\mathbf{v} : \quad v_1^o < v_1 \le v_1^o + \Delta v_1, \ldots, v_N^o < v_N \le v_N^o + \Delta v_N \tag{2.10}$$

where the Δv_i are small increments in the components of the random vector. From the relations (2.5) and (2.8) and the definition of the partial derivative it can be shown that

$$\Pr\left[\mathbf{v}^{\mathrm{o}} < \boldsymbol{v} \leq \mathbf{v}^{\mathrm{o}} + \Delta\mathbf{v}\right] = \int_{\mathbf{v}^{\mathrm{o}}}^{\mathbf{v}^{\mathrm{o}}+\Delta\mathbf{v}} f_{\boldsymbol{v}}(\mathbf{v})d\mathbf{v} \approx f_{\boldsymbol{v}}(\mathbf{v}^{\mathrm{o}})\Delta\mathrm{v}_1\Delta\mathrm{v}_2\ldots\Delta\mathrm{v}_N \tag{2.11}$$

From this it follows that the probability that $\mathbf{v}$ is in any *arbitrary* region $\mathcal{V}$ can be obtained by integrating the density function over that region. *Equation 2.11 is an extremely important result.* It states that the density function, evaluated at a point $\mathbf{v}^{\mathrm{o}}$, is proportional to the probability that the random vector $\boldsymbol{v}$ falls in a small region around $\mathbf{v}^{\mathrm{o}}$ represented by (2.10). Thus when the density function at a given point $\mathbf{v}^{\mathrm{o}}$ is large, there is high probability that the random vector will take on a value close to $\mathbf{v}^{\mathrm{o}}$. Note, however, that although the probability that $\boldsymbol{v}$ falls in the small region (2.10) may be nonzero, the probability that $\boldsymbol{v}$ is *equal* to the value $\mathbf{v}^{\mathrm{o}}$ is equal to zero for any continuous density function. This follows directly from (2.11) by taking the limit as the $\Delta\mathrm{v}_i$ approach zero.

Occasionally, it is necessary to represent density functions for random vectors that *do* take on specific values with nonzero probability. An example is when the v_k represent signal levels (say, $+1$ and -1) corresponding to a binary communication message. In fact, for this example the random vector can *only* take on (2^N) specific discrete values. Such cases can be included by allowing *impulses* to occur in the density function. For example, if the random vector $\boldsymbol{v}$ assumes the value $\mathbf{v}^{\mathrm{o}}$ with probability $\mathrm{P_o}$, then this discrete component of the density is represented formally by a term

$$\mathrm{P_o}\delta_c(\|\mathbf{v} - \mathbf{v}^{\mathrm{o}}\|)$$

where $\delta_c(\cdot)$ is the continuous unit impulse function. In this case the complete density has the form

$$f_{\boldsymbol{v}}(\mathbf{v}) = f'_{\boldsymbol{v}}(\mathbf{v}) + \mathrm{P_o}\delta_c(\|\mathbf{v} - \mathbf{v}^{\mathrm{o}}\|)$$

where $f'_{\boldsymbol{v}}(\mathbf{v})$ represents the continuous part of the density. The limiting value of the integral near the point $\mathbf{v}^{\mathrm{o}}$ would be

$$\lim_{\Delta\mathbf{v}\to\mathbf{0}} \int_{\mathbf{v}^{\mathrm{o}}}^{\mathbf{v}^{\mathrm{o}}+\Delta\mathbf{v}} f_{\boldsymbol{v}}(\mathbf{v})d\mathbf{v} = \lim_{\Delta\mathbf{v}\to\mathbf{0}} \int_{\mathbf{v}^{\mathrm{o}}}^{\mathbf{v}^{\mathrm{o}}+\Delta\mathbf{v}} \left[f'_{\boldsymbol{v}}(\mathbf{v}) + \mathrm{P_o}\delta_c(\|\mathbf{v} - \mathbf{v}^{\mathrm{o}}\|)\right] d\mathbf{v} = \mathrm{P_o} \tag{2.12}$$

Example 2.6 in Section 2.1.4 illustrates such a case.

Table 2.1 gives a summary of the definitions and relations for real-valued random vectors.

2.1.2 Distribution and Density Functions for a Complex Random Vector

The results of the preceding subsection can be generalized to the case of a random vector with complex components. For this purpose, define the *complex* random vector $\boldsymbol{x}$ with components $x_k = x_{\mathrm{r}k} + \jmath x_{\mathrm{i}k}$ as

$$\boldsymbol{x} = \boldsymbol{x}_{\mathrm{r}} + \jmath\boldsymbol{x}_{\mathrm{i}} \tag{2.13}$$

TABLE 2.1 DISTRIBUTION AND DENSITY FUNCTIONS OF A REAL RANDOM VECTOR

Probability interpretation	$F_{\boldsymbol{v}}(\mathbf{v}^o) = \Pr[\boldsymbol{v} \le \mathbf{v}^o]$	$f_{\boldsymbol{v}}(\mathbf{v}^o)\Delta\mathrm{v}_1\Delta\mathrm{v}_2 \ldots \Delta\mathrm{v}_N \approx \Pr\left[\mathbf{v}^o < \boldsymbol{v} \le \mathbf{v}^o + \Delta\mathbf{v}\right]$
Relations	$F_{\boldsymbol{v}}(\mathbf{v}^o) = \int_{-\infty}^{\mathbf{v}^o} f_{\boldsymbol{v}}(\mathbf{v})d\mathbf{v}$	$f_{\boldsymbol{v}}(\mathbf{v}) = \frac{\partial}{\partial \mathrm{v}_1} \cdots \frac{\partial}{\partial \mathrm{v}_N} F_{\boldsymbol{v}}(\mathbf{v})$
Properties	$F_{\boldsymbol{v}}(-\infty) = 0$ $F_{\boldsymbol{v}}(+\infty) = 1$	$\int_{-\infty}^{\infty} f_{\boldsymbol{v}}(\mathbf{v})d\mathbf{v} = 1$

where $\boldsymbol{x}_\mathrm{r}$ and $\boldsymbol{x}_\mathrm{i}$ are the real random vectors

$$\boldsymbol{x}_\mathrm{r} = \begin{bmatrix} x_{\mathrm{r}1} \\ x_{\mathrm{r}2} \\ \vdots \\ x_{\mathrm{r}N} \end{bmatrix}; \qquad \boldsymbol{x}_\mathrm{i} = \begin{bmatrix} x_{\mathrm{i}1} \\ x_{\mathrm{i}2} \\ \vdots \\ x_{\mathrm{i}N} \end{bmatrix} \tag{2.14}$$

The magnitude or norm of the complex random vector is defined by

$$\|\boldsymbol{x}\| \stackrel{\mathrm{def}}{=} \left(\sum_{k=1}^{N} |x_k|^2 \right)^{1/2} = (\boldsymbol{x}^{*T}\boldsymbol{x})^{1/2} \tag{2.15}$$

and satisfies

$$\|\boldsymbol{x}\|^2 = \|\boldsymbol{x}_\mathrm{r}\|^2 + \|\boldsymbol{x}_\mathrm{i}\|^2 \tag{2.16}$$

Let $\mathbf{x}^o$ correspondingly represent a particular value for the complex random vector, that is,

$$\mathbf{x}^o = \mathbf{x}_\mathrm{r}^o + \jmath\mathbf{x}_\mathrm{i}^o \tag{2.17}$$

where $\mathbf{x}_\mathrm{r}^o$ and $\mathbf{x}_\mathrm{i}^o$ are given specific values for the real and imaginary parts. Now define the event $\boldsymbol{x} \le \mathbf{x}^o$ to mean

$$\boldsymbol{x} \le \mathbf{x}^o: \quad \boldsymbol{x}_\mathrm{r} \le \mathbf{x}_\mathrm{r}^o \quad \text{and} \quad \boldsymbol{x}_\mathrm{i} \le \mathbf{x}_\mathrm{i}^o \tag{2.18}$$

The distribution function for the complex random vector $\boldsymbol{x}$ is then defined by

$$F_{\boldsymbol{x}}(\mathbf{x}^o) \stackrel{\mathrm{def}}{=} \Pr[\boldsymbol{x} \le \mathbf{x}^o] \stackrel{\mathrm{def}}{=} \Pr[\boldsymbol{x}_\mathrm{r} \le \mathbf{x}_\mathrm{r}^o, \boldsymbol{x}_\mathrm{i} \le \mathbf{x}_\mathrm{i}^o] \tag{2.19}$$

This function is in fact a joint distribution function for all of the real and imaginary parts of the vector components.

The probability density function of a complex random vector is likewise a joint density function for all real and imaginary parts of the vector components. Formally, it is defined as the derivative of the distribution function with respect to all real and imaginary parts:

$$f_{\boldsymbol{x}}(\mathbf{x}^o) = \frac{\partial}{\partial \mathrm{x}_{\mathrm{r}1}} \frac{\partial}{\partial \mathrm{x}_{\mathrm{i}1}} \cdots \frac{\partial}{\partial \mathrm{x}_{\mathrm{r}N}} \frac{\partial}{\partial \mathrm{x}_{\mathrm{i}N}} F_{\boldsymbol{x}}(\mathbf{x}) \Bigg|_{\mathbf{x}=\mathbf{x}^o} \tag{2.20}$$

The distribution function is then obtained by integrating the density over all real and imaginary parts:

$$\begin{aligned} F_{\boldsymbol{x}}(\mathbf{x}^{\mathrm{o}}) &= \int_{-\infty}^{\mathbf{x}^{\mathrm{o}}} f_{\boldsymbol{x}}(\mathbf{x})d\mathbf{x} \\ &\stackrel{\text{def}}{=} \int_{-\infty}^{\mathbf{x}_{\mathrm{r}}^{\mathrm{o}}}\int_{-\infty}^{\mathbf{x}_{\mathrm{i}}^{\mathrm{o}}} f_{\boldsymbol{x}}(\mathbf{x})d\mathbf{x}_{\mathrm{i}}d\mathbf{x}_{\mathrm{r}} \\ &= \int_{-\infty}^{\mathrm{x}_{\mathrm{r}1}^{\mathrm{o}}}\int_{-\infty}^{\mathrm{x}_{\mathrm{i}1}^{\mathrm{o}}}\cdots\int_{-\infty}^{\mathrm{x}_{\mathrm{r}N}^{\mathrm{o}}}\int_{-\infty}^{\mathrm{x}_{\mathrm{i}N}^{\mathrm{o}}} f_{\boldsymbol{x}}(\mathbf{x})d\mathrm{x}_{\mathrm{i}N}d\mathrm{x}_{\mathrm{r}N}\ldots d\mathrm{x}_{\mathrm{i}1}d\mathrm{x}_{\mathrm{r}1} \end{aligned} \tag{2.21}$$

Note in (2.21) that the single integral involving the complex random vector $\boldsymbol{x}$ is *defined* to mean the pair of integrals over the real and imaginary parts and should not be confused with a line integral in the complex plane. The line integral involves integration along a specific path and is totally different from what is being defined here.

The density function for a complex random vector has the interpretation that the quantity

$$f_{\boldsymbol{x}}(\mathbf{x}^{\mathrm{o}})\Delta\mathrm{x}_{\mathrm{r}1}\Delta\mathrm{x}_{\mathrm{i}1}\ldots\Delta\mathrm{x}_{\mathrm{r}N}\Delta\mathrm{x}_{\mathrm{i}N}$$

is the probability that $\mathbf{x}_{\mathrm{r}}^{\mathrm{o}} < \mathbf{x}_{\mathrm{r}} \leq \mathbf{x}_{\mathrm{r}}^{\mathrm{o}} + \Delta\mathbf{x}_{\mathrm{r}}$ *and* $\mathbf{x}_{\mathrm{i}}^{\mathrm{o}} < \mathbf{x}_{\mathrm{i}} \leq \mathbf{x}_{\mathrm{i}}^{\mathrm{o}} + \Delta\mathbf{x}_{\mathrm{i}}$. This can be written simply as

$$\Pr[\mathbf{x}^{\mathrm{o}} < \boldsymbol{x} \leq \mathbf{x}^{\mathrm{o}} + \Delta\mathbf{x}] \approx f_{\boldsymbol{x}}(\mathbf{x}^{\mathrm{o}})\Delta\mathrm{x}_{\mathrm{r}1}\Delta\mathrm{x}_{\mathrm{i}1}\ldots\Delta\mathrm{x}_{\mathrm{r}N}\Delta\mathrm{x}_{\mathrm{i}N} \tag{2.22}$$

As in the case of a real random vector, the probability of any event involving the random vector can be obtained by integrating over the corresponding regions of the vector space, that is,

$$\Pr[\boldsymbol{x} \in \mathcal{A}] = \int_{\mathcal{A}} f_{\boldsymbol{x}}(\mathbf{x})d\mathbf{x} = \int\int_{\mathcal{A}} f_{\boldsymbol{x}}(\mathbf{x})d\mathbf{x}_{\mathrm{i}}d\mathbf{x}_{\mathrm{r}} \tag{2.23}$$

It follows that the density function integrated over the entire vector space must be equal to 1. This can be written simply as

$$\int_{-\infty}^{\infty} f_{\boldsymbol{x}}(\mathbf{x})d\mathbf{x} = 1 \tag{2.24}$$

with the understanding that when the random vector is complex the integral is over both the real and imaginary parts. The relations for distributions and densities of complex random vectors are summarized in Table 2.2. You may want to compare that to the corresponding results for a real random vector in Table 2.1.

In the sections to follow it is assumed that random vectors are in general complex. In cases where a random vector is in fact real, the formulas to be derived will generally apply by ignoring any portion having to do explicitly with the imaginary part. Situations where this is not the case are pointed out explicitly. Two examples are now given illustrating the use of density functions for complex random vectors.

TABLE 2.2 DISTRIBUTION AND DENSITY FUNCTIONS OF A COMPLEX RANDOM VECTOR

Probability interpretation	$F_{\boldsymbol{x}}(\mathbf{x}^{\circ}) = \Pr[\boldsymbol{x} \leq \mathbf{x}^{\circ}] = \Pr[\boldsymbol{x}_{\mathrm{r}} \leq \mathbf{x}_{\mathrm{r}}^{\circ}, \boldsymbol{x}_{\mathrm{i}} \leq \mathbf{x}_{\mathrm{i}}^{\circ}]$	$f_{\boldsymbol{x}}(\mathbf{x}^{\circ})\Delta x_{r1}\Delta x_{i1}\ldots\Delta x_{rN}\Delta x_{iN} \approx \Pr[\mathbf{x}^{\circ} < \boldsymbol{x} \leq \mathbf{x}^{\circ} + \Delta\mathbf{x}]$
Relations	$F_{\boldsymbol{x}}(\mathbf{x}^{\circ}) = \int_{-\infty}^{\mathbf{x}^{\circ}} f_{\boldsymbol{x}}(\mathbf{x})d\mathbf{x} = \int_{-\infty}^{\mathbf{x}_{\mathrm{r}}^{\circ}}\int_{-\infty}^{\mathbf{x}_{\mathrm{i}}^{\circ}} f_{\boldsymbol{x}}(\mathbf{x})d\mathbf{x}_{\mathrm{i}}d\mathbf{x}_{\mathrm{r}}$	$f_{\boldsymbol{x}}(\mathbf{x}) = \dfrac{\partial}{\partial x_{r1}}\dfrac{\partial}{\partial x_{i1}}\cdots\dfrac{\partial}{\partial x_{rN}}\dfrac{\partial}{\partial x_{iN}}F_{\boldsymbol{x}}(\mathbf{x})$
Properties	$F_{\boldsymbol{x}}(-\infty - \jmath\infty) = 0$; $F_{\boldsymbol{x}}(+\infty + \jmath\infty) = 1$	$\int_{-\infty}^{\infty} f_{\boldsymbol{x}}(\mathbf{x})d\mathbf{x} = \int_{-\infty}^{\infty}\int_{-\infty}^{\infty} f_{\boldsymbol{x}}(\mathbf{x})d\mathbf{x}_{\mathrm{i}}d\mathbf{x}_{\mathrm{r}} = 1$

Example 2.2

The components of a two-dimensional complex random vector $\boldsymbol{x}$ are uniformly distributed over the region $\|\mathbf{x}\| \leq 1$, that is,

$$f_{\boldsymbol{x}}(\mathbf{x}) = \begin{cases} \dfrac{2}{\pi^2} & |x_1|^2 + |x_2|^2 \leq 1 \\ 0 & \text{otherwise} \end{cases}$$

It is desired to find the probability that $\boldsymbol{x}$ is in the region defined by

$$\|\mathbf{x}\| \leq \frac{\sqrt{2}}{2}$$

In order to do this, first consider the more general region

$$\mathcal{A}: \quad \|\mathbf{x}\| = \sqrt{\mathbf{x}^{*T}\mathbf{x}} = \sqrt{x_{r1}^2 + x_{i1}^2 + x_{r2}^2 + x_{i2}^2} \leq r$$

where r is a real number between 0 and 1. The probability that $\boldsymbol{x} \in \mathcal{A}$ is given by

$$\int_{\mathcal{A}} f_{\boldsymbol{x}}(\mathbf{x})d\mathbf{x} = \int_{-r}^{r}\int_{-\sqrt{r^2-x_{r1}^2}}^{+\sqrt{r^2-x_{r1}^2}}\int_{-\sqrt{r^2-x_{r1}^2-x_{i1}^2}}^{+\sqrt{r^2-x_{r1}^2-x_{i1}^2}}\int_{-\sqrt{r^2-x_{r1}^2-x_{i1}^2-x_{r2}^2}}^{+\sqrt{r^2-x_{r1}^2-x_{i1}^2-x_{r2}^2}} \left(\frac{2}{\pi^2}\right) dx_{i2}dx_{r2}dx_{i1}dx_{r1}$$

This set of integrals represents integration over a line segment, a circle, a sphere, and finally a hypersphere in four dimensions. Carrying out the integration is tedious but straightforward. The steps are

$$\int_{-r}^{r}\int_{-\sqrt{r^2-x_{r1}^2}}^{+\sqrt{r^2-x_{r1}^2}}\int_{-\sqrt{r^2-x_{r1}^2-x_{i1}^2}}^{+\sqrt{r^2-x_{r1}^2-x_{i1}^2}}\int_{-\sqrt{r^2-x_{r1}^2-x_{i1}^2-x_{r2}^2}}^{+\sqrt{r^2-x_{r1}^2-x_{i1}^2-x_{r2}^2}} \left(\frac{2}{\pi^2}\right) dx_{i2}dx_{r2}dx_{i1}dx_{r1}$$

$$= \frac{2}{\pi^2}\int_{-r}^{r}\int_{-\sqrt{r^2-x_{r1}^2}}^{+\sqrt{r^2-x_{r1}^2}}\int_{-\sqrt{r^2-x_{r1}^2-x_{i1}^2}}^{+\sqrt{r^2-x_{r1}^2-x_{i1}^2}} 2\sqrt{r^2 - x_{r1}^2 - x_{i1}^2 - x_{r2}^2}\, dx_{r2}dx_{i1}dx_{r1}$$

$$= \frac{2}{\pi^2}\int_{-r}^{r}\int_{-\sqrt{r^2-x_{r1}^2}}^{+\sqrt{r^2-x_{r1}^2}} \pi(r^2 - x_{r1}^2 - x_{i1}^2)\, dx_{i1}dx_{r1}$$

$$= \frac{2}{\pi^2}\int_{-r}^{r} \tfrac{4}{3}\pi(r^2 - x_{r1}^2)^{\frac{3}{2}} dx_{r1} = r^4$$

Evaluating this at $r = \frac{\sqrt{2}}{2}$ then yields

$$\Pr\left[\|\boldsymbol{x}\| \le \frac{\sqrt{2}}{2}\right] = \frac{1}{4}$$

Note, incidentally, that if r is considered to be a random variable defined to be the magnitude of the complex vector, that is, if

$$r = \|\boldsymbol{x}\| = \sqrt{x_{r1}^2 + x_{i1}^2 + x_{r2}^2 + x_{i2}^2}$$

then the probability that $\boldsymbol{x} \in \mathcal{A}$ is equal to the distribution function of r. That is,

$$F_r(\mathrm{r}) = \mathrm{r}^4; \qquad 0 \le \mathrm{r} \le 1$$

The density function for the magnitude of the complex vector is the derivative of this quantity, so

$$f_r(\mathrm{r}) = 4\mathrm{r}^3; \qquad 0 \le \mathrm{r} \le 1$$

The following example shows the advantage of changing coordinates to exploit a particular kind of symmetry.

Example 2.3

The density function for a two-dimensional complex random vector $\boldsymbol{x}$ is

$$f_{\boldsymbol{x}}(\mathbf{x}) = \frac{16}{\pi^2}|\mathrm{x}_1|^2|\mathrm{x}_2|^2 e^{-2(|\mathrm{x}_1|^2+|\mathrm{x}_2|^2)}$$

It is desired to find the probability of the event

$$\|\boldsymbol{x}\| \le \frac{\sqrt{2}}{2}$$

as in Example 2.2.

For this case it is convenient to convert to polar coordinates, that is, to write

$$\mathrm{x}_k = \rho_k e^{\jmath\phi_k}; \qquad k = 1, 2$$

The density then becomes

$$f_{\boldsymbol{x}}(\mathbf{x}) \equiv f_{\rho_1,\rho_2}(\rho_1, \rho_2) = \frac{16}{\pi^2}\rho_1^2\rho_2^2 e^{-2(\rho_1^2+\rho_2^2)}; \qquad 0 \le \rho_1, \rho_2 < \infty$$

From this it is clear that the density depends only on the *magnitude* of the vector components; it is constant (uniform) as a function of phase. If the region $\mathcal{A}$ is defined as

$$\mathcal{A}: \ \|\mathbf{x}\| \le \mathrm{r} \ \equiv \ \rho_1^2 + \rho_2^2 \le \mathrm{r}^2$$

then the probability that $\boldsymbol{x} \in \mathcal{A}$ is

$$\int_{\mathcal{A}} f_{\boldsymbol{x}}(\mathbf{x})d\mathbf{x} = \frac{16}{\pi^2}\int_0^{2\pi}\int_0^{2\pi}\int_0^{\mathrm{r}}\int_0^{\sqrt{\mathrm{r}^2-\rho_1^2}} \rho_1^3\rho_2^3 e^{-2(\rho_1^2+\rho_2^2)} d\rho_2 d\rho_1 d\phi_2 d\phi_1$$

The outer two integrals yield a factor of $4\pi^2$. To carry out the inner integrals, make the change of variables $\zeta_k = 2\rho_k^2$ and $d\zeta_k = 4\rho_k d\rho_k$. The expression then becomes

$$\int_0^{2\mathrm{r}^2} \zeta_1 e^{-\zeta_1} \int_0^{2\mathrm{r}^2-\zeta_1} \zeta_2 e^{-\zeta_2} d\zeta_2 d\zeta_1$$

Although this double integral is not trivial, it can be integrated by parts to obtain

$$\Pr[\|\boldsymbol{x}\| \leq \mathrm{r}] = 1 - e^{-2\mathrm{r}^2}\left(1 + 2\mathrm{r}^2 + 2\mathrm{r}^4 + \tfrac{4}{3}\mathrm{r}^6\right)$$

Evaluating this expression at $\mathrm{r} = \frac{\sqrt{2}}{2}$ yields

$$\Pr\left[\|\boldsymbol{x}\| \leq \tfrac{\sqrt{2}}{2}\right] = 0.0190$$

As in Example 2.2, the expression for $\Pr[\|\boldsymbol{x}\| \leq \mathrm{r}]$ represents the distribution function for the magnitude of the random vector. As an exercise at this point, you may want to derive an expression for the density function of $\|\boldsymbol{x}\|$ and show that it integrates to 1.

2.1.3 Joint Distribution and Density Functions

This subsection considers the joint distribution and density functions of two random vectors. It is assumed that in general both vectors may be complex, although most of the formulas can be interpreted for either the real or the complex case. Let $\boldsymbol{x}$ be the random vector defined by (2.13) and $\boldsymbol{y}$ be the random vector

$$\boldsymbol{y} = \boldsymbol{y}_{\mathrm{r}} + \jmath\boldsymbol{y}_{\mathrm{i}} \tag{2.25}$$

where

$$\boldsymbol{y}_{\mathrm{r}} = \begin{bmatrix} y_{\mathrm{r}1} \\ y_{\mathrm{r}2} \\ \vdots \\ y_{\mathrm{r}M} \end{bmatrix}; \quad \boldsymbol{y}_{\mathrm{i}} = \begin{bmatrix} y_{\mathrm{i}1} \\ y_{\mathrm{i}2} \\ \vdots \\ y_{\mathrm{i}M} \end{bmatrix} \tag{2.26}$$

and where the dimension M of $\boldsymbol{y}$ is possibly different from the dimension N of $\boldsymbol{x}$. The joint distribution function for random vectors $\boldsymbol{x}$ and $\boldsymbol{y}$ is defined by the probability of the joint event "$\boldsymbol{x} \leq \mathbf{x}^{\mathrm{o}}$ *and* $\boldsymbol{y} \leq \mathbf{y}^{\mathrm{o}}$," that is,

$$F_{\boldsymbol{xy}}(\mathbf{x}^{\mathrm{o}}, \mathbf{y}^{\mathrm{o}}) = \Pr\left[\boldsymbol{x} \leq \mathbf{x}^{\mathrm{o}}\ ,\ \boldsymbol{y} \leq \mathbf{y}^{\mathrm{o}}\right] \tag{2.27}$$

The joint density function is then defined as

$$f_{xy}(\mathbf{x}^o, \mathbf{y}^o) = \frac{\partial}{\partial \mathrm{x}_{r1}} \frac{\partial}{\partial \mathrm{x}_{i1}} \cdots \frac{\partial}{\partial \mathrm{x}_{iN}} \frac{\partial}{\partial \mathrm{y}_{r1}} \frac{\partial}{\partial \mathrm{y}_{i1}} \cdots \frac{\partial}{\partial \mathrm{y}_{iM}} F_{xy}(\mathbf{x}, \mathbf{y}) \Big|_{\substack{\mathbf{x}=\mathbf{x}^o \\ \mathbf{y}=\mathbf{y}^o}} \tag{2.28}$$

and it follows from this definition that

$$F_{xy}(\mathbf{x}^o, \mathbf{y}^o) = \int_{-\infty}^{\mathbf{x}^o} \int_{-\infty}^{\mathbf{y}^o} f_{xy}(\mathbf{x}, \mathbf{y}) d\mathbf{y} d\mathbf{x} \tag{2.29}$$

Equation 2.27 implies that[3]

$$\begin{aligned} F_{xy}(\mathbf{x}^o, -\infty) &= F_{xy}(-\infty, \mathbf{y}^o) = 0 & (a) \\ F_{xy}(\mathbf{x}^o, +\infty) &= F_x(\mathbf{x}^o) & (b) \\ F_{xy}(+\infty, \mathbf{y}^o) &= F_y(\mathbf{y}^o) & (c) \\ F_{xy}(+\infty, +\infty) &= 1 & (d) \end{aligned} \tag{2.30}$$

Because of (2.29) it follows from (2.30d) that

$$\int_{-\infty}^{\infty} \int_{-\infty}^{\infty} f_{xy}(\mathbf{x}, \mathbf{y}) d\mathbf{y} d\mathbf{x} = 1 \tag{2.31}$$

while from (2.30b) and (2.30c) it follows that

$$f_x(\mathbf{x}^o) = \int_{-\infty}^{\infty} f_{xy}(\mathbf{x}^o, \mathbf{y}) d\mathbf{y} \tag{2.32}$$

and

$$f_y(\mathbf{y}^o) = \int_{-\infty}^{\infty} f_{xy}(\mathbf{x}, \mathbf{y}^o) d\mathbf{x} \tag{2.33}$$

Equation 2.31 shows that the joint density function must integrate to 1 while (2.32) and (2.33) show that the densities for the individual vectors can be obtained from the joint density by integrating over the other variable. When the densities $f_x(\mathbf{x})$ and $f_y(\mathbf{y})$ are obtained in this way they are referred to as *marginal densities* for $\boldsymbol{x}$ and $\boldsymbol{y}$.

The interpretation of the joint density function as probability of events is similar to that discussed for single random variables. Specifically,

$$\begin{aligned} &\Pr[\mathbf{x}^o < \boldsymbol{x} \leq \mathbf{x}^o + \Delta\mathbf{x} \quad \textit{and} \quad \mathbf{y}^o < \boldsymbol{y} \leq \mathbf{y}^o + \Delta\mathbf{y}] \\ &\quad \approx f_{xy}(\mathbf{x}^o, \mathbf{y}^o) \Delta \mathrm{x}_{r1} \Delta \mathrm{x}_{i1} \ldots \Delta \mathrm{x}_{iN} \Delta \mathrm{y}_{r1} \Delta \mathrm{y}_{i1} \ldots \Delta \mathrm{y}_{iM} \end{aligned} \tag{2.34}$$

for small intervals $\Delta\mathbf{x}$ and $\Delta\mathbf{y}$. Equation 2.34 states that the probability that $\boldsymbol{x}$ is in a small rectangular region around $\mathbf{x}^o$, and $\boldsymbol{y}$ is simultaneously in a small rectangular region around

[3]Since $\boldsymbol{x}$ and $\boldsymbol{y}$ are in general complex, it would be more correct to use the expressions $-\infty - \jmath\infty$ and $\infty + \jmath\infty$ as arguments to the distribution function in (2.30). Since that notation would be rather cumbersome, however, it is deliberately avoided.

$\mathbf{y}^o$, is given by the density function evaluated at $\mathbf{x}^o$ and $\mathbf{y}^o$ multiplied by all of the increments in $\boldsymbol{x}$ and all of the increments in $\boldsymbol{y}$ defining the respective regions. The probability that either $\boldsymbol{x}$ or $\boldsymbol{y}$ takes on the specific value $\mathbf{x}^o$ or $\mathbf{y}^o$, however, is zero unless the joint density function contains impulses at those points.

The following examples illustrates some of the ideas involving the joint density function.

Example 2.4

The joint density function for a two-dimensional real random vector $\boldsymbol{x}$ and a one-dimensional real random vector $\boldsymbol{y}$ is

$$f_{\boldsymbol{xy}}(\mathbf{x}, \mathbf{y}) = \begin{cases} (\mathrm{x}_1 + 3\mathrm{x}_2)\mathrm{y}_1 & 0 \leq \mathrm{x}_1, \mathrm{x}_2, \mathrm{y}_1 \leq 1 \\ 0 & \text{otherwise} \end{cases}$$

It is desired to determine the probability of the event $\boldsymbol{y} \leq \left[\frac{1}{2}\right]$ and the marginal densities $f_{\boldsymbol{x}}(\mathbf{x})$ and $f_{\boldsymbol{y}}(\mathbf{y})$. (The brackets [ ] are placed around the value $\frac{1}{2}$ to show that it is the value of a *vector* although that vector is one-dimensional and has only a single component.) First observe from (2.29) and (2.30) that

$$\Pr\left[\boldsymbol{y} \leq \left[\tfrac{1}{2}\right]\right] = F_{\boldsymbol{y}}\left(\left[\tfrac{1}{2}\right]\right) = \int_{-\infty}^{\infty}\int_{-\infty}^{\left[\frac{1}{2}\right]} f_{\boldsymbol{xy}}(\mathbf{x}, \mathbf{y})d\mathbf{y}d\mathbf{x}$$

$$= \int_0^1\int_0^1\int_0^{1/2} (\mathrm{x}_1 + 3\mathrm{x}_2)\,\mathrm{y}_1 d\mathrm{y}_1 d\mathrm{x}_2 d\mathrm{x}_1 = \tfrac{1}{4}$$

To compute the marginal densities, apply (2.32) and (2.33):

$$f_{\boldsymbol{x}}(\mathbf{x}) = \int_0^1 (\mathrm{x}_1 + 3\mathrm{x}_2)\,\mathrm{y}_1 d\mathrm{y}_1 \quad 0 \leq \mathrm{x}_1, \mathrm{x}_2 \leq 1$$

$$= \begin{cases} \frac{1}{2}(\mathrm{x}_1 + 3\mathrm{x}_2) & 0 \leq \mathrm{x}_1, \mathrm{x}_2 \leq 1 \\ 0 & \text{otherwise} \end{cases}$$

$$f_{\boldsymbol{y}}(\mathbf{y}) = \int_0^1\int_0^1 (\mathrm{x}_1 + 3\mathrm{x}_2)\,\mathrm{y}_1 d\mathrm{x}_2 d\mathrm{x}_1 \quad 0 \leq \mathrm{y}_1 \leq 1$$

$$= \begin{cases} 2\mathrm{y}_1 & 0 \leq \mathrm{y}_1 \leq 1 \\ 0 & \text{otherwise} \end{cases}$$

2.1.4 Conditional Densities and Bayes' Rule

If A and B are two events, then the conditional probability of A given B is defined as

$$\Pr\left[\mathsf{A}|\mathsf{B}\right] = \frac{\Pr\left[\mathsf{AB}\right]}{\Pr\left[\mathsf{B}\right]} \tag{2.35}$$

If A is the event

$$\mathsf{A}: \quad \mathbf{x}^o < \boldsymbol{x} \leq \mathbf{x}^o + \Delta\mathbf{x} \tag{2.36}$$

and B is the event

$$\mathsf{B}: \quad \mathbf{y}^{\mathrm{o}} < \boldsymbol{y} \leq \mathbf{y}^{\mathrm{o}} + \Delta\mathbf{y} \tag{2.37}$$

then

$$\begin{aligned}
\Pr\left[\mathsf{A}|\mathsf{B}\right] &= \frac{f_{\boldsymbol{x}\boldsymbol{y}}(\mathbf{x}^{\mathrm{o}}, \mathbf{y}^{\mathrm{o}})\Delta\mathrm{x}_{\mathrm{r}1}\ldots\Delta\mathrm{x}_{\mathrm{i}N}\Delta\mathrm{y}_{\mathrm{r}1}\ldots\Delta\mathrm{y}_{\mathrm{i}M}}{f_{\boldsymbol{y}}(\mathbf{y}^{\mathrm{o}})\Delta\mathrm{y}_{\mathrm{r}1}\ldots\Delta\mathrm{y}_{\mathrm{i}M}} \\
&= \frac{f_{\boldsymbol{x}\boldsymbol{y}}(\mathbf{x}^{\mathrm{o}}, \mathbf{y}^{\mathrm{o}})}{f_{\boldsymbol{y}}(\mathbf{y}^{\mathrm{o}})}\Delta\mathrm{x}_{\mathrm{r}1}\ldots\Delta\mathrm{x}_{\mathrm{i}N}
\end{aligned} \tag{2.38}$$

(For real random vectors the terms $\Delta\mathrm{x}_{\mathrm{i}k}$ and $\Delta\mathrm{y}_{\mathrm{i}k}$ are not present.) The conditional density of $\boldsymbol{x}$ given $\boldsymbol{y}$ is *defined* as

$$\boxed{f_{\boldsymbol{x}|\boldsymbol{y}}(\mathbf{x}|\mathbf{y}) \stackrel{\text{def}}{=} \frac{f_{\boldsymbol{x}\boldsymbol{y}}(\mathbf{x}, \mathbf{y})}{f_{\boldsymbol{y}}(\mathbf{y})}} \tag{2.39}$$

which has the interpretation that $f_{\boldsymbol{x}|\boldsymbol{y}}(\mathbf{x}^{\mathrm{o}}|\mathbf{y}^{\mathrm{o}})\Delta\mathrm{x}_{\mathrm{r}1}\ldots\Delta\mathrm{x}_{\mathrm{i}N}$ is the probability that $\boldsymbol{x}$ is in the region $\mathbf{x}^{\mathrm{o}} < \boldsymbol{x} \leq \mathbf{x}^{\mathrm{o}} + \Delta\mathbf{x}$ given that $\boldsymbol{y}$ is in the region $\mathbf{y}^{\mathrm{o}} < \boldsymbol{y} \leq \mathbf{y}^{\mathrm{o}} + \Delta\mathbf{y}$. Note that

$$\int_{-\infty}^{\infty} f_{\boldsymbol{x}|\boldsymbol{y}}(\mathbf{x}|\mathbf{y})d\mathbf{x} = \int_{-\infty}^{\infty} \frac{f_{\boldsymbol{x}\boldsymbol{y}}(\mathbf{x}, \mathbf{y})}{f_{\boldsymbol{y}}(\mathbf{y})}d\mathbf{x} = \frac{f_{\boldsymbol{y}}(\mathbf{y})}{f_{\boldsymbol{y}}(\mathbf{y})} = 1 \tag{2.40}$$

That is, the conditional density $f_{\boldsymbol{x}|\boldsymbol{y}}$ integrated over $\boldsymbol{x}$ (for *any* value of $\boldsymbol{y}$) is equal to 1. This emphasizes that $f_{\boldsymbol{x}|\boldsymbol{y}}$ is density for $\boldsymbol{x}$ that depends on $\boldsymbol{y}$ almost as if it were a parameter. The integral of $f_{\boldsymbol{x}|\boldsymbol{y}}$ over $\boldsymbol{y}$ is meaningless.

Observe that if the events A and B defined above are independent for all values $\mathbf{x}^{\mathrm{o}}$ and $\mathbf{y}^{\mathrm{o}}$, then the conditional density $f_{\boldsymbol{x}|\boldsymbol{y}}(\mathbf{x}|\mathbf{y})$ is identical to the unconditional density $f_{\boldsymbol{x}}(\mathbf{x})$. In this case the *random vectors* are said to be statistically independent and it follows that

$$f_{\boldsymbol{x}\boldsymbol{y}}(\mathbf{x}, \mathbf{y}) = f_{\boldsymbol{x}}(\mathbf{x}) \cdot f_{\boldsymbol{y}}(\mathbf{y}) \tag{2.41}$$

for $\boldsymbol{x}$ and $\boldsymbol{y}$ independent. A short example shows how to check for independence.

Example 2.5

To check if the random vectors $\boldsymbol{x}$ and $\boldsymbol{y}$ in Example 2.4 are independent, form the product of the marginal densities

$$f_{\boldsymbol{x}}(\mathbf{x}) \cdot f_{\boldsymbol{y}}(\mathbf{y}) = \begin{cases} (\frac{1}{2}(\mathrm{x}_1 + 3\mathrm{x}_2)) \cdot (2\mathrm{y}_1) & 0 \leq \mathrm{x}_1, \mathrm{x}_2, \mathrm{y}_1 \leq 1 \\ 0 & \text{otherwise} \end{cases}$$

Since this is equal to the expression for $f_{\boldsymbol{x}\boldsymbol{y}}(\mathbf{x}, \mathbf{y})$ given in that example, the random vectors *are* statistically independent.

Bayes' rule is based on the fact that the joint probability of two events can be expressed in terms of either the conditional probability for the first event or the conditional probability of the second. Bayes' rule for events follows directly from (2.35) and has the form

$$\Pr\left[\mathsf{B}|\mathsf{A}\right] = \frac{\Pr\left[\mathsf{A}|\mathsf{B}\right]\Pr\left[\mathsf{B}\right]}{\Pr\left[\mathsf{A}\right]} \tag{2.42}$$

An analogous expression can be written for density functions. Specifically, since

$$f_{\boldsymbol{x}\boldsymbol{y}}(\mathbf{x},\mathbf{y}) = f_{\boldsymbol{x}|\boldsymbol{y}}(\mathbf{x}|\mathbf{y})f_{\boldsymbol{y}}(\mathbf{y}) = f_{\boldsymbol{y}|\boldsymbol{x}}(\mathbf{y}|\mathbf{x})f_{\boldsymbol{x}}(\mathbf{x}) \tag{2.43}$$

it follows that

$$\boxed{f_{\boldsymbol{y}|\boldsymbol{x}}(\mathbf{y}|\mathbf{x}) = \frac{f_{\boldsymbol{x}|\boldsymbol{y}}(\mathbf{x}|\mathbf{y})f_{\boldsymbol{y}}(\mathbf{y})}{f_{\boldsymbol{x}}(\mathbf{x})}} \tag{2.44}$$

This expression, which is Bayes' rule for density functions, can also be derived by using the events (2.36) and (2.37) in (2.42). Since the density $f_{\boldsymbol{x}}(\mathbf{x})$ can be obtained by integrating the joint density over $\mathbf{y}$, and since the joint density can be expressed by (2.43), one can write (2.44) as

$$f_{\boldsymbol{y}|\boldsymbol{x}}(\mathbf{y}|\mathbf{x}) = \frac{f_{\boldsymbol{x}|\boldsymbol{y}}(\mathbf{x}|\mathbf{y})f_{\boldsymbol{y}}(\mathbf{y})}{\int_{-\infty}^{+\infty} f_{\boldsymbol{x}|\boldsymbol{y}}(\mathbf{x}|\mathbf{y})f_{\boldsymbol{y}}(\mathbf{y})d\mathbf{y}} \tag{2.45}$$

which is another useful formula.

Equations (2.44) and (2.45) arise frequently in problems of statistical decision and estimation where $\boldsymbol{y}$ denotes a random vector that cannot be observed directly and $\boldsymbol{x}$ denotes a random vector related to $\mathbf{y}$ that *can* be observed. In this context the density $f_{\boldsymbol{y}}$ is called the *prior* density (i.e., the density *before* observation of $\boldsymbol{x}$) and $f_{\boldsymbol{y}|\boldsymbol{x}}$ is called the *posterior* density (density *after* observation of $\boldsymbol{x}$). Bayes' rule is used in this context to update the knowledge about $\boldsymbol{y}$ after the vector $\boldsymbol{x}$ is observed. The following example illustrates this application.

Example 2.6

A two-dimensional complex random vector $\boldsymbol{y}$ has the density function

$$f_{\boldsymbol{y}}(\mathbf{y}) = \mathrm{P}_1\delta_c(\|\mathbf{y}-\mathbf{m}_1\|) + \mathrm{P}_2\delta_c(\|\mathbf{y}-\mathbf{m}_2\|)$$

where $\mathrm{P}_1+\mathrm{P}_2 = 1$. A two-dimensional complex random vector $\boldsymbol{x}$ related to $\boldsymbol{y}$ has the conditional density function

$$f_{\boldsymbol{x}|\boldsymbol{y}}(\mathbf{x}|\mathbf{y}) = \frac{1}{\pi}e^{-\|\mathbf{x}-\mathbf{y}\|^2}$$

The joint density function is given by

$$\begin{aligned}
f_{\boldsymbol{x}\boldsymbol{y}}(\mathbf{x},\mathbf{y}) &= f_{\boldsymbol{x}|\boldsymbol{y}}(\mathbf{x}|\mathbf{y})f_{\boldsymbol{y}}(\mathbf{y}) \\
&= \frac{1}{\pi}e^{-\|\mathbf{x}-\mathbf{y}\|^2}\left[\mathrm{P}_1\delta_c(\|\mathbf{y}-\mathbf{m}_1\|) + \mathrm{P}_2\delta_c(\|\mathbf{y}-\mathbf{m}_2\|)\right] \\
&= \frac{1}{\pi}\left[\mathrm{P}_1e^{-\|\mathbf{x}-\mathbf{m}_1\|^2}\delta_c(\|\mathbf{y}-\mathbf{m}_1\|) + \mathrm{P}_2e^{-\|\mathbf{x}-\mathbf{m}_2\|^2}\delta_c(\|\mathbf{y}-\mathbf{m}_2\|)\right]
\end{aligned}$$

The marginal density for $\boldsymbol{x}$ is then given by

$$f_{\boldsymbol{x}}(\mathbf{x}) = \int_{-\infty}^{\infty} f_{\boldsymbol{xy}}(\mathbf{x}, \mathbf{y})d\mathbf{y} = \frac{1}{\pi}\left[\mathrm{P}_1 e^{-\|\mathbf{x}-\mathbf{m}_1\|^2} + \mathrm{P}_2 e^{-\|\mathbf{x}-\mathbf{m}_2\|^2}\right]$$

The posterior density for $\boldsymbol{y}$ is then

$$\begin{aligned} f_{\boldsymbol{y}|\boldsymbol{x}}(\mathbf{y}|\mathbf{x}) &= \frac{f_{\boldsymbol{xy}}(\mathbf{x}, \mathbf{y})}{f_{\boldsymbol{x}}(\mathbf{x})} \\ &= \mathrm{P}'_1(\mathbf{x})\delta_c(\|\mathbf{y} - \mathbf{m}_1\|) + \mathrm{P}'_2(\mathbf{x})\delta_c(\|\mathbf{y} - \mathbf{m}_2\|) \end{aligned}$$

where

$$\mathrm{P}'_i(\mathbf{x}) = \frac{\mathrm{P}_i e^{-\|\mathbf{x}-\mathbf{m}_i\|^2}}{\mathrm{P}_1 e^{-\|\mathbf{x}-\mathbf{m}_1\|^2} + \mathrm{P}_2 e^{-\|\mathbf{x}-\mathbf{m}_2\|^2}}; \qquad i = 1, 2$$

Note that in this particular example the prior density of $\boldsymbol{y}$ was such that $\boldsymbol{y}$ could take on only two values $\mathbf{m}_1$ and $\mathbf{m}_2$ with probabilities P_1 and P_2. After observation of $\boldsymbol{x}$, the density of $\boldsymbol{y}$ still allows it to take on only the two values, but with different probabilities P'_1 and P'_2 that depend on the observation $\mathbf{x}$. Note also that

$$\mathrm{P}'_1(\mathbf{x}) + \mathrm{P}'_2(\mathbf{x}) = 1$$

and thus the conditional density $f_{\boldsymbol{y}|\boldsymbol{x}}$ integrates (over $\boldsymbol{y}$) to 1.

2.2 EXPECTATION AND MOMENTS

In many cases, as you might expect, it is very difficult to work with the density function for a high-dimensional random vector. In fact, in many cases of practical interest, the density function may not be known at all. In these cases it is possible to work entirely with the moments of the distribution (usually, the first and second moments) and to perform many useful analyses based on these moments. Although the moments are defined theoretically in terms of the density function, it is not necessary to know the density function explicitly to compute estimates of the moments. This section defines moments of a random vector and derives some simple procedures for estimating the moments from data.

2.2.1 Expectations for a Single Random Vector

Let $\boldsymbol{x}$ be a random vector and let $\psi(\boldsymbol{x})$ represent any quantity derived from it. This quantity may be a scalar, vector, or matrix. The expectation of ψ is denoted by $\mathcal{E}\{\psi(\boldsymbol{x})\}$ and defined by the operation

$$\mathcal{E}\{\psi(\boldsymbol{x})\} = \int_{-\infty}^{\infty} \psi(\mathbf{x}) f_{\boldsymbol{x}}(\mathbf{x})d\mathbf{x} \tag{2.46}$$

If ψ is a vector or matrix, (2.46) is interpreted as the application of the expectation operation to every component of the vector or element of the matrix. Thus, the result of taking the expectation of a vector or matrix is another vector or matrix of the same size.

When the quantity $\psi(\boldsymbol{x})$ involves products of the components of $\boldsymbol{x}$, the resulting expectations are referred to as *moments* of the distribution or density. The most commonly considered moments are the first and second moments, which are most conveniently described in the forms discussed below. Higher-order moments can also be defined for random vectors but will not be specifically considered here.[4]

The *first moment* or *mean* of a random vector $\boldsymbol{x}$ is defined by

$$\mathbf{m}_{\boldsymbol{x}} = \mathcal{E}\{\boldsymbol{x}\} = \int_{-\infty}^{\infty} \mathbf{x} f_{\boldsymbol{x}}(\mathbf{x}) d\mathbf{x} \tag{2.47}$$

In line with the earlier discussion, since $\boldsymbol{x}$ is a vector, $\mathbf{m}_{\boldsymbol{x}}$ is a vector of the form

$$\mathbf{m}_{\boldsymbol{x}} = \begin{bmatrix} \mathrm{m}_1 \\ \mathrm{m}_2 \\ \vdots \\ \mathrm{m}_N \end{bmatrix} \tag{2.48}$$

The notation of (2.47) implies that the kth component m_k is given by

$$\begin{aligned} \mathrm{m}_k = \mathcal{E}\{x_k\} &= \int_{-\infty}^{\infty} \mathrm{x}_k f_{\boldsymbol{x}}(\mathbf{x}) d\mathbf{x} \\ &= \int_{-\infty}^{\infty} \int_{-\infty}^{\infty} \cdots \int_{-\infty}^{\infty} (\mathrm{x}_{\mathrm{r}k} + \jmath \mathrm{x}_{\mathrm{i}k}) f_{\boldsymbol{x}}(\mathbf{x}) d\mathrm{x}_{\mathrm{i}N} \ldots d\mathrm{x}_{\mathrm{i}1} d\mathrm{x}_{\mathrm{r}1} \end{aligned}$$

After integrating over all the other components this becomes

$$\mathrm{m}_k = \int_{-\infty}^{\infty} \mathrm{x}_{\mathrm{r}k} f_{x_{\mathrm{r}k}}(\mathrm{x}_{\mathrm{r}k}) d\mathrm{x}_{\mathrm{r}k} + \jmath \int_{-\infty}^{\infty} \mathrm{x}_{\mathrm{i}k} f_{x_{\mathrm{i}k}}(\mathrm{x}_{\mathrm{i}k}) d\mathrm{x}_{\mathrm{i}k} \tag{2.49}$$

where $f_{x_{\mathrm{r}k}}$ and $f_{x_{\mathrm{i}k}}$ are the marginal density functions for the real and imaginary parts of the kth component. In the case of a real random vector the second integral is not present.

The *correlation matrix* represents the complete set of second moments for the random vector and is defined by

$$\mathbf{R}_{\boldsymbol{x}} = \mathcal{E}\{\boldsymbol{x}\boldsymbol{x}^{*T}\} \tag{2.50}$$

For a real random vector this matrix has the form

$$\mathbf{R}_{\boldsymbol{x}} = \begin{bmatrix} \mathcal{E}\{(x_1)^2\} & \mathcal{E}\{x_1 x_2\} & \cdots & \mathcal{E}\{x_1 x_N\} \\ \mathcal{E}\{x_2 x_1\} & \mathcal{E}\{(x_2)^2\} & \cdots & \mathcal{E}\{x_2 x_N\} \\ \vdots & \vdots & & \vdots \\ \mathcal{E}\{x_N x_1\} & \mathcal{E}\{x_N x_2\} & \cdots & \mathcal{E}\{(x_N)^2\} \end{bmatrix} \tag{2.51}$$

[4]Higher-order moments for random processes are considered in Chapter 4 however.

and so contains second moments between all possible pairs of components of the random vector. For a complex random vector the matrix has the form

$$\mathbf{R}_x = \begin{bmatrix} \mathcal{E}\{|x_1|^2\} & \mathcal{E}\{x_1x_2^*\} & \cdots & \mathcal{E}\{x_1x_N^*\} \\ \mathcal{E}\{x_2x_1^*\} & \mathcal{E}\{|x_2|^2\} & \cdots & \mathcal{E}\{x_2x_N^*\} \\ \vdots & \vdots & & \vdots \\ \mathcal{E}\{x_Nx_1^*\} & \mathcal{E}\{x_Nx_2^*\} & \cdots & \mathcal{E}\{|x_N|^2\} \end{bmatrix} \tag{2.52}$$

Why products of the form $\mathcal{E}\{x_kx_l^*\}$ are useful is not so obvious at first. It would seem that since x_k and x_l are both complex, the expected value of the products between all real and imaginary parts (four numbers) would be needed to characterize the random variables. The answer lies in the fact that x_k and x_l represent samples of a random signal, and that the moments of the real and imaginary parts of the signal frequently have a special kind of symmetry that allows their characterization in terms of the single complex quantity $\mathcal{E}\{x_kx_l^*\}$. This can be explored in more general terms as follows. The complete set of moments for the real and imaginary parts of the random vector $\boldsymbol{x}$ can conveniently be expressed by forming the real random vector

$$\boldsymbol{v} = \begin{bmatrix} \boldsymbol{x}_\mathrm{r} \\ \boldsymbol{x}_\mathrm{i} \end{bmatrix} \tag{2.53}$$

and considering its real-valued correlation matrix

$$\mathcal{E}\{\boldsymbol{v}\boldsymbol{v}^T\} = \mathcal{E}\left\{\begin{bmatrix} \boldsymbol{x}_\mathrm{r} \\ \boldsymbol{x}_\mathrm{i} \end{bmatrix}\begin{bmatrix}\boldsymbol{x}_\mathrm{r}^T & \boldsymbol{x}_\mathrm{i}^T\end{bmatrix}\right\} = \begin{bmatrix} \mathcal{E}\{\boldsymbol{x}_\mathrm{r}\boldsymbol{x}_\mathrm{r}^T\} & \mathcal{E}\{\boldsymbol{x}_\mathrm{r}\boldsymbol{x}_\mathrm{i}^T\} \\ \mathcal{E}\{\boldsymbol{x}_\mathrm{i}\boldsymbol{x}_\mathrm{r}^T\} & \mathcal{E}\{\boldsymbol{x}_\mathrm{i}\boldsymbol{x}_\mathrm{i}^T\} \end{bmatrix} \tag{2.54}$$

For reasons that will become clear later in the study of random signals it is frequently possible to restrict the analysis of complex random vectors to those for which the symmetry conditions

$$\begin{aligned} \mathcal{E}\{\boldsymbol{x}_\mathrm{r}\boldsymbol{x}_\mathrm{r}^T\} &= \mathcal{E}\{\boldsymbol{x}_\mathrm{i}\boldsymbol{x}_\mathrm{i}^T\} = \mathbf{R}_x^\mathrm{E} \ (a) \\ \mathcal{E}\{\boldsymbol{x}_\mathrm{i}\boldsymbol{x}_\mathrm{r}^T\} &= -\mathcal{E}\{\boldsymbol{x}_\mathrm{r}\boldsymbol{x}_\mathrm{i}^T\} = \mathbf{R}_x^\mathrm{O} \ (b) \end{aligned} \tag{2.55}$$

hold. These properties state that the real and imaginary parts of the vector components satisfy the two conditions

$$\mathcal{E}\{x_{\mathrm{r}k}x_{\mathrm{r}l}\} = \mathcal{E}\{x_{\mathrm{i}k}x_{\mathrm{i}l}\}$$

and

$$\mathcal{E}\{x_{\mathrm{i}k}x_{\mathrm{r}l}\} = -\mathcal{E}\{x_{\mathrm{r}k}x_{\mathrm{i}l}\}$$

[which motivate the superscripts in (2.55): "E" for even and "O" for odd]. Now observe from (2.50) that the correlation matrix for $\boldsymbol{x}$ has the form

$$\begin{aligned} \mathbf{R}_x &= \mathcal{E}\{[\boldsymbol{x}_\mathrm{r} + \jmath\boldsymbol{x}_\mathrm{i}][\boldsymbol{x}_\mathrm{r} - \jmath\boldsymbol{x}_\mathrm{i}]^T\} \\ &= \left(\mathcal{E}\{\boldsymbol{x}_\mathrm{r}\boldsymbol{x}_\mathrm{r}^T\} + \mathcal{E}\{\boldsymbol{x}_\mathrm{i}\boldsymbol{x}_\mathrm{i}^T\}\right) + \jmath\left(\mathcal{E}\{\boldsymbol{x}_\mathrm{i}\boldsymbol{x}_\mathrm{r}^T\} - \mathcal{E}\{\boldsymbol{x}_\mathrm{r}\boldsymbol{x}_\mathrm{i}^T\}\right) \\ &= 2\mathbf{R}_x^\mathrm{E} + \jmath 2\mathbf{R}_x^\mathrm{O} \end{aligned} \tag{2.56}$$

where the last equality follows from (2.55). Thus all of the correlation terms present in (2.54) are also represented in the complex correlation matrix (2.56). In particular, the last equation implies that

$$\mathbf{R}_x^{\mathrm{E}} = \tfrac{1}{2}\mathrm{Re}\,[\mathbf{R}_x] \tag{2.57}$$

and

$$\mathbf{R}_x^{\mathrm{O}} = \tfrac{1}{2}\mathrm{Im}\,[\mathbf{R}_x] \tag{2.58}$$

The properties are summarized in Table 2.3.

TABLE 2.3 SYMMETRY RELATIONS ASSUMED IN DEFINING THE COMPLEX CORRELATION MATRIX

Symmetry relations	$\mathbf{R}_x^{\mathrm{E}} = \mathcal{E}\{\boldsymbol{x}_{\mathrm{r}}\boldsymbol{x}_{\mathrm{r}}^T\} = \mathcal{E}\{\boldsymbol{x}_{\mathrm{i}}\boldsymbol{x}_{\mathrm{i}}^T\}$	$\mathbf{R}_x^{\mathrm{O}} = -\mathcal{E}\{\boldsymbol{x}_{\mathrm{r}}\boldsymbol{x}_{\mathrm{i}}^T\} = \mathcal{E}\{\boldsymbol{x}_{\mathrm{i}}\boldsymbol{x}_{\mathrm{r}}^T\}$
Complex correlation matrix	$\mathbf{R}_x = \mathcal{E}\{\boldsymbol{x}\boldsymbol{x}^{*T}\} = 2\mathbf{R}_x^{\mathrm{E}} + \jmath 2\mathbf{R}_x^{\mathrm{O}}$	
Form of joint correlation matrix	$\mathcal{E}\left\{\begin{bmatrix} \boldsymbol{x}_{\mathrm{r}} \\ \boldsymbol{x}_{\mathrm{i}} \end{bmatrix}\begin{bmatrix} \boldsymbol{x}_{\mathrm{r}}^T & \boldsymbol{x}_{\mathrm{i}}^T \end{bmatrix}\right\} = \begin{bmatrix} \mathcal{E}\{\boldsymbol{x}_{\mathrm{r}}\boldsymbol{x}_{\mathrm{r}}^T\} & \mathcal{E}\{\boldsymbol{x}_{\mathrm{r}}\boldsymbol{x}_{\mathrm{i}}^T\} \\ \mathcal{E}\{\boldsymbol{x}_{\mathrm{i}}\boldsymbol{x}_{\mathrm{r}}^T\} & \mathcal{E}\{\boldsymbol{x}_{\mathrm{i}}\boldsymbol{x}_{\mathrm{i}}^T\} \end{bmatrix} = \begin{bmatrix} \mathbf{R}_x^{\mathrm{E}} & -\mathbf{R}_x^{\mathrm{O}} \\ \mathbf{R}_x^{\mathrm{O}} & \mathbf{R}_x^{\mathrm{E}} \end{bmatrix}$	

Incidentally, the need for the Hermitian transpose in the definition of the correlation matrix can now be seen by observing that for any complex random vector satisfying (2.55) the expectation of the outer product *without* the conjugate is

$$\boxed{\mathcal{E}\{\boldsymbol{x}\boldsymbol{x}^T\} = [\mathbf{0}] \qquad \text{(complex random vector)}} \tag{2.59}$$

The last expression follows from an expansion of terms similar to that in (2.56). In fact, (2.59) can be taken as a more concise statement of the symmetry conditions (2.55). To make (2.50) a useful definition *it is henceforth assumed that condition (2.59) holds for complex random vectors,* unless otherwise stated.

Now consider the quantity known as the *covariance matrix.* The covariance matrix is the set of second central moments (i.e., moments about the mean). It is defined by

$$\mathbf{C}_x = \mathcal{E}\{(\boldsymbol{x} - \mathbf{m}_x)(\boldsymbol{x} - \mathbf{m}_x)^{*T}\} \tag{2.60}$$

The form of the matrix is similar to (2.51) or (2.52) except that the elements are given by

$$\mathrm{c}_{kl} = \mathcal{E}\{(x_k - \mathrm{m}_k)(x_l - \mathrm{m}_l)^*\} \tag{2.61}$$

with

$$\mathrm{c}_{kk} = \mathcal{E}\left\{|x_k - \mathrm{m}_k|^2\right\} = \mathrm{Var}\,[x_k] \tag{2.62}$$

For reasons similar to those discussed for the correlation matrix, the terms in the complex covariance matrix represent all of the terms appearing in the real covariance matrix for the random vector $\mathbf{v}$ of (2.53).

Note from (2.50) and (2.60) that both the correlation matrix and the covariance matrix are Hermitian symmetric, that is,

$$\mathbf{R}_x = \mathbf{R}_x^{*T} \tag{2.63}$$

and

$$\mathbf{C}_x = \mathbf{C}_x^{*T} \tag{2.64}$$

The two matrices are also positive semidefinite; that is, they satisfy the relations

$$\mathbf{a}^{*T}\mathbf{R}_x\mathbf{a} \geq 0 \tag{2.65}$$

and

$$\mathbf{a}^{*T}\mathbf{C}_x\mathbf{a} \geq 0 \tag{2.66}$$

for *any* complex vector **a**. This follows because if **a** is any complex vector, the quadratic product in (2.65) can be written as

$$\mathbf{a}^{*T}\mathbf{R}_x\mathbf{a} = \mathbf{a}^{*T}\mathcal{E}\{\boldsymbol{x}\boldsymbol{x}^{*T}\}\,\mathbf{a} = \mathcal{E}\{|\boldsymbol{x}^{*T}\mathbf{a}|^2\}$$

which is always ≥ 0. In most practical situations the correlation matrix is strictly *positive definite*, which means that it satisfies (2.65) with strict inequality for any vector $\mathbf{a} \neq \mathbf{0}$. This is due in part to the unavoidable presence of noise, which adds at least a constant term to the diagonal of the matrix. In the theoretical analysis some correlation matrices are encountered that are positive semidefinite but not strictly positive definite. Similar statements can be made about the covariance matrix. To be technically correct we consistently use the term "positive semidefinite" when referring to the correlation or covariance matrix. You should realize, however, that in most cases the matrix is actually positive definite.

The covariance matrix and the correlation matrix can easily be related by noting that since the expectation E is a linear operator and $\mathbf{m}_x$ is a constant vector

$$\begin{aligned}
\mathcal{E}\{(\boldsymbol{x}-\mathbf{m}_x)(\boldsymbol{x}-\mathbf{m}_x)^{*T}\} &= \mathcal{E}\{\boldsymbol{x}\boldsymbol{x}^{*T} - \boldsymbol{x}\mathbf{m}_x^{*T} - \mathbf{m}_x\boldsymbol{x}^{*T} + \mathbf{m}_x\mathbf{m}_x^{*T}\} \\
&= \mathcal{E}\{\boldsymbol{x}\boldsymbol{x}^{*T}\} - \mathcal{E}\{\boldsymbol{x}\}\,\mathbf{m}_x^{*T} - \mathbf{m}_x\mathcal{E}\{\boldsymbol{x}^{*T}\} + \mathbf{m}_x\mathbf{m}_x^{*T} \\
&= \mathcal{E}\{\boldsymbol{x}\boldsymbol{x}^{*T}\} - \mathbf{m}_x\mathbf{m}_x^{*T}
\end{aligned}$$

Thus it follows from the definitions of the correlation and covariance matrices that

$$\mathbf{R}_x = \mathbf{C}_x + \mathbf{m}_x\mathbf{m}_x^{*T} \tag{2.67}$$

This relation is handy when the correlation matrix is known and we need to compute the covariance, or vice versa.

A fundamental property of the expectation is that it is independent of transformations applied to the random variables. That is, if ψ in (2.46) represents a vector-valued *function* of $\boldsymbol{x}$, for example, if

$$\boldsymbol{y} = \mathbf{g}(\boldsymbol{x}) \tag{2.68}$$

then it can be shown that

$$\int_{-\infty}^{\infty} \mathbf{y} f_y(\mathbf{y})d\mathbf{y} = \int_{-\infty}^{\infty} \mathbf{g}(\mathbf{x}) f_x(\mathbf{x})d\mathbf{x} \tag{2.69}$$

Thus, the notation $\mathcal{E}\{\boldsymbol{y}\}$ represents the same quantity as the notation $\mathcal{E}\{\mathbf{g}(\boldsymbol{x})\}$, although the former implies that the expectation is carried out using the density of $\boldsymbol{y}$ while the later implies that the expectation is carried out using the density of $\boldsymbol{x}$.

2.2.2 Joint and Conditional Expectations for Two Random Vectors

When a quantity ψ depends on *two* random vectors $\boldsymbol{x}$ and $\boldsymbol{y}$, its expectation is defined by

$$\mathcal{E}\{\psi(\boldsymbol{x},\boldsymbol{y})\} = \int_{-\infty}^{\infty}\int_{-\infty}^{\infty} \psi(\mathbf{x},\mathbf{y}) f_{\boldsymbol{xy}}(\mathbf{x},\mathbf{y})d\mathbf{y}d\mathbf{x} \tag{2.70}$$

The joint expectations that will be found to be most useful are the *cross-correlation* matrix

$$\mathbf{R}_{\boldsymbol{xy}} = \mathcal{E}\{\boldsymbol{x}\boldsymbol{y}^{*T}\} \tag{2.71}$$

and the *cross-covariance* matrix

$$\mathbf{C}_{\boldsymbol{xy}} = \mathcal{E}\{(\boldsymbol{x}-\mathbf{m}_{\boldsymbol{x}})(\boldsymbol{y}-\mathbf{m}_{\boldsymbol{y}})^{*T}\} \tag{2.72}$$

As in the case of a single random vector, these matrices can be shown to carry all the necessary information about the second moment properties of the random vectors if the appropriate symmetry conditions apply (see Problem 2.9). Neither $\mathbf{R}_{\boldsymbol{xy}}$ nor $\mathbf{C}_{\boldsymbol{xy}}$ is a square matrix unless the dimensions of $\boldsymbol{x}$ and $\boldsymbol{y}$ are the same. By a procedure analogous to the one used to derive (2.67) it can be shown that

$$\mathbf{R}_{\boldsymbol{xy}} = \mathbf{C}_{\boldsymbol{xy}} + \mathbf{m}_{\boldsymbol{x}}\mathbf{m}_{\boldsymbol{y}}^{*T} \tag{2.73}$$

Although the matrices $\mathbf{R}_{\boldsymbol{xy}}$ and $\mathbf{C}_{\boldsymbol{xy}}$ in general exhibit no symmetry, it is true from the definitions that

$$\mathbf{R}_{\boldsymbol{xy}} = \mathbf{R}_{\boldsymbol{yx}}^{*T} \tag{2.74}$$

and

$$\mathbf{C}_{\boldsymbol{xy}} = \mathbf{C}_{\boldsymbol{yx}}^{*T} \tag{2.75}$$

The random vectors $\boldsymbol{x}$ and $\boldsymbol{y}$ are said to be *uncorrelated* if

$$\mathbf{R}_{\boldsymbol{xy}} = \mathcal{E}\{\boldsymbol{x}\boldsymbol{y}^{*T}\} = \mathcal{E}\{\boldsymbol{x}\}\,\mathcal{E}\{\boldsymbol{y}^{*T}\} = \mathbf{m}_{\boldsymbol{x}}\mathbf{m}_{\boldsymbol{y}}^{*T} \tag{2.76}$$

Because of (2.73), it is equivalent to say that the random vectors are uncorrelated if

$$\mathbf{C}_{\boldsymbol{xy}} = \mathcal{E}\{(\boldsymbol{x}-\mathbf{m}_{\boldsymbol{x}})(\boldsymbol{y}-\mathbf{m}_{\boldsymbol{y}})^{*T}\} = [\mathbf{0}] \tag{2.77}$$

(i.e., if their cross-*covariance* is zero). This is the usual definition. Two vectors are said to be *orthogonal* if

$$\mathbf{R}_{\boldsymbol{xy}} = \mathcal{E}\{\boldsymbol{x}\boldsymbol{y}^{*T}\} = [\mathbf{0}] \tag{2.78}$$

(i.e., if the cross-*correlation* is zero). The use of the term "uncorrelated" when the cross-*covariance* is zero is unfortunate but conventional. However (2.73) shows that when the mean of *either* random vector is zero, the cross-correlation and the cross-covariance matrices are identical. In this case orthogonal random vectors are also uncorrelated. Because of the definition (2.76) it is easy to show that if $\boldsymbol{x}$ and $\boldsymbol{y}$ are *independent* random vectors, then $\boldsymbol{x}$ and $\boldsymbol{y}$ are uncorrelated. *The converse of this statement is generally not true.*

The concepts of orthogonal and uncorrelated are defined in an analogous way for the *components* x_k of a random vector $\boldsymbol{x}$. Specifically, the components are said to be orthogonal if

$$\mathcal{E}\{x_k x_l^*\} = 0; \qquad k \neq l \tag{2.79}$$

and they are uncorrelated if

$$\mathcal{E}\{(x_k - \mathrm{m}_k)(x_l - \mathrm{m}_l)^*\} = 0; \qquad k \neq l \tag{2.80}$$

Given a random vector $\boldsymbol{x}$ with any distribution, linear transformations can be found to a space where the vector components are orthogonal and/or uncorrelated. Procedures for doing this are developed in Section 2.6.

The correlation matrix of the *sum* of two random vectors $\boldsymbol{x}$ and $\boldsymbol{y}$ that have the same dimension is given by

$$\begin{aligned}
\mathbf{R}_{x+y} &= \mathcal{E}\{(\boldsymbol{x}+\boldsymbol{y})(\boldsymbol{x}+\boldsymbol{y})^{*T}\} \\
&= \mathcal{E}\{\boldsymbol{x}\boldsymbol{x}^{*T} + \boldsymbol{x}\boldsymbol{y}^{*T} + \boldsymbol{y}\boldsymbol{x}^{*T} + \boldsymbol{y}\boldsymbol{y}^{*T}\} \\
&= \mathbf{R}_x + \mathbf{R}_{xy} + \mathbf{R}_{yx} + \mathbf{R}_y
\end{aligned} \tag{2.81}$$

In a similar manner the covariance matrix for a sum of random vectors is given by

$$\mathbf{C}_{x+y} = \mathbf{C}_x + \mathbf{C}_{xy} + \mathbf{C}_{yx} + \mathbf{C}_y \tag{2.82}$$

If $\boldsymbol{x}$ and $\boldsymbol{y}$ are orthogonal so $\mathbf{R}_{xy} = \mathbf{R}_{yx} = [\mathbf{0}]$, then

$$\mathbf{R}_{x+y} = \mathbf{R}_x + \mathbf{R}_y \qquad (\boldsymbol{x}, \boldsymbol{y} \text{ orthogonal}) \tag{2.83}$$

Likewise if $\boldsymbol{x}$ and $\boldsymbol{y}$ are uncorrelated so $\mathbf{C}_{xy} = \mathbf{C}_{yx} = [\mathbf{0}]$, then

$$\mathbf{C}_{x+y} = \mathbf{C}_x + \mathbf{C}_y \qquad (\boldsymbol{x}, \boldsymbol{y} \text{ uncorrelated}) \tag{2.84}$$

This last equation is the generalization of the result for random variables that states that

$$\mathrm{Var}\,[x+y] = \mathrm{Var}\,[x] + \mathrm{Var}\,[y] \tag{2.85}$$

when x and y are uncorrelated.

A summary of some of the most important results about correlation and covariance matrices is given in Table 2.4. Note that since the covariance of data is the correlation with the mean removed, all properties true for the correlation matrix are also true for the covariance matrix.

Before leaving this section it is worthwhile to explore one other form of expectation, namely conditional expectation. The conditional expectation $\mathcal{E}\{\psi(\boldsymbol{x},\boldsymbol{y})|\boldsymbol{x}\}$ is defined as

$$\mathcal{E}\{\psi(\boldsymbol{x},\boldsymbol{y})|\boldsymbol{x}\} = \int_{-\infty}^{\infty} \psi(\mathbf{x},\mathbf{y}) f_{\boldsymbol{y}|\boldsymbol{x}}(\mathbf{y}|\mathbf{x}) d\mathbf{y} \tag{2.86}$$

TABLE 2.4 DEFINITIONS AND PROPERTIES FOR CORRELATION AND COVARIANCE MATRICES

(Auto)correlation and covariance	$\mathbf{R}_x = \mathcal{E}\{\boldsymbol{x}\boldsymbol{x}^{*T}\}$	$\mathbf{C}_x = \mathcal{E}\{(\boldsymbol{x}-\mathbf{m})(\boldsymbol{x}-\mathbf{m})^{*T}\}$
Symmetry	$\mathbf{R}_x = \mathbf{R}_x^{*T}$	$\mathbf{C}_x = \mathbf{C}_x^{*T}$
Positive (semi)definite property	$\mathbf{a}^{*T}\mathbf{R}_x\mathbf{a} \geq 0$ for any vector $\mathbf{a}$	$\mathbf{a}^{*T}\mathbf{C}_x\mathbf{a} \geq 0$ for any vector $\mathbf{a}$
Interrelation	$\mathbf{R}_x = \mathbf{C}_x + \mathbf{m}_x\mathbf{m}_x^{*T}$	
Cross-correlation and cross-covariance	$\mathbf{R}_{xy} = \mathcal{E}\{\boldsymbol{x}\boldsymbol{y}^{*T}\}$	$\mathbf{C}_{xy} = \mathcal{E}\{(\boldsymbol{x}-\mathbf{m}_x)(\boldsymbol{y}-\mathbf{m}_y)^{*T}\}$
Relation to $\mathbf{R}_{yx}$ and $\mathbf{C}_{yx}$	$\mathbf{R}_{xy} = \mathbf{R}_{yx}^{*T}$	$\mathbf{C}_{xy} = \mathbf{C}_{yx}^{*T}$
Interrelation	$\mathbf{R}_{xy} = \mathbf{C}_{xy} + \mathbf{m}_x\mathbf{m}_y^{*T}$	
Orthogonal and uncorrelated	$\boldsymbol{x}, \boldsymbol{y}$ orthogonal: $\mathbf{R}_{xy} = [\mathbf{0}]$	$\boldsymbol{x}, \boldsymbol{y}$ uncorrelated: $\mathbf{C}_{xy} = [\mathbf{0}]$
Sum of $\boldsymbol{x}$ and $\boldsymbol{y}$	$\mathbf{R}_{x+y} = \mathbf{R}_x + \mathbf{R}_y$ if $\boldsymbol{x}, \boldsymbol{y}$ orthogonal	$\mathbf{C}_{x+y} = \mathbf{C}_x + \mathbf{C}_y$ if $\boldsymbol{x}, \boldsymbol{y}$ uncorrelated

Note that this quantity is a *function* of $\boldsymbol{x}$, and since $\boldsymbol{x}$ is a random vector, the following chain rule applies

$$\boxed{\mathcal{E}\{\psi(\boldsymbol{x},\boldsymbol{y})\} = \mathcal{E}\{\mathcal{E}\{\psi(\boldsymbol{x},\boldsymbol{y})|\boldsymbol{x}\}\}} \tag{2.87}$$

Equation (2.87) holds because the right side can be written as

$$\int_{-\infty}^{\infty}\left(\int_{-\infty}^{\infty}\psi(\mathbf{x},\mathbf{y})f_{y|x}(\mathbf{y}|\mathbf{x})d\mathbf{y}\right)f_x(\mathbf{x})d\mathbf{x}$$

$$= \int_{-\infty}^{\infty}\int_{-\infty}^{\infty}\psi(\mathbf{x},\mathbf{y})f_{xy}(\mathbf{x},\mathbf{y})d\mathbf{y}d\mathbf{x} \tag{2.88}$$

It will be of particular interest later when $\boldsymbol{x}$ and $\boldsymbol{y}$ are jointly distributed to compute specific conditional expectations such as $\mathcal{E}\{\boldsymbol{y}|\boldsymbol{x}\}$. In fact, this last expression is a quantity that can be used to *estimate* $\boldsymbol{y}$ from $\boldsymbol{x}$.

2.2.3 Estimating Expected Values from Data

The computation of statistical moments and related quantities from data is the subject of estimation theory and is treated at some length in Chapter 6. For the present, however, it is useful to know how to estimate simple expected values when the density is not known but instead there are a number of samples $\mathbf{x}^{(1)}, \mathbf{x}^{(2)}, \ldots, \mathbf{x}^{(K)}$ of the random vector $\boldsymbol{x}$. To illustrate how this is done, consider the simple case of a real random vector in one dimension and assume that it is desired to estimate the expected value of the the quantity $\psi(\mathrm{x})$.

Figure 2.3(a) shows a histogram constructed from the samples with intervals $\Delta\mathrm{x}$ wide and Fig. 2.3(b) shows the actual (unknown) probability density $f_x(\mathrm{x})$. Let K_{x^o} be the number of samples in the interval $\mathrm{x}^o < \mathrm{x} \leq \mathrm{x}^o + \Delta\mathrm{x}$. If the total number of samples K is sufficiently large and the interval size $\Delta\mathrm{x}$ is sufficiently small, then both K_{x^o}/K and $f_x(\mathrm{x}^o)\Delta\mathrm{x}$ are good estimates of the probability that x is in the given interval. (This fact can be shown rigorously.) Thus it follows that

$$f_x(\mathrm{x}^o)\Delta\mathrm{x} \approx \frac{K_{\mathrm{x}^o}}{K} \tag{2.89}$$

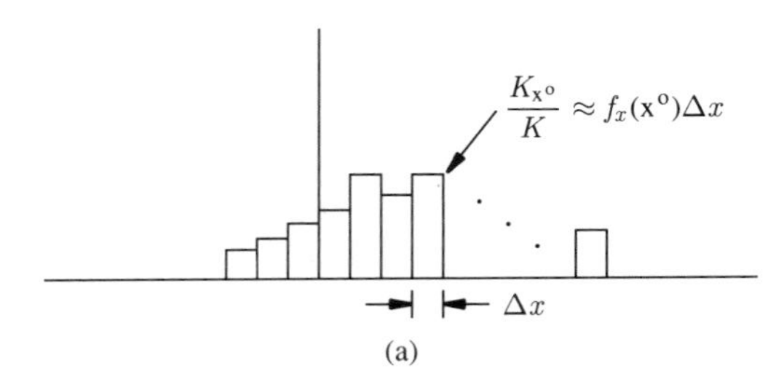

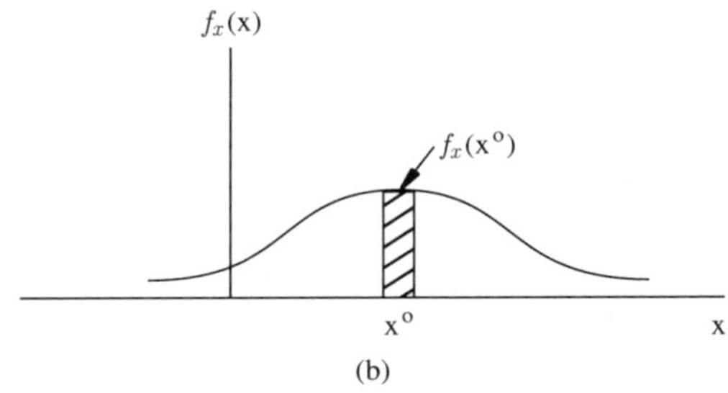

Figure 2.3 Comparison of histogram of random variable to probability density function. (a) Normalized histogram. (b) Corresponding density function. (From Charles W. Therrien, *Decision Estimation and Classification*, copyright © John Wiley & Sons, New York, 1989. Reproduced by permission.)

If both $f_x(\mathrm{x})$ and $\psi(\mathrm{x})$ do not vary appreciably over each interval, then the expectation (2.46) can be approximated by

$$\mathcal{E}\{\psi(x)\} \simeq \sum_{\text{all } \mathrm{x}^o} \psi(\mathrm{x}^o) f_x(\mathrm{x}^o)\Delta\mathrm{x} \simeq \frac{1}{K}\sum_{\text{all } \mathrm{x}^o} \psi(\mathrm{x}^o) K_{\mathrm{x}^o} \tag{2.90}$$

Next note that the last summation has a contribution equal to $K_{\mathrm{x}^o}\psi(\mathrm{x}^o)$ corresponding to the K_{x^o} samples in the interval $\mathrm{x}^o < \mathrm{x} \leq \mathrm{x}^o + \Delta\mathrm{x}$. This contribution can be replaced simply by summing $\psi(\mathrm{x}^{(i)})$ over all samples in the interval, assuming that for these samples $\psi(\mathrm{x}^{(i)}) \simeq \psi(\mathrm{x}^o)$. If this procedure is followed for all such intervals, then (2.90) can be written as

$$\mathcal{E}\{\psi(\boldsymbol{x})\} \simeq \frac{1}{K}\sum_{k=1}^{K} \psi(\mathbf{x}^{(k)}) \tag{2.91}$$

which has now been expressed in vector form since the general result for vectors follows from analogous arguments.

A similar result holds when estimating the expectation of a quantity involving two random vectors from samples $\mathbf{x}^{(1)}, \mathbf{x}^{(2)}, \ldots, \mathbf{x}^{(K)}$ and $\mathbf{y}^{(1)}, \mathbf{y}^{(2)}, \ldots, \mathbf{y}^{(K)}$. The estimate for the expectation takes the form

$$\mathcal{E}\{\psi(\boldsymbol{x}, \boldsymbol{y})\} \simeq \frac{1}{K}\sum_{k=1}^{K} \psi(\mathbf{x}^{(k)}, \mathbf{y}^{(k)}) \tag{2.92}$$

The following example illustrates these computations.

Example 2.7

Given four real sample vectors

$$\mathbf{x}^{(1)} = \begin{bmatrix} 1 \\ 0 \end{bmatrix}, \quad \mathbf{x}^{(2)} = \begin{bmatrix} -2 \\ 1 \end{bmatrix}, \quad \mathbf{x}^{(3)} = \begin{bmatrix} 2 \\ -2 \end{bmatrix}, \quad \mathbf{x}^{(4)} = \begin{bmatrix} 0 \\ 2 \end{bmatrix}$$

an estimate of the mean vector can be computed as

$$\hat{\mathbf{m}} = \frac{1}{K}\sum_{k=1}^{K} \mathbf{x}^{(k)} = \frac{1}{4}\left\{ \begin{bmatrix} 1 \\ 0 \end{bmatrix} + \begin{bmatrix} -2 \\ 1 \end{bmatrix} + \begin{bmatrix} 2 \\ -2 \end{bmatrix} + \begin{bmatrix} 0 \\ 2 \end{bmatrix} \right\} = \begin{bmatrix} \frac{1}{4} \\ \frac{1}{4} \end{bmatrix}$$

The estimated correlation matrix is

$$\begin{aligned}
\hat{\mathbf{R}}_x &= \frac{1}{K}\sum_{k=1}^{K} \mathbf{x}^{(k)}(\mathbf{x}^{(k)})^T \\
&= \frac{1}{4}\left\{ \begin{bmatrix} 1 \\ 0 \end{bmatrix}\begin{bmatrix} 1 & 0 \end{bmatrix} + \begin{bmatrix} -2 \\ 1 \end{bmatrix}\begin{bmatrix} -2 & 1 \end{bmatrix} + \begin{bmatrix} 2 \\ -2 \end{bmatrix}\begin{bmatrix} 2 & -2 \end{bmatrix} + \begin{bmatrix} 0 \\ 2 \end{bmatrix}\begin{bmatrix} 0 & 2 \end{bmatrix} \right\} \\
&= \begin{bmatrix} \frac{9}{4} & -\frac{3}{2} \\ -\frac{3}{2} & \frac{9}{4} \end{bmatrix}
\end{aligned}$$

Further, by computing an estimated covariance matrix for this example, it can be shown that the *estimates* also satisfy (2.67) (see Problem 2.10).

The estimate for a correlation matrix can be computed in a somewhat more convenient way by defining the matrix of sample vectors

$$\mathbf{X} = \begin{bmatrix} - & \mathbf{x}^{(1)*T} & - \\ - & \mathbf{x}^{(2)*T} & - \\ & \vdots & \\ & \vdots & \\ - & \mathbf{x}^{(K)*T} & - \end{bmatrix} \tag{2.93}$$

The matrix $\mathbf{X}$ will be referred to as the *data matrix*. Then since

$$\mathbf{X}^{*T}\mathbf{X} = \sum_{k=1}^{K} \mathbf{x}^{(k)}\mathbf{x}^{(k)*T}$$

the estimate for the correlation matrix can be written as

$$\hat{\mathbf{R}}_x = \frac{1}{K}\mathbf{X}^{*T}\mathbf{X} \tag{2.94}$$

The *columns* of the data matrix also play an important role in its use. If the matrix is partitioned along columns and written as

$$\mathbf{X} = \begin{bmatrix} | & | & & | \\ | & | & & | \\ \mathbf{x}_1 & \mathbf{x}_2 & \cdots & \mathbf{x}_N \\ | & | & & | \\ | & | & & | \end{bmatrix} \tag{2.95}$$

then a typical element $\hat{\mathrm{r}}_{kl}$ of the estimated correlation matrix is given by

$$\hat{\mathrm{r}}_{kl} = \frac{1}{K}\mathbf{x}_k^{*T}\mathbf{x}_l \tag{2.96}$$

In the subsequent discussions it is assumed that the correlation matrix is always estimated according to (2.94). The estimate for the covariance matrix can be defined similarly by first removing the mean for each sample vector. The following example illustrates the computation.

Example 2.8

The correlation matrix for Example 2.7 is computed here using (2.94). The data matrix is first formed as

$$\mathbf{X} = \begin{bmatrix} 1 & 0 \\ -2 & 1 \\ 2 & -2 \\ 0 & 2 \end{bmatrix}$$

Then the estimated correlation matrix is computed as

$$\hat{\mathbf{R}}_x = \frac{1}{4}\begin{bmatrix} 1 & -2 & 2 & 0 \\ 0 & 1 & -2 & 2 \end{bmatrix}\begin{bmatrix} 1 & 0 \\ -2 & 1 \\ 2 & -2 \\ 0 & 2 \end{bmatrix} = \frac{1}{4}\begin{bmatrix} 9 & -6 \\ -6 & 9 \end{bmatrix} = \begin{bmatrix} \frac{9}{4} & -\frac{3}{2} \\ -\frac{3}{2} & \frac{9}{4} \end{bmatrix}$$

This procedure is a bit simpler and more direct than the procedure of Example 2.7.

2.3 THE MULTIVARIATE GAUSSIAN DENSITY FUNCTION

Gaussian random sequences and Gaussian random vectors play a very important role in the design and analysis of signal processing systems. A Gaussian random vector is one characterized by a multivariate Normal or Gaussian density function. For a *real* random vector this density function has the form

$$\boxed{f_{\boldsymbol{x}}(\mathbf{x}) = \frac{1}{(2\pi)^{N/2}|\mathbf{C}_{\boldsymbol{x}}|^{1/2}} e^{-\frac{1}{2}(\mathbf{x}-\mathbf{m}_{\boldsymbol{x}})^T \mathbf{C}_{\boldsymbol{x}}^{-1}(\mathbf{x}-\mathbf{m}_{\boldsymbol{x}})} \quad \text{(real random vector)}} \tag{2.97}$$

where N is the dimension of $\boldsymbol{x}$. This density function is completely characterized by the mean vector $\mathbf{m}_{\boldsymbol{x}}$ and the covariance matrix $\mathbf{C}_{\boldsymbol{x}}$; it is the generalization of the one-dimensional Gaussian density for real random variables

$$f_x(\mathrm{x}) = \frac{1}{\sqrt{2\pi\sigma_x^2}} e^{-\frac{(\mathrm{x}-\mathrm{m})^2}{2\sigma_x^2}} \tag{2.98}$$

For a complex random vector the Gaussian density has the slightly different form

$$\boxed{f_{\boldsymbol{x}}(\mathbf{x}) = \frac{1}{\pi^N|\mathbf{C}_{\boldsymbol{x}}|} e^{-(\mathbf{x}-\mathbf{m}_{\boldsymbol{x}})^{*T} \mathbf{C}_{\boldsymbol{x}}^{-1}(\mathbf{x}-\mathbf{m}_{\boldsymbol{x}})} \quad \text{(complex random vector)}} \tag{2.99}$$

In the scalar form the complex Gaussian density is therefore

$$f_x(\mathrm{x}) = \frac{1}{\pi\sigma_x^2} e^{-\frac{(\mathrm{x}-\mathrm{m})^2}{\sigma_x^2}} \tag{2.100}$$

where in this case x is assumed to be a *complex* random variable.

Note that (2.97) is not just a special case of (2.99) with $\mathrm{Im}[\boldsymbol{x}] = \mathbf{0}$. In fact, it is just the other way around; (2.97) is more fundamental. As seen earlier the density function for a complex random vector is actually a joint density between the real and imaginary parts. For the Gaussian case the density of the real random vector $\boldsymbol{v}$ defined by (2.53) can be written in the form (2.97). Then when symmetry conditions analogous to (2.55) are met that permit the definition (2.60) to hold, it turns out that (2.99) is an equivalent expression. Problem 2.16 explores this point in detail.

When two random vectors $\boldsymbol{x}$ and $\boldsymbol{y}$ of dimensions N and M are *jointly* Gaussian, their joint density can be written by forming the larger vector

$$\boldsymbol{u} = \begin{bmatrix} \boldsymbol{x} \\ \boldsymbol{y} \end{bmatrix} \tag{2.101}$$

and the corresponding mean vector and covariance matrix

$$\mathbf{m}_{\boldsymbol{u}} = \mathcal{E}\left\{\begin{bmatrix} \boldsymbol{x} \\ \boldsymbol{y} \end{bmatrix}\right\} = \begin{bmatrix} \mathbf{m}_{\boldsymbol{x}} \\ \mathbf{m}_{\boldsymbol{y}} \end{bmatrix} \tag{2.102}$$

$$\mathbf{C}_u = \mathcal{E}\left\{ \begin{bmatrix} x - \mathbf{m}_x \\ y - \mathbf{m}_y \end{bmatrix} \begin{bmatrix} (x - \mathbf{m}_x)^{*T} & (y - \mathbf{m}_y)^{*T} \end{bmatrix} \right\}$$
$$= \begin{bmatrix} \mathbf{C}_x & \mathbf{C}_{xy} \\ \mathbf{C}_{xy}^{*T} & \mathbf{C}_y \end{bmatrix} \tag{2.103}$$

and then writing

$$f_{xy}(\mathbf{x}, \mathbf{y}) = f_u(\mathbf{u}) = \frac{1}{(2\pi)^{(N+M)/2}|\mathbf{C}_u|^{1/2}} e^{-\frac{1}{2}(\mathbf{u}-\mathbf{m}_u)^T \mathbf{C}_u^{-1}(\mathbf{u}-\mathbf{m}_u)} \tag{2.104}$$

(real random vectors)

and

$$f_{xy}(\mathbf{x}, \mathbf{y}) = f_u(\mathbf{u}) = \frac{1}{\pi^{N+M}|\mathbf{C}_u|} e^{-(\mathbf{u}-\mathbf{m}_u)^{*T} \mathbf{C}_u^{-1}(\mathbf{u}-\mathbf{m}_u)} \tag{2.105}$$

(complex random vectors)

From this it can be shown (see Problem 2.20) that the marginal densities f_x and f_y and the conditional densities $f_{x|y}$ and $f_{y|x}$ are also Gaussian. The latter have the form

$$f_{y|x}(\mathbf{y}|\mathbf{x}) = \frac{1}{(2\pi)^{M/2}|\mathbf{C}_{y|x}|^{1/2}} e^{-\frac{1}{2}(\mathbf{y}-\mathbf{m}_{y|x})^T \mathbf{C}_{y|x}^{-1}(\mathbf{y}-\mathbf{m}_{y|x})} \quad \text{(real random vectors)} \tag{2.106}$$

and

$$f_{y|x}(\mathbf{y}|\mathbf{x}) = \frac{1}{\pi^{M}|\mathbf{C}_{y|x}|} e^{-(\mathbf{y}-\mathbf{m}_{y|x})^{*T} \mathbf{C}_{y|x}^{-1}(\mathbf{y}-\mathbf{m}_{y|x})} \quad \text{(complex random vectors)} \tag{2.107}$$

where

$$\mathbf{m}_{y|x} = \mathbf{m}_y + \mathbf{C}_{xy}^{*T}\mathbf{C}_x^{-1}(\mathbf{x} - \mathbf{m}_x) \tag{2.108}$$

and

$$\mathbf{C}_{y|x} = \mathbf{C}_y - \mathbf{C}_{xy}^{*T}\mathbf{C}_x^{-1}\mathbf{C}_{xy} \tag{2.109}$$

in both cases.

2.4 LINEAR TRANSFORMATIONS OF RANDOM VECTORS

Since linear systems represent such an important class of signal processing systems, linear transformations of random vectors representing those signals are correspondingly important. This section describes how the first and second moment properties of random vectors are affected by a linear transformation. It further specifies the general effect that a linear transformation has on the density function. For Gaussian random vectors it turns out that the *form* of the density is unchanged. That is, *a linear transformation of a Gaussian random vector remains Gaussian.*

To begin, consider a random vector $\boldsymbol{y}$ defined by a linear transformation of the random vector $\boldsymbol{x}$:

$$\boldsymbol{y} = \mathbf{A}\boldsymbol{x} \tag{2.110}$$

The matrix $\mathbf{A}$ need not be a square matrix; that is, $\boldsymbol{x}$ can have any dimension N and $\boldsymbol{y}$ can have any dimension M. It is occasionally necessary to work with the norm of the matrix $\mathbf{A}$. The Euclidean norm, induced by the Euclidean norm of the vector, is defined by

$$\|\mathbf{A}\| \stackrel{\text{def}}{=} \max_{\|\mathbf{x}\|=1} \|\mathbf{Ax}\| \tag{2.111}$$

while the Frobenius norm is defined by

$$\|\mathbf{A}\|_F \stackrel{\text{def}}{=} \left(\sum_{i=1}^{M} \sum_{j=1}^{N} |a_{ij}|^2 \right)^{1/2} = \left(\text{tr } \mathbf{A}\mathbf{A}^{*T}\right)^{1/2} \tag{2.112}$$

Both forms are used in later chapters.

Given the transformation (2.110), the first and second moment quantities transform in a simple way. In particular, since the expectation is a linear operator, it follows that

$$\mathcal{E}\{\boldsymbol{y}\} = \mathcal{E}\{\mathbf{A}\boldsymbol{x}\} = \mathbf{A}\mathcal{E}\{\boldsymbol{x}\}$$

or

$$\boxed{\mathbf{m}_y = \mathbf{A}\mathbf{m}_x} \tag{2.113}$$

In a similar fashion,

$$\mathcal{E}\{\boldsymbol{y}\boldsymbol{y}^{*T}\} = \mathcal{E}\{(\mathbf{A}\boldsymbol{x})(\mathbf{A}\boldsymbol{x})^{*T}\} = \mathbf{A}\mathcal{E}\{\boldsymbol{x}\boldsymbol{x}^{*T}\}\,\mathbf{A}^{*T}$$

or

$$\boxed{\mathbf{R}_y = \mathbf{A}\mathbf{R}_x\mathbf{A}^{*T}} \tag{2.114}$$

By using (2.67), (2.113), and (2.114), a corresponding relation can be derived for the covariance matrix, namely

$$\boxed{\mathbf{C}_y = \mathbf{A}\mathbf{C}_x\mathbf{A}^{*T}} \tag{2.115}$$

Then from properties of the determinant and trace it follows that

$$|\mathbf{R}_y| = |\mathbf{A}||\mathbf{R}_x||\mathbf{A}^{*T}| = |\mathbf{A}^{*T}||\mathbf{A}||\mathbf{R}_x| = |\mathbf{A}|^*|\mathbf{A}||\mathbf{R}_x| \tag{2.116}$$

and

$$\text{tr } \mathbf{R}_y = \text{tr } \mathbf{A}\mathbf{R}_x\mathbf{A}^{*T} = \text{tr } \mathbf{A}^{*T}\mathbf{A}\mathbf{R}_x \tag{2.117}$$

The expression that appears on the right of (2.116) is the magnitude squared of the determinant. Since $\mathbf{A}$ is possibly complex, it is important not to write the product $|\mathbf{A}|^*|\mathbf{A}|$ as

$|\mathbf{A}|^2$ or (worse) as $||\mathbf{A}||^2$, which looks like the squared *norm*. Similar relations exist for the covariance matrix, namely

$$|\mathbf{C}_y| = |\mathbf{A}|^*|\mathbf{A}||\mathbf{C}_x| \tag{2.118}$$

and

$$\text{tr } \mathbf{C}_y = \text{tr } \mathbf{A}^{*T}\mathbf{A}\mathbf{C}_x \tag{2.119}$$

These relations lead to an important invariance property when $\mathbf{A}$ is a *unitary* matrix[5]. For a unitary matrix the product $\mathbf{A}^{*T}\mathbf{A}$ is equal to the identity matrix (i.e., $\mathbf{A}^{*T} = \mathbf{A}^{-1}$) and the determinant $|\mathbf{A}|$ is equal to 1. In this case *the determinant and trace of the correlation and covariance matrices remain unchanged.*

Linear transformations of random vectors also lead to relatively simple transformations of the probability density function if the transformation is nonsingular. To show this, assume first that $\boldsymbol{x}$ and $\boldsymbol{y}$ are both real and that $\boldsymbol{x}$, which lies in the space $\mathcal{X}$, is mapped to $\boldsymbol{y}$ in $\mathcal{Y}$ according to the general transformation (2.68). Assume further that the inverse transformation

$$\boldsymbol{x} = \mathbf{g}^{-1}(\boldsymbol{y}) \tag{2.120}$$

exists and that $\boldsymbol{x}$ lies in a small rectangular region depicted in Fig. 2.4 (for $N = 2$) with volume

$$\Delta V_x = \Delta \mathrm{x}_1 \Delta \mathrm{x}_2 \ldots \Delta \mathrm{x}_N \tag{2.121}$$

The rectangular region in $\mathcal{X}$ then maps into a generally nonrectangular region in $\mathcal{Y}$ whose volume is given by

$$\Delta V_y = J(\mathbf{y}^o, \mathbf{x}^o)\Delta V_x \tag{2.122}$$

where $J(\mathbf{y}^o, \mathbf{x}^o)$ is the Jacobian of the transformation

$$J(\mathbf{y}^o, \mathbf{x}^o) \stackrel{\text{def}}{=} \text{abs} \left| \begin{matrix} \frac{\partial g_1(\mathbf{x})}{\partial \mathrm{x}_1} & \frac{\partial g_2(\mathbf{x})}{\partial \mathrm{x}_1} & \cdots & \frac{\partial g_N(\mathbf{x})}{\partial \mathrm{x}_1} \\ \frac{\partial g_1(\mathbf{x})}{\partial \mathrm{x}_2} & \frac{\partial g_2(\mathbf{x})}{\partial \mathrm{x}_2} & \cdots & \frac{\partial g_N(\mathbf{x})}{\partial \mathrm{x}_2} \\ \vdots & \vdots & & \vdots \\ \frac{\partial g_1(\mathbf{x})}{\partial \mathrm{x}_N} & \frac{\partial g_2(\mathbf{x})}{\partial \mathrm{x}_N} & \cdots & \frac{\partial g_N(\mathbf{x})}{\partial \mathrm{x}_N} \end{matrix} \right|_{\mathbf{x}=\mathbf{x}^o, \mathbf{y}=\mathbf{y}^o} \tag{2.123}$$

("abs" denotes absolute value of the determinant, which could be negative) and the $g_i(\mathbf{x})$ are components of the vector function $\mathbf{g}(\mathbf{x})$. Since the event that $\boldsymbol{x}$ is in the small region of $\mathcal{X}$ is the same as the event that $\boldsymbol{y}$ is in the corresponding small region of $\mathcal{Y}$, the two probabilities can be equated to obtain

$$f_x(\mathbf{x}^o)\Delta V_x = f_y(\mathbf{y}^o)\Delta V_y = f_y(\mathbf{y}^o)J(\mathbf{y}^o, \mathbf{x}^o)\Delta V_x \tag{2.124}$$

Since the region with volume ΔV_x is small but arbitrary, this result must be independent of ΔV_x and therefore it follows that

$$f_y(\mathbf{y}) = \frac{1}{J(\mathbf{y}, \mathbf{x})} f_x(\mathbf{g}^{-1}(\mathbf{y})) \tag{2.125}$$

[5]When $\mathbf{A}$ is real it is said to be an *orthogonal* matrix. We use the single term *unitary* throughout to refer to both cases.

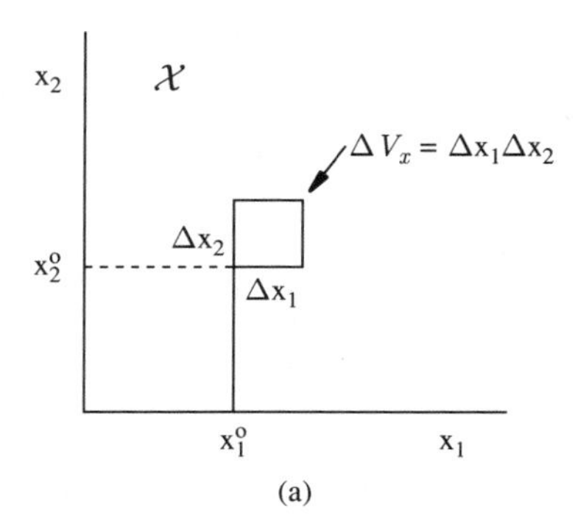

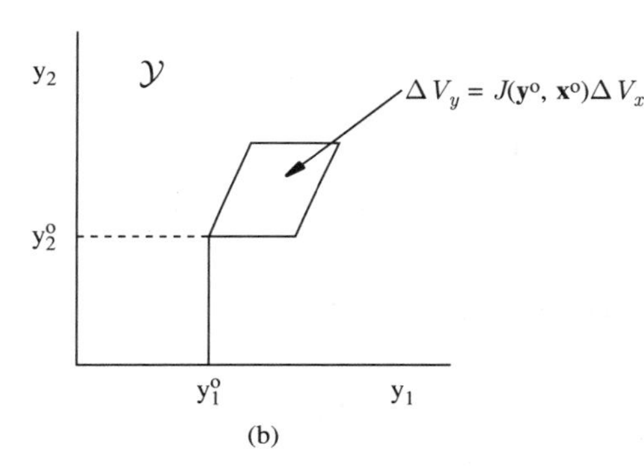

Figure 2.4 Mapping of small region of $\mathcal{X}$ space to $\mathcal{Y}$ space. (a) Small region of $\mathcal{X}$ space. (b) Corresponding region in $\mathcal{Y}$ space. (From Charles W. Therrien, *Decision Estimation and Classification*, copyright © John Wiley & Sons, New York, 1989. Reproduced by permission.)

When $\mathbf{g}$ is a nonsingular *linear* transformation, in the form (2.110), the Jacobian is equal to the absolute value of the determinant of $\mathbf{A}$. Therefore, the transformed density function is given by

$$f_{\boldsymbol{y}}(\mathbf{y}) = \frac{1}{\text{abs } |\mathbf{A}|} f_{\boldsymbol{x}}(\mathbf{A}^{-1}\mathbf{y}) \qquad \text{(real random vector)} \tag{2.126}$$

When $\boldsymbol{x}$ is complex, the same argument applied to both the real and imaginary parts leads to the relation

$$f_{\boldsymbol{y}}(\mathbf{y}) = \frac{1}{|\mathbf{A}|^*|\mathbf{A}|} f_{\boldsymbol{x}}(\mathbf{A}^{-1}\mathbf{y}) \qquad \text{(complex random vector)} \tag{2.127}$$

These last results can be used to show that a linear transformation applied to a Gaussian random vector results in another Gaussian random vector. This fact will be shown explicitly here for real random vectors; however, an essentially identical argument applies to complex random vectors.

Let us assume that $\boldsymbol{y}$ has the same dimension as $\boldsymbol{x}$ (i.e., that $\mathbf{A}$ is a square matrix.) This results in no loss of generality because if $\boldsymbol{y}$ were of lower dimension than $\boldsymbol{x}$, we could always form a vector $\boldsymbol{y}'$ by appending some rows to the matrix $\mathbf{A}$ and show that the vector $\boldsymbol{y}'$ is a Gaussian random vector. Then since $\boldsymbol{y}$ consists of selected components of $\boldsymbol{y}'$, it too would be Gaussian. On the other hand, if $\boldsymbol{y}$ were larger than $\boldsymbol{x}$, we could form the vector

$$\tilde{x} = \begin{bmatrix} x \\ x \end{bmatrix}$$

which is also Gaussian and proceed as just described.

It will further be assumed that $\mathbf{A}$ is nonsingular. If this were not the case, the distribution of $\boldsymbol{y}$ would not be defined except in a subspace of $\mathcal{Y}$. Under these circumstances (2.126) can be applied to (2.99) and (2.113) can be used to obtain

$$\begin{aligned} f_{\boldsymbol{y}}(\mathbf{y}) &= \frac{1}{\text{abs } |\mathbf{A}|} \cdot \frac{1}{(2\pi)^{N/2}|\mathbf{C}_{\boldsymbol{x}}|^{1/2}} e^{-\frac{1}{2}(\mathbf{A}^{-1}\mathbf{y}-\mathbf{m}_x)^T \mathbf{C}_x^{-1}(\mathbf{A}^{-1}\mathbf{y}-\mathbf{m}_x)} \\ &= \frac{1}{(2\pi)^{N/2}(|\mathbf{A}|^2|\mathbf{C}_{\boldsymbol{x}}|)^{1/2}} e^{-\frac{1}{2}(\mathbf{y}-\mathbf{m}_y)^T(\mathbf{A}^{-1})^T \mathbf{C}_x^{-1}\mathbf{A}^{-1}(\mathbf{y}-\mathbf{m}_y)} \end{aligned}$$

Then in view of (2.115) and (2.118), this can be written as

$$f_{\boldsymbol{y}}(\mathbf{y}) = \frac{1}{(2\pi)^{N/2}|\mathbf{C}_{\boldsymbol{y}}|^{1/2}} e^{-\frac{1}{2}(\mathbf{y}-\mathbf{m}_y)^T \mathbf{C}_y^{-1}(\mathbf{y}-\mathbf{m}_y)} \tag{2.128}$$

which proves the result.

2.5 REVERSAL OPERATION FOR RANDOM VECTORS

In the analysis of sequences and systems it is frequently necessary to reverse the order of the points in a sequence. For example, in carrying out convolution, the sequence representing the impulse response of the system has to be turned around before computing its product with the input sequence, or vice versa. When a sequence is represented by a random vector, the corresponding operation is to reorder the components of the vector so as to turn the vector "upside-down." The need for this operation is so frequent that it is worthwhile to define a special operation on vectors and matrices called *reversal* and denoted by a tilde (˜) over the variable. For a vector the reversal is defined as the vector whose components are in reverse order. For example consider the random vector

$$\boldsymbol{x} = \begin{bmatrix} x_1 \\ x_2 \\ \vdots \\ x_N \end{bmatrix} \tag{2.129}$$

Its reversal is defined by[6]

$$\tilde{\boldsymbol{x}} = \begin{bmatrix} x_N \\ x_{N-1} \\ \vdots \\ x_1 \end{bmatrix} \tag{2.130}$$

[6]The reversal of a vector can also be expressed as a matrix product $\mathbf{J}\boldsymbol{x}$ where $\mathbf{J}$ is a permutation matrix containing 1's on the reverse diagonal and 0's elsewhere. The new notation is defined because reversals are used frequently in this book and the permutation matrix notation becomes terribly cumbersome.

Now suppose that another random vector $\boldsymbol{y}$ is defined by the linear transformation

$$\boldsymbol{y} = \mathbf{A}\boldsymbol{x} \tag{2.131}$$

If both vectors are reversed, then the matrix **A** has to be reversed about *both its vertical axis and its horizontal axis* in order to preserve the equality. For example, consider the simple equation

$$\begin{bmatrix} y_1 \\ y_2 \\ y_3 \end{bmatrix} = \begin{bmatrix} a_{11} & a_{12} & a_{13} \\ a_{21} & a_{22} & a_{23} \\ a_{31} & a_{32} & a_{33} \end{bmatrix} \begin{bmatrix} x_1 \\ x_2 \\ x_3 \end{bmatrix}$$

(No special symmetry of the matrix is assumed.) If both of the vectors are reversed, we have to write

$$\begin{bmatrix} y_3 \\ y_2 \\ y_1 \end{bmatrix} = \begin{bmatrix} a_{33} & a_{32} & a_{31} \\ a_{23} & a_{22} & a_{21} \\ a_{13} & a_{12} & a_{11} \end{bmatrix} \begin{bmatrix} x_3 \\ x_2 \\ x_1 \end{bmatrix}$$

where the matrix has also been reversed. Notice that for a square matrix, reversing the matrix about both the vertical and horizontal axes is equivalent to flipping it about the main *and* reverse diagonals. If the reversal of the matrix **A** is denoted by $\tilde{\mathbf{A}}$, then the last equation can be written as

$$\tilde{\boldsymbol{y}} = \tilde{\mathbf{A}}\tilde{\boldsymbol{x}} \tag{2.132}$$

Since there will be multiple occasions to apply the reversal operation, it will be used in this book exactly like a vector or matrix transpose. When the operation is necessary we will simply add the $\tilde{}$ over the variable as we would attach the superscript ^{T}. A few properties of the reversal are given in Table 2.5. Note that the reversal of a product is the product of the reversals (*not* in reversed order) and that the operations of matrix inversion, transposition, conjugation (and, of course, conjugate transposition) all commute with the operation of reversal.

TABLE 2.5 PROPERTIES OF REVERSAL

	Quantity	Reversal
Matrix product	$\mathbf{AB}$	$\tilde{\mathbf{A}}\tilde{\mathbf{B}}$
Matrix inverse	$\mathbf{A}^{-1}$	$(\tilde{\mathbf{A}})^{-1}$
Matrix conjugate	$\mathbf{A}^*$	$(\tilde{\mathbf{A}})^*$
Matrix transpose	$\mathbf{A}^T$	$(\tilde{\mathbf{A}})^T$

In subsequent parts of this book there is a need to deal with the first and second moment properties of a reversed random vector. These are easily derived and related to the corresponding quantities for the unreversed vector. The mean is

$$\mathbf{m}_{\tilde{x}} = \mathcal{E}\{\tilde{\boldsymbol{x}}\} = \tilde{\mathbf{m}}_x \tag{2.133}$$

The correlation matrix is given by

$$\mathbf{R}_{\tilde{x}} = \mathcal{E}\{\tilde{x}\tilde{x}^{*T}\} = \tilde{\mathbf{R}}_x \tag{2.134}$$

Similarly from the two results above and the relation (2.67) it follows that

$$\mathbf{C}_{\tilde{x}} = \tilde{\mathbf{C}}_x \tag{2.135}$$

These results can be summarized by observing that as a general rule reversal can be interchanged with expectation.

2.6 DIAGONALIZATION OF THE CORRELATION OR COVARIANCE MATRIX BY UNITARY TRANSFORMATION

Sometimes it is desirable to transform a random vector x to another random vector x' whose components have the orthogonality property:

$$\mathcal{E}\{x'_k x'^*_l\} = 0; \qquad k \neq l$$

This provides the advantages and simplicity of working with orthogonal random varibles. Since the components are orthogonal the correlation matrix for the new random vector is a *diagonal* matrix. A similar procedure can be applied to diagonalize the covariance matrix. In this case the components of the random vector are *uncorrelated.*

There are two fundamentally different ways to perform this transformation or, equivalently, to diagonalize the original correlation or covariance matrix. One is based on a unitary transformation and is discussed here. The other is based on triangular decomposition and is discussed in the next section.

2.6.1 Diagonalizing the Correlation Matrix

The eigenvectors of the correlation matrix for the random vector x can be used to construct a unitary transformation. It turns out that this transformation provides the desired diagonalization of the correlation matrix and that the transformation so constructed can be thought of as a rotation of the coordinate system in which the vector x is represented. Although the transformation can be applied to any random vector, it is most useful when the mean is zero.

If $\mathbf{R}_x$ is the correlation matrix, an eigenvector $\mathbf{e}$ and an eigenvalue λ satisfy the relation

$$\mathbf{R}_x\mathbf{e} = \lambda\mathbf{e} \tag{2.136}$$

That is, $\mathbf{R}_x$, when regarded as a linear transformation, maps the eigenvector $\mathbf{e}$ into a scaled version of itself. Because the matrix $\mathbf{R}_x$ is Hermitian symmetric it is possible to find N *orthonormal* eigenvectors $\mathbf{e}_1, \mathbf{e}_2, \ldots, \mathbf{e}_N$ and a corresponding set of eigenvalues $\lambda_1, \lambda_2, \ldots, \lambda_N$ which are all *real.* (Some of these may have the same value.) Let $\mathbf{e}_l$ and $\mathbf{e}_k$ be any two such eigenvectors from this orthonormal set. Then it follows that

$$\mathbf{e}_l^{*T}\mathbf{R}_x\mathbf{e}_k = \lambda_k\mathbf{e}_l^{*T}\mathbf{e}_k = \begin{cases} \lambda_k & \text{if } l = k \\ 0 & \text{if } l \neq k \end{cases} \tag{2.137}$$

Now define a matrix $\mathbf{E}$ whose columns are the eigenvectors

$$\mathbf{E} = \begin{bmatrix} | & | & & | \\ \mathbf{e}_1 & \mathbf{e}_2 & \cdots & \mathbf{e}_N \\ | & | & & | \end{bmatrix} \tag{2.138}$$

From (2.137) it follows that

$$\mathbf{E}^{*T}\mathbf{R}_x\mathbf{E} = \begin{bmatrix} -- & \mathbf{e}_1^{*T} & -- \\ -- & \mathbf{e}_2^{*T} & -- \\ & \vdots & \\ -- & \mathbf{e}_N^{*T} & -- \end{bmatrix} \mathbf{R}_x \begin{bmatrix} | & | & & | \\ \mathbf{e}_1 & \mathbf{e}_1 & \cdots & \mathbf{e}_N \\ | & | & & | \end{bmatrix} = \begin{bmatrix} \lambda_1 & & & 0 \\ & \lambda_2 & & \\ & & \ddots & \\ 0 & & & \lambda_N \end{bmatrix} = \mathbf{\Lambda} \tag{2.139}$$

It can be seen from (2.110) and (2.114) that $\mathbf{\Lambda}$ is the correlation matrix for a random vector $\boldsymbol{x}'$ defined by

$$\boldsymbol{x}' = \mathbf{E}^{*T}\boldsymbol{x} = \begin{bmatrix} -- & \mathbf{e}_1^{*T} & -- \\ -- & \mathbf{e}_2^{*T} & -- \\ & \vdots & \\ -- & \mathbf{e}_N^{*T} & -- \end{bmatrix} \boldsymbol{x} \tag{2.140}$$

and since $\mathbf{\Lambda}$ is diagonal, the components of $\boldsymbol{x}'$ are orthogonal. Further, *the eigenvalues,* which are second moments of the components x'_k, *are always positive* (or possibly zero).

The last point is worth elaborating upon. A matrix is *positive semidefinite* if and only if all of its eigenvalues are *nonnegative* (see Problem 2.35). The matrix is *positive definite* if and only if the eigenvalues are all *positive.* Since the theoretical correlation matrix was previously shown to be positive semidefinite, the eigenvalues are guaranteed to be nonnegative. However, a check of the eigenvalues is useful when *estimating* a correlation matrix from data to ensure that the matrix has the required property. Recall that in most practical situations the matrix is (strictly) positive definite.

Since the columns of $\mathbf{E}$ are orthonormal, the matrix $\mathbf{E}^{*T}$ represents a unitary transformation, i.e.,

$$\mathbf{E}^{*T}\mathbf{E} = \mathbf{E}\mathbf{E}^{*T} = \mathbf{I} \tag{2.141}$$

The mean of the transformed vector is given by

$$\mathbf{m}_{x'} = \mathbf{E}^{*T}\mathbf{m}_x \tag{2.142}$$

and the correlation matrix is

$$\mathbf{R}_{x'} = \mathbf{E}^{*T}\mathbf{R}_x\mathbf{E} = \mathbf{\Lambda} \tag{2.143}$$

Let us mention in passing that the eigenvalue problem (2.136) can be written in the alternative matrix form

$$\mathbf{R}_x\mathbf{E} = \mathbf{\Lambda}\mathbf{E} \tag{2.144}$$

whereupon (2.143) follows directly.

TABLE 2.6 DIAGONALIZATION OF THE CORRELATION MATRIX BY ORTHOGONAL TRANSFORMATION

Transformation	$\boldsymbol{x}' = \mathbf{E}^{*T}\boldsymbol{x}$
Mean vector	$\mathbf{m}_{x'} = \mathbf{E}^{*T}\mathbf{m}_x$
Correlation matrix	$\mathbf{R}_{x'} = \mathbf{E}^{*T}\mathbf{R}_x\mathbf{E} = \mathbf{\Lambda}$

The results of this section are summarized in Table 2.6. Note that since

$$\mathbf{C}_{x'} = \mathbf{R}_{x'} - \mathbf{m}_{x'}\mathbf{m}_{x'}^{*T} = \mathbf{\Lambda} - \mathbf{m}_{x'}\mathbf{m}_{x'}^{*T}$$

the resulting covariance matrix is generally *not* diagonal.

Before proceeding, let us just note a few additional facts about the linear transformation (2.140). First, (2.143) can be pre- and post-multiplied by $\mathbf{E}$ and $\mathbf{E}^{*T}$; since $\mathbf{E}$ is a unitary matrix the result is

$$\boxed{\mathbf{R}_x = \mathbf{E}\mathbf{\Lambda}\mathbf{E}^{*T}} \tag{2.145}$$

This is a kind of cannonical form for the correlation matrix in terms of its eigenvector and eigenvalue matrices. Further, taking the inverse of this equation and noting that $\mathbf{E}^{-1} = \mathbf{E}^{*T}$ leads to the relation

$$\boxed{\mathbf{R}_x^{-1} = \mathbf{E}\mathbf{\Lambda}^{-1}\mathbf{E}^{*T}} \tag{2.146}$$

This equation is useful for inverting the correlation matrix since $\mathbf{\Lambda}^{-1}$ is simply a real diagonal matrix with diagonal elements $1/\lambda_j$. In addition, since $\mathbf{E}$ is orthonormal, the determinant and trace of $\mathbf{R}_x$ are the same as the determinant and trace of $\mathbf{\Lambda}$ (see Section 2.4). Thus it follows that

$$\boxed{|\mathbf{R}_x| = |\mathbf{\Lambda}| = \prod_{j=1}^{N} \lambda_j} \tag{2.147}$$

and

$$\boxed{\text{tr } \mathbf{R}_x = \text{tr } \mathbf{\Lambda} = \sum_{j=1}^{N} \lambda_j} \tag{2.148}$$

In most cases a computer program will be used to find the eigenvectors and eigenvalues. However it is worthwhile to review a procedure for doing it manually in simple cases[7]. To find the eigenvectors of a matrix, notice that (2.136) can be written in the equivalent form

$$(\mathbf{R}_x - \lambda\mathbf{I})\mathbf{e} = \mathbf{0} \tag{2.149}$$

[7]The procedure described here is very basic. For practical problems involving large matrices, numerically more efficient methods are used.

In order for nontrivial solutions of this equation to exist, it is required that

$$|\mathbf{R}_x - \lambda \mathbf{I}| = 0 \tag{2.150}$$

Equation (2.150) is the *characteristic equation* corresponding to (2.136); it is a polynomial in λ whose roots are the eigenvalues of $\mathbf{R}_x$. Once these eigenvalues are found, (2.149) can be used to determine the corresponding eigenvectors. This is demonstrated in a simple example.

Example 2.9

Consider the real correlation matrix

$$\mathbf{R}_x = \begin{bmatrix} 2 & -2 \\ -2 & 5 \end{bmatrix}$$

The characteristic equation for the eigenvalues is

$$\begin{vmatrix} 2-\lambda & -2 \\ -2 & 5-\lambda \end{vmatrix} = \lambda^2 - 7\lambda + 6 = 0$$

which has the two roots[8]

$$\lambda_1 = 6$$
$$\lambda_2 = 1$$

To find the first eigenvector, substitute $\lambda_1 = 6$ in (2.149):

$$\begin{bmatrix} -4 & -2 \\ -2 & -1 \end{bmatrix} \mathbf{e}_1 = \mathbf{0}$$

Since this system is underdetermined many solutions exist. However, if it is required that $\mathbf{e}_1^{*T}\mathbf{e}_1 = 1$, the unique solution is

$$\mathbf{e}_1 = \begin{bmatrix} \frac{1}{\sqrt{5}} \\ -\frac{2}{\sqrt{5}} \end{bmatrix}$$

Similarly, to find the second eigenvector, substitute $\lambda_2 = 1$ in (2.149) and solve

$$\begin{bmatrix} 1 & -2 \\ -2 & 4 \end{bmatrix} \mathbf{e}_2 = \mathbf{0}$$

to find

$$\mathbf{e}_2 = \begin{bmatrix} \frac{2}{\sqrt{5}} \\ \frac{1}{\sqrt{5}} \end{bmatrix}$$

[8]It is conventional to number the eigenvalues either from largest to smallest or from smallest to largest. We follow the first convention.

Thus the desired linear transformation is given by

$$x' = \mathbf{E}^{*T}x = \begin{bmatrix} \frac{1}{\sqrt{5}} & -\frac{2}{\sqrt{5}} \\ \frac{2}{\sqrt{5}} & \frac{1}{\sqrt{5}} \end{bmatrix} x$$

and the correlation matrix of x' is given by

$$\mathbf{\Lambda} = \begin{bmatrix} 6 & 0 \\ 0 & 1 \end{bmatrix}$$

2.6.2 Using Singular Value Decomposition

When the correlation matrix is estimated in the form (2.94), then the eigenvalues and eigenvectors of the estimated correlation matrix can frequently be found with better numerical precision by performing the *singular value decomposition* (SVD) of the data matrix. The enhanced numerical precision results because in SVD the products involved in forming the correlation matrix never have to be computed. In a machine with finite precision, this is a major advantage. Methods such as the SVD that deal directly with $\mathbf{X}$ instead of the product $\mathbf{X}^{*T}\mathbf{X}$ are sometimes called "square root" methods. More information on SVD can be found in [4, 5, 8, 12] and in Section 9.3.4 of Chapter 9; however, a brief introduction to the topic is provided here.

The singular value decomposition theorem states that any $K \times N$ matrix $\mathbf{X}$ can be decomposed and written as the product of matrices

$$\mathbf{X} = \mathbf{U}\mathbf{\Sigma}\mathbf{V}^{*T} \tag{2.151}$$

where $\mathbf{U}$ is the $K \times K$ unitary matrix of so-called left singular vectors,

$$\mathbf{U} = \begin{bmatrix} | & | & & | \\ \mathbf{u}_1 & \mathbf{u}_2 & \cdots & \mathbf{u}_K \\ | & | & & | \end{bmatrix} \tag{2.152}$$

$\mathbf{V}$ is the $N \times N$ unitary matrix of right singular vectors,

$$\mathbf{V} = \begin{bmatrix} | & | & & | \\ \mathbf{v}_1 & \mathbf{v}_2 & \cdots & \mathbf{v}_N \\ | & | & & | \end{bmatrix} \tag{2.153}$$

and $\mathbf{\Sigma}$ is the $K \times N$ matrix of nonnegative real singular values,

$$\mathbf{\Sigma} = \begin{bmatrix} \sigma_1 & 0 & \cdots & 0 \\ 0 & \sigma_2 & \cdots & 0 \\ \vdots & \vdots & \ddots & \vdots \\ 0 & 0 & \cdots & \sigma_N \\ 0 & 0 & \cdots & 0 \\ \vdots & \vdots & & \vdots \\ 0 & 0 & \cdots & 0 \end{bmatrix} \tag{2.154}$$

In writing this last matrix it is assumed that $K \geq N$ (the usual case for data matrices), that $\sigma_1 \geq \sigma_2 \geq \cdots \geq \sigma_K$, and that some of the lower-numbered σ_k may be zero. In general, there may be r nonzero singular values, where r is the rank of $\mathbf{X}$ (the number of independent columns). Let us reemphasize that regardless of whether $\mathbf{X}$ is real or complex, *the nonzero singular values are always real and positive.*

If the number of rows K is less than N, then $\boldsymbol{\Sigma}$ has the form

$$\boldsymbol{\Sigma} = \begin{bmatrix} \sigma_1 & 0 & \cdots & 0 & 0 & \cdots & 0 \\ 0 & \sigma_2 & \cdots & 0 & 0 & \cdots & 0 \\ \vdots & \vdots & \ddots & \vdots & \vdots & & \vdots \\ 0 & 0 & \cdots & \sigma_K & 0 & \cdots & 0 \end{bmatrix} \tag{2.155}$$

where again some of the higher-numbered singular values may be zero. In this case the rank r, and the number of nonzero singular values, is equal to the number of independent rows of $\mathbf{X}$.

Given (2.151) it is not hard to show that the eigenvectors of the matrix

$$\mathbf{X}^{*T}\mathbf{X}$$

are the right singular vectors of $\mathbf{X}$ and that the eigenvalues of $\mathbf{X}^{*T}\mathbf{X}$ are the squared singular values of $\mathbf{X}$. To show this, simply use (2.151) to write

$$\mathbf{X}^{*T}\mathbf{X} = \mathbf{V}\boldsymbol{\Sigma}^{*T}(\mathbf{U})^{*T}\mathbf{U}\boldsymbol{\Sigma}\mathbf{V}^{*T} = \mathbf{V}(\boldsymbol{\Sigma}^{T}\boldsymbol{\Sigma})\mathbf{V}^{*T}$$

where the last step follows because $\mathbf{U}$ is unitary. Now upon substituting (2.154) and carrying out the matrix multiplication, the last equation becomes

$$\mathbf{X}^{*T}\mathbf{X} = \mathbf{V} \begin{bmatrix} \sigma_1^2 & 0 & \cdots & 0 \\ 0 & \sigma_2^2 & \cdots & 0 \\ \vdots & \vdots & \ddots & \vdots \\ 0 & 0 & \cdots & \sigma_N^2 \end{bmatrix} \mathbf{V}^{*T}$$

This represents a decomposition of $\mathbf{X}^{*T}\mathbf{X}$ is the form (2.145). Since this decomposition is unique, the matrix $\mathbf{V}$ is the matrix of eigenvectors, and the σ_k^2 are the eigenvalues. Therefore, when the correlation matrix is estimated according to (2.94) the eigenvectors and eigenvalues are given by

$$\mathbf{e}_k = \mathbf{v}_k \qquad k = 1, 2, \ldots, N \tag{2.156}$$

and

$$\lambda_k = \frac{1}{K}\sigma_k^2 \qquad k = 1, 2, \ldots, N \tag{2.157}$$

It can similarly be shown that the left singular vectors $\mathbf{u}_k$ are the eigenvectors of the matrix

$$\mathbf{X}\mathbf{X}^{*T}$$

and that the σ_k^2 are eigenvalues of this matrix as well. Since $\mathbf{XX}^{*T}$ is a $K \times K$ matrix it has a total of K eigenvalues. If $K > N$, then there are at most N singular values, so the other $K - N$ eigenvalues are zero. (You may want to check this out as an exercise. The proof is virtually identical to that above.)

To demonstrate how the SVD can be used in place of solving an eigenvalue problem, let us apply these results to the correlation matrix of Example 2.8.

Example 2.10

By performing the SVD of the data matrix of Example 2.8 we find that it can be written as

$$\begin{bmatrix} 1 & 0 \\ -2 & 1 \\ 2 & -2 \\ 0 & 2 \end{bmatrix} = \begin{bmatrix} \frac{1}{\sqrt{30}} & -\frac{1}{\sqrt{6}} & \frac{-8+2\sqrt{5}}{15} & \frac{-4-4\sqrt{5}}{15} \\ -\sqrt{\frac{3}{10}} & \frac{1}{\sqrt{6}} & \frac{4+2\sqrt{5}}{15} & \frac{2-4\sqrt{5}}{15} \\ 2\sqrt{\frac{2}{15}} & 0 & \frac{8+\sqrt{5}}{15} & \frac{4-2\sqrt{5}}{15} \\ -\sqrt{\frac{2}{15}} & -\sqrt{\frac{2}{3}} & \frac{2}{5} & \frac{1}{5} \end{bmatrix} \begin{bmatrix} \sqrt{15} & 0 \\ 0 & \sqrt{3} \\ 0 & 0 \\ 0 & 0 \end{bmatrix} \begin{bmatrix} \frac{\sqrt{2}}{2} & -\frac{\sqrt{2}}{2} \\ -\frac{\sqrt{2}}{2} & -\frac{\sqrt{2}}{2} \end{bmatrix}$$

(It is not necessary to know how to perform the SVD since in almost all situations a computer algorithm will be used.) The right singular vectors

$$\mathbf{v}_1 = \begin{bmatrix} \frac{\sqrt{2}}{2} \\ -\frac{\sqrt{2}}{2} \end{bmatrix} \quad \text{and} \quad \mathbf{v}_2 = \begin{bmatrix} -\frac{\sqrt{2}}{2} \\ -\frac{\sqrt{2}}{2} \end{bmatrix}$$

obtained from the SVD are the eigenvectors of the estimated correlation matrix

$$\begin{bmatrix} \frac{9}{4} & -\frac{3}{2} \\ -\frac{3}{2} & \frac{9}{4} \end{bmatrix}$$

and the eigenvalues [from (2.157)] are

$$\lambda_1 = \frac{1}{4}(\sqrt{15})^2 = \frac{15}{4} \qquad \lambda_2 = \frac{1}{4}(\sqrt{3})^2 = \frac{3}{4}$$

The result can be verified by computing the eigenvectors and eigenvalues of the correlation matrix directly (see Problem 2.26).

2.6.3 Diagonalizing the Covariance Matrix

The covariance matrix can be diagonalized by a procedure similar to that used for the correlation matrix. In this case the transformation leads to a vector whose components are *uncorrelated* rather than orthogonal.

Since the covariance matrix like the correlation matrix is Hermitian symmetric, it is always possible to find N eigenvalues $\check{\lambda}_k$ and N corresponding unitary eigenvectors $\check{\mathbf{e}}_k$ satisfying the relation[9]

$$\mathbf{C}_x\check{\mathbf{e}}_k = \check{\lambda}_k\check{\mathbf{e}}_k\,; \qquad k = 1, 2, \ldots, N \tag{2.158}$$

[9]Note that in general the eigenvectors and eigenvalues of the covariance matrix are *not* the same as those of the correlation matrix.

Since the $\check{\mathbf{e}}_k$ are unitary, it follows from (2.158) that

$$\check{\mathbf{e}}_l^{*T}\mathbf{C}_{\boldsymbol{x}}\check{\mathbf{e}}_k = \begin{cases} \check{\lambda}_k & \text{if } l = k \\ 0 & \text{if } l \neq k \end{cases} \tag{2.159}$$

Consequently, if the matrix of eigenvectors is defined as

$$\check{\mathbf{E}} = \begin{bmatrix} | & | & & | \\ \check{\mathbf{e}}_1 & \check{\mathbf{e}}_2 & \cdots & \check{\mathbf{e}}_N \\ | & | & & | \end{bmatrix} \tag{2.160}$$

then by analogy with (2.139) it follows that

$$\check{\mathbf{E}}^{*T}\mathbf{C}_{\boldsymbol{x}}\check{\mathbf{E}} = \check{\boldsymbol{\Lambda}} \tag{2.161}$$

where

$$\check{\boldsymbol{\Lambda}} = \begin{bmatrix} \check{\lambda}_1 & & & 0 \\ & \check{\lambda}_2 & & \\ & & \ddots & \\ 0 & & & \check{\lambda}_N \end{bmatrix} \tag{2.162}$$

Thus if $\check{\boldsymbol{x}}$ is defined by

$$\check{\boldsymbol{x}} = \check{\mathbf{E}}^{*T}\boldsymbol{x} \tag{2.163}$$

then

$$\mathbf{m}_{\check{\boldsymbol{x}}} = \check{\mathbf{E}}^{*T}\mathbf{m}_{\boldsymbol{x}} \tag{2.164}$$

and

$$\mathbf{C}_{\check{\boldsymbol{x}}} = \check{\mathbf{E}}^{*T}\mathbf{C}_{\boldsymbol{x}}\check{\mathbf{E}} = \check{\boldsymbol{\Lambda}} \tag{2.165}$$

The components $\check{x}_k$ of $\check{\boldsymbol{x}}$ are therefore uncorrelated. Further, the eigenvalues, which are the variances of the components of $\check{\boldsymbol{x}}$, are always *real and positive* (or possibly zero).

The following useful relations then also follow for the covariance matrix.

$$\mathbf{C}_x = \check{\mathbf{E}}\check{\mathbf{\Lambda}}\check{\mathbf{E}}^{*T} \tag{2.166}$$

$$\mathbf{C}_x^{-1} = \check{\mathbf{E}}\check{\mathbf{\Lambda}}^{-1}\check{\mathbf{E}}^{*T} \tag{2.167}$$

$$|\mathbf{C}_x| = |\check{\mathbf{\Lambda}}| = \prod_{j=1}^{N} \check{\lambda}_j \tag{2.168}$$

$$\text{tr } \mathbf{C}_x = \text{tr } \check{\mathbf{\Lambda}} = \sum_{j=1}^{N} \check{\lambda}_j \tag{2.169}$$

The last two relations express the determinant and trace of $\mathbf{C}_x$ respectively as a product and sum of the variances of the transformed variables.

The procedure described here leads to vectors with uncorrelated components regardless of the form of the density characterizing the random vectors in the original coordinate frame. However, for *Gaussian* random vectors, it is easy to show that the density of the transformed vector takes the form of a product of one-dimensional Gaussian densities (see Problem 2.32) and therefore the components of $\check{x}$ are statistically independent.

The results just discussed can also be used to plot the contours of the Gaussian density function. The contours are defined by the relation

$$f_x(\mathbf{x}) = \text{const.} \tag{2.170}$$

where f_x is defined by (2.97) or (2.99). Since only the exponential term is a function of $\mathbf{x}$, the contours are equivalently defined by

$$(\mathbf{x} - \mathbf{m}_x)^{*T}\mathbf{C}_x^{-1}(\mathbf{x} - \mathbf{m}_x) = C \tag{2.171}$$

where C is a positive constant. By substituting (2.167) and using (2.163) and (2.164), the last equation can be written as

$$\begin{aligned}(\mathbf{x} - \mathbf{m}_x)^{*T}\,\mathbf{C}_x^{-1}\,(\mathbf{x} - \mathbf{m}_x) &= (\mathbf{x} - \mathbf{m}_x)^{*T}\,\check{\mathbf{E}}\check{\mathbf{\Lambda}}^{-1}\check{\mathbf{E}}^{*T}(\mathbf{x} - \mathbf{m}_x) \\ &= (\check{\mathbf{x}} - \mathbf{m}_{\check{x}})^{*T}\check{\mathbf{\Lambda}}^{-1}(\check{\mathbf{x}} - \mathbf{m}_{\check{x}}) = C\end{aligned} \tag{2.172}$$

This can then be written in the expanded form

$$\frac{|\check{\mathrm{x}}_1 - \check{\mathrm{m}}_1|^2}{\check{\lambda}_1} + \frac{|\check{\mathrm{x}}_2 - \check{\mathrm{m}}_2|^2}{\check{\lambda}_2} + \cdots + \frac{|\check{\mathrm{x}}_N - \check{\mathrm{m}}_N|^2}{\check{\lambda}_N} = C \tag{2.173}$$

which is the equation of a N-dimensional ellipsoid with center at $\mathbf{m}_{\check{x}}$ (see Fig. 2.5). The principal axes of the ellipsoid are aligned with the eigenvectors, and their sizes are given

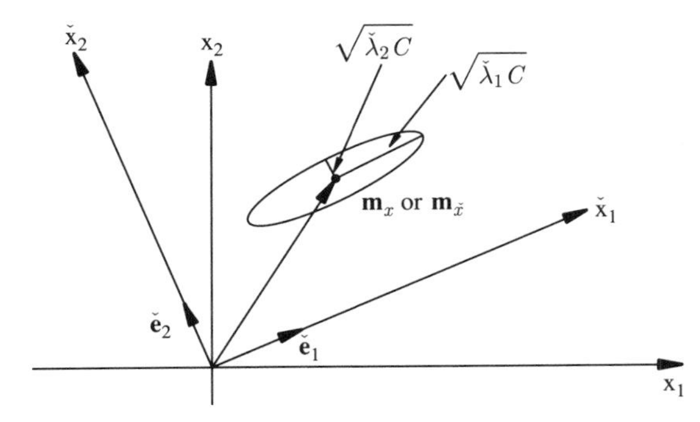

Figure 2.5 Typical contour of a Gaussian density function. (Concentration ellipse). (From Charles W. Therrien, *Decision Estimation and Classification*, copyright © John Wiley & Sons, New York, 1989. Reproduced by permission.)

by $2\sqrt{\breve{\lambda}_j C}$, $j = 1, 2, \ldots, N$. The fact that the ellipse is tilted with respect to the original coordinate system shows that there is correlation between the original vector components. In the case of two dimensions, a positive slope for the major axis of the ellipse shows positive correlation and a negative slope indicates negative correlation. When the axes of the ellipse are parallel to the coordinate axes, the vector components are uncorrelated. The contours of the Gaussian density are sometimes referred to as *concentration ellipsoids* since they indicate regions where the density is most concentrated. The concentration ellipsoids are useful even when the density is not Gaussian to give a rough idea of the spread of the density.

The following example illustrates plotting the concentration ellipsoids corresponding to a simple covariance matrix.

Example 2.11

Consider a real two-dimensional Gaussian random vector with covariance matrix

$$\mathbf{C}_x = \begin{bmatrix} 2 & -1 \\ -1 & 2 \end{bmatrix}$$

and mean vector

$$\mathbf{m}_x = \begin{bmatrix} 2 \\ 1 \end{bmatrix}$$

The eigenvalues and eigenvectors of the matrix are found to be

$$\breve{\lambda}_1 = 3, \quad \breve{\lambda}_2 = 1, \quad \breve{\mathbf{e}}_1 = \begin{bmatrix} \frac{1}{\sqrt{2}} \\ -\frac{1}{\sqrt{2}} \end{bmatrix}, \quad \breve{\mathbf{e}}_2 = \begin{bmatrix} \frac{1}{\sqrt{2}} \\ \frac{1}{\sqrt{2}} \end{bmatrix}$$

A contour of the density function is shown in Fig. EX2.11. The ellipse is centered at the coordinates of the mean vector and oriented with its axes parallel to the direction of the eigenvectors. The major and minor axes are proportional to the square root of the eigenvalues ($\sqrt{3}$ and 1), and the negative slope of the major axis shows that the vector components are negatively correlated.

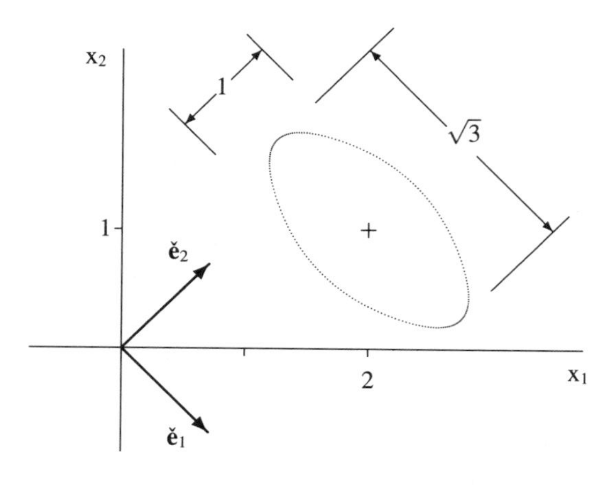

Figure EX2.11

2.6.4 Simultaneous Diagonalization

Sometimes it is necessary to diagonalize two correlation or covariance matrices at the same time. A typical case where this occurs is when one matrix represents a random signal and the other represents colored noise (see Chapter 4). If the observations are the sum of signal and noise, the only way to develop observations that are orthogonal or uncorrelated is to diagonalize the correlation or covariance matrices of the signal and noise simultaneously.

It is clearest to describe the procedure in terms of covariance matrices, which are denoted by $\mathbf{C}_{\boldsymbol{x}(A)}$ and $\mathbf{C}_{\boldsymbol{x}(B)}$. The contours corresponding to these covariance matrices are depicted in Fig. 2.6(a). It is obvious from the picture that rotating the axes to diagonalize one matrix will not diagonalize the other. The two matrices can be simultaneously diagonalized by combinations of rotation and scaling of the coordinate system, however. These are the steps in the procedure:

1. Find the eigenvalues and eigenvectors of $\mathbf{C}_{\boldsymbol{x}(B)}$ and write
$$\mathbf{C}_{\boldsymbol{x}(B)} = \check{\mathbf{E}}_B \check{\boldsymbol{\Lambda}}_B \check{\mathbf{E}}_B^{*T} \tag{2.174}$$
2. Apply the transformation
$$\boldsymbol{y} = \check{\boldsymbol{\Lambda}}_B^{-1/2} \check{\mathbf{E}}_B^{*T} \boldsymbol{x} \tag{2.175}$$
to the data. The matrix $\check{\boldsymbol{\Lambda}}_B^{-1/2}$ is the inverse of the diagonal matrix $\check{\boldsymbol{\Lambda}}_B^{1/2}$, whose elements are the square roots of the eigenvalues. The transformation in (2.175) is known as a *whitening transformation*; it produces the transformed covariance matrices
$$\begin{aligned}\mathbf{C}_{\boldsymbol{y}(B)} &= \left(\check{\boldsymbol{\Lambda}}_B^{-1/2} \check{\mathbf{E}}_B^{*T}\right) \mathbf{C}_{\boldsymbol{x}(B)} \left(\check{\boldsymbol{\Lambda}}_B^{-1/2} \check{\mathbf{E}}_B^{*T}\right)^{*T} \\ &= \check{\boldsymbol{\Lambda}}_B^{-1/2} \check{\mathbf{E}}_B^{*T} \cdot \check{\mathbf{E}}_B \check{\boldsymbol{\Lambda}}_B \check{\mathbf{E}}_B^{*T} \cdot \check{\mathbf{E}}_B \check{\boldsymbol{\Lambda}}_B^{-1/2} = \mathbf{I}\end{aligned} \tag{2.176}$$
and
$$\mathbf{C}_{\boldsymbol{y}(A)} = \left(\check{\boldsymbol{\Lambda}}_B^{-1/2} \check{\mathbf{E}}_B^{*T}\right) \mathbf{C}_{\boldsymbol{x}(A)} \left(\check{\boldsymbol{\Lambda}}_B^{-1/2} \check{\mathbf{E}}_B^{*T}\right)^{*T} = \check{\boldsymbol{\Lambda}}_B^{-1/2} \check{\mathbf{E}}_B^{*T} \mathbf{C}_{\boldsymbol{x}(A)} \check{\mathbf{E}}_B \check{\boldsymbol{\Lambda}}_B^{-1/2} \tag{2.177}$$

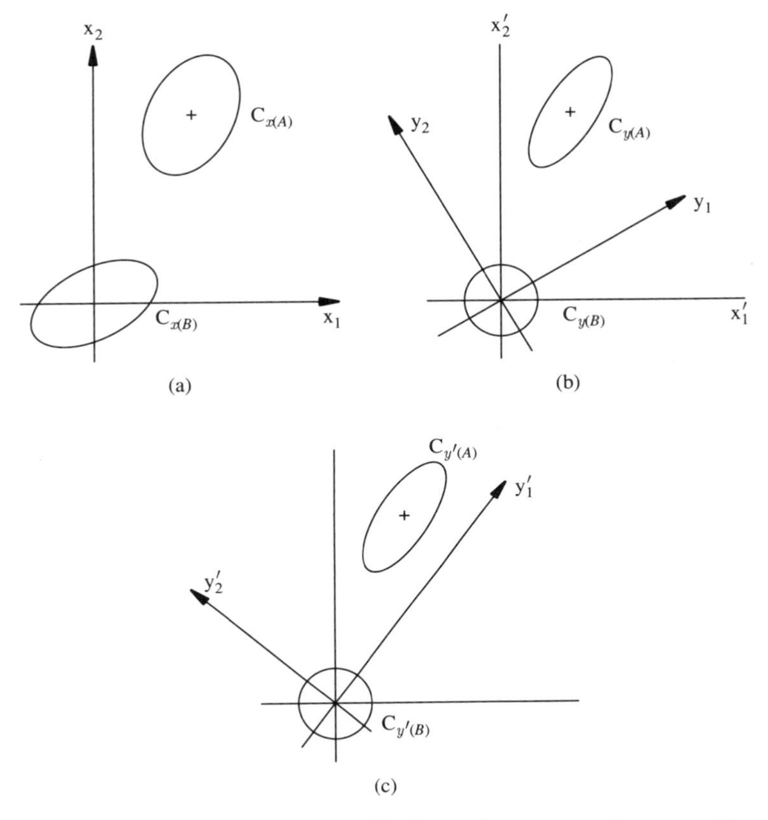

Figure 2.6 Simultaneous diagonalization of two covariance matrices. (a) Concentration ellipses in the original coordinate system. (b) Concentration ellipses and coordinate system after diagonalization and whitening of covariance B. (c) Concentration ellipses and coordinate system after simultaneous diagonalization.

The effect of this transformation is to rotate the coordinate system as shown in Fig. 2.6(b) and then scale the covariance matrix to the identity matrix. This is indicated by the circular contour for $\mathbf{C}_{\boldsymbol{y}(B)}$ in Fig. 2.6(b).

3. Find the eigenvalues and eigenvectors of $\mathbf{C}_{\boldsymbol{y}(A)}$ in the new coordinate system and diagonalize this covariance matrix. If the matrices of eigenvalues and eigenvectors are denoted by $\check{\mathbf{\Lambda}}_A$ and $\check{\mathbf{E}}_A$, then the transformation applied to the vector is

$$\boldsymbol{y}' = \check{\mathbf{E}}_A^{*T}\boldsymbol{y} \tag{2.178}$$

and the transformed covariance matrices are

$$\mathbf{C}_{\boldsymbol{y}'(A)} = \check{\mathbf{E}}_A^{*T}\mathbf{C}_{\boldsymbol{y}(A)}\check{\mathbf{E}}_A = \check{\mathbf{\Lambda}}_A \tag{2.179}$$

and

$$\mathbf{C}_{\boldsymbol{y}'(B)} = \check{\mathbf{E}}_A^{*T}\,\mathbf{I}\,\check{\mathbf{E}}_A = \mathbf{I} \tag{2.180}$$

This rotates the coordinate system again as shown in Fig. 2.6(c) and completes the simultaneous diagonalization.

The total effect of all of these steps is to transform the original data according to

$$\boldsymbol{y}' = (\check{\mathbf{E}}_{A/B})^{*T}\boldsymbol{x} \tag{2.181}$$

where

$$\check{\mathbf{E}}_{A/B} = \check{\mathbf{E}}_B\check{\mathbf{\Lambda}}_B^{-1/2}\check{\mathbf{E}}_A \tag{2.182}$$

The diagonalized covariance matrix corresponding to $\mathbf{C}_{\boldsymbol{x}(B)}$ is the identity matrix and the diagonalized covariance matrix corresponding to $\mathbf{C}_{\boldsymbol{x}(A)}$ is $\check{\mathbf{\Lambda}}_A$.

The entire procedure just described is equivalent to solving a generalized eigenvalue problem. This can be shown as follows. The last two parts of (2.179) can be written as

$$\mathbf{C}_{\boldsymbol{y}(A)}\check{\mathbf{E}}_A = \check{\mathbf{E}}_A\check{\mathbf{\Lambda}}_A \tag{2.183}$$

Substituting (2.177) for $\mathbf{C}_{\boldsymbol{y}(A)}$ then yields

$$\left(\check{\mathbf{\Lambda}}_B^{-1/2}\check{\mathbf{E}}_B^{*T}\mathbf{C}_{\boldsymbol{x}(A)}\check{\mathbf{E}}_B\check{\mathbf{\Lambda}}_B^{-1/2}\right)\check{\mathbf{E}}_A = \check{\mathbf{E}}_A\check{\mathbf{\Lambda}}_A$$

and rearranging this produces

$$\mathbf{C}_{\boldsymbol{x}(A)}\check{\mathbf{E}}_B\check{\mathbf{\Lambda}}_B^{-1/2}\check{\mathbf{E}}_A = \check{\mathbf{E}}_B\check{\mathbf{\Lambda}}_B^{1/2}\check{\mathbf{E}}_A\check{\mathbf{\Lambda}}_A = \check{\mathbf{E}}_B\,\check{\mathbf{\Lambda}}_B\check{\mathbf{E}}_B^{*T}\check{\mathbf{E}}_B\check{\mathbf{\Lambda}}_B^{-1/2}\,\check{\mathbf{E}}_A\check{\mathbf{\Lambda}}_A \tag{2.184}$$

where the identity

$$\check{\mathbf{\Lambda}}_B^{1/2} = \check{\mathbf{\Lambda}}_B\check{\mathbf{E}}_B^{*T}\check{\mathbf{E}}_B\check{\mathbf{\Lambda}}_B^{-1/2}$$

was used to obtain the rightmost side of (2.184). Finally, in view of (2.174) and (2.182), (2.184) can be written as

$$\boxed{\mathbf{C}_{\boldsymbol{x}(A)}\check{\mathbf{E}}_{A/B} = \mathbf{C}_{\boldsymbol{x}(B)}\check{\mathbf{E}}_{A/B}\check{\mathbf{\Lambda}}_A} \tag{2.185}$$

This is the generalized eigenvalue problem. Its solution yields the simutaneous diagonalizing transformation $\check{\mathbf{E}}_{A/B}$ and the diagonal matrix $\check{\mathbf{\Lambda}}_A$ whose elements are the variances of the data pertaining to $\mathbf{C}_{\boldsymbol{x}(A)}$ in the transformed coordinate system. The generalized eigenvalue problem can be written in component form as

$$\boxed{\mathbf{C}_{\boldsymbol{x}(A)}\check{\mathbf{e}}_{A/B} = \check{\lambda}_A\mathbf{C}_{\boldsymbol{x}(B)}\check{\mathbf{e}}_{A/B}} \tag{2.186}$$

where $\check{\mathbf{e}}_{A/B}$ is a generalized eigenvector (one of the columns of $\check{\mathbf{E}}_{A/B}$) and $\check{\lambda}_A$ is a generalized eigenvalue (the corresponding diagonal element of $\check{\mathbf{\Lambda}}_A$).

A concept that arises naturally in the discussion of simultaneous diagonalization is the whitening transformation. In the foregoing discussion the whitening transformation involved the matrix $\check{\mathbf{\Lambda}}_B^{-1/2}\check{\mathbf{E}}_B^{*T}$ [see (2.175)]. An alternative is to employ the transformation matrix

$$\mathbf{C}_{\boldsymbol{x}(B)}^{-1/2} \stackrel{\text{def}}{=} \check{\mathbf{E}}_B \check{\mathbf{\Lambda}}_B^{-1/2} \check{\mathbf{E}}_B^{*T} \tag{2.187}$$

which differs from the previous one by the additional factor $\check{\mathbf{E}}_B$. In terms of the geometric interpretation of Fig. 2.6 this transformation involves an additional rotation of the coordinate system in (b) back to its original position in (a) once the scaling of the first covariance matrix has been achieved. This is denoted by the scaled axes x_1' and x_2' in the figure. From there we can proceed to rotate the coordinate system to the position in Fig. 2.6(c) as before. The transformation involving (2.187) is called the Mahalanobis transformation. The inverse matrix

$$\mathbf{C}_{\boldsymbol{x}(B)}^{1/2} = (\mathbf{C}_{\boldsymbol{x}(B)}^{-1/2})^{-1} = \check{\mathbf{E}}_B \check{\mathbf{\Lambda}}_B^{1/2} \check{\mathbf{E}}_B^{*T} \tag{2.188}$$

is called the Hermitian square root of $\mathbf{C}_{\boldsymbol{x}(B)}$. It is a Hermitian symmetric matrix and satisfies the equation

$$\mathbf{C}_{\boldsymbol{x}(B)} = \left(\mathbf{C}_{\boldsymbol{x}(B)}^{1/2}\right)\left(\mathbf{C}_{\boldsymbol{x}(B)}^{1/2}\right)^{*T} \tag{2.189}$$

A procedure identical the the one above can be applied when it is desired to simultaneously diagonalize two correlation matrices $\mathbf{R}_{\boldsymbol{x}(A)}$ and $\mathbf{R}_{\boldsymbol{x}(B)}$. The needed transformation involves the matrix $\mathbf{E}_{A/B}$, which satisfies the generalized eigenvalue problem

$$\boxed{\mathbf{R}_{\boldsymbol{x}(A)}\mathbf{E}_{A/B} = \mathbf{R}_{\boldsymbol{x}(B)}\mathbf{E}_{A/B}\mathbf{\Lambda}_A} \tag{2.190}$$

or

$$\boxed{\mathbf{R}_{\boldsymbol{x}(A)}\mathbf{e}_{A/B} = \lambda_A \mathbf{R}_{\boldsymbol{x}(B)}\mathbf{e}_{A/B}} \tag{2.191}$$

Here $\mathbf{e}_{A/B}$ is a generalized eigenvector (column of $\mathbf{E}_{A/B}$) and λ_A is a generalized eigenvalue (diagonal element of $\mathbf{\Lambda}_A$) corresponding to the two *correlation* matrices. In the case where the means are not zero, the transformation analogous to the Mahalanobis transformation results in random vectors whose components are *orthogonal* rather than uncorrelated.

2.7 DIAGONALIZATION BY TRIANGULAR DECOMPOSITION

The preceding section developed a way to diagonalize the correlation or covariance matrix via a unitary transformation. Another way to diagonalize the matrix is by triangular or modified Cholesky decomposition. Both this procedure and the diagonalization using eigenvectors are useful in different situations. The method involving triangular decomposition however leads to a transformation that can be interpreted (depending on the exact form of the decomposition) as causal or anticausal linear filtering of the associated signal sequence. This transformation thus turns out to be very important for signal processing.

2.7.1 Lower–Upper Decomposition

Matrices that satisfy certain conditions of their principal minors [8] can be expressed in a form known as the LU (lower–upper) decomposition. Since the correlation matrix satisfies these conditions and is Hermitian symmetric, its LU decomposition can be written further as a product of three matrices in the form

$$\mathbf{R}_x = \mathbf{L}\mathbf{D}_L\mathbf{L}^{*T} \tag{2.192}$$

The matrix $\mathbf{L}$ is *unit lower triangular*, that is, all of the main diagonal elements are equal to 1 and all elements above the main diagonal are zero; the matrix $\mathbf{D}_L$ is diagonal. The matrix $\mathbf{L}^{*T}$ is *unit upper triangular* since all of its diagonal elements are 1 and the terms *below* the diagonal are zero. Now if (2.192) is premultiplied by $\mathbf{L}^{-1}$ and postmultiplied by $(\mathbf{L}^{*T})^{-1}$ the result is

$$\mathbf{L}^{-1}\mathbf{R}_x(\mathbf{L}^{-1})^{*T} = \mathbf{D}_L \tag{2.193}$$

From this form it can be recognized that $\mathbf{D}_L$ is the correlation matrix for a random vector $\boldsymbol{x}''$ defined by

$$\boldsymbol{x}'' = \mathbf{L}^{-1}\boldsymbol{x} \tag{2.194}$$

Since $\mathbf{D}_L$ is a diagonal matrix, the components of $\boldsymbol{x}''$ are orthogonal, and the diagonal elements of $\mathbf{D}_L$ are their second moments.

Unlike the eigenvector transformation, (2.194) cannot be interpreted as a simple rotation of the coordinate axes; however, (2.194) has the important interpretation that it represents a causal transformation. To show this let us first show that since $\mathbf{L}$ is unit lower triangular, $\mathbf{L}^{-1}$ has the same property. One way to see this is by induction. Suppose that the unit lower triangular property holds for inverse matrices of dimension $N-1$. Then for a unit lower triangular matrix of dimension N, we can write

$$\underbrace{\left[\begin{array}{c|c} \mathbf{L}_{N-1}^{-1} & \beta \\ \hline \gamma_N^T & \alpha \end{array}\right]}_{\mathbf{L}_N^{-1}} \underbrace{\left[\begin{array}{c|c} \mathbf{L}_{N-1} & \begin{matrix} 0 \\ 0 \\ \vdots \\ 0 \end{matrix} \\ \hline \boldsymbol{l}_N^T & 1 \end{array}\right]}_{\mathbf{L}_N} = \underbrace{\left[\begin{array}{cccc|c} 1 & & 0 & & 0 \\ & 1 & & & 0 \\ \times & \times & \cdots & & \times \\ & & & & \\ 0 & & & & 0 \\ \hline 0 & 0 & & & 1 \end{array}\right]}_{\mathbf{I}} \tag{2.195}$$

Clearly, in order for the product to be the identity matrix, the element α must be 1. Further, if any element in the last column of $\mathbf{L}_N^{-1}$ such as β is nonzero, then the elements in the corresponding row of the product (denoted by $\times$'s) are also nonzero. Since $\mathbf{I}$ is the identity matrix and cannot have nonzero elements except on its diagonal, the inverse matrix $\mathbf{L}_N^{-1}$ cannot have nonzero elements (except α) in the last column and must be unit lower triangular. Since the unit lower triangular property is trivially true for $N=1$, it is therefore true for all N.

Because $\mathbf{L}^{-1}$ is lower triangular, the kth component x''_k of the new random vector $\boldsymbol{x}''$ is a function of only the variables x_l for $l \leq k$ (see below).

$$\begin{bmatrix} x''_1 \\ x''_2 \\ \vdots \\ x''_k \\ \vdots \\ x''_N \end{bmatrix} \begin{bmatrix} 1 & 0 & \cdots & 0 & \cdots & 0 \\ \times & 1 & \cdots & 0 & \cdots & 0 \\ \vdots & & \ddots & & & \vdots \\ \times & \times & \cdots & 1 & \cdots & 0 \\ \vdots & \vdots & & & \ddots & \vdots \\ \times & \times & \cdots & \times & \cdots & 1 \end{bmatrix} \begin{bmatrix} x_1 \\ x_2 \\ \vdots \\ x_k \\ \vdots \\ x_N \end{bmatrix}$$

If the elements of $\boldsymbol{x}$ are consecutive samples of a random sequence, then $\mathbf{L}^{-1}$ represents a *causal* transformation of the sequence which leads to orthogonal random variables. Such transformations lie at the heart of many of the operations of optimal filtering and prediction.

As a final observation, note that since $\mathbf{L}$ is unit lower triangular, its determinant is equal to 1. Thus it follows from (2.192) that

$$|\mathbf{R}_{\boldsymbol{x}}| = |\mathbf{L}|\ |\mathbf{D}_L|\ |\mathbf{L}^{*T}| = |\mathbf{D}_L| = \prod_{k=1}^{N} d_k \tag{2.196}$$

where d_k are the diagonal elements of $\mathbf{D}_L$. Equation 2.196 is analogous to (2.147). Since the d_k are second moments of the vector components x''_k, (2.196) shows that the determinant of the correlation matrix can be expressed as a product of these second moments.

Diagonalization of the covariance matrix by triangular decomposition follows an identical procedure. The covariance matrix $\mathbf{C}_{\boldsymbol{x}}$ can simply be substituted for the correlation matrix $\mathbf{R}_{\boldsymbol{x}}$ in the equations of this section. In this case the components of the random vector $\boldsymbol{x}''$ defined by (2.194) are *uncorrelated* and the diagonal elements of the matrix $\mathbf{D}_L$ are their *variances*. In many situations it is common to remove the mean from the data but still to use the correlation matrix of the resulting zero-mean random process in the analysis. In this case diagonalizing the correlation matrix is identical to diagonalizing the covariance matrix of the original data.

Standard computer algorithms for LU decomposition usually produce a unit lower triangular factor and an upper triangular factor that is generally not *unit* upper triangular. These factors can easily be used to obtain the required form (2.192) and the transformation (2.194). The following example illustrates the procedure.

Example 2.12

A real two-dimensional random vector has mean zero and correlation matrix

$$\mathbf{R}_{\boldsymbol{x}} = \begin{bmatrix} 2 & -1 \\ -1 & 2 \end{bmatrix}$$

The standard LU decomposition of the matrix is found to be

$$\begin{bmatrix} 2 & -1 \\ -1 & 2 \end{bmatrix} = \begin{bmatrix} 1 & 0 \\ -\frac{1}{2} & 1 \end{bmatrix} \begin{bmatrix} 2 & -1 \\ 0 & \frac{3}{2} \end{bmatrix}$$

The elements of the diagonal matrix $\mathbf{D}_L$ appear as the terms on the main diagonal of the upper triangular term. Therefore, the required triangular decomposition can be written as

$$\begin{bmatrix} 2 & -1 \\ -1 & 2 \end{bmatrix} = \underbrace{\begin{bmatrix} 1 & 0 \\ -\frac{1}{2} & 1 \end{bmatrix}}_{\mathbf{L}} \underbrace{\begin{bmatrix} 2 & 0 \\ 0 & \frac{3}{2} \end{bmatrix}}_{\mathbf{D}_L} \underbrace{\begin{bmatrix} 1 & -\frac{1}{2} \\ 0 & 1 \end{bmatrix}}_{\mathbf{L}^T}$$

This also yields

$$\mathbf{L}^{-1} = \begin{bmatrix} 1 & 0 \\ -\frac{1}{2} & 1 \end{bmatrix}^{-1} = \begin{bmatrix} 1 & 0 \\ \frac{1}{2} & 1 \end{bmatrix}$$

as the required diagonalizing transformation. Since the mean was assumed equal to zero, the diagonal elements of $\mathbf{D}_L$ are the variances. The variances of the transformed variables are thus equal to 2 and $\frac{3}{2}$.

Although in most cases the triangular decomposition is obtained by a computer algorithm, in simple cases it is worth knowing a procedure for performing the decomposition manually. A simple procedure is demonstrated in the next example.

Example 2.13

It is desired to find the lower–upper triangular decomposition of the correlation matrix

$$\mathbf{R}_x = \begin{bmatrix} 4 & -2 & 1 \\ -2 & 4 & -2 \\ 1 & -2 & 4 \end{bmatrix}$$

The basic method for doing this is based on Gauss reduction of the matrix to upper triangular form[4]. If the computations are organized so that the same operations are performed on the identity matrix, the transformation $\mathbf{L}^{-1}$ can be computed at the same time. For this it is easiest to put the two matrices side by side in the form of one larger partitioned matrix something like

$$\left[\begin{array}{ccc|ccc} 4 & -2 & 1 & 1 & 0 & 0 \\ -2 & 4 & -2 & 0 & 1 & 0 \\ 1 & -2 & 4 & 0 & 0 & 1 \end{array}\right]$$

The first step is to reduce all the elements in the first column beneath the top element to zero. This is done by multiplying the top row by an appropriate number, called the "pivot," and subtracting it from a lower row. For this matrix, there are two such operations, which are shown below.

$$\left[\begin{array}{ccc|ccc} 4 & -2 & 1 & 1 & 0 & 0 \\ -2 & 4 & -2 & 0 & 1 & 0 \\ 1 & -2 & 4 & 0 & 0 & 1 \end{array}\right] \xrightarrow{-1/2} \left[\begin{array}{ccc|ccc} 4 & -2 & 1 & 1 & 0 & 0 \\ 0 & 3 & -\frac{3}{2} & \frac{1}{2} & 1 & 0 \\ 1 & -2 & 4 & 0 & 0 & 1 \end{array}\right]$$

$$\xrightarrow{1/4} \left[\begin{array}{ccc|ccc} 4 & -2 & 1 & 1 & 0 & 0 \\ 0 & 3 & -\frac{3}{2} & \frac{1}{2} & 1 & 0 \\ 0 & -\frac{3}{2} & \frac{15}{4} & -\frac{1}{4} & 0 & 1 \end{array}\right]$$

The pivot values are shown above the arrows in the foregoing steps. The next step is to reduce the elements in the second column beneath the main diagonal to zero. For this matrix there is only one element to reduce. This step is

$$\left[\begin{array}{ccc|ccc} 4 & -2 & 1 & 1 & 0 & 0 \\ 0 & 3 & -\frac{3}{2} & \frac{1}{2} & 1 & 0 \\ 0 & -\frac{3}{2} & \frac{15}{4} & -\frac{1}{4} & 0 & 1 \end{array}\right] \xrightarrow{-1/2} \left[\begin{array}{ccc|ccc} 4 & -2 & 1 & 1 & 0 & 0 \\ 0 & 3 & -\frac{3}{2} & \frac{1}{2} & 1 & 0 \\ 0 & 0 & 3 & 0 & \frac{1}{2} & 1 \end{array}\right]$$

where the right-hand block is $\mathbf{L}^{-1}$.

This produces the upper triangular form and the *inverse* lower triangular matrix as shown. The lower triangular matrix $\mathbf{L}$ is formed by starting with the identity matrix and *inserting* the pivots in the corresponding locations. For example, the first pivot $(-1/2)$ was used to reduce the element in row 2 of the first column to zero. Therefore this pivot goes into position (2,1) of $\mathbf{L}$. The complete result is

$$\mathbf{L} = \begin{bmatrix} 1 & 0 & 0 \\ -\frac{1}{2} & 1 & 0 \\ \frac{1}{4} & -\frac{1}{2} & 1 \end{bmatrix}$$

The combination of these results gives the basic LU decomposition of the correlation matrix.

$$\begin{bmatrix} 4 & -2 & 1 \\ -2 & 4 & -2 \\ 1 & -2 & 4 \end{bmatrix} = \begin{bmatrix} 1 & 0 & 0 \\ -\frac{1}{2} & 1 & 0 \\ \frac{1}{4} & -\frac{1}{2} & 1 \end{bmatrix} \begin{bmatrix} 4 & -2 & 1 \\ 0 & 3 & -\frac{3}{2} \\ 0 & 0 & 3 \end{bmatrix}$$

The required three-factor triangular decomposition can be found by noticing that the diagonal elements of the upper triangular matrix above are the same as those of the diagonal matrix $\mathbf{D}_L$ required in (2.192). Therefore, the complete triangular decomposition is

$$\mathbf{R}_x = \mathbf{L}\mathbf{D}_L\mathbf{L}^T = \begin{bmatrix} 1 & 0 & 0 \\ -\frac{1}{2} & 1 & 0 \\ \frac{1}{4} & -\frac{1}{2} & 1 \end{bmatrix} \begin{bmatrix} 4 & 0 & 0 \\ 0 & 3 & 0 \\ 0 & 0 & 3 \end{bmatrix} \begin{bmatrix} 1 & -\frac{1}{2} & \frac{1}{4} \\ 0 & 1 & -\frac{1}{2} \\ 0 & 0 & 1 \end{bmatrix}$$

2.7.2 Use of QR Factorization

The QR factorization is another important "square root" method from computational linear algebra. When applied to the data matrix $\mathbf{X}$ it has approximately the same relation to triangular decomposition as SVD has to eigenvector decomposition and has similar computational advantages. That is, the QR factorization of the *data matrix* provides the factors in the triangular decomposition of the *correlation matrix*.

To begin, consider the data matrix (2.95)

$$\mathbf{X} = \begin{bmatrix} | & | & & | \\ | & | & & | \\ \mathbf{x}_1 & \mathbf{x}_2 & \cdots & \mathbf{x}_N \\ | & | & & | \\ | & | & & | \end{bmatrix}$$

It is assumed that the number of rows K is at least equal to the number of columns N and that the columns are independent. In other words, the matrix is of full rank. Since the columns are independent, they can be used to derive a set of N *orthonormal* vectors $\mathbf{q}_1, \mathbf{q}_2, \ldots, \mathbf{q}_N$ by the Gram-Schmidt orthogonalization procedure. The procedure is as follows. The first orthonormal vector is taken as a normalized version of the first column.

$$\mathbf{q}'_1 = \mathbf{x}_1 \,; \qquad \mathbf{q}_1 = \mathbf{q}'_1 / \|\mathbf{q}'_1\|$$

The next vector is formed by taking the second column, subtracting off its component in the $\mathbf{q}_1$ direction, and normalizing.

$$\mathbf{q}'_2 = \mathbf{x}_2 - (\mathbf{x}_2^{*T}\mathbf{q}_1)\mathbf{q}_1 \,; \qquad \mathbf{q}_2 = \mathbf{q}'_2 / \|\mathbf{q}'_2\|$$

In general, the lth orthonormal vector is formed from the lth column by subtracting its components in the directions of the previously formed orthonormal vectors, and normalizing.

$$\mathbf{q}'_l = \mathbf{x}_l - \sum_{i=1}^{l-1} (\mathbf{x}_l^{*T}\mathbf{q}_i)\mathbf{q}_i \,; \qquad \mathbf{q}_l = \mathbf{q}'_l / \|\mathbf{q}'_l\|$$

This is done for all $l \leq N$. Notice by the nature of the construction that $\mathbf{q}_1$ and $\mathbf{q}_2$ lie in the plane defined by $\mathbf{x}_1$ and $\mathbf{x}_2$. Likewise, $\mathbf{q}_1, \mathbf{q}_2, \ldots, \mathbf{q}_l$ lie in the subspace defined by $\mathbf{x}_1, \mathbf{x}_2, \ldots, \mathbf{x}_l$. Therefore, it is possible to write

$$\underbrace{\begin{bmatrix} | & | & & | \\ | & | & & | \\ \mathbf{q}_1 & \mathbf{q}_2 & \cdots & \mathbf{q}_N \\ | & | & & | \\ | & | & & | \end{bmatrix}}_{\mathbf{Q}_1} = \underbrace{\begin{bmatrix} | & | & & | \\ | & | & & | \\ \mathbf{x}_1 & \mathbf{x}_2 & \cdots & \mathbf{x}_N \\ | & | & & | \\ | & | & & | \end{bmatrix}}_{\mathbf{X}} \underbrace{\begin{bmatrix} \rho_{11} & \rho_{12} & \cdots & \rho_{1N} \\ 0 & \rho_{22} & \cdots & \rho_{2N} \\ \vdots & & \ddots & \vdots \\ 0 & 0 & \cdots & \rho_{NN} \end{bmatrix}}_{\mathbf{R}_1^{-1}}$$

where the matrix $\mathbf{R}_1^{-1}$ is *upper triangular*.[10] This last equation can be rewritten as

$$\mathbf{X} = \mathbf{Q}_1\mathbf{R}_1 = \underbrace{\begin{bmatrix} | & | & & | \\ | & | & & | \\ \mathbf{q}_1 & \mathbf{q}_2 & \cdots & \mathbf{q}_N \\ | & | & & | \\ | & | & & | \end{bmatrix}}_{\mathbf{Q}_1} \underbrace{\begin{bmatrix} r_{11} & r_{12} & \cdots & r_{1N} \\ 0 & r_{22} & \cdots & r_{2N} \\ \vdots & & \ddots & \vdots \\ 0 & 0 & \cdots & r_{NN} \end{bmatrix}}_{\mathbf{R}_1} \tag{2.197}$$

where the matrix $\mathbf{R}_1$ is again upper triangular. This is one possible form of the QR factorization; the matrix $\mathbf{X}$ is expressed as the product of a rectangular matrix whose columns are

[10]We hesitate to use the symbol $\mathbf{R}_1$ since this matrix is *not a correlation matrix*, but it seems peculiar to use a letter other than "R" in the QR decomposition. Since the correlation matrix is always denoted by a bold R with a random vector as subscript (e.g., $\mathbf{R}_x$), use of the symbol $\mathbf{R}_1$ in this context will hopefully not be confusing.

orthonormal[11] and a square upper triangular matrix. This equation has the interpretation that $\mathbf{x}_1$ lies in the direction defined by $\mathbf{q}_1$, that $\mathbf{x}_1, \mathbf{x}_2$ lie in the plane determined by $\mathbf{q}_1, \mathbf{q}_2$, and that $\mathbf{x}_1, \mathbf{x}_2, \ldots, \mathbf{x}_l$ lie in the subspace defined by $\mathbf{q}_1, \mathbf{q}_2, \ldots, \mathbf{q}_l$ for $l \leq N$. The coefficients r_{ij} are given by

$$r_{ij} = \mathbf{q}_i^{*T}\mathbf{x}_j \tag{2.198}$$

The QR factorization is usually written in terms of larger matrices as

$$\begin{aligned} \mathbf{X} &= \begin{bmatrix} \mathbf{Q}_1 & \mathbf{Q}_2 \end{bmatrix} \begin{bmatrix} \mathbf{R}_1 \\ \mathbf{0} \end{bmatrix} \\ &= \underbrace{\begin{bmatrix} | & & | \\ | & & | \\ \mathbf{q}_1 & \cdots & \mathbf{q}_N \\ | & & | \\ | & & | \end{bmatrix}}_{\mathbf{Q}_1} \underbrace{\begin{bmatrix} | & & | \\ | & & | \\ \mathbf{q}_{N+1} & \cdots & \mathbf{q}_K \\ | & & | \\ | & & | \end{bmatrix}}_{\mathbf{Q}_2} \begin{bmatrix} r_{11} & r_{12} & \cdots & r_{1N} \\ 0 & r_{22} & \cdots & r_{2N} \\ \vdots & & \ddots & \vdots \\ 0 & 0 & \cdots & r_{NN} \\ 0 & 0 & \cdots & 0 \\ \vdots & \vdots & & \vdots \\ 0 & 0 & \cdots & 0 \end{bmatrix} \end{aligned} \tag{2.199}$$

where the columns $\mathbf{q}_{N+1} \cdots \mathbf{q}_K$ are a set of $K - N$ orthonormal vectors orthogonal to the original set $\mathbf{q}_1 \cdots \mathbf{q}_N$. The first term in the product is thus a unitary matrix, while the second term is an upper (or right) triangular rectangular matrix.

The QR decomposition can be used to obtain the triangular decomposition of an estimated correlation matrix which is in the form (2.94). The product $\mathbf{X}^{*T}\mathbf{X}$ can be written using (2.197) and the orthogonality of the columns of $\mathbf{Q}_1$ as

$$\mathbf{X}^{*T}\mathbf{X} = \mathbf{R}_1^{*T}\mathbf{Q}_1^{*T}\mathbf{Q}_1\mathbf{R}_1 = \mathbf{R}_1^{*T}\mathbf{R}_1 \tag{2.200}$$

This product of a lower triangular matrix and its Hermitian transpose is the Cholesky decomposition. Now if $\hat{\mathbf{R}}_x$ is defined by (2.94), then from (2.200),

$$\hat{\mathbf{R}}_x = \frac{1}{K}\mathbf{R}_1^{*T}\mathbf{R}_1 \tag{2.201}$$

Putting $\hat{\mathbf{R}}_x$ in the form

$$\hat{\mathbf{R}}_x = \hat{\mathbf{L}}\hat{\mathbf{D}}_L\hat{\mathbf{L}}^{*T} \tag{2.202}$$

requires that

$$\mathbf{R}_1^{*T} = \sqrt{K}\hat{\mathbf{L}}\hat{\mathbf{D}}_L^{1/2} \tag{2.203}$$

which leads to the relations

$$\hat{\mathbf{D}}_L^{1/2} = \frac{1}{\sqrt{K}}\operatorname{diag}(\mathbf{R}_1) \tag{2.204}$$

and

$$\hat{\mathbf{L}} = \frac{1}{\sqrt{K}}\mathbf{R}_1^{*T}\hat{\mathbf{D}}_L^{-1/2} \tag{2.205}$$

[11] The matrix is known as a "slice" of a unitary matrix.

Then of course $\hat{\mathbf{D}}_L$ is given by

$$\hat{\mathbf{D}}_L = (\hat{\mathbf{D}}_L^{1/2})^2 \tag{2.206}$$

Note that in computing $\hat{\mathbf{L}}$ from (2.205) it is important to use $\hat{\mathbf{D}}_L^{1/2}$ as computed from (2.204). If you compute $\hat{\mathbf{D}}_L$ first from (2.206) and take the square root, you could end up with the wrong sign! The following example illustrates the computation.

Example 2.14

A simple data matrix has the form

$$\mathbf{X} = \begin{bmatrix} \sqrt{3} & 0 \\ -\sqrt{3} & \sqrt{3} \\ 0 & -\sqrt{3} \end{bmatrix}$$

Note that the correlation matrix corresponding to this data matrix is

$$\hat{\mathbf{R}}_x = \frac{1}{3}\mathbf{X}^T\mathbf{X} = \begin{bmatrix} 2 & -1 \\ -1 & 2 \end{bmatrix}$$

which is the same as the correlation matrix in Example 2.12. The QR decomposition of the data matrix produces

$$\mathbf{X} = \begin{bmatrix} -\frac{1}{\sqrt{2}} & -\frac{1}{\sqrt{6}} & \frac{1}{\sqrt{3}} \\ \frac{1}{\sqrt{2}} & -\frac{1}{\sqrt{6}} & \frac{1}{\sqrt{3}} \\ 0 & \sqrt{\frac{2}{3}} & \frac{1}{\sqrt{3}} \end{bmatrix} \begin{bmatrix} -\sqrt{6} & \sqrt{\frac{3}{2}} \\ 0 & -\frac{3}{\sqrt{2}} \\ 0 & 0 \end{bmatrix}$$

Thus

$$\mathbf{R}_1 = \begin{bmatrix} -\sqrt{6} & \sqrt{\frac{3}{2}} \\ 0 & -\frac{3}{\sqrt{2}} \end{bmatrix}$$

This yields

$$\hat{\mathbf{D}}_L^{1/2} = \frac{1}{\sqrt{3}}\text{diag}\,(\mathbf{R}_1) = \begin{bmatrix} -\sqrt{2} & 0 \\ 0 & -\sqrt{\frac{3}{2}} \end{bmatrix}$$

$$\begin{aligned} \hat{\mathbf{L}} &= \frac{1}{\sqrt{3}}\mathbf{R}_1^T\hat{\mathbf{D}}_L^{-1/2} \\ &= \frac{1}{\sqrt{3}}\begin{bmatrix} -\sqrt{6} & 0 \\ \sqrt{\frac{3}{2}} & -\frac{3}{\sqrt{2}} \end{bmatrix}\begin{bmatrix} -\frac{1}{\sqrt{2}} & 0 \\ 0 & -\sqrt{\frac{2}{3}} \end{bmatrix} = \begin{bmatrix} 1 & 0 \\ -\frac{1}{2} & 1 \end{bmatrix} \end{aligned}$$

and finally,

$$\hat{\mathbf{D}}_L = (\hat{\mathbf{D}}_L^{1/2})^2 = \begin{bmatrix} 2 & 0 \\ 0 & \frac{3}{2} \end{bmatrix}$$

These are the same factors obtained in Example 2.12.

2.7.3 Upper–Lower Decomposition

The correlation or covariance matrix can be factored into an alternative upper–lower triangular decomposition. This decomposition is of the form

$$\mathbf{R}_x = \mathbf{U}_1 \mathbf{D}_U \mathbf{U}_1^{*T} \tag{2.207}$$

where the matrix $\mathbf{U}_1$ is unit upper triangular and the matrix $\mathbf{D}_U$ is diagonal. It will become clear from what follows that in general $\mathbf{U}_1$ is *not* equal to $\mathbf{L}^{*T}$ and $\mathbf{D}_U$ is *not* equal to $\mathbf{D}_L$.

Since (2.207) implies that

$$\mathbf{U}_1^{-1} \mathbf{R}_x (\mathbf{U}_1^{-1})^{*T} = \mathbf{D}_U \tag{2.208}$$

the transformation

$$\boldsymbol{x}''' = \mathbf{U}_1^{-1} \boldsymbol{x} \tag{2.209}$$

also leads to a vector with orthogonal components whose second moments are the elements of $\mathbf{D}_U$. Since $\mathbf{U}_1$ is an upper triangular matrix, this particular transformation represents an *anti*causal transformation if $\boldsymbol{x}$ is a signal vector. The fact that this transformation is anticausal does not minimize its importance. In later chapters it is shown that this transformation has equal importance to the causal transformation in the formulation of optimal filtering and prediction.

Some simple relations come to light when the triangular decomposition of both the correlation matrix $\mathbf{R}_x$ and its reversal $\tilde{\mathbf{R}}_x$ is considered. Suppose that we take the reversal of (2.207). This yields

$$\tilde{\mathbf{R}}_x = \tilde{\mathbf{U}}_1 \tilde{\mathbf{D}}_U \tilde{\mathbf{U}}_1^{*T} \tag{2.210}$$

Note that $\tilde{\mathbf{U}}_1$ is a unit *lower* triangular matrix and $\tilde{\mathbf{U}}_1^{*T}$ is unit upper triangular. $\tilde{\mathbf{D}}_U$ is a diagonal matrix whose elements are the elements of $\mathbf{D}_U$ in reverse order. Since the lower–upper decomposition of a correlation matrix is unique, (2.210) is the *lower–upper* decomposition of the reversed correlation matrix.

Now consider the reversal of (2.192):

$$\tilde{\mathbf{R}}_x = \tilde{\mathbf{L}} \tilde{\mathbf{D}}_L \tilde{\mathbf{L}}^{*T} \tag{2.211}$$

Since $\tilde{\mathbf{L}}$ is unit upper triangular and $\tilde{\mathbf{D}}_L$ is diagonal, this is the unique *upper–lower* decomposition of $\tilde{\mathbf{R}}_x$. Thus the upper–lower decomposition of a correlation matrix is the lower–upper decomposition of its reversal, and vice versa. The relations are summarized in Table 2.7.

It was seen how the QR factorization can provide the lower–upper triangular decomposition of the estimated correlation matrix directly from the data matrix. The previous relations provide a way to obtain the upper–lower triangular decomposition by using QR factorization. The data matrix $\mathbf{X}$ is reversed and the QR factorization is performed. The triangular factors then obtained are the reversals of the upper–lower triangular factors of $\hat{\mathbf{R}}_x$.

TABLE 2.7 TRIANGULAR DECOMPOSITION RELATIONS FOR THE CORRELATION MATRIX

Matrix	Lower-upper decomposition	Upper-lower decomposition
$\mathbf{R}_x$	$\mathbf{R}_x = \mathbf{L}\mathbf{D}_L\mathbf{L}^{*T}$	$\mathbf{R}_x = \mathbf{U}_1\mathbf{D}_U\mathbf{U}_1^{*T}$
$\tilde{\mathbf{R}}_x$	$\tilde{\mathbf{R}}_x = \tilde{\mathbf{U}}_1\tilde{\mathbf{D}}_U\tilde{\mathbf{U}}_1^{*T}$	$\tilde{\mathbf{R}}_x = \tilde{\mathbf{L}}\tilde{\mathbf{D}}_L\tilde{\mathbf{L}}^{*T}$

2.8 CHAPTER SUMMARY

This chapter develops important relations for random vectors. Random vectors are frequently used to represent a set of samples from a random sequence.

Random vectors are described by the joint distribution and density functions of their components. For complex random vectors this is a joint distribution or density function between the real and imaginary parts of all of the components. Joint and conditional distributions and densities for two random vectors and a form of Bayes' rule for densities are other important concepts.

When it is not practical to deal with the entire density function for a random vector, the concept of moments is of great importance. The mean vector, correlation matrix, and covariance matrix are defined in this chapter for both real and complex random vectors. It is shown that the definitions of correlation and covariance matrices for a complex random vector assume a certain symmetry in the second moment properties of the real and imaginary components of the vectors. This permits the moments of all components of the random vector to be specified by a single (complex-valued) matrix. Joint and conditional forms of the moments are also defined. When sample vectors are available, simple procedures described here permit estimation of the moments.

The multivariate Gaussian density function is introduced in this chapter. This density is completely specified by only the mean and covariance matrix of the random vector and is important for a number of reasons. The multivariate Gaussian density has a slightly different form if the random vector is real than if it is complex. Joint and conditional forms of this density are also discussed.

The effect of linear transformations on random vectors is also investigated. Linear transformations of random vectors lead to simple transformations of their moments. For Gaussian random vectors, linear transformations lead to random vectors that are still Gaussian.

The reversal of a vector is a reordering of its components where the last component becomes the first and the first component becomes the last. The reversal of a matrix is a reordering of its elements in *both* the rows and columns. Reversal of a matrix equation generally requires a reversal of all the vectors and matrices comprising it. Reversals are important for representing operations on vectors that are common in signal processing.

Finally, two fundamentally different ways of transforming a random vector to one with orthogonal or uncorrelated components are described. The first type of transformation uses an eigenvector decomposition of the correlation or covariance matrix, while the second involves a triangular decomposition.

The eigenvector transformation involves solving the eigenvalue problem for the correlation or covariance matrix. The singular value decomposition can be used as an efficient numerical method for computing this transformation when the correlation or covariance matrix is estimated from a data matrix in a standard form. The right singular vectors of the data matrix are the eigenvectors of the correlation matrix, and the singular values are the square roots of the eigenvalues. The eigenvector transformation can be interpreted as a rotation of the coordinate system in which the data is represented. For Gaussian random vectors this transformation provides a convenient way to plot the contours of the density, which are known as concentration ellipsoids. It is also possible to perform simultaneous diagonalization of two correlation or covariance matrices. The procedure is equivalent to solving a *generalized* eigenvalue problem.

Triangular decomposition is an alternative way to diagonalize the correlation or covariance matrix. Here two possibilities arise: lower–upper triangularization and upper–lower triangularization. The former leads to a "causal" transformation, while the latter leads to an "anticausal" transformation of the data. The QR decomposition provides an efficient way to compute the factors in the triangular decomposition of an estimated correlation matrix directly from the data matrix. Since product terms do not have to be computed from the data before factoring, this technique can provide better numerical precision.

REFERENCES

1. Athanasios Papoulis. *Probability and Statistics.* Prentice Hall, Inc., Englewood Cliffs, New Jersey, 1990.
2. Alvin W. Drake. *Fundamentals of Applied Probability Theory.* McGraw-Hill, New York, 1967. (Reissued 1991.)
3. Howard Anton and Chris Rorres. *Elementary Linear Algebra*, 6th ed. John Wiley & Sons, New York, 1991.
4. Gilbert Strang. *Linear Algebra and Its Applications*, 3rd ed. Harcourt Brace Jovanovich, San Diego, 1976.
5. G. W. Stewart. *Introduction to Matrix Computations.* Academic Press, New York, 1973.
6. J. H. Wilkinson. *The Algebraic Eigenvalue Problem.* Oxford University Press, New York, 1965.
7. Beresford N. Parlett. *The Symmetric Eigenvalue Problem.* Prentice Hall, Inc., Englewood Cliffs, New Jersey, 1980.
8. Gene H. Golub and Charles F. Van Loan. *Matrix Computations*, 2nd ed. Johns Hopkins University Press, Baltimore, Maryland, 1989.
9. K. V. Mardia, J. T. Kent, and J. M. Bibby. *Multivariate Analysis.* Academic Press, New York, 1979.
10. Donald F. Morrison. *Multivariate Statistical Methods.* McGraw-Hill, New York, 1967.
11. Kenneth S. Miller. *Complex Stochastic Processes.* Addison-Wesley, Reading, Massachusetts, 1974.
12. Charles L. Lawson and Richard J. Hanson. *Solving Least Squares Problems.* Prentice Hall, Inc., Englewood Cliffs, New Jersey, 1974.

PROBLEMS

2.1. A probability density function for a two-dimensional random vector $\boldsymbol{x}$ is defined by

$$f_{\boldsymbol{x}}(\mathbf{x}) = \begin{cases} A\mathrm{x}_1^2\mathrm{x}_2 & \mathrm{x}_1, \mathrm{x}_2 \geq 0 \text{ and } \mathrm{x}_1 + \mathrm{x}_2 \leq 1 \\ 0 & \text{otherwise} \end{cases}$$

(a) What is the distribution function $F_{\boldsymbol{x}}(\mathbf{x})$? Use this result to find the numerical value of the constant A.

(b) What is the marginal density $f_{x_2}(\mathrm{x}_2)$?

2.2. Let x and y be random variables (one-dimensional random vectors) with density functions

$$f_{x|y}(\mathrm{x}|\mathrm{y}) = \begin{cases} e^{-(\mathrm{x}-\mathrm{y})} & \mathrm{y} \leq \mathrm{x} < \infty \\ 0 & \mathrm{x} < \mathrm{y} \end{cases}$$

and

$$f_y(\mathrm{y}) = \begin{cases} 1 & 0 \leq \mathrm{y} \leq 1 \\ 0 & \text{otherwise} \end{cases}$$

(a) What is $f_{xy}(\mathrm{x}, \mathrm{y})$? Specify the region where the joint density is nonzero and sketch this region in the xy plane.

(b) What is $f_x(\mathrm{x})$? Don't forget to specify regions of definition.

2.3. The joint density function for the two-dimensional random vectors $\boldsymbol{x}$ and $\boldsymbol{y}$ is

$$f_{\boldsymbol{xy}}(\mathbf{x}, \mathbf{y}) = \begin{cases} \mathrm{x}_1\mathrm{x}_2 + 3\mathrm{y}_1\mathrm{y}_2 & 0 \leq \mathrm{x}_1, \mathrm{x}_2, \mathrm{y}_1, \mathrm{y}_2 \leq 1 \\ 0 & \text{otherwise} \end{cases}$$

Are $\boldsymbol{x}$ and $\boldsymbol{y}$ statistically independent? Show why or why not.

2.4. The components x_1 and x_2 of a two-dimensional random vector $\boldsymbol{x}$ are statistically independent and have the marginal densities shown in Fig. PR2.4.

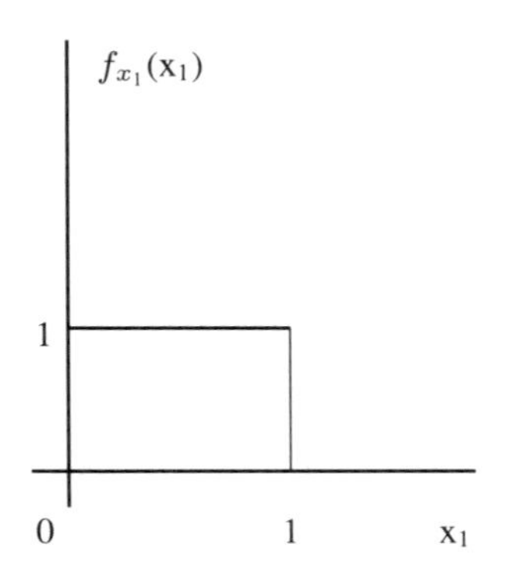

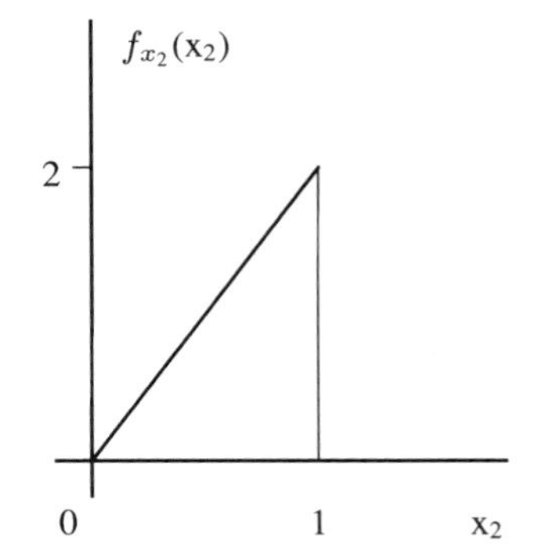

Figure PR2.4

Let A be the event $x_1 \leq x_2$.

(a) Find $f_x(\mathbf{x})$ and sketch it.

(b) What is $\Pr\left[\mathsf{A}\right]$?

(c) Define the density function conditioned on the event A as

$$f_{x|\mathsf{A}}(\mathbf{x}|\mathsf{A}) = \frac{f_x(\mathbf{x})}{\Pr\left[\mathsf{A}\right]} \quad \text{for } \mathrm{x}_1 \leq \mathrm{x}_2$$

Are the components of the random vector independent when conditioned on the event A? In other words, is it true that

$$f_{x|\mathsf{A}}(\mathbf{x}|\mathsf{A}) = f_{x_1|\mathsf{A}}(\mathrm{x}_1|\mathsf{A}) \cdot f_{x_2|\mathsf{A}}(\mathrm{x}_2|\mathsf{A})?$$

2.5. The following matrices are proposed to be correlation matrices. State if they are legitimate correlation matrices and if not, then why not.

(a)

$$\mathbf{R}_x = \begin{bmatrix} 3 & 2 & 1 \\ -2 & 3 & 2 \\ -1 & -2 & 3 \end{bmatrix}$$

(b)

$$\mathbf{R}_x = \begin{bmatrix} 2 & 4 \\ 4 & 2 \end{bmatrix}$$

(c)

$$\mathbf{R}_x = \begin{bmatrix} 4 & 2 & 1 \\ 2 & 4 & 2 \\ 1 & 2 & 4 \end{bmatrix}$$

(d)

$$\mathbf{R}_x = \begin{bmatrix} 1 & 1+\jmath \\ 1-\jmath & 1 \end{bmatrix}$$

(e)

$$\mathbf{R}_x = \begin{bmatrix} 2 & 2 & 3 \\ 2 & 4 & 2 \\ 3 & 2 & 2 \end{bmatrix}$$

(f)

$$\mathbf{R}_x = \begin{bmatrix} 1 & 1 & 1 \\ 1 & 1 & 1 \\ 1 & 1 & 1 \end{bmatrix}$$

(g)

$$\mathbf{R}_x = \begin{bmatrix} 1 & 1+\jmath & 0 \\ 1-\jmath & 2j & 1+\jmath \\ 0 & 1-\jmath & 1 \end{bmatrix}$$

2.6. **(a)** Calculate the mean vector, correlation matrix, and covariance matrix of the data given by

$$\mathbf{x}^{(1)} = \begin{bmatrix} 1 \\ 1 \end{bmatrix}, \ \mathbf{x}^{(2)} = \begin{bmatrix} 1 \\ 2 \end{bmatrix}, \ \mathbf{x}^{(3)} = \begin{bmatrix} 2 \\ 0 \end{bmatrix}, \ \mathbf{x}^{(4)} = \begin{bmatrix} 0 \\ 1 \end{bmatrix}$$

Show that the calculated values satisfy (2.67).

(b) If the data is known to be Gaussian, what is the density function ?

2.7. Repeat Problem 2.6 using the complex data

$$\mathbf{x}^{(1)} = \begin{bmatrix} 1-\jmath \\ 0 \end{bmatrix}, \ \mathbf{x}^{(2)} = \begin{bmatrix} 2+\jmath \\ 1-\jmath \end{bmatrix},$$

$$\mathbf{x}^{(3)} = \begin{bmatrix} -2+\jmath \\ 2+\jmath \end{bmatrix}, \ \mathbf{x}^{(4)} = \begin{bmatrix} 0 \\ -2+\jmath \end{bmatrix}$$

2.8. Estimate the mean vector, correlation matrix, and covariance matrix for the data

$$\mathbf{x}^{(1)} = \begin{bmatrix} 1+\jmath \\ 0 \\ 0 \end{bmatrix}, \ \mathbf{x}^{(2)} = \begin{bmatrix} 1-\jmath \\ 1+\jmath \\ 0 \end{bmatrix}, \ \mathbf{x}^{(3)} = \begin{bmatrix} -1+\jmath \\ 1-\jmath \\ 1+\jmath \end{bmatrix}$$

$$\mathbf{x}^{(4)} = \begin{bmatrix} -1-\jmath \\ -1+\jmath \\ 1-\jmath \end{bmatrix}, \ \mathbf{x}^{(5)} = \begin{bmatrix} 0 \\ -1-\jmath \\ -1+\jmath \end{bmatrix}, \ \mathbf{x}^{(6)} = \begin{bmatrix} 0 \\ 0 \\ -1-\jmath \end{bmatrix}$$

2.9. By following the development in Section 2.2.1 for the correlation matrix, show that if random vectors $\boldsymbol{x}$ and $\boldsymbol{y}$ have the properties

$$\mathcal{E}\{\boldsymbol{x}_r \boldsymbol{y}_r^T\} = \mathcal{E}\{\boldsymbol{x}_i \boldsymbol{y}_i^T\}$$

and

$$\mathcal{E}\{\boldsymbol{x}_i \boldsymbol{y}_r^T\} = -\mathcal{E}\{\boldsymbol{x}_r \boldsymbol{y}_i^T\}$$

then the cross-correlation matrix defined by (2.71) carries all the necessary information about the second moment properties of the random vectors. What do the conditions imply about the real and imaginary parts of the *components* of the random vectors?

2.10. Show that when the mean vector, correlation matrix, and covariance matrix are estimated from data according to the procedure of Section 2.2.3, the following are true:

(a) $\hat{\mathbf{R}}_x$ and $\hat{\mathbf{C}}_x$ are symmetric.

(b) $\hat{\mathbf{R}}_x$ and $\hat{\mathbf{C}}_x$ are always positive semidefinite and are positive definite as long as there are N linearly independent sample vectors where N is the dimension of the matrix.

(c) The estimates $\hat{\mathbf{m}}_x$, $\hat{\mathbf{R}}_x$, and $\hat{\mathbf{C}}_x$ satisfy (2.67).

2.11. State if the following pairs could represent the components of a complex correlation matrix. Give a reason for your answer.

(a)

$$\mathbf{R}_x^{\mathrm{E}} = \begin{bmatrix} 2 & 1 \\ 1 & 2 \end{bmatrix} \qquad \mathbf{R}_x^{\mathrm{O}} = \begin{bmatrix} 0 & 0.3 \\ -0.3 & 0 \end{bmatrix}$$

(b)

$$\mathbf{R}_x^{\mathrm{E}} = \begin{bmatrix} 2 & 2 \\ 2 & 1 \end{bmatrix} \qquad \mathbf{R}_x^{\mathrm{O}} = \begin{bmatrix} 0 & -0.5 \\ 0.5 & 0 \end{bmatrix}$$

(c)

$$\mathbf{R}_x^{\mathrm{E}} = \begin{bmatrix} 7 & 3 & 1 \\ 3 & 6 & 2 \\ 1 & 2 & 5 \end{bmatrix} \qquad \mathbf{R}_x^{\mathrm{O}} = \begin{bmatrix} 1 & 0.5 & 0.3 \\ -0.5 & 1 & 0.2 \\ -0.3 & -0.2 & 1 \end{bmatrix}$$

2.12. Show that the matrix

$$\mathbf{R}_v = \begin{bmatrix} \mathbf{R}_x^{\mathrm{E}} & -\mathbf{R}_x^{\mathrm{O}} \\ \mathbf{R}_x^{\mathrm{O}} & \mathbf{R}_x^{\mathrm{E}} \end{bmatrix}$$

is positive semidefinite if and only if the correlation matrix

$$\mathbf{R}_x = 2\mathbf{R}_x^{\mathrm{E}} + \jmath 2\mathbf{R}_x^{\mathrm{O}}$$

is positive semidefinite and that $\mathbf{R}_v$ is positive definite if and only if $\mathbf{R}_x$ is positive definite.

2.13. Show that if $\mathbf{C}_x^{-1} = \mathbf{F} + \jmath\mathbf{G}$, then the inverse covariance matrix $\mathbf{C}_v^{-1}$ corresponding to the real vector $\boldsymbol{v}$ of (2.53) has the form

$$\mathbf{C}_v^{-1} = \begin{bmatrix} 2\mathbf{F} & -2\mathbf{G} \\ 2\mathbf{G} & 2\mathbf{F} \end{bmatrix}$$

Find explicit expressions for $\mathbf{F}$ and $\mathbf{G}$ in terms of

$$\mathbf{C}_x^{\mathrm{E}} = \tfrac{1}{2}\mathrm{Re}\,\mathbf{C}_x \quad \text{and} \quad \mathbf{C}_x^{\mathrm{O}} = \tfrac{1}{2}\mathrm{Im}\,\mathbf{C}_x$$

2.14. The covariance matrix $\mathbf{C}_v$ corresponding to the real vector $\boldsymbol{v}$ of (2.53) and the covariance matrix $\mathbf{C}_x$ are related through the terms $\mathbf{C}_x^{\mathrm{E}}$ and $\mathbf{C}_x^{\mathrm{O}}$ in a manner similar to the relation of the correlation matrices in Table 2.3. Show that a concise, specific form of the relation between $\mathbf{C}_v$ and $\mathbf{C}_x$ is

$$\mathbf{C}_v = \tfrac{1}{2}\mathrm{Re}\begin{bmatrix} \mathbf{C}_x & \jmath\mathbf{C}_x \\ -\jmath\mathbf{C}_x & \mathbf{C}_x \end{bmatrix} = \tfrac{1}{2}\mathrm{Re}\left\{\begin{bmatrix} \mathbf{I} \\ -\jmath\mathbf{I} \end{bmatrix} \mathbf{C}_x \begin{bmatrix} \mathbf{I} & \jmath\mathbf{I} \end{bmatrix}\right\}$$

Further show that $\mathbf{C}_v^{-1}$ is related to $\mathbf{C}_x^{-1}$ by

$$\mathbf{C}_v^{-1} = 2\mathrm{Re}\left\{\begin{bmatrix} \mathbf{I} \\ -\jmath\mathbf{I} \end{bmatrix} \mathbf{C}_x^{-1} \begin{bmatrix} \mathbf{I} & \jmath\mathbf{I} \end{bmatrix}\right\}$$

2.15. Let $\check{\mathbf{e}}$ be an eigenvector of $\mathbf{C}_x$ corresponding to an eigenvalue $\check{\lambda}$. Show that the real covariance matrix $\mathbf{C}_v$ defined in Problem 2.14 has two real eigenvectors,

$$\mathbf{u_r} = \mathrm{Re}\begin{bmatrix} \check{\mathbf{e}} \\ -\jmath\check{\mathbf{e}} \end{bmatrix}; \qquad \mathbf{u_i} = \mathrm{Im}\begin{bmatrix} \check{\mathbf{e}} \\ -\jmath\check{\mathbf{e}} \end{bmatrix}$$

both corresponding to an eigenvalue of $\check{\lambda}/2$. Use this fact to show that the determinant of $\mathbf{C}_v$ is related to the determinant of $\mathbf{C}_x$ by

$$|\mathbf{C}_v| = \left(\tfrac{1}{2}\right)^{2N} |\mathbf{C}_x|^2$$

where N is the dimension of $\mathbf{C}_x$.

2.16. By forming the multivariate Gaussian density for the real random vector $\boldsymbol{v}$ of (2.53) and applying symmetry conditions for the covariance analogous to (2.55), show that the complex Gaussian density function has the form (2.99). You may want to use the results of Problems 2.14 and 2.15.

2.17. The entropy $\mathcal{H}$ for a random vector is defined by

$$\mathcal{H} = -\int_{-\infty}^{\infty} f_x(\mathbf{x}) \ln f_x(\mathbf{x}) d\mathbf{x} = -\mathcal{E}\{\ln f_x(\boldsymbol{x})\}$$

Beginning with the expression (2.99) for the density, show that the entropy for a complex Gaussian random vector is given by

$$\mathcal{H} = N(1 + \ln \pi) + \ln |\mathbf{C}_x|$$

What is the corresponding expression for a real Gaussian random vector?

2.18. By computing the product and showing that it is equal to the identity matrix, verify that the inverse of the matrix $\mathbf{C}_u$ in (2.103) is given by

$$\mathbf{C}_u^{-1} = \begin{bmatrix} \mathbf{C}_x^{-1} + \mathbf{C}_x^{-1}\mathbf{C}_{xy}\mathbf{C}_{y|x}^{-1}\mathbf{C}_{xy}^{*T}\mathbf{C}_x^{-1} & -\mathbf{C}_x^{-1}\mathbf{C}_{xy}\mathbf{C}_{y|x}^{-1} \\ -\mathbf{C}_{y|x}^{-1}\mathbf{C}_{xy}^{*T}\mathbf{C}_x^{-1} & \mathbf{C}_{y|x}^{-1} \end{bmatrix}$$

where $\mathbf{C}_{y|x}$ is given by (2.109).

2.19. Verify that the the matrix $\mathbf{C}_u$ in (2.103) can be written as

$$\mathbf{C}_u = \begin{bmatrix} \mathbf{I} & \mathbf{0} \\ \mathbf{C}_{xy}^{*T}\mathbf{C}_x^{-1} & \mathbf{I} \end{bmatrix} \begin{bmatrix} \mathbf{C}_x & \mathbf{C}_{xy} \\ \mathbf{0} & \mathbf{C}_{y|x} \end{bmatrix}$$

where $\mathbf{C}_{y|x}$ is given by (2.109). Use this to show that

$$|\mathbf{C}_u| = |\mathbf{C}_x| \cdot |\mathbf{C}_{y|x}|$$

2.20. By carrying out the matrix products and simplifying, show that the product of f_x and $f_{y|x}$ for Gaussian random vectors is equal to f_{xy} [see equations (2.97), (2.104), and (2.106) or (2.99), (2.105), and (2.107)]. Use the results of Problems 2.18 and 2.19 to help in showing this result.

2.21. **(a)** Given the correlation matrix and mean vector

$$\mathbf{R}_x = \begin{bmatrix} 9 & 3 & 1 \\ 3 & 9 & 3 \\ 1 & 3 & 9 \end{bmatrix} \qquad \mathbf{m}_x = \begin{bmatrix} 1 \\ -1 \\ 1 \end{bmatrix}$$

and the transformation

$$\boldsymbol{y} = \begin{bmatrix} 1.0 & 0.5 & 0.5 \\ 0.2 & 0.3 & 0.5 \\ 0.4 & 0.1 & 1.0 \end{bmatrix} \boldsymbol{x}$$

find the mean vector, correlation matrix, and covariance matrix of $\boldsymbol{y}$.

(b) Repeat this for the transformation

$$\boldsymbol{y} = \begin{bmatrix} 2.0 & 1.0 & 0.5 \\ 0.3 & 0.3 & 0.4 \end{bmatrix} \boldsymbol{x}$$

2.22. Let random variables x and y be defined by

$$x = 3u - 4v$$

$$y = 2u + v$$

where u and v are uncorrelated Gaussian random variables with mean values of 1 and variances of 1.

(a) Find the mean values of x and y.
(b) Find the variances of x and y.
(c) Write an expression for the joint density function of x and y.
(d) Write an expression for the conditional density of y given x.

Hint: What is the mean vector and covariance matrix for the vector whose components are u and v? Use these to find the corresponding quantites for x and y.

2.23. Let v_1, v_2, v_3, v_4 be a set of zero-mean independent random variables with variances equal to $1, 2, 3, 4$. Let x_1, x_2, x_3, x_4 be defined by

$$x_1 = v_1 + v_2 + v_3 + v_4$$

$$x_2 = -v_1 + v_2 + v_3 - v_4$$

$$x_3 = v_1 - v_2 + v_3 - v_4$$

$$x_4 = v_1 + v_2 - v_3 - v_4$$

Show that x_1 and x_2 are uncorrelated, and x_3 and x_4 are uncorrelated, but that x_2 and x_3 are *not* uncorrelated and x_1 and x_4 are *not* uncorrelated.

2.24. Given the correlation matrix

$$\mathbf{R}_x = \begin{bmatrix} 2 & 1 \\ 1 & 2 \end{bmatrix}$$

and the transformation

$$\boldsymbol{y} = \begin{bmatrix} 1 & -3 \\ 2 & -1 \\ 3 & 2 \end{bmatrix} \boldsymbol{x}$$

(a) Show that $\mathbf{R}_x$ is (strictly) positive definite.
(b) Is the matrix $\mathbf{R}_y$ (strictly) positive definite? Explain.
(c) Find the eigenvalues of the matrices $\mathbf{R}_x$ and $\mathbf{R}_y$. Is there anything about the eigenvalues of $\mathbf{R}_y$ that corroborates your answer to part (b)?

2.25. Show that all of the properties of the reversal listed in Table 2.5 are correct.

2.26. Show that the eigenvectors and eigenvalues of the correlation matrix

$$\begin{bmatrix} \frac{9}{4} & -\frac{3}{2} \\ -\frac{3}{2} & \frac{9}{4} \end{bmatrix}$$

are as given in Example 2.10.

2.27. Given a real *zero-mean* random vector $\boldsymbol{x}$ with components x_1 and x_2 having second moments

$$\mathcal{E}\{x_1^2\} = \mathcal{E}\{x_2^2\} = 2; \qquad \mathcal{E}\{x_1 x_2\} = 0$$

let the components of a new random vector $\boldsymbol{y}$ be defined by

$$y_1 = x_1 + 2x_2$$

$$y_2 = 2x_1 + x_2$$

(a) Find the mean vector and the covariance matrix for $\boldsymbol{y}$.
(b) Give an expression for the density function of $\boldsymbol{y}$ if $\boldsymbol{x}$ is known to be Gaussian. You can leave the result in a matrix form, but work out all determinants, matrix inverse, and so on.
(c) Plot the contours of the density function of $\boldsymbol{y}$.

2.28. Consider the orthogonal transformation of the correlated zero-mean random variables x_1 and x_2.

$$\begin{bmatrix} y_1 \\ y_2 \end{bmatrix} = \begin{bmatrix} \cos\theta & \sin\theta \\ -\sin\theta & \cos\theta \end{bmatrix} \begin{bmatrix} x_1 \\ x_2 \end{bmatrix}$$

Let $\mathcal{E}\{x_1^2\} = \sigma_1^2$, $\mathcal{E}\{x_2^2\} = \sigma_2^2$, and $\mathcal{E}\{x_1 x_2\} = \rho\sigma_1\sigma_2$. Determine the angle θ such that y_1 and y_2 are uncorrelated.

2.29. Say if the following statements are correct or incorrect and why.
(a) If the eigenvalues of a covariance matrix are all equal, then the components of the random vector are uncorrelated to begin with.
(b) If $\boldsymbol{x}$ is a Gaussian random vector and the *conditional* density function for $\boldsymbol{y}$ given $\boldsymbol{x}$ is Gaussian, then the *unconditional* density function for $\boldsymbol{y}$ is also Gaussian.
(c) The SVD of the correlation matrix is the same as its eigenvector decomposition.
(d) If a correlation matrix is positive semidefinite but not positive definite, then at least one of the terms on the diagonal of the matrix $\mathbf{D}_L$ in the lower–upper triangular decomposition is zero.

2.30. **(a)** Find the eigenvectors and eigenvalues of the correlation matrix in Problem 2.7.
(b) Find the vectors $\mathbf{x}^{(1)\prime}$, $\mathbf{x}^{(2)\prime}$, $\mathbf{x}^{(3)\prime}$, and $\mathbf{x}^{(4)\prime}$ using the eigenvector transformation and use this transformed data to estimate the mean vector, correlation matrix, and covariance matrix. Verify that the results are as expected.

2.31. **(a)** Find the lower–upper triangular decomposition of the correlation matrix estimated in Problem 2.6 and the corresponding linear transformation to diagonalize the correlation matrix.

(b) Find the vectors $\mathbf{x}^{(1)\prime\prime}$, $\mathbf{x}^{(2)\prime\prime}$, $\mathbf{x}^{(3)\prime\prime}$, and $\mathbf{x}^{(4)\prime\prime}$ using the transformation computed in part (a) and use this transformed data to estimate the mean vector, correlation matrix, and covariance matrix. Verify that the results are as expected.

2.32. By applying the eigenvector transformation (2.163) show that the transformed density $f_{\check{\boldsymbol{x}}}(\check{\mathbf{x}})$ for a Gaussian random vector $\boldsymbol{x}$ can be written as a product of one-dimensional Gaussian density functions.

2.33. The probability density function for a two-dimensional real random vector is given by

$$f_{\boldsymbol{x}}(\mathbf{x}) = \frac{1}{1.82\pi} e^{-\frac{1}{2}[\boldsymbol{x}-\mathbf{m}_x]^T \begin{bmatrix} 1 & 0.3 \\ 0.3 & 1 \end{bmatrix}^{-1} [\boldsymbol{x}-\mathbf{m}_x]}$$

where

$$\mathbf{m}_x = \begin{bmatrix} 2 \\ 1 \end{bmatrix}$$

Sketch the contours of this density function in the x_1, x_2 plane.

2.34. A covariance matrix for a certain real Gaussian density function is

$$\mathbf{C}_x = E\left[(\boldsymbol{x} - \mathbf{m}_x)(\boldsymbol{x} - \mathbf{m}_x)^T\right] = \begin{bmatrix} 1 & 0.5 \\ 0.5 & 1 \end{bmatrix}$$

$$\text{with } \mathbf{m}_x = \begin{bmatrix} 1 \\ 0 \end{bmatrix}$$

(a) What are its eigenvalues and eigenvectors?

(b) Sketch the concentration ellipse representing the contours of the density function in the $\mathbf{x}$ plane.

Repeat parts (a) and (b) if the off-diagonal terms are equal to -0.5

2.35. **(a)** Show that a necessary and sufficient condition for a hermitian symmetric matrix to be positive definite is that all the eigenvalues are positive. Also show that a necessary and sufficient condition for such a matrix to be positive semidefinite is that all the eigenvalues are nonnegative.

(b) Using properties of the eigenvalues, show that the determinant and trace of a matrix must be positive if the matrix is positive definite. Further show that these are necessary but *not* sufficient conditions.

2.36. Find the simplest possible expression for the eigenvalues of the two-dimensional covariance matrix

$$\mathbf{C}_x = \begin{bmatrix} \sigma_1^2 & \rho\sigma_1\sigma_2 \\ \rho^*\sigma_1\sigma_2 & \sigma_2^2 \end{bmatrix}$$

Also use the results of Problem 2.28 to obtain an expression for the eigenvectors in the real case. Specialize the result to the case $\sigma_1^2 = \sigma_2^2 = \sigma^2$. Express this last result in terms of σ^2 and ρ. What are the eigenvectors in this special case when ρ is real?

2.37. A Gaussian density function for a two-dimensional random vector $\boldsymbol{x}$ has mean zero and covariance matrix

$$\begin{bmatrix} 3 & -1 \\ -1 & 3 \end{bmatrix}$$

(a) Plot a contour for this density function in the x_1, x_2 plane. Label it completely and clearly.

(b) Show how the results of part (a) change if the covariance matrix is given by

$$\begin{bmatrix} 3 & +1 \\ +1 & 3 \end{bmatrix}$$

Draw the new contour with dashed lines.

2.38. Given the correlation matrix for the zero-mean random vector $\boldsymbol{x}$

$$\mathbf{R}_x = \begin{bmatrix} 1 & 0.2 \\ 0.2 & 1 \end{bmatrix}$$

find two different transformations of the form

$$\boldsymbol{y} = \mathbf{A}\boldsymbol{x}$$

such that the components of $\boldsymbol{y}$ are uncorrelated.

2.39. By following the procedure in Example 2.13 show that the matrix

$$\mathbf{R}_x = \begin{bmatrix} 3 & 2 & 1 \\ 2 & 3 & 2 \\ 1 & 2 & 3 \end{bmatrix}$$

can be expressed as

$$\mathbf{R}_x = \begin{bmatrix} 1 & 0 & 0 \\ \frac{2}{3} & 1 & 0 \\ \frac{1}{3} & \frac{4}{5} & 1 \end{bmatrix} \begin{bmatrix} 3 & 2 & 1 \\ 0 & \frac{5}{3} & \frac{4}{3} \\ 0 & 0 & \frac{8}{5} \end{bmatrix} = \begin{bmatrix} 1 & 0 & 0 \\ \frac{2}{3} & 1 & 0 \\ \frac{1}{3} & \frac{4}{5} & 1 \end{bmatrix} \begin{bmatrix} 3 & 0 & 0 \\ 0 & \frac{5}{3} & 0 \\ 0 & 0 & \frac{8}{5} \end{bmatrix} \begin{bmatrix} 1 & \frac{2}{3} & \frac{1}{3} \\ 0 & 1 & \frac{4}{5} \\ 0 & 0 & 1 \end{bmatrix}$$

and thus is of the form $\mathbf{L}\mathbf{D}_L\mathbf{L}^T$. Show also that the matrix $\mathbf{L}^{-1}$ has the lower triangular form

$$\mathbf{L}^{-1} = \begin{bmatrix} 1 & 0 & 0 \\ -\frac{2}{3} & 1 & 0 \\ \frac{1}{5} & -\frac{4}{5} & 1 \end{bmatrix}$$

and can be obtained by performing the same multiplication and subtraction operations that are used to reduce $\mathbf{R}$ on rows of the identity matrix. What are the factors in the decomposition

$$\mathbf{R}_x = \mathbf{U}_1\mathbf{D}_U\mathbf{U}_1^T$$

for this matrix?

2.40. **(a)** Find the lower–upper and upper–lower triangular decomposition of the matrix

$$\mathbf{R}_x = \begin{bmatrix} 2 & \jmath \\ -\jmath & 3 \end{bmatrix}$$

Is there any obvious relation between the terms in the lower–upper and upper–lower decomposition?

(b) Repeat part (a) for the matrix

$$\mathbf{R}_x = \begin{bmatrix} 2 & \jmath \\ -\jmath & 2 \end{bmatrix}$$

(c) Using the relations in Table 2.7, show what properties of the two matrices above account for the difference in your answers to parts (a) and (b).

COMPUTER ASSIGNMENTS

2.1. Using the procedures of Section 2.2.3, estimate the correlation matrix corresponding to the real vectors

$$\begin{aligned}
\mathbf{x}^{(1)} &= [0, 1, -1, 0, 1, -1]^T \\
\mathbf{x}^{(2)} &= [1, -1, 0, 0, -1, 0]^T \\
\mathbf{x}^{(3)} &= [-1, 1, 2, -2, 0, 0]^T \\
\mathbf{x}^{(4)} &= [0, -1, -1, 2, 0, 1]^T \\
\mathbf{x}^{(5)} &= [-1, 1, -1, 1, 0, -1]^T \\
\mathbf{x}^{(6)} &= [-1, 2, 0, -1, 0, 0]^T \\
\mathbf{x}^{(7)} &= [2, -2, 0, 0, -2, 2]^T \\
\mathbf{x}^{(8)} &= [0, -1, 1, 0, 2, -1]^T \\
\mathbf{x}^{(9)} &= [-1, 2, -1, 0, 2, -1]^T \\
\mathbf{x}^{(10)} &= [0, -1, -1, 2, 0, -1]^T
\end{aligned}$$

The data is provided on the diskette packaged with this book as file VEC1_10 for convenience. Find the eigenvalues of the correlation matrix. How do you interpret the result?
Add the following vector to the set (VEC11 on the diskette)

$$\mathbf{x}^{(11)} = [0, -1, +1, 2, 0, 1]^T$$

and recompute the correlation matrix and its eigenvalues. What accounts for the change in the number of nonzero eigenvalues?

2.2. Using all 11 of the sample vectors in Computer Assignment 2.1, estimate the covariance matrix and find the diagonalizing transformation.

$$\check{\boldsymbol{x}} = \check{\mathbf{E}}^T \boldsymbol{x}$$

Apply it to each of the data vectors. Estimate the covariance matrix for the transformed data vectors and verify that it is diagonal with elements equal to the eigenvalues of the original covariance matrix. Recompute the transformation and the eigenvalues of the original covariance matrix by using the SVD. (Don't forget to subtract the mean from the data.)

2.3. Using all 11 of the sample vectors in Computer Assignment 2.1, compute the factors $\mathbf{L}$ and $\mathbf{D}_L$ corresponding to the triangular decomposition of the correlation matrix. Find the transformation

$$\boldsymbol{x}'' = \mathbf{L}^{-1}\boldsymbol{x}$$

and apply it to each of the data vectors. Estimate the correlation matrix for the transformed data vectors and verify that it is equal to the matrix $\mathbf{D}_L$. Recompute the factors $\mathbf{L}$ and $\mathbf{D}_L$ directly from the data matrix by using QR factorization. Check your results with the previous computation.

2.4. The files X01, X02, and X03 on the diskette each contain 256 two-dimensional sample vectors (each line of the file is one vector).

(a) Estimate the covariance matrix for each data set and compute its eigenvectors and eigenvalues. How do the data sets differ?

(b) Make a scatter plot for each of the data sets; that is, plot the first component of each sample vector versus the second component for all the vectors. Superimpose on these a plot of the concentration ellipse for the covariance matrix. Observe how most of the data tend to be distributed within the ellipse.

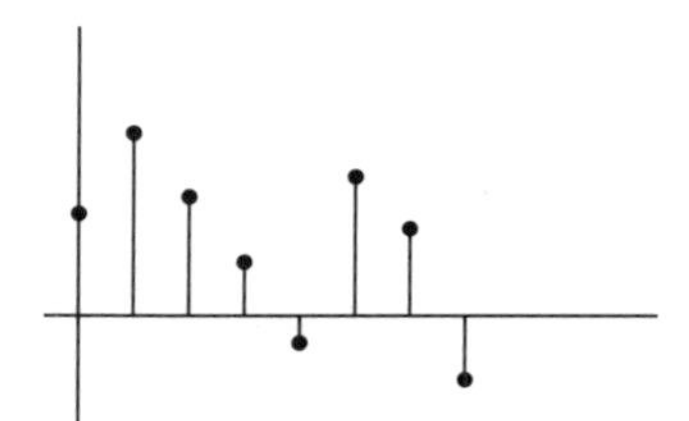

3

Random Processes

Random signals and sequences are represented mathematically by the concept of a random process. This chapter discusses random processes in general and introduces a number of specific random processes that are important in signal processing and other engineering applications. Most of the random processes described here are inherently discrete since they represent events that occur naturally at discrete points in time. A frequently encountered class of random sequences, however, results from sampling continuous random signals. Therefore, the concept of a continuous random process is defined and related to the discrete random process that can be derived from it by sampling. Since a large and important class of these random signals is the class of Gaussian random signals, we introduce the concept of a Gaussian white noise process and discuss Gaussian random processes in general.

This chapter is unique both in the scope of different types of random processes that it addresses and in the generality with which it approaches the analysis. Further chapters concentrate on the analysis of random sequences primarily in terms of their first and second moment properties. This chapter not only describes some specific types of random processes, but also develops some important general principles for dealing with random processes that serve through the remainder of the book.

3.1 RANDOM SIGNALS AND SEQUENCES

This section introduces the concept of a random process and defines a number of properties related to its statistical description. Although several examples of random processes are presented here, there is no attempt made to categorize different types of random processes or to investigate specific properties of these processes in detail. That more specific analysis is carried out in the sections that follow.

To begin the discussion, consider a sequence $x[n]$ such that its value for any choice of the parameter n is a random variable. Then $x[n]$ is called a discrete random signal or a *random sequence*. If the parameter n represents time, then the random sequence is sometimes referred to as a "time series." The underlying model that represents the random sequence is known as a *random process* or a stochastic process. Four examples of random sequences are given in Fig. 3.1. Figure 3.1(a) shows a sampled noise sequence. The values of the sequence may be any real or complex numbers and may or may not be independent from sample to sample.[1] Figure 3.1(b) shows a random sequence representing binary-coded data. This random process can take on only values of $+1$ and -1. If the samples of the sequence at each value of n are independent and equally likely, the process is referred to as binary white noise[2] and can be thought of as representing the ordered set of outcomes of flips of a coin.

Figure 3.1(c) shows a sequence consisting of a sinusoid with random amplitude and phase. This sequence could represent the sampled voltage on an alternating current (ac) power line. It is also a type of random sequence common in many radar, sonar, or communication problems. In the case of the ac power line the frequency is known to be constant but the amplitude can be undependable, due to unknown load conditions in the building or at the generating plant and so be considered a random variable. The phase is also a random variable because observation of this voltage could have started at any time. For any particular instance when we observe the voltage, however, the amplitude and phase are *constant*, and the sequence appears to be deterministic. Although this example may not at first seem to fit the definition of a random process, a little thought will convince you that it really does. Since the amplitude and phase are random variables, the signal at any point n prior to the observations is also a random variable. Once observed, the process is completely predictable, however. In fact, any two samples of the sequence (as long as they are not both at zero crossings) permit the other values to be exactly predicted for all time. It will be seen in subsequent chapters that random processes that are predictable in this manner form a very important class with properties distinct from random processes such as those of Fig. 3.1(a) and (b). In fact, an important theorem, presented in Chapter 7, provides for the decomposition of an arbitrary random process into a sum of predictable and nonpredictable parts.

Figure 3.1(d) shows a sequence representing a constant direct-current (dc) voltage, such as that of a battery. Such a constant sequence may also constitute a random process

[1]Complex-valued random sequences are discussed in greater detail in the next chapter. Complex sequences can arise as a representation of sampled bandpass real signals (see Appendix B).

[2]The term "white noise" refers to the spectral characteristics of the random process. This is discussed in detail in the next chapter.

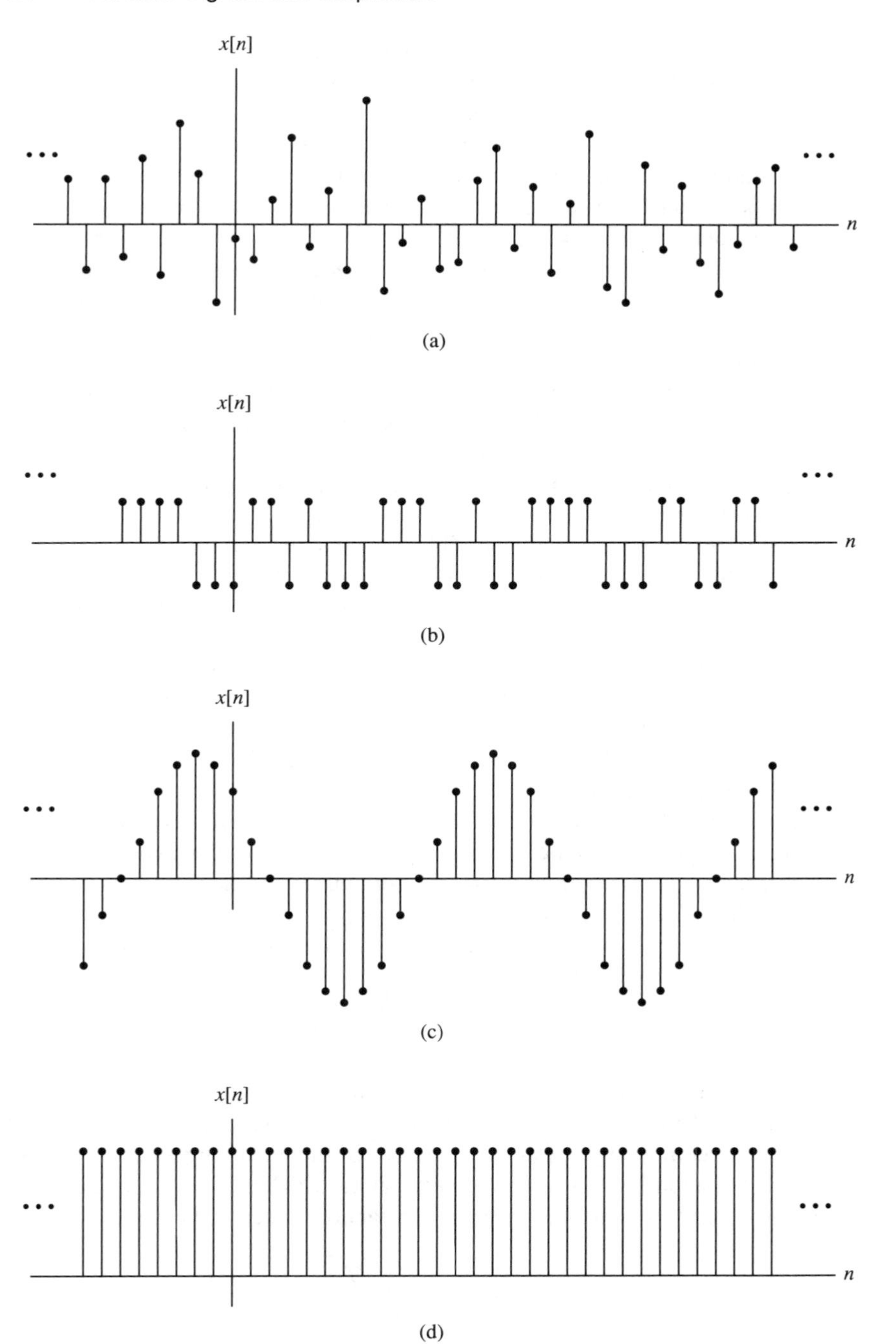

Figure 3.1 Examples of random sequences. (a) Noise. (b) Binary coded data. (c) Random sinusoid. (d) Battery voltage.

in that the value that the constant sequence takes on before it is observed is unknown (e.g., automobile batteries can vary over some given range around 12 volts). Once a single sample of a constant random sequence is observed, however, all of its remaining samples are known. It is the ultimate predictable random process.

A random sequence can be thought of in much the same way as a random variable. Both can be thought of as arising from an experiment that has many possible outcomes. In one case the outcome results in a single number (the value of the random variable), while in the other case the outcome results in an entire sequence. As an example, consider the case where we measure the rms voltage of the power line with a voltmeter. Since the exact value of the voltage that is observed cannot be predicted ahead of time, the measured voltage is a random variable. Alternatively, consider sampling and plotting this voltage as a function of time. Since the amplitude and phase at the time of sampling is not known beforehand, the result is a random sequence. Further, if this power line has superimposed some noise due to operation of heavy equipment in the building, the sequence so measured will be even less predictable and exhibit more random variations each time the experiment is performed.

The measurement of a random variable or a random sequence can be viewed in the following way. Imagine that it is possible to enumerate and list all the possible outcomes of the experiment as shown in Fig. 3.2. (Whether or not it is actually possible to construct such a listing does not really matter since this is primarily a way of *thinking* about random events.) Such a list of possible outcomes is called an *ensemble*. When the random experiment is performed, one member of the ensemble is selected randomly and presented. The result is called a *realization* of the random variable or the random process. While it is not usual to speak about an ensemble for a random variable, it is common to consider the ensemble for a random process. Mathematically, the set of all possible outcomes of a given experiment can be thought of as comprising the sample space $\mathcal{S}$. Then a random variable is a mapping from an element of $\mathcal{S}$ to the real line (or the complex plane if it is complex), while a random sequence is a mapping from $\mathcal{S}$ to some appropriate space of sequences.

Suppose that a random variable x is continuous and has a probability density function f_x. Since the values of the random variable within some small interval $\mathrm{x}^{\mathrm{o}} < x \leq \mathrm{x}^{\mathrm{o}} + \Delta x$, occur in the ensemble with a representation that is proportional to the value of the density function $f_x(\mathrm{x}^{\mathrm{o}})$, the expectation

$$\mathcal{E}\{x\} = \int_{-\infty}^{\infty} \mathrm{x} f_x(\mathrm{x}) d\mathrm{x} \tag{3.1}$$

which represents the average value of x, can be computed conceptually by averaging over the members of the ensemble. In a similar manner consider averaging over the ensemble of sequences comprising a random process as shown in Fig. 3.2. Since the average is computed by summing *down the ensemble* at each value of n, the result is a deterministic *sequence* which represents the mean of the random process. Formally, the mean is defined as the sequence

$$\mathrm{m}[n] = \mathcal{E}\{x[n]\} = \int_{-\infty}^{\infty} \mathrm{x}_n f_{x[n]}(\mathrm{x}_n) d\mathrm{x}_n \tag{3.2}$$

where for any fixed value of n, $f_{x[n]}$ represents the density function of the random variable $x[n]$. (Note that in general the density $f_{x[n]}$ is different for each value of n.) In a similar

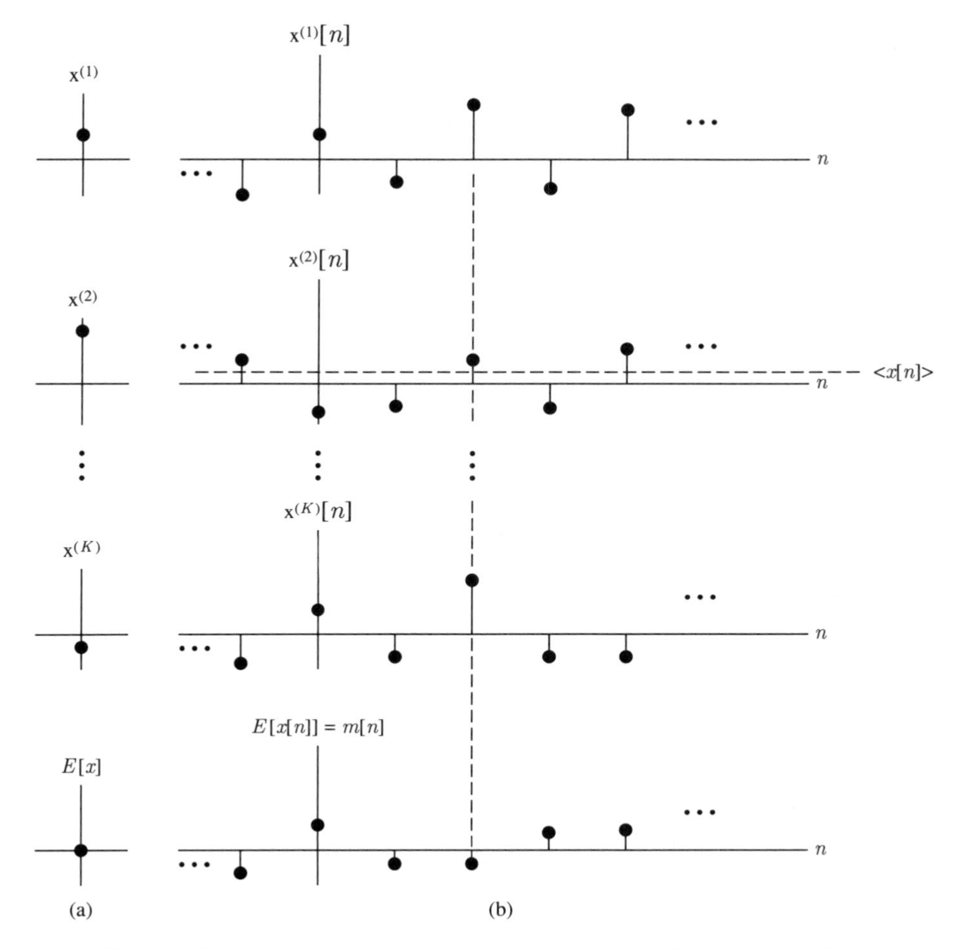

Figure 3.2 Members of an ensemble. (a) Random variable. (b) Random process.

manner the second moment of the random process can be defined as the expected value of $|x[n]|^2$, the variance of the process as the expected value of $|x[n] - \mathrm{m}[n]|^2$, and so forth. All are sequences (i.e., functions of n) and all can be considered to be obtained by averaging the corresponding quantities "down the ensemble." Thus the expectation of a random process is often referred to as an *ensemble average*.

In the practical analysis of random signals it may be difficult or impossible to determine the probability density function and thus to compute the theoretical expectation. However, if it *is* possible to repeat the experiment many times, then it may in fact be possible to generate some number of members of the ensemble and thus estimate the expectation by averaging over the members. In this case the concept of an ensemble is useful for more than just thought experiments and becomes a truly practical tool.

Two further concepts that are of importance in dealing with random processes are those of stationarity and ergodicity. To motivate these ideas, suppose that the random process of Fig. 3.2 were such that the mean is the same for every value of n. That is, the function $\mathrm{m}[n]$ is a constant (perhaps 0). Such a process exhibits a kind of uniformity that can simplify its statistical description. Now consider any member of the ensemble. Each such sequence is assumed to exist for all n from $-\infty$ to $+\infty$. Let the angle brackets $\langle\ \rangle$ denote the following averages computed along that *particular* ensemble member:

$$\langle x[n]\rangle_{n_0,n_1} \stackrel{\text{def}}{=} \frac{1}{n_1 - n_0 + 1} \sum_{n=n_0}^{n_1} x[n] \tag{3.3}$$

$$\langle x[n]\rangle \stackrel{\text{def}}{=} \lim_{M\to\infty} \langle x[n]\rangle_{-M,M} = \lim_{M\to\infty} \frac{1}{2M+1} \sum_{n=-M}^{M} x[n] \tag{3.4}$$

These averages will be referred to as "signal averages." Note that they are random variables. The signal average for any particular ensemble member using either a finite interval as in (3.3) or an infinite interval as in (3.4) may or may not result in the same value obtained from averaging down the ensemble. If the result were likely to be the same however, then the ensemble average (i.e., expectation) could possibly be replaced by the average along a single ensemble member.

Returning to the examples of Fig. 3.1 may be helpful at this point. Suppose that the random process in Fig. 3.1(a) represents noise measured in the front end of a radio receiver resulting from random disturbances in the atmosphere. If the experiment were repeated many times, then at each time point n we would expect to find equal numbers of occurrences of positive and negative values among the sequences in the ensemble; so the ensemble mean would be zero. Further, if any particular ensemble member were observed for a long enough time, it could be expected to find all values of the noise represented in about the same proportions; so too the signal average for the sequence would be zero. Similar arguments could be made about the binary-coded data sequence of Fig. 3.1(b), although the values of the sequence are always $+1$ and -1. On the other hand, consider the random process generated by measuring battery voltage as a function of time [Fig. 3.1(d)]. The ensemble average at any point n is constant and equal to 12 volts, so the mean $\mathrm{m}[n]$ is a constant as in the other examples. However, the average along most ensemble members will *not* be equal to the ensemble average. A voltage measured to be 11.75 volts will remain so for the duration of the sequence and will never average to 12 volts.

The foregoing examples are manifestations of the concepts of stationarity and ergodicity. A random process is said to be *stationary* (in the *strict* sense) if its statistical description is not a function of n. In particular, the joint probability density function $f_{x[n_0],x[n_1],\ldots x[n_L]}$ for any set of $L+1$ samples must be the same as that for any other set of $L+1$ samples with the same spacing, and this must be true for any value of L (see Fig. 3.3). The latter is equivalent to requiring that all the moments

$$\mathcal{E}\{x^{k_0}[n_0]\cdot x^{k_1}[n_1]\ldots x^{k_L}[n_L]\}$$

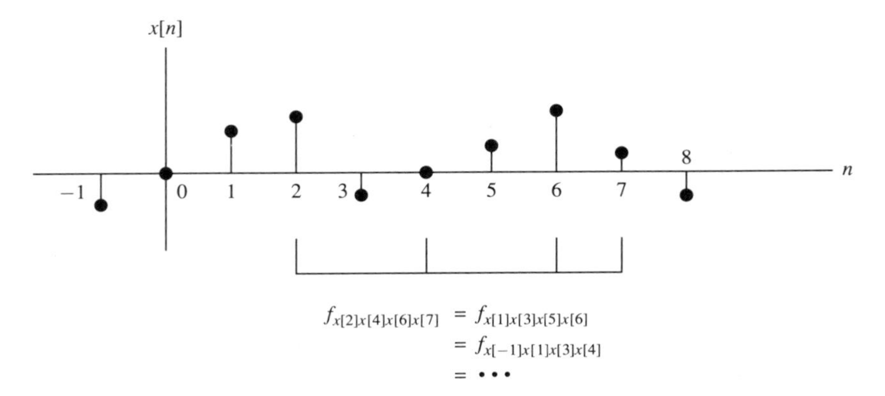

$$f_{x[2]x[4]x[6]x[7]} = f_{x[1]x[3]x[5]x[6]} = f_{x[-1]x[1]x[3]x[4]} = \cdots$$

Figure 3.3 Stationary random process. (Any set of samples with the same spacing has the same density function.)

for any selection of the powers $k_0, k_1, \ldots, k_L$ and any value of L are the same for all samples with the same intersample spacing, or that the moments written as

$$\mathcal{E}\{x^{k_0}[n] \cdot x^{k_1}[n+l_1] \ldots x^{k_L}[n+l_L]\}$$

where the l_i represent spacings between samples, are *not* a function of n. A random process is said to be *ergodic* if, in addition, such moments computed as signal averages are equal with probability 1 to the corresponding ensemble averages. In particular, a process is said to be strictly ergodic if it satisfies the property

$$\langle x^{k_0}[n]x^{k_1}[n+l_1] \ldots x^{k_L}[n+l_L]\rangle \doteq \mathcal{E}\{x^{k_0}[n]x^{k_1}[n+l_1] \ldots x^{k_L}[n+l_L]\} \tag{3.5}$$

for any choices of the spacings $l_1, l_2, \ldots, l_L$, and any choice of the powers $k_0, k_1, \ldots, k_L$. The notation "$\doteq$" means that the *event*

$$\langle x^{k_0}[n]x^{k_1}[n+l_1] \ldots x^{k_L}[n+l_L]\rangle = \mathcal{E}\{x^{k_0}[n]x^{k_1}[n+l_1] \ldots x^{k_L}[n+l_L]\}$$

has probability 1. A random process that satisfies only the condition

$$\langle x[n]\rangle \doteq \mathcal{E}\{x[n]\} \tag{3.6}$$

is sometimes said to be "ergodic in the mean," while one that satisfies

$$\langle x[n]x[n+l]\rangle \doteq \mathcal{E}\{x[n]x[n+l]\} \tag{3.7}$$

is said to be "ergodic in correlation." These last two conditions are sufficient for many analyses.

Note from the foregoing definitions that in order to be (strictly) ergodic, a random process *must* first be stationary. (The signal average cannot be equal to the ensemble average, if the ensemble average is a function of n.) However, the converse of that statement is not true. Consider the previous example of the random sinusoid of Fig. 3.1(c). If the phase is uniformly distributed between 0 and 2π, the random process is stationary; the moments of the density itself are independent of where the samples are taken as long as the spacing remains the same. The mean computed as both an ensemble average and as a signal average is zero, so the process is mean ergodic. The variance of any particular ensemble member is a constant (the squared rms voltage), which in general is not equal to the ensemble variance. Thus the random process is not strictly ergodic or even ergodic in correlation.

The concept of ergodicity is an important one for applications when it is impossible or impractical to generate an ensemble by repeating the experiment. For example, in seismic explorations it is very expensive to take many independent records of data. Therefore, analysts take advantage of ergodicity to draw inferences from a single recorded data sequence. In fact, in most areas of signal processing it is common to deal with only one (perhaps very long) record of data. If this record contains all the statistical variation that would be present in examining other members of the ensemble (i.e., if the process is ergodic), then meaningful analyses can be made using statistics computed from only the single data record.

The sections that follow describe several specific types of random processes and their statistical characterization. Some of these are stationary while others are not. Further, some of the stationary processes but not all are ergodic. The properties are usually apparent or can easily be shown from their definition.

3.2 SIMPLE DISCRETE RANDOM PROCESSES

Two simple related random processes are the Bernoulli process and the random walk. Two special cases of these are called binary white noise and the discrete Wiener process. In spite of their simplicity, these processes are quite useful and appear in many practical signal processing applications.

3.2.1 The Bernoulli Process and Binary White Noise

One of the simplest types of random process is typified by the binary data sequence that occurs in some communications problems. The sequence $x[n]$, $-\infty < n < \infty$, is a sequence of independent random variables taking on the two values $+1$ and -1 with probability P and $1 - \mathrm{P}$ respectively and is known as a *Bernoulli process*.[3] In the special case where $\mathrm{P} = \frac{1}{2}$ the process will be referred to as *binary white noise*. Although the Bernoulli process is very simple, it can be a reasonable model for certain types of discrete random signals. For example, in the linear-predictive coding of speech and image data, only the difference between the input sequence and a predicted value of the sequence based on past data is transmitted. The difference, which is the prediction error, is frequently coded with one bit and represents a sequence of binary-valued random variables. If the linear prediction is effective, then the samples of the sequence can be regarded as independent. Thus, the Bernoulli process is a suitable model for the transmitted coded sequence.

[3]The two values that the Bernoulli process is defined to take on could be *any* two values. Frequently, the values are taken to be 1 and 0.

It is fairly obvious that the Bernoulli process is stationary and ergodic. To demonstrate the stationarity notice that the probability for *any* subsequence of samples can be written easily. For example, consider the subsequence $\{+1, +1, -1, +1\}$ beginning at some arbitrary point n_o. The probability of this subsequence is[4]

$$\begin{aligned}\Pr[(x[n_o] = +1), (x[n_o+1] = +1), (x[n_o+2] = -1), (x[n_o+3] = +1)] \\ = \mathrm{P} \cdot \mathrm{P} \cdot (1-\mathrm{P}) \cdot \mathrm{P} \\ = \mathrm{P}^3(1-\mathrm{P})\end{aligned}$$

and this probability of four such consecutive values is clearly independent of where it is found in the sequence. To convince yourself of the ergodicity, imagine an ensemble of sequences of a Bernoulli process depicted as in Fig. 3.2. Since the samples are all independent random variables with the same distribution, a moment's thought should convince you that averages taken down the ensemble will be the same as averages taken along any sequence.

To further characterize the Bernoulli process, let us compute the first two (central) moments. The mean and variance are given by

$$\mathcal{E}\{x[n]\} = (+1) \cdot \mathrm{P} + (-1) \cdot (1-\mathrm{P}) = 2\mathrm{P} - 1 \tag{3.8}$$

and

$$\begin{aligned}\mathrm{Var}[x[n]] &= \mathcal{E}\{x^2[n]\} - (\mathcal{E}\{x[n]\})^2 \\ &= \left[(+1)^2 \cdot \mathrm{P} + (-1)^2 \cdot (1-\mathrm{P})\right] - (2\mathrm{P}-1)^2 \\ &= 1 - (2\mathrm{P}-1)^2 = 4\mathrm{P}(1-\mathrm{P})\end{aligned} \tag{3.9}$$

For the binary white noise process ($\mathrm{P} = \frac{1}{2}$) the mean is 0 and the variance is 1. Since the process is stationary, these quantities are not a function of n. The properties are summarized in Table 3.1.

TABLE 3.1 PROPERTIES OF THE BERNOULLI PROCESS AND BINARY WHITE NOISE

	Bernoulli process	Binary white noise
Distribution	$\Pr[x[n] = +1] = \mathrm{P}$	$\Pr[x[n] = +1] = \frac{1}{2}$
	$\Pr[x[n] = -1] = 1-\mathrm{P}$	$\Pr[x[n] = -1] = \frac{1}{2}$
Mean	$2\mathrm{P} - 1$	0
Variance	$4\mathrm{P}(1-\mathrm{P})$	1

The following two examples demonstrate some additional properties of the Bernoulli process.

[4]Parentheses are used around events such as $(x[n_o] = +1)$ only to improve readability.

Example 3.1

At a certain time n_o a Bernoulli process is observed to have the value $x[n_o] = +1$. It is desired to compute the distribution for the waiting time to the next "+1" in the sequence. Suppose that the time at which the next +1 occurs is at $n = n_o + l$. Then there must be $l - 1$ consecutive -1's occurring before the +1. Thus

$$\Pr[\text{time to the next } +1 \text{ is } l] = (1 - \mathrm{P})^{l-1}\mathrm{P}$$

for $l = 1, 2, 3, \ldots$. Note that the same answer would hold if $x[n_o]$ were equal to -1. This shows that *the Bernoulli process has no memory*. Further, the result is independent of the value of n_o. This again demonstrates the stationarity.

The next example deals with another random process derived from the Bernoulli process.

Example 3.2

A *counting process* is defined as a random process that counts the number of positive values in a Bernoulli process after a given time n_o up to any time $n > n_o$. Taking $n_o = 0$ yields

$$y[n] = \sum_{l=1}^{n} \tfrac{1}{2}(x[l] + 1)$$

Each time $x[l]$ takes on a value of +1 the corresponding term in the sum is equal to 1, and each time $x[l]$ is -1, the corresponding term is zero. Thus $y[n]$ represents the number of positive values in the sequence from $l = 1$ to $l = n$.

The counting process can be used to model the evolution of observations in a nonparametric signal detection procedure known as a sign test [1]. In this test a positive-valued signal is observed in zero-mean noise as seen in Fig. EX3.2a.

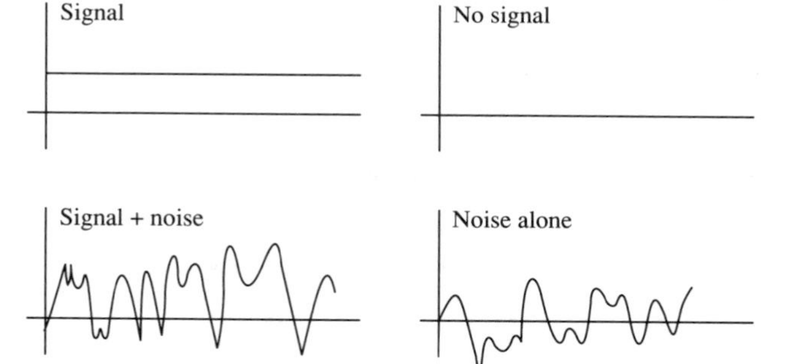

Figure EX3.2a

To give the detection procedure robust performance for noise processes with a wide variety of statistical characteristics, only the *sign* of the observations is used in the detector. Intuitively, for any fixed number of observations of the received sequence, we would expect to find that there are more positive values if the signal is present, and about equal numbers of positive and negative values if the signal is not present. The number of positive values, which can be modeled as a counting process, is therefore compared to a threshold to make a detection decision.

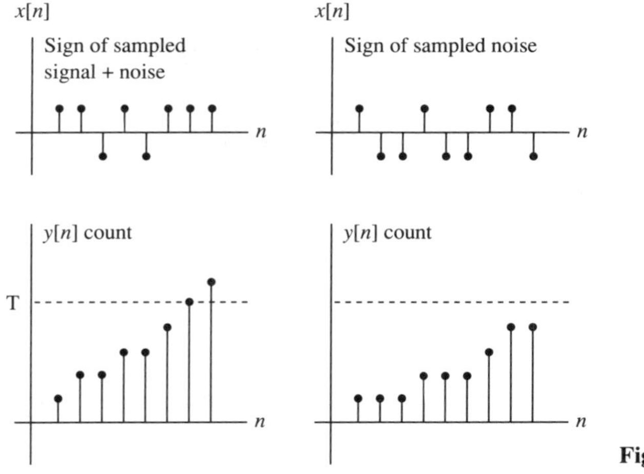

Figure EX3.2b

Notice that the counting process looks like a set of steps occurring at random times. Observe that at time n, $y[n]$ can take on any integer values between 0 and n. To compute the probability distribution of this random process, suppose that $y[n]$ is equal to some value r. This means that there are r positive values and $n - r$ negative values from $l = 1$ to $l = n$. These can occur in *any* order. The number of combinations of positive and negative values is given by the binomial coefficient

$$\binom{n}{r} = \frac{n!}{r!(n-r)!}$$

Thus it follows that

$$\Pr[y[n] = r] = \begin{cases} \binom{n}{r} \mathrm{P}^r (1-\mathrm{P})^{n-r} & r = 0, 1, \ldots, n \\ 0 & \text{otherwise} \end{cases}$$

This is the binomial distribution.

3.2.2 The Random Walk and the Discrete Wiener Process

It was seen earlier that the Bernoulli process can be a model for the one-bit coded error sequence in linear predictive coding. In this application the received error sequence is decoded and applied to a filter to construct the original waveform. While it is premature at this point to discuss most general types of filtering of random processes, a special case involving summation of a Bernoulli process is worth describing. The resulting process can be used to model the signal reconstruction in the simplest of predictive coding schemes known as delta modulation.

In delta modulation the waveform to be coded is represented by a sequence of discrete equilevel upward or downward steps as shown in Fig. 3.4. (The figure depicts continuous signals for clarity; however, since the waveforms are sampled in the coding, the signals are actually discrete.) The waveform can thus be represented by a binary sequence whose

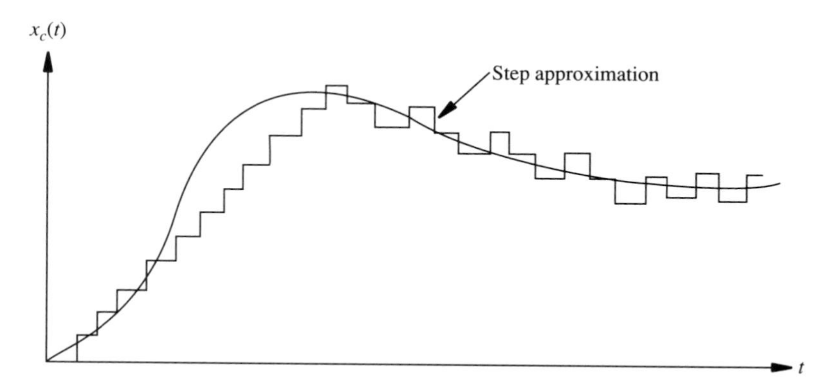

Figure 3.4 Step approximation to a waveform used in delta modulation.

values are $+1$ if an upward step is taken and -1 if a downward step is taken. If the time interval between steps is such that the steps can be considered to be statistically independent, then the resulting sequence is a Bernoulli process.

In the reconstruction procedure the sequence of $+1$'s and -1's are summed to approximately reproduce the waveform. If the steps are independent, the reconstruction can be modeled by a process known as a random walk.

To begin the study of the random walk, consider the random process defined by

$$x[n] = \sum_{k=-\infty}^{n} \xi[k] \tag{3.10}$$

where $\xi[n]$ is a Bernoulli process taking on values of $+1$ and -1 with probability P and $1 - \mathrm{P}$. Further, from the definition of the process, the differences $x[n] - x[n-1]$, $x[n-1] - x[n-2], \ldots, x[k] - x[k-1]$, ..., are all *independent*. A random process that has this property is said to have *independent increments*. Such a process has the important property that *events defined on non-overlapping intervals are independent*.

Consider now the behavior of the difference $x[n] - x[n_o]$ for $n > n_o$. For any fixed value of n_o the sequence of these differences defines another random process, namely

$$x_{n_o}[n] \stackrel{\text{def}}{=} x[n] - x[n_o] = \sum_{k=n_o+1}^{n} \xi[k]\ ; \qquad n > n_o \tag{3.11}$$

(It can be assumed that $x_{n_o}[n_o] = 0$.) This process is known as a *random walk* and is usually defined with $n_o = 0$ or $n_o = -1$. For the special case of $\mathrm{P} = \frac{1}{2}$, $\xi[k]$ is binary white noise, and $x_{n_o}[n]$ will be called a *discrete Wiener process*. Note that the original process $x[n]$ has a certain kind of consistency in that if the random walk had been defined as $x[n + k_o] - x[n_o + k_o]$ for any integer k_o, the statistical characteristics would be the same.

The value of a random walk can be thought of as the position of a point on a line that is subject to a random displacement of $+1$ or -1 from its current position at each epoch of time. If the point started at the coordinate $x[n_o]$ at time n_o, then its coordinate at time n, after $n - n_o$ random displacements, would be $x[n]$. Figure 3.5 shows a typical segment of a random walk $x_{n_o}[n]$ with $n_o = 0$.

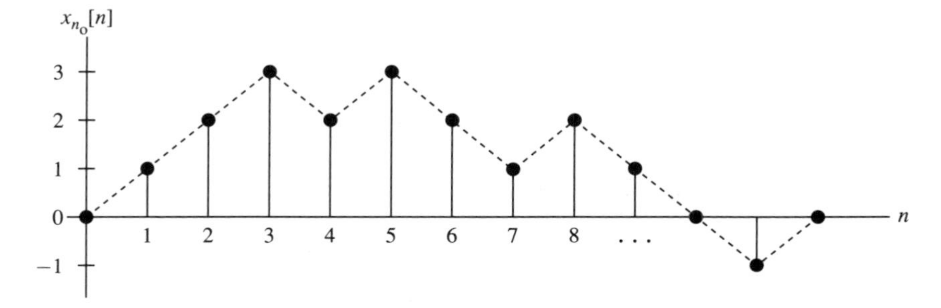

Figure 3.5 Segment of a random walk.

The first-order distribution of the random walk (i.e., $\Pr[x_{n_o}[n] = r]$, where r is an integer), is easily computed. To make the calculation easier, define the variable

$$l = n - n_o \tag{3.12}$$

Now suppose that the underlying Bernoulli sequence ξ takes on the value $+1$ a total of q times in this interval of length l. Then the sequence must take on the value -1 a total of $l - q$ times in this same interval. Thus if $x_{n_o}[n] = r$, then r is given by

$$r = q - (l - q) = 2q - l \tag{3.13}$$

It is thus seen that r can take on the values $-l, -l+2, -l+4, \ldots, l-2, l$. From (3.13) it is also seen that the number of positive values is given by

$$q = \frac{l+r}{2}$$

while the number of negative values is given by

$$l - q = \frac{l-r}{2}$$

The number of possible combinations of q positive values and $l - q$ negative values is given by the binomial coefficient

$$\binom{l}{q} = \frac{l!}{q!(l-q)!}$$

Therefore, from the last three equations, the probability that the random walk takes on a value r at $n = n_o + l$ is given by

$$\Pr[x_{n_o}[n_o + l] = r] = \begin{cases} \binom{l}{\frac{l+r}{2}} \mathrm{P}^{\frac{l+r}{2}}(1-\mathrm{P})^{\frac{l-r}{2}} & r = -l, -l+2, \ldots, l-2, l \\ 0 & \text{otherwise} \end{cases} \tag{3.14}$$

This distribution is similar to the one for the counting process of Example 3.2. However, it is nonzero either for only even values or for only odd values of r.

The mean and the variance for the random walk can be computed most easily by noting that the $\xi[k]$ are independent and using the previous results (3.8) and (3.9) for a Bernoulli process. The mean is thus

$$\mathcal{E}\{x_{n_o}[n]\} = \sum_{k=n_o+1}^{n} \mathcal{E}\{\xi[k]\} = (2\text{P}-1)(n-n_o) \tag{3.15}$$

while the variance is

$$\text{Var}\left[x_{n_o}[n]\right] = \sum_{k=n_o+1}^{n} \text{Var}\left[\xi[k]\right] = 4\text{P}(1-\text{P})(n-n_o) \tag{3.16}$$

For $\text{P} \neq \frac{1}{2}$ both the mean and the variance of the random walk are proportional to $n - n_o$. For $\text{P} = \frac{1}{2}$ (the discrete Wiener process) the mean is zero and the variance is equal to $n - n_o$. Thus in all cases the process is *nonstationary*. The properties are summarized in Table 3.2. Since the variance becomes unbounded as $n \to \infty$ or as $n_o \to -\infty$, the process can have wide deviation over long periods of time. Further, since the random walk changes value by only ± 1 with each consecutive sample, the sequences tend to have long runs of positive or negative values for large n. Figure 3.6 shows a typical realization of a discrete Wiener process exhibiting these properties. In the limiting case as the interval between consecutive samples becomes infinitesimal so that the sequence approaches a continuous signal, the random process becomes what is known as a Brownian motion process or a continuous

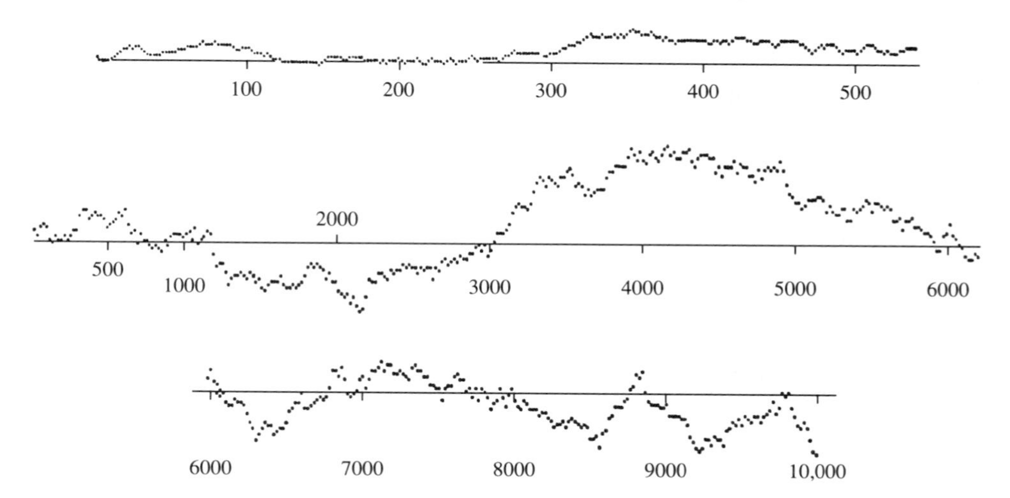

Figure 3.6 Sample sequences from a discrete Wiener process. (From William Feller, *An Introduction to Probability Theory and Its Applications,* Volume I, 2nd ed., copyright © John Wiley & Sons, New York, 1957; reproduced by permission.)

Wiener process. The continuous Wiener process is an extremely important process for representing noise in the analysis of continuous signals and systems. Its properties are discussed further in Section 3.6.

TABLE 3.2 PROPERTIES OF THE RANDOM WALK AND DISCRETE WIENER PROCESS

	Random walk	Discrete Wiener process
Distribution $\Pr[x[n] = r]$	$\binom{n-n_o}{\frac{n-n_o+r}{2}} \mathrm{P}^{\frac{n-n_o+r}{2}}(1-\mathrm{P})^{\frac{n-n_o-r}{2}}$ $\lvert r\rvert \leq n-n_o;\quad n-n_o+r$ even	$\binom{n-n_o}{\frac{n-n_o+r}{2}} \left(\frac{1}{2}\right)^{n-n_o}$ $\lvert r\rvert \leq n-n_o;\quad n-n_o+r$ even
Mean	$(2\mathrm{P}-1)(n-n_o)$	0
Variance	$4\mathrm{P}(1-\mathrm{P})(n-n_o)$	$n-n_o$

3.3 MARKOV PROCESSES

The Bernoulli process introduced in the preceding section is a random process with independent samples. While this model is appropriate for modeling some practical random sequences, it is obviously not suitable for modeling *all* binary data sequences. In most binary data sequences the samples are far from independent and are often highly correlated. As a result, the probability of occurrence of any particular value in the sequence is dependent upon the preceding values. The simplest kind of dependence arises when the probability of any sample depends only upon the value of the *immediately preceding* sample. This simple kind of dependency defines a type of random process known as a Markov process, which is a surprisingly good model for a number of practical signal processing, communication, and control problems.

Markov random processes do not have to be binary-valued. As an example of a Markov process whose samples can take on *any* continuous real value, consider the process generated by the difference equation

$$x[n] = \rho x[n-1] + w[n] \tag{3.17}$$

where $w[n]$ is a sequence of zero-mean independent identically distributed Gaussian random variables with density

$$f_w(\mathrm{w}) = \frac{1}{\sqrt{2\pi\sigma_o^2}} e^{-\frac{\mathrm{w}^2}{2\sigma_o^2}} \tag{3.18}$$

The conditional density of $x[n]$ given $x[n-1]$ is also Gaussian and is given by

$$f_{x[n]\,|\,x[n-1]}(\mathrm{x}_n|\mathrm{x}_{n-1}) = \frac{1}{\sqrt{2\pi\sigma_o^2}} e^{-\frac{(\mathrm{x}_n-\rho\mathrm{x}_{n-1})^2}{2\sigma_o^2}} \tag{3.19}$$

In fact, if $w[n]$ is independent with *any* density $f_w(\mathrm{w})$, the conditional density of $x[n]$ given $x[n-1]$ is $f_w(\mathrm{x}_n - \rho\mathrm{x}_{n-1})$. Note that $x[n-1]$ completely determines the distribution for

$x[n]$, and $x[n]$ completely determines the distribution for $x[n+1]$, and so forth. Thus the value of the sequence at any time n_o completely determines the distribution of $x[n]$ for any $n > n_o$. It will be seen presently that the value of the Markov process at n_o also determines the *joint* distribution for any set of samples at times greater than n_o.

The following serves as a formal definition of a Markov process.

Definition 3.1 (Markov Process). *A random process is a Markov process if the distribution of $x[n]$, given the infinite past, depends on only the previous sample $x[n-1]$, that is, if*

$$f_{x[n]\,|\,x[n-1],x[n-2],\ldots}(\mathrm{x}_n|\mathrm{x}_{n-1},\mathrm{x}_{n-2},\ldots) = f_{x[n]|x[n-1]}(\mathrm{x}_n\,|\,\mathrm{x}_{n-1}) \tag{3.20}$$

Equation 3.20 implies that a sequence of independent random variables is also a Markov process. Although this may be technically correct, such a sequence is not generally considered to be Markov. The definition also implies that the joint density for any set of samples of a Markov process is completely determined by the first-order densities $f_{x[n]}$ and the conditional densities $f_{x[n]\,|\,x[n-1]}$. To see this, consider, for example, the set of samples $x[0], x[1], \ldots, x[n]$. For *any* random process it is possible to write (leaving out the arguments to simplify notation)

$$f_{x[0]x[1]\ldots x[n]} = f_{x[n]\,|\,x[n-1]\ldots x[0]} \cdot f_{x[n-1]\,|\,x[n-2]\ldots x[0]} \cdots f_{x[1]\,|\,x[0]} \cdot f_{x[0]}$$

If the random process satisfies the Markov condition (3.20), then this expression reduces to

$$f_{x[0]x[1]\ldots x[n]} = f_{x[n]\,|\,x[n-1]} \cdot f_{x[n-1]\,|\,x[n-2]} \cdots f_{x[1]\,|\,x[0]} \cdot f_{x[0]} \tag{3.21}$$

The joint density for a set of samples starting at *any* point n_o in the random process can be written as a similar expression, that is,

$$\begin{aligned} &f_{x[n_o+1]x[n_o+2]\ldots x[n]\,|\,x[n_o]x[n_o-1]\ldots} \\ &\qquad = f_{x[n]\,|\,x[n-1]} \cdot f_{x[n-1]\,|\,x[n-2]} \cdots f_{x[n_o+1]\,|\,x[n_o]} \end{aligned} \tag{3.22}$$

In words, the joint density for samples at times greater than n_o, conditioned on the past, depends on $x[n_o]$ but does *not* depend on any values of the sequence prior to n_o. This is the essential property of a Markov process.

3.3.1 Markov Chains

When $x[n]$ takes on only a countable (discrete) set of values, a Markov random process is called a Markov chain. This will always be the case in *digital* signal processing since the values of the random sequence are represented with a finite number of bits. There is a tremendous volume of results on Markov chains; only some of the basic facts are given here. Further treatments on an introductory level with applications can be found in [2–7]. More advanced treatments can be found in [7–9].

Markov Processes with a Discrete Set of Possible Values. Let us begin by giving a formal definition of a Markov chain. Since the values of a Markov chain are discrete, its distribution is defined by a discrete set of probabilities.

Definition 3.2 (Markov Chain). A random process taking on only discrete values is a Markov chain if it satisfies the condition

$$\begin{aligned}\Pr\left[(x[n] = \mathrm{x}_n) \,|\, (x[n-1] = \mathrm{x}_{n-1}), (x[n-2] = \mathrm{x}_{n-2}), \ldots\right] \\ = \Pr\left[(x[n] = \mathrm{x}_n) \,|\, (x[n-1] = \mathrm{x}_{n-1})\right]\end{aligned} \tag{3.23}$$

Thus, in analogy with (3.21), the probability that any consecutive samples such as $x[0], x[1], \ldots, x[n]$ take on particular values is

$$\begin{aligned}\Pr\left[(x[0] = \mathrm{x}_0), (x[1] = \mathrm{x}_1), \ldots, (x[n] = \mathrm{x}_n)\right] \\ = \Pr\left[(x[n] = \mathrm{x}_n) \,|\, (x[n-1] = \mathrm{x}_{n-1})\right] \cdot \cdots \\ \cdot \Pr\left[(x[1] = \mathrm{x}_1) \,|\, (x[0] = \mathrm{x}_0)\right] \cdot \Pr\left[(x[0] = \mathrm{x}_0)\right]\end{aligned} \tag{3.24}$$

Let us now introduce some terminology that is common in the description of Markov chains. Let the possible values that $x[n]$ or x_n can take on be denoted by $\mathrm{S}_1, \mathrm{S}_2, \ldots, \mathrm{S}_Q$ (see Fig. 3.7). When $x[n] = \mathrm{S}_i$ the Markov chain is said to be in *state* i and the conditional probabilities

$$\mathrm{P}_{j|i}[n] = \Pr[(x[n] = \mathrm{S}_j) \,|\, (x[n-1] = \mathrm{S}_i)] \tag{3.25}$$

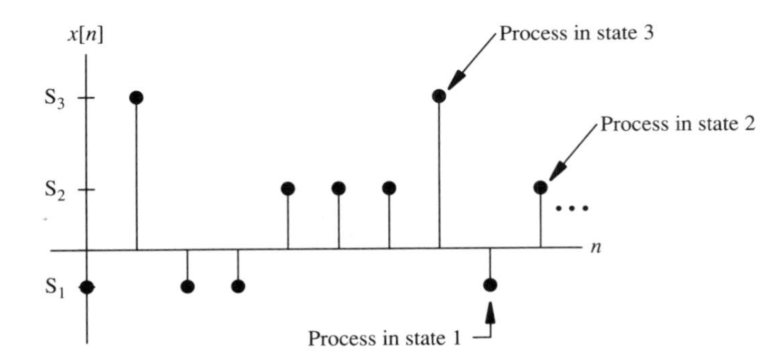

Figure 3.7 Identification of states in a Markov chain.

are known as state *transition probabilities*. A word about notation is in order here. The quantity x_n was previously used to represent a general value that $x[n]$ takes on. Thus the quantity

$$\Pr[(x[n] = \mathrm{x}_n) \,|\, (x[n-1] = \mathrm{x}_{n-1})]$$

represents the *conditional distribution* of $x[n]$ given $x[n-1]$. Equation 3.25, on the other hand, represents a *specific* probability when $x[n]$ and $x[n-1]$ take on *specific values* S_j

and S_i. Therefore, $P_{j|i}[n]$ is a *parameter* of the distribution, not the distribution itself. The usefulness of the state concept and state transition probabilities will become clear shortly. A Markov chain is stationary if the first-order probability distribution is independent of n and it has *constant* transition probabilities $P_{j|i}$ (independent of n).

Markov chains can represent many random processes quite well. Shannon gave this interesting example in terms of words of the English language [10, 11]. Words are chosen independently but with probabilities equal to their relative frequency of occurrence in English, and concatenated to form a "sentence." The result is:

> REPRESENTING AND SPEEDILY IS AN GOOD APT OR COME CAN DIFFERENT NATURAL HERE HE THE A IN CAME THE TO OF TO EXPERT GRAY COME TO FURNISHES THE LINE MESSAGE HAD BE THESE.

This sequence of words not only lacks meaning, but also seems strange because of peculiar juxtapositions of words such as "THE" and "TO" that would not occur in the natural language. On the other hand, if words are chosen with probabilities that depend on the previous word as in a Markov chain, the following "sentence" occurs:

> THE HEAD AND IN FRONTAL ATTACK ON AN ENGLISH WRITER THAT THE CHARACTER OF THIS POINT IS THEREFORE ANOTHER METHOD FOR THE LETTERS THAT THE TIME OF WHO EVER TOLD THE PROBLEM FOR AN UNEXPECTED.

While the result is still nonsensical, it sounds a good deal more like an English sentence. The studies here are not concerned with generating sequences of words to form sentences. However, the sequences of numbers that form a discrete signal can likewise be made to resemble many real-world measured signals if they are modeled by a Markov random process. That is the point in this example.

Let us return now to the formal consideration of Markov chains and assume that the transition probabilities are constant (i.e., independent of n) and of the form

$$P_{j|i} \stackrel{\text{def}}{=} \Pr[(x[n] = S_j) \,|\, (x[n-1] = S_i)] \tag{3.26}$$

The complete set of these transition probabilities is most conveniently represented by a matrix $\mathbf{P}$ whose elements are the $P_{j|i}$. For example, if the Markov chain has three states, the matrix of transition probabilities has the form

$$\mathbf{P} = \begin{bmatrix} P_{1|1} & P_{2|1} & P_{3|1} \\ P_{1|2} & P_{2|2} & P_{3|2} \\ P_{1|3} & P_{2|3} & P_{3|3} \end{bmatrix}$$

This type of matrix is known as a *stochastic matrix*. It has two salient properties: the elements are nonnegative, and the rows sum to 1. In what follows, it is generally more convenient to work with the transpose of this matrix, which for this example has the form

$$\mathbf{\Pi} = \mathbf{P}^T = \begin{bmatrix} \mathrm{P}_{1|1} & \mathrm{P}_{1|2} & \mathrm{P}_{1|3} \\ \mathrm{P}_{2|1} & \mathrm{P}_{2|2} & \mathrm{P}_{2|3} \\ \mathrm{P}_{3|1} & \mathrm{P}_{3|2} & \mathrm{P}_{3|3} \end{bmatrix}$$

The matrix $\mathbf{\Pi}$ will be called the *transition matrix*. The Markov chain such as the one depicted in Fig. 3.7 can also be represented graphically by a state transition diagram such as the one shown in Fig. 3.8. Branches in the diagram whose transition probabilities are zero are usually omitted and a branch such as the one labeled $\mathrm{P}_{1|1}$ that results in no change of state is called a "self-loop." The state transition diagram frequently provides a considerable amount of insight in the analysis of Markov chains.

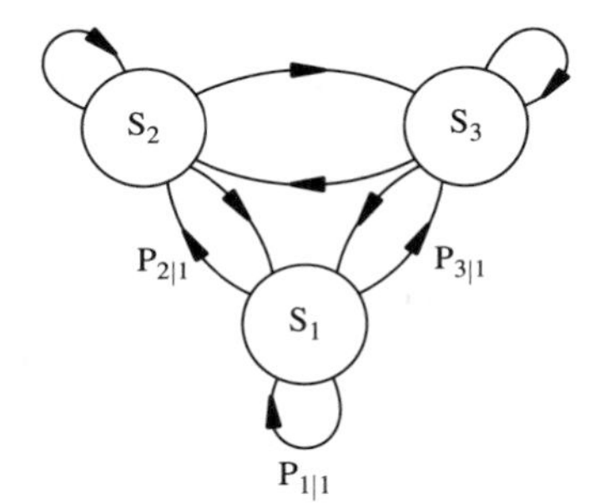

Figure 3.8 State transition diagram for a three-state Markov chain.

Higher-Order Transition Probabilities. In the analysis of Markov chains it is sometimes necessary to know the kth-order transition probabilities. These are defined as

$$\mathrm{P}_{j|i}^{(k)} \stackrel{\text{def}}{=} \Pr[(x[n] = \mathrm{S}_j) \,|\, (x[n-k] = \mathrm{S}_i)] \tag{3.27}$$

These higher-order transition probabilities can be computed very conveniently in terms of the ordinary (first-order) transition probabilities. To see how this is done, let us compute the term $\mathrm{P}_{2|1}^{(2)}$ for the three-state system considered above. From Fig. 3.8, the possible ways to get from state 1 to state 2 in exactly two transitions, and their probabilities, are

Possible paths	Probability
$1 \to 1 \to 2$	$\mathrm{P}_{1|1} \cdot \mathrm{P}_{2|1}$
$1 \to 2 \to 2$	$\mathrm{P}_{2|1} \cdot \mathrm{P}_{2|2}$
$1 \to 3 \to 2$	$\mathrm{P}_{3|1} \cdot \mathrm{P}_{2|3}$

The probability of starting in state 1 and arriving in state 2 after two transitions is therefore the sum

$$\mathrm{P}_{2|1}^{(2)} = \mathrm{P}_{1|1} \cdot \mathrm{P}_{2|1} \;+\; \mathrm{P}_{2|1} \cdot \mathrm{P}_{2|2} \;+\; \mathrm{P}_{3|1} \cdot \mathrm{P}_{2|3}$$

Note that this is just the "2|1" element in the product matrix

$$\mathbf{P}^2 = \mathbf{P} \cdot \mathbf{P} = \begin{bmatrix} \mathrm{P}_{1|1} & \mathrm{P}_{2|1} & \mathrm{P}_{3|1} \\ \mathrm{P}_{1|2} & \mathrm{P}_{2|2} & \mathrm{P}_{3|2} \\ \mathrm{P}_{1|3} & \mathrm{P}_{2|3} & \mathrm{P}_{3|3} \end{bmatrix} \begin{bmatrix} \mathrm{P}_{1|1} & \mathrm{P}_{2|1} & \mathrm{P}_{3|1} \\ \mathrm{P}_{1|2} & \mathrm{P}_{2|2} & \mathrm{P}_{3|2} \\ \mathrm{P}_{1|3} & \mathrm{P}_{2|3} & \mathrm{P}_{3|3} \end{bmatrix}$$

which is obtained by multiplying the first row in the first matrix by the second column in the second matrix. A similar analysis shows that all of the second-order transition probabilities are elements of the product matrix $\mathbf{P}^2$. You can further verify that the matrix $\mathbf{\Pi}^2$ also contains all of the second-order transition probabilities, but in the transposed arrangement.

The general result is that the matrices $\mathbf{P}^k$ or $\mathbf{\Pi}^k$ contain all of the kth-order transition probabilities. However, to show it by the previous argument would be quite tedious. To derive the result more easily and generally, assume that there are a total of Q states and define the vector $\mathbf{p}[n]$ as

$$\mathbf{p}[n] = \begin{bmatrix} \mathrm{p}_1[n] \\ \mathrm{p}_2[n] \\ \vdots \\ \mathrm{p}_Q[n] \end{bmatrix} = \begin{bmatrix} \Pr[(x[n] = \mathrm{S}_1)] \\ \Pr[(x[n] = \mathrm{S}_2)] \\ \vdots \\ \Pr[(x[n] = \mathrm{S}_Q)] \end{bmatrix} \tag{3.28}$$

In other words, the ith component of $\mathbf{p}[n]$ is just the probability that the process is in state i at time n. From the definition of $\mathbf{\Pi}$ it is then clear that $\mathbf{p}[n]$ satisfies the difference equation

$$\mathbf{p}[n] = \mathbf{\Pi}\mathbf{p}[n-1] \tag{3.29}$$

Iterating this k times yields

$$\mathbf{p}[n] = \mathbf{\Pi}^k \mathbf{p}[n-k] \tag{3.30}$$

Thus $\mathbf{\Pi}^k$ is the kth-order transition matrix, and its elements are the kth-order transition probabilities $\mathrm{P}^{(k)}_{j|i}$.

The previous analysis permits the easy establishment of another important result for Markov chains. Suppose that the matrix $\mathbf{\Pi}^k$ is factored as

$$\mathbf{\Pi}^k = \mathbf{\Pi}^l \mathbf{\Pi}^{(k-l)}$$

where l is any integer between 0 and k; then $\mathbf{\Pi}^l$ and $\mathbf{\Pi}^{(k-l)}$ are the lth-order and $(k-l)^{\text{th}}$-order transition matrices respectively. Writing out the matrix product in terms of its elements produces

$$\boxed{\mathrm{P}^{(k)}_{j|i} = \sum_{q=1}^{Q} \mathrm{P}^{(l)}_{j|q} \mathrm{P}^{(k-l)}_{q|i}; \qquad 0 < l < k} \tag{3.31}$$

This is the *Chapman-Kolmogorov equation*, which can be used as an alternative definition of a Markov chain. This equation states that the probability of moving from state i to j in k transitions is a sum of products of the probability of moving from state i to *each* of the other states in $k - l$ transitions and the probability of moving from each of these intermediate states to state j in l transitions. This principle is useful in many problems involving Markov chains.

Limiting-State Probabilities. When the set of transition probabilities for a Markov chain satisfies certain conditions, the Markov chain has a special limiting behavior for large numbers of observations. Consider the simple two-state Markov chain shown in Fig. 3.9.[5] The kth-order transition probabilities for several values of k are given in Table 3.3. It can be seen from the table that the higher-order transition probabilities apparently converge to a constant value, and that this value is *independent of the starting state*. Through (3.30) this implies that after a sufficient length of time the probability that the process is in any particular state approaches a constant value. Thus the process has only "limited memory" of the starting state. Let us assume the process is first observed at $n = n_o$. The vector $\mathbf{p}[n_o]$ will be called the "initial condition." The *limiting-state probability* vector for the Markov chain is defined as

$$\lim_{n\to\infty} \mathbf{p}[n] = \lim_{n\to\infty} \mathbf{\Pi}^{n-n_o}\mathbf{p}[n_o] \tag{3.32}$$

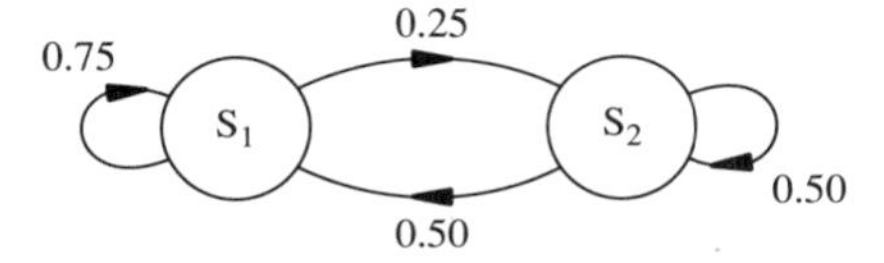

Figure 3.9 Simple two-state Markov chain that attains limiting-state probabilities.

TABLE 3.3 TRANSITION PROBABILITIES FOR THE TWO-STATE MARKOV CHAIN

	$k=1$	$k=2$	$k=3$	$k=4$	$k=5$	$k=6$	$k=7$	$k=8$
$P_{1\|1}^{(k)}$	0.750	0.688	0.672	0.668	0.667	0.667	0.667	0.667
$P_{2\|1}^{(k)}$	0.250	0.312	0.328	0.332	0.333	0.333	0.333	0.333
$P_{1\|2}^{(k)}$	0.500	0.625	0.656	0.664	0.666	0.667	0.667	0.667
$P_{2\|2}^{(k)}$	0.500	0.375	0.344	0.336	0.334	0.333	0.333	0.333

when the limit exists and is independent of $\mathbf{p}[n_o]$. Limiting-state probabilities do not exist for *all* Markov chains; Fig. 3.10 shows two cases typical of when they do not exist. In Fig. 3.10(a) there is a single "transient state" and two separate sets of "recurrent states" (states for which it is possible to transfer from and eventually back to again). Each such set is called a *class*. The asymptotic probabilities of the states given by (3.32) are *different* for different initial conditions since the chain always ends up in either one class or the other. Therefore, the limiting-state probabilities do not exist. In Fig. 3.10(a) the states are said to form a periodic class. The probability vector $\mathbf{p}[n]$ is periodic and will never reach a limiting constant value.

[5]Example taken from [2, pp. 166–167], © McGraw-Hill, 1967; quoted by permission.

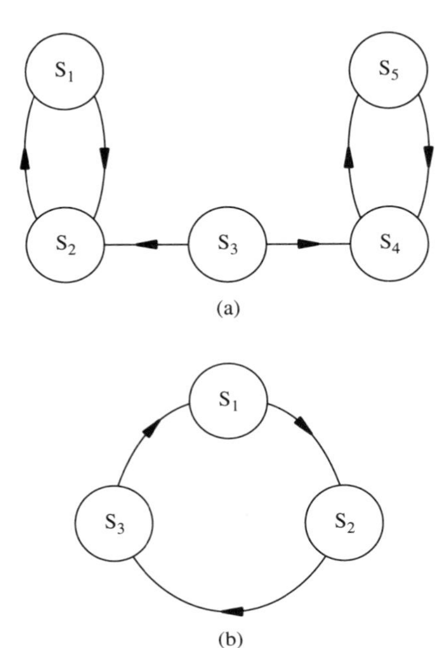

Figure 3.10 Markov chains for which no set of limiting-state probabilities exists. (a) Transient state and two recurrent classes of states. (b) Periodic class of states.

The existence of the limiting-state probabilities can be expressed algebraically as conditions on the eigenvalues of the transition matrix. It can be shown that the eigenvalues of $\mathbf{\Pi}$ (which may be complex) all have magnitudes less than or equal to 1.[6] In addition, the matrix always has *at least* one eigenvalue equal to 1. To see this, denote the eigenvector by $\bar{\mathbf{p}}$ and observe that the eigenvalue equation in this case

$$\mathbf{\Pi}\bar{\mathbf{p}} = \bar{\mathbf{p}} \tag{3.33}$$

has a solution if and only if the determinant $|\mathbf{\Pi} - \mathbf{I}|$ is equal to zero. This is always the case, however, since the columns of $\mathbf{\Pi}$ sum to 1. Consider the case of a three-state Markov chain, for example. In this case

$$\mathbf{\Pi} - \mathbf{I} = \begin{bmatrix} \mathrm{P}_{1|1} - 1 & \mathrm{P}_{1|2} & \mathrm{P}_{1|3} \\ \mathrm{P}_{2|1} & \mathrm{P}_{2|2} - 1 & \mathrm{P}_{2|3} \\ \mathrm{P}_{3|1} & \mathrm{P}_{3|2} & \mathrm{P}_{3|3} - 1 \end{bmatrix}$$

Since the columns all sum to zero, the rows are linearly dependent and the determinant is zero. Therefore, the solution to (3.33) exists and $\bar{\mathbf{p}}$ is an eigenvector corresponding to an eigenvalue of 1. It can be shown that the convergence of (3.32) to limiting state probabilities independent of the initial conditions occurs if and only if there is a *single* eigenvalue equal to 1 and there are no complex eigenvalues with magnitudes equal to 1. In this case, $\bar{\mathbf{p}}$,

[6]The proof of this fact is not elementary and depends on some results from the Frobenius theory of positive matrices [7, 8].

which solves (3.33), is that unique limiting-state value. The columns of the higher-order transition matrix then become identical, that is,

$$\lim_{n\to\infty} \mathbf{\Pi}^{n-n_o} = \begin{bmatrix} | & | & & | \\ \bar{\mathbf{p}} & \bar{\mathbf{p}} & \cdots & \bar{\mathbf{p}} \\ | & | & & | \end{bmatrix} \tag{3.34}$$

Problem 3.11 explores the proof of these statements.

To see how to solve (3.33) in the previous three-state case, write this equation as

$$\begin{bmatrix} P_{1|1}-1 & P_{1|2} & P_{1|3} \\ P_{2|1} & P_{2|2}-1 & P_{2|3} \\ P_{3|1} & P_{3|2} & P_{3|3}-1 \end{bmatrix} \begin{bmatrix} \bar{p}_1 \\ \bar{p}_2 \\ \bar{p}_3 \end{bmatrix} = \begin{bmatrix} 0 \\ 0 \\ 0 \end{bmatrix}$$

The solution of this equation determines the limiting-state probabilities to within a multiplicative constant. The condition

$$\bar{p}_1 + \bar{p}_2 + \bar{p}_3 = 1$$

then determines the constant. An example illustrates the entire procedure for a Markov chain with just two states.

Example 3.3

A certain binary sequence, with values of $+1$ and -1, is modeled by a Markov chain with transition matrix and state diagram shown in Fig. EX3.3.

$$\mathbf{\Pi} = \begin{bmatrix} 0.6 & 0.1 \\ 0.4 & 0.9 \end{bmatrix}$$

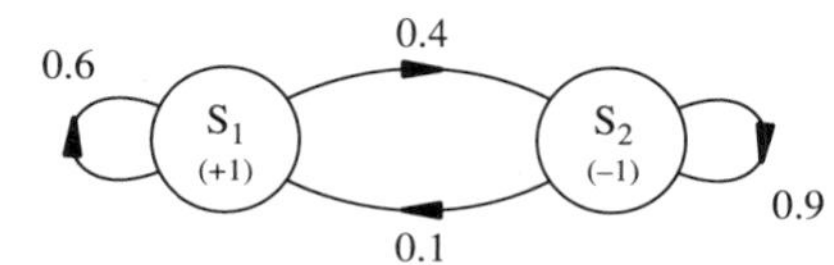

Figure EX3.3

It is desired to find the probability of a run of ten $+1$'s and a run of ten -1's.

It is first necessary to find the probabilities that when the sequence is observed at a random time, we see a $+1$ or a -1. These are the limiting-state probabilities, which can be found from (3.33):

$$\begin{bmatrix} -0.4 & 0.1 \\ 0.4 & -0.1 \end{bmatrix} \begin{bmatrix} \bar{p}_1 \\ \bar{p}_2 \end{bmatrix} = \begin{bmatrix} 0 \\ 0 \end{bmatrix}$$

This provides the relation

$$\bar{p}_2 = 4\bar{p}_1$$

The additional condition

$$\bar{p}_1 + \bar{p}_2 = 1$$

then yields the solution $\bar{p}_1 = 0.2$ and $\bar{p}_2 = 0.8$. The probability of a run of ten +1's is then

$$(0.2)(0.6)^9 = 0.0020$$

and the probability of a run of ten −1's is

$$(0.8)(0.9)^9 = 0.3099$$

Now compare these results to the probability of runs in a Bernoulli process with the same limiting-state probabilities. The probability of ten +1's for the Bernoulli process is

$$(0.2)^{10} = 1.024 \times 10^{-7}$$

while the probability of ten −1's is

$$(0.8)^{10} = 0.1074$$

These probabilities are significantly smaller than the probabilities for the Markov chain, as you might expect, since successive values of the Bernoulli process are independent.

The Random Walk as a Markov Chain. After seeing several examples of Markov chains it is not surprising to find that the random walk described in Section 3.2.2 is also a Markov chain, but with an infinite number of states. In fact, it can be shown that any process with independent increments is Markov, although the converse is not true.

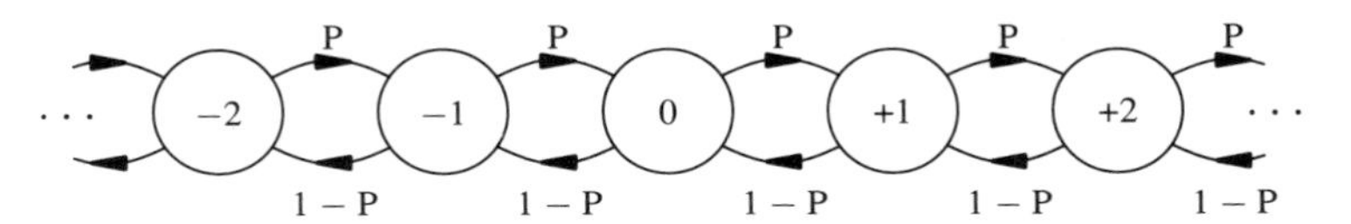

Figure 3.11 Random walk and its state transition diagram.

Figure 3.11 shows the state transition diagram for a random walk. This type of Markov chain is also known as a "birth and death" process, since it can serve as a model for population growth. Birth and death processes may also have self-loops associated with each state and a *finite* number of states. A typical state diagram is shown in Fig. 3.12. The transition matrix for a birth and death process is a tridiagonal (banded) matrix. For the example of Fig. 3.12 it has the form

$$\Pi = \begin{bmatrix} 1-b_1 & d_2 & 0 & 0 \\ b_1 & 1-b_2-d_2 & d_3 & 0 \\ 0 & b_2 & 1-b_3-d_3 & d_4 \\ 0 & 0 & b_3 & 1-d_4 \end{bmatrix}$$

Application of (3.33) to a birth and death process leads to the simple conditions

$$\bar{p}_{i+1} d_{i+1} = \bar{p}_i b_i\,; \qquad i = 1, 2, \ldots, Q-1 \tag{3.35}$$

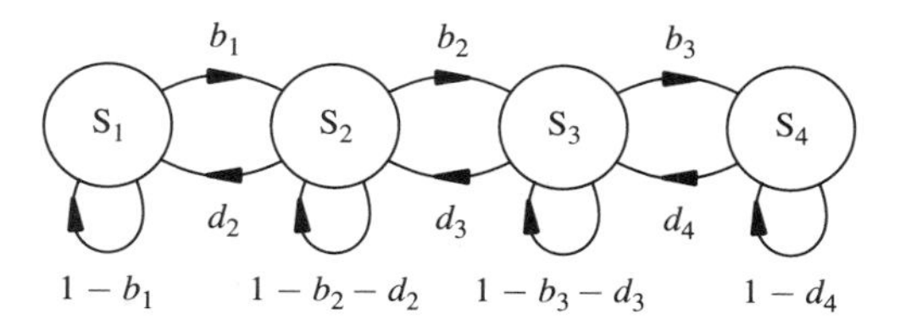

Figure 3.12 Birth and death process with four states.

This equation expresses the fact that at limiting-state conditions, each state is in a kind of equilibrium with the probability of an outward transition equal to the probability of an inward transition. Equation 3.35, together with the condition that the probabilities must sum to 1, provide enough simultaneous equations to solve for the limiting-state probabilities of a birth and death process.

Several variations of the random walk have been studied extensively (see, e.g., [4]). One of these variations is the random walk with "reflecting boundaries," where when a process enters one of the end-point states (state 1 or Q) it is forced to move into another state on the next transition. Another is the random walk with "absorbing boundaries" where upon entering an end-point state, the process remains there for ever after. All of these variations can be considered as special cases of the birth and death process.

3.3.2 Hidden Markov Models

An interesting class of random processes that has found applications in speech processing and other areas is known as a *hidden Markov model* [12, 13]. A similar model based on two-dimensional generalizations of the Markov chain known as Markov random fields has simultaneously found use in image processing [14, 15]. The basic idea is shown in Fig. 3.13. A random sequence shown in part (a) is generated as shown in part (b). Each of two systems generates a random process with some given statistical description. Which process is observed depends upon the position of a switch whose state is characterized by a Markov chain with the state transition diagram of Fig. 3.13(c). At each value of n the switch may either stay in its current position or change position according to the known transition probabilies. Figure 3.13(a) also shows the state of the switch for each value of n. In the analysis of speech, a model of this type (with more states) is quite useful because the observed speech sequence remains stationary for only brief intervals of time and may have significantly different statistical characteristics in adjacent time intervals.[7] Efficient procedures for estimating the model parameters or "training" the model are available [12,13], but here it is assumed that the parameters are already known.

Since this model deals with random processes on two different levels, it is said to be *doubly stochastic*. There is the random process $s[n]$ representing the switch position, which is a Markov chain, and the random process $x[n]$ which is formed by observing the output of the two systems according to the switch position. The underlying Markov chain $s[n]$ takes on a finite number of values (two in this example) but these values *cannot be observed directly*. Hence the Markov chain is "hidden." The observed process $x[n]$ may

[7]Usually in the speech application, the random process $x[n]$ is not the sampled speech waveform, but instead a set of measurements derived from it.

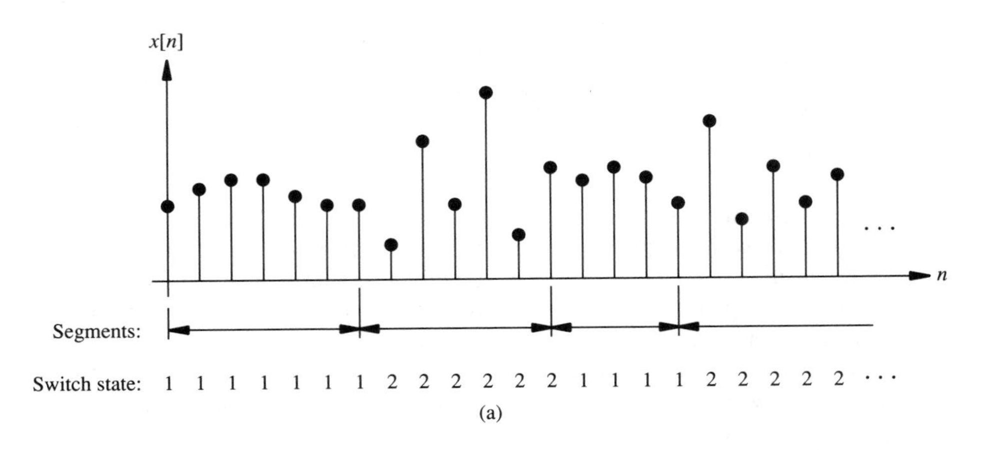

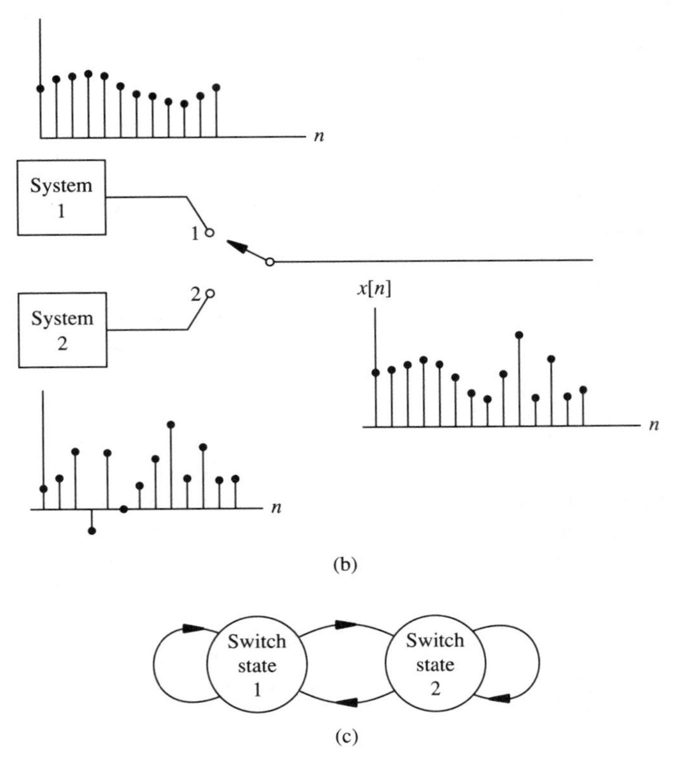

Figure 3.13 Hidden Markov model. (a) Sequence of observations. (b) Generation of the observation sequence. (c) State transition diagram. (From Charles W. Therrien, *Decision Estimation and Classification*, copyright © John Wiley & Sons, New York, 1989. Reproduced by permission.)

take on a continuous or discrete set of values according to what is produced by the two systems. The observed process $x[n]$ need not be Markov; it is the Markov character of the underlying process that gives the model its name.

One problem of importance with respect to hidden Markov models is to estimate the underlying state sequence $s[n]$ from the observed process $x[n]$. This problem occurs, for example, in the observation of a binary-valued Markov chain representing a communication message in additive noise. The received sequence $x[n]$ is the sum of the message sequence $s[n]$ (which is the "state") and the additive noise. While the message $s[n]$ takes on only two values, the received sequence $x[n]$ can take on a large range of other values due to the added noise. Let the value that $x[k]$ takes on be denoted by x_k for $0 \leq k \leq n$, and similarly let the value of $s[k]$ be denoted by s_k for k in this same range of values.[8] For simplicity, let us assume that the values that $s[k]$ or s_k can take on are simply the set of integers $1, 2, \ldots, Q$. In other words,

$$\mathrm{S}_i = i \qquad i = 1, 2, \ldots, Q \tag{3.36}$$

General procedures for estimating random variables such as the s_k from noisy observations are given in Chapter 6. These methods require the computation of the conditional probability

$$\Pr[(s[0] = \mathrm{s}_0), \ldots, (s[n] = \mathrm{s}_n) | (x[0] = \mathrm{x}_0), \ldots, (x[n] = \mathrm{x}_n)]$$

This conditional probability can be computed from Bayes' rule in the form

$$\Pr[\mathrm{s}_0, \ldots, \mathrm{s}_n | \mathrm{x}_0, \ldots, \mathrm{x}_n] = \frac{f_{\boldsymbol{x}|\boldsymbol{s}}(\mathrm{x}_0, \ldots, \mathrm{x}_n \mid \mathrm{s}_0, \ldots, \mathrm{s}_n) \Pr[\mathrm{s}_0, \ldots, \mathrm{s}_n]}{f_{\boldsymbol{x}}(\mathrm{x}_0, \ldots, \mathrm{x}_n)} \tag{3.37}$$

where in this equation a slightly abusive but less cumbersome notation is used for the conditional probability. One way to estimate the states is to choose those that maximize this conditional probability. The maximization is not trivial. In the simplest case it involves a form of dynamic programming known as the Viterbi algorithm [16]. The algorithm is discussed later in this subsection.

Another problem of interest involves computing the value of the unconditional density $f_{\boldsymbol{x}}(\mathrm{x}_0, \mathrm{x}_1, \ldots, \mathrm{x}_n)$ for a given set of observations from a hidden Markov model. This is of interest in classification problems. That is, suppose that there are *several* hidden Markov models representing different classes of data such as spoken words, and it is desired to decide which model has most likely produced a given sequence of observations. A reasonable procedure is to evaluate the density $f_{\boldsymbol{x}}(\mathrm{x}_0, \mathrm{x}_1, \ldots, \mathrm{x}_n)$ for each model and choose the class for which it is largest. When used in this context, the quantity $f_{\boldsymbol{x}}$ is called a "likelihood function."

Either of these two problems can be approached as follows. First of all, it is easy to write an expression for the probability of any given state sequence. In particular,

$$\Pr[\mathrm{s}_0, \mathrm{s}_1, \ldots, \mathrm{s}_n] = \bar{\mathrm{p}}_{\mathrm{s}_0} \left(\prod_{k=1}^{n} \mathrm{P}_{\mathrm{s}_k | \mathrm{s}_{k-1}} \right) \tag{3.38}$$

[8] To avoid excessively cumbersome notation, it is assumed throughout the discussion of hidden Markov models that observation of the process begins at $n_o = 0$.

where the term $\mathrm{P}_{\mathrm{s}_k|\mathrm{s}_{k-1}}$ represents $\mathrm{P}_{j|i}$ for $j = \mathrm{s}_k$ and $i = \mathrm{s}_{k-1}$. For example, if the underlying Markov model has $Q = 3$ states, then the probability of a particular state sequence such as $\{2, 3, 3, 1, 2\}$ is

$$\Pr[2, 3, 3, 1, 2] = \bar{\mathrm{p}}_2 \mathrm{P}_{3|2} \mathrm{P}_{3|3} \mathrm{P}_{1|3} \mathrm{P}_{2|3}$$

Now assume that when the system remains in any particular state, the observations $x[k]$ generated are independent. Further assume that the observations generated by *different* states are also independent.[9] Then the probability density for the observations, *given* the state sequence, can be written as

$$f_{x|s}(\mathrm{x}_0, \mathrm{x}_1, \ldots, \mathrm{x}_n \mid \mathrm{s}_0, \mathrm{s}_1, \ldots, \mathrm{s}_n) = \prod_{k=0}^{n} f_{x|s}(\mathrm{x}_k \mid \mathrm{s}_k) \tag{3.39}$$

Note that the term $f_{x|s}(\mathrm{x}_k \mid \mathrm{s}_k)$ really denotes a family of densities depending on the state. For example, in a three-state problem $f_{x|s}(\mathrm{x} \mid 1)$ may represent a Gaussian density for x with mean zero, $f_{x|s}(\mathrm{x} \mid 2)$ may represent a Gaussian density with a nonzero mean, and $f_{x|s}(\mathrm{x} \mid 3)$ could conceivably represent an entirely different form such as a uniform density. The densities simply correspond to the type of random process that is produced when the process is in a particular state; these densities are assumed to be known or to have been estimated. Equations 3.38 and 3.39 are the expressions needed to compute the two terms in the numerator of (3.37).

To find the unconditional density function $f_{x}(\mathrm{x}_0, \mathrm{x}_1, \ldots, \mathrm{x}_n)$, first combine (3.38) and (3.39) to write

$$f_{xs}(\mathrm{x}_0, \mathrm{x}_1, \ldots, \mathrm{x}_n, \mathrm{s}_0, \mathrm{s}_1, \ldots, \mathrm{s}_n) = \prod_{k=0}^{n} f_{x|s}(\mathrm{x}_k \mid \mathrm{s}_k) \mathrm{P}_{\mathrm{s}_k|\mathrm{s}_{k-1}} \tag{3.40}$$

where for convenience the term $\mathrm{P}_{\mathrm{s}_0|\mathrm{s}_{-1}}$ is defined to mean

$$\mathrm{P}_{\mathrm{s}_0|\mathrm{s}_{-1}} \stackrel{\text{def}}{=} \bar{\mathrm{p}}_{\mathrm{s}_0} \tag{3.41}$$

Then the probability density function for the observations alone is given by

$$f_{x}(\mathrm{x}_0, \mathrm{x}_1, \ldots, \mathrm{x}_n) = \sum_{\substack{\text{all sequences} \\ \text{of states}}} f_{xs}(\mathrm{x}_0, \mathrm{x}_1, \ldots, \mathrm{x}_n, \mathrm{s}_0, \mathrm{s}_1, \ldots, \mathrm{s}_n) \tag{3.42}$$

While this in theory solves the problem, (3.42) is computationally intensive. For a Markov chain with Q states it involves computation of the density and summation over Q^{n+1} different permutations of states. In other words, the computation increases *exponentially* with the length of the sequence. Fortunately, there exists a set of recursive procedures for computing the probabilities whose computation increases only linearly with n. These procedures, known as "forward" and "backward" procedures, are described below.

[9]The procedure can easily be generalized to the case where the observations are also Markov [e.g., observations of the form (3.17)].

Forward Procedure. The forward procedure begins by defining the *forward variable*

$$\alpha_i[k] \stackrel{\text{def}}{=} f_{\boldsymbol{x}s[k]}(\mathrm{x}_0, \mathrm{x}_1, \ldots, \mathrm{x}_k, i) \tag{3.43}$$

(This is the joint density function for the samples $x[0], x[1], \ldots, x[k]$ and the state $s[k]$ evaluated at $\mathrm{s}_k = i$.) The desired joint density function is related to the forward variables by

$$f_{\boldsymbol{x}}(\mathrm{x}_0, \mathrm{x}_1, \ldots, \mathrm{x}_n) = \sum_{i=1}^{Q} f_{\boldsymbol{x}s[n]}(\mathrm{x}_0, \mathrm{x}_1, \ldots, \mathrm{x}_n, i)$$

or

$$\boxed{f_{\boldsymbol{x}}(\mathrm{x}_0, \mathrm{x}_1, \ldots, \mathrm{x}_n) = \sum_{i=1}^{Q} \alpha_i[n]} \tag{3.44}$$

The forward variables are of interest because they can be used to develop an efficient recursion. In particular, note from definitions of conditional and marginal probability that

$$\begin{aligned} f_{\boldsymbol{x}s[k]}(\mathrm{x}_0, \ldots, \mathrm{x}_k, i) &= \sum_{j=1}^{Q} f_{\boldsymbol{x}s[k]s[k-1]}(\mathrm{x}_0, \ldots, \mathrm{x}_k, i, j) \\ &= \left[\sum_{j=1}^{Q} f_{\boldsymbol{x}s[k-1]}(\mathrm{x}_0, \ldots, \mathrm{x}_{k-1}, j) \Pr[(s[k] = i) \,|\, (s[k-1] = j)]\right] f_{x[k]\,|\,s[k]}(\mathrm{x}_k \,|\, i) \end{aligned}$$

This can be written as

$$\boxed{\alpha_i[k] = \left[\sum_{j=1}^{Q} \alpha_j[k-1] \cdot \mathrm{P}_{i|j}\right] f_{x|s}(\mathrm{x}_k|i) \qquad \begin{aligned} i &= 1, 2, \ldots, Q \\ k &= 1, 2, \ldots, n \end{aligned}} \tag{3.45}$$

with the initial conditions

$$\boxed{\alpha_i[0] = \bar{\mathrm{p}}_i\, f_{x|s}(\mathrm{x}_0|i); \qquad i = 1, 2, \ldots, Q} \tag{3.46}$$

The recursion can be described in terms of a lattice structure similar to that for an FFT algorithm and depicted in Fig. 3.14(a). There are $n+1$ columns of nodes corresponding to the $n+1$ observations and Q nodes in each column corresponding to the Q possible states. The forward variables $\alpha_i[k]$ are associated with the nodes as shown. Unlike the FFT, every node in the $k-1^{\text{st}}$ column is involved in the computation of a forward variable at a node in the kth column. A signal flow graph for the actual computation is shown in Fig. 3.14(b).

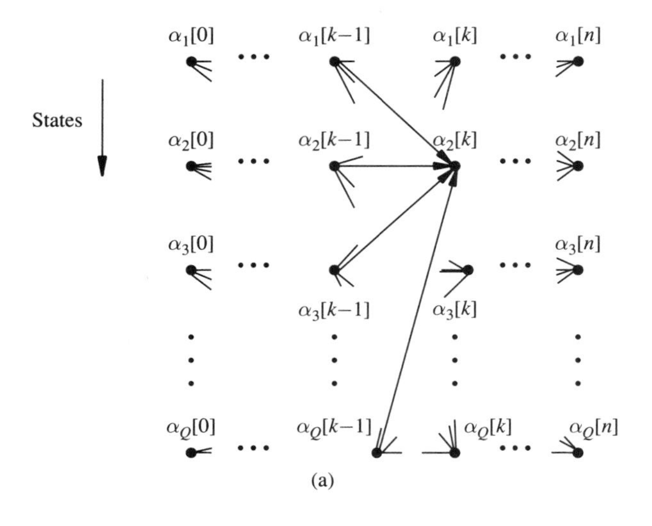

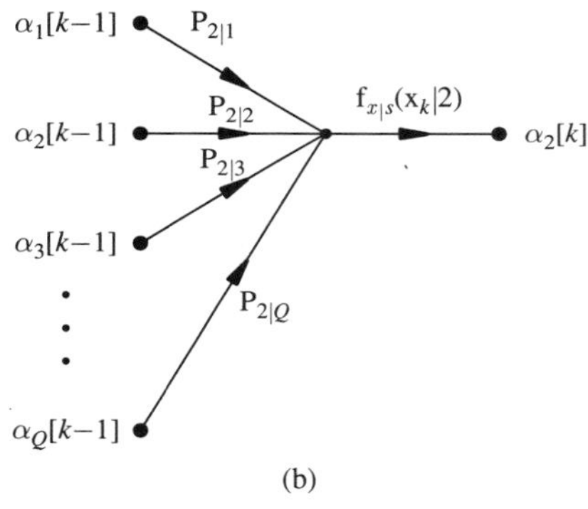

Figure 3.14 Computation of the likelihood function for a given set of observations for a hidden Markov model by the forward procedure. (a) Lattice structure showing dependence among the forward variables. (b) Signal flow graph for computing one of the forward variables.

The recursion is carried out by beginning in the leftmost column, applying (3.45) to compute the forward variables in the next column, and continuing. Once all of the forward variables are calculated, the joint density $f_{\boldsymbol{x}}(\mathrm{x}_0, \mathrm{x}_1, \ldots, \mathrm{x}_n)$ can be computed directly from (3.44). The amount of computation is proportional to $Q^2 \cdot (n+1)$, a considerable savings over evaluating all of the Q^{n+1} permutations cited earlier.

Backward Procedure. An equally efficient computational procedure to compute the value of the joint density function for a set of observations can be carried out by proceding backward along the random process. For this, define the backward variables as

$$\beta_i[k] \stackrel{\text{def}}{=} f_{\boldsymbol{x}\,|\,s[k]}(\mathrm{x}_{k+1}, \ldots, \mathrm{x}_n | i) \tag{3.47}$$

The joint density of the observations is related to the backward variables by noting that

$$f_{\boldsymbol{x}}(\mathrm{x}_0, \mathrm{x}_1, \ldots, \mathrm{x}_n) = \sum_{i=1}^{Q} f_{\boldsymbol{x}s[0]}(\mathrm{x}_0, \mathrm{x}_1, \ldots, \mathrm{x}_n, i)$$

$$= \sum_{i=1}^{Q} f_{\boldsymbol{x}\,|\,s[0]}(\mathrm{x}_1, \ldots, \mathrm{x}_n|i)\, f_{x[0]\,|\,s[0]}(\mathrm{x}_0|i)\, \Pr[(s[0] = i)]$$

or

$$f_{\boldsymbol{x}}(\mathrm{x}_0, \mathrm{x}_1, \ldots, \mathrm{x}_n) = \sum_{i=1}^{Q} \beta_i[0]\, f_{x|s}(\mathrm{x}_0|i)\, \bar{\mathrm{p}}_i \tag{3.48}$$

A recursion involving the backward variables can be set up by using the definition of conditional probability and summing over the states to write

$$f_{\boldsymbol{x}\,|\,s[k]}(\mathrm{x}_{k+1}, \ldots, \mathrm{x}_n|i) = \sum_{j=1}^{Q} f_{\boldsymbol{x}s[k+1]\,|\,s[k]}(\mathrm{x}_{k+1}, \ldots, \mathrm{x}_n, j|i)$$

$$= \sum_{j=1}^{Q} f_{\boldsymbol{x}\,|\,s[k+1]}(\mathrm{x}_{k+2}, \ldots, \mathrm{x}_n|j)$$

$$\times f_{x[k+1]\,|\,s[k+1]}(\mathrm{x}_{k+1}|j)\, \Pr[(s[k+1] = j)\,|\,(s[k] = i)]$$

This can be expressed using the definitions for the backward variables and transition probabilities as

$$\beta_i[k] = \sum_{j=1}^{Q} \beta_j[k+1] f_{x|s}(\mathrm{x}_{k+1}|j)\mathrm{P}_{j|i} \qquad i = 1, 2, \ldots, Q; \quad k = n-1, n-2, \ldots, 0 \tag{3.49}$$

with initial conditions

$$\beta_i[n] = 1; \qquad i = 1, 2, \ldots, Q \tag{3.50}$$

This recursion is carried out on a lattice structure of the same form as the one for the forward procedure. However, in this graph the variables $\beta_i[k]$ are associated with the nodes and the computation proceeds from right to left. The lattice and the basic computation (3.49) associated with the backward procedure are shown in Fig. 3.15. The backward procedure, like the forward procedure, requires computation proportional to $Q^2 \cdot (n+1)$ and the joint density of the observations is easily computed from (3.48) once all of the backward variables are known.

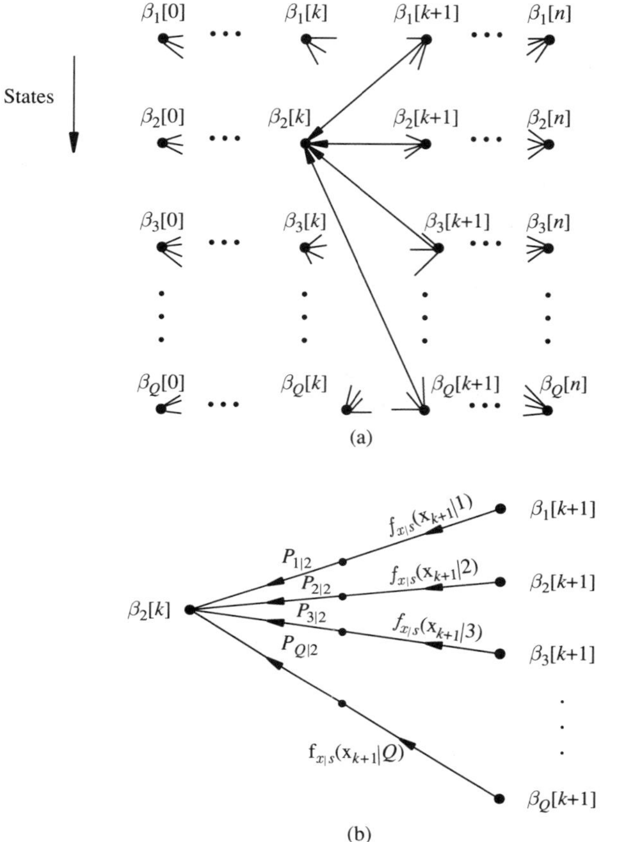

Figure 3.15 Computation of the likelihood function for a given set of observations for a hidden Markov model by the backward procedure. (a) Lattice structure showing dependence among the backward variables. (b) Signal flow graph for computing one of the backward variables.

In practical application of these procedures the terms in the computation for any resonably large values of n can become very small and cause underflow. As a result, practical algorithms use logarithms of the terms in (3.45) and (3.49) and employ suitable scaling procedures to ensure that correct results will be obtained. Some of these procedures are discussed in [13].

Estimating the State Sequence: The Viterbi Algorithm. When the joint probability density can be written in the form (3.40), a method similar to the forward and backward procedures can be applied to find a maximizing sequence of states. Since (3.40) is equivalent to the numerator in (3.37), this provides a way to estimate the underlying state sequence. The procedure, which is known as the Viterbi algorithm, begins by taking the logarithm of (3.40) and focusing on the equivalent problem of maximizing

$$\sum_{k=0}^{n} V_i[k] + B_{j,i}[k] \tag{3.51}$$

where $i = s_k$; $j = s_{k-1}$ and V_i and $B_{j,i}$ are defined by

$$V_i[k] \stackrel{\text{def}}{=} \log f_{x|s}(x_k \mid i) \tag{3.52}$$

and

$$B_{j,i}[k] \stackrel{\text{def}}{=} \log P_{i|j} \tag{3.53}$$

The maximization can be posed as a problem of finding a path in a graph which is similar to the lattice used in the forward and backward procedures. In the context of the Viterbi algorithm this graph is known as a "trellis." The trellis and a typical path from the leftmost column to the rightmost column is depicted in Fig. 3.16. The terms $V_i[k]$ are associated with the nodes and the terms $B_{j,i}[k]$ are associated with the branches of the trellis. *Note that the relation between the variables is not the same as in a signal flow graph.* If the *path weight* is defined as the sum of the node terms V_i and branch terms $B_{j,i}$ along the path, then the maximization of (3.51) can be interpreted as finding an *optimal* path through the trellis (i.e., one of maximum path weight).

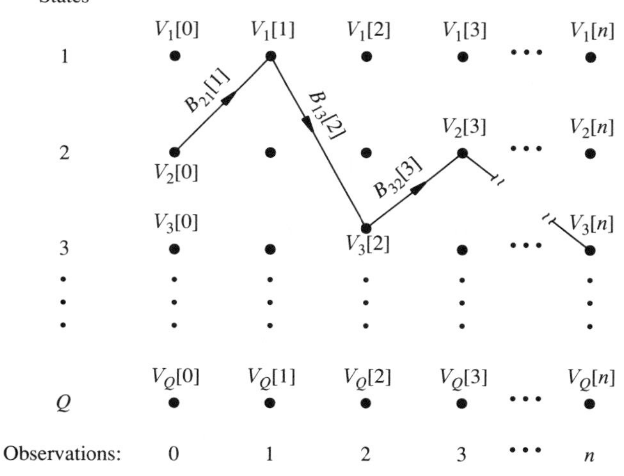

Figure 3.16 Path through the trellis in the Viterbi algorithm.

The search for the optimal path is based on the following principle. Suppose that an optimal path has been determined from column 0 to *each* node j in some column $k-1$. Denote the weights of these optimal paths by $W_j[k-1]$ and note that *one* of these paths must be part of the optimal path to any node in a subsequent column. In particular, if the optimal path to any specific node i in column k passes through the node j' in column $k-1$, then the total path weight is

$$W_{j'}[k-1] + B_{j',i}[k] + V_i[k]$$

If the path is truly optimal, then node j' is the node that maximizes this quantity. The optimal path weight for each node i in column k is thus given by

$$W_i[k] = \max_j \Big(W_j[k-1] + B_{j,i}[k] \Big) + V_i[k] \tag{3.54}$$

The maximizing node j' can be easily found by considering each node in column $k-1$ and evaluating this expression. The weight of the optimal path so computed and the node j' in the previous column from which the path extends is stored in a list with the node i. Thus at any stage in the algorithm only nodes in two adjacent columns need to be considered.

The procedure continues from column to column, finding the optimal paths to each node in each column. When the last column has been processed, the maximization is complete. The optimal path through the trellis is simply the optimal path for the node in column n that has the largest path weight. The optimal set of states can then be explicitly determined by tracing back from the chosen node in column n. Since the optimal path to the previous column has been stored with each node, this backtracing is easily done.

If, during the application of the algorithm, the optimal paths to all nodes in some column k happen to emanate from a *single* node j in column $k-1$, then the optimal path up to node j is automatically part of the optimal path to any succeeding columns. In problems involving a large number of observations (columns) and states (rows) this realization can be helpful since the columns up to $k-1$ can then be shifted out of memory to make room for more data.

3.4 MARTINGALES AND ABSOLUTELY FAIR PROCESSES

Another type of random process, which like the Markov process, has a limited dependence on past history is known as a *martingale*.[10] The martingale tends to be used more in the analysis of random processes than as a model for random signals. Nevertheless, the martingale can arise in certain statistical signal processing applications and it is instructive to examine its properties. This section provides a brief introduction to martingales.

Let us begin with a random process $x[n]$ with the property

$$\mathcal{E}\{x[n] \mid x[n-1], x[n-2], \ldots\} = 0 \tag{3.55}$$

Such a random process is said to be *absolutely fair*. The term arises from the use of such processes to describe gambling games where at each new round the expectation of the player's winnings is not influenced by his past winnings, and in fact is always zero [18]. Now define another random process $y[n]$ by the summation

$$y[n] = \sum_{k=n_o}^{n} x[k] + y[n_o] \tag{3.56}$$

where $y[n_o]$ is a constant initial condition. The process $y[n]$ is known as a *martingale* and has the property

$$\boxed{\mathcal{E}\{y[n] \mid y[n-1], y[n-2], \ldots, y[n_o]\} = y[n-1]} \tag{3.57}$$

Equation 3.57 is usually taken as the *definition* of a martingale.

[10] The term "martingale" comes from the Provençal name of the French community Martigues and is said to have long been used in a gambling context [17]. The Random House unabridged dictionary (2nd edition) defines *martingale* as "a system of gambling where stakes are doubled or otherwise raised after each loss."

To show that the sequence defined by (3.56) has the martingale property, observe that

$$y[n] = y[n-1] + x[n] \tag{3.58}$$

for $n > n_o$. Then taking the conditional expectation yields

$$\begin{aligned} \mathcal{E}\{y[n] \mid y[n-1], \ldots, y[n_o]\} &= \mathcal{E}\{y[n-1] \mid y[n-1], \ldots, y[n_o]\} \\ &\quad + \mathcal{E}\{x[n] \mid y[n-1], \ldots, y[n_o]\} \end{aligned} \tag{3.59}$$

The first term on the right side of this equation is just $y[n-1]$, while the second term on the right is equivalent to

$$\mathcal{E}\{x[n] \mid x[n-1], \ldots, x[n_o]\}$$

which by (3.55) is zero. Thus it follows from (3.59) that

$$\mathcal{E}\{y[n] \mid y[n-1], \ldots, y[n_o]\} = y[n-1] \tag{3.60}$$

and the sequence is a martingale. Since the summation of an absolutely fair process is a martingale it also follows that if $y[n]$ is a martingale, then the difference sequence

$$x[n] = y[n] - y[n-1]$$

is absolutely fair.

It is also true that any zero-mean process with independent increments, such as the discrete Wiener process, is a martingale. This follows because the difference process is a sequence of zero-mean independent random variables and so satisfies (3.55). The absolutely fair property is a much weaker condition than independence however. Any independent sequence with mean zero is absolutely fair, but the converse is not true. Consider the following example.

Example 3.4

The sequence $x[n]$ is a nonstationary random signal with the density shown in Fig. EX3.4.

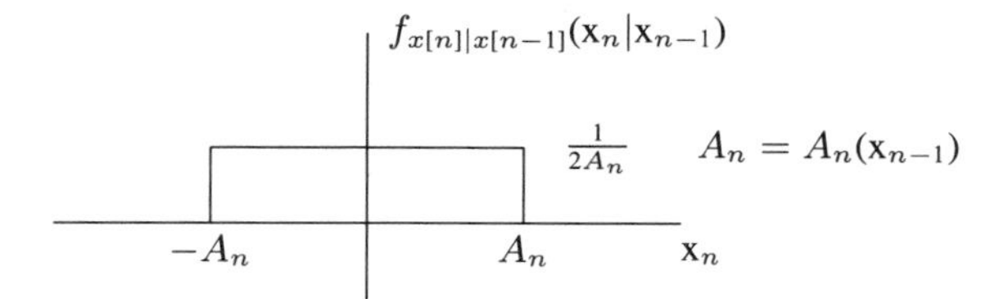

Figure EX3.4

The value A_n depends on $x[n-1]$ in some predetermined way; for example, $A_n = |\mathrm{x}_{n-1}|$. From the description of the random process it is clear that

$$\mathcal{E}\{x[n] \mid x[n-1], x[n-2], \ldots\} = 0$$

so that $x[n]$ is absolutely fair. However, since A_n depends on $x[n-1]$, it is clear that $x[n]$ and $x[n-1]$ are *not* independent.

It can be shown [18, 19] that if $y[n]$ is a martingale with bounded second moment $\mathcal{E}\{(y[n])^2\} < C < \infty$, then $y[n]$ converges to some limiting value y_o (a random variable) with probability 1, that is,

$$\lim_{n\to\infty} \left[\Pr[|y[n] - y_o| < \epsilon]\right] = 1$$

This result, known as the *martingale convergence theorem*, implies that all martingales with finite average power are inherently nonstationary.

A situation involving martingales arises in statistical hypothesis testing. This is discussed in the following example.

Example 3.5

In various signal classification problems one of two possible random processes is presented and the observer needs to decide which one of the two is actually present. For example, one of the random processes may represent a signal in noise, while the other may represent noise alone. The statistical theory that relates to such decisions is known as hypothesis testing. Let $f_n^{(1)}(x[0], x[1], \ldots, x[n])$ represent the density function for n samples of the first random process and $f_n^{(2)}(x[0], x[1], \ldots, x[n])$ represent the density function for n samples of the second random process. In sequential hypothesis testing [20] the *likelihood ratio*

$$\mathcal{L}[n] = \frac{f_n^{(1)}(x[0], x[1], \ldots, x[n])}{f_n^{(2)}(x[0], x[1], \ldots, x[n])}$$

is formed and compared to a pair of thresholds A and B as shown in Fig. EX3.5:

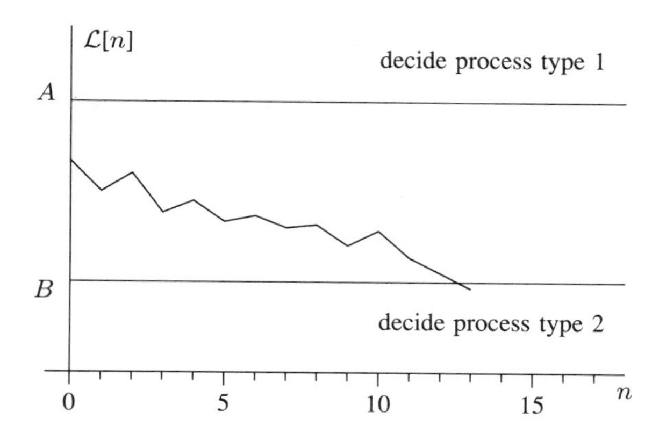

Figure EX3.5

The likelihood ratio is computed for increasing values of n as long as its value lies between A and B. When the value of $\mathcal{L}[n]$ crosses over one of the thresholds (falls above A or below B), a decision is made about the identity of the random process and the test stops.

Suppose now that the observed sequence is actually derived from a random process of type 2. It can then be shown that the likelihood ratio sequence $\mathcal{L}[n]$ is a martingale. To do this let us show, *given a process of type* 2, that

$$\mathcal{E}\{\mathcal{L}[n] \,|\, x[0], x[1], \ldots, x[n-1]\} = \mathcal{L}[n-1] \qquad \text{(I)}$$

If this is the case, then it must also be true that

$$\mathcal{E}\{\mathcal{L}[n] \mid \mathcal{L}[n-1], \ldots, \mathcal{L}[1], \mathcal{L}[0]\} = \mathcal{L}[n-1]$$

This follows from taking the conditional expectation of (I) given $\mathcal{L}[0], \ldots, \mathcal{L}[n-1]$.

To show (I), proceed as follows. The conditional density for $x[n]$ given $x[0], \ldots, x[n-1]$ is

$$\frac{f_n^{(2)}(x[0], \ldots, x[n-1], x[n])}{f_{n-1}^{(2)}(x[0], \ldots, x[n-1])}$$

Since $\mathcal{L}[n]$ depends on $x[n]$ it follows that

$$\begin{aligned}
&\mathcal{E}\{\mathcal{L}[n] \mid x[0], \ldots, x[n-1]\} \\
&\quad = \int_{-\infty}^{\infty} \mathcal{L}[n] \cdot \frac{f_n^{(2)}(x[0], \ldots, x[n-1], \mathrm{x}_n)}{f_{n-1}^{(2)}(x[0], \ldots, x[n-1])} \, d\mathrm{x}_n \\
&\quad = \int_{-\infty}^{\infty} \frac{f_n^{(1)}(x[0], \ldots, x[n-1], \mathrm{x}_n)}{f_n^{(2)}(x[0], \ldots, x[n-1], \mathrm{x}_n)} \cdot \frac{f_n^{(2)}(x[0], \ldots, x[n-1], \mathrm{x}_n)}{f_{n-1}^{(2)}(x[0], \ldots, x[n-1])} \, d\mathrm{x}_n \\
&\quad = \frac{1}{f_{n-1}^{(2)}(x[0], \ldots, x[n-1])} \int_{-\infty}^{\infty} f_n^{(1)}(x[0], \ldots, x[n-1], \mathrm{x}_n) \, d\mathrm{x}_n \\
&\quad = \frac{f_{n-1}^{(1)}(x[0], \ldots, x[n-1])}{f_{n-1}^{(2)}(x[0], \ldots, x[n-1])} = \mathcal{L}[n-1]
\end{aligned}$$

Thus the sequence $\mathcal{L}[n]$ is a martingale. The implication of this result is that the difference process $\mathcal{L}[n] - \mathcal{L}[n-1]$ is absolutely fair. This means that the change in the likelihood ratio at each observation is decoupled from its past values.

Martingales also arise in common but seemingly unlikely places. The following example shows how the martingale concept can be used to interpret the process of delta modulation described in Section 3.2.2.

Example 3.6

In the process of delta modulation the input waveform is approximated by a process that takes on only integer discrete values. This is the staircase-like function shown in Fig. 3.4. Denote this approximation by the random process $x[n]$ and note that the process never remains at the same level from $n-1$ to n; it always takes an upward or downward step.

Linear predictive coding of a sequence involves making an optimal prediction $\hat{x}[n]$ based on past values of the sequence and transmitting only the "prediction error": $x[n] - \hat{x}[n]$. It is shown in Chapter 6 that the optimal mean square prediction of the process is the conditional expectation

$$\hat{x}[n] = \mathcal{E}\{x[n] | x[n-1], x[n-2], \ldots\}$$

Suppose that the approximating function (i.e., the transmitted sequence) can be modeled by a martingale. Then the conditional expectation above is just equal to $x[n-1]$. In this case delta modulation is the optimum linear predictive coding scheme.

3.5 PERIODIC AND ALMOST PERIODIC RANDOM PROCESSES

A random process will be called *periodic* if there exists an integer P such that

$$f_{x[n_0],x[n_1],\ldots x[n_L]} = f_{x[n_0+k_0P],x[n_1+k_1P],\ldots x[n_L+k_LP]} \tag{3.61}$$

for all choices of the n_i, for any set of integers k_i, and for any value of L.[11] In most cases when a random process is periodic there is an explicit dependence on a sinusoid or complex exponential (i.e., a term of the form $e^{\jmath\omega n}$), as the following example illustrates.

Example 3.7

A sinusoidal random process has the form

$$x[n] = A\cos\omega_o n$$

where A is a zero-mean random variable with density $f_A(\mathrm{A})$. The probability density functions for the random process can be computed as follows. For any choice of the parameter $n = n_0$ the single sample $x[n_0]$ is just a known constant ($\cos\omega_o n_0$) times the random variable A. Therefore, the first-order density is simply

$$f_{x[n_0]}(\mathrm{x}_0) = \frac{1}{|\cos\omega_o n_0|} f_A\left(\frac{\mathrm{x}_0}{\cos\omega_o n_0}\right)$$

as long as $\cos\omega_o n_0 \neq 0$. For example, if A is uniformly distributed between -1 and $+1$, the densities appear as shown in Fig. EX3.7a:

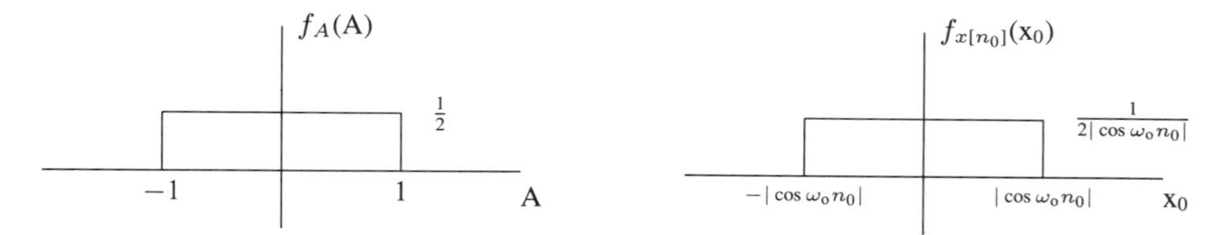

Figure EX3.7a

Note that as $\cos\omega_o n_0 \to 0$ the density for the sample approaches a unit impulse. This is as it should be since at a zero crossing the value of $x[n_0]$ is 0 with probability 1.

To find the joint density for two samples, we first find the conditional density $f_{x[n_1]\,|\,x[n_0]}$ and multiply this by $f_{x[n_0]}$ to get $f_{x[n_0]x[n_1]}$. To find the conditional density, assume that the value $x[n_0] = \mathrm{x}_0$ was observed. Then since $A = \mathrm{x}_0/\cos\omega_o n_0$, the value of $x[n_1]$ is precisely equal to

$$\mathrm{x}_1 = \left(\frac{\mathrm{x}_0}{\cos\omega_o n_0}\right)\cos\omega_o n_1$$

In other words, given $x[n_0]$, the random variable $x[n_1]$ is known with certainty. Its conditional density is therefore an impulse, as shown in Fig. EX3.7b.

[11]If (3.61) holds only for equal values of the integers $k_0 = k_1 = \cdots = k_L$, then the process is said to be *cyclostationary* [21, 22] in the discrete-time sense.

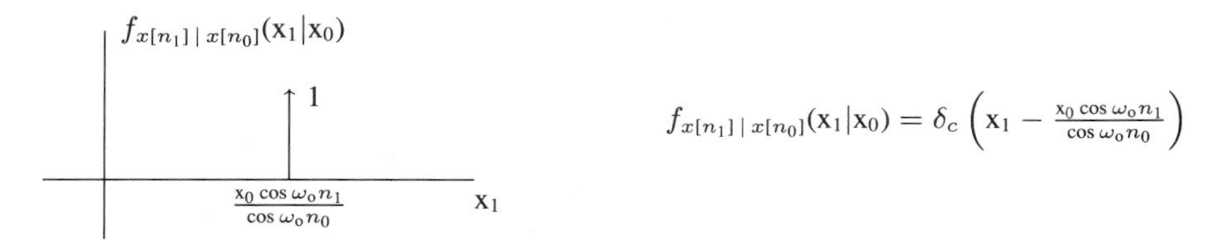

Figure EX3.7b

Therefore,

$$
\begin{aligned}
f_{x[n_0]x[n_1]}(\mathrm{x}_0, \mathrm{x}_1) &= f_{x[n_0]}(\mathrm{x}_0) f_{x[n_1]\,|\,x[n_0]}(\mathrm{x}_1|\mathrm{x}_0) \\
&= \frac{1}{|\cos \omega_o n_0|} f_A\left(\frac{\mathrm{x}_0}{\cos \omega_o n_0}\right) \delta_c\left(\mathrm{x}_1 - \frac{\mathrm{x}_0 \cos \omega_o n_1}{\cos \omega_o n_0}\right)
\end{aligned}
$$

Higher-order densities can be computed by multiplying by additional terms. For example, to compute $f_{x[n_0]x[n_1]x[n_2]}$ multiply the result above by

$$
f_{x[n_2]\,|\,x[n_0]x[n_1]}(\mathrm{x}_2|\mathrm{x}_0\mathrm{x}_1) = \delta_c\left(\mathrm{x}_2 - \frac{\mathrm{x}_1 \cos \omega_o n_2}{\cos \omega_o n_1}\right)
$$

By continuing in this way the joint density for any number of samples can be derived.

From the foregoing result and the definition (3.61), it can be seen that the *random process* $x[n]$ is periodic if and only if ω_o is a rational multiple of 2π, that is, if ω_o is of the form $\omega_o = 2\pi K/P$, where K and P are both integers.[12] In this case the random process is periodic with period equal to P.

Several other forms of periodic random processes can be mentioned that are generalizations of the random process in Example 3.7. For instance, the random process defined by

$$
x[n] = A\cos(\omega_o n + \phi) \tag{3.62}
$$

where A and ϕ are *both* random variables is also periodic if

$$
\omega_o = 2\pi K/P \tag{3.63}
$$

(If ϕ is uniformly distributed between 0 and 2π, then this random process is also stationary.) In addition, sums of sinusoidal random processes of the form

$$
x[n] = \sum_i A_i \cos(\omega_i n + \phi_i) \tag{3.64}
$$

where A_i and ϕ_i are real random variables, and sums of complex exponential random processes

$$
x[n] = \sum_i A_i e^{\jmath\omega_i n} \tag{3.65}
$$

[12]Any sequence from the ensemble is also periodic only under this condition.

where A_i is a complex random variable with random magnitude and phase, are periodic if the ω_i are of the form $2\pi K_i/P_i$.

If a random process is of the form (3.62), where ω does *not* satisfy the condition (3.63), the random process will be called *almost periodic*. Also, if a random process is of the form (3.64) or (3.65), where *any* of the ω_i are not of the form $2\pi K_i/P_i$, it is almost periodic. A continuous random signal of the form $x_c(t) = A_1 \cos 2\pi f_1 t + A_2 \cos 2\pi f_2 t$ is said to be almost periodic if f_1 and f_2 are not rational multiples of each other. Such a signal never repeats and any discrete sequence derived from it by sampling will also be almost periodic in the sense that we have defined it for sequences.

3.6 GAUSSIAN RANDOM PROCESSES: CONTINUOUS AND DISCRETE

Many random sequences that occur in signal processing applications are the result of sampling continuous random signals. This section begins by defining and briefly discussing the concept of a continuous random process. The continuous Wiener process is then defined as a limiting case of the discrete Wiener process and shown to be an example of a *Gaussian* random process. Other Gaussian processes are shown to arise in a similar way. The discussion of these continuous random processes provides sufficient motivation to consider a discrete Gaussian random process derived from a continuous Gaussian process by sampling. Continuous random processes are not all Gaussian processes and sampled continuous processes are consequently not always Gaussian. On the other hand, discrete Gaussian processes can arise in contexts other than as sampled continuous random processes. Nevertheless, discrete Gaussian processes that arise from sampling noise and other continuous signals are very common in signal processing applications and play an important role in the design of practical systems.

3.6.1 Continuous Random Processes

A discrete random signal is defined as a sequence $x[n]$ where for any value of the parameter n the quantity $x[n]$ is a random variable. A continuous random signal $x_c(t)$ is defined similarly. Like its deterministic counterpart, a continuous random signal is a function of a continuous parameter t (usually representing time). However, the value that the signal takes on at any selected value of t is a *random variable*. Continuous random signals are modeled by continuous random (or stochastic) processes.

To describe a continuous random process completely, it is necessary to specify the joint probability distribution or density function for *any* set of samples taken at *any* values $t_0, t_1, \ldots, t_N$ and any N (see Fig. 3.17). Since t is continuous and the set of values $t_0, t_1, \ldots, t_N$ may have any arbitrary spacing, specification of the joint distribution or density for a set of samples appears to be much more difficult than in the discrete case. However, there are a number of cases where this complete specification *is* possible.

A continuous random process may be real- or complex-valued. When it is complex, statistical characterization involves specifying the joint distribution or density of the real and imaginary parts.

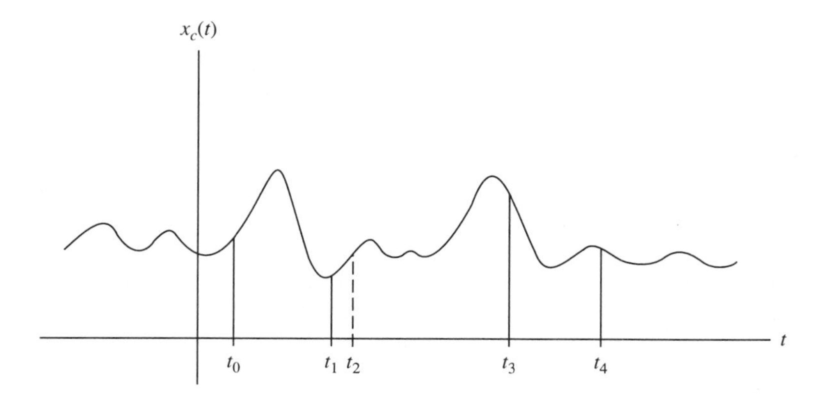

Figure 3.17 Continuous random signal and some of its samples.

A continuous random process is said to be *stationary* (in the strict sense) if the joint distribution and density for any set of samples depends only on the spacing between samples and not on the absolute locations $t_0, t_1, \ldots, t_N$ where the samples are taken. This implies that all the moments of the distribution (mean, correlation, and higher-order moments) are also functions of only the spacings between samples.

The concept of an *ensemble* applies to continuous as well as to discrete random processes. Statistical expectations such as $\mathcal{E}\{x_c(t)\}$ are interpreted as averages down the ensemble. Signal averages (also called temporal averages when t is time) denoted by brackets $\langle\ \rangle$ are defined as a limit, for example,

$$< x_c(t) >= \lim_{\Upsilon \to \infty} \frac{1}{2\Upsilon} \int_{-\Upsilon}^{\Upsilon} x_c(t)dt \tag{3.66}$$

A continuous random process is said to be *ergodic* if it is stationary, and signal averages are equal with probability 1 to the corresponding ensemble averages.

3.6.2 The Continuous Wiener Process

Most types of noise occurring in the natural world result from microscopic interactions at the molecular or submolecular level. Early in his career Norbert Wiener studied Brownian motion, the random motion that small particles suspended on the surface of a liquid undergo due to their continued bombardment by molecules of the liquid. This led to the formulation of a random process now known as the (continuous) Wiener process[13] and its role in a model for noise. The continuous Wiener process is defined to be the limit of a random walk as the time interval between samples approaches zero.

An alternative way to develop noise that will not be pursued here is through the concept of a Poisson process. The Poisson process deals with the occurrence of discrete events (e.g., electrons arriving at a positively charged plate) such that in any very small

[13]Also called a Wiener-Lévy process, a Brownian motion process, or a diffusion process.

increment of time Δt the probability of occurrence of an event is a constant, proportional to Δt. In this way the Poisson process is similar to a microscopic Bernoulli process with values 0 and 1. Since the Bernoulli process and the random walk are closely related, the continuous noise models that are developed from either the Poisson process or the Wiener process have very similar properties. The development of noise via the Poisson model can be found in a number of standard texts dealing with continuous random processes (see e.g., [5, 19, 23, 24]).

Let us begin the development of the continuous Wiener process by considering a discrete Wiener process (say, representing motion of a particle in one dimension) with a step size s. That is, at each time epoch the random process increases or decreases its value by $\pm s$ with probability $\mathrm{P} = \frac{1}{2}$. If the process is represented on a continuous time scale t with microscopic increments Δt, the motion is described by

$$x_c(k\Delta t) = \sum_{i=k_o+1}^{k} s\,\zeta[i]\;; \qquad k > k_o \tag{3.67}$$

where $\zeta[i]$ is a microscopic binary white noise process. From the previous analysis of the discrete Wiener process (see Table 3.2) the mean of this random process is zero and the variance is given by

$$\mathrm{Var}\,[x_c(k\Delta t)] = s^2(k - k_o) \tag{3.68}$$

The last equation can be written as

$$\mathrm{Var}\,[x_c(t)] = \frac{s^2(t - t_o)}{\Delta t}\;; \qquad \text{for } t = k\Delta t,\ t_o = k_o\Delta t \tag{3.69}$$

Now suppose that the interval Δt approaches zero and the step size s also approaches zero in such a way that the ratio $s^2/\Delta t$ approaches a constant ν_o:

$$\mathrm{Var}\,[x_c(t)] = \lim_{\Delta t \to 0} \frac{s^2(t - t_o)}{\Delta t} = \nu_o(t - t_o) \tag{3.70}$$

The resulting process in this limiting form is the *continuous Wiener process*. Since $x_c(t)$ is the sum of a large number of independent identically distributed random variables, the Central Limit Theorem applies and states that $x_c(t)$ has a Gaussian density function.

$$f_{x_c(t)}(\mathrm{x}) = \frac{1}{\sqrt{2\pi\nu_o(t - t_o)}} e^{-\frac{\mathrm{x}^2}{2\nu_o(t - t_o)}} \tag{3.71}$$

It is also fairly easy to see that any number of samples of this process are jointly Gaussian. For example, if there are two samples, say $x_c(t_1)$ and $x_c(t_2)$ with $t_1 < t_2$, they form the random variable

$$v = c_1 x_c(t_1) + c_2 x_c(t_2)$$

where c_1 and c_2 are any real numbers. The new random variable can be written using (3.67) as

$$\begin{aligned} v &= c_1 \sum_{i=k_o+1}^{k_1} s\zeta[i] + c_2 \sum_{i=k_o+1}^{k_2} s\zeta[i] \\ &= (c_1 + c_2) \sum_{i=k_o+1}^{k_1} s\zeta[i] + c_2 \sum_{i=k_1+1}^{k_2} s\zeta[i] \end{aligned}$$

where $k_1 = t_1/\Delta t$ and $k_2 = t_2/\Delta t$. Each of these last two terms is independent and Gaussian by the Central Limit Theorem (since $\Delta t \to 0$). Because v is the sum of two independent Gaussian random variables (recall that the Wiener process has independent increments), it too is Gaussian. Further, since the argument holds for *any* choice of the constants c_1 and c_2, it follows that that $x_c(t_1)$ and $x_c(t_2)$ are jointly Gaussian. A similar argument shows that any number of samples of the Wiener process are jointly Gaussian. Such a process is called a *Gaussian random process*. Since the variance of the Wiener process depends on t, however, the Wiener process is not stationary.

Although the Wiener process may not be a typical type of continuous random signal, the steps used in constructing the Wiener process imply something about random signals that *are* likely to be encountered in the real world. In particular, consider an impulsive process such as the microscopic binary white noise process or the Poisson process with spacing Δt applied to a linear filter. Since the response time of any real filter is long compared to the infinitesimal Δt, the effect of filtering is like integration. Therefore, the output of the filter, which is the sum of a large number of random variables, has a jointly Gaussian distribution. This argument can be made rigorous (see Problem 3.25) and so establishes the existence of continuous Gaussian random processes. *Complex* Gaussian processes can also occur as representations of narrowband real random processes (see Appendix B).

Since linear operations on Gaussian random processes are Gaussian, it becomes *mathematically convenient* to think of the input to such a filter as a Gaussian random process as well. This random process, which is formally the *derivative* of a Wiener process, is called *Gaussian white noise*. Such a random process, although fictitious, is commonly used in the analysis and modeling of continuous random signals.

3.6.3 Discrete Gaussian Processes

As was just seen, a continuous Gaussian random process can result as the response of a filter to a discrete process of microscopic amplitude whose sampling is at a microcopic interval Δt approaching zero. Now suppose that this continuous process is sampled at a much larger interval T to define a random sequence. Since the continuous process is Gaussian, the samples of the discrete process are jointly Gaussian.

Let us therefore *define* a discrete Gaussian process as a random process such that any set of its samples has a multivariate Gaussian distribution. Since the marginal densities for a multivariate Gaussian density are also Gaussian, it is equivalent to say that a discrete random process is Gaussian if the random vector of N contiguous samples

$$\boldsymbol{x} = \begin{bmatrix} x[n_o] \\ x[n_o+1] \\ \vdots \\ x[n_o+N-1] \end{bmatrix} \tag{3.72}$$

for any starting point n_o and any number of samples N, which has mean vector $\mathbf{m}_x$ and covariance matrix $\mathbf{C}_x$, is characterized by the Gaussian density function

$$f_x(\mathbf{x}) = \frac{1}{(2\pi)^{N/2}|\mathbf{C}_x|^{1/2}} e^{-\frac{1}{2}(\mathbf{x}-\mathbf{m}_x)^T \mathbf{C}_x^{-1}(\mathbf{x}-\mathbf{m}_x)} \quad \text{(real random vector)} \tag{3.73}$$

or

$$f_x(\mathbf{x}) = \frac{1}{\pi^N|\mathbf{C}_x|} e^{-(\mathbf{x}-\mathbf{m}_x)^{*T} \mathbf{C}_x^{-1}(\mathbf{x}-\mathbf{m}_x)} \quad \text{(complex random vector)} \tag{3.74}$$

Because linear transformations of Gaussian random vectors are also Gaussian, *Gaussian random processes remain Gaussian when processed by linear systems* (see Fig. 3.18). Thus Gaussian random processes and linear systems go hand in hand.

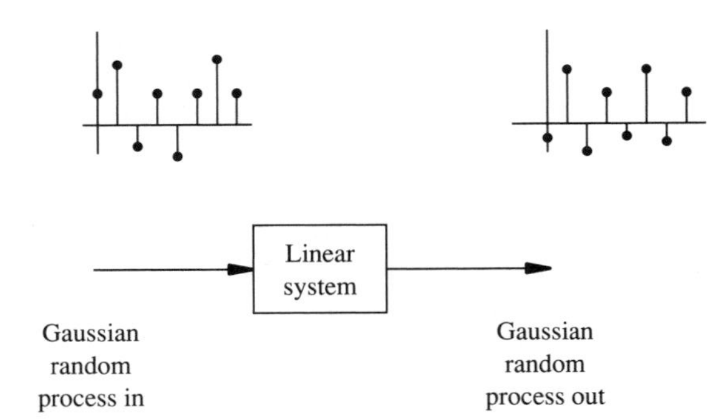

Figure 3.18 Transformation of a Gaussian random process by a linear system.

An especially important class of Gaussian random processes are *stationary* Gaussian random processes. These processes have a particularly simple statistical description and remain stationary if they are acted upon by linear *shift-invariant* systems. The following example illustrates how a simple Gaussian random process and the process resulting from its transformation by a simple linear system can be described statistically.

Example 3.8

A Gaussian "white noise" process consists of a sequence of zero-mean independent Gaussian random variables $w[n]$ with variance σ_o^2. Because a Gaussian random vector is defined by only its first two moments, a complete statistical description of this random process requires only being able to write the mean vector and covariance matrix for any set of samples of this random process. Since the mean is zero and the samples are independent, the white noise process has the simple characterization

$$\mathbf{m}_w = \mathbf{0}$$

and

$$\mathbf{C}_w = \begin{bmatrix} \sigma_o^2 & 0 & \cdots & 0 \\ 0 & \sigma_o^2 & \cdots & 0 \\ \vdots & \vdots & \ddots & \vdots \\ 0 & 0 & \cdots & \sigma_o^2 \end{bmatrix} = \sigma_o^2 \mathbf{I}$$

where the dimensions of the vector and matrix are determined by the number of samples taken.

The Gaussian white noise sequence is applied to a linear filter which produces an output random process according to the difference equation

$$x[n] = \rho x[n-1] + w[n]$$

with real coefficient ρ. From previous considerations, this output random process is known to be both a Gaussian random process and a Markov process. Such a random process is sometimes called a *Gauss-Markov process.* Now assume for convenience that the process is generated from the difference equation beginning at $n_o = 0$ with initial conditions $x[-1] = 0$. Then by carrying out successive steps of the recursion the output can be written in the convolution form

$$x[n] = \sum_{k=0}^{n} w[k]\rho^{n-k}$$

or in the matrix form

$$\begin{bmatrix} x[0] \\ x[1] \\ x[2] \\ \vdots \\ x[n] \end{bmatrix} = \begin{bmatrix} 1 & 0 & 0 & \cdots & 0 \\ \rho & 1 & 0 & \cdots & 0 \\ \rho^2 & \rho & 1 & \cdots & 0 \\ \vdots & \vdots & \vdots & \ddots & \vdots \\ \rho^n & \rho^{n-1} & & \cdots & 1 \end{bmatrix} \begin{bmatrix} w[0] \\ w[1] \\ w[2] \\ \vdots \\ w[n] \end{bmatrix}$$

Since this is in the form $\boldsymbol{x} = \mathbf{A}\boldsymbol{w}$ the mean vector and covariance matrix of the output process are given by $\mathbf{m}_x = \mathbf{A}\mathbf{m}_w$ and $\mathbf{C}_x = \mathbf{A}\mathbf{C}_w\mathbf{A}^T$. Thus

$$\mathbf{m}_x = \mathbf{0}$$

and

$$\mathbf{C}_x = \begin{bmatrix} 1 & 0 & 0 & \cdots & 0 \\ \rho & 1 & 0 & \cdots & 0 \\ \rho^2 & \rho & 1 & \cdots & 0 \\ \vdots & \vdots & \vdots & \ddots & \vdots \\ \rho^n & \rho^{n-1} & & \cdots & 1 \end{bmatrix} \begin{bmatrix} \sigma_o^2 & 0 & 0 & \cdots & 0 \\ 0 & \sigma_o^2 & 0 & \cdots & 0 \\ 0 & 0 & \sigma_o^2 & \cdots & 0 \\ \vdots & \vdots & \vdots & \ddots & \vdots \\ 0 & 0 & 0 & \cdots & \sigma_o^2 \end{bmatrix} \begin{bmatrix} 1 & \rho & \rho^2 & \cdots & \rho^n \\ 0 & 1 & \rho & \cdots & \rho^{n-1} \\ 0 & 0 & 1 & \cdots & \\ \vdots & \vdots & \vdots & \ddots & \vdots \\ 0 & 0 & 0 & \cdots & 1 \end{bmatrix}$$

After some algebraic manipulation this last expression can be put in the form

$$\mathbf{C}_x = \frac{\sigma_o^2}{1-\rho^2}\begin{bmatrix} 1 & \rho & \rho^2 & \cdots & \rho^n \\ \rho & 1 & \rho & \cdots & \rho^{n-1} \\ \rho^2 & \rho & 1 & \ddots & \vdots \\ \vdots & \vdots & \ddots & \ddots & \rho \\ \rho^n & \rho^{n-1} & \cdots & \rho & 1 \end{bmatrix} - \frac{\sigma_o^2\rho^2}{1-\rho^2}\begin{bmatrix} 1 & \rho & \rho^2 & \cdots & \rho^n \\ \rho & \rho^2 & \rho^3 & \cdots & \rho^{n+1} \\ \rho^2 & \rho^3 & \rho^4 & \cdots & \rho^{n+2} \\ \vdots & \vdots & \vdots & & \vdots \\ \rho^n & \rho^{n+1} & \rho^{n+2} & \cdots & \rho^{2n} \end{bmatrix}$$

The first matrix has the interesting property that the elements on each principal diagonal are equal. It will be seen later that this type of covariance matrix, known as a *Toeplitz* matrix, arises whenever a random process is stationary. The random process $x[n]$ is thus seen to have a stationary component that arises from applying the stationary sequence $w[n]$ to a linear *shift-invariant* system. The second matrix is a *Hankel* matrix and has the property that all elements on the reverse diagonals are equal. This component of the covariance represents the "transient response" for the system resulting from applying the noise at $n_o = 0$. If $|\rho|$ is less than 1, which is the condition for stability of the linear shift-invariant system, this transient component eventually disappears (one can observe that the terms in the lower right block of this matrix get closer and closer to zero). In the limit when the output sequence is observed at some point far removed from when the input was first applied, the resulting covariance has only the stationary (Toeplitz) component. The transient portion of the covariance matrix can be eliminated completely by changing the initial variance of the white noise (i.e., at $n = n_o$) to the value

$$\sigma_o'^2 = \frac{\sigma_o^2}{1-\rho^2} = \sigma_o^2 + \frac{\sigma_o^2\rho^2}{1-\rho^2}$$

The white noise covariance matrix then becomes

$$\mathbf{C}'_w = \begin{bmatrix} \sigma_o'^2 & 0 & \cdots & 0 \\ 0 & \sigma_o^2 & \cdots & 0 \\ \vdots & \vdots & \ddots & \vdots \\ 0 & 0 & \cdots & \sigma_o^2 \end{bmatrix}$$

and the product $\mathbf{AC}'_w\mathbf{A}^T$ contains an extra term that cancels the transient part of $\mathbf{C}_x$. The transient portion can be thought of as arising from a mismatch between the steady-state variance of the output process and that of the initial input when we force $\text{Var}[x[n_o]]$ to be equal to $\text{Var}[w[n_o]]$. Changing the the variance of $w[n_o]$ to match that of the steady-state response eliminates the transient.

Most of the remainder of this book focuses on the description of random processes in terms of their first and second moments (mean and correlation or covariance).[14] This is known as *second moment* characterization or second moment analysis. While for most random processes second moment characterization provides an adequate description, for Gaussian random processes second moment characterization provides a *complete* statistical description of the random process.

[14]A few brief sections in later chapters deal with higher-order moments.

3.7 CHAPTER SUMMARY

Random signals are described in terms of a mathematical model known as a random process. A discrete random process is comprised of a large (usually infinite) set of sequences representing possible outcomes of a random experiment. Such a set of possible outcomes is called an ensemble. A random process is said to be stationary if the density function or all of the moments are independent of where they are evaluated along the random process. That is, the statistics must depend only upon the *spacings* between samples and not upon their absolute locations. A random process is ergodic if it is stationary and signal averages, computed along any ensemble member, are with probability 1 equal to the corresponding ensemble averages.

Two simple types of random processes are the Bernoulli process and the random walk. The Bernoulli process is stationary and ergodic. The random walk is the summation of a Bernoulli process and is not stationary. An important property of the random walk, however, is the independence of its increments. Two special cases of the Bernoulli process and the random walk are known as binary white noise and the discrete Wiener process respectively. For both of these processes the parameter P is equal to $\frac{1}{2}$. Both of these processes have zero mean and have important applications in signal processing.

Markov processes are random processes that embody a simple type of dependency between successive samples. In particular, the conditional density and therefore all of the statistics, given a set of past values of the process depend on only the most recently observed past value. A Markov chain is a special type of Markov process that takes on only a discrete set of values which can be associated with a set of "states." The transition probabilities for a Markov chain can be represented either in a state transition diagram or in a stochastic or state transition matrix. Higher-order transition probabilities can be obtained by taking products of these matrices. The existence of limiting-state probabilities can be expressed as conditions on the matrix eigenvalues. The random walk can be thought of as a Markov chain with an infinite number of states.

A hidden Markov model results when a Markov chain generates another random process from within each state. The random process generated while in the state is directly observable, but the sequence of states is not. Such compound random processes are called "doubly stochastic." Efficient methods exist for computing the probability of a given set of outputs from a hidden Markov model and for estimating the underlying state sequence. The method to perform the latter task is known as the Viterbi algorithm.

A martingale is a special type of random process whose expected value given all of the past values in the sequence is just equal to the most recently observed past value. A martingale can be represented as the summation of an "absolutely fair" process, whose expected value given the previous values of the process is always zero. Martingales arise in a variety of situations involving the analysis of random processes and can have some limited use as models for random signals.

A periodic random process as described in this chapter is a random process whose joint density function does not change when any of the samples are displaced by one or more periods [see (3.61)]. Discrete sinusoidal or complex exponential processes which are not periodic are called "almost periodic." The sum of discrete periodic processes is also a periodic process and the sum of periodic and almost periodic processes is almost periodic.

Continuous random processes are the underlying mathematical models for continuous random signals. The complete statistical description of a continuous random process requires specification of the joint distribution or density function for any number of samples wherever they may be taken. Gaussian random processes arise when the joint density is multivariate Gaussian. The continuous Wiener process is derived as the limit of a discrete Wiener process as the increment between samples and the amplitude of the process shrink to zero in a prescribed way. The formal derivative of the continuous Wiener process is called Gaussian white noise.

A discrete Gaussian process can be described as a random process where all sets of successive samples are jointly Gaussian. A discrete Gaussian process can be derived from a continuous Gaussian random process by sampling.

REFERENCES

1. J. D. Gibson and J. L. Melsa. *Introduction to Nonparametric Detection with Applications*. Academic Press, New York, 1975.
2. Alvin W. Drake. *Fundamentals of Applied Probability Theory*. McGraw-Hill, New York, 1967.
3. Harold J. Larson and Bruno O. Schubert. *Probabilistic Models in Engineering Sciences*, Volume II. John Wiley & Sons, New York, 1979.
4. William Feller. *An Introduction to Probability Theory and Its Applications*, Volume I, 2nd ed. John Wiley & Sons, New York, 1957.
5. Athanasios Papoulis. *Probability, Random Variables, and Stochastic Processes*, 3rd ed. McGraw-Hill, New York, 1991.
6. Leo Breiman. *Probability and Stochastic Processes, with a View Toward Applications*, 2nd ed. Scientific Press, Palo Alto, California, 1986.
7. Samuel Karlin and Howard M. Taylor. *A First Course in Stochastic Processes*, 2nd ed. Academic Press, New York, 1975.
8. Samuel Karlin and Howard M. Taylor. *A Second Course in Stochastic Processes*. Academic Press, New York, 1981.
9. Joseph L. Doob. *Stochastic Processes*. John Wiley & Sons, New York, 1953.
10. Claude E. Shannon. A mathematical theory of communication. *Bell System Technical Journal*, 27(3), July 1948. (See also [11].)
11. Claude E. Shannon and Warren Weaver. *The Mathematical Theory of Communication*. University of Illinois Press, Urbana, Illinois, 1963.
12. L. R. Rabiner and B. H. Juang. An introduction to hidden Markov models. *IEEE ASSP Magazine*, 3(1):4–16, January 1986.
13. S. E. Levinson, L. R. Rabiner, and M. M. Sondhi. An introduction to the application of the theory of probabilistic functions of a Markov process to automatic speech recognition. *Bell System Technical Journal*, 62(4):1035–1075, April 1983.
14. H. Kaufman, J. W. Woods, V. K. Ingle, R. Mediavilla, and A. Radpour. Recursive image estimation: a multiple model approach. In *18th Conference on Decision and Control*, December 12–14, 1979. Fort Lauderdale, Florida.
15. C. W. Therrien. An estimation-theoretic approach to terrain image segmentation. *Computer Vision, Graphics, and Image Processing*, 22:313–326, 1983.

16. G. David Forney. The Viterbi algorithm. *Proceedings of the IEEE*, 6:268–278, 1973.

17. Peter Hall. Martingales. In *Encyclopedia of Statistical Sciences*, pages 278–285. John Wiley & Sons, New York, 1985.

18. William Feller. *An Introduction to Probability Theory and Its Applications,* Volume II, 2nd ed. John Wiley & Sons, New York, 1971.

19. Henry Stark and John W. Woods. *Probability, Random Processes, and Estimation Theory for Engineers.* Prentice Hall, Inc., Englewood Cliffs, New Jersey, 1986.

20. Abraham Wald. *Sequential Analysis.* John Wiley & Sons, New York, 1947.

21. William A. Gardner. Stationarizable random processes. *IEEE Transactions on Information Theory*, IT-24(1):8–22, January 1978.

22. William A. Gardner. *Representation and Estimation of Cyclostationary Processses.* Technical Report SIPL-82-1, University of California, Davis, California, April 1982.

23. Wilbur B. Davenport, Jr. and William L. Root. *An Introduction to the Theory of Random Signals and Noise.* McGraw-Hill, New York, 1958.

24. Carl W. Helstrom. *Probability and Stochastic Processes for Engineers*, 2nd ed. Macmillan, New York, 1991.

PROBLEMS

3.1. A random sequence is equal to a (constant) random variable A with density function $f_A(\text{A})$. Compute the joint density function for any three samples of this random process $x[i], x[j]$, and $x[k]$ with $i < j < k$. Is this random process stationary?

3.2. For each of the following random processes, state if the random process

i. has constant mean (independent of n).
ii. has constant variance (independent of n).
iii. is ergodic.
iv. is stationary.
v. is a first-order Markov process, or has independent samples.
vi. has independent increments.

(a) $x[n] = A\cos(\omega_o n)$ where A is a Gaussian random variable with mean m_A and variance σ_A^2.

(b) $x[n] = \cos(\omega_o n + \phi)$ where ϕ is a random variable uniformly distributed over the interval $[0, 2\pi]$.

(c) $x[n] = \xi[n]$, a Bernoulli process where

$$\xi[n] = \begin{cases} +1 & \text{with probability P} \\ -1 & \text{with probability } 1 - \text{P} \end{cases}$$

(d) $x[n] = \sum_{l=-\infty}^{n} \xi[l]$ where $\xi[n]$ is the Bernoulli process defined above.

(e) $x[n] = \xi[n] - \xi[n-1]$ where $\xi[n]$ is the Bernoulli process defined above.

3.3. A real random process $y[n]$ is defined by taking the difference between samples of a Bernoulli process, that is,

$$y[n] = x[n] - x[n-1]$$

where $x[n]$ is a Bernoulli process.

(a) What is the (first order) distribution of $y[n]$?

(b) Is this random process stationary?

(c) What is the correlation between two successive samples $\mathcal{E}\{y[n]y[n-1]\}$?

(d) What is the minimum spacing l such that the correlation between samples $\mathcal{E}\{y[n]y[n-l]\}$ is zero?

3.4. At a certain time n_o a sample from a Bernoulli process is observed to have a value of $+1$. What is the distribution for the waiting time l until *two* more $+1$'s occur in the sequence? Specialize your results to the case of a binary white noise process.

3.5. A binary white noise process is applied to a real linear shift-invariant filter with impulse response

$$h[n] = \rho^n u[n]$$

where $u[n]$ is the unit step function and the magnitude of the parameter ρ is less than 1.

(a) Assume that the processes have been going for a long period of time so that the output is in steady state. What are the maximum and minimum values of the output random process?

(b) Are all values of the output process that are between the minimum the maximum values equally likely? Show why or why not.

3.6. A common way to characterize the frequency content of a random process is in terms of the number of zero-crossings in a given interval of time. One can define *zero crossing* for a discrete signal or sequence as whenever the sequence changes from a positive to a negative value, or vice versa. Consider a Bernoulli sequence over some time interval $[0, N-1]$. What are the probabilities for 0 zero crossings and 1 zero crossing in this interval? Also write a recursive formula that expresses the probability of k zero crossings in terms of the probability of $k-1$ zero crossings in some interval of length L.

3.7. Given a random walk that starts with $x[0] = 0$, what is the distribution of the waiting time until the first return to zero? In other words, what is the probability that the next time that $x[n] = 0$ is at $n = k$?

3.8. A random walk starts with $x[0] = 0$. What is the probability that the rth return to zero occurs at $n = k$ for $r = 1, 2, 3$? Use the result of Problem 3.7. Your answer does not have to be in closed form, but you must give a specific formula by which the probability can be computed.

3.9. A two-state Markov chain $x[n]$ can take on the two values $S_1 = +1$ (state 1) and $S_2 = -1$ (state 2) and has the following state transition matrix:

$$\mathbf{\Pi} = \begin{bmatrix} \frac{3}{4} & \frac{1}{4} \\ \frac{1}{4} & \frac{3}{4} \end{bmatrix}$$

(a) Given that $x[n] = +1$, what is the probability that $x[n+3] = +1$? In other words, given that the random process is initially in state 1, what is the probability that it is found to be in state 1 after three transitions?

(b) If the process has been going on for a long period of time, what is the probability that if we observe the process at any time we find it in state 1?

(c) Suppose that we have a Bernoulli process with the same limiting-state probabilities as the Markov process. For the Bernoulli process, given that $x[n] = +1$, what is the probability that $x[n+3] = +1$?

3.10. Compute the probability of a run of 10 zeros in a binary (0,1) Markov chain with state transition matrix

$$\boldsymbol{\Pi} = \begin{bmatrix} 0.6 & 0.4 \\ 0.4 & 0.6 \end{bmatrix}$$

Also compute the probability of 10 zeros immediately followed by 10 ones.

3.11. The transition matrix for a Markov chain can be expressed in terms of its eigenvector matrix $\mathbf{E}$ and its eigenvalue matrix $\boldsymbol{\Lambda}$ as

$$\boldsymbol{\Pi} = \mathbf{E}\boldsymbol{\Lambda}\mathbf{E}^{-1}$$

where the eigenvectors are the columns of $\mathbf{E}$, and the eigenvalues are the diagonal elements of $\boldsymbol{\Lambda}$. Since $\boldsymbol{\Pi}$ is not necessarily symmetric, $\mathbf{E}$ is not necessarily orthogonal or unitary.

(a) Show that the rows of $\mathbf{E}^{-1}$ (call them $\mathbf{v}_i^T$) are eigenvectors of $\mathbf{P} = \boldsymbol{\Pi}^T$ and that $\boldsymbol{\Pi}$ and $\mathbf{P}$ have the same eigenvalues. The rows $\mathbf{v}_i^T$ are also called *left eigenvectors* of $\boldsymbol{\Pi}$ since they satisfy

$$\mathbf{v}_i^T\boldsymbol{\Pi} = \lambda_i\mathbf{v}_i^T$$

(b) It has already been observed that eigenvector $\bar{\mathbf{p}}$ of $\boldsymbol{\Pi}$ corresponding to an eigenvalue $\lambda = 1$ satisfies

$$\boldsymbol{\Pi}\bar{\mathbf{p}} = \bar{\mathbf{p}}$$

Show that a corresponding eigenvector of $\mathbf{P}$ corresponding to $\lambda = 1$ is

$$\mathbf{v}^T = \begin{bmatrix} 1 & 1 & \cdots & 1 \end{bmatrix}^T$$

(i.e., a vector of all 1's).

(c) Show that for any integer k, $\boldsymbol{\Pi}^k$ has the representation

$$\boldsymbol{\Pi}^k = \mathbf{E}\boldsymbol{\Lambda}^k\mathbf{E}^{-1}$$

and therefore, if $\boldsymbol{\Pi}$ has no complex eigenvalues with unit magnitude, the limiting form of $\boldsymbol{\Pi}^{n-n_o}$ is

$$\lim_{n\to\infty} \boldsymbol{\Pi}^{n-n_o} = \mathbf{E}\boldsymbol{\Lambda}^{\infty}\mathbf{E}^{-1}$$

where $\boldsymbol{\Lambda}^{\infty}$ is a diagonal matrix whose only nonzero diagonal terms are equal to 1. (Recall that all eigenvalues of $\boldsymbol{\Pi}$ have magnitudes ≤ 1.)

(d) Finally, show that if $\boldsymbol{\Pi}$ has only a single eigenvalue equal to 1 and no complex eigenvalues with unit magnitude, that

$$\lim_{n\to\infty} \boldsymbol{\Pi}^{n-n_o}\mathbf{p}[n_o] = \bar{\mathbf{p}}$$

independent of $\mathbf{p}[n_o]$, where $\bar{\mathbf{p}}$ is the eigenvector above. Show also in this case that the limiting value of $\boldsymbol{\Pi}^{n-n_o}$ is given by (3.34).

3.12. Show that the limiting-state probabilities for the simple two-state Markov chain discussed in Section 3.3.1 are $\bar{p}_1 = 2/3$ and $\bar{p}_2 = 1/3$.

3.13. Show that the random process discussed in Example 3.7 is a Markov process.

3.14. [15] A three-state Markov chain has the state transition matrix

$$\boldsymbol{\Pi} = \begin{bmatrix} 0.3 & 0.2 & 0.2 \\ 0.3 & 0.7 & 0.3 \\ 0.4 & 0.1 & 0.5 \end{bmatrix}$$

[15] A computer or programmable calculator is helpful for this problem.

(a) Does the system have a set of limiting-state probabilities? If so, what are the values of $\bar{p}_1$, $\bar{p}_2$, and $\bar{p}_3$?
(b) Compute the kth-order transition probabilities for $k = 2, 3, 4$.
(c) How many state transitions are needed to reach the limiting-state condition? (Assume that in this condition the kth-order transition probabilities remain constant to within three significant digits.)

3.15. A binary temporal data sequence consisting of $+1$'s and -1's is modeled by a two-state Markov chain with transition matrix

$$\mathbf{\Pi} = \begin{bmatrix} 0.7 & 0.5 \\ 0.3 & 0.5 \end{bmatrix}$$

State 1 corresponds to $S_1 = +1$; state 2 corresponds to $S_2 = -1$.
(a) Given that you have just observed a -1 at time n, what is the probability that you will observe a -1 at time $n + 2$?
(b) Given that you have just observed a $+1$ at time n, what is the probability that you will observe a -1 at time $n + 2$?
(c) Given that you start at a random time in the sequence, what is the probability that you observe 5 $+1$'s followed by 5 -1's? Is this the same as the probability that you observe 5 -1's followed by 5 $+1$'s? (Again assume that you start at a random time in the sequence.) State why or why not.

3.16. A Markov random process $x[n]$ can take on four values, namely 1, 2, 3, and 4 and can only increase or decrease by ± 1 or remain at the same value with each increment of the parameter n. Given that its value is 1, the process will increase with probability 0.7 and remain at 1 with probability 0.3. Given that its value is 2 or 3, the process will increase or decrease with probability 0.25 and remain the same with probability 0.5. Finally given that its value is 4, the process will decrease with probability 0.7 and remain at 4 with probability 0.3.
(a) Draw the state transition diagram.
(b) What is the transition matrix $\mathbf{\Pi}$?
(c) Given that $x[n] = 1$:
  i. What is the probability that the next three values are 1?
  ii. What is the probability that $x[n + 3] = 1$?
  iii. What is the probability that $x[n + 3] = 4$?

3.17. Say if the following statements about random processes are correct or incorrect and why.
(a) The conditional density involving any two samples of a periodic process is a periodic function of its arguments.
(b) The sum of two martingales is a martingale only if the two processes are statistically independent.
(c) A random process consisting of independent random variables with the same probability density function is always stationary.
(d) For a Markov random process, samples $x[n]$ and $x[n+l]$ for $l > 1$ are always statistically independent.
(e) Any random process that is a martingale is also a Markov process.
(f) Any weighted combination of a Bernoulli process will produce another random process that has independent increments.

3.18. A two-state hidden Markov model has the state transition matrix given in Example 3.3. For state 1, the observation is an independent random variable uniformly distributed between -1

and +1. For state 2, the observation is an independent random variable uniformly distributed between −2 and +2. The following sequence of three values of the process is observed:

$$0.5, \ -0.5, \ 1.2$$

Using the *forward procedure*, find the value of the density (the likelihood function) for this given set of observations.

3.19. Repeat Problem 3.18 using the *backward procedure*.

3.20. Using the Viterbi procedure, find the most likely state sequence for the model and the set of observations given in Problem 3.18.

3.21. Consider a two-state hidden Markov model. For state 1, the observations are independent zero-mean Gaussian random variables with variance σ_1^2. For state 2 the observations are independent zero-mean Gaussian random variables with variance σ_2^2. Is the unconditional density function

$$f_x(\mathrm{x}_0, \mathrm{x}_1, \ldots, \mathrm{x}_n)$$

Gaussian? Show why or why not.

3.22. Suppose that the first random process in Example 3.5 is a Gaussian random process with mean μ and variance σ_o^2 representing a constant signal in noise (hypothesis 1). Suppose that the second random process is also Gaussian but with mean zero and the same variance σ_o^2 representing the noise alone (hypothesis 2). Derive the simplest possible expression for the "log likelihood ratio"

$$\Lambda[n] = \ln \mathcal{L}[n]$$

Is this sequence a martingale under either hypothesis?

3.23. A Gaussian random process is defined by the relation

$$x[n] = x[n-1] + w[n]$$

where $w[n]$ is a sequence of independent random variables with density

$$f_w(\mathrm{w}) = \frac{1}{\sqrt{2\pi\sigma_w^2}} e^{-\frac{\mathrm{w}^2}{2\sigma_w^2}}$$

i.e., it is a Gaussian white noise process. Assume the initial condition $x[0] = x_o$ is also a Gaussian random variable with mean zero and variance σ_o^2 independent of w. Show that $x[n]$ is both a Markov random process and a martingale. Is every Markov random process also a martingale?

3.24. Find an expression for the correlation between any two samples of the random process of Example 3.7. That is, evaluate

$$\mathcal{E}\{x[n_1]x[n_0]\}$$

Is the correlation a function of only the spacing between samples? Repeat when the process contains an independent random phase uniformly distributed between 0 and 2π.

3.25. In this problem you show that when a continuous Wiener process is input to a continuous filter, the output is a continuous Gaussian random process. Assume that the Wiener process is given by the limiting form of (3.67) and that the linear filter has a continuous impulse response to an input $\zeta[i]$ occurring at $t_i = i\Delta t$ is given by

$$h_c(t - t_i) = e^{-\frac{t-t_i}{\tau}} u_c(t - t_i)$$

where $u_c(t)$ is the continuous unit step function.

(a) Assume the Wiener process begins at some time $t_o = k_o\Delta t$ and write an expression for the output $y_c(t)$ of the filter at some arbitrary time $t = k\Delta t$.

(b) Find a closed-form expression for the variance of $y_c(t)$ in terms of s, τ, Δt, t, and t_o. Take the limit as $\Delta t \to 0$ and $s \to 0$ to express this in terms of ν_o, τ, t, and t_o. Notice that $y_c(t)$ is Gaussian by the Central Limit Theorem.

(c) Finally, assume that the process has gone on for a very long time by letting $t_o \to -\infty$. Find a final expression for the variance of the Gaussian random process $y_c(t)$.

3.26. Consider the Gauss-Markov process defined in Example 3.8. Assume that the process has started a long time in the past so that the transient portion of the covariance has vanished. Then the covariance matrix for N successive samples of the random process is

$$\mathbf{C}_x = \frac{\sigma_o^2}{1-\rho^2}\begin{bmatrix} 1 & \rho & \rho^2 & \cdots & \rho^{N-1} \\ \rho & 1 & \rho & \cdots & \rho^{N-2} \\ \rho^2 & \rho & 1 & \ddots & \vdots \\ \vdots & \vdots & \ddots & \ddots & \rho \\ \rho^{N-1} & \rho^{N-2} & \cdots & \rho & 1 \end{bmatrix}$$

Verify that the inverse covariance matrix has the form

$$\mathbf{C}_x^{-1} = \frac{1}{\sigma_o^2}\begin{bmatrix} 1 & -\rho & 0 & \cdots & 0 \\ -\rho & 1+\rho^2 & -\rho & \cdots & 0 \\ 0 & -\rho & 1+\rho^2 & \ddots & \vdots \\ \vdots & \ddots & \ddots & \ddots & 0 \\ 0 & \ddots & \ddots & 1+\rho^2 & -\rho \\ 0 & 0 & \cdots & -\rho & 1 \end{bmatrix}$$

What is the density function of this random process if the mean is zero? (Note: The determinant of the covariance matrix is $|\mathbf{C}_x| = \frac{\sigma_o^2}{1-\rho^2}$. You may want to see if you can show that.)

COMPUTER ASSIGNMENTS

3.1. Compute and plot 50 samples each of the following random processes:

(a) a binary white noise process.

(b) a Bernoulli process with $P = 0.8$

(c) the 2-state Markov chain described in Example 3.3

To initiate the Markov chain, you can use the limiting-state probabilities computed in the example as the initial probabilities of a +1 and a −1. What differences do you and should you observe between these three random processes? Explain these differences.

Estimate the limiting-state probability of a +1 and a −1 for the Markov chain by computing the relative frequency of +1's and −1's in the sequence of 50 samples. Repeat this calculation for sequences of 100, 200, and 500 samples.

3.2. **(a)** Generate and plot some typical realizations (sequences) from a discrete Wiener process. Assume that the process starts at $n_o = 0$ and that $x[0] = 0$. Plot several members of the ensemble using at least 50 samples. Observe that the means computed along these sequences may be significantly different from zero due to long runs of positive or negative values.

(b) Now generate 100 realizations of a discrete Wiener process but do not plot them. Estimate the mean $\mathrm{m}[n] = \mathcal{E}\{x[n]\}$ by averaging "down the ensemble" and plot this function. Is the ensemble mean close to zero?

(c) Repeat part (b) for a random walk with $\mathrm{P} = 0.8$.

3.3. A hidden Markov model has two states and is described by the state transition matrix

$$\mathbf{\Pi} = \begin{bmatrix} 0.8 & 0.3 \\ 0.2 & 0.7 \end{bmatrix}$$

When the Markov chain is in state 1, the sequence $x[n]$ generated is a set of independent random variables uniformly distributed between −1 and +1. When in state 2 the sequence $x[n]$ is again a set of independent random variables but uniformly distributed between −2 and +2.

(a) Generate and plot several realizations of this random process each containing at least 30 samples. By comparing the random process to the corresponding state sequence, see if it is possible to tell by observing $x[n]$ at which points the underlying Markov model makes state transitions.

(b) Apply the Viterbi algorithm to the data and estimate the underlying state transition sequence. Compare the results to the true state sequence.

3.4. A random sequence $x[n]$ is generated according to the difference equation

$$x[n] = \rho x[n-1] + \xi[n]$$

where ρ is a constant and $\xi[n]$ is a binary white noise sequence taking on values −1 and +1 with equal probabilities.

Generate and plot 50 samples of the random sequence for $\rho = 0.95, 0.70$, and -0.95. What differences do you observe in these three different random sequences? What happens if $|\rho| > 1$?

4

Second Moment Analysis

Sometimes it is not possible to perform a complete analysis of a random process because the distribution or density function is not known and is too difficult to estimate. Still some very useful results can be obtained by working with the first few moments of the distribution. When the analysis is restricted to at most second moments it is called *second moment analysis*. For a *Gaussian* random process the first and second moments *completely* characterize the distribution; therefore, second moment analysis is all that is ever required. In other cases second moment analysis may not tell the full story about a random process, but it is frequently enough to describe the most important features.

The first moment of a random process is the mean sequence, which was already introduced in Chapter 3. The second moments are represented by the correlation and covariance functions, which are introduced in this chapter. These functions provide a way of characterizing the random process in the signal domain. When a random process is stationary, however, it can also be described in the frequency domain and the transform (z) domain. These other methods are a powerful set of tools that can provide a considerable amount of new insight about the random process, and they are the key to performing signal processing operations.

This chapter also develops the Karhunen-Loève representation for discrete random processes. This representation provides a kind of generalization of Fourier analysis to both stationary and nonstationary random process and it is closely related to the unitary transformations of random vectors involving the eigenvectors of the correlation matrix (see Chapter 2).

With the new tools in hand we revisit some specific types of random processes introduced in Chapter 3 and gain some new insights. For example, periodic random processes are examined and related to general nonperiodic random processes, and some further interpretations of Gaussian white noise are given. The chapter concludes with a brief introduction to the description of random processes in terms of their higher-order moments and cumulants. This area has become increasingly important in modern signal processing.

4.1 THE CORRELATION AND COVARIANCE FUNCTIONS

Second moment analysis of random processes in the signal domain involves the mean, correlation, and covariance functions. This section first defines these important quantities, and then proceeds to describe their fundamental properties.

4.1.1 Definitions

The mean of a random process was defined in Chapter 2 to be the ensemble average. Recall that the mean is a (deterministic) sequence defined by

$$\mathrm{m}_x[n] \stackrel{\text{def}}{=} \mathcal{E}\{x[n]\} \tag{4.1}$$

and that in general it does *not* have to be a constant function of n. Ordinarily, a subscript is attached to the function as it is in the definition above to show that the function pertains to the random process x.

The correlation betwen any two samples of a random process $x[n_1]$ and $x[n_0]$ is expressed by the *correlation function* (also called the *auto*correlation function)

$$\mathrm{R}_x[n_1, n_0] = \mathcal{E}\{x[n_1]x^*[n_0]\} \tag{4.2}$$

and usually has different values for different choices of n_1 and n_0 (see Fig. 4.1). The reason for the conjugate in the definition of the correlation function is related to certain symmetry properties assumed for the second moments of a complex random process and is discussed further in Section 4.5. It is similar to the reason for defining a correlation *matrix* with the conjugate transpose as in Chapter 2. For real random processes the conjugate can be ignored; for complex processes it will be seen that the expression without the conjugate is identically zero!

The usual interpretation of statistical correlation can be applied to random processes. When the correlation function $\mathrm{R}_x[n_1, n_0]$ is high and positive the values of the random process at the points n_1 and n_0 tend to "track" each other; large values of $x[n_0]$ tend to imply large values of $x[n_1]$ and small values of $x[n_0]$ tend to imply small values of $x[n_1]$. If $\mathrm{R}_x[n_1, n_0]$ is low, then $x[n_0]$ and $x[n_1]$ tend to have little to do with each other. If $\mathrm{R}_x[n_1, n_0]$ has a large *negative* value, then $x[n_1]$ and $x[n_0]$ tend to have *opposite* behavior; that is, large positive values of $x[n_0]$ tend to imply large negative values of $x[n_1]$; small

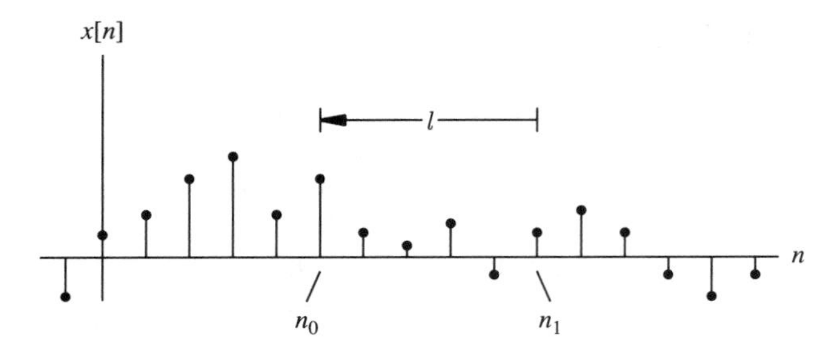

Figure 4.1 Definition of the correlation function for a random process.

positive values of $x[n_0]$ tend to imply small negative values of $x[n_1]$; and so on. When $\mathrm{R}_x[n_1, n_0]$ is identically zero the two samples of the random process $x[n_1]$ and $x[n_0]$ are *orthogonal*.

The covariance between any two samples of a random process is expressed by the *covariance function* (also called the *auto*covariance function)

$$\mathrm{C}_x[n_1, n_0] = \mathcal{E}\{(x[n_1] - m_x[n_1])(x[n_0] - m_x[n_0])^*\} \tag{4.3}$$

where again the conjugate on the second term is ignored for real-valued random processes. When $\mathrm{C}_x[n_1, n_0] = 0$, $x[n_1]$ and $x[n_0]$ are *uncorrelated*. Further, it is easy to show the general relation

$$\mathrm{R}_x[n_1, n_0] = \mathrm{C}_x[n_1, n_0] + \mathrm{m}_x[n_1]\mathrm{m}_x^*[n_0] \tag{4.4}$$

Since for a stationary random process the probability density is a function of only the *spacing* between samples and not of the absolute locations, it must be true that (for a stationary process)

$$\mathrm{m}_x[n] = m_x \quad \text{(a constant)} \tag{4.5}$$

and

$$\mathrm{R}_x[n_1, n_0] = R_x[n_1 - n_0] = R_x[l] \tag{4.6}$$

where R_x is a new function[1] that depends on only the single argument l which is the *difference* between the two indices n_1 and n_0 (see Fig. 4.1). The difference l is called the *lag*. In light of (4.4), equations 4.5 and 4.6 imply that for a stationary random process it is also true that the covariance is a function only of the lag

$$\mathrm{C}_x[n_1, n_0] = C_x[n_1 - n_0] = C_x[l] \tag{4.7}$$

These last properties are sufficiently important that they warrant recognition in the following definition:

Definition 4.1 (Wide-Sense Stationary). A random process $x[n]$ is said to be *wide-sense stationary* if the mean is a constant m_x and the correlation function is a function only of the *spacing* between the samples, $\mathrm{R}_x[n_1, n_0] = R_x[n_1 - n_0]$.

[1] A math italic font is used for the function that depends on the difference $n_1 - n_0$ while a roman font is used for the original function $\mathrm{R}_x[n_1, n_0]$.

From the foregoing discussion it is clear that strict-sense stationarity implies wide-sense stationarity. The converse is not true, however, unless the random process is Gaussian. In common usage, *the term "stationary" is usually taken to mean wide-sense stationary,* especially when the analysis deals with only second moment properties of the random process. That convention is also followed here.

The following example illustrates the calculation of correlation and covariance functions for a simple random process and the test for stationarity. Notice that in the example nothing is said about the *distribution or density* of the random variables; only their first and second moment properties are given.

Example 4.1

Let $v[n]$ be a real-valued process of independent random variables each with mean μ and variance σ^2. The correlation function for this random process is

$$\mathrm{R}_v[n_1, n_0] = \mathcal{E}\{v[n_1]v[n_0]\} = \begin{cases} \mu^2 & n_1 \neq n_0 \\ \sigma^2 + \mu^2 & n_1 = n_0 \end{cases}$$

This can be written as

$$\mathrm{R}_v[n_1, n_0] = \sigma^2\delta[n_1 - n_0] + \mu^2$$

Since the mean is constant, and the correlation function is a function only of the difference $n_1 - n_0$, this random process is wide-sense stationary.

Now consider the process $x[n]$ defined by

$$x[n] = nv[n-1]$$

Its correlation function is

$$\begin{aligned} \mathrm{R}_x[n_1, n_0] &= \mathcal{E}\{x[n_1]x[n_0]\} = \mathcal{E}\{n_1 v[n_1 - 1]n_0 v[n_0 - 1]\} \\ &= n_1 n_0 \mathcal{E}\{v[n_1 - 1]v[n_0 - 1]\} = n_1 n_0 \left(\sigma^2\delta[n_1 - n_0] + \mu^2\right) \end{aligned}$$

Since this is *not* purely a function of $n_1 - n_0$, this random process is not wide-sense stationary. The covariance function of this process can be found from either (4.3) or (4.4) to be

$$\mathrm{C}_x[n_1, n_0] = n_1 n_0 \sigma^2 \delta[n_1 - n_0]$$

which is also not a function of just $n_1 - n_0$.

A slightly more complicated example is given next.

Example 4.2

Let $x[n]$ be a random process defined by

$$x[n] = v[n] + \tfrac{1}{2}v[n-1]$$

where $v[n]$ is the process defined in Example 4.1. Assume for simplicity that $\mu = 0$ and $\sigma^2 = 1$. The correlation function of this process is

$$\begin{aligned}
\mathrm{R}_x[n_1, n_0] &= \mathcal{E}\{x[n_1]x[n_0]\} \\
&= \mathcal{E}\left\{\left(v[n_1] + \tfrac{1}{2}v[n_1 - 1]\right)\left(v[n_0] + \tfrac{1}{2}v[n_0 - 1]\right)\right\} \\
&= \mathcal{E}\{v[n_1]v[n_0]\} + \tfrac{1}{2}\mathcal{E}\{v[n_1 - 1]v[n_0]\} + \tfrac{1}{2}\mathcal{E}\{v[n_1]v[n_0 - 1]\} \\
&\quad + \tfrac{1}{4}\mathcal{E}\{v[n_1 - 1]v[n_0 - 1]\} \\
&= \tfrac{5}{4}\delta[n_1 - n_0] + \tfrac{1}{2}\delta[n_1 - n_0 - 1] + \tfrac{1}{2}\delta[n_1 - n_0 + 1]
\end{aligned}$$

Since R_x is a function of just the difference $n_1 - n_0$, the random process is stationary. Further, it is easy to see that since $\mu = 0$, the covariance function and the correlation function are identical.

For a stationary random process, the mean, correlation, and covariance functions can be defined as

$$\boxed{m_x = \mathcal{E}\{x[n]\}} \tag{4.8}$$

$$\boxed{R_x[l] = \mathcal{E}\{x[n]x^*[n-l]\}} \tag{4.9}$$

and

$$\boxed{C_x[l] = \mathcal{E}\{(x[n] - m_x)(x[n-l] - m_x)^*\}} \tag{4.10}$$

with the corresponding relation

$$\boxed{R_x[l] = C_x[l] + |m_x|^2} \tag{4.11}$$

Notice that

$$R_x[0] = \mathcal{E}\left\{|x[n]|^2\right\}$$

is the *average power* of the random process and that

$$C_x[0] = \mathcal{E}\left\{|x[n] - m_x|^2\right\}$$

is the *variance*. It is also clear from (4.9) and (4.10) why l is called the *lag*; it represents the shift of the sequence that appears in the definition of these functions. The lag will normally be denoted by the parameter l (or sometimes k) and it may take on both positive and negative values.

Here is one more example of computing a correlation function. This one involves a complex random process.

Example 4.3

Consider the complex random process defined by

$$x[n] = x_{\mathrm{r}}[n] + \jmath x_{\mathrm{i}}[n]$$

where $x_{\mathrm{r}}[n]$ and $x_{\mathrm{i}}[n]$ are *real* stationary random processes with mean zero and autocorrelation functions

$$R_{x_{\mathrm{r}}}[l] = R_{x_{\mathrm{i}}}[l] = \sigma^2\delta[l]$$

It is further assumed that the components x_r and x_i are orthogonal; that is,

$$\mathcal{E}\{x_r[n_1]x_i[n_0]\} = 0$$

for all values of n_1 and n_0.

Since the weighted sum of two orthogonal stationary random processes is stationary, the correlation function can be computed from (4.9) as

$$\begin{aligned} R_x[l] &= \mathcal{E}\{x[n]x^*[n-l]\} \\ &= \mathcal{E}\{(x_r[n] + \jmath x_i[n])(x_r[n-l] - \jmath x_i[n-l])\} \\ &= \mathcal{E}\{x_r[n]x_r[n-l]\} + \mathcal{E}\{x_i[n]x_i[n-l]\} - \jmath\mathcal{E}\{x_r[n]x_i[n-l]\} + \jmath\mathcal{E}\{x_i[n]x_r[n-l]\} \\ &= 2\sigma^2\delta[l] \end{aligned} \tag{4.12}$$

Since the means are zero, the covariance function is identical.

In this example the stationary form of the definition (4.9) was used to compute the correlation function. If the assumption about the stationarity of the random process had been wrong, the correlation function would have turned out to be a function of *both* l and n. (You may want to try this on the nonstationary process of Example 4.1.)

4.1.2 Properties

The correlation and covariance functions have a number of important properties. The two most fundamental from which all others can be derived are the symmetry and the positive semidefinite properties. These are defined below.

The symmetry property states that the correlation and covariance functions are *conjugate symmetric* functions of lag, that is,

$$\boxed{R_x[l] = R_x^*[-l]} \tag{4.13}$$

$$\boxed{C_x[l] = C_x^*[-l]} \tag{4.14}$$

These properties follow directly from the definitions and imply that the correlation between $x[n_1]$ and $x[n_0]$ is the *conjugate* of the correlation between and $x[n_1]$ and $x[n_0]$ (see Fig. 4.1). For real random processes the correlation in both directions is the same. Correspondingly, (4.13) and (4.14) show that for real random processes the correlation and covariance are *even* functions of lag.

The positive semidefinite property of the correlation *function* is analogous to the positive semidefinite property of the correlation *matrix* for random vectors (see Chapter 2) and can be interpreted in the following way. When a random process is transformed by a linear shift-invariant system, the result is another random process. The requirement that the average power in the output process is positive imposes a certain constraint on the correlation function of the input. This constraint is the positive semidefinite property. To see how this works, consider an arbitrary linear shift-invariant system with impulse response sequence $a[n]$ (see Fig. 4.2). If the input to the system is a random process x, then the output is a random process y given by

$$y[n] = \sum_{n_0=-\infty}^{\infty} a[n_0]x[n-n_0] \tag{4.15}$$

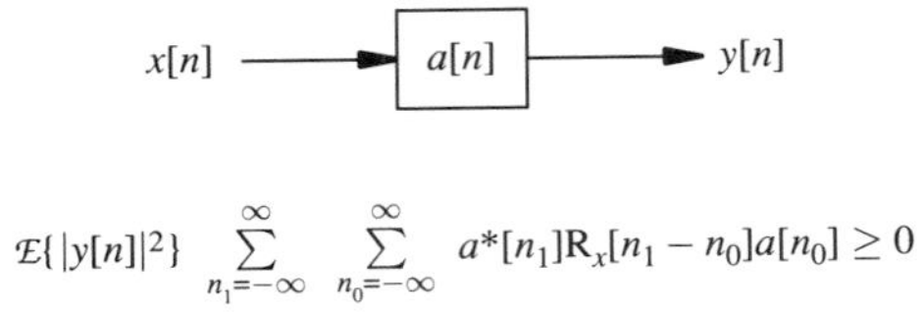

$x[n] \longrightarrow \boxed{a[n]} \longrightarrow y[n]$

$$\mathcal{E}\{|y[n]|^2\} \sum_{n_1=-\infty}^{\infty} \sum_{n_0=-\infty}^{\infty} a^*[n_1]R_x[n_1-n_0]a[n_0] \geq 0$$

Figure 4.2 Linear shift-invariant system used to show the positive definite property of the correlation function.

Now look at the average output power of this system:

$$\begin{aligned}
\mathcal{E}\{|y[n]|^2\} &= \mathcal{E}\{y[n]y^*[n]\} \\
&= \mathcal{E}\left\{\sum_{n_0=-\infty}^{\infty} a[n_0]x[n-n_0] \sum_{n_1=-\infty}^{\infty} a^*[n_1]x^*[n-n_1]\right\} \\
&= \sum_{n_1=-\infty}^{\infty} \sum_{n_0=-\infty}^{\infty} a^*[n_1]\mathcal{E}\{x[n-n_0]x^*[n-n_1]\}\, a[n_0] \\
&= \sum_{n_1=-\infty}^{\infty} \sum_{n_0=-\infty}^{\infty} a^*[n_1]R_x[n_1-n_0]a[n_0] \geq 0
\end{aligned}$$

where the last inequality follows since the average power $\mathcal{E}\{|y[n]|^2\}$ must be greater than or equal to zero.[2] This condition is the *positive semidefinite* property of the correlation function. An identical property holds for the covariance function.

From a purely mathematical point of view the positive semidefinite property can be stated as[3]

$$\boxed{\sum_{n_1=-\infty}^{\infty} \sum_{n_0=-\infty}^{\infty} a^*[n_1]R_x[n_1-n_0]a[n_0] \geq 0} \tag{4.16}$$

and

$$\boxed{\sum_{n_1=-\infty}^{\infty} \sum_{n_0=-\infty}^{\infty} a^*[n_1]C_x[n_1-n_0]a[n_0] \geq 0} \tag{4.17}$$

[2]Note that the conclusion does not depend on the interpretation of $|y[n]|^2$ as power but only on the fact that the expectation of a squared quantity is always ≥ 0.

[3]More generally, even a correlation function for a nonstationary random process is positive semidefinite. That is, $\sum_{n_1=-\infty}^{\infty} \sum_{n_0=-\infty}^{\infty} a^*[n_1]R_x[n_1, n_0]a[n_0] \geq 0$. However, this more general result will not be particularly useful in later sections.

where $a[n]$ is any real- or complex-valued sequence. *The condition* (4.16) [or (4.17)] *and the symmetry property* (4.13) [or (4.14)] *are necessary and sufficient conditions for any discrete sequence to qualify as a correlation* [or covariance] *function.* These two fundamental properties are summarized in Table 4.1.

TABLE 4.1 FUNDAMENTAL PROPERTIES OF THE CORRELATION FUNCTION

Conjugate symmetry	$R_x[l] = R_x^*[-l]$
Positive semidefinite property	$\sum_{n_1=-\infty}^{\infty} \sum_{n_0=-\infty}^{\infty} a^*[n_1]R_x[n_1 - n_0]a[n_0] \geq 0$ for an arbitrary sequence $a[n]$

Note: Identical properties pertain to the covariance function; substitute C_x for R_x.

It can be shown that (4.16) and (4.17) imply that

$$R_x[0] \geq |R_x[l]| \quad l \neq 0 \tag{4.18}$$

and

$$C_x[0] = \text{Var}\big[x[n]\big] \geq |C_x[l]| \quad l \neq 0 \tag{4.19}$$

so the magnitudes of the correlation and covariance functions assume their maximum values at the origin. These properties are quite simple to show for a real-valued sequence (see Problem 4.3). For a complex sequence the proof is slightly more complicated, but not unreasonably so. To show (4.18), let $a[n]$ be chosen as the specific sequence

$$a[n] = \begin{cases} b & \text{for } n = 0 \\ b^* & \text{for } n = l \\ 0 & \text{otherwise} \end{cases}$$

where b is a complex constant that is to be specified shortly. Then inserting this in (4.16) produces

$$b^*bR_x[0-0] + b^2R_x[l-0] + (b^*)^2R_x[0-l] + bb^*R_x[l-l] \geq 0$$

or (using the symmetry property of the correlation function)

$$2|b|^2R_x[0] + 2\text{Re}\left(b^2R_x[l]\right) \geq 0 \tag{4.20}$$

Now let $\phi[l]$ be the *phase* of $R_x[l]$, that is,

$$R_x[l] = |R_x[l]|e^{\jmath\phi[l]}$$

and take b to be

$$b = e^{-\jmath \frac{\phi[l]+\pi}{2}}$$

so that

$$b^2 = -e^{-\jmath\phi[l]} \text{ and } |b|^2 = 1$$

Then (4.20) becomes

$$2R_x[0] + 2\text{Re}\left(-|R_x[l]|\right) \geq 0$$

and since $|R_x[l]|$ is real, (4.18) follows.

When a random process is periodic, all of its moments are also periodic; thus the correlation and covariance become periodic functions. In this case the functions have an infinite number of maxima with the value $R_x[0]$ or $C_x[0]$ occurring at every integer multiple of the period.

So far in this section properties for the correlation function and similar properties for the covariance function were stated separately. Observe, however, that since the covariance function *is* a correlation function for the random process that has its mean removed, any property that is true for the correlation is also true for the covariance. (The converse is *not* true.) From here on properties stated for only the correlation function should be understood to apply also to the covariance.

Figure 4.3 is a sketch of some typical random sequences with zero mean and the corresponding forms of their correlation functions. When the sequence changes slowly the correlation function is positive and broad, indicating that successive samples of the sequence are highly correlated. If the sequence exhibits a more random appearance with no particular structure, then the correlation function tends to be narrow, indicating little or no correlation from sample to sample. Finally, if the sequence changes sign rapidly from sample to sample with large positive values followed by large negative values (and small positive values followed by small negative values), then the correlation function also changes sign but has a broad envelope indicating high *negative* sample-to-sample correlation.

4.2 CORRELATION AND COVARIANCE MATRICES

This section puts to use some of the ideas developed in Chapter 2 by defining the correlation and covariance *matrices* for a random process and showing how they relate to the correlation and covariance *functions* defined in the preceding section.

Let $\boldsymbol{x}$ be a random vector consisting of N samples of the random process x.

$$\boldsymbol{x} = \begin{bmatrix} x[0] \\ x[1] \\ \vdots \\ x[N-1] \end{bmatrix} \tag{4.21}$$

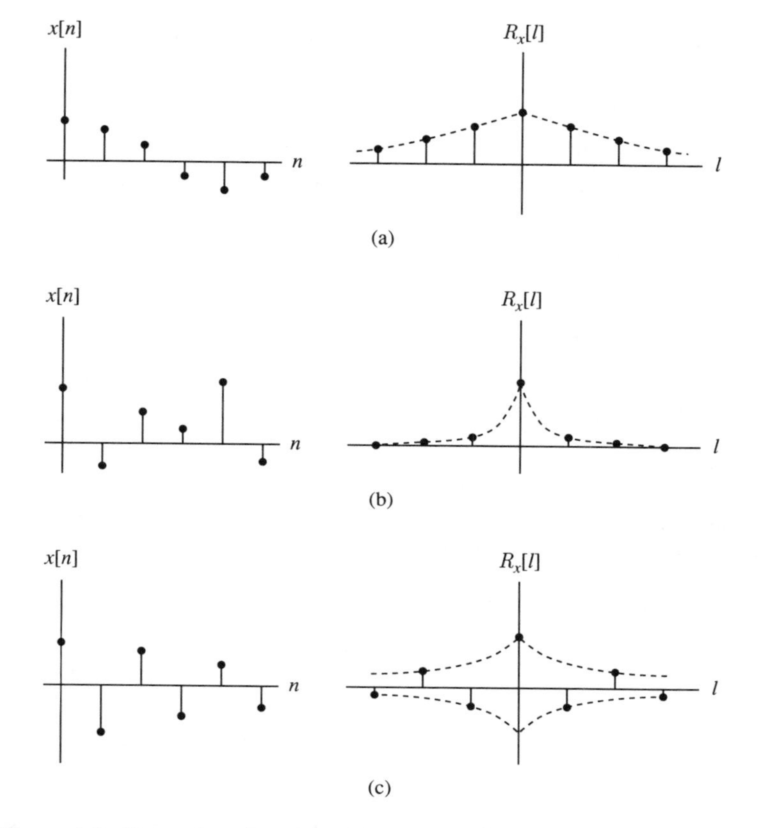

Figure 4.3 Examples of random sequences and their correlation functions. (a) High correlation. (b) Low correlation. (c) High negative correlation.

The *mean vector* is given by

$$\mathbf{m}_x = \mathcal{E}\{\boldsymbol{x}\} = \begin{bmatrix} \mathrm{m}_x[0] \\ \mathrm{m}_x[1] \\ \vdots \\ \mathrm{m}_x[N-1] \end{bmatrix} \tag{4.22}$$

and so is completely determined by the mean function $\mathrm{m}_x[n]$ for the random process. For a stationary random process the mean vector has all of its components equal to the same constant m_x.

The *correlation matrix* for the random process is defined by

$$
\begin{aligned}
\mathbf{R}_x &= \mathcal{E}\{\boldsymbol{x}\boldsymbol{x}^{*T}\} \\
&= \begin{bmatrix}
\mathcal{E}\{|x[0]|^2\} & \mathcal{E}\{x[0]x^*[1]\} & \cdots & \mathcal{E}\{x[0]x^*[N-1]\} \\
\mathcal{E}\{x[1]x^*[0]\} & \mathcal{E}\{|x[1]|^2\} & \cdots & \mathcal{E}\{x[1]x^*[N-1]\} \\
\vdots & \vdots & \vdots & \vdots \\
\mathcal{E}\{x[N-1]x^*[0]\} & \mathcal{E}\{x[N-1]x^*[1]\} & \cdots & \mathcal{E}\{|x[N-1]|^2\}
\end{bmatrix} \\
&= \begin{bmatrix}
\mathrm{R}_x[0,0] & \mathrm{R}_x[0,1] & \cdots & \mathrm{R}_x[0,N-1] \\
\mathrm{R}_x[1,0] & \mathrm{R}_x[1,1] & \cdots & \mathrm{R}_x[1,N-1] \\
\vdots & \vdots & \vdots & \vdots \\
\mathrm{R}_x[N-1,0] & \mathrm{R}_x[N-1,1] & \cdots & \mathrm{R}_x[N-1,N-1]
\end{bmatrix}
\end{aligned}
\tag{4.23}
$$

This matrix is completely specified by the correlation *function* for the random process. For a stationary random process the correlation function depends on only a single argument and the correlation matrix takes the special form

$$
\mathbf{R}_x = \begin{bmatrix}
R_x[0] & R_x[-1] & & & R_x[-N+1] \\
R_x[1] & R_x[0] & \ddots & & \\
& \ddots & \ddots & \ddots & \\
& & \ddots & R_x[0] & R_x[-1] \\
R_x[N-1] & & & R_x[1] & R_x[0]
\end{bmatrix}
\tag{4.24}
$$

Recall that this form, where all elements on each of the main diagonals are equal, is known as a *Toeplitz* matrix; it arises because all elements on a given diagonal represent correlations between terms of the random process with the *same* lag separation. Thus *the correlation matrix for any stationary random process is always a [Hermitian] symmetric Toeplitz matrix.* It will be seen in later chapters that this special property has many important implications for the analysis of stationary random processes.

The *covariance matrix* for a random process is defined by

$$
\begin{aligned}
\mathbf{C}_x &= \mathcal{E}\{(\boldsymbol{x}-\mathbf{m}_x)(\boldsymbol{x}-\mathbf{m}_x)^{*T}\} \\
&= \begin{bmatrix}
\mathrm{C}_x[0,0] & \mathrm{C}_x[0,1] & \cdots & \mathrm{C}_x[0,N-1] \\
\mathrm{C}_x[1,0] & \mathrm{C}_x[1,1] & \cdots & \mathrm{C}_x[1,N-1] \\
\vdots & \vdots & \vdots & \vdots \\
\mathrm{C}_x[N-1,0] & \mathrm{C}_x[N-1,2] & \cdots & \mathrm{C}_x[N-1,N-1]
\end{bmatrix}
\end{aligned}
\tag{4.25}
$$

and is completely specified by the covariance *function.* For the special case of a stationary random process it also has the symmetric Toeplitz form

$$\mathbf{C}_x = \begin{bmatrix} C_x[0] & C_x[-1] & & C_x[-N+1] \\ C_x[1] & C_x[0] & \ddots & \\ & \ddots & \ddots \quad \ddots & \\ & & \ddots \quad C_x[0] & C_x[-1] \\ C_x[N-1] & & C_x[1] & C_x[0] \end{bmatrix} \quad (4.26)$$

It should be clear that for a stationary random process, the correlation and covariance matrices do not change if N samples are taken anywhere in the random process. That is, if $\boldsymbol{x}$ is defined as

$$\boldsymbol{x} = \begin{bmatrix} x[n_\mathrm{o}] \\ x[n_\mathrm{o}+1] \\ \vdots \\ x[n_\mathrm{o}+N-1] \end{bmatrix} \quad (4.27)$$

where n_o is an arbitrary starting point, the correlation and covariance matrices still have the forms (4.24) and (4.26). This is in keeping with the meaning of stationarity, since for a stationary random process the second moment statistics do not depend on absolute location of the samples.

The positive semidefinite property of the correlation function can also be derived from the positive semidefinite property of the correlation matrix. To do so, assume that the dimension N is arbitrary but *odd* and form the vector

$$\mathbf{a} = \begin{bmatrix} a[-\frac{N-1}{2}] \\ a[1-\frac{N-1}{2}] \\ \vdots \\ a[0] \\ \vdots \\ a[-1+\frac{N-1}{2}] \\ a[\frac{N-1}{2}] \end{bmatrix}$$

where $a[n]$ is again any sequence. Since the correlation matrix of any order N is positive semidefinite, it follows that

$$\mathbf{a}^{*T}\mathbf{R}_x\mathbf{a} = \sum_{n_1=-\frac{N-1}{2}}^{\frac{N-1}{2}} \sum_{n_0=-\frac{N-1}{2}}^{\frac{N-1}{2}} a^*[n_1]R_x[n_1-n_0]a[n_0] \geq 0$$

Then taking the limit as $N \to \infty$ yields (4.16). Equation 4.17 can be derived in a similar way.

Let us conclude this section with a simple example illustrating the relation between the correlation function and the correlation matrix.

Example 4.4

The correlation function for a certain random process has the exponential form

$$R_x[l] = 4(-0.5)^{|l|}$$

The correlation matrix for $N = 3$ is

$$\begin{bmatrix} 4 & -2 & 1 \\ -2 & 4 & -2 \\ 1 & -2 & 4 \end{bmatrix}$$

which is clearly Toeplitz. The eigenvalues of this matrix are found to be $\lambda_1 = 7.4$, $\lambda_2 = 3.0$, and $\lambda_3 = 1.6$. Since the eigenvalues are all positive, the correlation matrix is *positive definite*.

For $N = 4$ the correlation matrix has the Toeplitz form

$$\begin{bmatrix} 4 & -2 & 1 & -0.5 \\ -2 & 4 & -2 & 1 \\ 1 & -2 & 4 & -2 \\ -0.5 & 1 & -2 & 4 \end{bmatrix}$$

The eigenvalues of this matrix turn out to be $\lambda_1 = 8.3$, $\lambda_2 = 4.0$, $\lambda_3 = 2.2$, and $\lambda_4 = 1.5$, which implies that the matrix is positive definite, as required.

It can be shown that the correlation *function* given above satisfies the positive semidefinite condition (4.16) with strict inequality (see Problem 4.4). This implies that the correlation matrix of *any* order is positive definite.

4.3 CROSS-CORRELATION AND COVARIANCE

It is often necessary to deal with correlations or covariances between two or more random processes. For this reason the cross-correlation and cross-covariance functions for a random process $x[n]$ and another random process $y[n]$ are defined as

$$\mathrm{R}_{xy}[n_1, n_0] = \mathcal{E}\{x[n_1]y^*[n_0]\} \tag{4.28}$$

and

$$\mathrm{C}_{xy}[n_1, n_0] = \mathcal{E}\{(x[n_1] - \mathrm{m}_x[n_1])(y[n_0] - \mathrm{m}_y[n_0])^*\} \tag{4.29}$$

These two functions satisfy the relation

$$\mathrm{R}_{xy}[n_1, n_0] = \mathrm{C}_{xy}[n_1, n_0] + \mathrm{m}_x[n_1]\mathrm{m}_y^*[n_0] \tag{4.30}$$

The two random processes are said to be *orthogonal* if their cross-correlation function is identically zero; they are *uncorrelated* if the cross-covariance is zero. It follows from (4.30) that for two uncorrelated random processes

$$\mathrm{R}_{xy}[n_1, n_0] = \mathrm{m}_x[n_1]\mathrm{m}_y^*[n_0]$$

This can be taken as an alternative definition of the term "uncorrelated" (as in Chapter 2).

The following definition applies to the stationarity of two random processes.

Definition 4.2 (Jointly Stationary). The random processes $x[n]$ and $y[n]$ are said to be *jointly stationary* (in the wide sense) if

1. $x[n]$ and $y[n]$ are each wide-sense stationary; and
2. the cross-correlation is a function of only the *spacing* of the samples, $\mathrm{R}_{xy}[n_1, n_0] = R_{xy}[n_1 - n_0]$.

Thus for jointly stationary random processes, the cross-correlation and the cross-covariance functions are functions only of the lag l and can be defined as

$$\boxed{R_{xy}[l] = \mathcal{E}\{x[n]y^*[n-l]\}} \tag{4.31}$$

and

$$\boxed{C_{xy}[l] = \mathcal{E}\{(x[n] - m_x)(y[n-l] - m_y)^*\}} \tag{4.32}$$

Equation (4.30) for jointly stationary random processes becomes

$$\boxed{R_{xy}[l] = C_{xy}[l] + m_x m_y^*} \tag{4.33}$$

The cross-correlation and cross-covariance functions have no special symmetry (they can even be one-sided). However, they have the following properties:

$$\boxed{R_{xy}[l] = R_{yx}^*[-l]} \tag{4.34}$$

and

$$\boxed{C_{xy}[l] = C_{yx}^*[-l]} \tag{4.35}$$

(Note the reversal of the subscripts in these last two equations.) Figure 4.4 illustrates the cross-correlation between two random processes. The correlation between $x[n_1]$ and $y[n_0]$ is the *conjugate* of the correlation between $y[n_0]$ and $x[n_1]$ (or the *same* if both sequences are *real*). However, *the correlation between* $x[n_1]$ *and* $y[n_0]$ *is not the same as the correlation between* $y[n_1]$ *and* $x[n_0]$. If this seems at all strange, then consider the following example. Suppose that $x[n]$ is a white noise process (such as the binary white noise process or the Gaussian white noise process considered in Chapter 3[4]). Suppose further that this process is the input to a causal IIR system and that $y[n]$ is the output of that same system as depicted in Fig. 4.5(a). Assume that $n_0 < n_1$, as shown. Then the input at n_0 affects the output at n_1, so $y[n_1]$ and $x[n_0]$ are correlated. The value of $y[n_0]$ does *not* affect $x[n_1]$, however. After all, y is the output, and x is the input. Further, since the

[4]Also see Section 4.9.

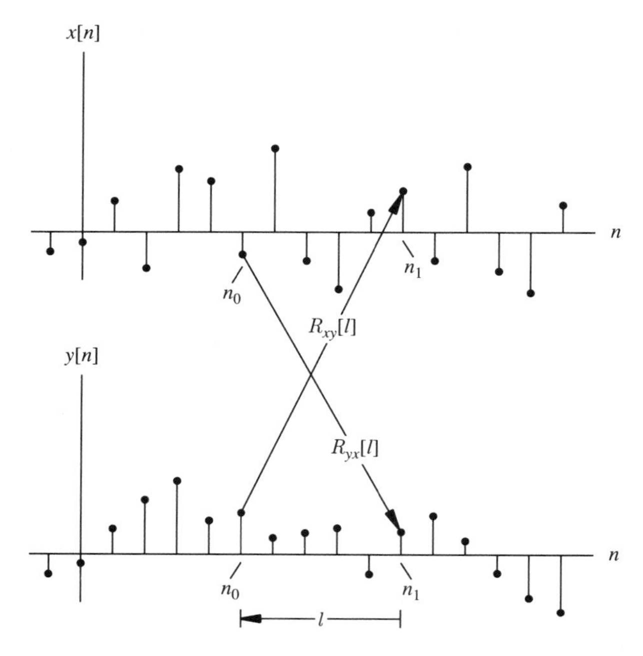

Figure 4.4 Illustration of cross-correlation: $R_{xy}[l] \neq R_{yx}[l]$

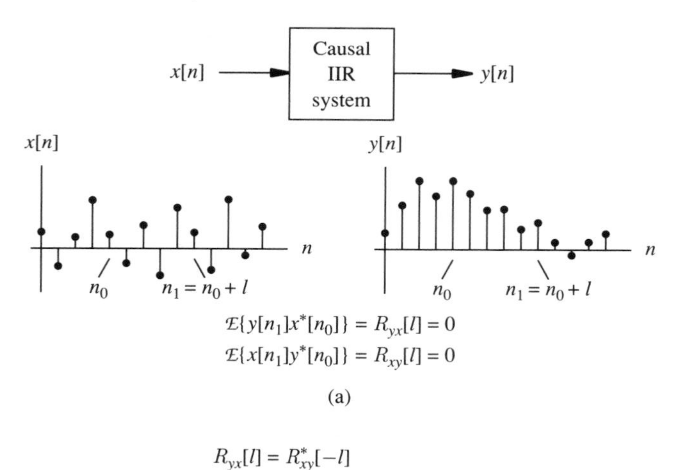

$\mathcal{E}\{y[n_1]x^*[n_0]\} = R_{yx}[l] = 0$

$\mathcal{E}\{x[n_1]y^*[n_0]\} = R_{xy}[l] = 0$

(a)

$R_{yx}[l] = R^*_{xy}[-l]$

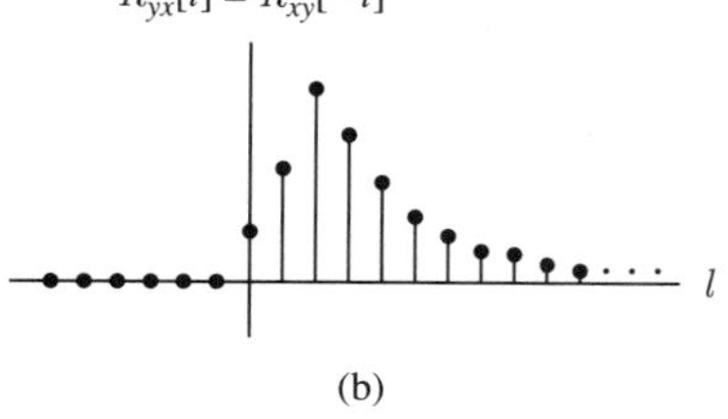

(b)

Figure 4.5 Cross-correlation between input and output of a causal linear system. (a) System with white noise input and random output. (b) Cross-correlation function for input and output.

system is causal, $x[n_1]$ does *not* affect $y[n_0]$. In other words, $x[n_1]$ and $y[n_0]$ *do not affect each other*. Therefore, the correlation between them is zero! Figure 4.5(b) shows a typical cross-correlation function for the input and output which clearly illustrates the asymmetry. Note also that the maximum value of the cross-correlation function does not necessarily occur at the origin. This is not too surprising since the cross-correlation function is *not*, in general, positive semidefinite. In fact, it is shown in Chapter 5 that when a system is driven with white noise, $R_{yx}[l]$ is proportional to the system impulse response. Therefore, the cross-correlation function can be almost anything!

One other fact that is worth mentioning with respect to the cross-correlation function is that its magnitude is bounded by both the geometric and the arithmetic mean of the autocorrelation functions at lag zero. That is,

$$|R_{xy}[l]| \leq \left(R_x[0]R_y[0]\right)^{\frac{1}{2}} \tag{4.36}$$

and

$$|R_{xy}[l]| \leq \tfrac{1}{2}\left(R_x[0] + R_y[0]\right) \tag{4.37}$$

These bounds follow from the corresponding definitions and are not too difficult to derive (see Problem 4.8).

Cross-correlation and cross-covariance *matrices* can be defined in a manner analogous to the definition of (auto-)correlation and covariance matrices in Section 4.2 and their elements can be written in terms of the cross-correlation and cross-covariance *functions*. For the case of jointly stationary random processes, the matrices exhibit Toeplitz structure in that elements on all of the diagonals are equal. If $\mathbf{R}_{xy}$ is a square matrix, then it will be a square Toeplitz matrix, but *it will not be Hermitian symmetric in general*. In other words, elements below the main diagonal are *not* equal to the conjugates of corresponding elements above the main diagonal.

As a simple example consider a random vector $\boldsymbol{x}$ consisting of three samples and a random vector $\boldsymbol{y}$ consisting of four samples of two jointly stationary random processes. The cross-correlation matrix has the form

$$\mathbf{R}_{xy} = \begin{bmatrix} R_{xy}[0] & R_{xy}[-1] & R_{xy}[-2] & R_{xy}[-3] \\ R_{xy}[1] & R_{xy}[0] & R_{xy}[-1] & R_{xy}[-2] \\ R_{xy}[2] & R_{xy}[1] & R_{xy}[0] & R_{xy}[-1] \end{bmatrix}$$

Since in general $R_{xy}[l] \neq R_{xy}^*[-l]$ the matrix has no particular symmetry. However, because the processes are jointly stationary, all elements along a diagonal are equal.

4.4 FREQUENCY- AND TRANSFORM-DOMAIN DESCRIPTION OF RANDOM PROCESSES

Frequency- and transform-domain methods are very powerful tools for the analysis of deterministic sequences. This section shows how the Fourier transform and the z-transform extend to the analysis of stationary random processes.

4.4.1 Power Density Spectrum and Cross-Spectrum

Frequency-domain characterizations of random processes relate to the average power in a random process and are therefore called power density spectra. These spectra give quantitative descriptions of the frequency components of a random process or, in the case of cross-spectra, the frequency components shared between two random processes.

Power Density Spectrum. The Fourier transform of the correlation function for a random process is called the *power spectral density function* and is denoted by $S_x(e^{\jmath\omega})$. The power spectral density function is defined by[5]

$$\boxed{S_x(e^{\jmath\omega}) = \sum_{l=-\infty}^{\infty} R_x[l]e^{-\jmath\omega l}} \tag{4.38}$$

and its inverse is given by

$$\boxed{R_x[l] = \frac{1}{2\pi}\int_{-\pi}^{\pi} S_x(e^{\jmath\omega})e^{\jmath\omega l}d\omega} \tag{4.39}$$

A plot of $S_x(e^{\jmath\omega})$ versus ω is called the *power density spectrum* (or simply the *power spectrum* or *spectrum*) of the random process, and its value at a given radian frequency[6] ω is called the power spectral density. The reason for the term "power density" becomes clear by noticing that the average "power" in the random process is given by

$$\text{avg. power} = \mathcal{E}\{|x[n]|^2\} = R_x[0] = \frac{1}{2\pi}\int_{-\pi}^{\pi} S_x(e^{\jmath\omega})d\omega \tag{4.40}$$

Since the integral of $S_x(e^{\jmath\omega})$ is power, $S_x(e^{\jmath\omega})$ itself must be *power density*. The power density $S_x(e^{\jmath\omega})$ can also be interpreted as representing the correlation of the random process *at the frequency* ω.

In the most general case, the power spectral density function can be written in the form

$$S_x(e^{\jmath\omega}) = S'_x(e^{\jmath\omega}) + \sum_i 2\pi \mathsf{P}_i \delta_c(e^{\jmath\omega} - e^{\jmath\omega_i}) \tag{4.41}$$

The term $S'_x(e^{\jmath\omega})$ represents the *continuous* part of the spectrum, while the sum of weighted impulses represents the discrete part or "lines" in the spectrum. The impulses or lines arise from periodic or almost periodic components of the random process. (Recall that when a

[5]The version of this definition involving signal averages is sometimes called the Wiener-Khintchine relation or (more recently) the Einstein-Wiener-Khintchine relation in acknowledgment of the connection between signal averages and harmonic content pointed out by these early investigators.

[6]Also called "digital frequency."

process is periodic, its correlation function is a periodic function of lag.) Note also that any process whose mean is nonzero has an impulse in the spectrum at $\omega = 0$ corresponding to the constant term $|m_x|^2$ in the correlation function. Thus *in order for a random process to have a purely continuous spectrum, its mean must be zero.* Since it is impossible to distinguish a random process with nonzero mean from one with a random zero frequency (dc) component using just the power spectrum or the correlation function, these strictly second moment quantities are ambiguous when the mean of the process is not specified separately. To avoid this ambiguity it is frequently assumed that the mean of the random process has been removed so that any impulse occurring at zero frequency represents a discrete random component.

A typical power density spectrum (for a *complex* random process) is shown in Fig. 4.6. Note that the power density spectrum (like the Fourier transform of any discrete sequence) is periodic with period 2π. This follows directly from (4.38) since (from that definition)

$$S_x(e^{j(\omega+2\pi k)}) = S_x(e^{j\omega})$$

for any integer k.

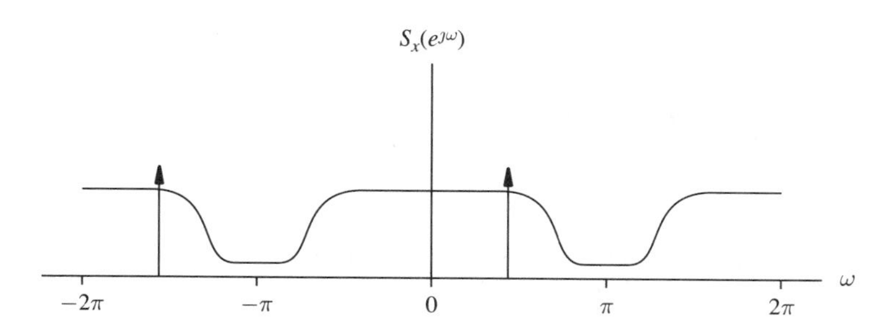

Figure 4.6 Typical power density spectrum for a complex random process, showing continuous and discrete components.

Some examples of correlation functions and their corresponding power density spectra are depicted in Fig. 4.7. When the correlation function is "broad," indicating small changes between samples of the random sequence, the power spectrum is narrow, indicating that there are only low-frequency components in the process. Conversely, when the correlation function is "narrow," indicating low correlation between samples of increasingly large separation, the power spectrum is broad. This shows that the sequence can have high-frequency components which permit it to change significantly (and irregularly) between successive samples. Finally, when the correlation function shows the rapidly varying characteristic corresponding to high negative correlation, the power spectrum contains *only* high frequencies, as shown in Fig. 4.7(c). In this case the corresponding sequence is virtually *guaranteed* to change rapidly between successive samples (compare Figs. 4.3 and 4.7).

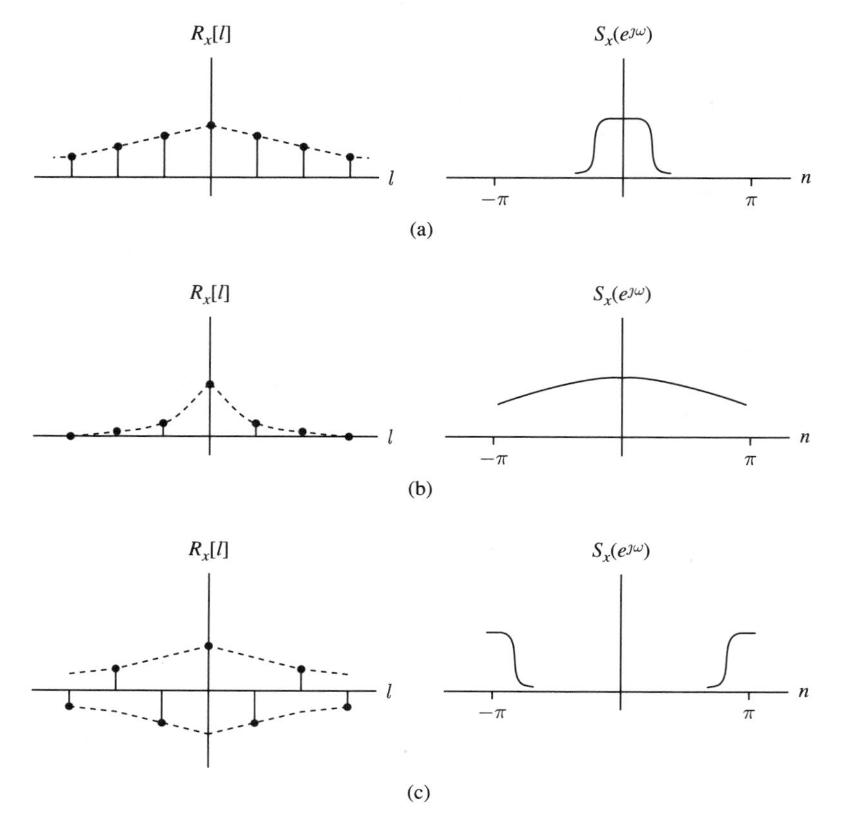

Figure 4.7 Comparison of correlation functions and power density spectra. (a) High correlation. (b) Low correlation. (c) High negative correlation.

The power density spectrum has two fundamental properties that derive from the two corresponding properties of the correlation function (see Table 4.1). These are

1. $S_x(e^{j\omega})$ is *real*.
2. $S_x(e^{j\omega})$ is *nonnegative*: $S_x(e^{j\omega}) \geq 0$

The first property corresponds to the fact that the $R_x[l]$ is conjugate symmetric and follows directly from a property of the Fourier transform; the second derives from the fact that $R_x[l]$ is positive semidefinite and will be proven shortly. For *real random processes* the power spectrum is also *even*, that is,

$$S_x(e^{j\omega}) = S_x(e^{-j\omega})$$

This follows from the fact that the autocorrelation function is real. The power spectral density function in Fig. 4.6 corresponds to a *complex* random process because the discrete component is not an even function of ω. Properties (1) and (2) show that the power density spectrum has truly zero phase. In most other contexts a "zero-phase" function usually means one whose phase can be either zero or π.

The nonnegative property of the power density spectrum can be shown as follows. Equation 4.16 can be written as

$$\sum_{n_1=-\infty}^{\infty} a^*[n_1] \sum_{n_0=-\infty}^{\infty} R_x[n_1 - n_0]a[n_0] = \sum_{n_1=-\infty}^{\infty} a^*[n_1]b[n_1] \geq 0 \tag{4.42}$$

where

$$b[n] = \sum_{n_0=-\infty}^{\infty} R_x[n - n_0]a[n_0] \tag{4.43}$$

Since the sequence $b[n]$ is recognized as the convolution of $R_x[n]$ with the sequence $a[n]$, it follows that

$$B(e^{\jmath\omega}) = S_x(e^{\jmath\omega})A(e^{\jmath\omega}) \tag{4.44}$$

where $B(e^{\jmath\omega})$ and $A(e^{\jmath\omega})$ are the Fourier transforms of $b[n]$ and $a[n]$ respectively. Now by using Parseval's theorem and applying (4.44) and (4.42) it follows that

$$\begin{aligned}\sum_{n_1=-\infty}^{\infty} a^*[n_1]b[n_1] &= \frac{1}{2\pi}\int_{-\pi}^{\pi} A^*(e^{\jmath\omega})B(e^{\jmath\omega})d\omega \\ &= \frac{1}{2\pi}\int_{-\pi}^{\pi} A^*(e^{\jmath\omega})S_x(e^{\jmath\omega})A(e^{\jmath\omega})d\omega \\ &= \frac{1}{2\pi}\int_{-\pi}^{\pi} S_x(e^{\jmath\omega})|A(e^{\jmath\omega})|^2 d\omega \geq 0 \end{aligned} \tag{4.45}$$

Now suppose that $S_x(e^{\jmath\omega})$ had any point ω_o where it were negative. Since $a[n]$ can be chosen arbitrarily, let us choose $a[n] = e^{\jmath\omega_o n}$ so that $A(e^{\jmath\omega})$ becomes an impulse in frequency at $\omega = \omega_o$. The result of (4.45) is then an impulse with negative area

$$2\pi S_x(e^{\jmath\omega_o})$$

which contradicts the inequality.[7] Since the inequality *must* hold, it follows that $S_x(e^{\jmath\omega})$ must be greater than or equal to zero for all values of ω. From the same argument it follows that if R_x is (strictly) *positive definite*, then $S_x(e^{\jmath\omega})$ must be a *strictly positive* function of ω.

A simple correlation function that occurs frequently in the remainder of the book is the *exponential correlation function* defined by

$$\boxed{R_x[l] = \begin{cases} \sigma^2\rho^l & l \geq 0 \\ \sigma^2(\rho^*)^{-l} & l < 0 \end{cases}} \tag{4.46}$$

[7]The argument here is mathematically "loose" but can be made rigorous through appropriate limiting arguments or via the theory of generalized functions [1], [2, App. I].

For the real case it is given by the more compact formula

$$R_x[l] = \sigma^2 \rho^{|l|} \qquad -\infty < l < \infty \tag{4.47}$$

The power spectral density function corresponding to (4.46) or (4.47) is

$$\boxed{S_x(e^{\jmath\omega}) = \frac{\sigma^2(1-|\rho|^2)}{1+|\rho|^2 - 2|\rho|\cos(\omega - \angle\rho)}} \tag{4.48}$$

where $\angle\rho$ is the angle of the complex parameter ρ (see Problem 4.10). For real vaues of ρ this becomes

$$S_x(e^{\jmath\omega}) = \frac{\sigma^2(1-\rho^2)}{1+\rho^2 - 2\rho\cos\omega}$$

Note that this function is strictly positive as long as the magnitude of ρ is less than 1. The exponential correlation function and the corresponding power spectral density function are depicted in Fig. 4.8 for a positive real parameter ρ. The spectrum has a low-pass character with bandwidth that *decreases* for increasing values of ρ. For $\rho < 0$ the functions appear more like those in Fig. 4.7(c); the correlation function has the same envelope but the samples alternate between positive and negative values, and the spectrum has a high-pass character.

Cross-Power Density Spectrum. It is also useful to give a frequency-domain description of the statistical relations that exist between *two* random processes. The *cross-power spectral density function* is defined as the Fourier transform of the cross-correlation function:

$$\boxed{S_{xy}(e^{\jmath\omega}) = \sum_{l=-\infty}^{\infty} R_{xy}[l]e^{-\jmath\omega l}} \tag{4.49}$$

Its value is called the *cross-power spectrum* (or simply the *cross-spectrum*) for the two random processes x and y.

The cross-spectrum does *not* have zero phase but is complex in general. It is also periodic with period 2π and satisfies the property

$$\boxed{S_{xy}(e^{\jmath\omega}) = S_{yx}^*(e^{\jmath\omega})} \tag{4.50}$$

(see Problem 4.21). This implies that S_{xy} and S_{yx} have the same magnitude and opposite phase.

For real random processes the cross-correlation function R_{xy} is real. Therefore, its Fourier transform satisfies the condition

$$S_{xy}(e^{\jmath\omega}) = S_{xy}^*(e^{-\jmath\omega})$$

Thus, *for real random processes*, the magnitude of S_{xy} is even and the phase is odd.

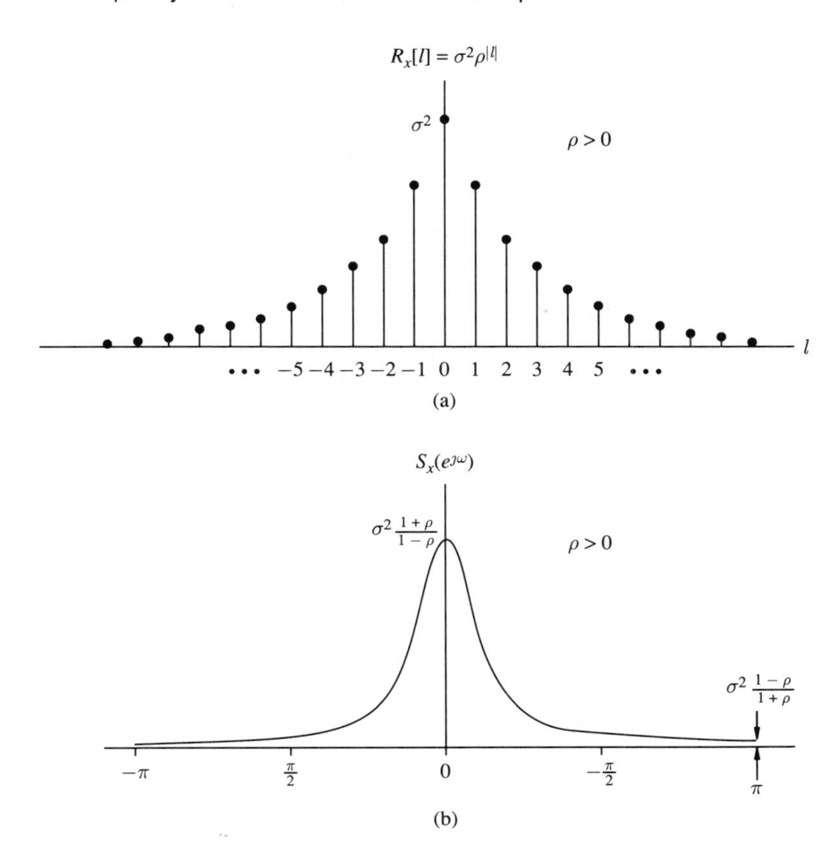

Figure 4.8 Real exponential correlation function and corresponding power spectral density ($\rho > 0$). (a) Correlation function. (b) Power spectral density function.

The cross-correlation function can be obtained from the cross-power density spectrum through the inverse relation

$$\boxed{R_{xy}[l] = \frac{1}{2\pi}\int_{-\pi}^{\pi} S_{xy}(e^{j\omega})e^{j\omega l}d\omega} \tag{4.51}$$

Since the integral is a kind of summation, an interpretation that can be given to the cross-spectrum is that $S_{xy}(e^{j\omega_o})$ measures the correlation between the two random processes *at a given frequency* ω_o.

The normalized cross-spectrum

$$\Gamma_{xy}(e^{j\omega}) \stackrel{\text{def}}{=} \frac{S_{xy}(e^{j\omega})}{\sqrt{S_x(e^{j\omega})}\sqrt{S_y(e^{j\omega})}} \tag{4.52}$$

is called the *coherence function.* Its squared magnitude

$$|\Gamma_{xy}(e^{\jmath\omega})|^2 = \frac{|S_{xy}(e^{\jmath\omega})|^2}{S_x(e^{\jmath\omega})S_y(e^{\jmath\omega})} \tag{4.53}$$

is called the *magnitude squared coherence (MSC)* and is often used instead of $|S_{xy}(e^{\jmath\omega})|$. The MSC has the important property that

$$0 \leq |\Gamma_{xy}(e^{\jmath\omega})|^2 \leq 1 \tag{4.54}$$

(see Problem 4.20).

4.4.2 Complex Spectral Density Functions

The analysis of stationary random processes through the z-transform is also extremely important. The second moment quantities that describe a random process in the transform domain are known as the complex spectral density function and the complex cross-spectral density function. The power spectral density function and cross-power spectral density function discussed in the preceding subsection can be regarded as special cases of the complex spectral density functions when the latter are evaluated on the unit circle.

Complex Spectral Density. The z-transform of the correlation function

$$\boxed{S_x(z) = \sum_{l=-\infty}^{\infty} R_x[l]z^{-l}} \tag{4.55}$$

is known as the complex spectral density function or simply the complex spectral density. Since the power density spectrum is simply the complex spectral density function evaluated on the unit circle, this explains the notation $S_x(e^{\jmath\omega})$.

The inverse of the complex spectral density is given by the contour integral

$$\boxed{R_x[l] = \frac{1}{2\pi j}\oint_C S_x(z)z^{l-1}dz} \tag{4.56}$$

where the contour of integration C is taken to be counterclockwise and in the region of convergence.[8] From the conjugate symmetry of the correlation function and the definition (4.55) it follows that

$$\boxed{S_x(z) = S_x^*(1/z^*)} \tag{4.57}$$

(see Problem 4.15). For the case when R_x is *real*, this specializes to

$$S_x(z) = S_x(z^{-1}) \qquad \text{(real random process)} \tag{4.58}$$

[8]If you are not familiar with the bilateral z-transform and its inversion through contour integration, you may want to review these topics in a suitable basic text in signal processing such as [3, Chap. 4].

The possible existence of lines in the power spectral density function [see (4.41)] poses some mathematical problems in defining the complex spectral density function. [In fact, it also poses difficulty in defining $S_x(e^{\jmath\omega})$, which is a special case of $S_x(z)$.] The trouble is that terms of the form $e^{\jmath\omega l}$ in the correlation function do not have *any* region of convergence; thus strictly speaking, neither their z-transform nor their Fourier transform exists. In the case of the power spectral density these terms are included formally as impulses in frequency, since this gives the correct inverse transform. We will do the same thing here and write the complex spectral density function as

$$S_x(z) = S'_x(z) + \sum_i 2\pi \mathsf{P}_i \delta_c(z - e^{\jmath\omega_i}) \tag{4.59}$$

where $S'_x(z)$ corresponds to the continuous part of the spectrum. This provides consistent notation with (4.41) and permits dealing with periodic and almost periodic processes. Mathematically, the definition (4.55) and its inverse through the contour integral (4.56) can only be used for the continuous part of the complex spectral density function.

The continuous part of the complex spectral density function is most frequently a rational polynomial. As a result of (4.57), factors of the numerator or denominator of $S'_x(z)$ must occur in pairs of the form

$$(1 - z_o z^{-1})(1 - z_o^* z)$$

This implies that for every pole or zero that occurs at

$$z_o = |z_o| e^{\jmath\phi_o}$$

there is a corresponding pole or zero that occurs at the point

$$1/z_o^* = \frac{1}{|z_o| e^{-\jmath\phi_o}}$$

In other words, the *poles and zeros of the complex spectral density occur at conjugate reciprocal locations.* Further, for real polynomials in z, the roots must occur in conjugate pairs. This implies that for real random processes the poles and zeros of the complex spectral density function not on the real axis occur in groups of four at locations

$$z_o, \quad 1/z_o, \quad z_o^*, \quad 1/z_o^*$$

For either real or complex random processes roots on the real axis occur in pairs at locations z_o and $1/z_o$.

When *zeros* occur *on the unit circle*, they *must occur in even multiplicities.* Otherwise, the power density spectrum $S(e^{\jmath\omega})$ could be negative or complex. Poles on the unit circle are not allowed to exist at all, for reasons that will become clear shortly. The types of terms in the correlation function that such poles would logically produce are accounted for by the sum of impulses in (4.59) (i.e., the *discrete* part of the function). Figure 4.9 shows some possible locations of the poles and zeros of a complex spectral density function.

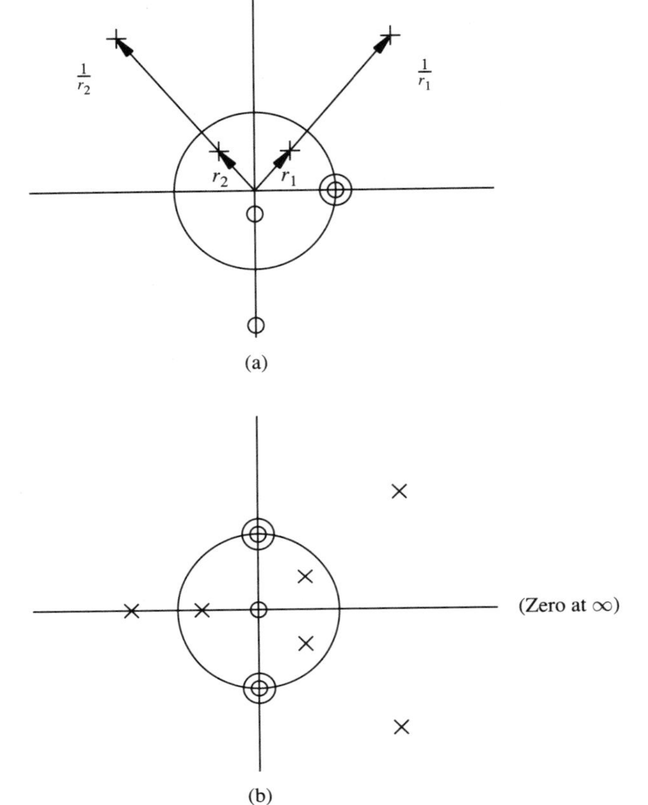

Figure 4.9 Possible pole and zero locations for a complex spectral density function. (a) Complex random process. (b) Real random process.

The symmetry properties (4.57) or (4.58) of the complex spectral density function imply specific things about the region of convergence. In particular, these properties imply that if $S_x(z)$ converges only for values in a region

$$|z| > r_R$$

where r_R is a positive number, then it also converges only for values of z such that

$$|z| < \frac{1}{r_R}$$

Thus the complex spectral density function of a random process always converges in an annular region of the form

$$r_R < |z| < \frac{1}{r_R} \tag{4.60}$$

where r_R is a positive number less than 1. Since poles can never occur in the region of convergence, it is clear why *poles <u>on</u> the unit circle are not allowed.* This form of region of convergence may be less familiar if you are used to dealing primarily with causal sequences; it requires special care when inverting the z-transform. An example illustrates region of convergence for a simple complex spectral density function and its inversion through contour integration.

Example 4.5

A real random process has the exponential correlation function

$$R_x(l) = \sigma^2 \rho^{|l|}$$

The complex spectral density function for the process can be computed by

$$S_x(z) = \sum_{l=-\infty}^{\infty} \sigma^2 \rho^{|l|} z^{-l} = \sum_{l=0}^{\infty} \sigma^2 \rho^l z^{-l} + \sum_{l=-\infty}^{-1} \sigma^2 \rho^{-l} z^{-l}$$

The first term can be put in closed form by using the formula for an infinite geometric series:

$$\sum_{l=0}^{\infty} \sigma^2 \rho^l z^{-l} = \sigma^2 \sum_{l=0}^{\infty} (\rho z^{-1})^l = \frac{\sigma^2}{1-\rho z^{-1}}; \qquad |z| > |\rho|$$

(The condition for convergence of the series is $|\rho z^{-1}| < 1$ or $|z| > |\rho|$.) The second term can be written as

$$\sum_{k=1}^{\infty} \sigma^2 \rho^k z^k = \sigma^2 \rho z \sum_{l=0}^{\infty} (\rho z)^l = \frac{\sigma^2 \rho z}{1-\rho z}; \qquad |z| < 1/|\rho|$$

Therefore, the complex power density spectrum is

$$\begin{aligned} S_x(z) &= \frac{\sigma^2}{1-\rho z^{-1}} + \frac{\sigma^2 \rho z}{1-\rho z} \\ &= \frac{\sigma^2(1-\rho^2)}{(1-\rho z^{-1})(1-\rho z)}; \qquad |\rho| < |z| < 1/|\rho| \end{aligned}$$

The poles of this function and the region of convergence are shown in Fig. EX4.5a. Clearly this z-transform exists only if $|\rho| < 1$.

The correlation function is recovered by integrating (4.56) over a contour in the region of convergence such as the one shown. The integral is evaluated using residues according to the formula

$$\begin{aligned} R_x(l) = \frac{1}{2\pi j} \oint S_x(z) z^{l-1} dz &= \sum \text{residues } \left[S_x(z) z^{l-1}\right] \\ &= \sum \text{residues } \left[\frac{\sigma^2(1-\rho^2) z^l}{(z-\rho)(1-\rho z)}\right] \end{aligned}$$

Note that for $l \geq 0$ there are no poles at the origin and the only pole enclosed by the contour is the one at $z = \rho$. Therefore, the inverse transform for $l \geq 0$ is

$$\begin{aligned} R_x(l) &= \text{Res}\left[\frac{\sigma^2(1-\rho^2) z^l}{(z-\rho)(1-\rho z)} \text{ at } z = \rho\right] \\ &= \left.\frac{\sigma^2(1-\rho^2) z^l}{(z-\rho)(1-\rho z)} \cdot (z-\rho)\right|_{z=\rho} = \sigma^2 \rho^l \end{aligned}$$

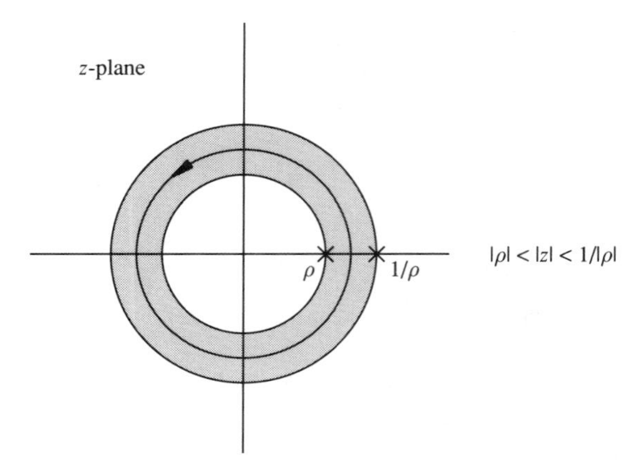

Figure EX4.5a

Since a real correlation function is known to be an even function of l, there is no actual need to carry out the inversion for $l < 0$. However, to show how we could proceed, let us make the transformation of variables $z = 1/w$ and write the inversion formula as

$$R_x(l) = \frac{1}{2\pi \jmath} \oint S_x(w^{-1}) w^{-l-1} dw$$

This avoids the difficulty of a high order pole depending on lag l that exists if we use the inversion formula in z. The function

$$S_x(w^{-1}) w^{-l-1} = \frac{\sigma^2(1-\rho^2) w^{-l}}{(w-\rho)(1-\rho w)}$$

has poles only at $w = \rho$ and $w = 1/\rho$ for $l < 0$ and converges in the annular region in between as shown in Fig. EX4.5b.

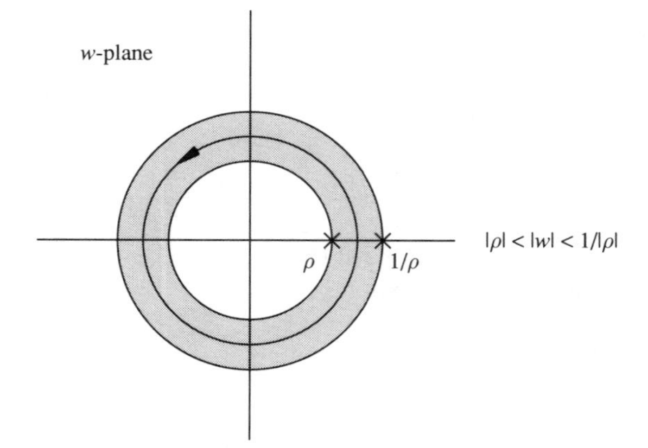

Figure EX4.5b

The integration in the w plane is thus similar to the previous integral in the z plane and yields

$$R_x(l) = \text{Res}\left[\frac{\sigma^2(1-\rho^2)w^{-l}}{(w-\rho)(1-\rho w)} \text{ at } w=\rho\right]$$

$$= \left.\frac{\sigma^2(1-\rho^2)w^{-l}}{(w-\rho)(1-\rho w)}\cdot(w-\rho)\right|_{w=\rho} = \sigma^2\rho^{-l}$$

for $l < 0$. The complete autocorrelation function is then

$$R_x(l) = \sigma^2\rho^{|l|} \qquad -\infty < l < \infty$$

Although contour integration is often a convenient way to evaluate the inverse z-transform of a complex spectral density function, it is not the only possible procedure. The correlation function can also be recovered by using the method of partial fraction expansion. Even if partial fraction expansion is used, however, region of convergence plays a vital role in determining how to associate the terms with "causal" and "anticausal" parts of the correlation function. A numerical example illustrates this best.

Example 4.6

Suppose that the value of ρ in Example 4.5 is 0.8 and the value of σ^2 is 2. Then the complex spectral density function is

$$S_x(z) = \frac{0.72}{(1-0.8z^{-1})(1-0.8z)} = \frac{-0.90z^{-1}}{(1-0.8z^{-1})(1-1.25z^{-1})}$$

The function can be expanded by partial fractions to obtain[9]

$$S_x(z) = \frac{2}{1-0.8z^{-1}} + \frac{-2}{1-1.25z^{-1}}$$

Now by noting the poles and the region of convergence (see Example 4.5) it can be seen that the first term corresponds to a *"causal"* sequence:

$$\frac{2}{1-0.8z^{-1}} \Leftrightarrow 2\cdot(0.8)^l; \qquad l \geq 0$$

The second term corresponds to an *"anticausal"* sequence:[10]

$$\frac{-2}{1-1.25z^{-1}} \Leftrightarrow 2\cdot(1.25)^l = 2\cdot(0.8)^{-l}; \qquad l < 0$$

Putting these results together yields

$$R_x[l] = 2\cdot(0.8)^{|l|}; \qquad -\infty < l < \infty$$

[9]The partial fraction expansion can also be written as a sum of terms of the form $\frac{z}{z-a}$. We follow the notation of Oppenheim and Schafer [3, 4], which is more common in digital signal processing.

[10]Note that a term of the form $\frac{1}{1-az^{-1}} = \frac{z}{z-a}$ may correspond to either a sequence a^l for $l \geq 0$ *or* a sequence $-a^l$ for $l < 0$, depending on whether the region of convergence is $|z| > |a|$ or $|z| < |a|$ respectively.

These last two examples deal with the real form of the exponential correlation function introduced earlier, and its z-transform. The exponential correlation function and its complex spectral density function form an important z-transform pair. In the next chapter it is shown that this particular type of correlation function describes the random process that results when a first-order (single-pole) linear system is driven by white noise. For convenience and later reference we list the general form of the exponential correlation function and its transform in Table 4.2. Some various equivalent forms for the complex spectral density function are given in the box below. In general, we prefer forms such as (4.61)(a) and (4.61)(b) involving mixed powers of z and z^{-1} in representing the complex spectral density function, since they clearly display the necessary symmetry.

TABLE 4.2 EXPONENTIAL CORRELATION FUNCTION AND ITS COMPLEX SPECTRAL DENSITY FUNCTION

Correlation function	Complex spectral density
$R_x[l] = \begin{cases} \sigma^2\rho^l & l \geq 0 \\ \sigma^2(\rho^*)^{-l} & l < 0 \end{cases}$	$S_x(z) = \dfrac{\sigma^2(1-\lvert\rho\rvert^2)}{(1-\rho z^{-1})(1-\rho^* z)}$

Note: See (4.48) on p. 160 for $S_x(e^{j\omega})$.

$$
\begin{aligned}
S_x(z) &= \frac{\sigma^2(1-|\rho|^2)}{(1-\rho z^{-1})(1-\rho^* z)} && (a) \\
S_x(z) &= \frac{\sigma^2(1-|\rho|^2)}{-\rho^* z + (1+|\rho|^2) - \rho z^{-1}} && (b) \\
S_x(z) &= \frac{\sigma^2(\rho - 1/\rho^*)z^{-1}}{1-(\rho+1/\rho^*)z^{-1} + (\rho/\rho^*)z^{-2}} && (c) \\
S_x(z) &= \frac{\sigma^2(\rho - 1/\rho^*)z}{z^2-(\rho+1/\rho^*)z + (\rho/\rho^*)} && (d)
\end{aligned}
\tag{4.61}
$$

Complex Cross-Spectral Density. The complex cross-spectral density, which is the z-transform of the cross-correlation function, is also needed on a number of occasions. The complex cross-spectral density function is defined by

$$
S_{xy}(z) = \sum_{l=-\infty}^{\infty} R_{xy}[l]z^{-l} \tag{4.62}
$$

Its inverse is given by the contour integral

$$\boxed{R_{xy}[l] = \frac{1}{2\pi j}\oint_C S_{xy}(z)z^{l-1}dz} \tag{4.63}$$

where the contour C is taken in the region of convergence. Evaluation of $S_{xy}(z)$ on the unit circle yields the cross-power density spectrum discussed in Section 4.4.1. Since the cross-correlation function has no special symmetry, the poles and zeros of the complex cross-spectral density function can occur anywhere in the z plane. For real R_{xy}, however (i.e., real processes x and y), the singularities must occur in complex conjugate pairs.

Although the complex cross-spectral density function *itself* does not have any special symmetry properties, the function S_{xy} does share a relation with S_{yx}. Specifically, it follows from the definition (4.62) and the relation (4.34) that

$$\boxed{S_{xy}(z) = S^*_{yx}(1/z^*)} \tag{4.64}$$

(see Problem 4.21). For the case where both x and y are *real*, this property reduces to

$$S_{xy}(z) = S_{yx}(z^{-1}) \tag{4.65}$$

4.5 SYMMETRY PROPERTIES OF CORRELATION AND SPECTRA FOR COMPLEX RANDOM PROCESSES

Let us summarize the properties and relations of the second moment quantities discussed so far in this chapter. Those related to the autocorrelation function are given in Table 4.3; those relating to the cross-correlation are given in Table 4.4. As mentioned before, the quantities defined in Table 4.3 provide all of the corresponding functions for the real and imaginary parts of the random process when certain conditions are met. These conditions are investigated here. Similar conditions, not specifically developed here, permit computation of all of the second moment functions of real and imaginary parts of *two* random processes from the quantities defined in Table 4.4 (i.e., those functions relating to cross-correlation).

TABLE 4.3 FUNCTIONS PERTAINING TO AUTOCORRELATION AND PROPERTIES

Function and definition	Properties
$R_x[l] = \mathcal{E}\{x[n]x^*[n-l]\}$	$R_x[l] = R^*_x[-l]$
	$\sum_{n_1=-\infty}^{\infty}\sum_{n_0=-\infty}^{\infty} a^*[n_1]R_x[n_1-n_0]a[n_0] \geq 0$
$S_x(e^{\jmath\omega}) = \sum_{l=-\infty}^{\infty} R_x[l]e^{-\jmath\omega l}$	$S_x(e^{\jmath\omega})$ is real.
	$S_x(e^{\jmath\omega}) \geq 0$
$S_x(z) = \sum_{l=-\infty}^{\infty} R_x[l]z^{-l}$	$S_x(z) = S^*_x(1/z^*)$

TABLE 4.4 FUNCTIONS PERTAINING TO CROSS-CORRELATION AND RELATIONS

Function and definition	Relations
$R_{xy}[l] = \mathcal{E}\{x[n]y^*[n-l]\}$	$R_{xy}[l] = R^*_{yx}[-l]$
$S_{xy}(e^{\jmath\omega}) = \sum_{l=-\infty}^{\infty} R_{xy}[l]e^{-\jmath\omega l}$	$S_{xy}(e^{\jmath\omega}) = S^*_{yx}(e^{\jmath\omega})$
$S_{xy}(z) = \sum_{l=-\infty}^{\infty} R_{xy}[l]z^{-l}$	$S_{xy}(z) = S^*_{yx}(1/z^*)$

To begin the discussion, let us write a complex random process in terms of its real and imaginary parts as

$$x[n] = x_{\rm r}[n] + \jmath x_{\rm i}[n] \tag{4.66}$$

When $x[n]$ is the sampled complex signal corresponding to a bandpass random process, then $x_{\rm r}$ and $x_{\rm i}$ are proportional to the real-valued quadrature components of the sampled bandpass signal (see Appendix B). The usual assumption about sequences derived from such bandpass random signals is that the samples at any points n_1 and n_0 satisfy the condition

$$\boxed{\mathcal{E}\{x[n_1]x[n_0]\} = 0 \qquad \text{(for } x[n] \text{ complex)}} \tag{4.67}$$

Note that (4.67) is written *without* the conjugate on $x[n_0]$. As shown in Appendix B, this condition is necessary and sufficient for the stationarity of bandpass random signals. For nonstationary bandpass signals it is an assumption that frequently holds if the second moment statistics of the random process are slowly varying (with respect to the carrier). Since (4.67) is a necessary and sufficient condition for the definitions (4.2) and (4.3) to provide a complete second moment description of the random process, (4.67) *will be required to hold for all complex random processes considered from here on.* Condition (4.67) is equivalent to requiring a particular kind of statistical symmetry between the real and imaginary parts of the complex process. With this symmetry, (4.2) or (4.9) provide all the needed second moment information for a complex random process.

To derive the symmetry conditions first substitute (4.66) in (4.67) to obtain

$$\begin{aligned}
&\mathcal{E}\{(x_{\rm r}[n_1] + \jmath x_{\rm i}[n_1])(x_{\rm r}[n_0] + \jmath x_{\rm i}[n_0])\} \\
&\quad = \left(\mathcal{E}\{x_{\rm r}[n_1]x_{\rm r}[n_0]\} - \mathcal{E}\{x_{\rm i}[n_1]x_{\rm i}[n_0]\}\right) \\
&\qquad + \jmath\left(\mathcal{E}\{x_{\rm i}[n_1]x_{\rm r}[n_0]\} + \mathcal{E}\{x_{\rm r}[n_1]x_{\rm i}[n_0]\}\right) \\
&\quad = \left(\mathrm{R}_{x_{\rm r}}[n_1,n_0] - \mathrm{R}_{x_{\rm i}}[n_1,n_0]\right) + \jmath\left(\mathrm{R}_{x_{\rm i}x_{\rm r}}[n_1,n_0] + \mathrm{R}_{x_{\rm r}x_{\rm i}}[n_1,n_0]\right) = 0
\end{aligned}$$

Since each term in parentheses must be zero, this implies that

$$\begin{aligned}
\mathrm{R}_{x_{\rm r}}[n_1,n_0] &= \mathrm{R}_{x_{\rm i}}[n_1,n_0] \; (a) \\
\mathrm{R}_{x_{\rm i}x_{\rm r}}[n_1,n_0] &= -\mathrm{R}_{x_{\rm r}x_{\rm i}}[n_1,n_0] \; (b)
\end{aligned} \tag{4.68}$$

The symmetry conditions (4.68) can be interpreted as follows. Suppose that $x[n]$ represents samples of the modulation of some bandpass random process with carrier frequency f_c. The bandpass process is then proportional to the real part of a complex signal

$$v_c(t) = x_c(t)e^{j2\pi f_c t}$$

where $x_c(t)$ is the modulation process and

$$x[n] = x_c(nT)$$

where T is the sampling interval. If $v_c(t)$ is represented by a vector (phasor) in the complex plane, then as t increases, $v_c(t)$ rotates about the origin (see Fig. 4.10). Now assume that at some particular sampling time the phase of the carrier is zero so that $v_c(nT) = x_c(nT)$. Thus $x_{\rm r}[n]$ lies along the positive real axis and $x_{\rm i}[n]$ lies along the positive imaginary axis as shown in Fig. 4.10(a). The real part of v_c is thus associated with $x_{\rm r}$ and the imaginary part is associated with $x_{\rm i}$. Now assume a short time later the phase of the carrier has changed to $\pi/2$ so that v_c has the orientation shown in Fig. 4.10(b). Thus the real and imaginary parts of v_c have changed roles; if the process is sampled at this time the *real* part of v_c is associated with $-x_{\rm i}$ and the *imaginary* part is associated with $x_{\rm r}$. It is clear therefore that if the statistical properties of $v_c(t)$ are to be time-invariant (or at least *slowing* changing), then $x_{\rm r}$ and $x_{\rm i}$ must have identical autocorrelation functions and the cross-correlation functions must be related as in (4.68)(b). These are the symmetry conditions that are generally assumed.

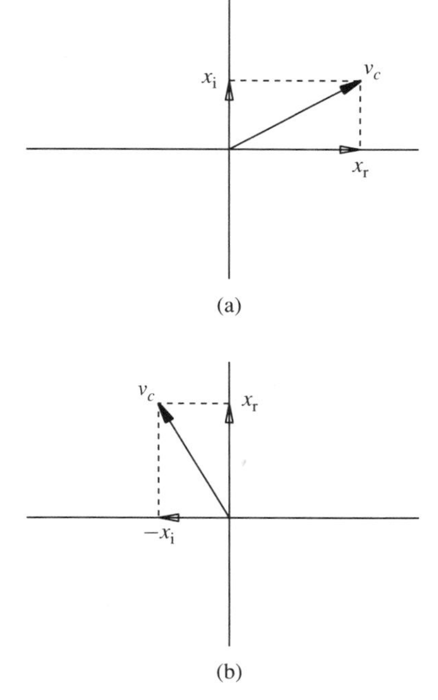

Figure 4.10 Rotating vector (phasor) in the complex plane representing a bandpass random process. (a) Carrier phase zero. (c) Carrier phase $\pi/2$.

Let us now see how these conditions relate to the correlation function for the complex process $x[n]$. Substituting (4.66) in the definition (4.2) yields

$$\begin{aligned}
\mathrm{R}_x[n_1, n_0] &= \mathcal{E}\{(x_\mathrm{r}[n_1] + \jmath x_\mathrm{i}[n_1])(x_\mathrm{r}[n_0] - \jmath x_\mathrm{i}[n_0])\} \\
&= \left(\mathcal{E}\{x_\mathrm{r}[n_1]x_\mathrm{r}[n_0]\} + \mathcal{E}\{x_\mathrm{i}[n_1]x_\mathrm{i}[n_0]\}\right) \\
&\quad + \jmath\left(\mathcal{E}\{x_\mathrm{i}[n_1]x_\mathrm{r}[n_0]\} - \mathcal{E}\{x_\mathrm{r}[n_1]x_\mathrm{i}[n_0]\}\right) \\
&= \left(\mathrm{R}_{x_\mathrm{r}}[n_1, n_0] + \mathrm{R}_{x_\mathrm{i}}[n_1, n_0]\right) + \jmath\left(\mathrm{R}_{x_\mathrm{i}x_\mathrm{r}}[n_1, n_0] - \mathrm{R}_{x_\mathrm{r}x_\mathrm{i}}[n_1, n_0]\right)
\end{aligned}$$

Then in view of (4.68) this can be written as

$$\mathrm{R}_x[n_1, n_0] = 2\mathrm{R}_x^\mathrm{E}[n_1, n_0] + \jmath 2\mathrm{R}_x^\mathrm{O}[n_1, n_0] \tag{4.69}$$

where

$$\begin{aligned}
\mathrm{R}_x^\mathrm{E}[n_1, n_0] &\stackrel{\text{def}}{=} \mathrm{R}_{x_\mathrm{r}}[n_1, n_0] = \mathrm{R}_{x_\mathrm{i}}[n_1, n_0] \ (a) \\
\mathrm{R}_x^\mathrm{O}[n_1, n_0] &\stackrel{\text{def}}{=} -\mathrm{R}_{x_\mathrm{r}x_\mathrm{i}}[n_1, n_0] = \mathrm{R}_{x_\mathrm{i}x_\mathrm{r}}[n_1, n_0] \ (b)
\end{aligned} \tag{4.70}$$

Thus it follows from (4.69) and (4.70) that

$$\boxed{\begin{aligned}
\mathrm{R}_{x_\mathrm{r}}[n_1, n_0] = \mathrm{R}_{x_\mathrm{i}}[n_1, n_0] &= \tfrac{1}{2}\mathrm{Re}[\mathrm{R}_x[n_1, n_0]] \quad (a) \\
\mathrm{R}_{x_\mathrm{r}x_\mathrm{i}}[n_1, n_0] = -\mathrm{R}_{x_\mathrm{i}x_\mathrm{r}}[n_1, n_0] &= -\tfrac{1}{2}\mathrm{Im}[\mathrm{R}_x[n_1, n_0]] \quad (b)
\end{aligned}} \tag{4.71}$$

In other words, all autocorrelation functions and cross-correlation functions for the real and imaginary parts of $x[n]$ can be derived from the (complex) autocorrelation function $\mathrm{R}_x[n_1, n_0]$.

When the complex random process is truly wide-sense stationary (4.69) and (4.70) can be written as

$$R_x[l] = 2R_x^\mathrm{E}[l] + \jmath 2R_x^\mathrm{O}[l] \tag{4.72}$$

where

$$\begin{aligned}
R_x^\mathrm{E}[l] &\stackrel{\text{def}}{=} R_{x_\mathrm{r}}[l] = R_{x_\mathrm{i}}[l] \quad (a) \\
R_x^\mathrm{O}[l] &\stackrel{\text{def}}{=} -R_{x_\mathrm{r}x_\mathrm{i}}[l] = R_{x_\mathrm{i}x_\mathrm{r}}[l] \quad (b)
\end{aligned} \tag{4.73}$$

From the conjugate symmetric property (4.13) and (4.72) it follows that $R_x^\mathrm{E}[l]$ is an *even* function and $R_x^\mathrm{O}[l]$ is an *odd* function of l. This is the reason for the superscripts "E" and "O." From (4.72) and (4.73) it then follows that

$$\boxed{\begin{aligned}
R_{x_\mathrm{r}}[l] = R_{x_\mathrm{i}}[l] &= \tfrac{1}{2}\mathrm{Re}[R_x[l]] \quad (a) \\
R_{x_\mathrm{r}x_\mathrm{i}}[l] = -R_{x_\mathrm{i}x_\mathrm{r}}[l] &= -\tfrac{1}{2}\mathrm{Im}[R_x[l]] \quad (b)
\end{aligned}} \tag{4.74}$$

TABLE 4.5 RELATIONS BETWEEN SECOND MOMENT QUANTITIES FOR A COMPLEX RANDOM PROCESS

Function pertaining to real and imaginary parts	Expression in terms of same function for the complex random process
$R_{x_r}[l],\ R_{x_i}[l]$	$\frac{1}{2}\mathrm{Re}[R_x[l]] = R_x^{\mathrm{E}}[l]$
$R_{x_r x_i}[l],\ -R_{x_i x_r}[l]$	$-\frac{1}{2}\mathrm{Im}[R_x[l]] = -R_x^{\mathrm{O}}[l]$
$S_{x_r}(e^{\jmath\omega}),\ S_{x_i}(e^{\jmath\omega})$	$\frac{1}{4}\left(S_x(e^{\jmath\omega}) + S_x(e^{-\jmath\omega})\right) = \frac{1}{2}\left[S_x(e^{\jmath\omega})\right]_{\mathrm{EVEN}}$
$S_{x_r x_i}(e^{\jmath\omega}),\ -S_{x_i x_r}(e^{\jmath\omega})$	$\jmath\frac{1}{4}\left(S_x(e^{\jmath\omega}) - S_x(e^{-\jmath\omega})\right) = \jmath\frac{1}{2}\left[S_x(e^{\jmath\omega})\right]_{\mathrm{ODD}}$
$S_{x_r}(z),\ S_{x_i}(z)$	$\frac{1}{4}\left(S_x(z) + S_x^*(z^*)\right) = \frac{1}{2}\left[S_x(z)\right]_{\mathrm{CONJ\ SYM}}$
$S_{x_r x_i}(z),\ -S_{x_i x_r}(z)$	$\jmath\frac{1}{4}\left(S_x(z) - S_x^*(z^*)\right) = \jmath\frac{1}{2}\left[S_x(z)\right]_{\mathrm{CONJ\ ANTISYM}}$

These relations and the corresponding relations for the spectra are listed in Table 4.5. Note that the spectra S_{x_r} and S_{x_i} are real (as they must be) and that the cross-spectra $S_{x_r x_i}$ and $S_{x_i x_r}$ are purely imaginary. The notation in the third and fourth lines of the table refers to the even and odd parts of the spectrum.

The complex spectral densities are of course neither purely real nor purely imaginary in general. However, they do exhibit certain symmetry properties, as the last two lines in Table 4.5 show. These last two items are proportional to the *conjugate symmetric part* and the *conjugate antisymmetric part* of the complex spectral density function respectively.

In summary, the considerations of this section show that the definitions (4.2) and (4.9) for the correlation function of a complex random process presume that there is a special kind of symmetry between the second moment statistics of the real and imaginary parts. This symmetry, which is implied by (4.67), is both necessary and sufficient for the stationarity of a bandpass random process; for a nonstationary process it is a condition that usually exists if the changes in the random process are slow with respect to the carrier frequency (i.e., if the process is "narrowband"). Under these circumstances all of the second moment functions (correlation and spectra) for the real and imaginary components can be derived from the corresponding second moment functions of the complex process.

4.6 THE DISCRETE KARHUNEN-LOÈVE TRANSFORM

Frequently in mathematics and engineering it is useful to expand a function as a linear combination of orthogonal basis functions. The orthogonality of the basis functions makes the representation efficient and mathematically convenient. In signal processing applications it also guarantees that the components of the signal with respect to the basis functions do not interfere with each other. For example, in communication problems the use of orthogonal functions permits several different messages to share the same transmission medium, such as a cable or fiber optic link.

The most familiar set of orthogonal basis functions in signal processing are those used in the Fourier transform or DFT. These are complex exponentials (i.e., terms of the form $e^{\jmath\omega n}$). Other orthogonal basis functions commonly appear in coding applications such as in the Hadamard and Harr transforms, slant transforms, and others.

In the analysis of random signals it is useful to have a set of basis functions for which the *components* of the signal are also orthogonal (in the statistical sense) or uncorrelated. It is shown in Section 4.7.2 that when a random process is stationary and the observation interval is long compared to the duration of the correlation function, the complex exponentials of the DFT provide such a basis. However, it is possible to find a potentially different transformation and associated set of basis functions that has uncorrelated or statistically orthogonal coefficients even if the observation interval is short and the random process is nonstationary. This transformation, which amounts to a generalization of the DFT for random processes, is called the discrete Karhunen-Loève transform (DKLT).[11] The procedure is related to the diagonalization of the correlation matrix by the unitary eigenvector transformation discussed in Chapter 2.

4.6.1 Development of the DKLT

To begin this topic consider a segment of a random sequence $\{x[n]; n = 0, 1, \ldots, N-1\}$. This segment can be expanded in any set of orthonormal basis functions $\varphi_i[n]$ as

$$x[n] = \kappa_1\varphi_1[n] + \kappa_2\varphi_2[n] + \cdots + \kappa_N\varphi_N[n] \tag{4.75}$$

where the κ_i are coefficients in the expansion and where by "orthonormal" it is meant that the functions satisfy the relation

$$\sum_{n=0}^{N-1} \varphi_i^*[n]\varphi_j[n] = \begin{cases} 1 & i = j \\ 0 & i \neq j \end{cases} \tag{4.76}$$

The procedure is illustrated in Fig. 4.11. It follows from (4.75) and (4.76) that the coefficients are given by

$$\boxed{\kappa_i = \sum_{n=0}^{N-1} \varphi_i^*[n]x[n]} \tag{4.77}$$

Now it is desired to find a *particular* orthonormal set of functions such that

$$\mathcal{E}\{\kappa_i\kappa_j^*\} = \begin{cases} \varsigma_j^2 & i = j \\ 0 & i \neq j \end{cases} \tag{4.78}$$

Therefore, if the random process has zero mean, which is the usual assumption, the coefficients are *uncorrelated.* Otherwise, they are statistically orthogonal [not to be confused with orthogonality of the basis functions as defined by (4.76)].

[11]The Karhunen-Loève transformation was originally developed for continuous random processes [5, 6] and is an extremely important tool for their analysis (see, e.g., [7, 8]).

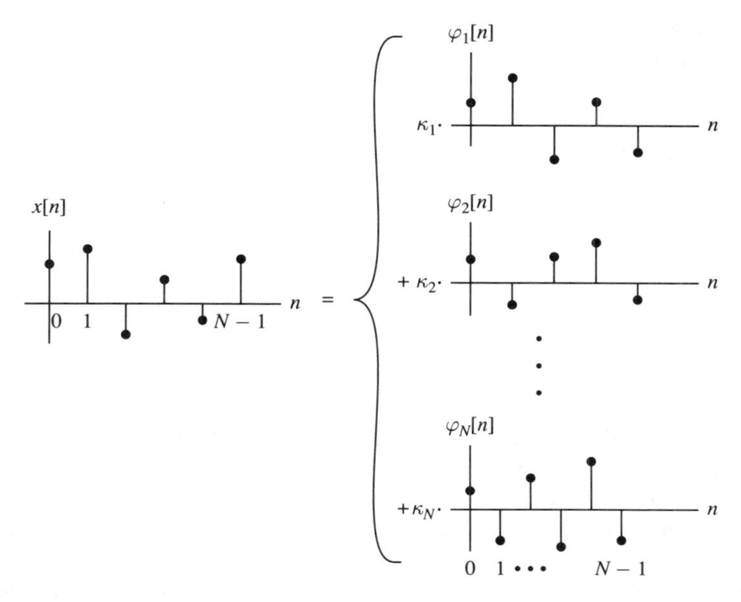

Figure 4.11 Expansion of a random sequence in a set of orthonormal basis functions.

The solution to this problem can be obtained deftly by restating it in terms of random vectors and applying the results from Chapter 2. Let $\boldsymbol{x}$ be the random vector defined by (4.21) and define the vector of coefficients

$$\boldsymbol{\kappa} = \begin{bmatrix} \kappa_1 \\ \kappa_2 \\ \vdots \\ \kappa_N \end{bmatrix} \tag{4.79}$$

and the matrix

$$\boldsymbol{\Phi} = \begin{bmatrix} | & | & & | \\ \boldsymbol{\varphi}_1 & \boldsymbol{\varphi}_2 & \cdots & \boldsymbol{\varphi}_N \\ | & | & & | \end{bmatrix}; \qquad \boldsymbol{\varphi}_i = \begin{bmatrix} \varphi_i[0] \\ \varphi_i[1] \\ \vdots \\ \varphi_i[N-1] \end{bmatrix}, \quad i = 1, 2, \ldots, N \tag{4.80}$$

Note from (4.80) and (4.76) that the columns of $\boldsymbol{\Phi}$ are a set of orthonormal vectors satisfying

$$\boldsymbol{\varphi}_i^{*T} \boldsymbol{\varphi}_j = \begin{cases} 1 & i = j \\ 0 & i \neq j \end{cases} \tag{4.81}$$

Therefore, $\mathbf{\Phi}$ is a *unitary* matrix. Equations 4.75 and 4.77 can now be written using this matrix formulation as

$$\boldsymbol{x} = \begin{bmatrix} | & | & & | \\ \boldsymbol{\varphi}_1 & \boldsymbol{\varphi}_2 & \cdots & \boldsymbol{\varphi}_N \\ | & | & & | \end{bmatrix} \begin{bmatrix} \kappa_1 \\ \kappa_2 \\ \vdots \\ \kappa_N \end{bmatrix} = \mathbf{\Phi}\boldsymbol{\kappa} \tag{4.82}$$

and

$$\boldsymbol{\kappa} = \begin{bmatrix} -- & \boldsymbol{\varphi}_1^{*T} & -- \\ -- & \boldsymbol{\varphi}_2^{*T} & -- \\ & \vdots & \\ -- & \boldsymbol{\varphi}_N^{*T} & -- \end{bmatrix} \boldsymbol{x} = \mathbf{\Phi}^{*T}\boldsymbol{x} \tag{4.83}$$

where since $\mathbf{\Phi}$ is unitary, $\mathbf{\Phi}^{*T}$ is the inverse of $\mathbf{\Phi}$.

These equations have the following interpretation. If the sequence $x[n]$ is thought of as a vector $\boldsymbol{x}$ in an N-dimensional space, then the κ_i can be regarded as components of the same vector with respect to a *rotated coordinate system*. Further, recall from Chapter 2 that if the $\boldsymbol{\varphi}_i$ are chosen as the *eigenvectors of the correlation matrix*, then the resulting κ_i satisfy (4.78). The desired set of basis functions are thus determined by the eigenvectors of the correlation matrix. These functions are called *eigenfunctions* of the random process and satisfy the equation

$$\sum_{k=0}^{N-1} R_x[l-k]\varphi_i[k] = \lambda_i \varphi_i[l] \qquad i = 1, 2, \cdots, N \tag{4.84}$$

This last equation, in fact, is just the component form of the matrix eigenvalue equation

$$\mathbf{R}_x \boldsymbol{\varphi}_i = \lambda_i \boldsymbol{\varphi}_i$$

With this choice of basis functions, the transformation (4.77) is the DKLT, and the representation (4.75) (which is the inverse transform) is called the *Karhunen-Loève expansion* for the random process [9].

The DKLT is also *unique* in that it is the *only* transformation that results in (4.78). To show this, consider *any* set of orthonormal basis functions (not necessarily those corresponding to the DKLT). Since the functions and the corresponding vectors are orthonormal, the requirement (4.78) can be written as

$$\mathcal{E}\{\kappa_i \kappa_j^*\} = \mathcal{E}\{\boldsymbol{\varphi}_i^{*T}\boldsymbol{x}\boldsymbol{x}^{*T}\boldsymbol{\varphi}_j\} = \boldsymbol{\varphi}_i^{*T}\mathbf{R}_x\boldsymbol{\varphi}_j = \begin{cases} \varsigma_j^2 & i = j \\ 0 & i \neq j \end{cases} \tag{4.85}$$

Now examine this condition more closely by writing it as

$$\boldsymbol{\varphi}_i^{*T}\mathbf{u}_j = \begin{cases} \varsigma_j^2 & i = j \\ 0 & i \neq j \end{cases} \tag{4.86}$$

where

$$\mathbf{u}_j = \mathbf{R}_x \boldsymbol{\varphi}_j$$

is whatever vector results from multiplying φ_j by the correlation matrix. Condition (4.86) states that $\mathbf{u}_j$ is orthogonal to all of the φ_i for $i \neq j$. But since the φ_i form an orthonormal set, $\mathbf{u}_j$ must be equal to a constant times φ_j, that is,

$$\mathbf{u}_j = \mathbf{R}_x \varphi_j = \lambda_j \, \varphi_j \tag{4.87}$$

This states that φ_j must be an eigenvector and λ_j must be the corresponding eigenvalue. Since this is true for any choice of j, it follows that *the unique set of* $\{\varphi_i\}$ *that result in* (4.78) *are the eigenvectors of the correlation matrix.* Then from (4.85) and (4.87) it also follows that

$$\varsigma_j^2 = \lambda_j$$

A plot of the eigenfunctions for a typical stationary random process is shown in Fig. 4.12. That these functions appear to be almost sinusoidal is not a coincidence but is related to the stationary property of the random process and the length of the observation interval as mentioned above. This point is developed further in Section 4.7.2.

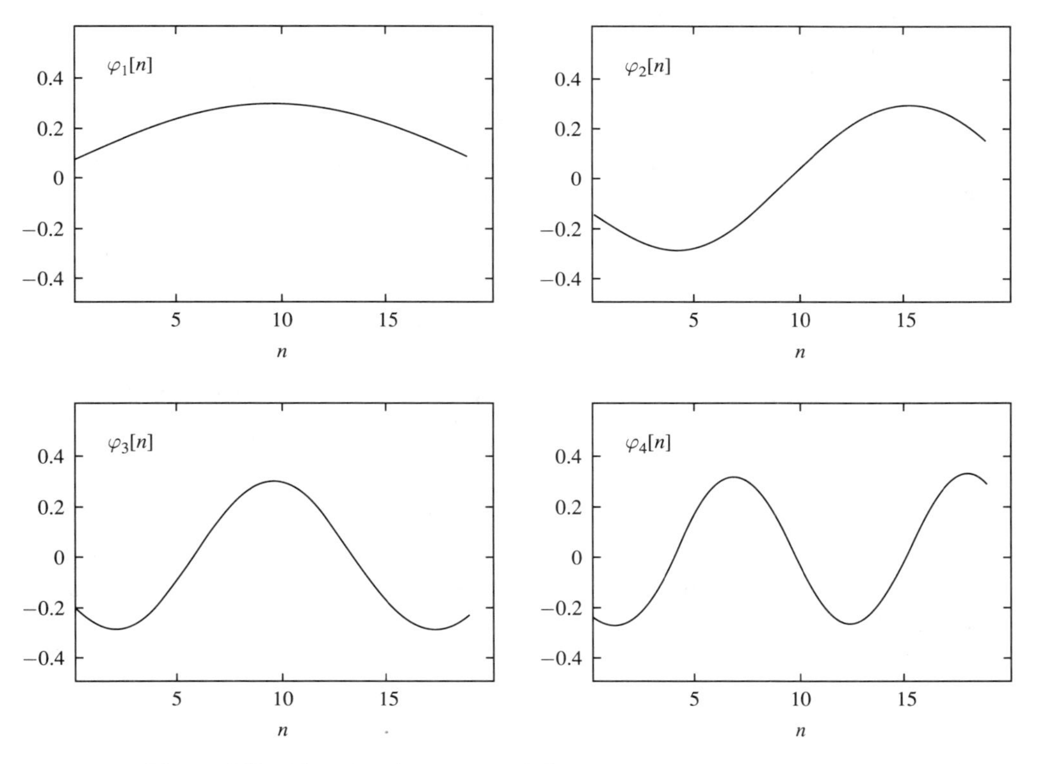

Figure 4.12 Eigenfunctions corresponding to a random process with correlation function $R_x[l] = (0.5)^{|l|}$.

4.6.2 Optimal Representation Property of the DKLT

A further property of the DKLT is that it is the most efficient representation of the random process if the expansion is truncated to use fewer than N orthonormal basis functions. One reason for wanting to truncate the expansion occurs if the random process $x[n]$ consists of a signal in additive noise. It can turn out that by using a truncated expansion, a significant part of the noise is eliminated while most of the signal is kept intact. This application is explored in more detail at the end of the section.

To develop the optimal representation property, consider approximating the random sequence in terms of some number $M < N$ of basis functions as

$$\hat{x}[n] = \sum_{i=1}^{M} \kappa_i \varphi_i[n]; \qquad M < N \tag{4.88}$$

and define the error sequence

$$\varepsilon[n] = x[n] - \hat{x}[n] \tag{4.89}$$

The problem is to find the appropriate set of basis functions to minimize the average energy in the error process

$$\mathcal{E} = \mathcal{E}\left\{ \sum_{n=0}^{N-1} \varepsilon^2[n] \right\} \tag{4.90}$$

This quantity is referred to as the *mean-square error*. To solve this problem, let us return to the vector representation of the process. From (4.82) we can write

$$\boldsymbol{x} = \sum_{i=1}^{N} \kappa_i \boldsymbol{\varphi}_i = \underbrace{\sum_{i=1}^{M} \kappa_i \boldsymbol{\varphi}_i}_{\hat{\boldsymbol{x}}} + \underbrace{\sum_{i=M+1}^{N} \kappa_i \boldsymbol{\varphi}_i}_{\boldsymbol{\varepsilon}}$$

where the vector quantities $\hat{\boldsymbol{x}}$ and $\boldsymbol{\varepsilon}$ defined here correspond to the sequences $\hat{x}[n]$ and $\varepsilon[n]$. Equation 4.90 can then be expressed as

$$\mathcal{E} = \mathcal{E}\{\boldsymbol{\varepsilon}^{*T}\boldsymbol{\varepsilon}\} = \mathcal{E}\left\{ \left(\sum_{i=M+1}^{N} \kappa_i^* \boldsymbol{\varphi}_i^{*T} \right) \left(\sum_{j=M+1}^{N} \kappa_j \boldsymbol{\varphi}_j \right) \right\} = \sum_{i=M+1}^{N} \mathcal{E}\{|\kappa_i|^2\} \tag{4.91}$$

where the last step follows because the $\boldsymbol{\varphi}_i$ satisfy the orthonormal condition (4.81). Finally, (4.85) can be used to write the mean-square error in compact form as

$$\mathcal{E} = \sum_{i=M+1}^{N} \boldsymbol{\varphi}_i^{*T} \mathbf{R}_x \boldsymbol{\varphi}_i \tag{4.92}$$

The problem is now to minimize (4.92) subject to the constraints

$$\boldsymbol{\varphi}_i^{*T} \boldsymbol{\varphi}_i = 1; \qquad i = M+1, M+2, \ldots, N \tag{4.93}$$

This problem can be attacked formally by using the methods based on Lagrange multipliers discussed in Appendix A.[12] Specifically, we consider minimizing the augmented quantity (known as the *Lagrangian*)

$$\mathcal{L} = \sum_{i=M+1}^{N} \boldsymbol{\varphi}_i^{*T}\mathbf{R}_x\boldsymbol{\varphi}_i + \sum_{i=M+1}^{N} \lambda_i(1 - \boldsymbol{\varphi}_i^{*T}\boldsymbol{\varphi}_i) \tag{4.94}$$

where λ_i are real Lagrange multipliers associated with each real constraint. When all constraints are satisfied, the additional terms in (4.94) are all equal to zero, so that minimizing $\mathcal{L}$ is equivalent to minimizing $\mathcal{E}$. A necessary condition for the minimum is that the complex gradient $\nabla_{\boldsymbol{\varphi}_i^*}\mathcal{L}$ be equal to zero. (See Appendix A for definition and formulas.) The result is equivalent to taking the derivative with respect to all real and imaginary parts of the $\boldsymbol{\varphi}_i$ and setting them simultaneously to zero. From Table A.2 in Appendix A, the gradient can be expressed as[13]

$$\nabla_{\boldsymbol{\varphi}_i^*}\mathcal{L} = \mathbf{R}_x\boldsymbol{\varphi}_i - \lambda_i\boldsymbol{\varphi}_i = \mathbf{0}$$

or

$$\mathbf{R}_x\boldsymbol{\varphi}_i = \lambda_i\boldsymbol{\varphi}_i \qquad i = M+1, \ldots, N \tag{4.95}$$

Therefore, each $\boldsymbol{\varphi}_i$ must be an eigenvector with corresponding eigenvalue λ_i. Equation 4.92 can then be reexpressed using (4.95) as

$$\mathcal{E} = \sum_{i=M+1}^{N} \boldsymbol{\varphi}_i^{*T}\mathbf{R}_x\boldsymbol{\varphi}_i = \sum_{i=M+1}^{N} \boldsymbol{\varphi}_i^{*T}(\lambda_i\boldsymbol{\varphi}_i) = \sum_{i=M+1}^{N} \lambda_i \tag{4.96}$$

The procedure is now clear. Since the eigenvalues of a correlation matrix are never negative, we should use the $N-M$ *smallest* eigenvalues of $\mathbf{R}_x$ in (4.96) in order to minimize $\mathcal{E}$. Since these smallest eigenvalues correspond to the terms *left out* of the expansion, *the optimal basis functions for the truncated expansion* (4.87) *correspond to the eigenvectors of* $\mathbf{R}_x$ *with the* M largest *eigenvalues*. An example illustrates this optimal property of the DKLT.

Example 4.7

Consider a real random process with the correlation matrix

$$\begin{bmatrix} 4.000 & 2.400 & 1.440 & 0.864 \\ 2.400 & 4.000 & 2.400 & 1.440 \\ 1.440 & 2.400 & 4.000 & 2.400 \\ 0.864 & 1.440 & 2.400 & 4.000 \end{bmatrix}$$

The eigenvalues of this matrix are found to be

$$\lambda_1 = 9.584, \quad \lambda_2 = 3.597, \quad \lambda_3 = 1.716, \quad \lambda_4 = 1.139$$

[12]More general discussion of the Lagrange multiplier technique can be found in several places (see, e.g., [10]).

[13]The expression when $\mathbf{R}_x$ and $\boldsymbol{\varphi}_i$ are real is slightly different (see Appendix A) but leads to the same conclusion.

with corresponding eigenvectors

$$\mathbf{e}_1 = \begin{bmatrix} 0.448 \\ 0.547 \\ 0.547 \\ 0.448 \end{bmatrix} \; ; \quad \mathbf{e}_2 = \begin{bmatrix} 0.637 \\ 0.306 \\ -0.306 \\ -0.637 \end{bmatrix} \; ; \quad \mathbf{e}_3 = \begin{bmatrix} 0.547 \\ -0.448 \\ -0.448 \\ 0.547 \end{bmatrix} \; ; \quad \mathbf{e}_4 = \begin{bmatrix} 0.306 \\ -0.637 \\ 0.637 \\ -0.306 \end{bmatrix}$$

Now consider a sequence consisting of four samples of the random process, and suppose it is desired to find an approximation of the form

$$\hat{x}[n] = \kappa_1\varphi_1[n] + \kappa_2\varphi_2[n] + \kappa_3\varphi_3[n] \qquad 0 \leq n \leq 3$$

The optimal approximation results by choosing $\varphi_i = \mathbf{e}_i$, $i = 1, 2, 3$, that is, by choosing $\varphi_i[n]$ as the components of the first three eigenvectors. The mean-square error in the approximation is then equal to λ_4 or 1.139. If it is desired to approximate $x[n]$ using only two orthonormal basis functions, then the optimal choice is $\varphi_1[n]$ and $\varphi_2[n]$ and the minimum mean-square error is $\lambda_3 + \lambda_4 = 2.855$.

To see how the DKLT can be useful in a practical application, consider the case of a zero-mean random signal $s[n]$ in additive white noise $\eta[n]$. The observation sequence is

$$x[n] = s[n] + \eta[n]$$

Suppose first that the signal and noise are both stationary and that their respective power spectral densities are as depicted in Fig. 4.13(a). Since the signal has a narrow bandwidth and the noise is white, a natural thing to do is to pass the received observation sequence through a linear filter to eliminate the out-of-band noise.[14] This is depicted in Fig. 4.13(b).

Suppose now that the random signal is not stationary. Then the power spectral density is not defined and the solution involving linear filtering is not so clear. The average energy in the signal, however, is given by

$$\begin{aligned} &\mathcal{E}\{|s[0]|^2 + |s[1]|^2 + \cdots + |s[N-1]|^2\} \\ &\qquad = \mathrm{R}_s[0,0] + \mathrm{R}_s[1,1] + \cdots + \mathrm{R}_s[N-1, N-1] \end{aligned}$$

The last expression is the *trace* of the correlation matrix, and since the trace is invariant under a unitary transformation, it is equal to the *sum of the eigenvalues.* Figure 4.13(c) shows the concentration ellipsoids for the signal and noise correlation matrices in a two-dimensional case. Since the signal ellipsoid is long and thin, most of the signal energy lies along the direction of the first eigenvector. Thus a good approximation to the signal can be obtained by using only a single eigenfunction in the DKLT. In using this truncated approximation the noise energy is reduced by a factor of 2. Therefore, the use of the truncated representation increases the signal-to-noise ratio and is counterpart for the nonstationary case to eliminating the out-of-band noise.

[14]Specific procedures for filtering random processes are discussed in the next chapter.

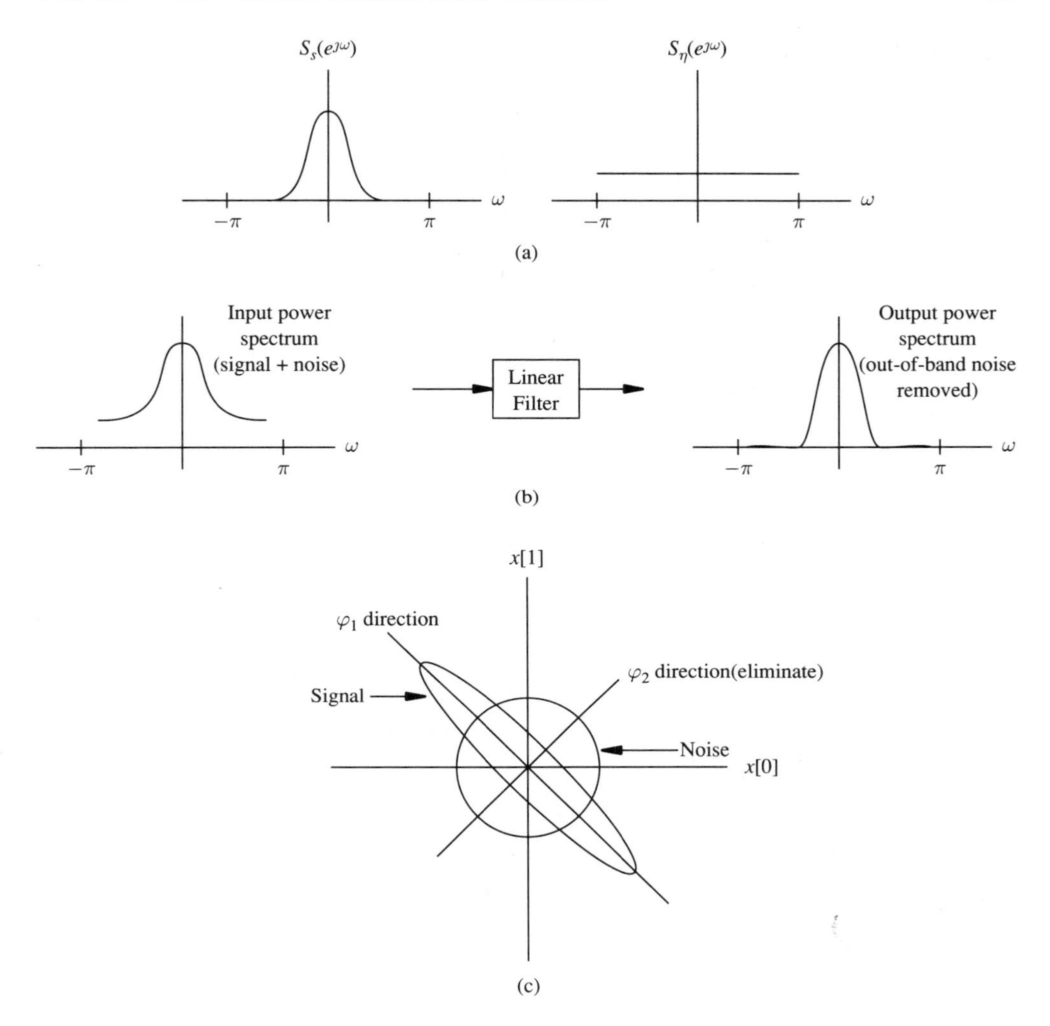

Figure 4.13 Comparison of linear filtering and DKLT methods to reduce noise. (a) Power spectral densities pertaining to a stationary random signal and noise. (b) Filtering to suppress out-of-band noise. (c) Concentration ellipses pertaining to a nonstationary random signal and noise.

The use of eigenvectors to characterize the signal and noise portions of the observation is one of the cornerstones of modern statistical signal processing. Several methods for signal modeling and spectrum analysis based on these ideas are encountered in later chapters of this book. In one important case the signals are found to lie entirely within a *subspace* of the vector space which is spanned by certain of the eigenvectors. This characterization leads to several important high-resolution methods for spectrum estimation.

4.7 PERIODIC AND ALMOST PERIODIC PROCESSES

This section discusses periodic random processes in terms of their second moment properties. It is shown here that the DKLT for a periodic stationary random process is identical to the DFT. It is further shown that when the DFT of the correlation function of any *general* (nonperiodic) random process is taken over a long enough interval, the result for all practical purposes is identical to the DKLT.

4.7.1 Second Moment Properties

In Chapter 3 a periodic random process with period P was defined as one whose joint density functions satisfy the condition

$$f_{x[n_0],x[n_1],\ldots x[n_L]} = f_{x[n_0+k_0P],x[n_1+k_1P],\ldots x[n_L+k_LP]} \tag{4.97}$$

for all choices of the n_i and for any set of integers k_i and any L. This condition implies among other things that the mean of the random process is a periodic function, and that the correlation function is periodic in each of its arguments. Let us examine this last point a little further. Since the density satisfies (4.97) the following two moments must be equal:

$$\mathcal{E}\{x[n_1]x^*[n_0]\} = \mathcal{E}\{x[n_1 + k_1P]x^*[n_0 + k_0P]\}$$

Thus it follows that

$$\mathrm{R}_x[n_1, n_0] = \mathrm{R}_x[n_1 + k_1P, n_0 + k_0P] \tag{4.98}$$

for all integers k_1 and k_0. Since the mean is periodic, it follows that the covariance function satisfies a similar relation.

$$\mathrm{C}_x[n_1, n_0] = \mathrm{C}_x[n_1 + k_1P, n_0 + k_0P] \tag{4.99}$$

Figure 4.14 shows a periodic random process and points where the process has identical correlation. The points n_0, n_0', n_0'' and n_1, n_1', n_1'' are separated by one or more multiples of the period.

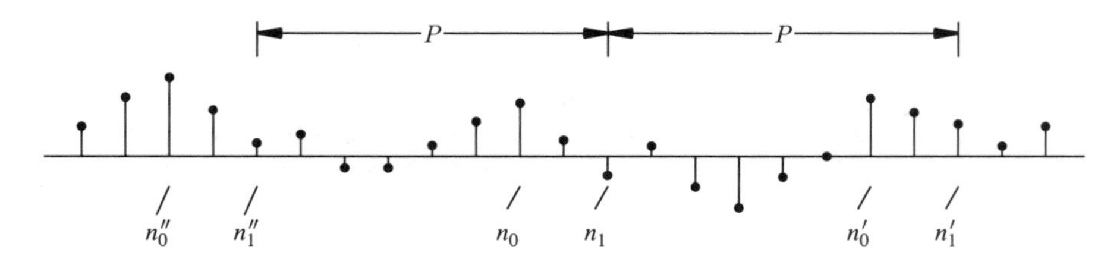

Figure 4.14 Periodic random process and points where it has identical correlation. [$\mathrm{R}_x[n_1, n_0] = \mathrm{R}_x[n_1', n_0] = \mathrm{R}_x[n_1'', n_0']$ etc.]

For a periodic *stationary* random process, the correlation and covariance functions depend only on the spacing between samples. Hence if

$$l = n_1 - n_0$$

and

$$k = k_1 - k_0$$

then (4.99) implies that

$$R_x[l] = R_x[l + kP] \tag{4.100}$$

and

$$C_x[l] = C_x[l + kP] \tag{4.101}$$

for all integers k. Thus the correlation function and covariance function are *periodic*.

Since any discrete periodic function has a Fourier transform consisting entirely of impulses at multiples of the fundamental frequency, the power density spectra of stationary periodic random processes consists entirely of impulses or "lines" that are harmonically related. The following example shows how to compute the correlation function and power spectral density function for a simple periodic random process and illustrates several very important points.

Example 4.8

A basic type of periodic random process is the random complex exponential with fixed frequency ω_o

$$x[n] = Ae^{\jmath\omega_o n}$$

where the complex amplitude A is a random variable with random magnitude and phase

$$A = |A|e^{\jmath\phi}$$

A sufficient condition for the random process to be stationary is that the amplitude and phase are independent and that the phase is uniformly distributed between $-\pi$ and π. To show this, first consider the mean, which is given by

$$\mathcal{E}\{x[n]\} = \mathcal{E}\{A\}\, e^{\jmath\omega_o n}$$

The only way this can be a constant is if $\mathcal{E}\{A\} = 0$. For the given random process, however,

$$\mathcal{E}\{A\} = \mathcal{E}\left\{|A|e^{\jmath\phi}\right\} = \mathcal{E}\{|A|\} \cdot \mathcal{E}\left\{e^{\jmath\phi}\right\} = 0$$

which follows because $\mathcal{E}\left\{e^{\jmath\phi}\right\} = 0$ when ϕ is uniformly distributed. (It is worthwhile to note here that since $|A|$ is never negative, the expectation $\mathcal{E}\{|A|\}$ *cannot* be zero, so $\mathcal{E}\left\{e^{\jmath\phi}\right\}$ *must* be zero if the process is to be stationary.)

A further condition that the complex random process must satisfy is (4.67) if either of the definitions (4.2) or (4.9) is to be useful relation. The expectation of $x[n_1]x[n_0]$ is

$$\mathcal{E}\{x[n_1]x[n_0]\} = \mathcal{E}\{Ae^{\jmath\omega_o n_1} \cdot Ae^{\jmath\omega_o n_0}\} = \mathcal{E}\left\{A^2\right\} e^{\jmath\omega_o(n_1+n_0)}$$

In order for this to be zero it is necessary that $\mathcal{E}\left\{A^2\right\} = 0$. For the given random process

$$\mathcal{E}\left\{A^2\right\} = \mathcal{E}\left\{|A|^2 e^{\jmath 2\phi}\right\} = \mathcal{E}\left\{|A|^2\right\} \cdot \underbrace{\mathcal{E}\left\{e^{\jmath 2\phi}\right\}}_{0} = 0$$

which follows again because ϕ is uniformly distributed. Since (4.67) is satisfied, the correlation function can be computed from

$$\begin{aligned} R_x[n_1, n_0] &= \mathcal{E}\{x[n_1]x^*[n_0]\} \\ &= \mathcal{E}\left\{Ae^{\jmath\omega_o n_1} \cdot A^* e^{-\jmath\omega_o n_0}\right\} = \mathcal{E}\left\{|A|^2\right\} e^{\jmath\omega_o(n_1-n_0)} \end{aligned}$$

From the form of this equation, it follows that the random process is stationary, and we can write

$$R_x[l] = \mathsf{P}_o e^{\jmath\omega_o l}$$

where

$$\mathsf{P}_o = \text{Var}\,[A] = \mathcal{E}\left\{|A|^2\right\}$$

is the power of the random process.

Recall from Chapter 3 that processes of the type considered in the example were noted to be periodic as long as ω_o satisfies the condition

$$\omega_o = 2\pi K/P \tag{4.102}$$

for some integers K and P. However, the results of the example do not depend on this condition being satisfied. Random processes in the form of the one in the example where ω_o does *not* satisfy the condition (4.102) were defined to be *almost periodic*. The results apply to these processes as well.

It is straightforward to expand the example to processes of the form

$$x[n] = \sum_i A_i e^{\jmath\omega_i n} \tag{4.103}$$

where the A_i are complex random variables satisfying the conditions of the example. Then, provided that (4.67) is met, random processes of the form (4.103) have correlation functions that are sums of the same complex exponentials (see Problem 4.24). Further, it is easy to show that the process $x[n]$ is stationary if and only if the A_i are orthogonal, that is, if

$$\mathcal{E}\left\{A_i A_j^*\right\} = 0; \qquad i \neq j \tag{4.104}$$

(see Problem 4.24). Since, as the example shows, the mean of each term is zero, (4.104) implies that the A_i are uncorrelated. The power spectra of such random processes consist only of "lines." If the random process is periodic, these lines occur at rational multiples of 2π and are harmonically related. If the random process contains any *almost periodic* components, then the corresponding lines occur at irrational multiples of 2π (see Fig. 4.15). In either case the characteristic line spectrum gives another interpretation of periodic or almost periodic stationary random processes.

Condition (4.104) holds in many cases, including when two complex exponentials combine to form a real sinusoid

$$Ae^{\jmath\omega_o n} + A^* e^{-\jmath\omega_o n} = 2|A|\cos(\omega_o n + \phi)$$

and when the terms are harmonically related as in the case of a random square wave. Problem 4.24 explores the conditions in more detail.

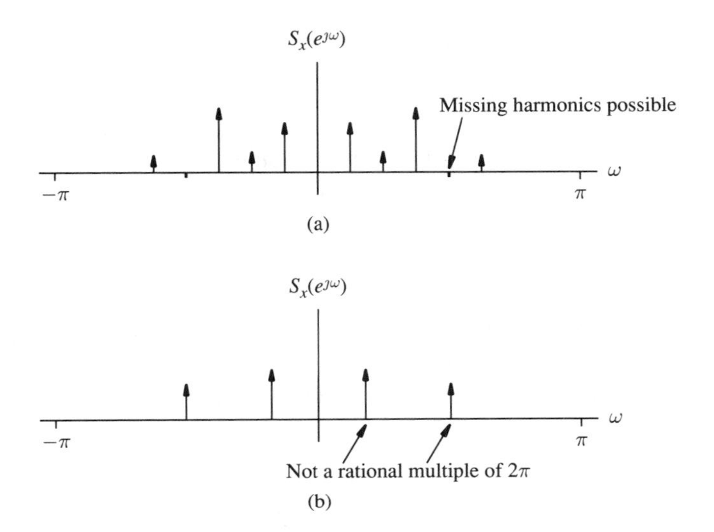

Figure 4.15 Line spectra. (a) Periodic process. (b) Almost periodic process.

Table 4.6 summarizes the results for this class of random processes. Note that there is no restriction on the density for $|A|$ except, of course, that it is nonzero only for $|A| \geq 0$. An important special case, however, is the Rayleigh density

$$f_r(\mathrm{r}) = \begin{cases} \frac{1}{\sigma_o^2}\mathrm{r}\, e^{-\frac{\mathrm{r}^2}{2\sigma_o^2}} & \mathrm{r} \geq 0 \\ 0 & \mathrm{r} < 0 \end{cases} \tag{4.105}$$

(random varable r represents $|A|$). In this case the real and imaginary parts of the random process are independent zero-mean Gaussian random variables with variance

$$\sigma_o^2 = \frac{2\mathsf{P}_o}{4 - \pi}$$

This in fact is the only case where the real and imaginary parts of the process are independent [11, pp. 134–135].

TABLE 4.6 CORRELATION FUNCTION FOR A STATIONARY RANDOM PROCESS CONSISTING OF A SUM OF COMPLEX EXPONENTIALS WITH INDEPENDENT AMPLITUDE AND PHASE AND UNIFORMLY DISTRIBUTED PHASE

Random process	$x[n] = \sum_i A_i e^{j\omega_i n}$;	$\mathcal{E}\{A_i A_j^*\} = 0, \quad i \neq j$
Correlation function	$R_x[l] = \sum_i \mathsf{P}_i e^{j\omega_i l}$;	$\mathsf{P}_i = \mathcal{E}\{\lvert A_i\rvert^2\} = \mathrm{Var}\,[A_i]$

4.7.2 Karhunen-Loève Representation

The eigenfunctions correponding to a periodic random process turn out to be the Fourier basis functions that appear in the DFT. This implies that for a periodic random process, the DFT and the DKLT are identical. While this fact is interesting in itself, it has further implications. Since the correlation function for a general random process can be approximated by one that is periodically extended, the eigenfunctions for most general random processes are very nearly equal to the Fourier basis functions if the observation interval is sufficiently long. These facts are explored here.

Eigenfunctions of Periodic Random Processes. It was seen that the power density spectrum of a periodic random process with period P consists entirely of impulses or lines. Now consider the DFT of the autocorrelation function for such a random process:

$$\mathrm{S}_x[k] = \sum_{l=0}^{P-1} R_x[l] W_P^{lk} \tag{4.106}$$

where

$$W_P \stackrel{\text{def}}{=} e^{-\jmath \frac{2\pi}{P}} \tag{4.107}$$

The DFT represents the values of the power density spectrum at the lines. The correlation function can be recovered from the $\mathrm{S}_x[k]$ via the inverse DFT

$$R_x[l] = \frac{1}{P} \sum_{l=0}^{P-1} \mathrm{S}_x[l] W_P^{-lk} \tag{4.108}$$

Now consider the $P \times P$ correlation matrix for this random process.

$$\mathbf{R}_x = \begin{bmatrix} R_x[0] & R_x[-1] & R_x[-2] & \cdots & R_x[-P+1] \\ R_x[1] & R_x[0] & R_x[-1] & \cdots & R_x[-P+2] \\ R_x[2] & R_x[1] & R_x[0] & \cdots & R_x[-P+3] \\ \vdots & \vdots & \vdots & \ddots & \vdots \\ R_x[P-1] & R_x[P-2] & R_x[P-3] & \cdots & R_x[0] \end{bmatrix} \tag{4.109}$$

The correlation function is depicted in Fig. 4.16. Because of the periodicity the correlation matrix has the special form

$$\mathbf{R}_x = \begin{bmatrix} R_x[0] & R_x[-1] & R_x[-2] & \cdots & R_x[-P+1] \\ R_x[-P+1] & R_x[0] & R_x[-1] & \cdots & R_x[-P+2] \\ R_x[-P+2] & R_x[-P+1] & R_x[0] & \cdots & R_x[-P+3] \\ \vdots & \vdots & \vdots & \ddots & \vdots \\ R_x[-1] & R_x[-2] & R_x[-3] & \cdots & R_x[0] \end{bmatrix} \tag{4.110}$$

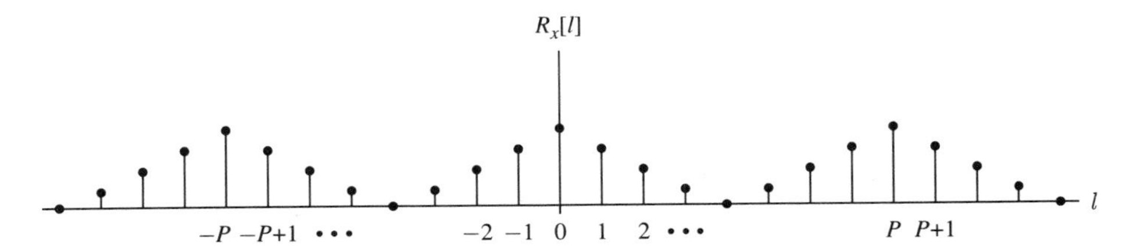

Figure 4.16 Typical correlation function for a periodic random process.

Each row is just a circularly rotated version of the row above it. Matrices of this form are called *circulant matrices*, and it is well known that the eigenvectors of such matrices are comprised of complex exponentials (see, e.g., [12] or [13, App. 2]). This can easily be shown as follows. To make the indexing correspond to what is normally used in the DFT, denote the eigenvectors of $\mathbf{R}_x$ by $\mathbf{w}_0, \mathbf{w}_1, \ldots, \mathbf{w}_{P-1}$. Then we claim that the eigenvectors have the specific form

$$\mathbf{w}_k = \frac{1}{\sqrt{P}} \begin{bmatrix} 1 \\ W_P^{-k} \\ W_P^{-2k} \\ \vdots \\ W_P^{-(P-1)k} \end{bmatrix}; \qquad k = 0, 1, \ldots, P-1 \tag{4.111}$$

(where the factor of $1/\sqrt{P}$ is for purposes of normalization) and the corresponding eigenvalues are equal to $\mathrm{S}_x[k]$. To check this out, substitute these quantities in the eigenvalue equation

$$\mathbf{R}_x \mathbf{w}_k = \mathrm{S}_x[k] \mathbf{w}_k \tag{4.112}$$

to obtain

$$\begin{bmatrix} R_x[0] & R_x[-1] & R_x[-2] & \cdots & R_x[-P+1] \\ R_x[-P+1] & R_x[0] & R_x[-1] & \cdots & R_x[-P+2] \\ R_x[-P+2] & R_x[-P+1] & R_x[0] & \cdots & R_x[-P+3] \\ \vdots & \vdots & \vdots & \ddots & \vdots \\ R_x[-1] & R_x[-2] & R_x[-3] & \cdots & R_x[0] \end{bmatrix} \cdot \frac{1}{\sqrt{P}} \begin{bmatrix} 1 \\ W_P^{-k} \\ W_P^{-2k} \\ \vdots \\ W_P^{-(P-1)k} \end{bmatrix}$$

$$\stackrel{?}{=} \mathrm{S}_x[k] \cdot \frac{1}{\sqrt{P}} \begin{bmatrix} 1 \\ W_P^{-k} \\ W_P^{-2k} \\ \vdots \\ W_P^{-(P-1)k} \end{bmatrix}$$

Conjugating both sides, canceling the common term $1/\sqrt{P}$, and observing that $\mathrm{S}_x[k]$ is real results in

$$\begin{bmatrix} R_x[0] & R_x[1] & R_x[2] & \cdots & R_x[P-1] \\ R_x[P-1] & R_x[0] & R_x[1] & \cdots & R_x[P-2] \\ R_x[P-2] & R_x[P-1] & R_x[0] & \cdots & R_x[P-3] \\ \vdots & \vdots & \vdots & \ddots & \vdots \\ R_x[1] & R_x[2] & R_x[3] & \cdots & R_x[0] \end{bmatrix} \begin{bmatrix} 1 \\ W_P^k \\ W_P^{2k} \\ \vdots \\ W_P^{(P-1)k} \end{bmatrix}$$

$$\stackrel{?}{=} \mathrm{S}_x[k] \cdot \begin{bmatrix} 1 \\ W_P^k \\ W_P^{2k} \\ \vdots \\ W_P^{(P-1)k} \end{bmatrix}$$

This represents a set of P linear equations. The first equation is just (4.106), the defining equation for the DFT. The second equation is just the DFT of the circularly rotated correlation sequence. Therefore, it represents W_P^k times $\mathrm{S}_x[k]$. The third equation is the DFT of the sequence rotated two places and so is equal to $W_P^{2k}\mathrm{S}_x[k]$, and so on. Since the matrix equation is clearly satisfied, this shows that the $\mathbf{w}_k$ *are eigenvectors with corresponding eigenvalues* $\mathrm{S}_x[k]$.

Now since the $\mathbf{w}_k$ are eigenvectors of the correlation matrix, the matrix $\mathbf{W}$ defined by

$$\mathbf{W} = \begin{bmatrix} | & | & & | \\ \mathbf{w}_0 & \mathbf{w}_1 & \cdots & \mathbf{w}_{P-1} \\ | & | & & | \end{bmatrix} \tag{4.113}$$

is unitary, that is,

$$\mathbf{W}^{*T}\mathbf{W} = \mathbf{W}\mathbf{W}^{*T} = \mathbf{I} \tag{4.114}$$

Further, since $\mathbf{W}$ is the matrix of eigenvectors, it diagonalizes the correlation matrix, that is,

$$\mathbf{W}^{*T}\mathbf{R}_x\mathbf{W} = \begin{bmatrix} \mathrm{S}_x[0] & 0 & \cdots & 0 \\ 0 & \mathrm{S}_x[1] & & 0 \\ \vdots & & \ddots & \vdots \\ 0 & 0 & \cdots & \mathrm{S}_x[P-1] \end{bmatrix} \tag{4.115}$$

The terms on the diagonal are of course the eigenvalues. Now consider the random sequence $x[n]$ and its normalized DFT[15]

$$X'[k] \stackrel{\text{def}}{=} \frac{X[k]}{\sqrt{P}} = \sum_{n=0}^{P-1} x[n]\frac{W_P^{-nk}}{\sqrt{P}} \tag{4.116}$$

[15]This is also called a unitary DFT.

If the vector $\boldsymbol{X}'$ is defined as the vector of normalized DFT coefficients, then (4.116) can be written in matrix notation as

$$\boldsymbol{X}' = \begin{bmatrix} X'[0] \\ X'[1] \\ \vdots \\ X'[P-1] \end{bmatrix} = \begin{bmatrix} -- & \mathbf{w}_0^{*T} & -- \\ -- & \mathbf{w}_1^{*T} & -- \\ & \vdots & \\ -- & \mathbf{w}_{P-1}^{*T} & -- \end{bmatrix} \boldsymbol{x} = \mathbf{W}^{*T}\boldsymbol{x} \tag{4.117}$$

The inverse DFT is

$$x[n] = \frac{1}{P}\sum_{k=0}^{P-1} X[k]W_P^{nk} = \sum_{k=0}^{P-1} X'[k]\frac{W_P^{nk}}{\sqrt{P}} \tag{4.118}$$

which can be expressed in matrix notation as

$$\boldsymbol{x} = \begin{bmatrix} | & | & & | \\ \mathbf{w}_0 & \mathbf{w}_1 & \cdots & \mathbf{w}_{P-1} \\ | & | & & | \end{bmatrix} \begin{bmatrix} X'[0] \\ X'[1] \\ \vdots \\ X'[P-1] \end{bmatrix} = \mathbf{W}\boldsymbol{X}' \tag{4.119}$$

Since $\mathbf{W}$ a is unitary matrix, (4.117) and (4.119) represent a transformation of the random vector $\boldsymbol{x}$ that corresponds to a rotation of the coordinate system. From the form of (4.117) it can be seen that the correlation matrix of $\boldsymbol{X}'$ is given by (4.115). Thus the random vector $\boldsymbol{X}'$ has orthogonal components

$$\mathcal{E}\{X'[l]X'^*[k]\} = \frac{1}{P}\mathcal{E}\{X[l]X^*[k]\} = 0 \quad \text{for } l \neq k \tag{4.120}$$

and the average power of these terms is

$$\mathcal{E}\{X'[k]X'^*[k]\} = \frac{1}{P}\mathcal{E}\{|X[k]|^2\} = \mathrm{S}_x[k] \tag{4.121}$$

Implications for General Random Processes. The development in the preceding subsection shows that for periodic random processes, the normalized DFT (4.116) and (4.118) is identical to the DKLT (4.77) and (4.75). In particular, the Fourier transform yields a representation with orthogonal coefficients and provides the optimal representation of the random process using just a subset of the basis functions.

The foregoing result has further implications, however, for general random processes that are not periodic or almost periodic. The argument to be presented is not rigorous, but it serves to indicate why the eigenfunctions for most general random processes appear sinusoidal. For most stationary random processes the samples become uncorrelated as the separation between them becomes large. Thus as the lag becomes sufficiently large, the correlation function approaches a constant equal to the squared magnitude of the mean [see (4.11)]. For purposes of the present argument it can be assumed that the mean is zero. Let us thus consider the DKLT for a set of N samples of a general random process. The correlation function for these samples is represented on an interval $-N \leq l \leq N$. Assume that N is sufficiently large that the correlation function becomes approximately zero at the ends of the interval. The situation is depicted in Fig. 4.17(a). Now consider taking a $2N$-point DFT of this correlation function. Taking the DFT implicitly assumes that the correlation function is periodically extended with period $P = 2N$ as shown in Fig. 4.17(b). Since we are now dealing with a periodic correlation function, the Karhunen-Loève basis functions are the Fourier basis functions appearing in (4.118). Nevertheless, observe that these Fourier basis functions are the eigenfunctions pertaining to the correlation matrix depicted in Fig. 4.17(c) and not to the correlation matrix of the original random process. This circulant correlation matrix has extra terms in the two corners due to the periodicity which are not present in the correlation matrix for the original random process. If N is sufficiently large, however,

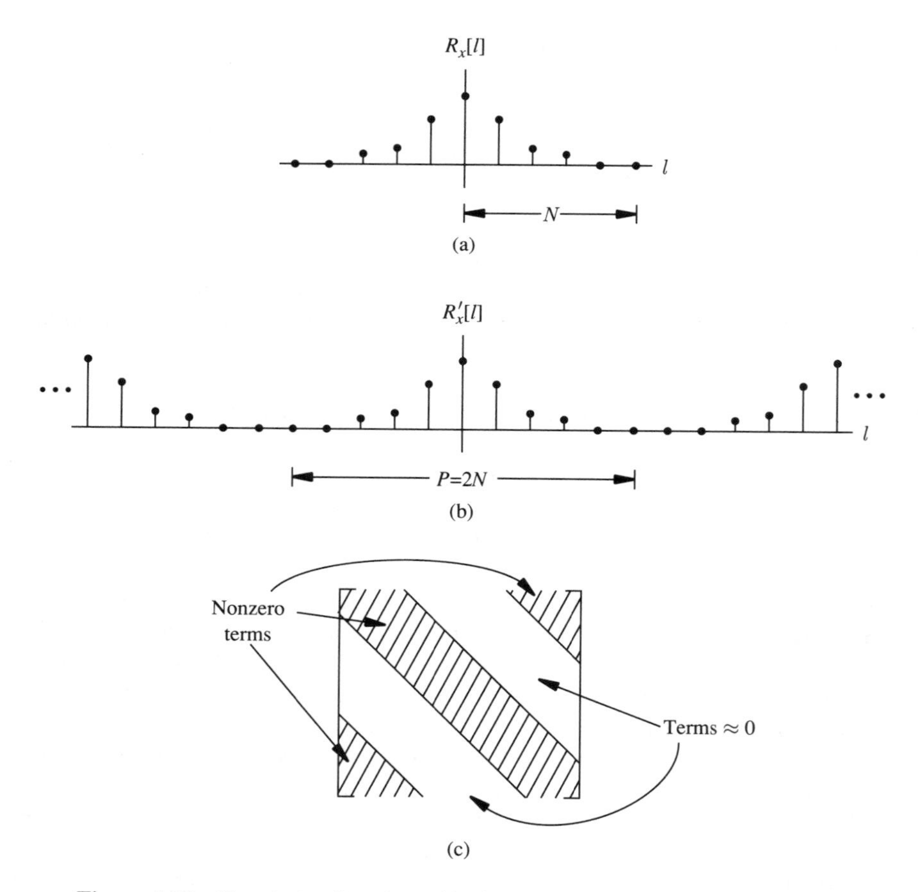

Figure 4.17 Correlation function with "long" observation interval. (a) Original correlation function. (b) Periodically extended correlation function. (c) Structure of the correlation matrix.

the contributions of these terms to the eigenvalue problem are small and the Fourier basis functions become a good approximation to the optimal basis functions of the DKLT.

In many practical situations, the magnitude of the correlation function becomes small within a relatively short observation interval. Thus when a segment of such a correlation function is used, the DKLT is practically equivalent (within a scale factor) to the ordinary DFT. This explains the sinusoidal behavior of the eigenfunctions observed in Fig. 4.12.

4.8 SAMPLED CONTINUOUS RANDOM PROCESSES

A great number of discrete random processes arise from sampling continuous random processes. The concept of a continuous random process was introduced in Chapter 3. Here the second moment properties of continuous random processes are briefly explored and related to those of the discrete random processes that arise through sampling. The next subsection summarizes the basic second moment properties. The following subsection develops a stochastic version of the Nyquist sampling theorem. The theorem gives explicit relations between the continuous and the sampled versions of the random signal and relates the radian frequency scale for the random sequence to the frequency scale of the continuous random signal in hertz.

4.8.1 Second Moment Characterization

As for the case of discrete random processes, complete statistical characterization of continuous random processes is not always possible. Most analyses of random signals are therefore based on the first and second moments of the distribution. These moments are most conveniently expressed as the mean, correlation, and covariance functions, which are defined for a general (nonstationary) continuous random process $x_c(t)$ as

$$\mathrm{m}_{x_c}^c(t) = \mathcal{E}\{x_c(t)\} \tag{4.122}$$

$$\mathrm{R}_{x_c}^c(t_1, t_0) = \mathcal{E}\{x_c(t_1)x_c^*(t_0)\} \tag{4.123}$$

$$\mathrm{C}_{x_c}^c(t_1, t_0) = \mathcal{E}\big\{(x_c(t_1) - \mathrm{m}_{x_c}^c(t_1))(x_c(t_0) - \mathrm{m}_{x_c}^c(t_0))^*\big\} \tag{4.124}$$

(The superscript "c" on these quantities indicates that they are the *continuous* versions of the moments.) These quantities are related as

$$\mathrm{R}_{x_c}^c(t_1, t_0) = \mathrm{C}_{x_c}^c(t_1, t_0) + \mathrm{m}_{x_c}^c(t_1)\mathrm{m}_{x_c}^{c*}(t_0) \tag{4.125}$$

As in the case of discrete random processes, the mean, correlation, and covariance functions defined here permit deriving the corresponding quantities for the real and imaginary parts of a complex random process if the necessary symmetry conditions hold. (See Appendix B.)

For a *stationary* random process the mean is a constant (independent of t) and the correlation and covariance functions depend only on the difference $\tau = t_1 - t_0$. Therefore, the moments for a stationary random process can be defined as

$$m_{x_c}^c = \mathcal{E}\{x_c(t)\} \tag{4.126}$$

$$R_{x_c}^c(\tau) = \mathcal{E}\{x_c(t)x_c^*(t-\tau)\} \tag{4.127}$$

$$C_{x_c}^c(\tau) = \mathcal{E}\{(x_c(t) - m_{x_c}^c)(x_c(t-\tau) - m_{x_c}^c)^*\} \tag{4.128}$$

with the relation

$$R_{x_c}^c(\tau) = C_{x_c}^c(\tau) + m_{x_c}^c m_{x_c}^{c*} \tag{4.129}$$

A continuous random process is *wide-sense stationary* if only the properties mentioned in this paragraph hold, even if the process is not strict-sense stationary.

The continuous correlation function has properties that are similar to those for the discrete correlation function. These are summarized in Table 4.7. Identical properties pertain to the covariance function. The first property in the table follows directly from the definition. The second property can be proven in a manner identical to the proof of the discrete version and can be shown to imply that the correlation has its largest magnitude at the origin (see Problem 4.26):

$$R_{x_c}^c(0) \geq |R_{x_c}^c(\tau)| \qquad \tau \neq 0 \tag{4.130}$$

[Note that $R_{x_c}^c(0) = \mathcal{E}\{|x(t)|^2\}$, which represents the average power, is always a positive real number.] Similar properties hold for the covariance function.

TABLE 4.7 PROPERTIES OF THE CORRELATION FUNCTION FOR A CONTINUOUS RANDOM PROCESS

Conjugate symmetry	$R_{x_c}^c(\tau) = R_{x_c}^{c*}(-\tau)$
Positive semidefinite property	$\int_{-\infty}^{\infty}\int_{-\infty}^{\infty} a_c^*(t_1)R_{x_c}^c(t_1 - t_0)a_c(t_0)dt_1dt_0 \geq 0$ for an arbitrary function $a_c(t)$

The *power spectral density function* for a stationary random process is defined as the Fourier transform of the correlation function

$$S_{x_c}^c(f) = \int_{-\infty}^{\infty} R_{x_c}^c(\tau)e^{-\jmath 2\pi f\tau}d\tau \tag{4.131}$$

where f is the frequency in *hertz*.[16] The correlation function can be recovered from the power spectral density function through the inverse transform

$$\boxed{R^c_{x_c}(\tau) = \int_{-\infty}^{\infty} S^c_{x_c}(f)e^{j2\pi f\tau}\,df} \tag{4.132}$$

The properties of the correlation function cited earlier result in two corresponding properties for the power density spectrum. These are essentially identical to the properties of the power density spectrum for discrete random processes, namely

1. $S^c_{x_c}(f)$ is a *real-valued* function of frequency.
2. $S^c_{x_c}(f)$ is *nonnegative*.

$$S^c_{x_c}(f) \geq 0; \qquad -\infty < f < \infty \tag{4.133}$$

For *real* random processes the power density spectrum is also an *even* function, that is,

$$S^c_{x_c}(f) = S^c_{x_c}(-f) \tag{4.134}$$

(real random process)

Note that the power density spectrum for continuous signals is *not* periodic. Typical power density spectra for real and complex random processes are shown in Fig. 4.18.

As in the discrete case, two random processes are said to be *jointly stationary* if they are each wide-sense stationary and if the correlation $\mathcal{E}\{x_c(t_1)y^*_c(t_0)\}$ depends only on the difference $\tau = t_1 - t_0$. The cross-correlation, cross-covariance, and cross-power spectral density functions can be defined for two jointly stationary continuous random processes as

$$R^c_{x_cy_c}(\tau) = \mathcal{E}\{x_c(t)y^*_c(t-\tau)\} \tag{4.135}$$

$$C^c_{x_cy_c}(\tau) = \mathcal{E}\{(x_c(t) - m^c_{x_c})(y_c(t-\tau) - m^c_{y_c})^*\} \tag{4.136}$$

$$S^c_{x_cy_c}(f) = \int_{-\infty}^{\infty} R^c_{x_cy_c}(\tau)e^{-j2\pi f\tau}\,d\tau \tag{4.137}$$

---

[16]This definition seems preferable to the alternative definition

$$S^c_{x_c}(\Omega) = \int_{-\infty}^{\infty} R^c_{x_c}(\tau)e^{-j\Omega\tau}\,d\tau$$

where Ω is the frequency in rad/s since the primary motivation in defining the power spectrum for continuous processes is to relate it to the power spectrum for discrete processes involving the frequency variable ω and in most practical engineering applications the spectrum of the continuous process is expressed in hertz.

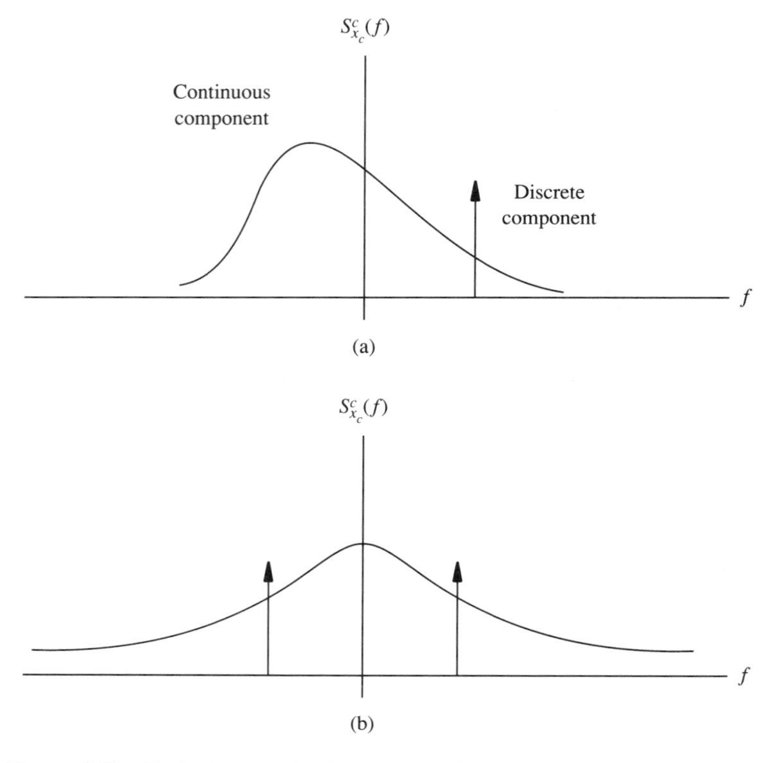

Figure 4.18 Typical power density spectrum for a continuous random process. (a) Complex random process. (b) Real random process.

From the definitions it is clear that

$$R^c_{x_c y_c}(\tau) = R^{c*}_{y_c x_c}(-\tau) \tag{4.138}$$

$$C^c_{x_c y_c}(\tau) = C^{c*}_{y_c x_c}(-\tau) \tag{4.139}$$

and

$$S^c_{x_c y_c}(f) = S^{c*}_{y_c x_c}(f) \tag{4.140}$$

Other than these, the cross-correlation, cross-covariance, and cross-power spectral density function have no special properties.

4.8.2 Sampling Theory

When a discrete random process represents a sampled continuous random process it is appropriate to ask if sampling relations akin to those for deterministic signals exist. This section answers the question by developing a version of the sampling theorem for continuous random processes.

Let the discrete random process $x[n]$ be defined by sampling a continuous random process $x_c(t)$ at uniform intervals with spacing T. That is,

$$x[n] = x_c(nT); \qquad n = \ldots, -1, 0, 1, 2, \ldots \tag{4.141}$$

The correlation function for the discrete process is given by

$$\begin{aligned} R_x[l] &= \mathcal{E}\{x[n]x^*[n-l]\} \\ &= \mathcal{E}\{x_c(nT)x_c^*(nT - lT)\} = R_{x_c}^c(lT) \end{aligned} \tag{4.142}$$

Thus the correlation function for the discrete random process is obtained by sampling the correlation function for the continuous random process at the sampling interval T.

We know that if the Fourier transform of a deterministic signal is bandlimited, then the signal can be reconstructed from samples taken at the Nyquist sampling rate. For a random signal it is resonable to suspect that a similar bandlimited condition may be needed for the power spectrum. Let us therefore assume that the power density spectrum $S_{x_c}^c(f)$ corresponding to a given random process $x_c(t)$ is nonzero only in a region $-W < f < W$. The correlation function and its power spectrum are shown in Fig. 4.19(a). The sampled correlation function with samples taken at $T = 1/2W$ and its Fourier transform are shown in Fig. 4.19(b). These correspond to the discrete correlation function and power density spectrum shown in Fig. 4.19(c), where to maintain the identity (4.142) between the samples of the continuous and the discrete correlation function, $S_x(e^{\jmath\omega})$ has been scaled by a factor $1/T$. Since $R_{x_c}^c(\tau)$ is a deterministic function it can evidently be recovered by applying an ideal lowpass filter of width $2W$ to the spectrum of the sampled correlation function. This corresponds to the signal-domain interpolation

$$R_{x_c}^c(\tau) = \sum_{l=-\infty}^{\infty} R_{x_c}^c(lT)\,\text{sinc}(2\pi W\tau - l\pi) \tag{4.143}$$

where

$$\text{sinc}(\xi) = \frac{\sin(\xi)}{\xi} \tag{4.144}$$

That the correlation function can be recovered from samples is an important precursor to the main result. The following shows that the random process itself can be recovered from samples in the following sense.

Theorem 4.1 (Sampling). Let $x_c(t)$ be a random process with power density spectrum that is zero outside of an interval $-W < f < W$. Let $x[n] = x_c(nT)$ be a sequence corresponding to samples of the continuous process taken at the sampling interval $T = 1/2W$. Define the random process $\check{x}_c(t)$ by

$$\check{x}_c(t) = \sum_{n=-\infty}^{\infty} x[n]\,\text{sinc}(2\pi Wt - n\pi)$$

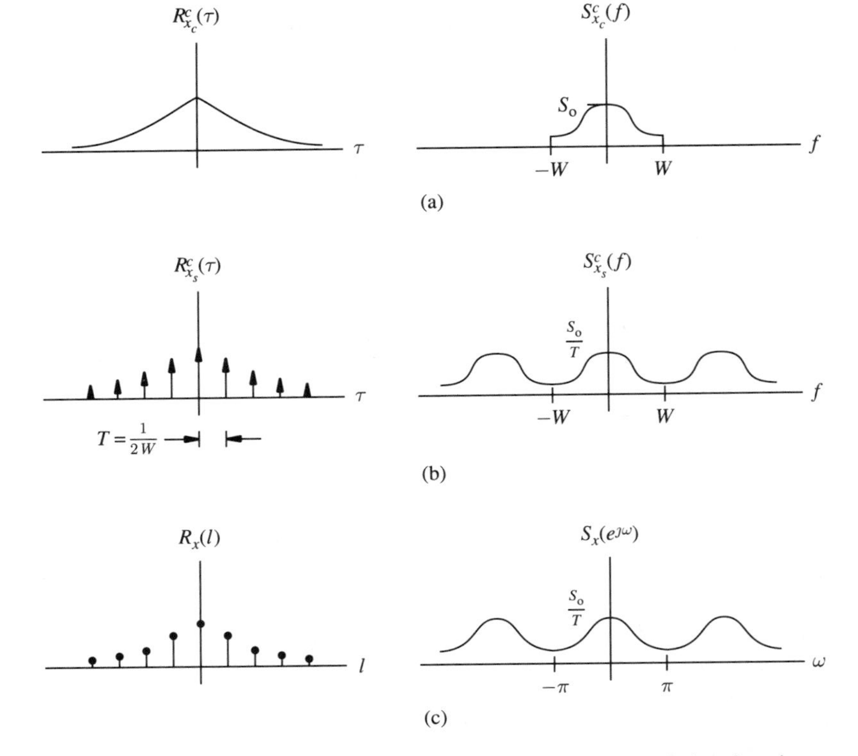

Figure 4.19 Relations between continuous and discrete correlation functions and spectra. (a) Continuous correlation function and corresponding spectrum. (b) Sampled correlation function and corresponding spectrum. (c) Discrete correlation function and corresponding power spectrum.

Then $\check{x}_c(t)$ is the reconstruction of $x_c(t)$ in the sense that

$$\mathcal{E}\{|x_c(t) - \check{x}_c(t)|^2\} = 0$$

When two signals are equivalent in the sense defined in this theorem they are said to be equal in the mean-square sense or to be equal *almost everywhere*. In order to more easily prove the theorem, the following intermediate result will first be shown.

Lemma. *Let the conditions of Theorem 4.1 apply and let $\check{x}_c(t)$ be defined as in the theorem. Then*

$$\mathcal{E}\{[x_c(t) - \check{x}_c(t)]x_c^*(\xi)\} = 0; \qquad -\infty < \xi < \infty$$

This is a kind of orthogonality principle. The proof is as follows.

First note that since $S^c_{x_c}(f)$ is bandlimited, $S^c_{x_c}(f)e^{-j2\pi f\tau_o}$ is also bandlimited. Therefore [see (4.143)],

$$R^c_{x_c}(\tau - \tau_o) = \sum_{l=-\infty}^{\infty} R^c_{x_c}(lT - \tau_o)\ \mathrm{sinc}(2\pi W\tau - l\pi)$$

This can be rewritten, using different variables, as

$$R^c_{x_c}(t - \xi) - \sum_{n=-\infty}^{\infty} R^c_{x_c}(nT - \xi)\ \mathrm{sinc}(2\pi Wt - n\pi) = 0 \tag{4.145}$$

which is more convenient for later use.

Now note that $\check{x}_c(t)$ can be written as

$$\check{x}_c(t) = \sum_{n=-\infty}^{\infty} x_c(nT)\ \mathrm{sinc}(2\pi Wt - n\pi) \tag{4.146}$$

and this can be used to write

$$\begin{aligned} &\mathcal{E}\{[x_c(t) - \check{x}_c(t)]x_c^*(\xi)\} \\ &\quad = \mathcal{E}\{x_c(t)x_c^*(\xi)\} - \sum_{n=-\infty}^{\infty} \mathcal{E}\{x_c(nT)x_c^*(\xi)\}\ \mathrm{sinc}(2\pi Wt - n\pi) \\ &\quad = R^c_{x_c}(t - \xi) - \sum_{n=-\infty}^{\infty} R^c_{x_c}(nT - \xi)\ \mathrm{sinc}(2\pi Wt - n\pi) = 0 \end{aligned}$$

where the last result is equal to zero because of (4.145). This proves the lemma.

The proof of the theorem now follows. First write the error between the original and reconstructed sequences as

$$\begin{aligned} &\mathcal{E}\{|x_c(t) - \check{x}_c(t)|^2\} \\ &\qquad = \mathcal{E}\{[x_c(t) - \check{x}_c(t)]x_c^*(t)\} - \mathcal{E}\{[x_c(t) - \check{x}_c(t)]\check{x}_c^*(t)\} \end{aligned}$$

From the lemma, the first term on the right side is equal to zero. The second term can be written using (4.146) as

$$\sum_{n=-\infty}^{\infty} \mathcal{E}\{[x_c(t) - \check{x}_c(t)]x_c^*(nT)\}\ \mathrm{sinc}(2\pi Wt - n\pi)$$

which is also equal to zero because of the lemma. This proves the theorem.

The sampling theorem for random signals states that a bandlimited random signal can be reconstructed from samples taken at the Nyquist rate and that the reconstruction is equal to the original random signal "almost everywhere." Equivalently, it states that the error signal $x_c(t) - \check{x}_c(t)$ has average power equal to zero. This then justifies representation of a bandlimited continuous random process by a discrete random process comprised of samples of the continuous process taken at the Nyquist rate.

4.9 WHITE NOISE

White noise is the term applied to any zero-mean random process whose power density spectrum is a constant. The name comes from an analogy to the idealized spectrum of white light which would contain power at all frequencies in equal proportions. (Physical white light does not really have a flat spectrum.) Since the power spectral density function is a constant, the correlation function for white noise is an impulse. In other words, *the samples of a white noise process are completely uncorrelated.*

Although discrete white noise processes can arise in a number of ways, a most common and important source of discrete white noise results from sampling a continuous white noise process. Continuous white noise is a kind of idealized random process that does not really exist in the physical world and is even difficult to describe mathematically. Nevertheless, it is an important conceptual tool in the analysis of continuous signals and systems. Therefore, when a discrete white noise process arises from sampling a continuous white noise process it is essential to understand the properties of both processes and their interrelation.

The organization of topics in this section is as follows. The next subsection introduces the concept of a continuous white noise process and develops some of its properties. It is shown that continuous white noise can be thought of as the formal derivative of the continuous Wiener process studied in Chapter 3. The section that follows examines the discrete process that results from sampling a bandlimited approximation to continuous white noise. It is pointed out that caution has to be exercised in defining the relation between continuous and discrete bandlimited white noise. Finally, the last subsection investigates discrete white noise processes in general and discusses their statistical characterization.

4.9.1 Continuous Gaussian White Noise

Let us begin the discussion in this subsection by reviewing the properties of the continuous Wiener process developed in Chapter 3. Recall that the continuous Wiener process was derived as the limiting case of a discrete Wiener process[17] as the step size and the interval between samples approaches zero. Because the Wiener process is the sum of an infinite number of identically distributed random steps, the Wiener process is a Gaussian random process. It is also nonstationary with mean zero and variance $\nu_o(t - t_o)$, where t_o is the time origin. The density for any single sample of the process is therefore

$$f_{x_c(t)}(\mathrm{x}) = \frac{1}{\sqrt{2\pi\nu_o(t - t_o)}} e^{-\frac{\mathrm{x}^2}{2\nu_o(t - t_o)}} \tag{4.147}$$

[17] The discrete Wiener process is also known as a symmetric random walk.

Without attempting to provide great mathematical rigor, we can argue that the correlation function for the Wiener process has the form

$$R_{x_c}^c(t_1, t_2) = \mathcal{E}\{x_c(t_1)x_c(t_2)\} = \nu_o \min(t_1 - t_o, t_2 - t_o) \tag{4.148}$$

To see this, suppose that $t_1 > t_2$ and write

$$\mathcal{E}\{x_c(t_1)x_c(t_2)\} = \mathcal{E}\{x_c(t_2)x_c(t_2)\} + \mathcal{E}\{\big(x_c(t_1) - x_c(t_2)\big)x_c(t_2)\}$$

The first term is just the variance $\nu_o(t_2 - t_o)$ while the second term is equal to zero since the process has independent increments.[18] When $t_1 \leq t_2$ the situation is just reversed, so that (4.148) follows.

It will be seen that the logical way to define a continuous white noise process is as the formal derivative of a Wiener process. The basic *mathematical* difficulty with such a definition however is that although the derivative of a continuous random process can be properly defined (see, e.g., [9, 14, 15]), the Wiener process by its very nature is not continuous *in any sense* and so does not really possess a proper derivative! Still, it is possible to consider a kind of limiting argument and define the white noise process to be what would logically result if it were possible to carry out the limiting operations as desired.

To begin the argument, consider a process w_c' defined by

$$w_c'(t) \stackrel{\text{def}}{=} \frac{x_c(t + \Delta t) - x_c(t)}{\Delta t} \tag{4.149}$$

where Δt is any very small but *finite* positive change in t. Now look at the correlation of this process at any two points t_1 and $t_2 > t_1$. The correlation is given by

$$\mathcal{E}\{w_c'(t_1)w_c'(t_2)\} = \mathcal{E}\left\{\frac{\big[x_c(t_1 + \Delta t) - x_c(t_1)\big]\big[x_c(t_2 + \Delta t) - x_c(t_2)\big]}{(\Delta t)^2}\right\} \tag{4.150}$$

$$= \frac{\mathcal{E}\{x_c(t_1 + \Delta t)x_c(t_2 + \Delta t) - x_c(t_1)x_c(t_2 + \Delta t) - x_c(t_1 + \Delta t)x_c(t_2) + x_c(t_1)x_c(t_2)\}}{(\Delta t)^2}$$

Now first consider the case where the two points are separated by more than the amount Δt (i.e., $t_1 > t_2 + \Delta t$). Then applying (4.148) to (4.150) yields

$$\mathcal{E}\{w_c'(t_1)w_c'(t_2)\} = \frac{\nu_o(t_2 + \Delta t - t_o) - \nu_o(t_2 + \Delta t - t_o) - \nu_o(t_2 - t_o) + \nu_o(t_2 - t_o)}{(\Delta t)^2}$$

$$= 0; \qquad t_1 > t_2 + \Delta t$$

In other words, any two samples of the process $w_c'(t)$ separated by more than Δt are uncorrelated.

[18] Recall that the independent increment property means that "events" such as $x_c(t_1) - x_c(t_2)$ and $x_c(t_2)$, which are defined on nonoverlapping intervals, are independent. The intervals are nonoverlapping because the second interval $(t_2, t_1]$ does not include the point t_2 (see Chapter 3).

Next consider the case where $t_2 \leq t_1 \leq t_2 + \Delta t$. Again from (4.148) and (4.150) we have

$$\mathcal{E}\{w_c'(t_1)w_c'(t_2)\} = \frac{\nu_o(t_2 + \Delta t - t_o) - \nu_o(t_1 - t_o) - \nu_o(t_2 - t_o) + \nu_o(t_2 - t_o)}{(\Delta t)^2}$$

$$= \frac{\nu_o \Delta t - \nu_o(t_1 - t_2)}{(\Delta t)^2} = \frac{\nu_o}{\Delta t}\left(1 - \frac{t_1 - t_2}{\Delta t}\right) \quad t_2 \leq t_1 \leq t_2 + \Delta t$$

Note that since this is a function only of the difference $t_1 - t_2$, the process $w_c'(t)$ is *wide-sense stationary*. The last two equations imply that the correlation function for $w_c'(t)$ has the form shown in Fig. 4.20(a). Notice that the variance $R^c_{w_c'}(0)$ (which follows from the last equation with $t_2 = t_1$) is equal to $\nu_o/\Delta t$. Now without attempting to take any formal limit, notice that smaller and smaller values of Δt yield larger and larger values for the variance, while points at which the process is uncorrelated move closer and closer together. As Δt decreases the correlation function becomes very tall and narrow, but the area (which is the total energy) remains constant.

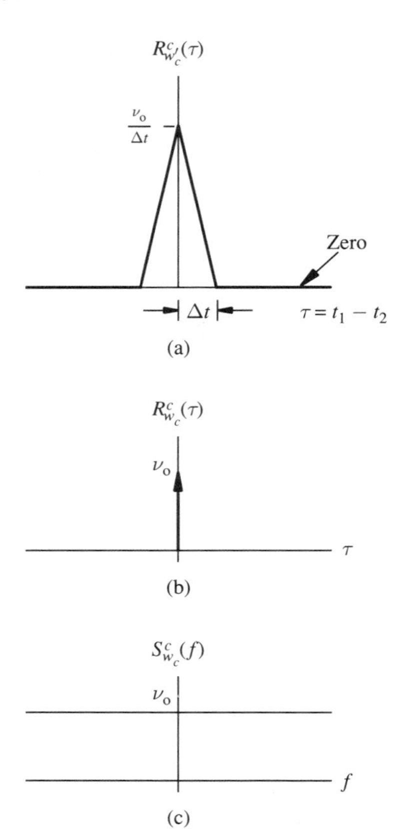

Figure 4.20 Second moment functions for continuous white noise. (a) Correlation function for the process $w_c'(t)$. (b) Correlation function for white noise. (c) Power spectral density function for white noise.

This behavior motivates *defining* a continuous white noise process as a random process that has a correlation function which is the limiting case of the correlation function depicted in Fig. 4.20(a): namely, a unit impulse with area ν_o. This process has *infinite* variance (or power) but finite energy, and points separated by any distance no matter how small are uncorrelated. This new process $w_c(t)$ is called a *continuous white noise process*. Since its correlation function is an impulse, its power spectral density is a constant ν_o and the process has *infinite* bandwidth. The correlation and power spectral density functions for continuous white noise are *defined* by

$$R^c_{w_c}(\tau) = \nu_o \delta_c(\tau) \tag{4.151}$$

and

$$S^c_{w_c}(f) = \nu_o \tag{4.152}$$

Both are depicted in Fig. 4.20 [(b) and (c)].

The continuous white noise process, although in most ways fictitious, is still a vital concept in the study of random processes. It serves to model the various types of noise encountered in the physical world. Since all of these physical noise processes are observed only after passing through electronic, optical, or other systems, so too white noise is used in analyses primarily with models for other parts of a system. In this context its impossible physical properties and intractible mathematical properties can be excused. Further, since we do not wish to contradict the principle that linear transformations of Gaussian random processes are Gaussian, and filtering of a physical noise process frequently results in a Gaussian output process (see Chapter 3), we *define* the continuous white noise process developed in this section to be *Gaussian!*

4.9.2 Sampling Bandlimited White Noise

While most deterministic and random processes can be sampled at some suitably high rate and then reconstructed from the samples, it is impossible to do this for continuous white noise. This follows because continuous white noise by definition has infinite bandwidth, i.e., *it is not bandlimited.* As a result it is necessary to restrict analysis for discrete systems to noise that is not truly white but has a constant spectrum over some limited finite band—say, $-W < f < +W$. This type of noise is called *bandlimited white noise*. The restriction to bandlimited white noise is not of great significance since this type of noise is more representative of what is likely to be encountered in the physical world anyway. For one thing, since the bandwidth is finite, this noise has finite rather than infinite power. Further, if the bandwidth of the noise is larger than that of any of the systems to which it is applied, then the output is the same as if white noise of infinite bandwidth had appeared at the input.

Care must be exercised when designing or analyzing a discrete system to perform processing of sampled continuous signals, however, if the true bandwidth of the noise exceeds twice the sampling frequency. This is one reason that continuous signals are usually passed through a sharp cutoff filter at the sampling bandwidth (sometimes called an "antialiasing" filter) before sampling.

To illustrate the procedure, consider a bandlimited continuous white noise process with the rectangular power spectrum shown in Fig. 4.21(a). The corresponding correlation

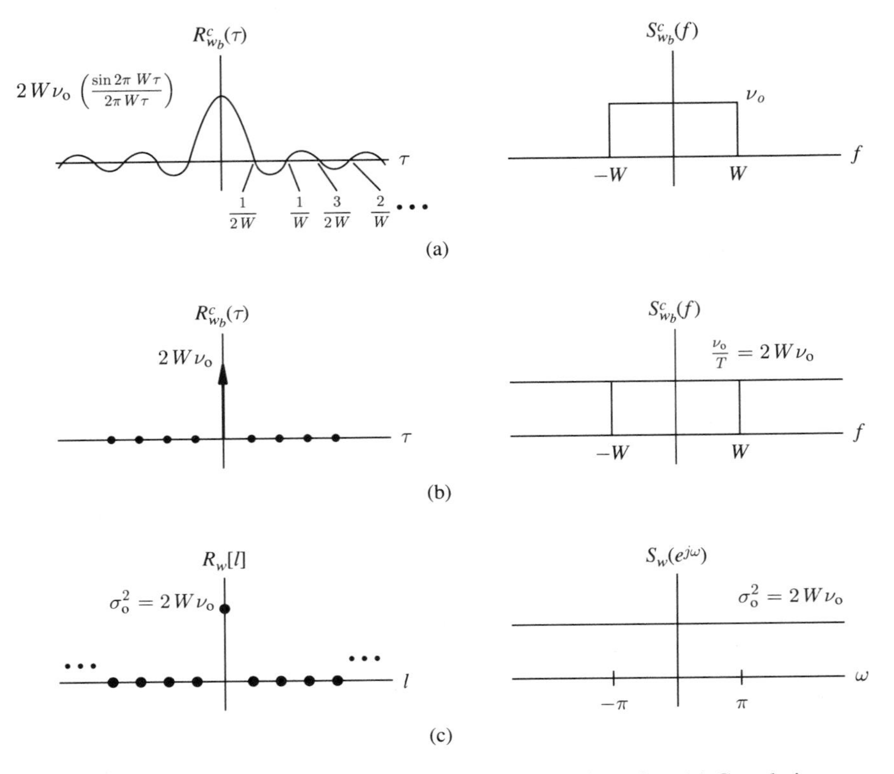

Figure 4.21 Sampling of bandlimited continuous white noise. (a) Correlation function and spectrum of bandlimited white noise. (b) Correlation function and spectrum for the sampled noise. (c) Discrete representation of the sampled noise.

function also shown has zero crossings at $\tau = l/2W, \quad l = \pm 1, \pm 2, \pm 3, \ldots$. If this random process is sampled by a periodic impulse train with spacing $T = 1/2W$, then the result is as shown in Fig. 4.21(b). The sampling impulse occurring at the origin has area $2W\nu_o$, and all others, which fall at zero crossings, have area zero. In the frequency domain the original spectrum is repeated periodically without aliasing to form a totally flat spectrum. Because of the sampling the amplitude is modified to $\nu_o/T = 2W\nu_o$. The sampled correlation function and spectrum then have the discrete representation shown in Fig. 4.21(c), where we have defined the noise power for the discrete system to be

$$\boxed{\sigma_o^2 = 2W\nu_o} \tag{4.153}$$

Since the correlation function of the discrete process is an impulse occurring at the origin, the discrete process consists of uncorrelated samples and has a constant power spectrum. This process is called *discrete white noise* and its relation to the bandlimited continuous white noise through (4.153) is clear.

If the noise bandwidth is greater than twice the sampling frequency, however, the relationship between the continuous and the discrete noise process can become ambiguous. For example, suppose that the bandlimited continuous white noise has a bandwidth twice as large as in the previous discussion, but perhaps because the other signals in the system have smaller bandwidth, the sampling is still taken at a spacing of $T = 1/2W$. The situation is depicted in Fig. 4.22. Because of the undersampling the noise is totally aliased and the noise power of the sampled noise is doubled. The discrete representation is still the same except that the noise power σ_o^2 now represents a continuous noise power of $4W\nu_o$ instead of $2W\nu_o$.

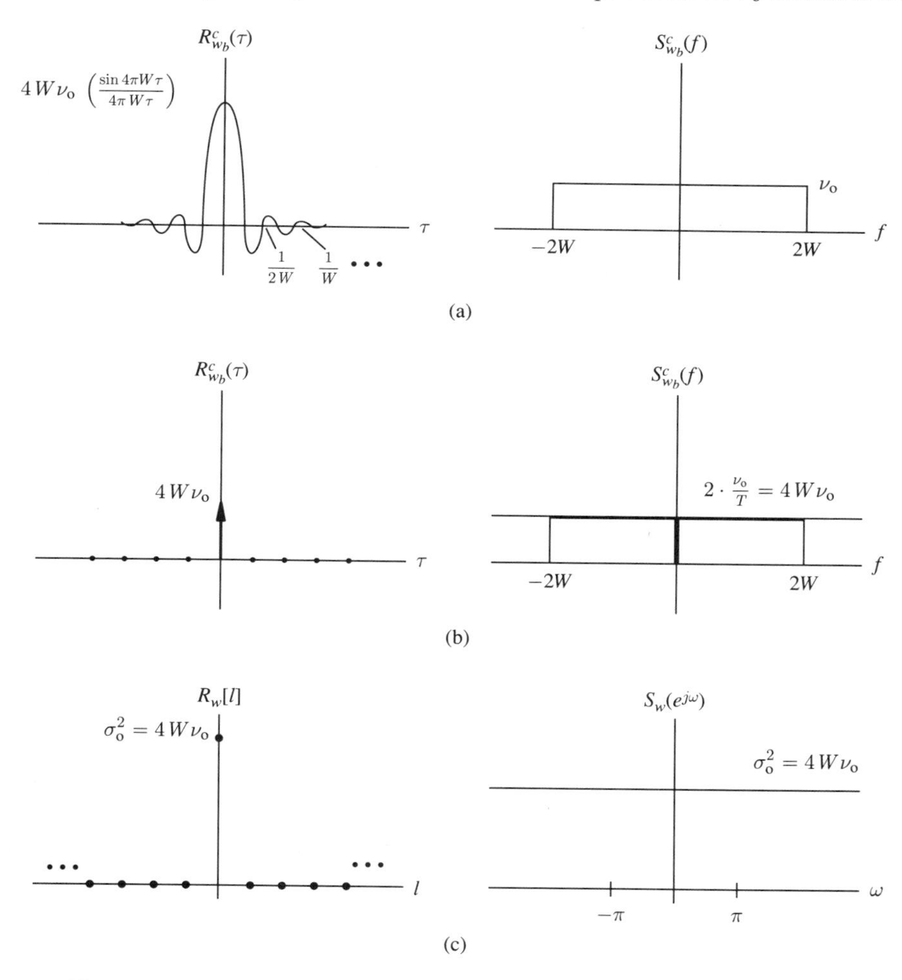

Figure 4.22 Undersampling of bandlimited continuous white noise. (a) Correlation function and spectrum of bandlimited white noise. (b) Correlation function and spectrum for the sampled noise. (c) Discrete representation of the sampled noise.

The point of this last example is to illustrate that some care must be taken when representing sampled continuous random processes by discrete random processes, and that the situation cannot be taken out of context. Normally, as mentioned earlier, the signals are filtered prior to sampling so that the out-of-band noise does not become folded in and degrade the signal-to-noise ratio.

4.9.3 Discrete White Noise

Any discrete zero-mean random process whose samples are uncorrelated will be called a "white" process or a *white noise process*.[19] The correlation function of a white noise process $w[n]$ has the form

$$R_w[n_1, n_0] = \sigma_o^2[n_1]\delta[n_1 - n_0] \tag{4.154}$$

Usually, a white noise process is taken to mean a *stationary* white noise process so that

$$\boxed{R_w[l] = \sigma_o^2\delta[l]} \tag{4.155}$$

The correlation function and corresponding power spectral density function

$$\boxed{S_w(e^{\jmath\omega}) = \sigma_o^2} \tag{4.156}$$

are shown in Fig. 4.23.

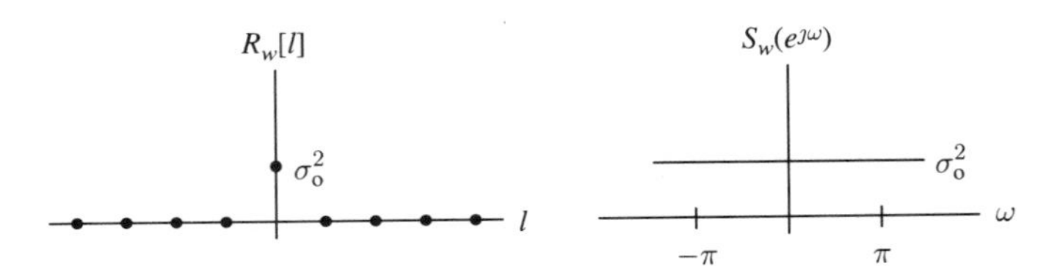

Figure 4.23 Correlation function and power density spectrum for white noise.

A very important example of white noise is the Gaussian white noise process that results from sampling continuous white noise. If $\boldsymbol{w}$ is a random vector of N consecutive samples from this process, then the process has a covariance matrix of the form

$$\mathbf{C}_w = \mathbf{R}_w = \begin{bmatrix} \sigma_o^2 & 0 & \cdots & 0 \\ 0 & \sigma_o^2 & \cdots & 0 \\ \vdots & \vdots & \ddots & \vdots \\ 0 & 0 & \cdots & \sigma_o^2 \end{bmatrix} = \sigma_o^2\mathbf{I} \tag{4.157}$$

[19]From here on the term "white noise" will be assumed to mean discrete white noise and the modifier "discrete" will not be used.

and the probability density function is given by

$$f_w(\mathbf{w}) = \frac{1}{(2\pi\sigma_o^2)^{N/2}} e^{-\frac{\|\mathbf{w}\|^2}{2\sigma_o^2}} = \prod_{n=0}^{N-1} \frac{1}{\sqrt{2\pi\sigma_o^2}} e^{-\frac{(w_n)^2}{2\sigma_o^2}} \tag{4.158}$$

(real random process)

Observe that the samples of a Gaussian white noise process are not only uncorrelated but also *independent.*

A random process that frequently occurs in analysis of radar, sonar, and communications systems is the complex Gaussian white noise process that results when a bandpass noise process is represented by a complex random process at baseband. Such a process has the complex density function

$$f_w(\mathbf{w}) = \frac{1}{(\pi\sigma_o^2)^N} e^{-\frac{\|\mathbf{w}\|^2}{\sigma_o^2}} = \prod_{n=0}^{N-1} \frac{1}{\pi\sigma_o^2} e^{-\frac{|w_n|^2}{\sigma_o^2}} \tag{4.159}$$

(complex random process)

and can also be thought of as generated by a sequence of independent complex random variables whose magnitude has the Rayleigh density (4.105) and whose phase is uniformly distributed between $-\pi$ and π [16]. The real and imaginary components of course have a jointly Gaussian density function and satisfy the symmetry conditions discussed in Section 4.5.

Although Gaussian white noise is a common occurrence in many signal processing systems, *a discrete white noise process need not be Gaussian.* The only requirement of discrete white noise in fact is that the mean is zero and the samples are uncorrelated. The binary white noise process was already encountered in Chapter 3. Although this process takes on only the two values $+1$ and -1, it satisfies the definition given here. In fact, any process with independent, identically distributed samples is a white noise process. A process for which the samples are independent, rather than just uncorrelated, is sometimes called a *strictly* white process [11]. Many discrete white noise processes are of this type.

4.10 HIGHER-ORDER MOMENTS

It was stated in the introduction to this chapter that second moment analysis is frequently sufficient to characterize the most important features of a random process, even if the process is not Gaussian. There are, however, cases involving non-Gaussian random processes where second moment analysis does *not* provide all of the needed information. An example is when phase of the random process is important. Second moment quantities (the autocorrelation function, power spectral density, and complex spectral density function) suppress phase information. Since computing the entire density function for the random process is usually impractical or impossible, a compromise is to deal with the higher-order (i.e., higher than second order) moments of the density. This approach not only provides tractible analysis methods in the signal domain, but also leads to frequency-domain methods through suitable extensions of Fourier analysis for stationary random processes.

This section provides a brief introduction to the characterization of random processes using higher-order moments. Although the field is currently an area of intense research and new results are constantly being reported, known results in higher order moments can already fill a volume of its own. The intent here is only to give a general description of the methods and to give some references to the literature where you can find more information (see, e.g., [17–24]). The next subsection gives some general facts about higher-order moments and related functions. The section following that then looks at the third moment functions in a little more detail.

4.10.1 Moments, Cumulants, and Higher-Order Spectra

Let us begin this topic by starting with the definition of moments for a stationary random process. The first four moments are defined by

$$M_x^{(1)} = \mathcal{E}\{x[n]\} = m_x \qquad (a)$$

$$M_x^{(2)}[l_1] = \mathcal{E}\{x^*[n]x[n+l_1]\} = R_x[l_1] \qquad (b)$$

$$M_x^{(3)}[l_1, l_2] = \mathcal{E}\{x^*[n]x[n+l_1]x[n+l_2]\} \qquad (c) \qquad (4.160)$$

$$M_x^{(4)}[l_1, l_2, l_3] = \mathcal{E}\{x^*[n]x^*[n+l_1]x[n+l_2]x[n+l_3]\} \qquad (d)$$

The first two moments are the equal to the mean and correlation function defined earlier. The third and fourth moments provide new information. For the complex case the expressions given in (4.160)(c) and (d) are only one way to define the moments. There is no compelling reason to associate the conjugate with one term or the other, and it is sometimes useful to define the complex moments differently for different problems [23].

Although the moments provide all of the needed information for higher-order analysis of a random process it is usually preferable to work with related quantities called *cumulants*, which more clearly exhibit the additional information brought in by the higher-order statistics. Their use is analogous to using the covariance instead of the correlation function in second moment analysis to remove the effect of the mean. Cumulants measure the departure of a random process from a Gaussian random process.

Cumulants can be defined as terms in the expansion of a *cumulant-generating function* (analogous to a moment generating function) or via certain partitions of the set of random variables $x[n], x[n+l_1], \ldots, x[n+l_k]$ [24]. A more intuitively satisfying definition, however, is as follows [17]. If $x[n]$ is a non-Gaussian random process, and $x'[n]$ is a *Gaussian* random process with same mean and correlation function as $x[n]$, then the first two cumulants of $x[n]$ are equal to the mean and covariance function of $x[n]$, while the cumulants $C_x^{(k)}$ of higher order satisfy

$$C_x^{(k)}[l_1, l_2, \ldots, l_k] = M_x^{(k)}[l_1, l_2, \ldots, l_k] - M_{x'}^{(k)}[l_1, l_2, \ldots, l_k] \qquad k = 3, 4 \tag{4.161}$$

Explicit expressions for the first four cumulants for the case of a zero-mean random process are

$$C_x^{(1)} = \mathcal{E}\{x[n]\} = m_x = 0 \qquad (a)$$

$$C_x^{(2)}[l_1] = \mathcal{E}\{x^*[n]x[n+l_1]\} = R_x[l_1] \qquad (b)$$

$$C_x^{(3)}[l_1, l_2] = \mathcal{E}\{x^*[n]x[n+l_1]x[n+l_2]\} \qquad (c)$$

(4.162)

$$\begin{aligned} C_x^{(4)}[l_1, l_2, l_3] = &\ \mathcal{E}\{x^*[n]x^*[n+l_1]x[n+l_2]x[n+l_3]\} \\ &- C_x^{(2)}[l_2]C_x^{(2)}[l_3 - l_1] - C_x^{(2)}[l_3]C_x^{(2)}[l_2 - l_1] \end{aligned} \qquad (d)$$

(complex random process)

$$\begin{aligned} C_x^{(4)}[l_1, l_2, l_3] = &\ \mathcal{E}\{x[n]x[n+l_1]x[n+l_2]x[n+l_3]\} - C_x^{(2)}[l_1]C_x^{(2)}[l_3 - l_2] \\ &- C_x^{(2)}[l_2]C_x^{(2)}[l_3 - l_1] - C_x^{(2)}[l_3]C_x^{(2)}[l_2 - l_1] \end{aligned} \qquad (e)$$

(real random process)

The expression for the fourth-order cumulant involves the subtraction of the fourth-order moment for the Gaussian process, which can be expressed in terms of second-order cumulants. That is why the expression is more complicated. In addition, the expressions differ by one term for the real and complex cases. The difference stems from the condition (4.67), which is assumed to hold for the complex case. Since the cumulants of order three and four satisfy (4.161), *the higher-order cumulants for a Gaussian random process are identically zero*.

The spectral quantities related to the cumulants are called *cumulant spectra*, *higher-order spectra*, or *polyspectra*. The second-order spectrum is the ordinary power spectrum

$$S_x^{(2)}(e^{\jmath\omega}) = \sum_{l_1=-\infty}^{\infty} C_x^{(2)}[l_1]e^{-\jmath\omega l_1} = S_x(e^{\jmath\omega}) \tag{4.163}$$

The third- and fourth-order spectra are called the *bispectrum* and *trispectrum* respectively. These quantities are the two-dimensional and three-dimensional Fourier transforms of the third- and fourth-order cumulants:

$$S_x^{(3)}(e^{\jmath\omega^{(1)}}, e^{\jmath\omega^{(2)}}) = \sum_{l_1=-\infty}^{\infty} \sum_{l_2=-\infty}^{\infty} C_x^{(3)}[l_1, l_2] e^{-\jmath(\omega^{(1)} l_1 + \omega^{(2)} l_2)} \tag{4.164}$$

(bispectrum)

$$S_x^{(4)}(e^{\jmath\omega^{(1)}}, e^{\jmath\omega^{(2)}}, e^{\jmath\omega^{(3)}}) = \sum_{l_1=-\infty}^{\infty} \sum_{l_2=-\infty}^{\infty} \sum_{l_3=-\infty}^{\infty} C_x^{(4)}[l_1, l_2, l_3] e^{-\jmath(\omega^{(1)} l_1 + \omega^{(2)} l_2 + \omega^{(3)} l_3)}$$

(trispectrum) (4.165)

Note that these higher-order spectra are periodic in *each* of their arguments ($\omega^{(1)}$, $\omega^{(2)}$, $\omega^{(3)}$) with period 2π. Since for a Gaussian random process the higher-order cumulants are zero, the bispectrum and trispectrum for a Gaussian process are identically zero.

The third- and fourth-order cumulants and corresponding spectral quantities have been used in a number of applications dealing with non-Gaussian processes. Cumulants and spectra of order higher than the fourth are difficult to compute reliably and so far have found limited practical use. The following subsection examines the third-order cumulant and the bispectrum a little more closely and shows one simple example of its application. It should be noted, however, that when the random process has a density that is symmetric about the mean, the third-order cumulant and the bispectrum are identically zero and it is necessary to consider fourth-order quantities to perform any higher-order statistical analysis.

4.10.2 Symmetry Properties of Third Moment Quantities

It is obvious from the definition that the third-order cumulant has the symmetry property

$$C_x^{(3)}[l_1, l_2] = C_x^{(3)}[l_2, l_1] \tag{4.166}$$

In addition, it can be shown from the definition that for a real random process the third-order cumulant has the following addition symmetry properties (see Problem 4.30):

$$\begin{aligned} C_x^{(3)}[l_1, l_2] &= C_x^{(3)}[-l_2, l_1 - l_2] = C_x^{(3)}[-l_1, l_2 - l_1] \\ &= C_x^{(3)}[l_2 - l_1, -l_1] = C_x^{(3)}[l_1 - l_2, -l_2] \end{aligned} \tag{4.167}$$

(real random process)

The symmetry properties are depicted in Fig. 4.24. For the complex cumulant, knowledge of $C_x^{(3)}[l_1, l_2]$ in a half plane is sufficient to specify all of its other values, while for the real cumulant, knowledge in the first *octant* [shown shaded in Fig. 4.24(b)] is sufficient to define it in all of the remaining five regions.

For convenience of notation, let us denote the bispectrum by the function $B_x(\omega^{(1)}, \omega^{(2)})$. Then (4.164) can be rewritten as

$$B_x(\omega^{(1)}, \omega^{(2)}) = \sum_{l_1=-\infty}^{\infty} \sum_{l_2=-\infty}^{\infty} C_x^{(3)}[l_1, l_2] e^{-\jmath(\omega^{(1)} l_1 + \omega^{(2)} l_2)} \tag{4.168}$$

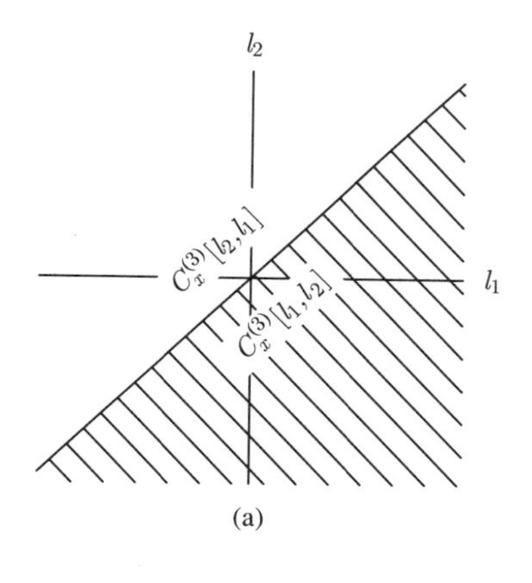

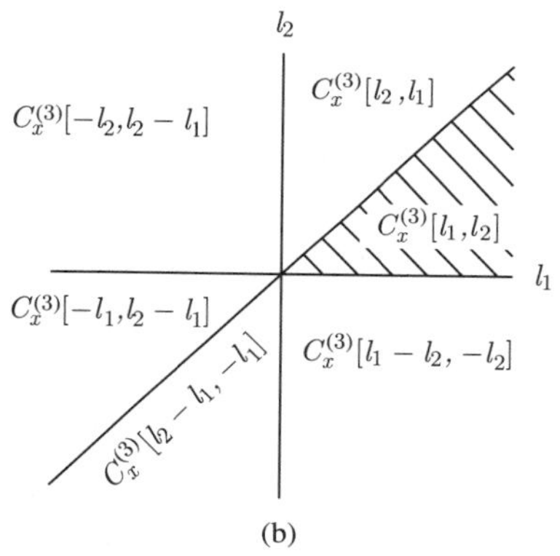

Figure 4.24 Regions of symmetry for the third-order cumulant. (a) The two regions of symmetry for the cumulant of a complex random process. (b) The six regions of symmetry for the cumulant of a real random process.

From (4.168) and (4.166) it follows that the bispectrum has the basic symmetry property

$$B_x(\omega^{(1)},\omega^{(2)}) = B_x(\omega^{(2)},\omega^{(1)}) \tag{4.169}$$

Thus in the general complex case the bispectrum, like the cumulant, has half-plane symmetry (see Fig. 4.25).

In the case of a real random process, it follows from (4.167) that

$$\begin{aligned} B_x(\omega^{(1)},\omega^{(2)}) &= B_x(-\omega^{(1)}-\omega^{(2)},\omega^{(1)}) = B_x(-\omega^{(1)}-\omega^{(2)},\omega^{(2)}) \\ &= B_x(\omega^{(2)},-\omega^{(1)}-\omega^{(2)}) = B_x(\omega^{(1)},-\omega^{(1)}-\omega^{(2)}) \\ &\qquad \text{(real random process)} \end{aligned} \tag{4.170}$$

If these are combined with the additional conjugate symmetric property

$$B_x(\omega^{(1)},\omega^{(2)}) = B_x^*(-\omega^{(1)},-\omega^{(2)}) \qquad \text{(real random process)} \tag{4.171}$$

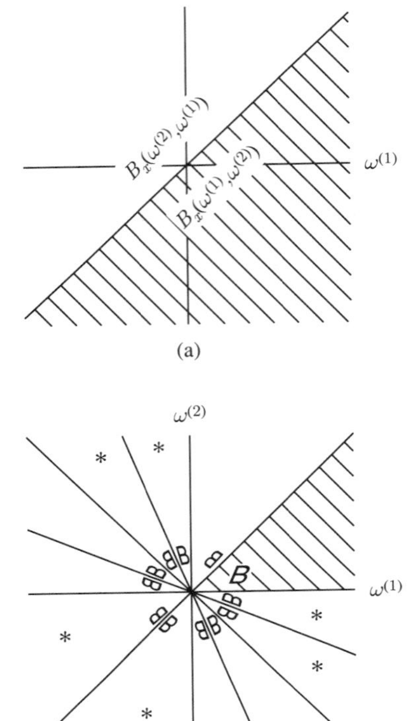

Figure 4.25 Regions of symmetry for the bispectrum. (a) The two regions of symmetry for the bispectrum of a complex process. (b) The 12 regions of symmetry for the bispectrum of a real process. Position of letter B shows how bispectrum features are reflected in regions of symmetry; $*$ shows regions where bispectrum is *conjugate* symmetric.

then it can be shown that there are 12 regions of symmetry depicted in Fig. 4.25(b) (see Problem 4.30). Thus the bispectrum needs only to be computed explicitly in the shaded region, not over the entire plane.

The bispectrum is useful in detecting phase relationships between harmonicaly related components of a random process known as *quadratic phase coupling* [20, 21]. The ordinary power spectrum is insensitive to such phase coupling. The following example illustrates this point.

Example 4.9

A complex periodic random signal consists of three components:

$$x[n] = \underbrace{e^{\jmath(\omega_a n+\phi_a)}}_{s_a[n]} + \underbrace{e^{\jmath(\omega_b n+\phi_b)}}_{s_b[n]} + \underbrace{e^{\jmath(\omega_c n+\phi_c)}}_{s_c[n]}$$

To simplify the discussion all of the components have been chosen to have unit magnitude, although this condition is certainly not required. The frequencies ω_a, ω_b, and ω_c are harmonically related so that

$$\omega_a = \omega_b + \omega_c$$

The phases ϕ_a and ϕ_b are independent and uniformly distributed between $-\pi$ and π. Two cases are considered for the phase ϕ_c.

Case 1. ϕ_c is independent of the other two phases and is also uniformly distributed between $-\pi$ and π. Note that $x[n]$ is of the form given in Table 4.6. For example, the component $s_a[n]$ has the form

$$s_a[n] = Ae^{\jmath\omega_a n}$$

where the complex amplitude is given by

$$A = 1 \cdot e^{\jmath\phi_a}$$

(i.e., the magnitude of A is fixed but the phase is random). It can easily be shown that the complex amplitudes satisfy the necessary orthogonality conditions (4.104). Therefore, the power density spectrum consists of three lines as shown in Fig. EX4.9a.

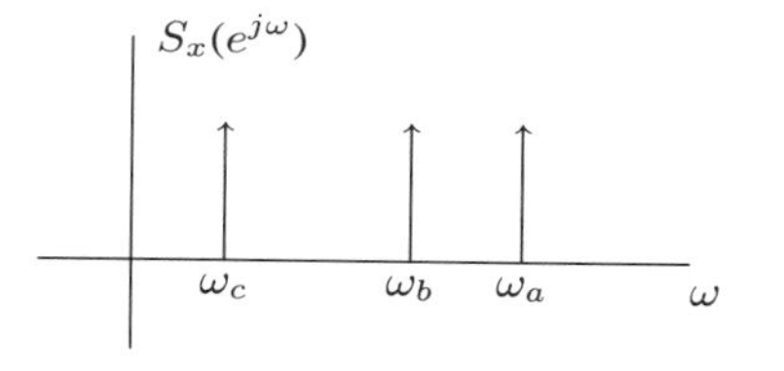

Figure EX4.9a

Now consider the cumulant

$$C_x^{(3)}[l_1, l_2] = \mathcal{E}\{[x[n]^* x[n+l_1]x[n+l_2]\}$$

This involves cross-products of terms such as

$$\mathcal{E}\{s_a^*[n]s_b[n+l_1]s_c[n+l_2]\}$$

Since all of the components are independent and have zero mean, all such cross-terms are zero. The terms involving a single component such as

$$\mathcal{E}\{s_a^*[n]s_a[n+l_1]s_a[n+l_2]\}$$

are also zero. Specifically,

$$\mathcal{E}\left\{e^{-\jmath(\omega_a n+\phi_a)}e^{\jmath(\omega_a(n+l_1)+\phi_a)}e^{\jmath(\omega_a(n+l_2)+\phi_a)}\right\} = e^{\jmath\omega_a(n+l_1+l_2)}\mathcal{E}\left\{e^{\jmath\phi_a}\right\} = 0$$

Thus, since $C_x^{(3)}$ is identically zero, the bispectrum is identically zero.

Case 2. ϕ_a is linearly related to the other two phases; in particular,

$$\phi_a = \phi_b + \phi_c$$

In this case the power density spectrum is exactly the same as before; the coupling between the phase of the components does not destroy the orthogonality between the complex amplitudes.

The cumulant involves expectations of products of the components as before, most of which are zero. However, consider the cross-term

$$\mathcal{E}\{s_a^*[n]s_b[n+l_1]s_c[n+l_2]\} = \mathcal{E}\{e^{-\jmath(\omega_a n+\phi_b+\phi_c)}e^{\jmath(\omega_b(n+l_1)+\phi_b)}e^{\jmath(\omega_c(n+l_2)+\phi_c)}\}$$
$$= e^{\jmath(\omega_b l_1+\omega_c l_2)}\mathcal{E}\{e^{\jmath(\phi_b+\phi_c-\phi_b-\phi_c}\} = e^{\jmath(\omega_b l_1+\omega_c l_2)}$$

where we used the fact that $\omega_a = \omega_b + \omega_c$. This term is nonzero and produces an impulse in the bispectrum at $\omega^{(1)} = \omega_b$ and $\omega^{(2)} = \omega_c$, indicating the phase coupling between these frequency components as shown in Fig. EX4.9b.

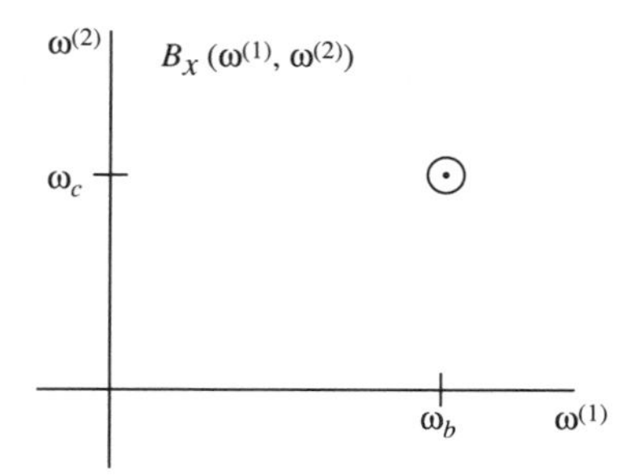

Figure EX4.9b

Further uses of cumulants and higher-order spectra become apparent when transformations of random processes are considered. Linear transformations of random processes produce prescribed effects on all the moments of a random process. If the processes are non-Gaussian, then cumulant analysis can be useful. Nonlinear transformations of even a Gaussian random process produce random processes with higher-order moments different from those of a Gaussian process and require analysis using higher-order moments or cumulants if the effect of the nonlinearity is to be fully accounted for. Thus cumulants and higher-order spectra will probably play an important role in future applications of statistical signal processing.

4.11 CHAPTER SUMMARY

This chapter concentrates on representing random processes in terms of their second moment properties. It begins by defining the mean, correlation, and covariance functions for random processes. A random process is defined to be *wide-sense stationary* if the mean is constant and the correlation and covariance functions are dependent on just the *difference* between the locations of the samples. Strict-sense stationary random processes are always wide-sense stationary, but wide-sense stationary processes are not strict-sense stationary unless they are Gaussian. A correlation or covariance function for a stationary random process is and must be both conjugate symmetric and positive semidefinite.

The correlation and covariance matrices introduced in Chapter 2 can be defined for a vector of samples from a random process; the components of the matrices are then determined by the correlation and covariance functions. When a random process is stationary the matrices have Toeplitz structure; the $N \times N$ Toeplitz matrix is completely specified by just N instead of N^2 correlation or covariance parameters.

Cross-correlation and cross-covariance functions specify the second moments between two random processes. These functions are not positive semidefinite in general and have no required symmetry properties. Cross-correlation and cross-covariance matrices for random processes are comprised of terms from these functions. Two random processes are jointly stationary (in the wide sense) if they are each wide-sense stationary and their cross-correlation and cross-covariance functions depend only on the difference between the locations of the samples.

Stationary random processes are characterized in the frequency and transform domains by the power spectral density function and the complex spectral density function. The former is the Fourier transform of the correlation function; the latter is the z-transform of the correlation function. Both satisfy special symmetry properties. Moreover, the power spectral density is always real and never negative. Roots of complex spectral density functions that are rational polynomials always occur in conjugate reciprocal pairs. The Fourier and z-transforms of the cross-correlation function are called the cross-power spectral density function and the complex cross-spectral density function. Neither is required to have any special symmetry properties.

The definitions of the correlation (and covariance) function for complex random processes assume that certain symmetry conditions exist between the second moments of the real and imaginary parts of the random process. These conditions are derived from conditions on a bandpass random process that are required for its stationarity and permit the autocorrelation and cross-correlation functions for the components of the random process to be derived from the real and imaginary parts of the complex autocorrelation function. Corresponding relations exist for the power spectral density and the complex spectral density functions.

The discrete Karhunen-Loève transform (DKLT) is a representation of a random process in an orthonormal basis set with orthogonal coefficients. The DKLT basis functions are known as eigenfunctions of the random process and correspond to eigenvectors of the correlation matrix. When a random process is stationary and periodic the DKLT is identical to the discrete Fourier transform (DFT). For nonperiodic stationary random processes the DFT is approximately the same as the DKLT if the observation interval is taken to be sufficiently long. The DKLT basis functions in this case are harmonically related complex exponentials.

Many discrete random processes are representations of sampled continuous random processes. Correlation functions and power spectral density functions for continuous random processes are defined analogously to those for discrete random processes. If a continuous random process is stationary and bandlimited, then a stochastic version of the sampling theorem guarantees that the continuous random process can be reconstructed "almost everywhere" from samples taken at the Nyquist rate.

White noise (continuous or discrete) is a random process with a flat or constant power density spectrum. Continuous white noise can be *thought of* as the formal derivative of a continuous Wiener process. Discrete Gaussian white noise can be derived by sampling a bandlimited approximation to continuous white noise. Its average power is proportional to the bandwidth of the continuous noise. More generally, discrete white noise is any random process whose correlation function is an impulse and whose power spectral density function is a constant. It need not be Gaussian.

Finally, non-Gaussian discrete random processes can be described in terms of higher-order moments and cumulants when second moment characterization is not sufficient. Corresponding higher-order spectra such as the bispectrum and trispectrum can be defined for stationary random processes as multidimensional Fourier transforms of the cumulants. These higher-order spectra can exhibit phase information that relates to properties of the random processes that are not capable of being inferred from the ordinary power density spectrum. An example of this is phase coupling between two random signals. Higher-order moment analysis of random processes has only recently started to attract more research interest, and procedures based on higher-order moments can be expected to become more predominant as these methods are developed further.

REFERENCES

1. M. J. Lighthill. *An Introduction to Fourier Analysis and Generalized Functions.* Cambridge University Press, New York, 1959.
2. Athanasios Papoulis. *The Fourier Integral and Its Applications.* McGraw-Hill, New York, 1962.
3. Alan V. Oppenheim and Ronald W. Schafer. *Discrete-Time Signal Processing.* Prentice Hall, Inc., Englewood Cliffs, New Jersey, 1989.
4. Alan V. Oppenheim and Ronald W. Schafer. *Digital Signal Processing.* Prentice Hall, Inc., Englewood Cliffs, New Jersey, 1975.
5. K. Karhunen. Über linearen methoden in der wahrscheinlichkeitsrechnung. *Ann. Acad. Sci. Fennical, Series A*, 1(2), 1947. English translation by I. Selin, "On linear methods in probability theory," The Rand Corp., Doc. T-131, August 11, 1960.
6. M. Loève. Sur les fonctions aléatoires stationnaires de second order. *Rev. Sci.*, 83:297–310, 1945.
7. Wilbur B. Davenport, Jr. and William L. Root. *An Introduction to the Theory of Random Signals and Noise.* McGraw-Hill, New York, 1958.
8. Harry L. Van Trees. *Detection, Estimation, and Modulation Theory*, Part I. John Wiley & Sons, New York, 1968.
9. Michel Loève. *Probability Theory.* Van Nostrand, Princeton, New Jersey, 1955.
10. Gilbert Strang. *Introduction to Applied Mathematics.* Wellesley Cambridge Press, Wellesley, Massachusetts, 1986.
11. Athanasios Papoulis. *Probability, Random Variables, and Stochastic Processes*, 3rd ed. McGraw-Hill, New York, 1991.
12. Douglas F. Elliott and K. Ramamohan Rao. *Fast Transforms: Algorithms, Analyses, Applications.* Academic Press, New York, 1982.
13. Alan S. Willsky. *Digital Signal Processing and Control and Estimation Theory: Points of Tangency, Areas of Intersection, and Parallel Directions.* MIT Press, Cambridge, Massachusetts, 1979.

14. Henry Stark and John W. Woods. *Probability, Random Processes, and Estimation Theory for Engineers*. Prentice Hall, Inc., Englewood Cliffs, New Jersey, 1986.

15. Harold J. Larson and Bruno O. Schubert. *Probabilistic Models in Engineering Sciences*, Volume II. John Wiley & Sons, New York, 1979.

16. Kenneth S. Miller. *Complex Stochastic Processes*. Addison-Wesley, Reading, Massachusetts, 1974.

17. Jerry M. Mendel. Tutorial on higher-order statistics (spectra) in signal processing and system theory: theoretical results and some applications. *Proceedings of the IEEE*, 79(3):278–305, March 1991.

18. Jerry M. Mendel. Use of higher order statistics in signal processing and system theory: a short perspective. In *Proceedings of 8th International Conference on Mathematical Theory of Networks and Systems*, Phoenix, Arizona, June 1987.

19. Jerry M. Mendel. Use of higher order statistics in signal processing and system theory: an update. In *Proceedings of SPIE Conference on Advanced Algorithms and Architectures for Signal Processing*, San Diego, California, August 1988.

20. Chrysostomos L. Nikias. Higher-order spectral analysis. In Simon Haykin, editor, *Advances in Spectrum Analysis and Array Processing*, pages 326–365. Prentice Hall, Inc., Englewood Cliffs, New Jersey, 1992.

21. Chrysostomos L. Nikias and Mysore R. Raghuveer. Bispectrum estimation: a digital signal processing framework. *Proceedings of the IEEE*, 75(7):869–891, July 1987.

22. D. R. Brillinger. An introduction to polyspectra. *Annals of Mathematical Statistics*, 36:1351–1374, 1965.

23. Chrysostomos L. Nikias. *Higher Order Moments in Digital Signal Processing*. Prentice Hall, Inc., Englewood Cliffs, New Jersey.

24. Murray Rosenblatt. *Stationary Sequences and Random Fields*. Birkhäuser, Cambridge, Massachusetts, 1985.

PROBLEMS

4.1. **(a)** Determine the mean and autocorrelation function for the random process

$$x[n] = v[n] + 3v[n-1]$$

where $v[n]$ is a sequence of independent random variables with mean μ and variance σ^2.

(b) Is $x[n]$ stationary?

4.2. Random processes $x[n]$ and $y[n]$ are defined by

$$x[n] = v_1[n] + 3v_2[n-1]$$

and

$$y[n] = v_2[n+1] + 3v_1[n-1]$$

where $v_1[n]$ and $v_2[n]$ are independent white noise processes each with variance equal to 0.5.

(a) What are the autocorrelation functions of x and y? Are these processes wide-sense stationary?

(b) What is the cross-correlation function $\mathrm{R}_{xy}[n_1, n_0]$? Are these processes jointly stationary (in the wide sense)?

4.3. Beginning with (4.16), give an alternative simpler proof of property (4.18) when R_x is a real-valued correlation function.

4.4. **(a)** Show that if a correlation function can be decomposed and written as

$$R_x[l] = h[l] * h^*[-l]$$

for some suitable nonzero sequence h, then it satisfies condition (4.16) with strict inequality, i.e., it is *positive definite*.

(b) Show that the correlation function in Example 4.4 can be written in the form above where

$$h[l] = \sqrt{3}(-0.5)^l u[l]$$

where $u[l]$ is the unit step function. Therefore prove that the correlation function in the example is positive definite.

4.5. Random processes $v_1[n]$ and $v_2[n]$ are independent and have the same correlation function

$$\mathrm{R}_v[n_1, n_0] = 0.5\delta[n_1 - n_0]$$

(a) What is the correlation function of the random process

$$x[n] = v_1[n] + 2v_1[n+1] + 3v_2[n-1]?$$

Is this random process wide-sense stationary?

(b) Find the correlation matrix for a random vector consisting of eight consecutive samples of $x[n]$.

4.6. In the following, $x[n]$ is a sample of a real stationary zero-mean random process. The random vector $\boldsymbol{x}$ is the vector of samples

$$\boldsymbol{x} = [x[0]\ x[2]\ x[3]\ x[4]\ x[5]]^T$$

Note that $x[1]$ is missing. Say if the following statements are correct or incorrect and why.

(a) The covariance matrix for this random vector is a *Toeplitz* matrix.

(b) The covariance matrix for this random vector is a *symmetric* matrix.

(c) If the covariance matrix of the vector $\boldsymbol{x}$ is given, we can determine from it the covariance matrix for the vector

$$\boldsymbol{x}' = [x[0]\ x[1]\ x[2]\ x[3]\ x[4]\ x[5]]^T$$

where $x[1]$ is not missing.

4.7. The density function for a stationary Gaussian random process is described by its mean vector and covariance matrix. Since the process is stationary, the covariance matrix is Toeplitz. It may occur to you that if the inverse of this Toeplitz matrix is also Toeplitz, then the computation of the quadratic product in the exponent of the Gaussian density function would be simplified. Is the inverse of a symmetric Toeplitz matrix Toeplitz? If the answer is "yes," then you must prove it. If the answer is "no," then show it by counterexample.

4.8. By considering appropriate combinations of $x[n]$ and $y[n-l]$, prove the bounds (4.36) and (4.37) on the cross-correlation function.

4.9. State whether or not each of the following sequences could represent a legitimate correlation function for a random process. Show why or why not.

(a)

$$R_x[l] = e^{-l^2}$$

(b)

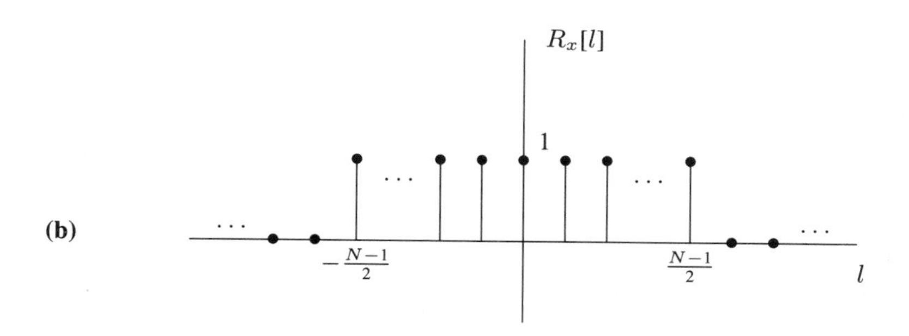

(c)

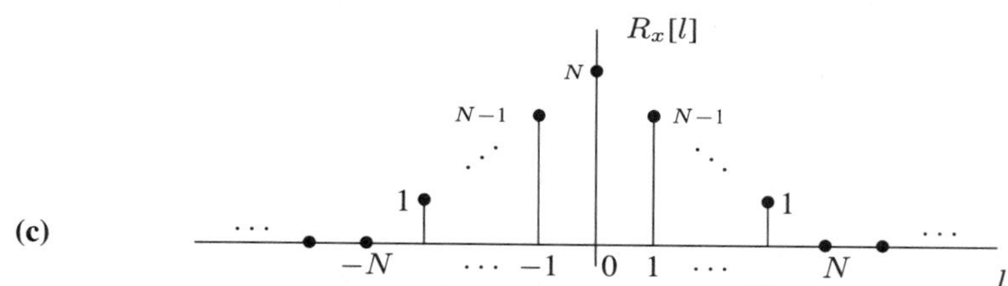

4.10. By direct use of the Fourier transform, show that the power spectral density function of the exponential correlation function (4.46) is given by (4.48). Show also that the formula can be derived by evaluating the complex spectral density function in Table 4.2 on the unit circle.

4.11. Find the power spectral density functions corresponding to the following correlation functions and verify that they are real and nonnegative.

(a)

$$R_x[l] = \begin{cases} 3 - |l| & -3 \leq l \leq 3 \\ 0 & \text{otherwise} \end{cases}$$

(b)

$$R_x[l] = 2(-0.6)^{|l|} + \delta[l]$$

4.12. Find the complex spectral density functions corresponding to the correlation functions in Problem 4.11 and their regions of convergence.

4.13. Find the correlation function corresponding to the following complex spectral density functions using contour integration.

(a)

$$S_x(z) = \frac{1.5}{z + 2.5 + z^{-1}}$$

(b)

$$S_x(z) = \frac{40z}{-99z^2 + 202z - 99}$$

(c)

$$S_x(z) = \frac{0.6z - 2 + 0.6z^{-1}}{0.6z - 1.36 + 0.6z^{-1}}$$

(d)

$$S_x(z) = \frac{5}{z^2 + \frac{5}{2} + z^{-2}}$$

(e)

$$S_x(z) = -\frac{z - \frac{26}{5} + z^{-1}}{z^2 + \frac{5}{2} + z^{-2}}$$

4.14. Repeat Problem 4.13 using partial fraction expansion.

4.15. Show that the complex spectral density function $S_x(z)$ satisfies the property (4.57). Show further that when R_x is real, $S_x(z)$ satisfies the property (4.58).

4.16. A nonstationary random process $x[n]$ has the correlation matrix

$$\mathbf{R}_x = \begin{bmatrix} 4 & 1 & 0 & 0 \\ 1 & 4 & 0 & 0 \\ 0 & 0 & 2 & 1 \\ 0 & 0 & 1 & 2 \end{bmatrix}$$

(a) What are the DKLT basis functions?

(b) What is the mean-square error in representing the random process as

$$\hat{x}[n] = \kappa_1 \varphi_1[n]$$

where $\varphi_1[n]$ is the first eigenfunction of the process?

(c) What is the mean-square error in representing the process by

$$\hat{x}[n] = \kappa_1 \varphi_1[n] + \kappa_2 \varphi_2[n]$$

where φ_1 and φ_2 are the first two eigenfunctions? Is there more than one choice for the basis functions that would give the same mean-square error?

4.17. Consider representing a segment of a stationary random process with $N = 3$ time samples. The first three values of the correlation function for this random process are given by

$$R[0] = 3, \quad R[1] = 2, \quad R[2] = 1$$

What is the mean-square error in representing the random sequence as

$$\hat{x}[n] = \kappa_1 \varphi_1[n]$$

where $\varphi_1[n]$ is the eigenvector basis function corresponding to the largest eigenvalue of the correlation matrix? Is this larger or smaller than the mean-square error that would be encountered in representing the sequence by

$$\hat{x}[n] = \kappa_2 \varphi_2[n] + \kappa_3 \varphi_3[n]$$

where $\varphi_2[n]$ and $\varphi_3[n]$ are the remaining two eigenvector basis functions?

4.18. A random process $x[n]$ is defined by

$$x[n] = \mathrm{s}[n] + w[n]$$

where $\mathrm{s}[n]$ is a *deterministic* signal and $w[n]$ is a zero-mean white noise sequence with variance σ^2. Consider taking N samples of the random process and defining the vectors

$$\boldsymbol{x} = \begin{bmatrix} x[0] \\ x[1] \\ \vdots \\ x[N-1] \end{bmatrix} ; \quad \mathbf{s} = \begin{bmatrix} \mathrm{s}[0] \\ \mathrm{s}[1] \\ \vdots \\ \mathrm{s}[N-1] \end{bmatrix} ; \quad \boldsymbol{w} = \begin{bmatrix} w[0] \\ w[1] \\ \vdots \\ w[N-1] \end{bmatrix}$$

(a) What is the correlation matrix $\mathbf{R}_x$?

(b) What are the eigenvectors and eigenvalues of the correlation matrix?

(c) What is the mean-square error if $x[n]$ is represented as a truncated Karhunen-Loève expansion using only the first orthonormal basis function?

4.19. State if the following statements are correct or incorrect and why.

(a) The eigenfunctions corresponding to a stationary random process are always sinusoidal.

(b) The mean-square error involved in the DKLT for data with nonzero mean can always be reduced if the mean is first removed.

(c) The complex cross-spectral density function is always nonnegative on the unit circle.

(d) The third and higher moments of a white noise process are always zero.

4.20. In this problem you will show that the magnitude squared coherence satisfies the upper bound in (4.54).

(a) Define a random process

$$v[n] = a_1^* x[n] + a_2^* y[n]$$

where a_1 and a_2 are any arbitrary complex numbers. Show that the correlation function $R_v[l]$ is given by

$$R_v[l] = \mathbf{a}^{*T} \begin{bmatrix} R_x[l] & R_{xy}[l] \\ R_{yx}[l] & R_y[l] \end{bmatrix} \mathbf{a} \quad \text{where} \quad \mathbf{a} = \begin{bmatrix} a_1 \\ a_2 \end{bmatrix}$$

(b) By taking the Fourier transform of this equation, and observing that $S_v(e^{\jmath\omega}) \geq 0$ for any choice of the a_1 and a_2, show that the power spectral density matrix

$$\begin{bmatrix} S_x(e^{\jmath\omega}) & S_{xy}(e^{\jmath\omega}) \\ S_{yx}(e^{\jmath\omega}) & S_y(e^{\jmath\omega}) \end{bmatrix}$$

is positive semidefinite.

(c) Since the power spectral density matrix is positive semidefinite, show that this implies that

$$|\Gamma(e^{\jmath\omega})|^2 = \frac{|S_{xy}(e^{\jmath\omega})|^2}{S_x(e^{\jmath\omega})S_y(e^{\jmath\omega})} \leq 1$$

4.21. Show that the complex cross-spectral density function satisfies the property (4.64). Thus when $z = e^{\jmath\omega}$ the cross-power spectral density satisfies (4.50). Show further that when R_{xy} is *real*, (4.64) becomes equivalent to (4.65).

4.22. What are the correlation functions for the real and imaginary parts of the exponential correlation function in Table 4.2? What is the cross-correlation function? Determine also the complex spectral density function for the real and imaginary parts of the process and sketch its poles and zeros.

4.23. A certain real random process is defined by

$$x[n] = A\cos\omega_o n + w[n]$$

where A is a Gaussian random variable with mean zero and variance σ_A^2 and $w[n]$ is a white noise process with variance σ_w^2 independent of A.

(a) What is the correlation function of $x[n]$?

(b) Can the power spectrum of $x[n]$ be defined? If so, what is the power spectral density function?

(c) Repeat parts (a) and (b) in the case when the cosine has an independent random phase uniformly distributed between $-\pi$ and π.

4.24. **(a)** Derive a general expression for the correlation function of the random process defined by (4.103) and show that the process is wide-sense stationary if and only if the amplitudes satisfy the orthogonality condition (4.104).

(b) By decomposing a real sinusoid with real random amplitude and uniform phase into the sum of two complex sinusoids, show that such a random process satisfies the orthogonality condition above and is therefore a stationary random process.

(c) A sampled random square wave with discrete period P has the form

$$x[n] = A\,\mathrm{sqr}(nT - \tau)$$

where A is a real random variable, T is the sampling interval, and τ is a random delay parameter uniformly distributed over $[0, PT]$ and independent of A. This signal is comprised of the fundamental frequency and only odd harmonics. Show that this periodic random process is also stationary.

(d) Give an example of a random process in the form of (4.103) that does *not* satisfy the orthogonality conditions (4.104). Recall that each component of the process is assumed to have uniform phase and amplitude distributed independent of phase.

4.25. A sufficient condition for the random process

$$x[n] = Ae^{\jmath\omega n} = |A|e^{\jmath(\omega n+\phi)}$$

to be wide-sense stationary is that A and ϕ be independent and ϕ be uniformly distributed. This condition also guarrantees that the essential requirement

$$\mathcal{E}\{x[n_1]x[n_0]\} = 0$$

is satisfied for the complex random process. Show that the foregoing condition is only *sufficient* for the stationarity, but not *necessary*. In other words, show by counterexample that $|A|$ and ϕ need not be independent, and that if they are independent, the phase need not be uniformly distributed in order for the random process to be stationary.

4.26. By following a procedure similar to that in Section 4.1.2, prove the positive semidefinite property for the correlation function of a continuous random process. Then by taking the function $a_c(t)$ in Table 4.7 to be an appropriate combination of continuous-time impulses, show that the property (4.130) holds.

4.27. By following an argument similar to that which was given in Section 4.4 for discrete random processes, show that the positive semi-definite property of the correlation function for continuous random processes implies that the power spectrum is always ≥ 0.

4.28. Find an expression for the power spectral density function of a continuous real random process with correlation function

$$R_{x_c}^{c}(\tau) = \sigma_x^2 e^{-\frac{|\tau|}{\tau_0}}$$

4.29. A bandlimited random signal is sampled at T = 1 ms.

(a) If this signal is used to form a discrete random sequence, what is the highest frequency in hertz that can be represented in the power density spectrum $S_x(e^{\jmath\omega})$?

(b) What must be the maximum bandwidth of the process if the continuous-time signal is to be reconstructed from the discrete-time sequence?

(c) If the continuous random process has a correlation function of the form

$$R_{x_c}^{c}(\tau) = \sigma_x^2 e^{-\frac{|\tau|}{\tau_0}}$$

what is the smallest value of the parameter τ_0 such that the continuous spectrum at the bandlimit corresponding to the sampling interval above is 40 dB below its value at $f = 0$? This gives a measure of the "correlation time" for a random process that can be adequately represented by samples at the given sampling interval.

4.30. **(a)** Starting with the definition (4.162)(c), show that the real third-order cumulant satisfies the symmetry properties (4.167) and has the symmetry regions depicted in Fig. 4.24(b).

(b) Given the symmetry properties (4.167), show that the bispectrum for a real random process has the symmetry properties (4.170) and (4.171) and that when these are combined, it leads to the 12 regions of symmetry depicted in Fig. 4.25(b).

COMPUTER ASSIGNMENTS

4.1. On the enclosed diskette you will find four real-valued data sets called

S00.DAT
S01.DAT
S02.DAT
S03.DAT

Each of these represents a random sequence with 512 samples. The format of each file is similar; you can look at the file with an editor. The first line contains the integer 512, representing the number of data points. Successive lines of the file contain the floating-point values of the sequence, four numbers per line. (The first four numbers are the first four values of the sequence, the next four numbers are the next four values, and so on.)

(a) Plot each of the sequences and generate hard copies of the plots. Tell whether you think each of these sequences is positively correlated, negatively correlated, or more-or-less uncorrelated.

(b) Compute the mean of each sequence as a signal average. Do the results of your computation seem reasonable based on the plots?

(c) Compute and plot the correlation function for the range of values $0 \leq l < 100$ using the estimate

$$\hat{R}_x[l] = \frac{1}{512 - l} \sum_{n=0}^{511-l} x[n+l]x[n]$$

Was your guess about correlation of these sequences in part (a) correct? You may want to look at expanded plots of just the first few lags to determine the correlation.

4.2. Form a Toeplitz correlation matrix using the first 20 lag values of the estimated correlation function for the data sets S01 and S03 on the enclosed diskette. Compute and plot the first four eigenfunctions for the data sets. Do they appear to be sinusoidal? What are the prominent differences in these sets of eigenfunctions? How do these differences relate to the differences observed in plots of the sequences?

4.3. Form a Toeplitz correlation matrix using the first eight lag values of the estimated correlation function for the data sets S00, S01, S02, and S03 on the enclosed diskette. For each random process compute the eigenvalues and eigenvectors of the correlation matrix. Plot the first eight values of each random sequence and the result of approximating each sequence by a truncated Karhunen-Loève expansion using one, two, and four basis functions corresponding to the eigenvectors of its correlation matrix. For each truncated representation also plot the error sequence $\varepsilon[n] = \mathrm{x}[n] - \hat{\mathrm{x}}[n]$. Observe that the average squared error decreases as more basis functions are used.

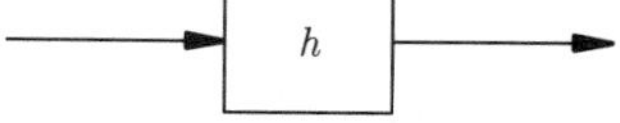

5

Linear Transformations

Signal processing involves transformation of signals to enhance certain characteristics, to suppress noise, and usually to extract meaningful information. This chapter deals with the processing of random sequences by linear systems. For the most part it deals with linear *shift-invariant* systems and their various representations. Linear shift-invariant systems produce transformations of the first and second moment functions of a random process that are similar to the transformations that the system applies to the signal itself. These transformations are easily described in the signal domain or the frequency and transform (z) domains. In the signal domain linear transformations of the moments can be expressed using both the convolution and the difference equation representation of linear systems.

Some other topics related to the processing of random processes by linear systems are also discussed in this chapter. An application of major importance to radar, sonar, and communication systems is the matched filter. This topic is explored for both deterministic and random signals. In another development it is shown that the continuous part of the complex spectral density function can be factored into causal and anticausal components and that this leads to a canonical representation for random processes as the output of a linear filter driven by white noise. This representation has applications in spectral analysis, coding, and in many other areas. It is also an important conceptual tool. The last section of the chapter provides a brief look at higher order moments of a random process and develops explicit relations that show how the moments, cumulants, and higher-order spectra are transformed when a general random process is applied to a linear system.

5.1 TRANSFORMATION BY LINEAR SYSTEMS

Transformations by linear systems are extremely important and are probably the most common class of signal processing operations. This section deals with linear transformations in the signal domain. General linear transformations are considered first; then results are specialized to the case of linear shift-invariant systems. It turns out for the latter case, in particular, that the mean, correlation, and covariance functions for the output can be computed quite neatly from the system impulse response and the corresponding first and second moment functions of the input.

5.1.1 Random Processes and General Linear Systems

Consider the general linear system shown in Fig. 5.1, which is not necessarily shift-invariant. The system input and output are related by the summation

$$y[n] = \sum_{k=-\infty}^{\infty} \mathrm{h}[n,k]x[k] \tag{5.1}$$

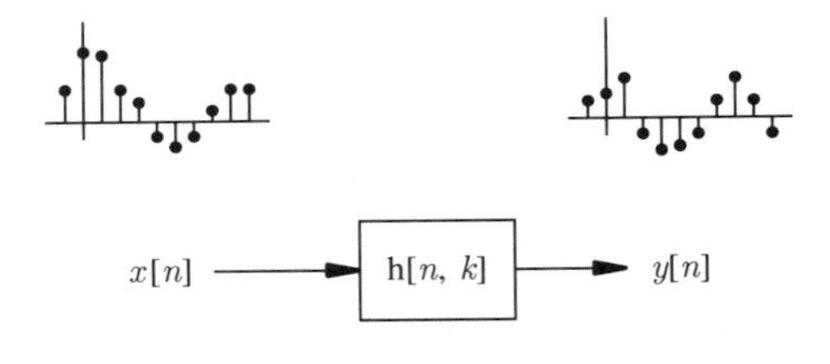

Figure 5.1 General linear system with random process input and output. $\mathrm{h}[n,k]$ is the response at point n to an impulse occurring at point k.

where $\mathrm{h}[n,k]$ is the response at a point n to an impulse occurring at a point k. It is straightforward to see how the means of these two random processes are related by taking the expectation on both sides of the equation. This results in

$$\mathcal{E}\{y[n]\} = \sum_{k=-\infty}^{\infty} \mathrm{h}[n,k]\mathcal{E}\{x[k]\}$$

or

$$\mathrm{m}_y[n] = \sum_{k=-\infty}^{\infty} \mathrm{h}[n,k]\mathrm{m}_x[k] \tag{5.2}$$

Thus the mean transforms in exactly the same way as the random process itself.

The correlation function of y can be obtained almost as easily by multiplying (5.1) by y^* and taking the expectation. In particular this yields

$$\mathcal{E}\{y[n_1]y^*[n_0]\} = \mathcal{E}\left\{\left(\sum_{k=-\infty}^{\infty} \mathrm{h}[n_1,k]x[k]\right) y^*[n_0]\right\}$$

$$= \sum_{k=-\infty}^{\infty} \mathrm{h}[n_1,k]\mathcal{E}\{x[k]y^*[n_0]\}$$

or

$$\mathrm{R}_y[n_1,n_0] = \sum_{k=-\infty}^{\infty} \mathrm{h}[n_1,k]\mathrm{R}_{xy}[k,n_0] \tag{5.3}$$

Note that the autocorrelation of the output R_y is expressed in terms of the impulse response and the *cross*-correlation R_{xy}. This is not exactly what we wanted. However, it is possible to compute the cross-correlation function from the autocorrelation function of the input by following a similar procedure. It is easiest to compute R_{yx} and then use the relation

$$\mathrm{R}_{xy}[n_1,n_0] = \mathrm{R}_{yx}^*[n_0,n_1] \tag{5.4}$$

to find R_{xy}. To compute R_{yx} multiply (5.1) by x^* (instead of y^*) and take the expectation as before. This yields

$$\mathcal{E}\{y[n_1]x^*[n_0]\} = \sum_{k=-\infty}^{\infty} \mathrm{h}[n_1,k]\mathcal{E}\{x[k]x^*[n_0]\}$$

or

$$\mathrm{R}_{yx}[n_1,n_0] = \sum_{k=-\infty}^{\infty} \mathrm{h}[n_1,k]\mathrm{R}_x[k,n_0] \tag{5.5}$$

The last three equations applied in the order (5.5), (5.4), (5.3), or combined into a single equation can then be used to determine the correlation function for the output process. Note that when the area of support for the correlation functions and the impulse reponse is *finite*, those functions can be represented as matrices $\mathbf{R}_x$, $\mathbf{R}_y$, and $\mathbf{H}$. In this case (5.3) and (5.5) represent matrix products and the combination of (5.3) through (5.5) is just the familiar relation

$$\mathbf{R}_y = \mathbf{H}\mathbf{R}_x\mathbf{H}^{*T}$$

It is easy to show that the covariance functions satisfy equations identical to those satisfied by the correlation functions. For example, using the relation

$$\mathrm{C}_y[n_1,n_0] = \mathrm{R}_y[n_1,n_0] - \mathrm{m}_y[n_1]\mathrm{m}_y^*[n_0] \tag{5.6}$$

and applying (5.3) and (5.2) leads to

$$\mathrm{C}_y[n_1, n_0] = \sum_{k=-\infty}^{\infty} \mathrm{h}[n_1, k]\mathrm{R}_{xy}[k, n_0] - \sum_{k=-\infty}^{\infty} \mathrm{h}[n_1, k]\mathrm{m}_x[k]\mathrm{m}_y^*[n_0]$$

$$= \sum_{k=-\infty}^{\infty} \mathrm{h}[n_1, k](\mathrm{R}_{xy}[k, n_0] - \mathrm{m}_x[k]\mathrm{m}_y^*[n_0])$$

or

$$\mathrm{C}_y[n_1, n_0] = \sum_{k=-\infty}^{\infty} \mathrm{h}[n_1, k]\mathrm{C}_{xy}[k, n_0] \tag{5.7}$$

Now combining this with the relation

$$\mathrm{C}_{xy}[n_1, n_0] = \mathrm{C}_{yx}^*[n_0, n_1] \tag{5.8}$$

and the result

$$\mathrm{C}_{yx}[n_1, n_0] = \sum_{k=-\infty}^{\infty} \mathrm{h}[n_1, k]\mathrm{C}_x[k, n_0] \tag{5.9}$$

derived in a similar manner, provides a sequence of equations that can be used to compute the covariance function of the output given the covariance function of the input. These equations apply to any random process, whether stationary or not, and to any transformation that qualifies as linear through the general representation (5.1).

5.1.2 Stationary Processes and Linear Shift-Invariant Systems

When the system input is stationary *or* the system is shift-invariant, no particularly interesting simplifications arise for the results of the preceding subsection. However, when *both* of these conditions hold, the equations simplify and the output of the linear system is also found to be stationary. This can be shown as follows.

When the system is shift-invariant, the impulse response $\mathrm{h}[n, k]$ is of the form $h[n-k]$ and the output is given by the convolution

$$y[n] = \sum_{k=-\infty}^{\infty} h[n-k]x[k] = \sum_{l=-\infty}^{\infty} h[l]x[n-l] \tag{5.10}$$

The mean of the output is therefore

$$\mathcal{E}\{y[n]\} = \sum_{k=-\infty}^{\infty} h[k]\mathcal{E}\{x[n-k]\}$$

or, since the mean of x is constant,

$$\boxed{m_y = m_x \cdot \sum_{k=-\infty}^{\infty} h[k]} \tag{5.11}$$

The cross-correlation between output and input can be found by a procedure similar to that of the preceding subsection. Alternatively, we can apply (5.5) directly, noting that since the input is stationary and the system is shift-invariant, both h and R_x are functions of just the difference of their arguments. This yields

$$\begin{aligned}\mathrm{R}_{yx}[n_1, n_0] &= \sum_{k=-\infty}^{\infty} h[n_1 - k]R_x[k - n_0] \\ &= \sum_{k'=-\infty}^{\infty} h[k']R_x[n_1 - n_0 - k']\end{aligned}$$

where the sum was converted by making the change of variables $k' = n_1 - k$. This shows that R_{yx} is also a function of just the difference $n_1 - n_0$. Substituting the variable $l = n_1 - n_0$ produces

$$R_{yx}[l] = \sum_{k'=-\infty}^{\infty} h[k']R_x[l - k']$$

or

$$R_{yx}[l] = h[l] * R_x[l] \tag{5.12}$$

where "$*$" denotes convolution. Since $R_{yx}[l] = R^*_{xy}[-l]$ and $R^*_x[-l] = R_x[l]$, it follows that

$$R_{xy}[l] = h^*[-l] * R_x[l] \tag{5.13}$$

for any impulse response sequence h. Now by a procedure paralleling that used to derive (5.12), it can be shown that

$$R_y[l] = h[l] * R_{xy}[l] \tag{5.14}$$

Then substituting (5.13) in (5.14) results in

$$\boxed{R_y[l] = h[l] * h^*[-l] * R_x[l]} \tag{5.15}$$

Equation 5.15 gives an explicit relation between the input and output correlation functions; it can also be written in the double summation form

$$R_y[l] = \sum_{k_0=-\infty}^{\infty} \sum_{k_1=-\infty}^{\infty} R_x[l - k_1 + k_0]h[k_1]h^*[k_0] \tag{5.16}$$

but the double convolution (5.15) is probably simpler and easier to remember. Alternatively, the output correlation function can be found by a three-step procedure involving (5.12), (5.14), and the symmetry relation

$$R_{yx}[l] = R^*_{xy}[-l] \tag{5.17}$$

Figure 5.2 depicts the relations (5.11), (5.12), and (5.14) graphically. If the mean of x, considered as a constant sequence, is applied to the system, the output is the mean of

y. Also, if the correlation function R_x is applied as the input sequence to the system, the output sequence is the cross-correlation function R_{yx}. If we then form the cross-correlation R_{xy} and apply *this* sequence to the system, the output sequence is R_y. Figure 5.2 is useful not so much for actually *computing* the mean and correlation functions as for *representing* the relations between them. *Think of it as an important mnemonic device.* The same relations occur in other forms when linear transformations are represented through difference equations and in the frequency and transform domains.

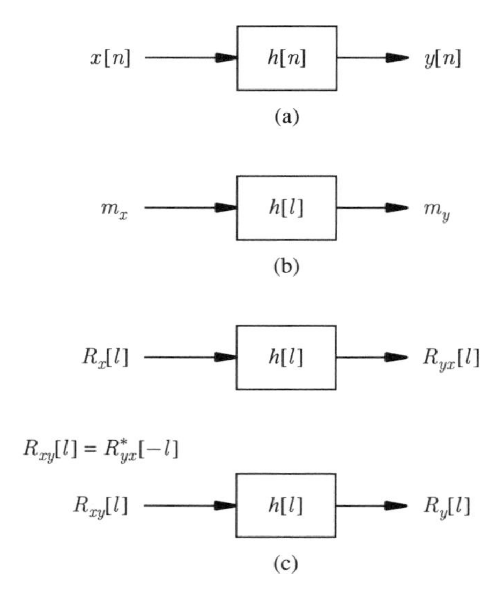

Figure 5.2 Transformation of the mean and correlation function by a linear shift-invariant system. (a) System with input and output sequences. (b) Transformation of mean. (c) Transformation of correlation function.

Equation 5.12 is the basis for a classical method of identification of an unknown linear system. The procedure is depicted in Fig. 5.3. The unknown system is excited with a white noise input with correlation function $R_x[l] = \delta[l]$. The output is cross-correlated with the input and the result from (5.12) is

$$R_{yx}[l] = h[l] * \delta[l] = h[l]$$

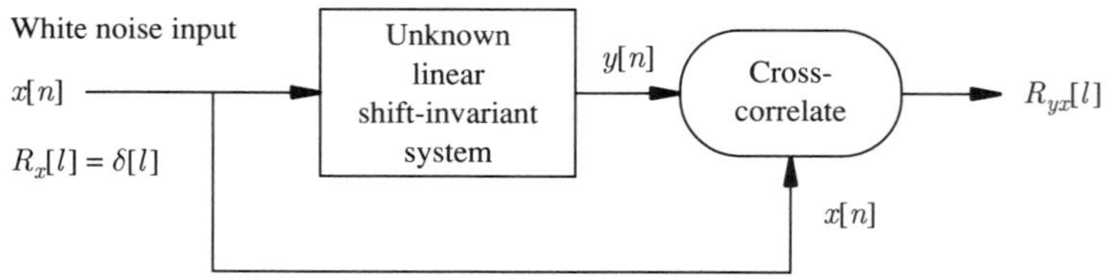

White noise input

$x[n]$

$R_x[l] = \delta[l]$

Unknown linear shift-invariant system

$y[n]$

Cross-correlate

$R_{yx}[l]$

$x[n]$

Figure 5.3 System identification by cross-correlation.

If the discrete system represents a sampled continuous system, this procedure may be a better way to compute the impulse response than actually driving the system with an impulse for two reasons. First, it may be easier to generate an approximation to white noise than to generate an approximation to an impulse since the latter must have finite energy in an almost zero-width pulse. Second, application of the impulse to a physical system requires driving it *very* hard (although for a very short time) and may cause damage, while driving it with white noise is much less traumatic.

Before concluding this section, let us observe that since the covariance function is actually a correlation function for the random process with the mean removed, the covariance also satisfies a set of equations analogous to (5.12)–(5.15). For completeness these are listed below.

$$C_{yx}[l] = h[l] * C_x[l] \tag{5.18}$$

$$C_{xy}[l] = h^*[-l] * C_x[l] \tag{5.19}$$

$$C_y[l] = h[l] * C_{xy}[l] \tag{5.20}$$

$$C_y[l] = h[l] * h^*[-l] * C_x[l] \tag{5.21}$$

The following example illustrates the application of these results.

Example 5.1

The linear shift-invariant system shown in Fig. EX5.1a is driven by a process with mean m_o and covariance function $C_x[l] = \sigma_o^2\delta[l]$. (This is white noise with an added nonzero mean.)

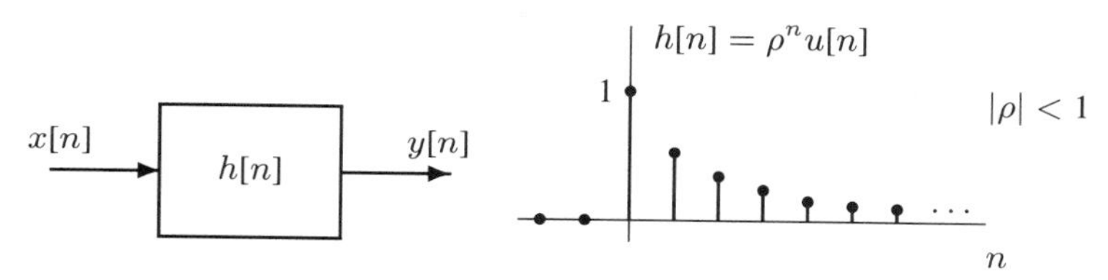

Figure EX5.1a

It is desired to compute the mean, correlation function, and covariance function of the output, and the cross-correlation and cross-covariance functions between input and output.

The mean of the output is very simple to compute. Substituting the given impulse response in (5.11) produces

$$m_y = m_o \sum_{k=0}^{\infty} \rho^k = \frac{m_o}{1-\rho}$$

Since the input and output have nonzero mean, it is easiest to compute first the auto- and cross-*covariance* functions from (5.18) through (5.21). Then the corresponding correlation functions can be found easily.

From (5.18), the cross-covariance of the output is

$$C_{yx}[l] = h[l] * C_x[l] = \left(\rho^l u[l]\right) * \left(\sigma_o^2\delta[l]\right) = \sigma_o^2\rho^l u[l]$$

and therefore

$$C_{xy}[l] = C_{yx}^*[-l] = \sigma_o^2(\rho^*)^{-l}u[-l]$$

as shown in Fig. EX.5.1b.

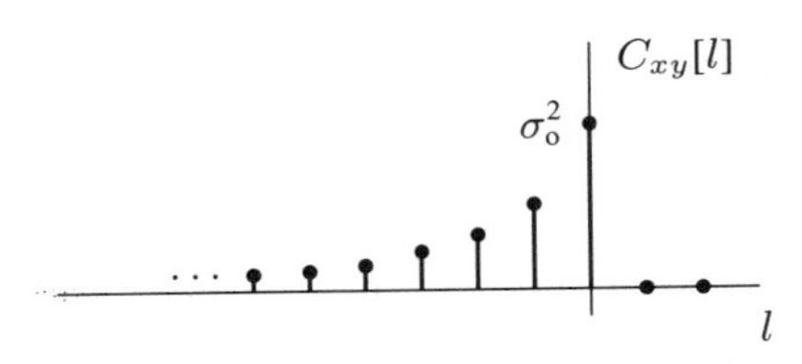

Figure EX5.1b

Then from (5.21) we have

$$C_y[l] = h[l] * C_{xy}[l] = \sum_{k=-\infty}^{\infty} h[k] C_{xy}[l-k]$$

To help in carrying out the convolution, the terms in the summation are depicted in Fig. EX5.1c for a typical value of $l > 0$.

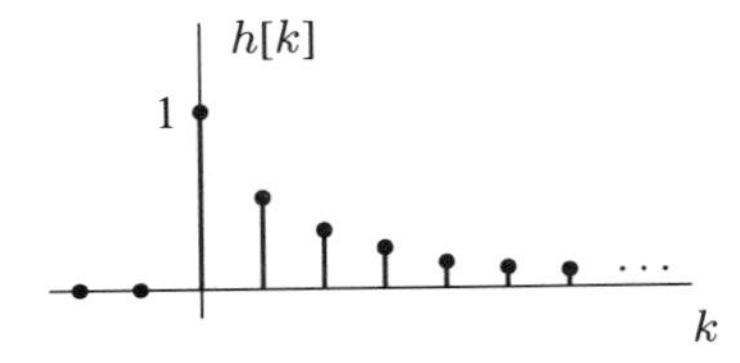

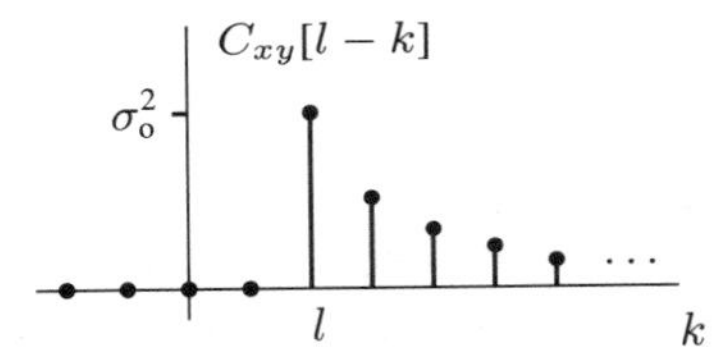

Figure EX5.1c

Thus for $l > 0$ the summation is

$$C_y[l] = \sum_{k=l}^{\infty} \sigma_o^2 \rho^k \cdot (\rho^*)^{-(l-k)}$$

Upon making the substitution $i = k - l$, this becomes

$$C_y[l] = \sigma_o^2 \sum_{i=0}^{\infty} \rho^{i+l} (\rho^*)^i = \sigma_o^2 \rho^l \sum_{i=0}^{\infty} \left(|\rho|^2\right)^i = \frac{\sigma_o^2 \rho^l}{1 - |\rho|^2}; \qquad l > 0$$

In a similar manner, for $l \leq 0$ we find that

$$C_y[l] = \frac{\sigma_o^2 (\rho^*)^{-l}}{1 - |\rho|^2}; \qquad l \leq 0$$

The cross-correlation function can now be computed as

$$
\begin{aligned}
R_{xy}[l] &= C_{xy}[l] + m_x m_y^* \\
&= \sigma_o^2(\rho^*)^{-l}u[-l] + m_o \cdot \left(\frac{m_o}{1-\rho}\right)^* \\
&= \sigma_o^2(\rho^*)^{-l}u[-l] + \frac{|m_o|^2}{1-\rho^*}
\end{aligned}
$$

and the autocorrelation function of the output is

$$
\begin{aligned}
R_y[l] &= C_y[l] + |m_y|^2 \\
&= \begin{cases} \frac{\sigma_o^2}{1-|\rho|^2}\rho^l + \left|\frac{m_o}{1-\rho}\right|^2 & l > 0 \\ \frac{\sigma_o^2}{1-|\rho|^2}(\rho^*)^{-l} + \left|\frac{m_o}{1-\rho}\right|^2 & l \leq 0 \end{cases}
\end{aligned}
$$

Observe that when the mean is zero, this is the exponential correlation function encountered in Chapter 4. This shows that a process with the exponential correlation function can always be generated by applying white noise to a stable first-order system. The variance parameter σ^2 of the process is given by

$$
\sigma^2 = \frac{\sigma_o^2}{1-|\rho|^2}
$$

In the real case (with zero mean) the correlation function has the simpler form

$$
R_y[l] = C_y[l] = \frac{\sigma_o^2}{1-\rho^2}\rho^{|l|}; \qquad \forall l
$$

5.1.3 Effect of Linear Transformations on Cross-correlation

Another situation that sometimes occurs with respect to random processes and linear transformations is depicted in Fig. 5.4. A random process x is transformed by a linear system to produce another random process v. The process x is related to a third random process y and the cross-correlation function for x and y is known. It is desired to find the cross-correlation function for v and y.

Let us first consider the case of a general linear system. The output $v[n]$ is given by

$$
v[n] = \sum_{k=-\infty}^{\infty} \mathrm{h}[n,k]x[k] \tag{5.22}
$$

To find the cross-correlation between v and y, multiply both sides of this equation by y^* and take the expectation. This yields

$$
\mathcal{E}\{v[n_1]y^*[n_0]\} = \sum_{k=-\infty}^{\infty} \mathrm{h}[n_1,k]\mathcal{E}\{x[k]y^*[n_0]\}
$$

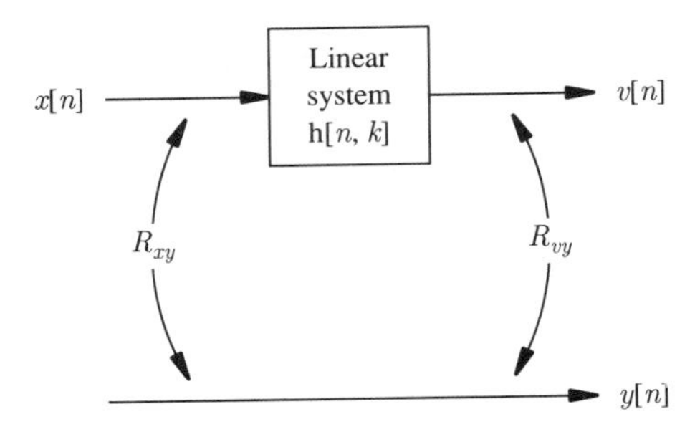

Figure 5.4 Cross-correlation with respect to a third random process.

or

$$\mathrm{R}_{vy}[n_1, n_0] = \sum_{k=-\infty}^{\infty} \mathrm{h}[n, k]\mathrm{R}_{xy}[k, n_0] \tag{5.23}$$

If the linear system is shift-invariant, then the input-output relation can be written as

$$v[n] = \sum_{k=-\infty}^{\infty} h[k]x[n-k] \tag{5.24}$$

Multiplying by $y^*[n-l]$ and taking the expectation as above then produces

$$R_{vy}[l] = \sum_{k=-\infty}^{\infty} h[k]R_{xy}[l-k] \tag{5.25}$$

or

$$\boxed{R_{vy}[l] = h[l] * R_{xy}[l]} \tag{5.26}$$

This last equation is a handy relation that finds use in later parts of the book. The covariance functions, of course, satisfy similar relations. By taking the z-transform of (5.26), it follows that

$$S_{vy}(z) = H(z)S_{xy}(z) \tag{5.27}$$

5.2 DIFFERENCE EQUATION REPRESENTATION OF LINEAR SHIFT-INVARIANT TRANSFORMATIONS

Let us now look at linear shift-invariant systems that can be represented by linear difference equations. Although the results of the preceding section also apply to these systems, the difference equation offers an alternative representation of the results that is sometimes quite important. To begin, consider a linear shift-invariant system represented by the difference equation

$$y[n] + \mathrm{a}_1 y[n-1] + \cdots + a_P y[n-P] = \mathrm{b}_0 x[n] + \cdots + \mathrm{b}_Q x[n-Q] \tag{5.28}$$

Taking the expectation of both sides of this equation leads to

$$\mathcal{E}\{y[n] + a_1 y[n-1] + \cdots + a_P y[n-P]\} = \mathcal{E}\{b_0 x[n] + \cdots + b_Q x[n-Q]\}$$

or

$$m_y(1 + a_1 + \cdots + a_P) = m_x(b_0 + \cdots + b_Q) \tag{5.29}$$

Thus the mean of the output has the specific form

$$\boxed{m_y = \frac{\sum_{j=0}^{Q} b_j}{1 + \sum_{i=1}^{P} a_i}\, m_x} \tag{5.30}$$

Figure 5.2(c) implies that the correlation functions must satisfy difference equations identical to that of the random process itself. To see this explicitly, multiply (5.28) by $y^*[n-l]$ and form the expectation.

$$\begin{aligned}\mathcal{E}\{&\big(y[n] + a_1 y[n-1] + \cdots + a_P y[n-P]\big) y^*[n-l]\} \\ &= \mathcal{E}\{\big(b_0 x[n] + \cdots + b_Q x[n-Q]\big) y^*[n-l]\}\end{aligned}$$

After multiplying through and taking the expectation, this yields

$$\boxed{R_y[l] + a_1 R_y[l-1] + \cdots + a_P R_y[l-P] = b_0 R_{xy}[l] + \cdots + b_Q R_{xy}[l-Q]} \tag{5.31}$$

In a similar manner, multiplying (5.28) by $x^*[n-l]$ and taking the expectation results in

$$\boxed{R_{yx}[l] + a_1 R_{yx}[l-1] + \cdots + a_P R_{yx}[l-P] = b_0 R_x[l] + \cdots + b_Q R_x[l-Q]} \tag{5.32}$$

Thus the correlation functions satisfy the same difference equations as the input and output of the system. This fact can be used to determine R_{xy} and R_y.

To show that the covariance function satisfies a similar set of equations, let us first subtract (5.28) and (5.29). A typical term on the left-hand side is therefore

$$a_i(y[n-i] - m_y)$$

while a typical term on the right is

$$b_j(x[n-j] - m_x)$$

If both sides of the equation are now multiplied by $(y[n-l] - m_y)^*$ and the expectation is taken, then the typical term on the left becomes

$$a_i \mathcal{E}\{(y[n-i] - m_y)(y[n-l] - m_y)^*\} = a_i C_y[l-i]$$

while the typical term on the right becomes

$$b_j \mathcal{E}\{(x[n-j] - m_x)(y[n-l] - m_y)^*\} = b_j C_{xy}[l-j]$$

Thus it follows that the covariance functions satisfy the difference equation

$$\boxed{C_y[l] + a_1 C_y[l-1] + \cdots + a_P C_y[l-P] = b_0 C_{xy}[l] + \cdots + b_Q C_{xy}[l-Q]} \quad (5.33)$$

Proceding in a similar manner by subtracting (5.29) from (5.28), multiplying the result by $(x[n-l]-m_x)^*$, and taking the expectation leads to the analogous result

$$\boxed{C_{yx}[l] + a_1 C_{yx}[l-1] + \cdots + a_P C_{yx}[l-P] = b_0 C_x[l] + \cdots + b_Q C_x[l-Q]} \quad (5.34)$$

These last two difference equations can be used to solve for C_{xy} and C_y. The following example illustrates how this is done.

Example 5.2

Consider the difference equation corresponding to the causal first order linear system of Example 5.1

$$y[n] = \rho y[n-1] + x[n]$$

where $x[n]$ is a white noise process with added mean m_o and variance σ_o^2. To be general it is assumed here that the parameter ρ is complex; later simplifications are noted that arise when ρ is real.

The mean satisfies (5.29), which in this case takes the form

$$m_y(1-\rho) = m_o$$

or

$$m_y = \frac{m_o}{1-\rho}$$

The cross-covariance satisfies (5.34), which takes the form

$$C_{yx}[l] - \rho C_{yx}[l-1] = \sigma_o^2 \delta[l]$$

The solution of this difference equation (the impulse response) is then

$$C_{yx}[l] = \sigma_o^2 \rho^l u[l]$$

where $u[l]$ is the unit step function.

To find the autocovariance function of the $y[n]$ use (5.33) in the form

$$C_y[l] - \rho C_y[l-1] = C_{xy}[l]$$

where

$$C_{xy}[l] = C_{yx}^*[-l] = \sigma_o^2 (\rho^*)^{-l} u[-l]$$

Thus the input to the difference equation is an exponential for $l \leq 0$ and zero for $l > 0$. The difference equation can be solved in two steps. For $l \leq 0$ assume a solution of the form

$$C_y[l] = c_1 (\rho^*)^{-l}; \qquad l \leq 0$$

where c_1 is a constant to be determined. Substituting this in the difference equation produces

$$c_1(\rho^*)^{-l} - \rho c_1(\rho^*)^{-(l-1)} = \sigma_o^2(\rho^*)^{-l}$$

Then canceling the common term $(\rho^*)^{-l}$ yields

$$c_1(1 - \rho\rho^*) = \sigma_o^2$$

or

$$c_1 = \frac{\sigma_o^2}{1 - |\rho|^2}$$

Thus

$$C_y[l] = \frac{\sigma_o^2}{1 - |\rho|^2}(\rho^*)^{-l}; \qquad l \leq 0$$

For $l > 0$ the input is zero, so only the homogeneous solution of the difference equation exists. This homogeneous solution is of the form

$$C_y[l] = c_2\rho^l\ ; \qquad l > 0$$

For $l = 1$ the difference equation has the form

$$C_y[1] - \rho C_y[0] = 0$$

which provides the boundary condition. Substituting the assumed form of the solution yields

$$c_2\rho - \rho C_y[0] = 0$$

and

$$c_2 = C_y[0] = \frac{\sigma_o^2}{1 - |\rho|^2}$$

or

$$C_y[l] = \frac{\sigma_o^2}{1 - |\rho|^2}\rho^l; \qquad l > 0$$

This completes the solution.

The cross-correlation function and autocorrelation function of the output can then be computed as in Example 5.1 to yield

$$R_{xy}[l] = \sigma_o^2(\rho^*)^{-l}u[-l] + \frac{|m_o|^2}{1 - \rho^*}$$

and

$$R_y[l] = \begin{cases} \frac{\sigma_o^2}{1-|\rho|^2}\rho^l + \left|\frac{m_o}{1-\rho}\right|^2 & l > 0 \\ \frac{\sigma_o^2}{1-|\rho|^2}(\rho^*)^{-l} + \left|\frac{m_o}{1-\rho}\right|^2 & l \leq 0 \end{cases}$$

As noted in Example 5.2, if the mean is zero, this is the exponential correlation function. Further, in the real case (with zero mean) we have the simpler expression

$$R_y[l] = \frac{\sigma_o^2}{1 - \rho^2}\rho^{|l|}; \qquad \forall l$$

5.3 SPECTRAL REPRESENTATION OF LINEAR SHIFT-INVARIANT TRANSFORMATIONS

Let us now consider how a linear shift-invariant transformation affects the power density spectra and complex power density spectra of a stationary random process. The needed relations are derived quite easily from (5.12)–(5.15) and properties of the Fourier transform. Specifically, since R_{yx} is the convolution of h and R_x, it follows from (5.12) that

$$S_{yx}(e^{\jmath\omega}) = H(e^{\jmath\omega})S_x(e^{\jmath\omega}) \tag{5.35}$$

where $H(e^{\jmath\omega})$ is the system frequency response (the Fourier transform of h). Similarly, from (5.13), (5.14), and (5.15) it follows that

$$S_{xy}(e^{\jmath\omega}) = H^*(e^{\jmath\omega})S_x(e^{\jmath\omega}) \tag{5.36}$$

$$S_y(e^{\jmath\omega}) = H(e^{\jmath\omega})S_{xy}(e^{\jmath\omega}) \tag{5.37}$$

and

$$S_y(e^{\jmath\omega}) = H(e^{\jmath\omega})H^*(e^{\jmath\omega})S_x(e^{\jmath\omega}) \tag{5.38}$$

or

$$\boxed{S_y(e^{\jmath\omega}) = |H(e^{\jmath\omega})|^2 S_x(e^{\jmath\omega})} \tag{5.39}$$

The last equation is a very important and frequently used formula. It can be used to give an alternative proof of the fact that the power spectral density function of *any* random process must be nonnegative; the argument is probably more intuitive than the one presented in Chapter 4. In particular, suppose that a random process $x[n]$ had a region of its power spectral density that is negative, as shown in Fig. 5.5(a). Then the process could be applied to a very narrowband filter whose squared magnitude is depicted in Fig. 5.5(b). The spectrum of the output process would consist only the negative regions and thus the average output power would be negative [see Fig. 5.5(c)]. Since it is impossible for any random process to have negative average power,[1] it follows that the input process cannot have any negative regions of its spectrum.

A set of relations similar to (5.36)–(5.38) can be derived for the complex spectral density function. Specifically, since $h^*[-l]$ corresponds to $H^*(1/z^*)$, it follows from (5.13)–(5.15) that

$$S_{yx}(z) = H(z)S_x(z) \tag{5.40}$$

$$S_{xy}(z) = H^*(1/z^*)S_x(z) \tag{5.41}$$

$$S_y(z) = H(z)S_{xy}(z) \tag{5.42}$$

and

$$S_y(z) = H(z)H^*(1/z^*)S_x(z) \tag{5.43}$$

[1] Average power is the expectation of the squared magnitude $\mathcal{E}\left\{|y[n]|^2\right\}$, which cannot be negative.

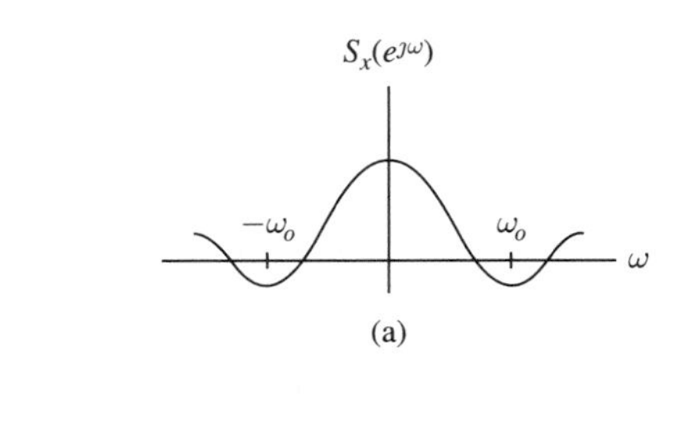

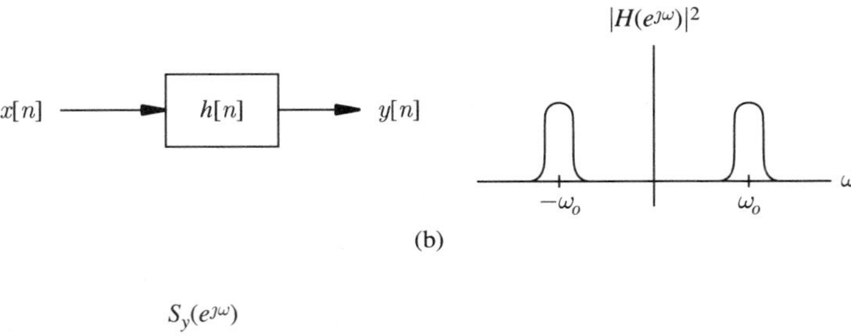

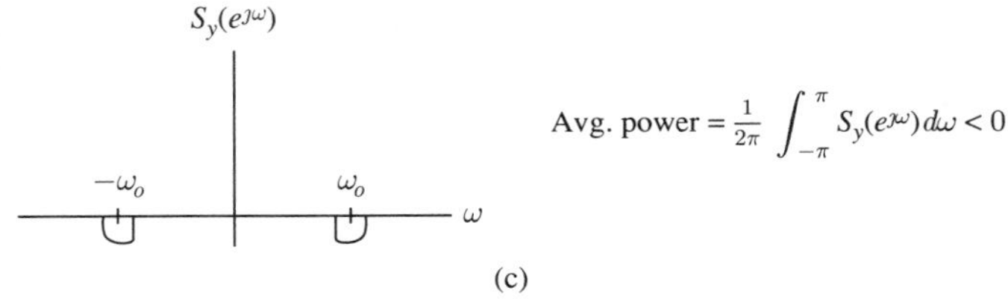

Figure 5.5 Demonstration that the power spectral density function cannot be negative. (a) Input power spectrum. (b) Narrowband filter. (c) Output power spectrum.

Note that S_y satisfies the required property for a complex spectral density function, namely $S_y(z) = S_y^*(1/z^*)$.

For a real filter $H^*(1/z^*)$ is equal to $H(z^{-1})$, so (5.41) and (5.43) simplify accordingly. If, in addition, the input process is real, then $S_x(z) = S_x(z^{-1})$; it then follows that S_y has this same property.

The results on transformation of stationary random signals by linear shift-invariant systems are summarized in Table 5.1. (Remember that all of these relations are implied by Fig. 5.2.) Let us continue consideration of the first-order system with an example that illustrates application of the new results in the table.

Example 5.3

The first-order system of Examples 5.1 and 5.2 has the system transfer function

$$H(z) = \frac{1}{1 - \rho z^{-1}}$$

The complex spectral density function for the white noise with added mean is

$$S_x(z) = \sigma_o^2 + 2\pi|m_o|^2\delta_c(z-1)$$

TABLE 5.1 LINEAR TRANSFORMATION RELATIONS

System defined by: $y[n] = h[n] * x[n]$		
$R_{yx}[l] = h[l] * R_x[l]$	$S_{yx}(e^{j\omega}) = H(e^{j\omega})S_x(e^{j\omega})$	$S_{yx}(z) = H(z)S_x(z)$
$R_{xy}[l] = h^*[-l] * R_x[l]$	$S_{xy}(e^{j\omega}) = H^*(e^{j\omega})S_x(e^{j\omega})$	$S_{xy}(z) = H^*(1/z^*)S_x(z)$
$R_y[l] = h[l] * R_{xy}[l]$	$S_y(e^{j\omega}) = H(e^{j\omega})S_{xy}(e^{j\omega})$	$S_y(z) = H(z)S_{xy}(z)$
$R_y[l] = h[l] * h^*[-l] * R_x[l]$	$S_y(e^{j\omega}) = \lvert H(e^{j\omega})\rvert^2 S_x(e^{j\omega})$	$S_y(z) = H(z)H^*(1/z^*)S_x(z)$

Note: For real $h[n]$, $H^*(1/z^*) = H(z^{-1})$.

The cross-power density spectrum is therefore

$$S_{xy}(e^{j\omega}) = H^*(e^{j\omega})S_x(e^{j\omega}) = \left(\frac{1}{1-\rho e^{-j\omega}}\right)^* \left(\sigma_o^2 + 2\pi|m_o|^2\delta_c(e^{j\omega}-1)\right)$$

$$= \frac{\sigma_o^2}{1-\rho^* e^{j\omega}} + \frac{2\pi|m_o|^2}{1-\rho^*}\delta_c(e^{j\omega}-1)$$

The second term, which arises whenever the mean is nonzero, gives impulses at $\omega = 0$ and at integer multiples of 2π.

The power density spectrum of the output is

$$S_y(e^{j\omega}) = |H(e^{j\omega})|^2 S_x(e^{j\omega}) = \frac{\sigma_o^2}{|1-\rho e^{-j\omega}|^2} + \frac{2\pi|m_o|^2}{|1-\rho|^2}\delta_c(e^{j\omega}-1)$$

The continuous term on the right can be simplified to

$$\frac{\sigma_o^2}{(1-\rho e^{-j\omega})(1-\rho^* e^{j\omega})} = \frac{\sigma_o^2}{1+|\rho|^2 - 2|\rho|\cos(\omega - \angle\rho)}$$

Thus

$$S_y(e^{j\omega}) = \frac{\sigma_o^2}{1+|\rho|^2 - 2|\rho|\cos(\omega - \angle\rho)} + \frac{2\pi|m_o|^2}{|1-\rho|^2}\delta_c(e^{j\omega}-1)$$

For real values of ρ and m_o this reduces to

$$S_y(e^{j\omega}) = \frac{\sigma_o^2}{1+\rho^2-2\rho\cos\omega} + \frac{2\pi m_o^2}{(1-\rho)^2}\delta_c(e^{j\omega}-1)$$

Again, the impulses or "lines" in the spectrum occur whenever the mean is nonzero.

The complex spectral density functions are also easily computed. These are

$$S_{xy}(z) = H^*(1/z^*)S_x(z) = \left(\frac{1}{1-\rho z^*}\right)^* \left(\sigma_o^2 + 2\pi|m_o|^2\delta_c(z-1)\right)$$

$$= \frac{\sigma_o^2}{1-\rho^* z} + \frac{2\pi|m_o|^2}{1-\rho^*}\delta_c(z-1)$$

and

$$S_y(z) = H(z)H^*(1/z^*)S_x(z) = \left(\frac{1}{1-\rho z^{-1}}\right)\left(\frac{1}{1-\rho z^*}\right)^* \left(\sigma_o^2 + 2\pi|m_o|^2\delta_c(z-1)\right)$$

The product of the first two terms can be written as

$$\frac{1}{1-\rho z^{-1}} \cdot \frac{1}{1-\rho^* z} = \frac{1}{-\rho^* z + (1+|\rho|^2) - \rho z^{-1}}$$

Thus the complex spectral density is

$$S_y(z) = \frac{\sigma_o^2}{-\rho^* z + (1+|\rho|^2) - \rho z^{-1}} + \frac{2\pi|m_o|^2}{|1-\rho|^2}\delta_c(z-1)$$

The functions computed in this example can be recognized as those pertaining to the exponential correlation function. The results of the last three examples for the case of real random processes are summarized in Table 5.2.

TABLE 5.2 FIRST AND SECOND MOMENT QUANTITIES FOR OUTPUT OF FIRST-ORDER LINEAR SYSTEM WITH WHITE NOISE INPUT

System and white noise input (with added mean)	$x[n]$ → $h[n]$ → $y[n]$ $m_x = m_o$ $R_x[l] = \sigma_o^2\delta[l] + m_o^2$ $y[n] = \rho y[n-1] + x[n]$ $h[n] = \rho^n u[n]$
Mean and correlation function	$m_y = \frac{m_o}{1-\rho}$ $R_y[l] = \frac{\sigma_o^2}{1-\rho^2}\rho^{\lvert l\rvert} + \frac{m_o^2}{(1-\rho)^2}$
Power spectral density function	$S_y(e^{\jmath\omega}) = \frac{\sigma_o^2}{1+\rho^2-2\rho\cos\omega} + \frac{2\pi m_o^2}{(1-\rho)^2}\delta_c(e^{\jmath\omega}-1)$
Complex spectral density function	$S_y(z) = \frac{\sigma_o^2}{-\rho z+(1+\rho^2)-\rho z^{-1}} + \frac{2\pi m_o^2}{(1-\rho)^2}\delta_c(z-1)$

Note: Real random processes shown here; see examples 5.1–5.3 for corresponding results pertaining to complex processes. Also see Table 4.2 and equation 4.48, Chapter 4.

5.4 THE MATCHED FILTER

An important application involving linear transformations of random processes is the *matched filter*, which is used to detect a signal in additive noise. The matched filter can be developed from at least two different points of view. One is based on statistical detection theory (see, e.g., [1–3]), which is not dealt with specifically in this book. The basic premises are explored in Problem 5.23, however. The other deals with optimizing a signal-to-noise ratio. This is the approach taken here. We begin with the case of a known, deterministic signal in noise, then generalize the result to the case of a random signal in noise. The solution involves a generalized eigenvalue problem which is further interpreted in terms of a whitening transformation (for the noise) applied to the received sequence.

5.4.1 Deterministic Signal in Noise

Let us begin by considering a discrete finite-length deterministic signal which is observed in additive noise. The situation is depicted in Fig. 5.6. Assume that the received signal $\mathrm{s}[n]$ contains P nonzero values extending from $n = n_{\mathrm{o}}$ to $n = n_{\mathrm{p}} = n_{\mathrm{o}} + P - 1$. The observed sequence is

$$x[n] = \mathrm{s}[n] + \eta[n] \tag{5.44}$$

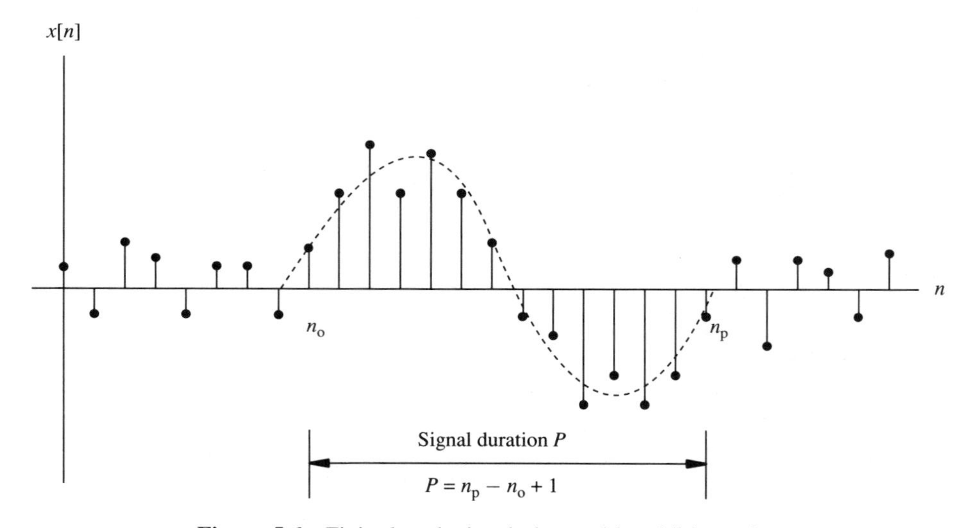

Figure 5.6 Finite-length signal observed in additive noise.

where $\eta[n]$ is the noise. The observation interval $0 \leq n < N$ is assumed large enough to contain all the nonzero values of the signal (i.e., $N > n_{\mathrm{p}} \geq P - 1$). The observation sequence is to be processed by a causal FIR filter. The goal is to design this filter so that at $n = n_{\mathrm{p}}$, when the nonzero part of the signal has completely passed through it, the filter will show a definite peak in its response. If design of such a filter is possible, then the filter will indicate the presence of the signal and its location on the n axis. This situation is shown in Fig. 5.7.

Intuitively, it would seem that the performance of this filter depends on its relative sensitivity to the signal and the noise; that is, it should pass most of the signal while simultaneously rejecting most of the noise. This idea can be formalized as follows. Let

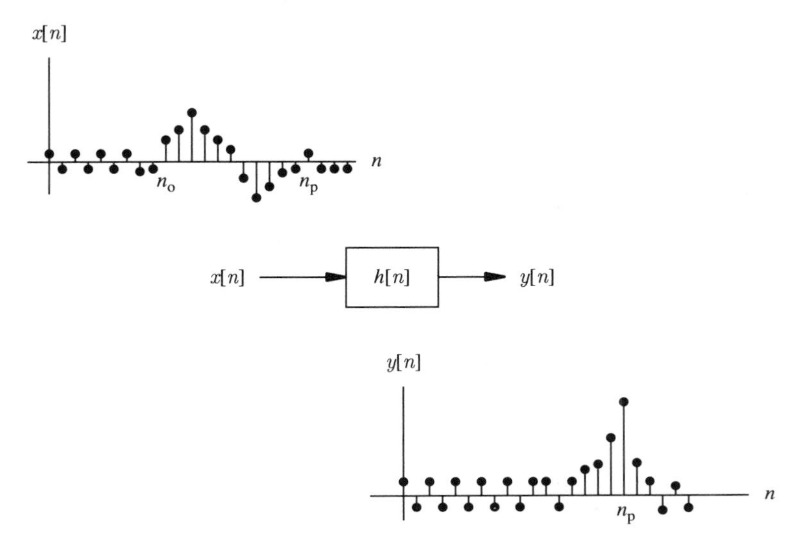

Figure 5.7 Matched filter and its response to signal in noise.

$h[n]$ represent the filter impulse response and $y[n]$ represent the output resulting from the input (5.44). Then $y[n]$ can be written as

$$y[n] = y_s[n] + y_\eta[n] \tag{5.45}$$

where $y_s[n]$ is the response due to the signal alone and $y_\eta[n]$ is the response due to the noise. The goal then is to design the filter so that the signal-to-noise ratio at $n = n_p$,

$$\text{SNR} \stackrel{\text{def}}{=} \frac{|y_s[n_p]|^2}{\mathcal{E}\{|y_\eta[n_p]|^2\}} \tag{5.46}$$

is maximized. Let us take the duration of the filter to be equal to P so that the filter can process the entire length of the signal. Then define the P-dimensional vectors of samples:

$$\mathbf{h} = \begin{bmatrix} h[0] \\ h[1] \\ \vdots \\ h[P-1] \end{bmatrix} \qquad \boldsymbol{x} = \begin{bmatrix} x[n_o] \\ x[n_o+1] \\ \vdots \\ x[n_p] \end{bmatrix}$$

$$\mathbf{s} = \begin{bmatrix} s[n_o] \\ s[n_o+1] \\ \vdots \\ s[n_p] \end{bmatrix} \qquad \boldsymbol{\eta} = \begin{bmatrix} \eta[n_o] \\ \eta[n_o+1] \\ \vdots \\ \eta[n_p] \end{bmatrix}$$

It follows that

$$y[n_p] = \sum_{k=0}^{P-1} h[k]x[n_p - k] = \mathbf{h}^T \tilde{\boldsymbol{x}} \tag{5.47}$$

where the ˜ denotes the reversal of the vector, and correspondingly,

$$y_s[n_p] = \mathbf{h}^T \tilde{\mathbf{s}} \tag{5.48}$$

and

$$y_\eta[n_p] = \mathbf{h}^T \tilde{\boldsymbol{\eta}} \tag{5.49}$$

These last two relations can be used to express the signal-to-noise ratio in terms of the vectors and thus make it simpler to perform the maximization. From (5.48) and (5.49) it follows that

$$|y_s[n_p]|^2 = (\mathbf{h}^T \tilde{\mathbf{s}})^* (\tilde{\mathbf{s}}^T \mathbf{h}) = \mathbf{h}^{*T} \tilde{\mathbf{s}}^* \tilde{\mathbf{s}}^T \mathbf{h}$$

and

$$\begin{aligned} \mathcal{E}\left\{|y_\eta[n_p]|^2\right\} &= \mathcal{E}\{(\mathbf{h}^T \tilde{\boldsymbol{\eta}})^* (\tilde{\boldsymbol{\eta}}^T \mathbf{h})\} \\ &= \mathbf{h}^{*T} \tilde{\mathbf{R}}_\eta^* \mathbf{h} = \mathbf{h}^{*T} \mathbf{R}_\eta \mathbf{h} \end{aligned}$$

where the last equality ($\tilde{\mathbf{R}}_\eta^* = \mathbf{R}_\eta$) follows because $\mathbf{R}_\eta$ is Toeplitz. Substitution of these results in (5.46) then yields

$$\text{SNR} = \frac{\mathbf{h}^{*T} \tilde{\mathbf{s}}^* \tilde{\mathbf{s}}^T \mathbf{h}}{\mathbf{h}^{*T} \mathbf{R}_\eta \mathbf{h}}$$

The problem now is to maximize SNR by choosing the vector **h**. It is clear from the form of the expression, however, that if a value for **h** is found that maximizes SNR, then any constant times **h** will also maximize SNR. Therefore, a constraint needs to be introduced on the magnitude of **h**. Although this can be done in any number of ways, a convenient method is to constrain the denominator to be a fixed value, specifically,

$$\mathbf{h}^{*T} \mathbf{R}_\eta \mathbf{h} = 1 \tag{5.50}$$

and maximize the numerator. With this constraint the signal-to-noise ratio can be written simply as

$$\text{SNR} = \mathbf{h}^{*T} \tilde{\mathbf{s}}^* \tilde{\mathbf{s}}^T \mathbf{h} \tag{5.51}$$

and the problem can be formulated as one of choosing **h** to maximize the quantity

$$\mathcal{L} = \mathbf{h}^{*T} \tilde{\mathbf{s}}^* \tilde{\mathbf{s}}^T \mathbf{h} + \lambda(1 - \mathbf{h}^{*T} \mathbf{R}_\eta \mathbf{h}) \tag{5.52}$$

where since the constraint is real, λ is a real Lagrange multiplier (see Appendix A). A necessary condition for the minimum can then be obtained by taking the gradient with respect to $\mathbf{h}^*$ using the formulas in Table A.2, Appendix A, and setting it to zero. This produces

$$\nabla_{\mathbf{h}^*} \mathcal{L} = \tilde{\mathbf{s}}^* \tilde{\mathbf{s}}^T \mathbf{h} - \lambda \mathbf{R}_\eta \mathbf{h} = \mathbf{0}$$

or

$$(\tilde{\mathbf{s}}^* \tilde{\mathbf{s}}^T) \mathbf{h} = \lambda \mathbf{R}_\eta \mathbf{h} \tag{5.53}$$

This equation can be recognized as a generalized eigenvalue problem involving the matrices $(\tilde{\mathbf{s}}^* \tilde{\mathbf{s}}^T)$ and $\mathbf{R}_\eta$. Thus **h** is required to be a generalized eigenvector of (5.53). The generalized eigenvalue problem was encountered in Chapter 2 and plays a further important role here. Note that when the noise is white, $\mathbf{R}_\eta$ is a diagonal matrix $\sigma_o^2\mathbf{I}$, so (5.53) becomes an ordinary eigenvalue problem.

Before proceeding with the solution of (5.53), notice that if $\mathbf{h}$ is a solution to (5.53) (i.e., a generalized eigenvector), then

$$\mathbf{h}^{*T}\tilde{\mathbf{s}}^*\tilde{\mathbf{s}}^T\mathbf{h} = \lambda\mathbf{h}^{*T}\mathbf{R}_\eta\mathbf{h} = \lambda \tag{5.54}$$

where the last equality follows because of (5.50). A comparison of this last equation to (5.51) shows that λ is the SNR corresponding to the eigenvector $\mathbf{h}$. Thus it follows that the filter that maximizes the signal-to-noise ratio corresponds to the generalized eigenvector with the *largest* eigenvalue.

Let us now proceed to solve (5.53). Since the matrix $\tilde{\mathbf{s}}^*\tilde{\mathbf{s}}^T$ is of unit rank, it is possible to obtain an explicit expression for the solution. The key is to notice that since it is possible to find $P-1$ linearly independent vectors each of which is orthogonal to the P-dimensional vector $\tilde{\mathbf{s}}^*$, each of these is an eigenvector with eigenvalue 0. Since SNR is equal to the eigenvalue, these eigenvectors produce an SNR of zero. Clearly, none of these is the eigenvector we are looking for. Now it can easily be verified that the remaining eigenvector is proportional to $\mathbf{R}_\eta^{-1}\tilde{\mathbf{s}}^*$. In particular, substitution of this quantity into (5.53) yields

$$\tilde{\mathbf{s}}^*\underbrace{\tilde{\mathbf{s}}^T\mathbf{R}_\eta^{-1}\tilde{\mathbf{s}}^*} = \lambda\mathbf{R}_\eta\mathbf{R}_\eta^{-1}\tilde{\mathbf{s}}^* = \lambda\tilde{\mathbf{s}}^*$$

The bracketed quantity on the left is a scalar, which must be equal to the eigenvalue λ. (If this were not the case, then the substitution of $\mathbf{R}_\eta^{-1}\tilde{\mathbf{s}}^*$ would have produced a contradiction.)

It is easy to show that since $\mathbf{R}_\eta$ is a Toeplitz matrix, the quadratic form $\tilde{\mathbf{s}}^T\mathbf{R}_\eta^{-1}\tilde{\mathbf{s}}^*$ can be rewritten as $\mathbf{s}^{*T}\mathbf{R}_\eta^{-1}\mathbf{s}$ (see Problem 5.16). Therefore, the maximum SNR can be written as

$$\boxed{\text{SNR}_{MAX} = \lambda_{MAX} = \mathbf{s}^{*T}\mathbf{R}_\eta^{-1}\mathbf{s}} \tag{5.55}$$

Since $\mathbf{h}$ needs to satisfy the condition (5.50), the optimal filter is given by

$$\boxed{\mathbf{h} = \frac{1}{\sqrt{\mathbf{s}^{*T}\mathbf{R}_\eta^{-1}\mathbf{s}}}\mathbf{R}_\eta^{-1}\tilde{\mathbf{s}}^*} \tag{5.56}$$

(also see Problem 5.16).

Let us just mention in passing that the condition (5.50) is actually a special kind of orthonormal condition. The eigenvectors $\{\mathbf{e}_i\}$ of the generalized eigenvalue problem satisfy the condition

$$\mathbf{e}_i^{*T}\mathbf{R}_\eta\mathbf{e}_j = \begin{cases} 1 & i=j \\ 0 & i \neq j \end{cases} \tag{5.57}$$

and are said to be orthonormal *relative to the matrix* $\mathbf{R}_\eta$[4, 5].

In the commonly occurring case where the noise is white with variance σ_o^2, the inverse covariance matrix in (5.56) is

$$\mathbf{R}_\eta^{-1} = \frac{1}{\sigma_o^2}\mathbf{I}$$

so that (5.55) and (5.56) become

$$\text{SNR} = \frac{\|\mathbf{s}\|^2}{\sigma_o^2} \tag{5.58}$$

and

$$\mathbf{h} = \frac{1}{\sigma_o\|\mathbf{s}\|}\tilde{\mathbf{s}}^* \tag{5.59}$$

Therefore, the optimal filter is a *reversed, conjugated replica of the signal.* This property inspires the term "matched filter." In the general case where the noise is not white, the solution can be derived by applying a whitening transformation to all sequences and designing a filter that is a reversed replica of the *transformed* signal. More will be said about that interpretation later.

The following example illustrates the steps involved in constructing a matched filter.

Example 5.4

A simple real transient signal has the form shown in Fig. EX5.4a.

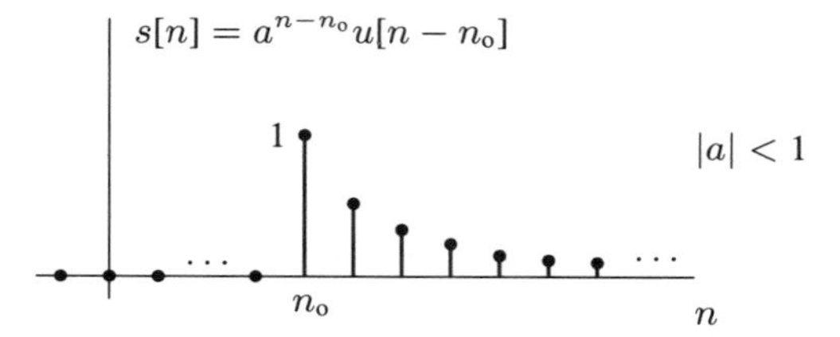

Figure EX5.4a

where $u[n]$ is the unit step function. It is desired to design an FIR matched filter to detect this signal in noise.

If the signal is regarded as having some finite length P, after which it is essentially zero, then the impulse response of the matched filter, from (5.59), is proportional to the reversed truncated signal

$$h[n] = \frac{1}{\sigma_o\|\mathbf{s}\|}\cdot a^{P-1-n} \qquad 0 \le n \le P-1$$

The normalizing constant is given by

$$\sigma_o\|\mathbf{s}\| = \sigma_o\left(\sum_{k=0}^{P-1}(a^k)^2\right)^{\frac{1}{2}} = \sigma_o\left(\frac{1-a^{2P}}{1-a^2}\right)^{\frac{1}{2}}$$

The impulse response is depicted in Fig. EX5.4b. The signal-to-noise ratio is given by

$$\text{SNR} = \frac{\|\mathbf{s}\|^2}{\sigma_o^2} = \frac{1}{\sigma_o^2}\left(\frac{1-a^{2P}}{1-a^2}\right)$$

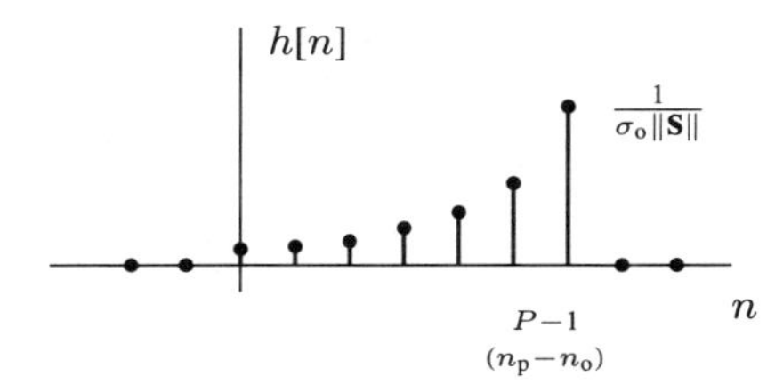

Figure EX5.4b

Now consider the case where the noise is not white but has the exponential correlation function.

$$R_\eta[l] = \sigma\rho^{|l|} = \frac{\sigma_o^2}{1-\rho^2}\rho^{|l|}$$

Problem **3.26** in Chapter 3 shows that the inverse correlation matrix corresponding to this correlation function has a particularly simple banded form. This is depicted below for the case of $P = 5$.

$$\mathbf{R}_\eta^{-1} = \frac{1}{\sigma_o^2}\begin{bmatrix} 1 & -\rho & 0 & 0 & 0 \\ -\rho & 1+\rho^2 & -\rho & 0 & 0 \\ 0 & -\rho & 1+\rho^2 & -\rho & 0 \\ 0 & 0 & -\rho & 1+\rho^2 & -\rho \\ 0 & 0 & 0 & -\rho & 1 \end{bmatrix}$$

The matched filter thus has the form [see (5.55) and (5.56)]

$$\mathbf{h} = \frac{1}{\sqrt{\text{SNR}}}\frac{1}{\sigma_o^2}\begin{bmatrix} 1 & -\rho & 0 & 0 & 0 \\ -\rho & 1+\rho^2 & -\rho & 0 & 0 \\ 0 & -\rho & 1+\rho^2 & -\rho & 0 \\ 0 & 0 & -\rho & 1+\rho^2 & -\rho \\ 0 & 0 & 0 & -\rho & 1 \end{bmatrix}\begin{bmatrix} a^4 \\ a^3 \\ a^2 \\ a \\ 1 \end{bmatrix}$$

where SNR represents the maximum signal-to-noise ratio, achieved by the matched filter. The terms of the matched filter are

$$h[0] = \frac{1}{\sigma_o^2\sqrt{\text{SNR}}}\left[a^4 - \rho a^3\right]$$

$$h[n] = \frac{1}{\sigma_o^2\sqrt{\text{SNR}}}\left[(1+\rho^2)a^{4-n} - \rho a^{3-n} - \rho a^{5-n}\right] \qquad 1 \le n \le 3$$

$$h[4] = \frac{1}{\sigma_o^2\sqrt{\text{SNR}}}\left[1 - \rho a\right]$$

To evaluate SNR, observe that the inverse correlation matrix can be factored as

$$\mathbf{R}_\eta^{-1} = \frac{1}{\sigma_o^2}\begin{bmatrix} 1 & 0 & 0 & 0 & 0 \\ -\rho & 1 & 0 & 0 & 0 \\ 0 & -\rho & 1 & 0 & 0 \\ 0 & 0 & -\rho & 1 & 0 \\ 0 & 0 & 0 & -\rho & 1 \end{bmatrix}\begin{bmatrix} 1 & -\rho & 0 & 0 & 0 \\ 0 & 1 & -\rho & 0 & 0 \\ 0 & 0 & 1 & -\rho & 0 \\ 0 & 0 & 0 & 1 & -\rho \\ 0 & 0 & 0 & 0 & 1 \end{bmatrix}$$

Therefore, SNR can be written as

$$\text{SNR} = (\mathbf{s}')^T \mathbf{s}'$$

where

$$\mathbf{s}' = \frac{1}{\sigma_o} \begin{bmatrix} 1 & -\rho & 0 & 0 & 0 \\ 0 & 1 & -\rho & 0 & 0 \\ 0 & 0 & 1 & -\rho & 0 \\ 0 & 0 & 0 & 1 & -\rho \\ 0 & 0 & 0 & 0 & 1 \end{bmatrix} \begin{bmatrix} 1 \\ a \\ a^2 \\ a^3 \\ a^4 \end{bmatrix} = \frac{1}{\sigma_o} \begin{bmatrix} 1-\rho a \\ a(1-\rho a) \\ a^2(1-\rho a) \\ a^3(1-\rho a) \\ a^4 \end{bmatrix}$$

SNR is then given by

$$\text{SNR} = \frac{1}{\sigma_o^2}(1-\rho a)^2 \left[1 + a^2 + a^4 + a^6 + a^8/(1-\rho a)^2\right]$$

The filter impulse response is depicted in Fig. EX5.4c for the parameter values $a = 0.95$, $\sigma_o^2 = 0.25$, and $\rho = -0.40$ (negatively correlated noise). Note that when the noise is not white, $h[n]$ does not necessarily resemble the signal.

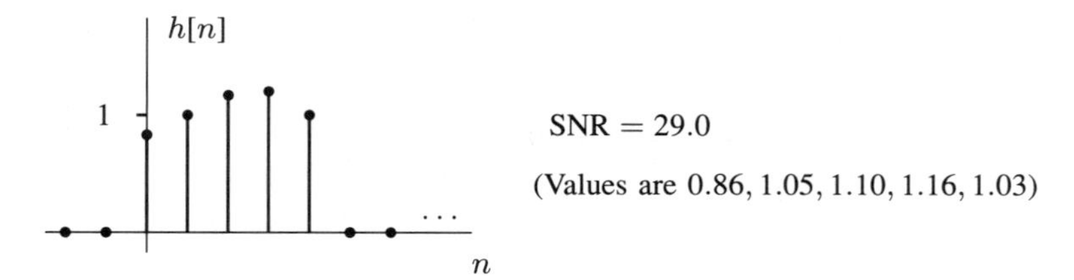

Figure EX5.4c

It is not too difficult to generalize the formulas above for the special case of $P = 5$ to an arbitrary value of P. The results are

$$h[0] = \frac{1}{\sigma_o^2\sqrt{\text{SNR}}}(a-\rho)a^{P-2}$$

$$h[n] = \frac{1}{\sigma_o^2\sqrt{\text{SNR}}}\left((1+\rho^2)a - \rho(1+a^2)\right)a^{P-2-n} \qquad 1 \le n \le P-2$$

$$h[P-1] = \frac{1}{\sigma_o^2\sqrt{\text{SNR}}}(1-\rho a)$$

where the value of SNR is

$$\text{SNR} = \frac{1}{\sigma_o^2}(1-\rho a)^2 \left[\frac{1-a^{2(P-1)}}{1-a^2} + \frac{a^{2(P-1)}}{(1-\rho a)^2}\right]$$

5.4.2 Random Signal in Noise

The results of the preceding section can be extended to the case of a random signal; that is, the case where both the signal $s[n]$ and the noise $\eta[n]$ are random processes. It is assumed that the signal and noise are independent and that P observations are taken as before, culminating at $n = n_{\text{p}}$. The signal-to-noise ratio now has the form

$$\text{SNR} \stackrel{\text{def}}{=} \frac{\mathcal{E}\{|y_{\text{s}}[n_{\text{p}}]|^2\}}{\mathcal{E}\{|y_\eta[n_{\text{p}}|^2\}} = \frac{\mathbf{h}^{*T}\mathbf{R}_s\mathbf{h}}{\mathbf{h}^{*T}\mathbf{R}_\eta\mathbf{h}}$$

where $\mathbf{R}_s$ is the correlation matrix for the signal. Now by constraining $\mathbf{h}$ to satisfy (5.50) and following the steps that led to (5.53), it follows that $\mathbf{h}$ must satisfy the generalized eigenvalue problem

$$\mathbf{R}_s\mathbf{h} = \lambda\mathbf{R}_\eta\mathbf{h} \tag{5.60}$$

Premultiplying both sides of (5.60) by $\mathbf{h}^{*T}$ and employing the constraint (5.50) then yields

$$\lambda = \text{SNR} = \mathbf{h}^{*T}\mathbf{R}_s\mathbf{h} \tag{5.61}$$

Therefore, it is seen that the matched filter again corresponds to the generalized eigenvector with the *largest* eigenvalue. In this case there is no simple analytic expression for the matched filter as there is when the signal is deterministic; a formula such as (5.56) would not make sense since the signal is random. The generalized eigenvalue problem can be solved with a computer, however, so the solution is easy to obtain in practice. Some further interpretation is also given below.

In the case where the noise is white, the correlation matrix is $\mathbf{R}_\eta = \sigma_{\text{o}}^2\mathbf{I}$. In this case (5.60) becomes an ordinary eigenvalue problem

$$\mathbf{R}_s\mathbf{h} = \lambda\sigma_{\text{o}}^2\mathbf{h} = \lambda'\mathbf{h}$$

and thus the solution is the eigenvector corresponding to the largest eigenvalue of $\mathbf{R}_s$ (see Fig. 5.8). Recall from the results on the DKLT that this vector provides the best mean-square representation of the signal.[2] The corresponding eigenfunction $\varphi_1[n]$, scaled and reversed, becomes the impulse response of the matched filter.

The case of nonwhite (or *colored*) noise can also be interpreted geometrically through the procedure of simultaneous diagonalization discussed in Chapter 2. Let $\mathbf{E}_\eta$ and $\mathbf{\Lambda}_\eta$ represent the eigenvector and eigenvalue matrices of $\mathbf{R}_\eta$, so $\mathbf{R}_\eta$ can be represented as

$$\mathbf{R}_\eta = \mathbf{E}_\eta\mathbf{\Lambda}_\eta{\mathbf{E}_\eta}^{*T} \tag{5.62}$$

The Mahalanobis transformation

$$x' = \mathbf{R}_\eta^{-1/2}x = (\mathbf{E}_\eta\mathbf{\Lambda}_\eta^{1/2}{\mathbf{E}_\eta}^{*T})x \tag{5.63}$$

[2]From an analogy with the deterministic case, you might expect the solution to be the largest eigenvector of the correlation matrix for the vector $\tilde{\mathbf{s}}^*$. This is true, in fact, since $\mathbf{R}_{\tilde{\mathbf{s}}^*}$ is equal to $\mathbf{R}_s$.

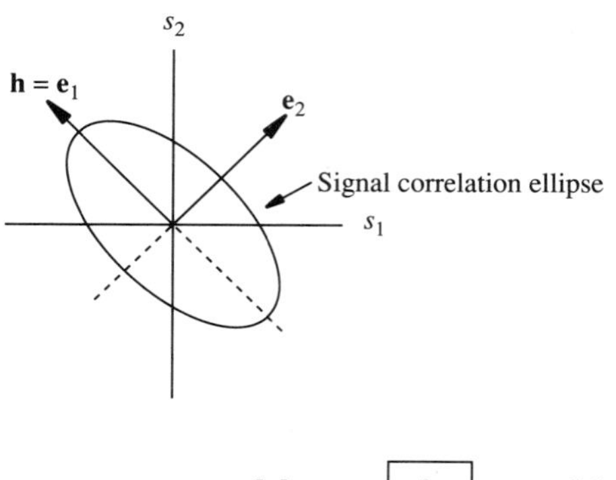

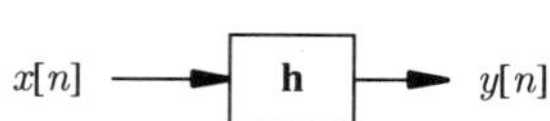

Figure 5.8 Geometrical interpretation of matched filter for random signal in white noise. Impulse response of matched filter corresponds to largest eigenvector and lies along major axis of the ellipse.

introduced in Chapter 2 (Section 2.6.4) whitens the noise and converts the problem to one of a random signal in white noise. Geometrically, this transformation represents a rotation and scaling of the coordinate system in which the vectors are represented (see Fig. 5.9). The noise correlation matrix becomes equal to the identity matrix. In the new coordinate system the random signal is also affected by the transformation (5.63), so the correlation matrix of the signal becomes

$$\mathbf{R}_{s'} = \mathbf{R}_\eta^{-1/2}\mathbf{R}_s\mathbf{R}_\eta^{-1/2} \tag{5.64}$$

[Observe that the transformation is Hermitian symmetric, so $(\mathbf{R}_\eta^{-1/2})^{*T} = \mathbf{R}_\eta^{-1/2}$.] With this transformation, the generalized eigenvalue problem (5.60) is converted to the ordinary eigenvalue problem

$$\mathbf{R}_{s'}\mathbf{h}' = \lambda\mathbf{h}'$$

The matched filter in this new coordinate system is then represented by the eigenvector $\mathbf{h}'$ of $\mathbf{R}_{s'}$, corresponding to its largest eigenvalue. The output of the matched filter can be expressed as

$$y[n_\text{p}] = (\mathbf{h}')^T\tilde{\boldsymbol{x}}' = (\mathbf{h}')^T\tilde{\mathbf{R}}_\eta^{-1/2}\tilde{\boldsymbol{x}} = \mathbf{h}^T\tilde{\boldsymbol{x}} \tag{5.65}$$

where

$$\mathbf{h} = \mathbf{R}_\eta^{-1/2}\mathbf{h}' \tag{5.66}$$

The Mahalanobis transformation (5.63) is also the transformation needed to convert the matched filter problem for a *deterministic* signal in colored noise to one involving white noise. In other words, the problem can be solved by prewhitening and constructing the replica of the *transformed* signal (see Problem 5.22).

In application of the matched filter to a problem involving colored noise, the prewhitening step may or may not be part of the actual implementation. Generally speaking, it is not necessary to prewhiten the noise explicitly since that operation is implicit in the filter derived by solution of the generalized eigenvalue problem. In any case, if the implementation is to include an explicit prewhitening step, the Mahalanobis transformation (5.63) is not convenient to perform since it is noncausal and not shift-invariant. An alternative approach, therefore, is to perform the prewhitening by triangular decomposition of the noise correlation matrix. This leads to a whitening transformation that can be applied to the signal in the

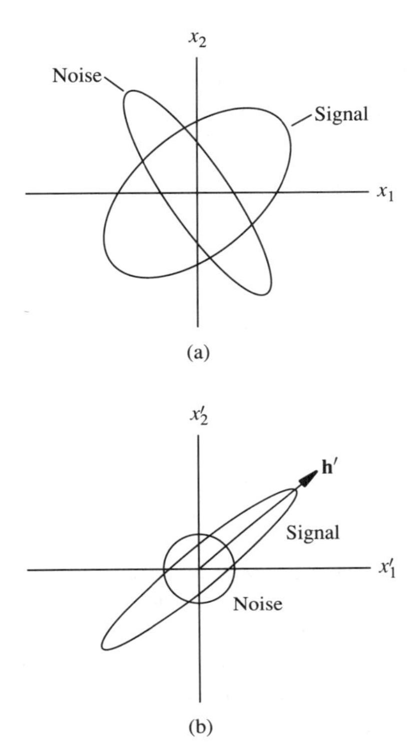

Figure 5.9 Geometrical interpretation of matched filter for random signal in colored noise. (a) Original signal and noise concentration ellipses. (b) Concentration ellipses after whitening the noise.

form of a causal linear filter (see, e.g., [2]). This technique is not pursued here; however, there is much more discussion of causal whitening filters in Chapters 7 and 8.

5.5 SPECTRAL FACTORIZATION AND INNOVATIONS REPRESENTATION OF RANDOM PROCESSES

In Section 5.3 it was seen that a random process with exponential correlation function can be represented as the output of a first-order linear system driven by white noise. This representation of a random process is not just something that applies to a first-order system; it is much more general. It turns out that most random processes with a continuous power density spectrum can be generated as the output of a *causal* linear filter driven by white noise. The filter in this representation also has a causal inverse, so that given the filter, the white noise is completely equivalent to the random process, and vice versa. This white noise-driven model is called the *innovations representation* for the random process and is a very powerful analytical tool.

In order to find the innovations representation for a given random process, it is necessary to factor the complex spectral density function into causal and anticausal components. This is most easily done for rational complex spectral density functions where the components correspond to minimum- and maximum-phase systems; but it is also possible in general for any process satisfying an integral relation known as the Paley-Wiener condition. This section begins by discussing minimum- and maximum-phase systems since they play an important role in the sequel. It then discusses the Paley-Wiener condition, which is the basis of spectral factorization for general linear shift-invariant systems. Finally, it shows how to perform the spectral factorization for rational spectral density functions and to find the innovations representation.

The treatment of minimum-phase and maximum-phase systems is in the form of a review which is sufficient for the purpose here. Some of the references such as [6–8] provide a more complete discussion of these systems.

5.5.1 Minimum- and Maximum-Phase Systems

A *minimum-phase polynomial* is one that has all of its zeros *strictly inside* the unit circle. A *minimum-phase system* is a *causal* linear shift-invariant system with rational transfer function

$$H_{min}(z) = \frac{B(z)}{A(z)} \tag{5.67}$$

where both $B(z)$ and $A(z)$ are minimum-phase polynomials. Thus a minimum-phase system is a *causal stable system with a causal stable inverse*. When a system is not minimum-phase, it is referred to as a *nonminimum-phase* system. Figure 5.10 shows pole-zero plots for typical minimum-phase and nonminimum-phase systems. Note that according to the definition, the system of Fig. 5.10(a) would not be minimum phase if it had poles or zeros *on* the unit circle. Since the inverse of a system has its poles and zeros interchanged, the causal inverse of the minimum-phase system in Fig. 5.10(a) also has its poles inside the unit circle and is stable. The inverse of the nonminimum-phase system in Fig. 5.10(b), however, has poles *outside* the unit circle and is *not* stable.

A *maximum-phase system* is the antithesis of a minimum-phase system. It is represented by a ratio of two *maximum-phase polynomials* whose zeros all lie *strictly outside* the unit circle.[3] It is an anticausal stable sytem with an anticausal stable inverse.

A nonminimum-phase system that has no zeros or poles *on* the unit circle can be made minimum-phase by moving zeros (or poles) which are outside the unit circle to their conjugate reciprocal locations *inside* the unit circle (see Fig. 5.11). It will be seen shortly that this leaves the *magnitude* of the frequency response the same, and only alters the phase. The operation of moving a *zero* inside the unit circle can be represented formally by multiplying the nonminimum-phase transfer function by a term of the form

$$\frac{z^{-1} - z_o^*}{1 - z_o z^{-1}} = (-z_o^*)\, \frac{1 - (1/z_o^*)z^{-1}}{1 - z_o z^{-1}} \tag{5.68}$$

The denominator of this term cancels the zero in the transfer function at $z = z_o$ and the numerator replaces it with a zero at $z = 1/z_o^*$. Moving a *pole* to inside the unit circle involves multiplying by a term that has the reciprocal of this form.

[3]Note that neither the definition of a minimum-phase system nor that of a maximum-phase system allows poles or zeros *on* the unit circle.

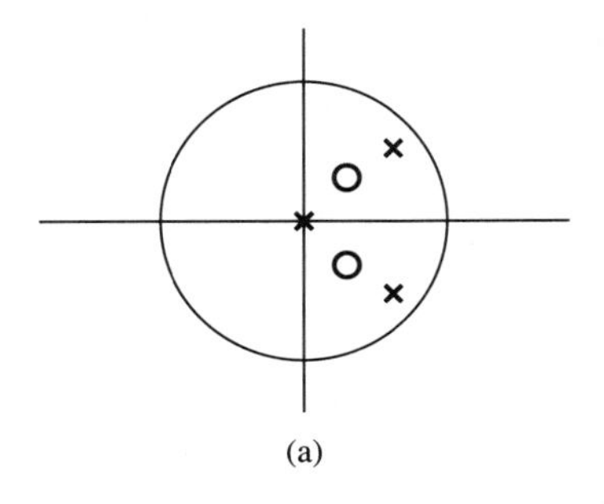

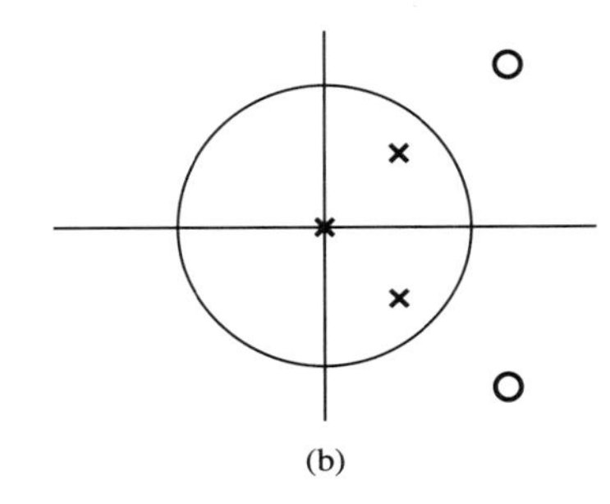

Figure 5.10 Pole-zero plots of typical minimum-phase and nonminimum-phase systems. (a) Minimum-phase system. (b) Nonminimum-phase system.

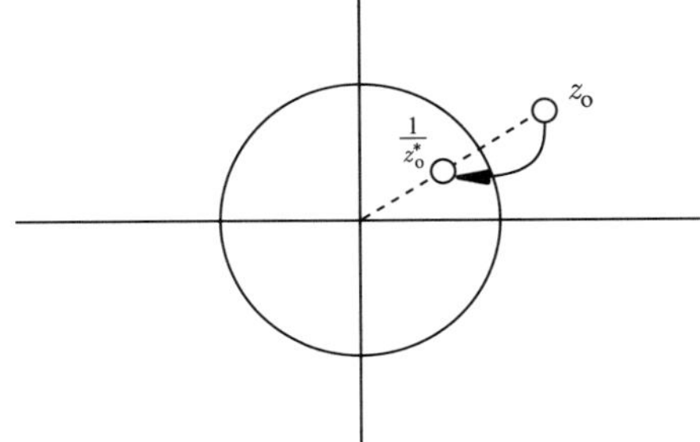

Original system:

$H(z) = H_1(z) \cdot (1 - z_o z^{-1})$

New system:

$H'(z) = H(z)\left(\frac{z^{-1} - z_o^*}{1 - z_o z^{-1}}\right) = H_1(z)(z^{-1} - z_o^*)$

Figure 5.11 Changing a nonminimum-phase system to a minimum-phase system by moving singularities to inside the unit circle.

Let us now examine the change in frequency response that occurs when a transfer function is multiplied by a term of the form (5.68). The frequency response corresponding to (5.68) is found by substituting $z = e^{\jmath\omega}$, specifically,

$$\frac{e^{-\jmath\omega} - z_o^*}{1 - z_o e^{-\jmath\omega}} = e^{-\jmath\omega}\frac{1 - z_o^* e^{\jmath\omega}}{1 - z_o e^{-\jmath\omega}} \tag{5.69}$$

The expression on the right has a magnitude of 1 for all values of ω because the numerator and denominator of the fraction are complex conjugates of each other. Thus multiplying by terms of the form (5.68) alters the phase but not the magnitude of a transfer function. A filter consisting of a product of terms like (5.68) or their reciprocals is called an *allpass* filter because it passes all frequencies without change in magnitude. The phase may of course be altered.

Provided that a system never has any poles or zeros precisely *on* the unit circle, then it follows from the previous discussion that any *nonminimum-phase system* can be converted to a minimum-phase system with the same magnitude frequency response by cascading with an appropriate allpass system as shown in Fig. 5.12(a). It follows further that any nonminimum-phase system has a canonical representation as the cascade of a minimum-phase system and an allpass system as shown in Fig. 5.12(b). The allpass system in this case serves to move zeros and poles from *inside* the unit circle to *outside* the unit circle.

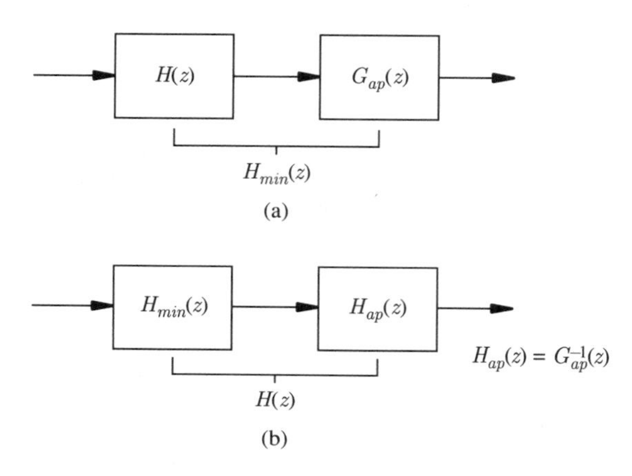

Figure 5.12 Relations between minimum-phase and nonminimum-phase systems. (a) Conversion of a nonminimum-phase system to a minimum-phase system by cascading with an allpass system. (b) Representation of a nonminimum-phase system as the cascade of a minimum-phase and an allpass system.

The term "mimimum-phase system" is actually inappropriate, since for positive values of frequency the phase is technically not minimum but *maximum* over all filters with the same magnitude response [6]. Specifically, if the *phase lag* of a system is defined as

$$\text{phase lag} \stackrel{\text{def}}{=} -\angle H(e^{j\omega}) \tag{5.70}$$

then the phase lag of a minimum-phase system for any value of ω is smaller than that for any other system with the same overall magnitude response (see Fig. 5.13). A minimum-phase system should probably be called a "minimal phase-lag system." The former term, like many others, seems to be etched in history, however, and is unlikely to change.

Minimum-phase systems have two other related properties that add to their characterization. The first is that the group delay (defined as the derivative of the phase lag) is minimal among all filters with the same magnitude response. The second is that the partial energy, defined as the sum of the magnitude squared impulse response from $-\infty$ to n, is larger than that of any other system with the same magnitude frequency response. Let $h_{min}[k]$ correspond to $H_{min}(e^{j\omega})$ and $h[k]$ correspond to $H(e^{j\omega})$ with $|H(e^{j\omega})| = |H_{min}(e^{j\omega})|$ for all ω. Then this last property, expressed mathematically as

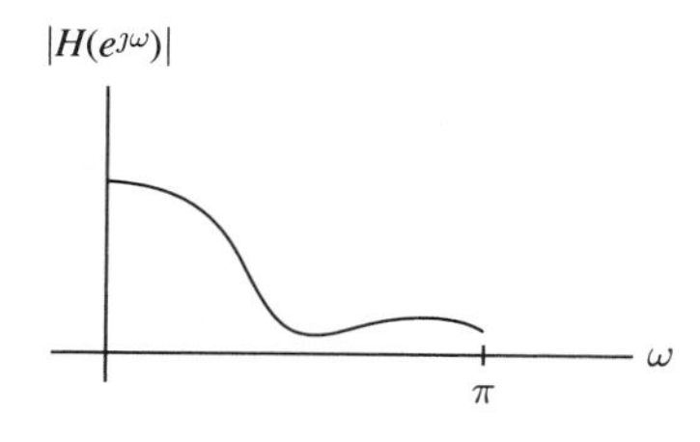

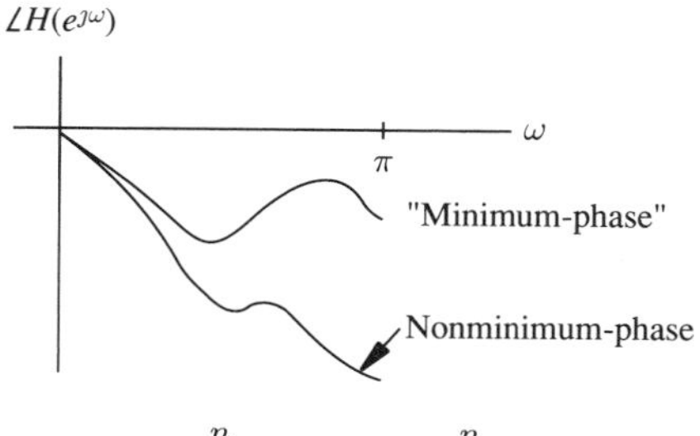

Figure 5.13 Magnitude and phase of the frequency response for a minimum-phase and a nonminimum-phase system.

$$\sum_{k=0}^{n} |h[k]|^2 \leq \sum_{k=0}^{n} |h_{min}[k]|^2 \qquad n \geq 0 \tag{5.71}$$

implies that the impulse response of the minimum-phase system is more compressed toward the origin than that of any other system with the same magnitude frequency response (see Fig. 5.14).

5.5.2 The Paley–Wiener Condition

Let us now turn to the problem of spectral factorization. Recall from Chapter 4 that a general complex spectral density function can be written in the form

$$S_x(z) = S'_x(z) + \sum_i 2\pi \mathsf{P}_i \delta_c(z - e^{j\omega_i}) \tag{5.72}$$

where $S'_x(z)$ represents the continuous part of the spectrum, and the sum of impulses represents the discrete part or "lines" in the spectrum. This section focuses on random processes with only a continuous part. It is shown here that when the following condition is met,

$$\boxed{\int_{-\pi}^{\pi} |\ln S_x(e^{j\omega})| d\omega < \infty} \tag{5.73}$$

then the complex spectral density function can be factored as

$$\boxed{S_x(z) = \mathcal{K}_o \cdot H_{ca}(z) H^*_{ca}(1/z^*)} \tag{5.74}$$

where $\mathcal{K}_o$ is a positive constant, and $H_{ca}(z)$ represents a causal stable system with a causal stable inverse. When $S_x(z)$ is a rational polynomial these factors represent minimum-phase and maximum-phase systems respectively.

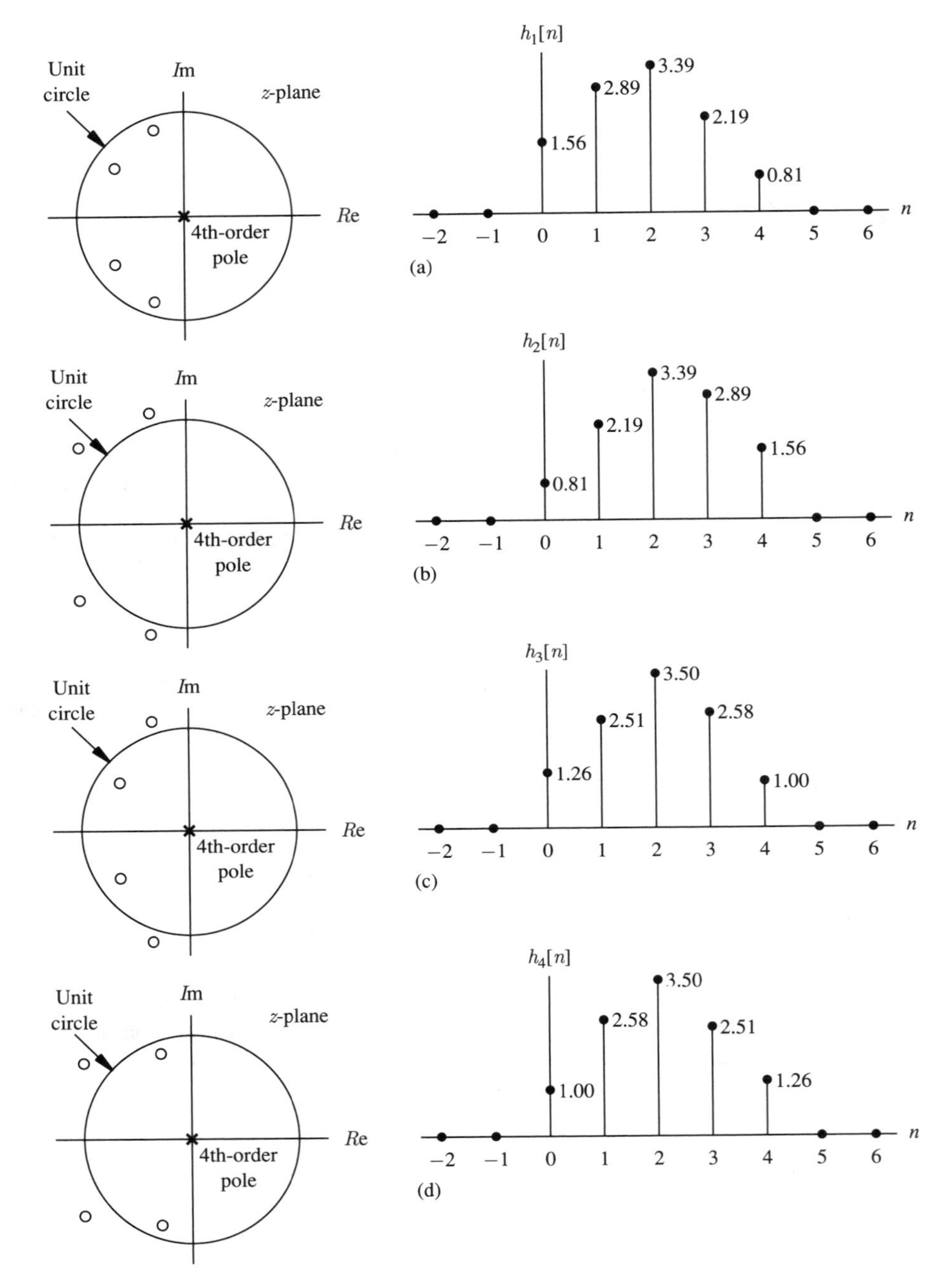

Figure 5.14 Compression of energy in the impulse response of a minimum-phase system. (a) Minimum-phase system. (b)–(d) Nonminimum-phase systems. (From A. V. Oppenheim and R. W. Schafer, *Discrete-Time Signal Processing*, ©1989, pp. 248–249; reprinted by permission of Prentice-Hall, Inc., Englewood Cliffs, New Jersey.)

Equation 5.73 is known as the Paley–Wiener condition [9, 10].[4] Although isolated zeros at single points are allowed, the Paley–Wiener condition does not permit $S_x(e^{\jmath\omega})$ to have any extended regions along the ω axis where it is zero. This implies that $S_x(e^{\jmath\omega})$ *cannot be bandlimited.* Spectral density functions that are rational polynomials always satisfy the Paley–Wiener condition, although the condition allows for more general types of complex spectral density functions.

Any process whose complex spectral density function satisfies (5.73), and therefore can be factored as in (5.74), is called a *regular process.* A regular process can therefore be realized as shown in Fig. 5.15(a) [i.e., as the output of a causal linear filter $H_{ca}(z)$ driven by white noise with variance $\mathcal{K}_o$]. This follows because the complex spectral density of the output process in Fig. 5.15(a) is given by

$$\begin{aligned} S_x(z) &= S_w(z)H_{ca}(z)H_{ca}^*(1/z^*) \\ &= \mathcal{K}_o \cdot H_{ca}(z)H_{ca}^*(1/z^*) \end{aligned}$$

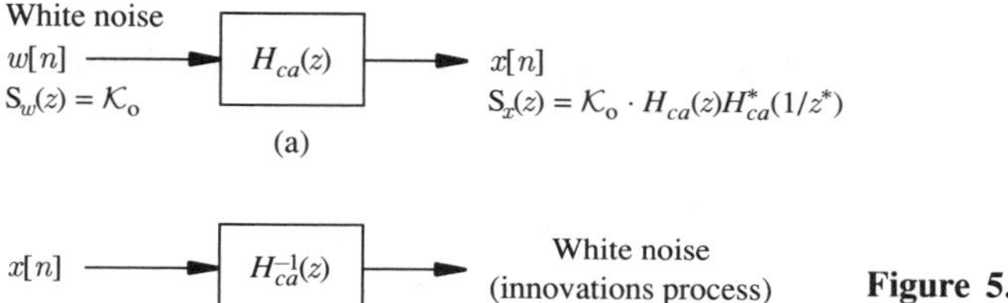

Figure 5.15 Innovations representation of a random process. (a) Signal model. (b) Inverse filter.

This is the *innovations representation* of the random process. By definition, any process that is regular has an innovations representation; conversely, if a process is not regular, the innovations representation does not exist. Further, the Paley–Wiener condition is necessary and sufficient to define a regular process.

The inverse filter $H_{ca}^{-1}(z)$ depicted in Fig. 5.15(b) is a whitening filter for the random process. When $x[n]$ is applied to the filter, the resulting white noise output is called the *innovations process.* (The reason for the name is discussed later in this section.) Since the two processes are related by an invertible linear transformation, they are equivalent representations of one another.

Let us now consider the proof of the Paley–Wiener condition. It will be shown in particular that the condition implies the factorization (5.74). This proves the sufficiency of the condition. The necessity of the condition is considered in Problem 5.26. Equation 5.73 implies that $\ln S_x(e^{\jmath\omega})$ has a convergent Fourier series of the form

$$\ln S_x(e^{\jmath\omega}) = \sum_{k=-\infty}^{\infty} c_k e^{-\jmath k\omega} \tag{5.75}$$

[4]Paley and Wiener developed a form of this condition for continuous processes. This form for discrete processes was known and used by Wold, Kolmogorov, and others [11, 12].

where since $\ln S_x(e^{\jmath\omega})$ is real,

$$c_{-k} = c_k^* \tag{5.76}$$

Now consider the function

$$\ln S_x(z) = \sum_{k=-\infty}^{\infty} c_k z^{-k} \tag{5.77}$$

Since this function converges on the unit circle, it must converge in an annular region

$$r'_R < |z| < 1/r'_R$$

no matter how small. Then (5.77) can be written as

$$\begin{aligned}\ln S_x(z) &= c_0 + \sum_{k=1}^{\infty} c_k z^{-k} + \sum_{k=-\infty}^{-1} c_k z^{-k} \\ &= c_0 + \sum_{k=1}^{\infty} c_k z^{-k} + \sum_{k'=1}^{\infty} c_{k'}^* z^{k'}\end{aligned}$$

where the change of variables $k = -k'$ was made and (5.76) was used in the second line. The complex spectral density function can thus be represented as

$$S_x(z) = e^{c_0} \cdot \left(e^{\sum_{k=1}^{\infty} c_k z^{-k}}\right) \left(e^{\sum_{k=1}^{\infty} c_k (1/z^*)^{-k}}\right)^* \tag{5.78}$$

which is in the form (5.74). The first term is a positive constant; the second term is the z-transform of a causal system. The causal system is stable because its region of convergence includes the unit circle. The last term correspondingly represents an anticausal stable system.

Since the Paley–Wiener condition involves the absolute value of the logarithm, it is clear that if the factorization is possible for $S_x(z)$, it is also possible for $1/S_x(z)$. Therefore, the Paley–Wiener condition not only guarantees that the causal stable factor exists, but it also guarantees that its inverse is causal and stable.

The following example illustrates the factorization of a spectral density function that is not a rational polynomial. The next subsection treats the more common case of a rational complex spectral density in depth.

Example 5.5

A complex spectral density function has the form

$$S_x(z) = e^{\frac{1+z^2}{z}} = e^{z^{-1}+z}$$

This function satisfies the required condition $S_x(z) = S_x^*(1/z^*)$. The power spectral density function is

$$S_x(e^{\jmath\omega}) = e^{2\cos\omega}$$

which is positive for all values of ω. The function satisfies the Paley–Wiener condition since

$$\int_{-\pi}^{\pi} |\ln S_x(e^{\jmath\omega})| d\omega = \int_{-\pi}^{\pi} |2\cos\omega| d\omega < \infty$$

The factorization can be done by inspection to obtain

$$S_x(z) = 1 \cdot e^{z^{-1}} \cdot e^{z}$$

where the causal factor is seen to be

$$H_{ca}(z) = e^{z^{-1}}$$

which converges everywhere except at $z = 0$. The impulse response of the filter is given by

$$h_{ca}[n] = \frac{1}{n!}u[n]$$

where $u[n]$ is the unit step function. This follows because

$$H_{ca}(z) = e^{z^{-1}} = \sum_{n=0}^{\infty} \frac{1}{n!} z^{-n}$$

5.5.3 Spectral Factorization for Rational Functions

In most practical problems the continuous part of the complex spectral density function is, or can be approximated by, a rational polynomial. Such complex spectral density functions always satisfy the Paley–Wiener condition; however, it is not necessary to show this fact explicitly since a procedure is developed here for performing the factorization directly.

The causal and anticausal factors of a rational polynomial correspond to minimum-phase and maximum-phase systems. To see this, consider a random process $x[n]$ with rational complex spectral density function $S_x(z)$. Recall that $S_x(z)$ can have no poles on the unit circle since otherwise there would be no region of convergence of the required form $r_R < |z| < 1/r_R$. Although zeros *may* occur on the unit circle, let us at first assume that they do not, and discuss the implications of this assumption later. Recall also that since $S_x(z)$ satisfies the symmetry property

$$S_x(z) = S_x^*(1/z^*) \tag{5.79}$$

zeros and poles of $S_x(z)$ occur in conjugate reciprocal pairs. That is, for every zero say at location z_o, there is another corresponding zero at location $1/z_o^*$. Likewise, for every pole there is a corresponding pole in the conjugate reciprocal location. Now since it is desired to obtain a filter that is causally stable and has a causal stable inverse, let us factor the numerator and denominator of $S_x(z)$ and group the poles and zeros that are *inside* the unit circle into one term denoted by $H_{ca}(z)$. This term by construction is minimum-phase. The remaining poles and zeros are also grouped together to form another term, $H_{ac}(z)$. Since this term contains only poles and zeros that are outside the unit circle, it is clearly maximum-phase. As a demonstration of this procedure, consider the function

$$S_x(z) = \frac{-12z^2 + 30 - 12z^{-2}}{6z^2 + 20 + 6z^{-2}} = \frac{-12 + 30z^{-2} - 12z^{-4}}{6 + 20z^{-2} + 6z^{-4}}$$

This can be factored as

$$S_x(z) = \frac{-2(1+\sqrt{2}z^{-1})(1+(1/\sqrt{2})z^{-1})(1-\sqrt{2}z^{-1})(1-(1/\sqrt{2})z^{-1})}{(1+\jmath\sqrt{3}z^{-1})(1+(\jmath/\sqrt{3})z^{-1})(1-\jmath\sqrt{3}z^{-1})(1-(\jmath/\sqrt{3})z^{-1})}$$

The pole-zero plot is shown in Fig. 5.16. Grouping the poles and zeros into those inside and outside the unit circle permits this function to be expressed as

$$\begin{aligned} S_x(z) &= -2\cdot\left(\frac{(1+(1/\sqrt{2})z^{-1})(1-(1/\sqrt{2})z^{-1})}{(1+(\jmath/\sqrt{3})z^{-1})(1-(\jmath/\sqrt{3})z^{-1})}\right)\left(\frac{(1+\sqrt{2}z^{-1})(1-\sqrt{2}z^{-1})}{(1-\jmath\sqrt{3}z^{-1})(1+\jmath\sqrt{3}z^{-1})}\right) \\ &= \frac{4}{3}\cdot\left(\frac{(1+(1/\sqrt{2})z^{-1})(1-(1/\sqrt{2})z^{-1})}{(1+(\jmath/\sqrt{3})z^{-1})(1-(\jmath/\sqrt{3})z^{-1})}\right)\left(\frac{(1+(1/\sqrt{2})z)(1-(1/\sqrt{2})z)}{(1+(\jmath/\sqrt{3})z)(1-(\jmath/\sqrt{3})z)}\right) \end{aligned} \tag{5.80}$$

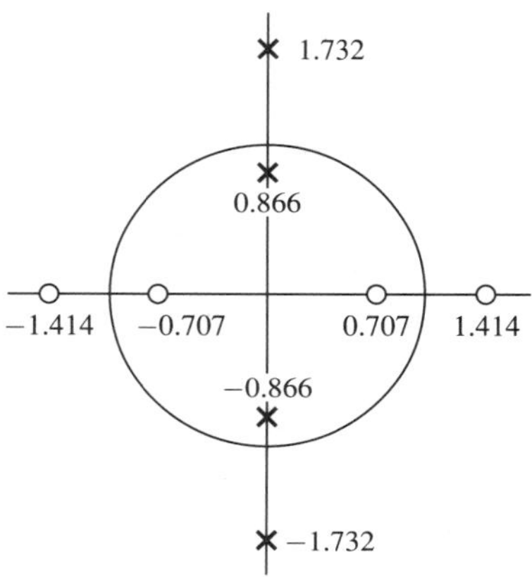

Figure 5.16 Pole-zero plot of an example complex spectral density function.

From the foregoing discussion it is apparent that every rational $S_x(z)$ has a factored representation of the form

$$S_x(z) = \mathcal{K}_o \cdot H_{ca}(z)H_{ac}(z) \tag{5.81}$$

where $\mathcal{K}_o$ is a normalizing constant, and (provided there are no roots on the unit circle) $H_{ca}(z)$ is minimum-phase, and $H_{ac}(z)$ is maximum-phase. Although there are different ways to perform this factorization, as evident from (5.80), the factors can always be chosen such that $\mathcal{K}_o$ is positive, and

$$H_{ac}(z) = H^*_{ca}(1/z^*) \tag{5.82}$$

If it were not possible to do this, then there would be no way that (5.79) could be satisfied. If it is further required that $H_{ca}(z)$ be the ratio of two *comonic* polynomials in z^{-1} (i.e., the constant term in numerator and denominator are both 1), then the decomposition (5.81) is *unique* and corresponds to (5.74). The power spectral density function thus has the form

$$S_x(e^{\jmath\omega}) = \mathcal{K}_o \cdot H_{ca}(e^{\jmath\omega})H^*_{ca}(e^{\jmath\omega}) = \mathcal{K}_o \cdot |H_{ca}(e^{\jmath\omega})|^2 \tag{5.83}$$

from which it is evident that $\mathcal{K}_o > 0$.

To summarize, when the complex spectral density function is a rational polynomial, its spectral factorization is a factorization into minimum- and maximum-phase components. The filter $H_{ca}(z)$ that appears in the innovations representation of Fig. 5.15 is thus a minimum-phase filter. Other similar representations of the random process are theoretically possible, including noncausal filters and causal filters with the same magnitude function that are not minimum-phase. These all produce processes with the *same* correlation function. The minimum-phase innovations representation, however, has the unique property that its inverse is also a causal stable filter [see Fig. 5.15(b)].

To find the innovations representation for a random process, it is necessary that the forms of the factors correspond *exactly to* (5.74). This is the factorization in the second line of (5.80). If the equation is not put in this form, it is usually not possible to identify the constant $\mathcal{K}_o$ correctly. For example, in the first line of (5.80) the first factor is $H_{ca}(z)$ but the second factor is not equal to $H^*_{ca}(1/z^*) = H_{ca}(z^{-1})$. The constant multiplying the two factors is thus not $\mathcal{K}_o$. In fact, the constant in this expression is *negative!*

It is also important when performing the spectral factorization to ensure that the components are truly minimum and maximum phase. Simply finding two factors that satisfy the condition (5.82) is not enough. This is illustrated in the following simple example.

Example 5.6

A complex spectral density function for a certain real random process is

$$S_x(z) = \frac{-(1/a)}{z - (a + 1/a) + z^{-1}}$$

This can be written in the equivalent form

$$S_x(z) = \frac{1}{-az + (1 + a^2) - az^{-1}} = \frac{1}{1 - az^{-1}} \cdot \frac{1}{1 - az}$$

which leads to the correct identification $\mathcal{K}_o = 1$ and

$$H_{ca}(z) = \frac{1}{1 - az^{-1}}$$

Notice a possible pitfall here. Suppose that the function had been factored as

$$S_x(z) = \frac{1}{-az + (1 + a^2) - az^{-1}} = \frac{1}{(z - a)(z^{-1} - a)}$$

Then it might be tempting to take

$$H_{ca}(z) = \frac{1}{z - a} = \frac{z^{-1}}{1 - az^{-1}} \qquad \text{(I)}$$

since it satisfies the symmetry condition

$$H^*_{ca}(1/z^*) = H_{ca}(z^{-1}) = \frac{1}{z^{-1} - a}$$

However, the term (I) is *not minimum-phase*. It has a zero at $z = \infty$ for one thing. The inverse z-transform is

$$a^{n-1}u[n-1]$$

where $u[n]$ is the unit step function, so the partial energy is not smaller than that of the impulse response

$$a^n u[n]$$

which *is* minimum-phase. Also, the inverse

$$H_{ca}^{-1}(z) = z - a$$

is not causal. This problem can be resolved by supplying an extra pole and zero at the origin, that is, by writing $S_x(z)$ in the equivalent form

$$S_x(z) = \frac{z \cdot z^{-1}}{-az + (1 + a^2) - az^{-1}} = \frac{z}{(z - a)} \frac{z^{-1}}{(z^{-1} - a)}$$

This is now a correct spectral factorization with

$$H_{ca}(z) = \frac{z}{z - a} = \frac{1}{1 - az^{-1}}$$

which is truly minimum-phase.

Before leaving this section let us make a few remarks about the continuous part of the complex spectral density function, its relation to the discrete part, and the assumption that there are no zeros on the unit circle. It is shown in Chapter 7 that the innovations representation of a random process is related to the predictability of the process. If the filter $H_{ca}^{-1}(z)$ is applied to the random process, its output represents the *difference* between the given random process $x[n]$ and its optimal prediction using past values of the process. The output of $H_{ca}^{-1}(z)$ thus represents only the *new* information present in the process; this is the basis of the term "innovations."[5] It is also shown there that the discrete spectral components (or "lines"), which do not admit an innovations representation, relate to the components of the process that can be predicted with zero error. It was already seen that poles cannot exist on the unit circle. Such singularities would correspond to undamped sinusoids or complex exponentials, which are predictable with zero error and are represented instead by the discrete components of the spectrum. Zeros on the unit circle, while theoretically possible if they occur in even multiplicities, require careful treatment. Since they become poles of the inverse filter, definition of the prediction filter for the process in this case requires some special considerations beyond the scope of this book (see [13]). In practice, zeros that are precisely *on* the unit circle can never occur because the inevitable noise, no matter how small, adds at least a constant term to the power density spectrum and bounds it away from zero.

The relations between regular processes, for which an innovations representation exists, and (perfectly) predictable processes that correspond to the discrete terms of the spectrum, are formalized in the theory of the Wold decomposition [11, 14], which is discussed

[5]Term due to Wiener (1894–1964).

in Chapter 7. Spectral factorization and the innovations representation are also related to the triangular decomposition of the correlation matrix discussed in Chapter 2. This point is investigated further in Chapters 7 and 8.

5.6 TRANSFORMATION OF HIGHER-ORDER MOMENTS

This chapter has so far concentrated on linear transformations of first and second moment quantities. Here the transformation of the higher-order moments, cumulants, and higher-order spectra is explored briefly. These transformations are of interest when a non-Gaussian random process is applied to a linear shift-invariant system. It will be assumed, as it was in the previous discussion of higher-order moments, that the random process is stationary and has zero mean.

Consider a linear system with a random process $x[n]$ as input and a random process $y[n]$ as output as in Fig. 5.1. Assume, however, that the linear system is shift-invariant with an impulse response $h[n]$ and input-output relation (5.10). Let the moments and cumulants of order k for the input and the output be denoted by $M_x^{(k)}$, $M_y^{(k)}$, $C_x^{(k)}$, and $C_y^{(k)}$.

Let us first begin with the moments of the input and the output. Since the second-order moment is just the correlation function, it follows from (5.16) that

$$M_y^{(2)}[l_1] = \sum_{k_0=-\infty}^{\infty} \sum_{k_1=-\infty}^{\infty} M_x^{(2)}[l_1 - k_1 + k_0] h[k_1] h^*[k_0] \tag{5.84}$$

The third-order moment can be found directly from the definition

$$M_y^{(3)}[l_1, l_2] = \mathcal{E}\{y^*[n] y[n+l_1] y[n+l_2]\} \tag{5.85}$$

and the transformation (5.10). Combination of these two equations produces

$$\begin{aligned} M_y^{(3)}[l_1, l_2] \\ &= \mathcal{E}\left\{ \sum_{k_0=-\infty}^{\infty} h^*[k_0] x^*[n-k_0] \sum_{k_1=-\infty}^{\infty} h[k_1] x[n+l_1-k_1] \sum_{k_2=-\infty}^{\infty} h[k_2] x[n+l_2-k_2] \right\} \\ &= \sum_{k_0=-\infty}^{\infty} \sum_{k_1=-\infty}^{\infty} \sum_{k_2=-\infty}^{\infty} \mathcal{E}\{x^*[n-k_0] x[n+l_1-k_1] x[n+l_2-k_2]\}\, h[k_2] h[k_1] h^*[k_0] \end{aligned}$$

or

$$\begin{aligned} M_y^{(3)}[l_1, l_2] \\ &= \sum_{k_0=-\infty}^{\infty} \sum_{k_1=-\infty}^{\infty} \sum_{k_2=-\infty}^{\infty} M_x^{(3)}[l_1 - k_1 + k_0, l_2 - k_2 + k_0] h[k_2] h[k_1] h^*[k_0] \end{aligned} \tag{5.86}$$

The fourth-order moment can be found by an essentially identical procedure. The result is

$$M_y^{(4)}[l_1, l_2, l_3] = \sum_{k_0=-\infty}^{\infty} \sum_{k_1=-\infty}^{\infty} \sum_{k_2=-\infty}^{\infty} \sum_{k_3=-\infty}^{\infty} M_x^{(4)}[l_1 - k_1 + k_0, l_2 - k_2 + k_0, l_3 - k_3 + k_0] \cdot h[k_3]h[k_2]h^*[k_1]h^*[k_0] \tag{5.87}$$

The transformation results for the moments can now be used to find the relations between the cumulants of the input and output. Since the input and output have mean zero, the cumulants of order less than four are equal to the correponding moments (see Chapter 4). Therefore, it follows from (5.84) and (5.86) that

$$\boxed{C_y^{(2)}[l_1] = \sum_{k_0=-\infty}^{\infty} \sum_{k_1=-\infty}^{\infty} C_x^{(2)}[l_1 - k_1 + k_0]h[k_1]h^*[k_0]} \tag{5.88}$$

and

$$\boxed{C_y^{(3)}[l_1, l_2] = \sum_{k_0=-\infty}^{\infty} \sum_{k_1=-\infty}^{\infty} \sum_{k_2=-\infty}^{\infty} C_x^{(3)}[l_1 - k_1 + k_0, l_2 - k_2 + k_0]h[k_2]h[k_1]h^*[k_0]} \tag{5.89}$$

The result for the fourth-order cumulant is less direct but also straightforward to derive. By beginning with the relations

$$\begin{aligned} C_y^{(4)}[l_1, l_2, l_3] = M_y^{(4)}[l_1, l_2, l_3] &- C_y^{(2)}[l_1]C_y^{(2)}[l_3 - l_2] \\ &- C_y^{(2)}[l_2]C_y^{(2)}[l_3 - l_1] - C_y^{(2)}[l_3]C_y^{(2)}[l_2 - l_1] \quad (a) \\ &\text{(real random process)} \\ C_y^{(4)}[l_1, l_2, l_3] = M_y^{(4)}[l_1, l_2, l_3]& \\ &- C_y^{(2)}[l_2]C_y^{(2)}[l_3 - l_1] - C_y^{(2)}[l_3]C_y^{(2)}[l_2 - l_1] \quad (b) \\ &\text{(complex random process)} \end{aligned} \tag{5.90}$$

and applying (5.87) and (5.88), it can be shown that the cumulants satisfy a relation identical to that of the moments, that is,

$$\boxed{\begin{aligned} &C_y^{(4)}[l_1, l_2, l_3] = \\ &\sum_{k_0=-\infty}^{\infty} \sum_{k_1=-\infty}^{\infty} \sum_{k_2=-\infty}^{\infty} \sum_{k_3=-\infty}^{\infty} C_x^{(4)}[l_1 - k_1 + k_0, l_2 - k_2 + k_0, l_3 - k_3 + k_0] \\ &\qquad \cdot h[k_3]h[k_2]h^*[k_1]h^*[k_0] \end{aligned}} \tag{5.91}$$

(This formula applies to the complex case and also to the real case if the conjugates are dropped.) The details are left as a problem (Problem 5.30).

Although (5.88), (5.89), and (5.91) are explicit relations for the transformations of the cumulants, it is worthwhile to interpret these equations further. Since $C_x^{(2)}$ and $C_y^{(2)}$ are just the correlation functions, we know from Section 5.1.2 that (5.88) can be expressed as

$$C_y^{(2)}[l] = C_x^{(2)}[l] * h[l] * h^*[-l] \tag{5.92}$$

Let us see if something like this can be derived for the third-order cumulant. Begin by writing (5.89) as

$$C_y^{(3)}[l_1, l_2] = \sum_{k_0=-\infty}^{\infty} \sum_{k_1=-\infty}^{\infty} G_1[l_1 - k_1 + k_0, l_2 + k_0] h[k_1] h^*[k_0] \tag{5.93}$$

where

$$\boxed{G_1[l_1', l_2'] = \sum_{k_2=-\infty}^{\infty} C_x^{(3)}[l_1', l_2' - k_2] h[k_2]} \tag{5.94}$$

Note that for any fixed value of the first argument, G_1 is the convolution of $C_x^{(3)}$ with the impulse response h. Now further write (5.93) as

$$C_y^{(3)}[l_1, l_2] = \sum_{k_0=-\infty}^{\infty} G_2[l_1 + k_0, l_2 + k_0] h^*[k_0] \tag{5.95}$$

where

$$\boxed{G_2[l_1', l_2'] = \sum_{k_1=-\infty}^{\infty} G_1[l_1' - k_1, l_2'] h[k_1]} \tag{5.96}$$

and note that for any fixed value of the second argument, G_2 is the convolution of G_1 with h. Finally, by making the simple change of variables $k' = -k_0$, rewrite (5.95) as

$$\boxed{C_y^{(3)}[l_1, l_2] = \sum_{k'=-\infty}^{\infty} G_2[l_1 - k', l_2 - k'] h^*[-k']} \tag{5.97}$$

This can also be interpreted as a convolution, but it is a convolution of the data with $h^*[-l]$ along a line at 45 degrees in the (l_1, l_2) plane. The steps involved in computing $C_y^{(3)}$ from $C_x^{(3)}$ in a finite rectangular region are depicted in Fig. 5.17. G_1 is first computed from (5.94) by convolving each *column* of $C_x^{(3)}$ with the impulse response. Next, G_2 is computed from (5.96) by convolving each *row* of G_1 with h. Finally, $C_y^{(3)}$ is computed from (5.97) by

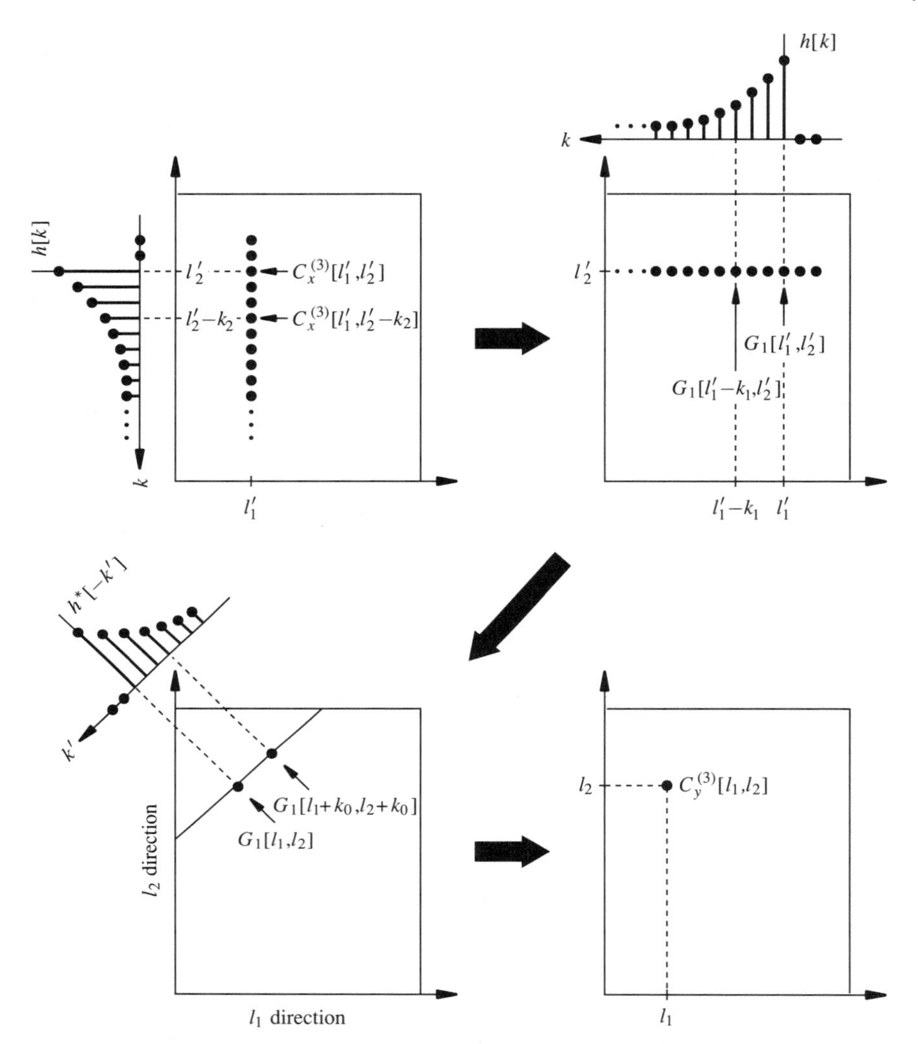

Figure 5.17 Steps involved in computing third-order cumulant of output from cumulant of input of a linear shift-invariant system.

convolving along the *diagonals* of G_2 with $h^*[-l]$. [This step may be easier to visualize using (5.95) instead of (5.97).] It could similarly be shown that if the values of the fourth-order input cumulant are represented as points in a cube, then the output cumulant can be computed by successively convolving along each of the axes with $h[l]$ or $h^*[l]$ and then convolving along the major diagonal of the cube with $h^*[-l]$.

Let us now turn to the transformation of the higher-order spectra. The bispectrum of the output is defined by the two-dimensional Fourier transform

$$B_y(\omega^{(1)},\omega^{(2)}) = S_y^{(3)}(e^{\jmath\omega^{(1)}},e^{\jmath\omega^{(2)}}) = \sum_{l_1=-\infty}^{\infty}\sum_{l_2=-\infty}^{\infty} C_y^{(3)}[l_1,l_2]e^{-\jmath(\omega^{(1)}l_1+\omega^{(2)}l_2)} \tag{5.98}$$

If (5.89) is now substituted for $C_y^{(3)}$ and the order of summation is interchanged, the right side of this equation becomes

$$\sum_{l_1=-\infty}^{\infty}\sum_{l_2=-\infty}^{\infty}\sum_{k_0=-\infty}^{\infty}\sum_{k_1=-\infty}^{\infty}\sum_{k_2=-\infty}^{\infty} C_x^{(3)}[l_1-k_1+k_0,l_2-k_2+k_0] \cdot h[k_2]h[k_1]h^*[k_0]e^{-\jmath(\omega^{(1)}l_1+\omega^{(2)}l_2)}$$

$$= \sum_{k_0=-\infty}^{\infty} h^*[k_0]e^{+\jmath(\omega^{(1)}+\omega^{(2)})k_0} \cdot \sum_{k_1=-\infty}^{\infty} h[k_1]e^{-\jmath\omega^{(1)}k_1} \cdot \sum_{k_2=-\infty}^{\infty} h[k_2]e^{-\jmath\omega^{(2)}k_2}$$

$$\cdot \sum_{l_1=-\infty}^{\infty}\sum_{l_2=-\infty}^{\infty} C_x^{(3)}[l_1-k_1+k_0,l_2-k_2+k_0]e^{-\jmath\omega^{(1)}(l_1-k_1+k_0)}e^{-\jmath\omega^{(2)}(l_2-k_2+k_0)}$$

Thus it follows that

$$\boxed{B_y(\omega^{(1)},\omega^{(2)}) = H^*(e^{\jmath(\omega^{(1)}+\omega^{(2)})})H(e^{\jmath\omega^{(1)}})H(e^{\jmath\omega^{(2)}})B_x(\omega^{(1)},\omega^{(2)})} \tag{5.99}$$

By a similar procedure, it can be shown that the trispectra satisfy the relation

$$\boxed{\begin{aligned}&S_y^{(4)}(e^{\jmath\omega^{(1)}},e^{\jmath\omega^{(2)}},e^{\jmath\omega^{(3)}})\\ &=H^*(e^{\jmath(\omega^{(1)}+\omega^{(2)}+\omega^{(3)})})H^*(e^{-\jmath\omega^{(1)}})H(e^{\jmath\omega^{(2)}})H(e^{\jmath\omega^{(3)}})S_x^{(4)}(e^{\jmath\omega^{(1)}},e^{\jmath\omega^{(2)}},e^{\jmath\omega^{(3)}})\end{aligned}} \tag{5.100}$$

Notice that unlike the ordinary power spectral density function, these quantities are sensitive to the *phase* as well as the magnitude of the filter. For example, for the bispectrum

$$\angle B_y(\omega^{(1)},\omega^{(2)}) = -\angle H(e^{\jmath(\omega^{(1)}+\omega^{(2)})}) + \angle H(e^{\jmath\omega^{(1)}}) + \angle H(e^{\jmath\omega^{(2)}}) + \angle B_x(\omega^{(1)},\omega^{(2)})$$

This property can be useful in identifying the magnitude as well as the phase of a system by driving it with non-Gaussian white noise, whose bispectrum is a constant. Figure 5.18 shows three linear systems with the same magnitude transfer function [15]. When each is driven by non-Gaussian white noise, the same autocorrelation function and the same power spectral density function results. The bispectra provide a way of identifying the three systems, however, because the third moments, shown in Table 5.3, and therefore the cumulants, are not identical.

Procedures based on higher-order moments can be used to form models for random processes analogous to the innovations representation [15–18]. In these models a linear filter is driven by non-Gaussian white noise in order to match the higher-order cumulants or higher-order spectra of the process. The exact procedures are beyond the scope of our

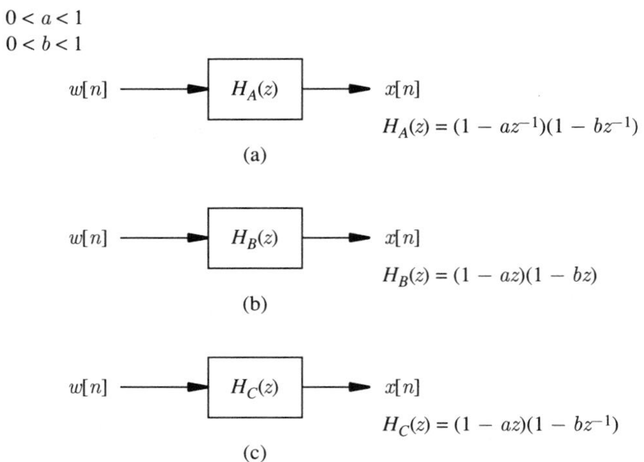

Figure 5.18 Three real linear systems with the same output power spectral density function but different higher-order spectra. (a) Minimum-phase system. (b) Maximum-phase system. (c) Mixed-phase system.

TABLE 5.3 SECOND AND THIRD MOMENT QUANTITIES FOR LINEAR SYSTEMS DRIVEN BY WHITE NOISE

		Minimum-phase	Maximum-phase	Mixed-phase
$H(z)$		$(1-ax^{-1})(1-bz^{-1})$	$(1-az)(1-bz)$	$(1-az)(1-bz^{-1})$
$x[n]$		$w[n]-(a+b)w[n-1] + abw[n-2]$	$w[n]-(a+b)w[n+1] + abw[n+2]$	$-aw[n+1]+(1+ab)w[n] - bw[n-1]$
Third moments or cumulants	$C_x^{(3)}[0,0]$	$1-(a+b)^3+a^3b^3$	$1-(a+b)^3+a^3b^3$	$(1+ab)^3-a^3-b^3$
	$C_x^{(3)}[1,1]$	$(a+b)^2-(a+b)a^2b^2$	$-(a+b)+ab(a+b)^2$	$-a(1+ab)^2+(1+ab)b^2$
	$C_x^{(3)}[2,2]$	a^2b^2	ab	$-ab^2$
	$C_x^{(3)}[1,0]$	$-(a+b)+ab(a+b)^2$	$(a+b)^2-(a+b)a^2b^2$	$a^2(1+ab)-(1+ab)^2b$
	$C_x^{(3)}[2,0]$	ab	a^2b^2	$-a^2b$
	$C_x^{(3)}[2,1]$	$-(a+b)ab$	$-(a+b)ab$	$ab(1+ab)$
Correlation function	$R_x[0]$	←	$1+a^2b^2+(a+b)^2$	→
	$R_x[1]$	←	$-(a+b)(1+ab)$	→
	$R_x[2]$	←	ab	→

Note: Minimum, maximum, and mixed-phase process with identical power spectra (or autocorrelations): $0 < a < 1$, $0 < b < 1$.
Source: C. L. Nikias and M. R. Raghuveer, Bispectrum Estimation: A Digital Signal Processing Framework, *Proc. IEEE*, 75(7), July 1987, © IEEE 1987; reproduced by permission.

present discussion. However, it can be noted that this normally results in a filter *different* from the one required for the innovations representation.

5.7 CHAPTER SUMMARY

This chapter deals with transformations of random processes by linear systems. The primary focus is on linear shift-invariant transformations and their effect on the first and second moment quantities. Although the effects of these transformations are explored in many forms, just a few simple principles emerge. First, the system performs the same transformation on the mean as it performs on the signal itself. That is, if $y[n]$ is the response to $x[n]$, then m_y is the response to m_x. Similar relations exist for the correlation function. Specifically, if the sequence $R_x[l]$ were the input to the system, then the sequence $R_{yx}[l]$ would be the output; if the sequence $R_{xy}[l]$ were the input, the sequence $R_y[l]$ would be the output. These input-output relations lead to corresponding results for the power spectral density function and the complex spectral density function. Although the mean and correlation functions are never actually used as the input to a linear system, thinking about them in this manner is a convenient way to summarize their transformations.

The matched filter is an optimal linear filter that provides maximum output signal-to-noise ratio when the input consists of a signal in additive noise. For the case of a deterministic signal in white noise the impulse response of the matched filter is proportional to the reversed signal waveform. For a random signal in white noise the impulse response is derived from the principal eigenvector of the signal correlation matrix. Specifically, it is equal to the first DKLT eigenfunction scaled and reversed. When the noise is not white, the FIR matched filter results in a generalized eigenvalue problem. The solution can be thought of as a prewhitening of the noise followed by a solution of the problem for white noise.

When the continuous part of the complex spectral density function satisfies a certain integral constraint known as the Paley-Wiener condition, it can be factored into purely causal and purely anticausal components with causal and anticausal inverses. When the complex spectral density is a rational polynomial, the factorization is achieved by simply grouping the terms corresponding to individual poles and zeros into a set that is inside and another set that is outside the unit circle. Poles and zeros outside the unit circle are in the conjugate reciprocal locations of corresponding poles and zeros inside the unit circle. The factors thus correspond to minimum phase and maximum phase components. The representation of the complex spectral density function in factored form implies that the random process can be modeled as the output of a causal linear filter driven by white noise. This is called the innovations representation. The random process and the white noise input are equivalent in that there is a causal and causally invertible linear transformation from one to the other.

Finally, the higher-order moments and cumulants of a zero-mean stationary random process are transformed in relatively simple ways when the random process is applied to a linear shift-invariant system. The transformation can be expressed as a series of convolutions with the impulse response along various directions over which the moments and cumulants are defined. The higher-order spectra are also transformed in a relatively simple way. Since the higher-order spectra of a non-Gaussian random process contain phase information, it is possible to model such a random process as a nonminimum-phase linear filter driven by non-Gaussian white noise.

REFERENCES

1. Louis L. Scharf. *Statistical Signal Processing: Detection, Estimation, and Time Series Analysis*. Addison-Wesley, Reading, Massachusetts, 1991.
2. Anthony D. Whalen. *Detection of Signals in Noise*. Academic Press, New York, 1971.
3. Harry L. Van Trees. *Detection, Estimation, and Modulation Theory*, Part I. John Wiley & Sons, New York, 1968.
4. Gilbert Strang. *Linear Algebra and Its Applications*, 3rd ed. Harcourt Brace Jovanovich, San Diego, California, 1976.
5. Francis B. Hildebrand. *Methods of Applied Mathematics*, 2nd ed. Prentice Hall, Inc., Englewood Cliffs, New Jersey, 1965.
6. Alan V. Oppenheim and Ronald W. Schafer. *Discrete-Time Signal Processing*. Prentice Hall, Inc., Englewood Cliffs, New Jersey, 1989.
7. Sophocles J. Orfanidis. *Optimum Signal Processing: An Introduction*, 2nd ed. Macmillan, New York, 1988.
8. Athanasios Papoulis. *Probability, Random Variables, and Stochastic Processes*, 2nd ed. McGraw-Hill, New York, 1991.
9. Raymond E. A. C. Paley and Norbert Wiener. Fourier transforms in the complex domain. *American Mathematical Society Colloquium Publication* 19, New York, 1934.
10. Norbert Wiener. *Extrapolation, Interpolation, and Smoothing of Stationary Time Series*. MIT Press (formerly Technology Press), Cambridge, Massachusetts, 1949.
11. H. Wold. *The Analysis of Stationary Time Series*, 2nd ed. Almquist and Wicksell, Uppsala, Sweden, 1954. (Originally published in 1938 as *A Study in the Analysis of Stationary Time Series*.)
12. A. N. Kolmogorov. Interpolation and extrapolation of stationary random sequences. *Bulletin of the Academy of Sciences, USSR*, 5:3–14, 1941. (In Russian with a short form in German.)
13. Peter Whittle. *Prediction and Regulation by Linear Least-Squares Methods*, 2nd ed. University of Minnesota Press, Minneapolis, Minnesota, 1983.
14. A. Papoulis. Predictable processes and Wold's decomposition: a review. *IEEE Transactions on Acoustics, Speech, and Signal Processing*, ASSP-33(4):933–938, August 1985.
15. Chrysostomos L. Nikias and Mysore R. Raghuveer. Bispectrum estimation: a digital signal processing framework. *Proceedings of the IEEE*, 75(7):869–891, July 1987.
16. Ananthram Swami. *System Identification Using Cumulants*. Ph.D. thesis, University of Southern California, University Park/MC-0272, Los Angeles, California, December 1988. Also USC SIPI Report 140.
17. Georgios B. Giannakis and Jerry M. Mendel. Identification of nonminimum phase systems using higher order statistics. *IEEE Transactions on Acoustics, Speech, and Signal Processing*, 37(3):360–377, March 1989.
18. Georgios B. Giannakis. On the identifiablity of non-Gaussian ARMA models using cumulants. *IEEE Transactions on Automatic Control*, 35(1):18–26, January 1990.

PROBLEMS

5.1. A linear shift-invariant system has the impulse response shown in Fig. PR5.1.

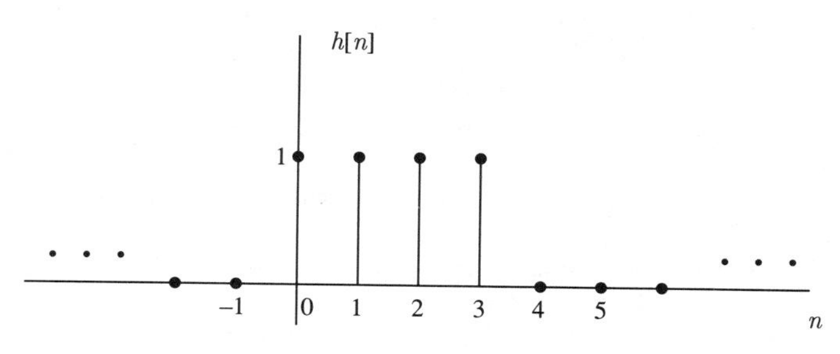

Figure PR5.1

If the system is driven with stationary white noise with variance σ_o^2,
(a) What is the mean of the output?
(b) What is the cross-correlation function between input and output?
(c) What is the correlation function of the output?

5.2. A random process with correlation function

$$R_x[l] = \rho^{|l|}$$

is passed through a linear shift-invariant system with impulse response

$$h[n] = \delta[n] - \delta[n-1]$$

(a) Compute and sketch the cross-correlation function $R_{yx}[l]$.
(b) Compute and sketch the output correlation function $R_y[l]$.

5.3. A white noise process is passed through an ideal lowpass filter whose frequency response is sketched in Fig. PR5.3.

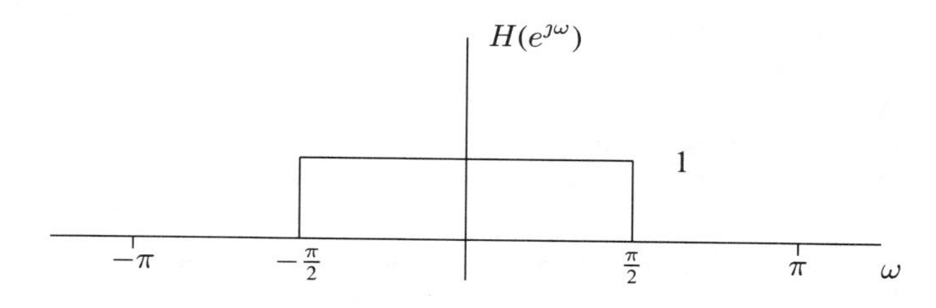

Figure PR5.3

(a) What is the mean of the output of the filter?

(b) What is the correlation function of the output?

5.4. A linear system is defined by

$$y[n] = 0.7y[n-1] + x[n] - x[n-1]$$

(a) Compute the first four values of $R_{yx}[l]$ if it is known that $R_x[l] = \delta[l]$.
(b) What is $R_{xy}[l]$ for $-3 \leq l \leq 3$?
(c) What is the power spectral density function $S_y(e^{\jmath\omega})$?

5.5. A causal linear shift-invariant system is described by the difference equation

$$y[n] - \frac{5}{6}y[n-1] + \frac{1}{6}y[n-2] = x[n-1]$$

Observe that the correlation and cross-correlation functions satisfy the difference equations

$$R_{yx}[l] - \frac{5}{6}R_{yx}[l-1] + \frac{1}{6}R_{yx}[l-2] = R_x[l-1]$$

$$R_y[l] - \frac{5}{6}R_y[l-1] + \frac{1}{6}R_y[l-2] = R_{yx}[1-l]$$

(a) Show that if the input x is a white noise process with unit variance, then the solution to the first equation is

$$R_{yx}[l] = 6\left(\left(\frac{1}{2}\right)^l - \left(\frac{1}{3}\right)^l\right)u[l]$$

where $u[l]$ is the unit step function.

(b) The function above is now used as an input to the second equation. Since the equation is driven with the sum of two exponentials, for $l < 0$ it is reasonable to assume that the response will be of the form

$$R_y[l] = c_1\left(\frac{1}{2}\right)^{-l} + c_2\left(\frac{1}{3}\right)^{-l}; \qquad l < 0$$

Since there is no input for $l > 0$ it is reasonable to assume that the only response will be the transient response, which has a similar form:

$$R_y[l] = c_1'\left(\frac{1}{2}\right)^{l} + c_2'\left(\frac{1}{3}\right)^{l}; \qquad l > 0$$

With these considerations, find the solution to the second equation.

(c) What is the system function $H(z)$ of the original system? Use this to find the z-transform of the correlation function $R_y[l]$.

(d) What is the power spectrum $S_y(e^{\jmath\omega})$?

(e) Find the impulse response of the original system and use the convolution relationships (5.13)–(5.15) to find the output correlation function $R_y[l]$. Check your answer with part (c).

5.6. Find a general closed-form expression for the correlation function of the random process $y[n]$ described by the first-order difference equation

$$y[n] + ay[n-1] = x[n] + bx[n-1]$$

when the input $x[n]$ is white noise with variance σ_o^2.

5.7. A signal with correlation function $R[l] = \left(\frac{1}{2}\right)^{|l|}$ is applied to a linear shift-invariant system with impulse response $h[n] = \delta[n] + \delta[n-1]$.
(a) Compute the correlation function of the output.
(b) Compute the power spectrum of the *input*.
(c) Compute the power spectrum of the *output*.

5.8. A random signal $x[n]$ is passed through a linear system with impulse response

$$h[n] = \delta[n] - 2\delta[n-1]$$

(a) Find the cross-correlation function between input and output $R_{xy}[l]$ if the input is white noise with variance σ_o^2.
(b) Find the correlation function of the output $R_y[l]$.
(c) Find the output power spectral density $S_y(e^{j\omega})$.

5.9. The impulse response of a linear shift-invariant system is given by

$$h[n] = \begin{cases} (-1)^n & 0 \leq n \leq 3 \\ 0 & \text{otherwise} \end{cases}$$

A white noise process with variance $\sigma_o^2 = 1$ is applied to this system. Call the input to the system $x[n]$ and the output $y[n]$.
(a) What is the cross-correlation function $R_{xy}[l]$? Sketch this function.
(b) What is the correlation function $R_y[l]$ of the output? Sketch this neatly.

5.10. A real random process with correlation function

$$R_x[l] = 2(0.8)^{|l|}$$

is applied to a linear shift-invariant system whose difference equation is

$$y[n] = 0.5y[n-1] + x[n]$$

(a) What is the complex spectral density function of the output $S_y(z)$?
(b) What is the correlation function of the output $R_y[l]$?

5.11. A random process $x[n]$ consists of independent random variables each with uniform density

$$f_x(\mathrm{x}) = \begin{cases} \frac{1}{2} & -1 \leq \mathrm{x} \leq +1 \\ 0 & \text{otherwise} \end{cases}$$

This process is applied to a linear shift-invariant system with impulse response

$$h[n] = \begin{cases} (\frac{1}{2})^n & n \geq 0 \\ 0 & n < 0 \end{cases}$$

Let the output process be denoted by $y[n]$.
(a) Compute $R_{yx}[l]$.
(b) What is $R_y[l]$?
(c) What is $S_y(z)$? Use this to compute $S_y(e^{j\omega})$.

5.12. A linear shift-invariant system has the impulse response

$$h[n] = 2\delta[n] + \delta[n-1] - \delta[n-2]$$

(a) What is the correlation function of the output if the correlation function of the input is $\delta[l]$? Sketch it versus l.
(b) What is the complex spectral density function of the output?

5.13. A random process is defined by

$$x[n] = s[n] + \eta[n]$$

where $\eta[n]$ is a unit-variance white noise process and $s[n]$ is defined by

$$s[n] = \rho s[n-1] + w[n]$$

where $w[n]$ is another unit-variance white noise process independent of $\eta[n]$.
(a) What is the correlation function $R_x[l]$?
(b) What is $S_x(z)$?

5.14. A linear shift-invariant FIR filter has impulse response

$$h[n] = \delta[n] - \frac{1}{2}\delta[n-1] + \frac{1}{4}\delta[n-2]$$

The filter is driven by a white noise process $x[n]$ with variance σ_o^2. Call the output process $y[n]$.
(a) Determine the cross-correlation function $R_{xy}[l]$.
(b) Determine the correlation function of the output $R_y[l]$.
(c) Determine the output spectral density function $S_y(e^{j\omega})$.

5.15. You are given a causal linear shift-invariant system described by the difference equation

$$y[n] = 0.2y[n-1] + x[n]$$

What is:
(a) the correlation function of the output $R_y[l]$ when the input is white noise with unit variance?
(b) the complex spectral density function of the output $S_y(z)$?
(c) the power spectral density of the output $S_y(e^{j\omega})$ when the input is a zero-mean random process with correlation function

$$R_x[l] = 2(0.5)^{|l|}$$

5.16. By using the properties of reversal, the fact that the eigenvalue is real, and the fact that $\mathbf{R}_\eta$ is a Toeplitz matrix, show that the eigenvalue $\lambda = \tilde{\mathbf{s}}^T\mathbf{R}_\eta^{-1}\tilde{\mathbf{s}}^*$ can be written as $\mathbf{s}^{*T}\mathbf{R}_\eta^{-1}\mathbf{s}$. Also show that the matched filter (5.56) satisfies (5.50) and when substituted in (5.54) results in the eigenvalue given by (5.55).

5.17. A waveform frequently used in radar and sonar is the CW pulse

$$\cos(\omega_o n)(u[n] - u[n-P])$$

where $u[n]$ is the unit step function.
(a) Find the matched filter corresponding to this signal for the case of white noise with variance σ_o^2.
(b) Find the form of the matched filter for noise with the exponential correlation function

$$R_\eta[l] = \frac{\sigma_o^2}{1-\rho^2}\rho^{|l|}$$

(It is not necessary to evaluate the SNR explicitly for this part.)

5.18. Show that the expressions for the impulse response and SNR for the colored noise matched filter in Example 5.4 generalize to the closed-form expressions given in that example for an arbitrary value of P.

5.19. Compute and sketch the impulse response of the matched filter in Example 5.4 for the following values of a and ρ. Assume that $\sigma_o^2 = 0.25$.

(a) $a = 0.7,\ \rho = 0$
(b) $a = 0.7,\ \rho = 0.2$
(c) $a = 0.7,\ \rho = 0.9$
(d) $a = 0.9,\ \rho = 0.7$
(e) $a = 0.9,\ \rho = -0.4$

5.20. The matched filter for a deterministic signal can be formulated in the frequency domain as follows. Let $S_\eta(e^{\jmath\omega})$ represent the power spectral density function of the noise. Let $\mathrm{S}(e^{\jmath\omega})$ and $H(e^{\jmath\omega})$ represent the Fourier transforms of the signal and the matched filter respectively. Now in order to maximize SNR, impose the constraint $y_s[n_p] = 1$ and *minimize* $\mathcal{E}\{|y_\eta[n_p]|^2\}$.

(a) By using the method for constrained minimization of a complex quantity (see Appendix A), show that this leads to minimizing the complex Lagrangian

$$\mathcal{L} = \frac{1}{2\pi}\int_{-\pi}^{\pi} \Big\{|H(e^{\jmath\omega})|^2 S_\eta(e^{\jmath\omega}) + \lambda\left(1 - H(e^{\jmath\omega})\mathrm{S}(e^{\jmath\omega})e^{\jmath\omega n_p}\right) + \lambda^*\left(1 - H^*(e^{\jmath\omega})\mathrm{S}^*(e^{\jmath\omega})e^{-\jmath\omega n_p}\right)\Big\}\, d\omega$$

where λ is a complex Lagrange multiplier.

(b) To find the matched filter it is necessary to choose $H(e^{\jmath\omega})$ to minimize the expression in part (a). Some results from the calculus of variations show that it is possible to take the gradient inside the integral, treating $H(e^{\jmath\omega})$ as if it were an ordinary variable instead of a function. By forming ∇_{H^*} of the integrand and setting it to zero, show that the expression for the optimal filter is given by

$$H(e^{\jmath\omega}) = \frac{\lambda^* \mathrm{S}^*(e^{\jmath\omega})e^{-\jmath\omega n_p}}{S_\eta(e^{\jmath\omega})}$$

Since SNR is independent of any scale factor, we can argue that a suitable filter is obtained if we ignore the constant λ^*, that is,

$$H(e^{\jmath\omega}) = \frac{\mathrm{S}^*(e^{\jmath\omega})e^{-\jmath\omega n_p}}{S_\eta(e^{\jmath\omega})}$$

Note that the resulting filter may be IIR and/or noncausal.

5.21. **(a)** Using the frequency domain expression for the matched filter developed in Problem 5.20, develop the form of the matched filter in colored noise for the simple transient signal of Example 5.4.

(b) Express the matched filter of part (a) in the signal domain and compare it to the matched filter derived in Example 5.4. Since the filter is not constrained to be FIR, the two results are not necessarily identical.

5.22. In this problem we explore an alternative derivation of the matched filter for a deterministic signal starting with the general expression

$$\text{SNR} = \frac{\mathbf{h}^{*T}\tilde{\mathbf{s}}^{*}\tilde{\mathbf{s}}^{T}\mathbf{h}}{\mathbf{h}^{*T}\mathbf{R}_{\eta}\mathbf{h}}$$

(a) For the case of white noise, $\mathbf{R}_{\eta} = \sigma_{o}^{2}\mathbf{I}$. The expression for SNR thus reduces to

$$\text{SNR} = \frac{\mathbf{h}^{*T}\tilde{\mathbf{s}}^{*}\tilde{\mathbf{s}}^{T}\mathbf{h}}{\sigma_{o}^{2}\mathbf{h}^{*T}\mathbf{h}}$$

By application of the Schwartz inequality, show that the value of $\mathbf{h}$ that maximizes SNR is given by

$$\mathbf{h} = K\tilde{\mathbf{s}}^{*}$$

where K is an arbitrary constant.

(b) For colored noise $\left(\mathbf{R}_{\eta} \neq \sigma_{o}^{2}\mathbf{I}\right)$, show by using the transformation

$$\boldsymbol{x}' = \mathbf{R}_{\eta}^{-1/2}\boldsymbol{x}$$

that the colored noise problem reduces to the white noise problem and that the resulting filter for the colored noise case is given by

$$\mathbf{h} = K\mathbf{R}_{\eta}^{-1}\tilde{\mathbf{s}}^{*}$$

5.23. The matched filter can be derived from the point of view of optimal detection theory for Gaussian random processes. Assume that the signal is deterministic and that the noise is a zero-mean complex Gaussian random process. Further assume that the exact starting point n_{o} of the signal is known and that the signal has length P. Then there are two possible hypotheses, namely:

$$\begin{array}{lll} \mathrm{H}_1: & x[n] = \mathrm{s}[n] + \eta[n] & \text{(signal present)} \\ \mathrm{H}_0: & x[n] = \eta[n] & \text{(signal not present)} \end{array}$$

Let $\boldsymbol{x}$ be the vector of the P samples $x[n_{o}], x[n_{o}+1], \ldots, x[n_{p}]$. Then optimal detection theory states that the optimal test for the presence of the signal has the form

$$\ln \frac{f_{\boldsymbol{x}}^{(1)}(\mathbf{x})}{f_{\boldsymbol{x}}^{(0)}(\mathbf{x})} \begin{cases} \geq \vartheta & \text{decide H}_1 \\ < \vartheta & \text{decide H}_0 \end{cases}$$

where $f_{\boldsymbol{x}}^{(i)}$ is the density function for $\boldsymbol{x}$ under hypothesis i, and ϑ is some appropriate threshold value. The quantity on the left is called the *log likelihood ratio.*

(a) Find the densities $f_{\boldsymbol{x}}^{(1)}$ and $f_{\boldsymbol{x}}^{(0)}$ assuming that the signal $\mathrm{s}[n]$ has a known deterministic form.

(b) Evaluate the log likelihood ratio and express the test in simplest form, bringing any constant terms to the right side of the equation and incorporating them into a new threshold ϑ'.

(c) Show that the optimal decision rule has the general form

$$\text{Re}\left[\tilde{\mathbf{h}}^{T}\mathbf{x}\right] \geq \vartheta'$$

which can be implemented with a matched filter. What is the formula for the filter $\mathbf{h}$?

(d) Now assume that s is a zero-mean Gaussian random process, independent of the noise, with correlation matrix $\mathbf{R}_s$. Show that in this case the likelihood ratio test *cannot* be implemented directly as a matched filter.

5.24. Following the derivations that were given in Section 5.1 for discrete random processes, show that for continuous random processes, if $y_c(t) = h_c(t) * x_c(t)$, where $*$ denotes continuous convolution, then

$$m_{y_c}^c = m_{x_c}^c \int_{-\infty}^{\infty} h_c(\tau) d\tau$$

and

$$R_{y_c x_c}^c(\tau) = h_c(\tau) * R_{x_c}^c(\tau)$$
$$R_{x_c y_c}^c(\tau) = h_c^*(-\tau) * R_{x_c}^c(\tau)$$
$$R_{y_c}^c(\tau) = h_c(\tau) * R_{x_c y_c}^c(\tau) = h_c(\tau) * h_c^*(-\tau) * R_{x_c}^c(\tau)$$

Also show that

$$S_{y_c}^c(f) = |H_c(f)|^2 S_{x_c}^c(f)$$

where $H_c(f)$ is the frequency response of the continuous filter $h_c(t)$.

5.25. A causal linear shift-invariant system is described by the difference equation

$$y[n] - 0.6y[n-1] = x[n] + 1.25x[n-1]$$

The input to this system has a power spectral density function

$$S_x(e^{\jmath\omega}) = \frac{1}{1.64 + 1.6\cos\omega}$$

(a) What is the power spectral density function of the output $S_y(e^{\jmath\omega})$?

(b) Is the system represented by the difference equation minimum-phase? If not, give the difference equation for the system with the same magnitude frequency response that *is* minimum-phase.

5.26. In Section 5.5.2 it was shown that the Paley–Wiener condition (5.73) is sufficient to define a regular process. This problem investigates the *necessity* of the condition. In other words, it shows that if a spectral density function can be factored as

$$S_x(z) = \mathcal{K}_o H_{ca}(z) H_{ca}^*(1/z^*)$$

which implies that

$$S_x(e^{\jmath\omega}) = \mathcal{K}_o |H_{ca}(e^{\jmath\omega})|^2$$

then it satisfies the Paley–Wiener condition. Note that if the last equation is written for simplicity as

$$S_x(e^{\jmath\omega}) = |H'_{ca}(e^{\jmath\omega})|^2$$

where $H'_{ca}(e^{\jmath\omega}) = \sqrt{\mathcal{K}_o} H_{ca}(e^{\jmath\omega})$, then the Paley–Wiener condition can be written as

$$\int_{-\pi}^{\pi} |\ln S_x(e^{\jmath\omega})| \, d\omega = 2\int_{-\pi}^{\pi} |\ln |H'_{ca}(e^{\jmath\omega})|| \, d\omega < \infty$$

Thus if $\gamma(\omega)$ is defined as

$$\gamma(\omega) = \ln |H'_{ca}(e^{\jmath\omega})|$$

then it is equivalent to show that

$$\int_{-\pi}^{\pi} |\gamma(\omega)| \, d\omega < \infty$$

(a) Define the function

$$\gamma^{+}(\omega) = \begin{cases} \gamma(\omega) & \geq 0 \\ 0 & \text{otherwise} \end{cases}$$

Show that

$$\int_{-\pi}^{\pi} e^{2\gamma(\omega)} d\omega > 2 \int_{-\pi}^{\pi} \gamma^{+}(\omega) d\omega$$

Further show that since

$$\int_{-\pi}^{\pi} S_x(e^{\jmath\omega}) d\omega < \infty$$

this implies that

$$\int_{-\pi}^{\pi} \gamma^{+}(\omega) < \infty$$

(b) Show also that

$$\int_{-\pi}^{\pi} \gamma(\omega) d\omega < \infty$$

[Assume that $H'_{ca}(z)$ has no zeros on the unit circle.]

(c) Finally, show from parts (a) and (b) that

$$2 \int_{-\pi}^{\pi} |\gamma(\omega)| d\omega = 2 \int_{-\pi}^{\pi} |\ln |H'_{ca}(e^{\jmath\omega})|| d\omega < \infty$$

5.27. Factor the following complex spectral density functions into minimum- and maximum-phase components.

(a)

$$S_x(z) = \frac{-16}{12z - 25 + 12z^{-1}}$$

(b)

$$S_x(z) = -\frac{3 - 10z^{-2} + 3z^{-4}}{3 + 10z^{-2} + 3z^{-4}}$$

5.28. A random process has the complex spectral density function

$$S_x(z) = \frac{z - 2.5 + z^{-1}}{z - 2.05 + z^{-1}}$$

(a) Factor this function into minimum- and maximum-phase terms. What are the poles and zeros?

(b) Find the innovations representation of the random process.

5.29. Find the innovations representation for the random processes whose complex spectral density functions are given by

(a)

$$S_x(z) = \frac{54z - 180 + 54z^{-1}}{72z - 145 + 72z^{-1}}$$

(b)

$$S_x(z) = \frac{-5z^{-2}}{1 - \frac{5}{2}z^{-2} + z^{-4}}$$

(c)

$$S_x(z) = \frac{121}{110z^4 + 221 + 110z^{-4}}$$

5.30. Combine (5.87) and (5.88) with (5.90) and simplify to derive the relation (5.91). You will need to be careful about your choice of summation indices when combining these equations.

5.31. A random process whose third-order cumulant is

$$C_x^{(3)}[l_1, l_2] = \beta_o \delta[l_1]\delta[l_2]$$

is input to the linear shift-invariant system of Problem 5.1.

(a) What is the third-order cumulant of the output? Check to see that it has the proper symmetry.

(b) What is the bispectrum of the output process? Specify magnitude and phase.

5.32. Consider the first-order difference equation

$$y[n] - \rho y[n-1] = x[n]$$

By generalizing the procedure in Section 5.2, show how to compute the third-order cumulant $C_y^{(3)}[l_1, l_2]$ from $C_x^{(3)}[l_1, l_2]$ by solving a series of three two-dimensional difference equations. Define any cross-cumulants such as $C_{xyy}^{(3)}[l_1, l_2]$ that you need in order to do this.

COMPUTER ASSIGNMENTS

5.1. The sequence S00.DAT on the enclosed disk is a realization of a white noise process with $\sigma_o^2 = 1$. Call this sequence $x[n]$. Generate another sequence $y[n]$ by applying this sequence to a first-order filter that is described by the difference equation

$$y[n] = \rho y[n-1] + x[n]$$

Generate three separate filtered sequences using each of the values $\rho = 0.95$, $\rho = 0.7$, and $\rho = -0.95$. Plot each of these sequences and the white noise sequence. What differences do you observe?

Compute the autocorrelation function for each of the three new sequences using the formula in Computer Assignment 4.1 and define the estimated correlation coefficient $\hat{\rho}$ as

$$\hat{\rho} = \frac{\hat{R}_y[1]}{\hat{R}_y[0]}$$

Compute $\hat{\rho}$ for each of the generated sequences. How well does it compare with the theoretical value?

5.2. Simulate a short delayed transient signal consisting of 32 points as follows:

$$\mathrm{s}[n] = \begin{cases} 0 & 0 \leq n \leq 7 \\ (0.7)^n & 8 \leq n \leq 23 \\ 0 & 24 \leq n \leq 31 \end{cases}$$

(a) Assume a matched filter is to be used to detect this signal in additive white Gaussian noise with $\sigma_o^2 = 1$. Determine the matched filter for the signal. Note that the normalization of the filter is such that $||\mathbf{h}|| = 1$. Convolve this filter with the signal (with no added noise) and plot the response.

(b) Determine two other filters $h_{\text{hi}}[n]$ and $h_{\text{lo}}[n]$ *not* matched to the signal whose impulse responses are proportional to $(0.5)^{(15-n)}, \quad 0 \leq n \leq 15$ and $(0.9)^{(15-n)}, \quad 0 \leq n \leq 15$ respectively. Normalize these filters so that $||\mathbf{h}_{\text{hi}}|| = ||\mathbf{h}_{\text{lo}}|| = 1$. Plot the response of each of these "unmatched" filters to the signal s$[n]$ (with no noise) and compare the responses to that of the matched filter.

(c) Divide the sequence S00.DAT, on the enclosed disk into 16 independent realizations of white Gaussian noise with $\sigma_o^2 = 1$ and 32 points in each sequence. For each of the realizations, add the noise sequence to the signal and filter the signal; plot the response of the three filters on the same scale. Make a table showing the maximum value of the filter output (for each of the three filters) and the corresponding estimated starting point of the signal $\hat{n}_o$. Also tabulate the output of each of the filters at $n = 23$, the point at which the expected value of the *matched* filter is largest. Average these values over the 16 trials and compare the results. Note that although the expected value of the matched filter is theoretically largest at $n = 23$, nothing has been shown about the properties of the estimated value $\hat{n}_o$.

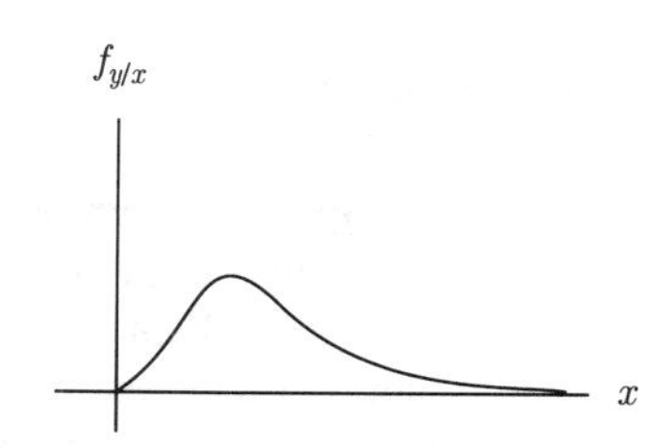

6

Estimation

The topic of estimation deals with inferring the values of unknown or random quantities from a set of related observations that are random variables. For example, given a set of observations from a Gaussian distribution with unknown mean, determining the mean from the observations is a problem in estimation. Likewise, when we try to infer the input to a system from samples of the output that are corrupted by noise, this is also a problem in estimation. There are fundamentally two classes of estimation problems to consider. The first is when the quantity to be estimated is fixed but unknown, as in the first example. The second is when the quantity to be estimated is a random variable, as in the second example. The latter problem is a little more difficult and requires some additional information. It is like trying to shoot a moving target instead of a fixed target.

The first type of estimation problem frequently arises in attempts to determine parameters related to the distribution of a random process. This topic is therefore referred to as *parameter estimation.* Examples are estimation of the mean, correlation function, or power spectral density function from one or more realizations of the process. The second type of estimation problem typically arises in problems related to filtering, prediction, and estimation of the random process itself. This topic can be called *random variable estimation.* Although we make a fundamental distinction between the two types of estimation problems, the two topics have many things in common and even their domain of application can overlap. For example, it is shown that random variable estimation is sometimes useful for estimating the parameters of a density function.

This chapter begins with the problem of parameter estimation, then continues to develop classical estimation methods for random variables. It concludes with a discussion of *linear mean-square estimation*, which forms the basis for most methods of statistical signal processing and provides a suitable lead-in for the topic of optimal filtering, taken up in Chapter 7.

6.1 ESTIMATION OF PARAMETERS

This section deals with the general topic of parameter estimation. It begins by introducing what is known as a maximum likelihood estimate and its formulation. It then describes a particular iterative technique for computing a maximum likelihood estimate when there is missing information, and continues to discuss properties that pertain to maximum likelihood and other types of estimates.

6.1.1 Maximum Likelihood Estimation

Let us initially approach the topic of parameter estimation by considering a set of observations denoted by the random vector $\boldsymbol{x}$. The vector $\boldsymbol{x}$ could represent a set of scalar observations

$$\boldsymbol{x} = \begin{bmatrix} x_1 & x_2 & \cdots & x_N \end{bmatrix}^T$$

or a set of vector observations

$$\boldsymbol{x} = \begin{bmatrix} \boldsymbol{v}^{(1)T} & \boldsymbol{v}^{(2)T} & \cdots & \boldsymbol{v}^{(N)T} \end{bmatrix}^T$$

The individual observations may or may not be independent. It is assumed that the density for the set of observations is of some known form (e.g., Gaussian, uniform, etc.) and depends on some fixed but unknown parameter θ. The density for the observations is thus denoted by

$$f_{\boldsymbol{x};\theta}(\mathbf{x};\theta) \tag{6.1}$$

When a particular set of observations $\mathbf{x}^{\mathrm{o}}$ is given, the *maximum likelihood estimate* of the parameter, denoted by $\hat{\theta}_{ml}$, is defined as that value of θ that maximizes $f_{\boldsymbol{x};\theta}(\mathbf{x}^{\mathrm{o}};\theta)$.

To see how this works in a simple case and why the definition is reasonable, consider a single real random variable x having a Gaussian density function with known variance σ_{o}^2 but unknown mean m. Assume further that just a single observation of the random variable is taken. Thus the density (6.1) is of the form

$$f_{x;m}(\mathrm{x};m) = \frac{1}{\sqrt{2\pi\sigma_{\mathrm{o}}^2}} e^{-\frac{(\mathrm{x}-m)^2}{2\sigma_{\mathrm{o}}^2}} \tag{6.2}$$

Figure 6.1 shows an observed value x^{o} for the random variable and density functions corresponding to several possible choices for the mean. Since the probability that the

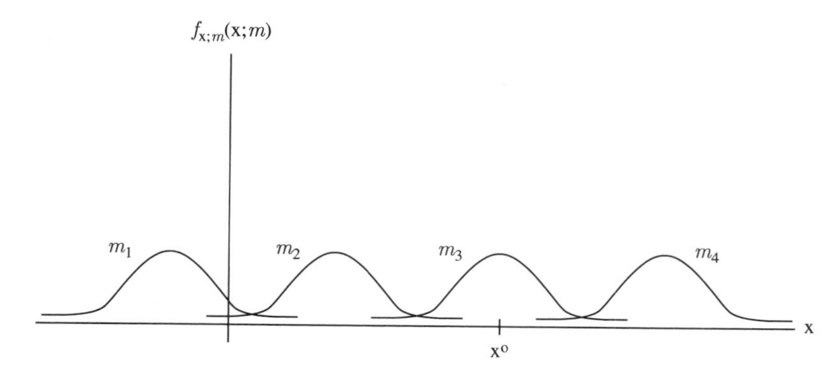

Figure 6.1 Maximum likelihood estimation of the mean.

random variable x is in a very small interval around the value x^o is proportional to the density function at the point x^o, choosing m_1, m_2, or m_4 as an estimate for the mean would imply that the given observation x_0 is *unlikely*. The density function that results in the largest probability for the given observation is the one with mean $m = m_3 = x^o$. This value is a much better choice. The value $\hat{m} = m_3$ is called the maximum likelihood estimate for the mean m, that is,

$$\hat{m}_{ml} = x^o$$

The principle is the same for all maximum likelihood estimation problems. The maximum likehood estimate for a parameter θ is that estimate that makes the *given* value of the observation vector the *most likely value*. In other words, *it is the value of θ that maximizes $f_{x;\theta}$*. The density function *when viewed as a function of θ*, for *fixed* values of the observations, is known as the *likelihood function*. This last point cannot be overemphasized. We have been used to thinking of $f_{x;\theta}$ as a function of x. Now it is necessary to turn this thinking around and view $f_{x;\theta}$ as a function of θ. Since the observation vector x is random, we write the likelihood function as $f_{x;\theta}(x;\theta)$ using the italic x as the argument instead of the roman **x** that is used when $f_{x;\theta}$ is regarded as a density function. The maximum likelihood estimate for θ is thus defined by

$$\boxed{\hat{\theta}_{ml}(x) = \underset{\theta}{\text{argmax}} \quad f_{x;\theta}(x;\theta)} \tag{6.3}$$

The notation emphasizes that $\hat{\theta}_{ml}$ depends on the random observation vector x and so *is itself a random variable*. This is a very important point that is discussed further in the next subsection.

As a more typical example of maximum likelihood estimation, consider again the real random variable with density function (6.2). In this case, however, assume that several independent observations of the random variable are available so $x = [x_1 x_2 \cdots x_N]^T$. Since the observations are independent, their joint density function is the product of the densities for each of the x_i. The likelihood function is therefore

$$f_{\boldsymbol{x};m}(\boldsymbol{x};m) = f_{x_1,x_2,\ldots,x_N;m}(x_1,x_2,\ldots,x_N;m) = \prod_{i=1}^{N} \frac{1}{\sqrt{2\pi\sigma_o^2}} e^{-\frac{(x_i-m)^2}{2\sigma_o^2}} \tag{6.4}$$

Since the logarithm is a monotonic function, maximizing (6.4) is equivalent to maximizing the *log likelihood function*

$$\ln f_{\boldsymbol{x};m}(\boldsymbol{x};m) = -N\ln(\sqrt{2\pi\sigma_o^2}) - \sum_{i=1}^{N} \frac{(x_i-m)^2}{2\sigma_o^2} \tag{6.5}$$

To clearly show the dependence on the parameter m, let us expand this and write it in the form

$$\ln f_{\boldsymbol{x};m}(\boldsymbol{x};m) = \left(-\frac{N}{2\sigma_o^2}\right)m^2 + \left(\sum_{i=1}^{N}\frac{x_i}{\sigma_o^2}\right)m - \left(N\ln(\sqrt{2\pi\sigma_o^2}) + \sum_{i=1}^{N}\frac{x_i^2}{2\sigma_o^2}\right) \tag{6.6}$$

This is a quadratic function of m which is convex upward as shown in Fig. 6.2.[1] Figure 6.2 shows that the likelihood function has a single maximum that occurs at the point where the derivative is zero. Taking the derivative with respect to m and setting the result to zero yields

$$\frac{\partial \ln f_{\boldsymbol{x};m}(\boldsymbol{x};m)}{\partial m} = -\frac{N}{\sigma_o^2}m + \frac{1}{\sigma_o^2}\sum_{i=1}^{N} x_i = 0 \tag{6.7}$$

or

$$\hat{m}_{ml} = \frac{1}{N}\sum_{i=1}^{N} x_i \tag{6.8}$$

This estimate is called the *sample mean*. Since the density function for complex random variables has a form similar to (6.2), the estimate above for the sample mean also applies to complex random variables. Let us consider this result in the context of a specific signal processing example.

Example 6.1

A constant but unknown signal is observed in additive Gaussian white noise. That is,

$$x[n] = \mathrm{A} + w[n]$$

where A is the unknown signal, $w[n]$ is the noise, and $x[n]$ is the observed random process. If A and $w[n]$ are complex, then the noise at any point n has the complex Gaussian density function

$$f_w(\mathrm{w}) = \frac{1}{\pi\sigma_w^2} e^{-\frac{|\mathrm{w}|^2}{\sigma_w^2}}$$

[1]Note that $f_{\boldsymbol{x};\theta}$ or its logarithm can in general exhibit a dependence on the parameter *which is entirely different from its dependence on the random vector*. In the present case the dependence on both is quadratic.

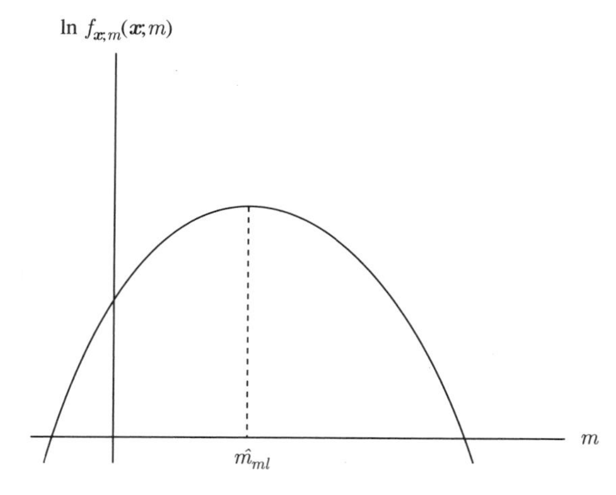

Figure 6.2 Plot of the likelihood function for the mean of a Gaussian random variable.

If $\boldsymbol{x}$ represents the vector of samples $x[0], x[1], \ldots, x[N_s - 1]$, then the observations for this example are independent and the joint density for the observations is a product of marginal densities. The likelihood function is therefore

$$f_{\boldsymbol{x};\mathrm{A}}(\boldsymbol{x};\mathrm{A}) = \prod_{n=0}^{N_s-1} \frac{1}{\pi\sigma_w^2} e^{-\frac{|x[n]-\mathrm{A}|^2}{\sigma_w^2}}$$

and the log likelihood function is

$$\ln f_{\boldsymbol{x};\mathrm{A}}(\boldsymbol{x};\mathrm{A}) = -N_s \ln(\pi\sigma_w^2) - \sum_{n=0}^{N_s-1} \frac{|x[n]-\mathrm{A}|^2}{\sigma_w^2}$$

Instead of rearranging this to show the explicit dependence on A, we can simply observe that the log likelihood function is continuous and set the complex gradient to zero. Applying the formulas in Table A.3 of Appendix A yields

$$\nabla_{\mathrm{A}^*} \ln f_{\boldsymbol{x};\mathrm{A}}(\boldsymbol{x};\mathrm{A}) = -\sum_{n=0}^{N_s-1} \frac{(x[n]-\mathrm{A})}{\sigma_w^2} = 0$$

Solving this for A produces the maximum likelihood estimate for the signal:

$$\hat{\mathrm{A}}_{ml} = \frac{1}{N_s} \sum_{n=0}^{N_s-1} x[n]$$

This is the sample mean of the random process.

To state a general principle, whenever the likelihood function is continuous *and the maximum does not occur at a boundary,* the maximum likelihood estimate for a real parameter θ can be obtained as a solution to either of the equations

$$\begin{array}{|ll|}\hline \left.\frac{\partial f_{\boldsymbol{x};\theta}(\boldsymbol{x};\theta)}{\partial\theta}\right|_{\theta=\hat{\theta}_{ml}} = 0 & (a) \\ \left.\frac{\partial \ln f_{\boldsymbol{x};\theta}(\boldsymbol{x};\theta)}{\partial\theta}\right|_{\theta=\hat{\theta}_{ml}} = 0 & (b) \\ \hline \end{array} \tag{6.9}$$

Usually, one or the other of these equations is most convenient for a particular problem. For complex parameters the equivalent equations are

$$\begin{array}{|ll|}\hline \nabla_{\theta^*} f_{\boldsymbol{x};\theta}(\boldsymbol{x};\theta)\big|_{\theta=\hat{\theta}_{ml}} = 0 & (a) \\ \nabla_{\theta^*} \ln f_{\boldsymbol{x};\theta}(\boldsymbol{x};\theta)\big|_{\theta=\hat{\theta}_{ml}} = 0 & (b) \\ \hline \end{array} \tag{6.10}$$

(When the true derivative with respect to the complex parameter θ exists, these two sets of equations are identical.) Equations 6.9(a) and (b) or 6.10(a) and (b) are called the *likelihood equation* and the *log likelihood equation* respectively. One further example demonstrates their use.

Example 6.2

The time T between requests for service on a computer bus is a random variable with exponential density function

$$f_T(\mathrm{T}) = \begin{cases} \alpha e^{-\alpha \mathrm{T}} & \mathrm{T} \geq 0 \\ 0 & \text{otherwise} \end{cases}$$

The real parameter α is known as the *arrival rate*. A total of $N+1$ requests are observed with independent[2] interarrival times $T_1, T_2, \ldots T_N$. It is desired to estimate the arrival rate α. The likelihood function for this problem is

$$f_{T_1,T_2,\ldots,T_N;\alpha}(T_1, T_2, \ldots, T_N; \alpha) = \prod_{i=1}^{N} \alpha e^{-\alpha T_i} = \alpha^N e^{-\alpha \sum_{i=1}^{N} T_i}$$

To find the maximum likelihood estimate for α, take the logarithm of this expression and set the derivative of the log likelihood function to zero. This yields

$$\frac{\partial}{\partial \alpha}\left[N \ln \alpha - \alpha \sum_{i=1}^{N} T_i\right] = \frac{N}{\alpha} - \sum_{i=1}^{N} T_i = 0$$

Solving this for α produces the maximum likelihood estimate

$$\hat{\alpha}_{ml} = \frac{1}{\frac{1}{N}\sum_{i=1}^{N} T_i}$$

The estimate is the reciprocal of the average interarrival time.

[2]The underlying model for the arrival of requests is a Poisson point process [1, 2]. Such a process has independent increments, with the result that events defined on nonoverlapping intervals are always independent.

6.1.2 The EM Algorithm

A general iterative approach to compute maximum likelihood estimates is known as the EM algorithm [3]. (EM stands for "expectation/maximization" or "estimate maximize.") The algorithm applies when the given observations are either incomplete *or can be viewed as* incomplete. Let us explain what is meant by this in a little more detail. Let the given observations be denoted by $\boldsymbol{y}$ and be dependent on a parameter θ. The likelihood function for the given observations is thus

$$f_{y;\theta}(\boldsymbol{y};\theta)$$

The observations are considered to be derived from another set of data $\boldsymbol{x}$ which for some reason cannot be observed directly. Mathematically, $\boldsymbol{y}$ is the result of a noninvertible (many-to-one) transformation of $\boldsymbol{x}$, that is,

$$\boldsymbol{y} = \boldsymbol{T}(\boldsymbol{x})$$

The data $\boldsymbol{x}$ is dependent on the same parameter θ in such a way that maximizing the likelihood function

$$f_{x;\theta}(\boldsymbol{x};\theta)$$

is (relatively) easy, while maximizing the likelihood function $f_{y;\theta}(\boldsymbol{y};\theta)$ is (relatively) hard. The data $\boldsymbol{y}$ is called the "incomplete data," while the data $\boldsymbol{x}$ is referred to as the "complete data."

To give a simple example of this situation,[3] consider an array of five adjacent sensors arranged as shown in Fig. 6.3, whose outputs are quantized to 7 bits, representing discrete values between 0 and 127. Assume that the outputs $\{x_0, x_1, x_2, x_3, x_4\}$ of the array are described by a multinomial distribution with a single parameter P of the specific form

$$\frac{(x_0+x_1+x_2+x_3+x_4)!}{x_0!x_1!x_2!x_3!x_4!}\left(\tfrac{1}{2}\right)^{x_0}\left(\tfrac{1}{4}\mathrm{P}\right)^{x_1}\left(\tfrac{1}{4}-\tfrac{1}{4}\mathrm{P}\right)^{x_2}\left(\tfrac{1}{4}-\tfrac{1}{4}\mathrm{P}\right)^{x_3}\left(\tfrac{1}{4}\mathrm{P}\right)^{x_4} \tag{6.11}$$

It can easily be shown (see Problem 6.3) that the maximum likelihood estimate for P requires solution of the simple linear equation

$$(x_1+x_2+x_3+x_4)\mathrm{P} - (x_1+x_4) = 0 \tag{6.12}$$

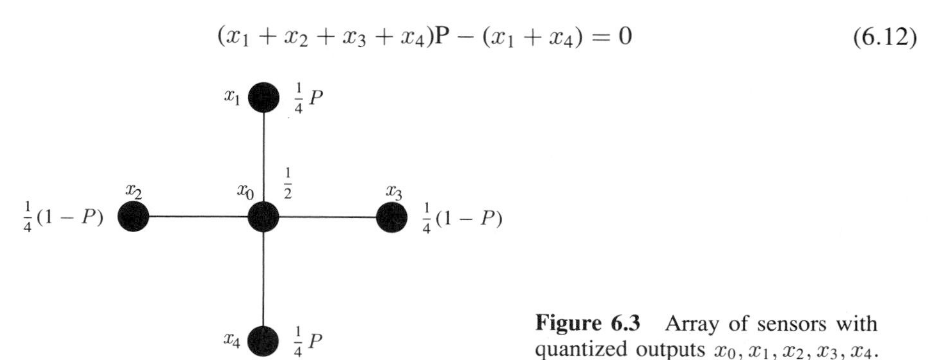

Figure 6.3 Array of sensors with quantized outputs x_0, x_1, x_2, x_3, x_4.

[3]Example adapted from the original paper by Dempster et al. [3].

which yields

$$\hat{P}_{ml} = \frac{x_1 + x_4}{x_1 + x_2 + x_3 + x_4} \tag{6.13}$$

Now assume that the array has a defect such that the output of the two sensors 0 and 1 are added together. Thus the only data that can be observed is the "incomplete" data

$$y_1 = x_0 + x_1$$
$$y_2 = x_2$$
$$y_3 = x_3$$
$$y_4 = x_4$$

This incomplete data is characterized by the multinomial distribution

$$\frac{(y_1 + y_2 + y_3 + y_4)!}{y_1!y_2!y_3!y_4!}\left(\tfrac{1}{2} + \tfrac{1}{4}\mathrm{P}\right)^{y_1}\left(\tfrac{1}{4} - \tfrac{1}{4}\mathrm{P}\right)^{y_2}\left(\tfrac{1}{4} - \tfrac{1}{4}\mathrm{P}\right)^{y_3}\left(\tfrac{1}{4}\mathrm{P}\right)^{y_4} \tag{6.14}$$

It turns out that the maximum likelihood estimate for the parameter P using the incomplete data requires solution of the quadratic equation

$$(y_1 + y_2 + y_3 + y_4)\mathrm{P}^2 - (y_1 - 2y_2 - 2y_3 - y_4)\mathrm{P} - 2y_4 = 0 \tag{6.15}$$

This estimate is of course different from the estimate (6.13); but using the EM algorithm it is possible to find the maximum likelihood estimate based on the incomplete data through an iterative procedure by repetitively estimating the missing data in (6.12) and solving the *linear* equation instead of solving the quadratic equation directly. Of course, solving the quadratic equation in this example is not a "big deal" and we may prefer doing that to iterating. However, this example illustrates the trade-off between more difficult computation and iteration; in many cases finding a direct maximum likelihood solution to the problem with incomplete data can be extremely difficult.

Let us now develop the precise form of the EM algorithm. The development follows along the lines of that in [4]. Begin by writing the likelihood function for the complete data as

$$f_{x;\theta}(\boldsymbol{x};\theta) = f_{x|y;\theta}(\boldsymbol{x}|\boldsymbol{y};\theta) f_{y;\theta}(\boldsymbol{y};\theta) \tag{6.16}$$

Taking the logarithm of this equation and rearranging yields

$$\ln f_{y;\theta}(\boldsymbol{y};\theta) = \ln f_{x;\theta}(\boldsymbol{x};\theta) - \ln f_{x|y;\theta}(\boldsymbol{x}|\boldsymbol{y};\theta) \tag{6.17}$$

It is important for the next step to remember that the terms in (6.17) are *likelihood functions*, that is, known functions of the parameter θ that are also dependent on the random vectors $\boldsymbol{x}$ and $\boldsymbol{y}$. Let us then evaluate (6.17) at some arbitrary value of the parameter θ' and take the expectation of this entire equation using the conditional density $f_{x|y;\theta}(\boldsymbol{x}|\boldsymbol{y};\hat{\theta})$, where $\hat{\theta}$ is the current estimate for the parameter. Denoting this expectation by $\mathcal{E}\left\{\,\cdot\,|\boldsymbol{y};\hat{\theta}\right\}$, we obtain

$$\ln f_{y;\theta}(\boldsymbol{y};\theta') = U(\theta',\hat{\theta}) - V(\theta',\hat{\theta}) \tag{6.18}$$

where

$$U(\theta', \hat{\theta}) = \mathcal{E}\Big\{\ln f_{x;\theta}(\boldsymbol{x}; \theta')\big|\boldsymbol{y}; \hat{\theta}\Big\} \tag{6.19}$$

and

$$V(\theta', \hat{\theta}) = \mathcal{E}\Big\{\ln f_{x|y;\theta}(\boldsymbol{x}|\boldsymbol{y}; \theta')\big|\,\boldsymbol{y}; \hat{\theta}\Big\} \tag{6.20}$$

and where the conditional expectation of the term on the left of (6.17) is just equal to the term itself.

Now it can be shown (see Problem 6.4) that the function V satisfies the condition

$$V(\theta', \hat{\theta}) \leq V(\hat{\theta}, \hat{\theta}) \tag{6.21}$$

In other words, if $\hat{\theta}$ were the true value of the parameter, then the expected value of the log likelihood function is maximized when the value θ' is also equal to the true value. As a result, if θ' is chosen such that

$$U(\theta', \hat{\theta}) > U(\hat{\theta}, \hat{\theta})$$

then it follows from (6.18) that

$$\ln f_{y;\theta}(\boldsymbol{y}; \theta') > \ln f_{y;\theta}(\boldsymbol{y}; \hat{\theta})$$

This last observation is the basis of the EM algorithm and provides the method by which the term $\ln f_{y;\theta}$ can be maximized. The steps in the algorithm are summarized in the box below.

EM ALGORITHM

Start with an arbitrary initial estimate $\hat{\theta}^{(0)}$.
For $i = 0, 1, 2, \ldots$

E-step: Compute

$$U(\theta', \hat{\theta}^{(i)}) = \mathcal{E}\{\ln f_{x;\theta}(\boldsymbol{x}; \theta')|\,\boldsymbol{y}; \hat{\theta}^{(i)}\}$$

as a function of the parameter θ'.

M-step: Find $\hat{\theta}^{(i+1)}$ as

$$\hat{\theta}^{(i+1)} = \underset{\theta'}{\operatorname{argmax}}\ U(\theta', \hat{\theta}^{(i)})$$

The last step (M-step) computes a maximum likelihood estimate for the *complete* data given the observed incomplete data. At each iteration the value of the log likelihood function can only increase. Therefore, if U is continuous in θ' and $\hat{\theta}$, the algorithm converges to a stationary point of the function. The algorithm may converge to only a local maximum, however, in which case it may be necessary to try another starting value and repeat.

Let us illustrate the EM algorithm by using it to find a solution to the array problem introduced earlier. The distributions in this case are discrete but the same procedure applies. Assume that the given (incomplete) observations are

$$\{x_1, x_2, x_3, x_4\} = \{125, 18, 20, 34\} \tag{6.22}$$

The estimate for P can be found directly by solving (6.15); this results in the value

$$\hat{\mathrm{P}} = \frac{15 + \sqrt{53809}}{394} = 0.626821498 \tag{6.23}$$

To solve the problem using the EM algorithm the function U is first formed using the discrete distribution (6.11). The log of the distribution is

$$\ln \frac{(x_0 + x_1 + x_2 + x_3 + x_4)!}{x_0!x_1!x_2!x_3!x_4!} + x_0 \ln\left(\tfrac{1}{2}\right) + x_1 \ln\left(\tfrac{1}{4}\mathrm{P}\right) + x_2 \ln\left(\tfrac{1}{4} - \tfrac{1}{4}\mathrm{P}\right) + x_3 \ln\left(\tfrac{1}{4} - \tfrac{1}{4}\mathrm{P}\right) + x_4 \ln\left(\tfrac{1}{4}\mathrm{P}\right)$$

Substituting the known values for x_2, x_3, and x_4 and taking the conditional expectation yields

$$\begin{aligned} U(\mathrm{P}', \hat{\mathrm{P}}^{(i)}) &= \mathcal{E}\left\{ \ln \frac{(x_0 + x_1 + 18 + 20 + 34)!}{x_0!x_1!18!\,20!\,34!} \,\middle|\, \hat{\mathrm{P}}^{(i)} \right\} \\ &+ \mathcal{E}\{x_0|\hat{\mathrm{P}}^{(i)}\} \ln\left(\tfrac{1}{2}\right) + \mathcal{E}\{x_1|\hat{\mathrm{P}}^{(i)}\} \ln\left(\tfrac{1}{4}\mathrm{P}'\right) \\ &+ 18 \ln\left(\tfrac{1}{4} - \tfrac{1}{4}\mathrm{P}'\right) + 20 \ln\left(\tfrac{1}{4} - \tfrac{1}{4}\mathrm{P}'\right) + 34 \ln\left(\tfrac{1}{4}\mathrm{P}'\right) \end{aligned} \tag{6.24}$$

Now the two simple expectations can be expressed in terms of the current estimate $\hat{\mathrm{P}}^{(i)}$ as

$$\begin{aligned} x_0^{(i)} &\stackrel{\text{def}}{=} \mathcal{E}\{x_0|\hat{\mathrm{P}}^{(i)}\} = 125 \cdot \frac{\frac{1}{2}}{\frac{1}{2} + \frac{1}{4}\hat{\mathrm{P}}^{(i)}} \quad (a) \\ x_1^{(i)} &\stackrel{\text{def}}{=} \mathcal{E}\{x_1|\hat{\mathrm{P}}^{(i)}\} = 125 \cdot \frac{\frac{1}{4}\hat{\mathrm{P}}^{(i)}}{\frac{1}{2} + \frac{1}{4}\hat{\mathrm{P}}^{(i)}} \quad (b) \end{aligned} \tag{6.25}$$

Fortunately, an explicit expression for the expectation of the factorial term is not needed in the remaining steps since it does not involve the parameter P′. Let us therefore denote this term by $G(\hat{\mathrm{P}}^{(i)})$ and write (6.24) as

$$\begin{aligned} U(\mathrm{P}', \hat{\mathrm{P}}^{(i)}) &= G(\hat{\mathrm{P}}^{(i)}) + x_0^{(i)} \ln \tfrac{1}{2} \\ &+ x_1^{(i)} \ln\left(\tfrac{1}{4}\mathrm{P}'\right) + 38 \ln\left(\tfrac{1}{4} - \tfrac{1}{4}\mathrm{P}'\right) + 34 \ln\left(\tfrac{1}{4}\mathrm{P}'\right) \end{aligned} \tag{6.26}$$

where $x_0^{(i)}$ and $x_1^{(i)}$ are given by (6.25). This constitutes the E-step. To apply the M-step, (6.26) is maximized with respect to P′ by taking the derivative and setting it to zero. This yields a linear equation identical to (6.12) and the corresponding estimate

$$\hat{\mathrm{P}}^{(i+1)} = \frac{x_1^{(i)} + x_4}{x_1^{(i)} + x_2 + x_3 + x_4} \tag{6.27}$$

The EM algorithm for this problem thus reduces to two simple equations: (6.25b) and (6.27). These can be combined into a single equation if desired. Table 6.1 shows the results of the EM algorithm for eight iterations, starting with an initial estimate of $\hat{\mathrm{P}}^{(0)} = 0.5$. The general theory implies the exponential type of convergence exhibited in this example [3].

TABLE 6.1 APPLICATION OF THE EM ALGORITHM IN ESTIMATING THE PARAMETER P OF THE MULTINOMIAL DISTRIBUTION

Iteration i	Current estimate $\hat{\mathrm{P}}^{(i)}$	Deviation from ML estimate $\hat{\mathrm{P}}^{(i)} - \hat{\mathrm{P}}$	Relative deviation $\frac{\hat{\mathrm{P}}^{(i+1)} - \hat{\mathrm{P}}}{\hat{\mathrm{P}}^{(i)} - \hat{\mathrm{P}}}$
0	0.500000000	0.126821498	0.1465
1	0.608247423	0.018574075	0.1346
2	0.624321050	0.002500448	0.1330
3	0.626488879	0.000332619	0.1328
4	0.626777322	0.000044176	0.1328
5	0.626815632	0.000005866	0.1328
6	0.626820719	0.000000779	0.1328
7	0.626821394	0.000000103	0.1328
8	0.626821484	0.000000014	—

Note: A. P. Dempster, N. M. Laird, and D. B. Rubin, "Maximum Likelihood from Incomplete Data via the EM Algorithm," *Journal of the Royal Statistical Society, Series B, Vol. 39, No. 1*, 1977. Reprinted by permission.

The EM algorithm was demonstrated here for only a simple case. The benefits are greatly magnified when dealing with problems where there are multiple parameters and highly nonlinear likelihood functions. The true power of the EM algorithm, however, lies in the diversity of situations in which it can be applied. It applies not only to problems where there are missing data but to a much larger class of problems that can be cast in a form to fit the assumptions of the method. Given a difficult maximum likelihood problem, if we can somehow find a related underlying problem with an easier maximum likelihood solution and regard the given data as "incomplete" with respect to the underlying problem, then the EM algorithm yields an iterative solution to the original problem. This approach has been applied to obtain a maximum likelihood solution to a number of difficult signal processing problems (see, e.g., [5]). The EM algorithm inherently allows for great flexibility in choosing the underlying problem involving the "complete" data. The key to an effective solution however is a clever choice of this problem.

6.1.3 Properties of Estimates

The maximum likelihood estimate is only one of many possible estimates for a parameter of a density. For example, in Section 6.2.1 various estimates of the correlation function are discussed, none of which are necessarily maximum likelihood estimates. Any quantity that is used as an *estimate* for a parameter is a *function* of the observations. Using somewhat abusive notation, we can write

$$\hat{\theta}_N = \hat{\theta}_N(\boldsymbol{x})$$

where the subscript N has been added to indicate that the estimate also depends on the number of observations. Since the observations are random variables, *the estimate is also a random variable* and it is important to understand its statistical properties. For example, it is generally desirable for the expected value of an estimate to be equal to the true value of the quantity being estimated. This would imply that "on the average" the value of the estimate is close to the actual parameter value. If the expected value of the estimate is not equal to the true parameter value, then the difference between the two values is called the *bias*. Clearly, it is useful to know if an estimate has a nonzero bias and the value of the bias.

Another important statistic is the *variance* of the estimate. If an estimate is known to have a low variance, then there is high probability that its value will be close to the mean. On the other hand, if the variance is large, then the estimate can easily take on a wide range of values, and so is not very reliable. The following statistical properties of estimates can be used to quantify some of the foregoing ideas and serve as a minimal description for purposes of comparing different estimates.

1. An estimate $\hat{\theta}_N$ is *unbiased* if

$$\mathcal{E}\{\hat{\theta}_N\} = \theta$$

Otherwise, the estimate is *biased* with bias $\mathrm{b}(\theta) = \mathcal{E}\{\hat{\theta}_N\} - \theta$. An estimate is *asymptotically unbiased* if

$$\lim_{N\to\infty} \mathcal{E}\{\hat{\theta}_N\} = \theta$$

2. An estimate $\hat{\theta}_N$ is *consistent* if

$$\lim_{N\to\infty} \Pr\left[|\hat{\theta}_N - \theta| < \epsilon\right] = 1$$

for any arbitrarily small number ϵ. The sequence of estimates $\{\hat{\theta}_N\}$ for increasing values of N is said to *converge in probability* to the true value of the parameter θ.[4]

3. An estimate is said to be *efficient* with respect to another estimate if it has a lower variance. Thus if $\hat{\theta}_N$ is unbiased and efficient with respect to $\hat{\theta}_{N-1}$ for all N, then $\hat{\theta}_N$ is a consistent estimate.

[4] In general, an estimate is consistent if it converges *in some sense* to the true value of the parameter. Other possible (stronger) ways to define consistency is that the estimate converges with probability 1 or that it converges in mean square. (See, e.g., [1] or [6] for a discussion of different types of convergence.)

This last statement follows from the Tchebycheff inequality [1], which for this case states that

$$\Pr\left[|\hat{\theta}_N - \theta| \geq \epsilon\right] \leq \frac{\text{Var}\left[\hat{\theta}_N\right]}{\epsilon^2}$$

Thus if the variance of $\hat{\theta}_N$ decreases with N, the probability that $|\hat{\theta}_N - \theta| \geq \epsilon$ approaches zero as $N \to \infty$. In other words, the probability that $|\hat{\theta}_N - \theta| < \epsilon$ approaches 1. This last property is illustrated in Fig. 6.4.

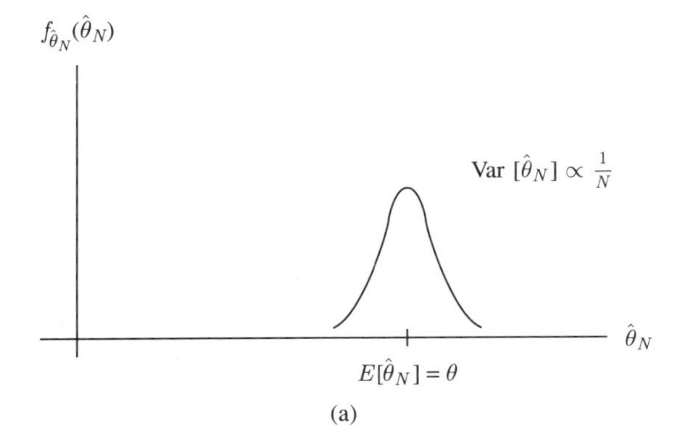

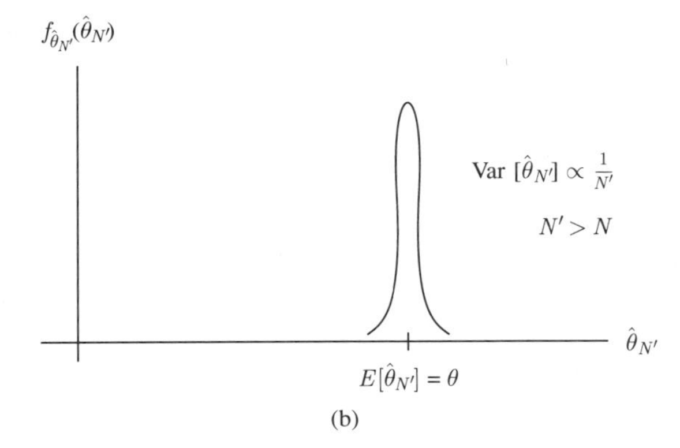

Figure 6.4 Density function for an unbiased estimate whose variance decreases with N. (a) Density function of the estimate $\hat{\theta}_N$. (b) Density function of the estimate $\hat{\theta}_{N'}$ with $N' > N$.

The following example examines these properties for the sample mean (6.8).

Example 6.3

Assume that the set of observations in an experiment $x_1, x_2, \ldots, x_N$ are independent and that the random variables x_i each have mean m and variance σ_o^2. The sample mean is given by

$$\hat{m} = \frac{1}{N}\sum_{i=1}^{N} x_i$$

The mean of this estimate is

$$\mathcal{E}\{\hat{m}\} = \frac{1}{N}\sum_{i=1}^{N}\mathcal{E}\{x_i\} = \frac{1}{N}Nm = m$$

Thus the sample mean is unbiased. The variance of the estimate is given by

$$\text{Var}\,[\hat{m}] = \frac{1}{N^2}\sum_{i=1}^{N}\text{Var}\,[x_i] = \frac{1}{N^2}N\sigma_o^2 = \frac{\sigma_o^2}{N}$$

Since the estimate is unbiased and the variance decreases with N, the estimate is consistent. This implies that as the number of samples gets very large, the probability that the estimate differs from the *true* value of the mean approaches zero.

6.1.4 The Cramér-Rao Bound

One of the most powerful results in estimation theory states that the variance of *any* unbiased estimate, regardless of its form, is greater than or equal to a certain number that depends on the probability density. The discussion in this subsection applies only to a *real* parameter θ or a complex parameter if and only if the necessary derivatives exist; more general results follow in the next subsection. Specifically, if $\hat{\theta}$ is an unbiased estimate of the parameter θ, its variance satisfies the inequality[5]

$$\boxed{\text{Var}\left[\hat{\theta}\right] \geq \frac{1}{\mathcal{E}\left\{\left(\frac{\partial \ln f_{\boldsymbol{x};\theta}(\boldsymbol{x};\theta)}{\partial\theta}\right)^2\right\}}} \tag{6.28}$$

This result is known as the Cramér-Rao bound and holds as long as the partial derivative of the log likelihood function exists and is absolutely integrable.[6] An estimate that satisfies the bound with equality is said to be *most efficient* and is called a *minimum-variance estimate.* It will be shown that the bound is satisfied with equality *if and only if* the estimate satisfies the condition

$$\hat{\theta}(\boldsymbol{x}) - \theta = K(\theta)\cdot\frac{\partial \ln f_{\boldsymbol{x};\theta}(\boldsymbol{x};\theta)}{\partial\theta} \tag{6.29}$$

where $K(\theta)$ is some quantity that may be a function of θ but *is not* a function of $\hat{\theta}$.

[5] A generalization of this bound for biased estimates is

$$\text{Var}\left[\hat{\theta}\right] \geq \frac{\left(1+\frac{\partial \text{b}(\theta)}{\partial\theta}\right)^2}{\mathcal{E}\left\{\left(\frac{\partial \ln f_{\boldsymbol{x};\theta}(\boldsymbol{x};\theta)}{\partial\theta}\right)^2\right\}}$$

where $\text{b}(\theta)$ is the bias

$$\text{b}(\theta) = \mathcal{E}\left\{\hat{\theta}\right\} - \theta$$

(See [7, p. 147, Prob. 2.4.17].)

[6] The bound is named after Cramér [8, Chap. 32] and Rao [9], although Fisher had actually shown a form of this result some 23 years earlier [10].

A second form of the Cramér-Rao bound can be written as

$$\boxed{\operatorname{Var}\left[\hat{\theta}\right] \geq \frac{1}{-\mathcal{E}\left\{\frac{\partial^2 \ln f_{\boldsymbol{x};\theta}(\boldsymbol{x};\theta)}{\partial\theta^2}\right\}}} \tag{6.30}$$

and is valid as long as the second partial derivative of the log likelihood function exists and is absolutely integrable. Both forms of the bound are equivalent in that the right-hand sides of (6.28) and (6.30) produce the same value for the lower bound, and both require (6.29) to hold for equality.

If an unbiased minimum-variance estimate exists and the maximum likelihood estimate does not occur at a boundary, then the maximum likelihood estimate *is* the minimum-variance estimate. To show this, note that if a minimum-variance estimate exists, it must satisfy (6.29), while if the maximum likelihood estimate does not occur at a boundary, it satisfies the log likelihood equation (6.9b). Combining these two equations then yields

$$\hat{\theta}(\boldsymbol{x}) - \hat{\theta}_{ml} = K(\hat{\theta}_{ml}) \cdot \left.\frac{\partial \ln f_{\boldsymbol{x};\theta}(\boldsymbol{x};\theta)}{\partial\theta}\right|_{\theta=\hat{\theta}_{ml}} = 0$$

or

$$\hat{\theta}(\boldsymbol{x}) = \hat{\theta}_{ml} \tag{6.31}$$

Let us now provide the proof of the Cramér-Rao bound. The first form (6.28) and the condition for equality (6.29) can be derived by the following procedure. Since $\hat{\theta}$ is unbiased we have

$$\mathcal{E}\left\{\hat{\theta}(\boldsymbol{x})\right\} = \theta$$

or

$$\mathcal{E}\left\{(\hat{\theta}(\boldsymbol{x}) - \theta)\right\} = \int_{-\infty}^{\infty} f_{\boldsymbol{x};\theta}(\mathbf{x};\theta)(\hat{\theta}(\mathbf{x}) - \theta)d\mathbf{x} = 0 \tag{6.32}$$

Now if the partial derivative $\frac{\partial f_{\boldsymbol{x};\theta}}{\partial\theta}$ exists and is absolutely integrable, then (6.32) can be differentiated inside the integral using the product rule to obtain

$$\int_{-\infty}^{\infty} \left(\frac{\partial f_{\boldsymbol{x};\theta}}{\partial\theta}(\hat{\theta}(\mathbf{x}) - \theta) - f_{\boldsymbol{x};\theta}(\mathbf{x};\theta)\right) d\mathbf{x} = 0$$

or

$$\int_{-\infty}^{\infty} \frac{\partial f_{\boldsymbol{x};\theta}}{\partial\theta}(\hat{\theta}(\mathbf{x}) - \theta)d\mathbf{x} = 1 \tag{6.33}$$

Then by using the formula for the derivative of the logarithm, (6.33) can be rewritten as

$$\int_{-\infty}^{\infty} \frac{\partial \ln f_{\boldsymbol{x};\theta}}{\partial\theta} f_{\boldsymbol{x};\theta} \cdot (\hat{\theta}(\mathbf{x}) - \theta)d\mathbf{x} = 1 \tag{6.34}$$

Finally, if the integrand in (6.34) is written as a product of two factors and the Schwartz inequality for functions is applied,[7] the result is

$$1 = \left(\int_{-\infty}^{\infty} \left(\frac{\partial \ln f_{x;\theta}}{\partial \theta} f_{x;\theta}^{1/2}\right) \left((\hat{\theta}(\mathbf{x}) - \theta) f_{x;\theta}^{1/2}\right) d\mathbf{x}\right)^2 \tag{6.35}$$

$$\leq \left(\int_{-\infty}^{\infty} \left(\frac{\partial \ln f_{x;\theta}}{\partial \theta}\right)^2 f_{x;\theta} d\mathbf{x}\right) \cdot \left(\int_{-\infty}^{\infty} (\hat{\theta}(\mathbf{x}) - \theta)^2 f_{x;\theta} d\mathbf{x}\right)$$

The last integral on the right is recognized as $\text{Var}[\hat{\theta}]$ and (6.9) follows. The Schwartz inequality becomes an equality if and only if the two integrands are related by a constant multiplier, independent of **x**, that is, when

$$(\hat{\theta}(\mathbf{x}) - \theta) f_{x;\theta}^{1/2} = K(\theta) \cdot \left(\frac{\partial \ln f_{x;\theta}}{\partial \theta} f_{x;\theta}^{1/2}\right)$$

which yields condition (6.29).

The second form (6.30) of the Cramér-Rao bound can be proven as follows. Start with the simple condition

$$\int_{-\infty}^{\infty} f_{x;\theta}(\mathbf{x};\theta) d\mathbf{x} = 1 \tag{6.36}$$

and differentiate this once to obtain

$$\int_{-\infty}^{\infty} \frac{\partial f_{x;\theta}}{\partial \theta} d\mathbf{x} = \int_{-\infty}^{\infty} \frac{\partial \ln f_{x;\theta}}{\partial \theta} f_{x;\theta} d\mathbf{x} = 0 \tag{6.37}$$

and then again to obtain[8]

$$\int_{-\infty}^{\infty} \left(\frac{\partial^2 \ln f_{x;\theta}}{\partial \theta^2} f_{x;\theta} + \left(\frac{\partial \ln f_{x;\theta}}{\partial \theta}\right)^2 f_{x;\theta}\right) d\mathbf{x} = 0 \tag{6.38}$$

or

$$\mathcal{E}\left\{\left(\frac{\partial \ln f_{x;\theta}}{\partial \theta}\right)^2\right\} = -\mathcal{E}\left\{\left(\frac{\partial^2 \ln f_{x;\theta}}{\partial \theta^2}\right)\right\} \tag{6.39}$$

Substitution of (6.39) into (6.28) then establishes (6.30).

As a simple application of the Cramér-Rao bound consider the following example which relates to the sample mean.

[7] The Schwartz inequality states that

$$\left(\int_{-\infty}^{\infty} g_1(x) g_2(x) dx\right)^2 \leq \left(\int_{-\infty}^{\infty} g_1^2(x) dx\right) \left(\int_{-\infty}^{\infty} g_2^2(x) dx\right)$$

for any two functions g_1 and g_2 (see, e.g., [11] or [12]).

[8] The condition that the second partial derivative of the log likelihood function exists and is absolutely integrable permits this second differentiation.

Example 6.4

Consider a set of observations $\boldsymbol{x} = [x_1 x_2 \cdots x_N]^T$ from the Gaussian density function (6.2). The likelihood function and the log likelihood function for the mean are given by (6.4) and (6.5) respectively. The derivative of the log likelihood function is

$$\frac{\partial \ln f_{\boldsymbol{x};m}(\boldsymbol{x}; m)}{\partial m} = \sum_{i=1}^{N} \frac{(x_i - m)}{\sigma_o^2}$$

Since the x_i are uncorrelated, the cross terms are zero and the expectation of this squared quantity is

$$\mathcal{E}\left\{\left(\frac{\partial \ln f_{\boldsymbol{x};m}(\boldsymbol{x}; m)}{\partial m}\right)^2\right\} = \sum_{i=1}^{N} \frac{\mathcal{E}\{(x_i - m)^2\}}{\sigma_o^4} = \frac{N\sigma_o^2}{\sigma_o^4} = \frac{N}{\sigma_o^2}$$

Therefore, by (6.28), the bound on the variance of *any* unbiased estimate of the mean of the Gaussian density is

$$\text{Var}\,[\hat{m}] \geq \frac{\sigma_o^2}{N}$$

This result can also be achieved using (6.30) since

$$-\frac{\partial^2 \ln f_{\boldsymbol{x};m}(\boldsymbol{x}; m)}{\partial m^2} = -\frac{\partial}{\partial m}\left(\sum_{i=1}^{N} \frac{(x_i - m)}{\sigma_o^2}\right) = \sum_{i=1}^{N} \frac{1}{\sigma_o^2} = \frac{N}{\sigma_o^2}$$

Note from the variance expression in Example 6.3 that the sample mean satisfies this bound with equality. This result is not really surprising, since it was shown in Section 6.1.1 that the sample mean is the maximum likelihood estimate for the mean of the Gaussian density.

To confirm that the sample mean satisfies condition (6.29), note from above that the derivative of the log likelihood function can be expressed as

$$\frac{\partial \ln f_{\boldsymbol{x};m}(\boldsymbol{x}; m)}{\partial m} = \frac{1}{\sigma_o^2} \sum_{i=1}^{N} (x_i - m) = \frac{N}{\sigma_o^2}\left(\left[\frac{1}{N}\sum_{i=1}^{N} x_i\right] - m\right)$$

or

$$(\hat{m}_{ml} - m) = \frac{\sigma_o^2}{N} \cdot \frac{\partial \ln f_{\boldsymbol{x};m}(\boldsymbol{x}; m)}{\partial m}$$

where $\hat{m}_{ml}$ is the sample mean (6.8). This last equation is in the form of (6.29).

Note in Example 6.4 that the multiplying factor $K(\theta)$ that appears in (6.29) turns out to be equal to the bound itself. As seen in the next subsection and shown in Problem 6.9, this fact is not a mere coincidence.

6.1.5 Estimating Multiple Parameters

The ideas of the previous subsections can be extended to the estimation of multiple parameters. Multiple parameters occur in estimating the mean vector of a nonstationary random process, in estimating the correlation function for a stationary random process at multiple lags, or in estimating the elements of the correlation matrix for a stationary or nonstationary random process. In describing the procedure it is most convenient to formulate the problem as estimating the elements of some vector parameter $\boldsymbol{\theta}$ of the density function, regardless of whether or not the parameters to be estimated actually appear as a vector in the expression for the density.

Maximum likelihood estimation for a vector parameter follows along the same lines as estimation for a scalar parameter. If the density function for the observations is $f_{\boldsymbol{x};\boldsymbol{\theta}}(\mathbf{x};\boldsymbol{\theta})$, then this same quantity, viewed as a function of the vector parameter $\boldsymbol{\theta}$, is the likelihood function. The maximum likelihood estimate is therefore defined by

$$\hat{\boldsymbol{\theta}}_{ml}(\boldsymbol{x}) = \underset{\boldsymbol{\theta}}{\operatorname{argmax}} \quad f_{\boldsymbol{x};\boldsymbol{\theta}}(\boldsymbol{x};\boldsymbol{\theta}) \tag{6.40}$$

When the likelihood function is a continuous function of $\boldsymbol{\theta}$, then the maximum likelihood estimate can be determined from the vector likelihood equation or log likelihood equation

$$\begin{aligned} \nabla_{\boldsymbol{\theta}^*} f_{\boldsymbol{x};\boldsymbol{\theta}}(\boldsymbol{x};\boldsymbol{\theta}) &= \mathbf{0} \qquad (a) \\ \nabla_{\boldsymbol{\theta}^*} \ln f_{\boldsymbol{x};\boldsymbol{\theta}}(\boldsymbol{x};\boldsymbol{\theta}) &= \mathbf{0} \quad (b) \end{aligned} \tag{6.41}$$

where the (real or complex) gradient is as defined in Appendix A. Either of these expressions represents a system of scalar equations that can be solved for the vector parameter. The equations are usually *nonlinear*.

Properties of estimates for a vector parameter are defined in an analogous way. Specifically, if

$$\hat{\boldsymbol{\theta}}_N = \hat{\boldsymbol{\theta}}_N(\boldsymbol{x})$$

then the following definitions apply.

1. An estimate $\hat{\boldsymbol{\theta}}_N$ is *unbiased* if

$$\mathcal{E}\{\hat{\boldsymbol{\theta}}_N\} = \boldsymbol{\theta}$$

Otherwise, the estimate is *biased* with bias $\mathbf{b}(\boldsymbol{\theta}) = \mathcal{E}\{\hat{\boldsymbol{\theta}}_N\} - \boldsymbol{\theta}$. An estimate is *asymptotically unbiased* if

$$\lim_{N\to\infty} \mathcal{E}\{\hat{\boldsymbol{\theta}}_N\} = \boldsymbol{\theta}$$

2. An estimate is $\hat{\boldsymbol{\theta}}_N$ is *consistent* if

$$\lim_{N\to\infty} \Pr\big[\|\hat{\boldsymbol{\theta}}_N - \boldsymbol{\theta}\| < \epsilon\big] = 1$$

for any arbitrarily small number ϵ. The sequence of estimates $\{\hat{\boldsymbol{\theta}}_N\}$ is then said to *converge in probability* to the parameter $\boldsymbol{\theta}$.

3. An estimate $\hat{\boldsymbol{\theta}}$ will be said to be *efficient* with respect to another estimate $\hat{\boldsymbol{\theta}}'$ if the difference of their covariance matrices $\mathbf{C}_{\hat{\theta}'} - \mathbf{C}_{\hat{\theta}}$ is positive definite. This implies that the variance of every component of $\hat{\boldsymbol{\theta}}$ must be smaller than the variance of the corresponding component of $\hat{\boldsymbol{\theta}}'$. If $\hat{\boldsymbol{\theta}}_N$ is unbiased and efficient with respect to $\hat{\boldsymbol{\theta}}_{N-1}$ for all N, then $\hat{\boldsymbol{\theta}}_N$ is a consistent estimate.

The proof of this last statement is similar to the proof in the scalar case.

For real-valued vector parameters there is a generalized version of the Cramér-Rao bound which is as follows. Define the matrix

$$\mathbf{J}(\boldsymbol{\theta}) = \mathcal{E}\{\boldsymbol{s}(\boldsymbol{x};\boldsymbol{\theta})\boldsymbol{s}^T(\boldsymbol{x};\boldsymbol{\theta})\} \tag{6.42}$$

where $\boldsymbol{s}(\boldsymbol{x};\boldsymbol{\theta})$ is the gradient of the log likelihood function

$$\boldsymbol{s}(\boldsymbol{x};\boldsymbol{\theta}) \stackrel{\text{def}}{=} \nabla_{\boldsymbol{\theta}} \ln f_{\boldsymbol{x};\boldsymbol{\theta}}(\boldsymbol{x};\boldsymbol{\theta}) \tag{6.43}$$

(real parameter $\boldsymbol{\theta}$)

The parenthetical remark about the real parameter is added to emphasize that the complex gradient cannot necessarily be substituted in (6.43) in all cases to obtain a complex version of the results (see Problem 6.10). The real vector $\boldsymbol{s}$ is called the *score* for $\boldsymbol{\theta}$ based on $\boldsymbol{x}$ [13] and has mean zero (see Problem 6.8). If $\hat{\boldsymbol{\theta}}$ is substituted for $\boldsymbol{\theta}$, the score is a measure of the optimality of the estimate with scores near $\mathbf{0}$ being more desirable; the maximum likelihood estimate has a score of exactly $\mathbf{0}$. The matrix $\mathbf{J}$, which is the covariance matrix of $\boldsymbol{s}$, is known as the *Fisher information matrix*; it is assumed here to be nonsingular. If $\hat{\boldsymbol{\theta}}$ is any unbiased estimate and $\mathbf{C}_{\hat{\theta}}$ is its covariance matrix, then the Cramér-Rao bound can be stated as

$$\boxed{\mathbf{C}_{\hat{\theta}} \geq \mathbf{J}^{-1} \quad \text{(real parameter } \boldsymbol{\theta})} \tag{6.44}$$

where the notation $\geq$ means that the difference matrix $\mathbf{C}_{\hat{\theta}} - \mathbf{J}^{-1}$ is positive semidefinite.[9] This result implies that if $\hat{\theta}_i$ is the ith component of the vector $\hat{\boldsymbol{\theta}}$, then its variance is lower bounded by

$$\text{Var}\left[\hat{\theta}_i\right] \geq j_{ii}^{(-1)} \tag{6.45}$$

where $j_{ii}^{(-1)}$ is the ith diagonal element of the inverse matrix $\mathbf{J}^{-1}$. This gives a bound on the variance of the individual parameters. The bound (6.45) is satisfied with equality if and only if the estimate satisfies an equation of the form

$$\hat{\theta}_i(\boldsymbol{x}) - \theta_i = \boldsymbol{k}_i^T(\boldsymbol{\theta}) \cdot \boldsymbol{s}(\boldsymbol{x};\boldsymbol{\theta}) \tag{6.46}$$

[9]For a *biased* estimate the bound is modified to

$$\mathbf{C}_{\hat{\theta}} \geq \left(\mathbf{I} + \nabla_{\boldsymbol{\theta}}\mathbf{b}^T(\boldsymbol{\theta})\right)^T \mathbf{J}^{-1} \left(\mathbf{I} + \nabla_{\boldsymbol{\theta}}\mathbf{b}^T(\boldsymbol{\theta})\right)$$

where $\mathbf{b}(\boldsymbol{\theta})$ is the bias $\mathbf{b}(\boldsymbol{\theta}) = \mathcal{E}\left\{\hat{\boldsymbol{\theta}}\right\} - \boldsymbol{\theta}$ and the quantity $\nabla_{\boldsymbol{\theta}}\mathbf{b}^T(\boldsymbol{\theta})$ is a square matrix obtained by performing the gradient on every component of $\mathbf{b}$. (See [14, p. 272, Prob. 6.31].)

where this notation means that the vector $\boldsymbol{k}_i$ may be a function of $\boldsymbol{\theta}$ but *must not be* a function of $\hat{\boldsymbol{\theta}}$ (see Problem 6.9). It can further be shown that the overall bound is satisfied with equality (i.e., $\mathbf{C}_{\hat{\theta}} = \mathbf{J}^{-1}$) when all of the conditions (6.46) are satisfied for $i = 1, 2, \ldots, N$, that is, if the estimate is of the form

$$\hat{\boldsymbol{\theta}}(\boldsymbol{x}) - \boldsymbol{\theta} = \mathbf{K}(\boldsymbol{\theta})\boldsymbol{s}(\boldsymbol{x}; \boldsymbol{\theta}) \tag{6.47}$$

and further that $\mathbf{K}$ is uniquely defined by

$$\mathbf{K}(\boldsymbol{\theta}) = \mathbf{J}^{-1}(\boldsymbol{\theta}) \tag{6.48}$$

(This fact is also explored in Problem 6.9.)

The bound (6.44) can of course be applied to complex parameters by separating them into real and imaginary parts and including those parts separately in the real vector $\boldsymbol{\theta}$. The essential difficulty in developing a direct complex version of this bound is that the symmetry conditions (see Chapter II) that are required for defining the complex covariance matrices $\mathbf{C}_{\hat{\theta}}$ and $\mathbf{J}$ may not hold even if they hold for the observation vector. In cases where the symmetry conditions hold, however, a complex version of the bound identical in form to (6.44) can be obtained by substituting the complex gradient $\nabla_{\boldsymbol{\theta}^*}$ for the real gradient in (6.43). The complex version of the bound is explored in detail in Problem 6.10.

Let us now give a brief proof of the Cramér-Rao bound for vector parameters. Consider the quantity

$$\nabla_{\boldsymbol{\theta}} \mathcal{E}\left\{(\hat{\boldsymbol{\theta}} - \boldsymbol{\theta})^T\right\} = \nabla_{\boldsymbol{\theta}} \int_{-\infty}^{\infty} f_{x;\theta}(\mathbf{x}; \boldsymbol{\theta})(\hat{\boldsymbol{\theta}} - \boldsymbol{\theta})^T d\mathbf{x} = [\mathbf{0}]$$

This quantity is defined to be a matrix where every column represents the gradient of an element of the transposed vector; since $\hat{\boldsymbol{\theta}}$ is unbiased, the result is a matrix of zeros. Using the product rule for differentiation, we can write

$$\int_{-\infty}^{\infty} \left((\nabla_{\boldsymbol{\theta}} f_{x;\theta})(\hat{\boldsymbol{\theta}} - \boldsymbol{\theta})^T - f_{x;\theta}(\mathbf{x}; \boldsymbol{\theta}) \cdot \mathbf{I}\right) d\mathbf{x} = [\mathbf{0}]$$

or

$$\int_{-\infty}^{\infty} f_{x;\theta}(\mathbf{x}; \boldsymbol{\theta}) \left(\nabla_{\boldsymbol{\theta}} \ln f_{x;\theta}\right) (\hat{\boldsymbol{\theta}} - \boldsymbol{\theta})^T d\mathbf{x} \; - \; \mathbf{I} = [\mathbf{0}]$$

[The introduction of $\ln f_{x;\theta}$ follows by a similar procedure to its introduction in the scalar case; see (6.37).] This last equation simply states that

$$\mathcal{E}\left\{\boldsymbol{s} \cdot (\hat{\boldsymbol{\theta}} - \boldsymbol{\theta})^T\right\} = \mathbf{I} \tag{6.49}$$

Now consider the real vector

$$\begin{bmatrix} \hat{\boldsymbol{\theta}}(\boldsymbol{x}) \\ \boldsymbol{s}(\boldsymbol{x}; \boldsymbol{\theta}) \end{bmatrix}$$

and form its covariance matrix

$$\mathcal{E}\left\{\begin{bmatrix} (\hat{\boldsymbol{\theta}} - \boldsymbol{\theta}) \\ \boldsymbol{s} \end{bmatrix} \begin{bmatrix} (\hat{\boldsymbol{\theta}} - \boldsymbol{\theta})^T & \boldsymbol{s}^T \end{bmatrix}\right\} = \begin{bmatrix} \mathbf{C}_{\hat{\theta}} & \mathbf{I} \\ \mathbf{I} & \mathbf{J} \end{bmatrix}$$

It is easy to verify that this covariance matrix can be written in factored form as

$$\begin{bmatrix} \mathbf{C}_{\hat{\theta}} & \mathbf{I} \\ \mathbf{I} & \mathbf{J} \end{bmatrix} = \begin{bmatrix} \mathbf{I} & \mathbf{J}^{-1} \\ \mathbf{0} & \mathbf{I} \end{bmatrix} \begin{bmatrix} \mathbf{C}_{\hat{\theta}} - \mathbf{J}^{-1} & \mathbf{0} \\ \mathbf{0} & \mathbf{J} \end{bmatrix} \begin{bmatrix} \mathbf{I} & \mathbf{0} \\ \mathbf{J}^{-1} & \mathbf{I} \end{bmatrix}$$

Since the overall covariance matrix is positive semidefinite, each of the diagonal blocks in the center matrix must be positive semidefinite. This yields (6.44).

The interpretation of the Cramér-Rao bound in terms of concentration ellipsoids is depicted in Fig. 6.5. If the *deviation* in the estimate is defined as[10]

$$\boldsymbol{\delta}(\boldsymbol{x};\boldsymbol{\theta}) \stackrel{\text{def}}{=} \hat{\boldsymbol{\theta}}(\boldsymbol{x}) - \boldsymbol{\theta}$$

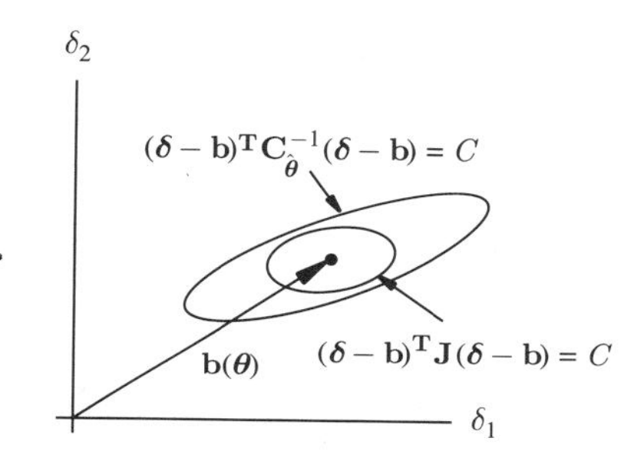

Figure 6.5 Concentration ellipses for the deviation of the estimate of a vector parameter; geometric interpretation of the Cramér-Rao bound.

then the bias of the estimate $\mathbf{b}(\boldsymbol{\theta})$ is the mean deviation (i.e., its expected value). The concentration ellipse for the deviation, with covariance $\mathbf{C}_{\hat{\theta}}$ is shown in the figure. If the deviation is Gaussian, then this represents a contour of the density. The minimum deviation covariance of the Cramér-Rao bound is represented by the smaller ellipse with covariance $\mathbf{J}^{-1}$. Geometrically, the bound states that the $\mathbf{J}^{-1}$ ellipsoid lies entirely within the $\mathbf{C}_{\hat{\theta}}$ ellipsoid. In the best case (when $\hat{\boldsymbol{\theta}}$ is the maximum likelihood estimate), the two ellipsoids coincide.

Let us conclude by giving an example of some of the concepts discussed in this subsection for a vector parameter.

Example 6.5

A d-dimensional real random vector $\boldsymbol{v}$ is described by a Gaussian density function

$$f_{\boldsymbol{v}}(\mathbf{v}) = \frac{1}{(2\pi)^{\frac{d}{2}}|\mathbf{C}_v|^{\frac{1}{2}}} e^{-\frac{1}{2}(\mathbf{v}-\mathbf{m})^T\mathbf{C}_v^{-1}(\mathbf{v}-\mathbf{m})}$$

Given N independent samples of the random vector $\boldsymbol{v}^{(1)}, \boldsymbol{v}^{(2)}, \ldots, \boldsymbol{v}^{(N)}$ it is desired to form a maximum likelihood estimate for the mean vector $\mathbf{m}$. The likelihood function for this problem is

$$f_{\boldsymbol{v}^{(1)},\boldsymbol{v}^{(2)},\ldots,\boldsymbol{v}^{(N)};\mathbf{m}}(\boldsymbol{v}^{(1)}, \boldsymbol{v}^{(2)}, \ldots, \boldsymbol{v}^{(N)}; \mathbf{m}) = \prod_{i=1}^{N} \frac{1}{(2\pi)^{\frac{d}{2}}|\mathbf{C}_v|^{\frac{1}{2}}} e^{-\frac{1}{2}(\boldsymbol{v}^{(i)}-\mathbf{m})^T\mathbf{C}_v^{-1}(\boldsymbol{v}^{(i)}-\mathbf{m})}$$

[10]In this book we use the term "deviation" to mean the difference between an estimate and the true value of a quantity, and the term "error" to mean the difference between the true value and the estimate.

The log likelihood function is thus

$$\ln f_{\boldsymbol{v}^{(1)},\boldsymbol{v}^{(2)},\ldots,\boldsymbol{v}^{(N)};\mathbf{m}}(\boldsymbol{v}^{(1)},\boldsymbol{v}^{(2)},\ldots,\boldsymbol{v}^{(N)};\mathbf{m}) = \text{const.} \;-\frac{1}{2}\sum_{i=1}^{N}(\boldsymbol{v}^{(i)}-\mathbf{m})^T\mathbf{C}_v^{-1}(\boldsymbol{v}^{(i)}-\mathbf{m})$$

Taking the gradient with respect to $\mathbf{m}$ (see Table A.1, Appendix A) and setting it to zero yields

$$\nabla_{\mathbf{m}}\ln f_{\boldsymbol{v}^{(1)},\boldsymbol{v}^{(2)},\ldots,\boldsymbol{v}^{(N)};\mathbf{m}}(\boldsymbol{v}^{(1)},\boldsymbol{v}^{(2)},\ldots,\boldsymbol{v}^{(N)};\mathbf{m}) = \sum_{i=1}^{N}\mathbf{C}_v^{-1}(\boldsymbol{v}^{(i)}-\mathbf{m}) = \mathbf{0}$$

or, since $\mathbf{C}_v^{-1}$ is a constant,

$$\hat{\boldsymbol{m}}_N = \frac{1}{N}\sum_{i=1}^{N}\boldsymbol{v}^{(i)}$$

This is the sample mean vector.

Let us check the properties of this estimate. The expected value is

$$\mathcal{E}\{\hat{\boldsymbol{m}}_N\} = \frac{1}{N}\sum_{i=1}^{N}\mathcal{E}\left\{\boldsymbol{v}^{(i)}\right\} = \frac{1}{N}(N\mathbf{m}) = \mathbf{m}$$

so the estimate is unbiased. The covariance matrix of the estimate is

$$\mathcal{E}\left\{(\hat{\boldsymbol{m}}_N-\mathbf{m})(\hat{\boldsymbol{m}}_N-\mathbf{m})^T\right\} = \frac{1}{N^2}\mathcal{E}\left\{\left(\sum_{i=1}^{N}(\boldsymbol{v}^{(i)}-\mathbf{m})\right)\left(\sum_{j=1}^{N}(\boldsymbol{v}^{(j)}-\mathbf{m})^T\right)\right\}$$

Since the $\boldsymbol{v}^{(i)}$ are independent, this reduces to

$$\frac{1}{N^2}\sum_{i=1}^{N}\mathcal{E}\left\{(\boldsymbol{v}^{(i)}-\mathbf{m})(\boldsymbol{v}^{(i)}-\mathbf{m})^T\right\} = \frac{1}{N}\mathbf{C}_v$$

Since the estimate is unbiased and the covariance of the estimate decreases with N, the estimate is consistent.

Finally, let us check the bound (6.45). The foregoing analysis shows that

$$\boldsymbol{s} = \frac{\partial\ln f_{\boldsymbol{v}^{(1)},\boldsymbol{v}^{(2)},\ldots,\boldsymbol{v}^{(N)};\mathbf{m}}(\boldsymbol{v}^{(1)},\boldsymbol{v}^{(2)},\ldots,\boldsymbol{v}^{(N)};\mathbf{m})}{\partial\mathbf{m}} = \sum_{i=1}^{N}\mathbf{C}_v^{-1}(\boldsymbol{v}^{(i)}-\mathbf{m})$$

Since this can be put in the form (6.46), $\hat{\boldsymbol{m}}_N$ is evidently a minimum-variance estimate[11]. However, let us proceed to check the bound explicitly. The Fisher information matrix is

$$\begin{aligned}\mathbf{J} = \mathcal{E}\left\{\boldsymbol{s}\boldsymbol{s}^T\right\} &= \mathcal{E}\left\{\left(\sum_{i=1}^{N}\mathbf{C}_v^{-1}(\boldsymbol{v}^{(i)}-\mathbf{m})\right)\left(\sum_{j=1}^{N}\mathbf{C}_v^{-1}(\boldsymbol{v}^{(j)}-\mathbf{m})\right)^T\right\}\\ &= \mathbf{C}_v^{-1}\sum_{i=1}^{N}\mathcal{E}\left\{(\boldsymbol{v}^{(i)}-\mathbf{m})(\boldsymbol{v}^{(i)}-\mathbf{m})^T\right\}\mathbf{C}_v^{-1}\\ &= \mathbf{C}_v^{-1}(N\mathbf{C}_v)\mathbf{C}_v^{-1} = N\mathbf{C}_v^{-1}\end{aligned}$$

[11]The expression can be written as
$\boldsymbol{s} = \sum_{i=1}^{N}\mathbf{C}_v^{-1}(\boldsymbol{v}^{(i)}-\mathbf{m}) = N\mathbf{C}_v^{-1}\left(\frac{1}{N}\left(\sum_{i=1}^{N}\boldsymbol{v}^{(i)}\right)-\mathbf{m}\right)$ or $\hat{\boldsymbol{m}}_N-\mathbf{m} = (1/N)\mathbf{C}_v\boldsymbol{s}$

Thus

$$\mathbf{J}^{-1} = \frac{1}{N}\mathbf{C}_v$$

and it is seen that the variance of the estimate and the bound (6.45) are identical.

6.2 ESTIMATION OF FIRST AND SECOND MOMENTS FOR A RANDOM PROCESS

In most practical problems the parameters of a random process are not known and must be estimated from the data. The estimation of the second moment quantities for a stationary random process are especially worthy of consideration since these quantities are needed so frequently. This section begins by examining methods for estimating the mean and correlation function and then continues to describe two common methods for estimating the correlation matrix.

6.2.1 Estimating the Mean and Correlation Function

The mean and correlation or covariance functions are among the most important parameters associated with the probability density for a random process. In the case of a Gaussian random process these parameters appear explicitly in the density as elements of the mean vector and covariance matrix. In other cases these parameters may appear only implicitly in that any density function can be expanded as a power series in its moments. In either case estimating the mean and correlation function is often a prerequisite to any further analysis.

To begin, assume that N_s samples of a stationary, ergodic random process are observed. Denote these samples by $x[0], x[1], \ldots, x[N_s - 1]$ and note that they are ordinarily *not independent*. The usual estimate of the mean of $x[n]$ is the *sample mean* defined by

$$\boxed{\hat{m}_x = \frac{1}{N_s}\sum_{n=0}^{N_s-1} x[n]} \tag{6.50}$$

It is easy to see that as in the case of random variables, the sample mean for a random process is unbiased. This follows since

$$\mathcal{E}\{\hat{m}_x\} = \frac{1}{N_s}\sum_{n=0}^{N_s-1} \mathcal{E}\{x[n]\} = \frac{1}{N_s}\sum_{n=0}^{N_s-1} m_x = m_x \tag{6.51}$$

The expression for the variance of the sample mean is slightly more complicated since as we noted above the observations $x[n]$ are generally not independent. If we define a N_s-dimensional constant vector of 1's as

$$\mathbf{1} = [1, 1, \ldots, 1]^T \tag{6.52}$$

and the vector $\boldsymbol{x}$ as

$$\boldsymbol{x} = \begin{bmatrix} x[0] \\ x[1] \\ \vdots \\ x[N_s - 1] \end{bmatrix} \tag{6.53}$$

then the sample mean can be written as

$$m_x = \frac{1}{N_s} \mathbf{1}^T \boldsymbol{x} \tag{6.54}$$

In this case the variance can be written as

$$\begin{aligned} \text{Var}\,[\hat{m}_x] &= \mathcal{E}\{|\hat{m}_x - m_x|^2\} \\ &= \frac{1}{N_s^2} \mathbf{1}^T \mathcal{E}\{(\boldsymbol{x} - \mathbf{m}_x)(\boldsymbol{x} - \mathbf{m}_x)^{*T}\}\,\mathbf{1} \\ &= \frac{1}{N_s^2} \mathbf{1}^T \mathbf{C}_x \mathbf{1} \end{aligned} \tag{6.55}$$

where $\mathbf{m}_x$ is the mean and $\mathbf{C}_x$ is the covariance of the vector $\boldsymbol{x}$. Note that the quadratic product in (6.55) simply serves to sum up all terms in the covariance matrix. Since the matrix is Toeplitz there will be N_s terms with value $C_x[0]$, $N_s - 1$ terms with value $C_x[+1]$ and $C_x[-1]$, and so on. Therefore, the last equation can be written in the form

$$\text{Var}\,[\hat{m}_x] = \frac{1}{N_s^2} \sum_{l=-N_s+1}^{N_s-1} [N_s - |l|] C_x[l] = \frac{1}{N_s} \sum_{l=-N_s+1}^{N_s-1} \left(1 - \frac{|l|}{N_s}\right) C_x[l] \tag{6.56}$$

where C_x is the covariance function for the random process. This has the form of a summation of the terms of $C_x[l]$ weighted by a triangular window. Since the variance of $\hat{m}_x$ is seen to approach zero as N_s gets large, and the sample mean is unbiased, the sample mean is a consistent estimate.

An estimate for the correlation function, known as the *sample correlation function*, can be defined in two different ways. The first form, denoted by $\hat{R}'_x$, is defined by

$$\begin{aligned} \hat{R}'_x[l] &= \frac{1}{N_s - l} \sum_{n=0}^{N_s-1-l} x[n+l]x^*[n]; \quad 0 \le l < N_s \\ \hat{R}'_x[l] &= \frac{1}{N_s - |l|} \sum_{n=0}^{N_s-1-|l|} x[n]x^*[n+|l|]; \quad -N_s < l < 0 \end{aligned} \tag{6.57}$$

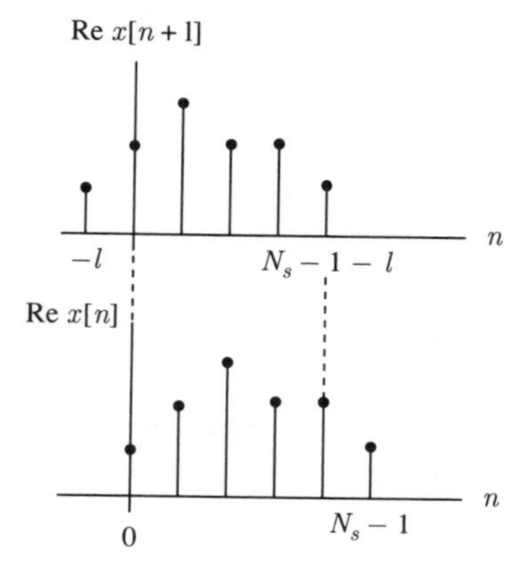

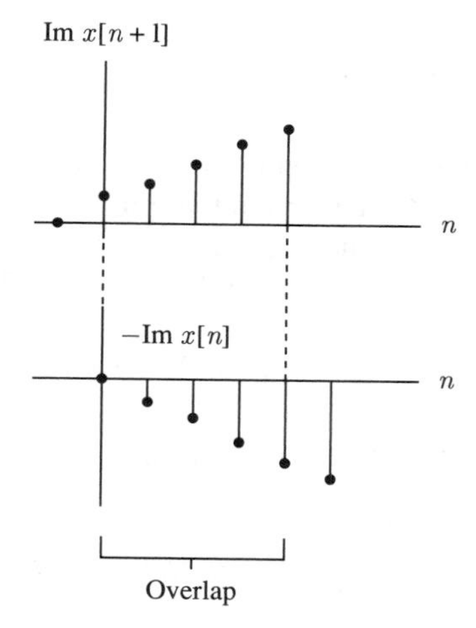

Figure 6.6 Estimation of the sample correlation function ($l \geq 0$).

Since the estimate satisfies the property $\hat{R}'_x[-l] = \hat{R}'^*_x[l]$, it is not necessary to use the second equation to perform the actual computation. As illustrated in Fig. 6.6, the sample correlation is computed from (6.57) by conjugating and shifting $x[n]$ with respect to itself, computing the sum of products in the overlap region, and dividing by the number of terms $(N_s - l)$. For real random processes the estimate for both positive and negative values of lag can be written compactly in a single formula as

$$\hat{R}'_x[l] = \frac{1}{N_s - |l|} \sum_{n=0}^{N_s-1-|l|} x[n+|l|]x[n] \; ; \quad |l| < N_s \tag{6.58}$$

(real random process)

Since the sum of products is divided by the number of terms in the sum, the estimate defined by (6.57) or (6.58) is intuitively reasonable. Further, the estimate is statistically unbiased since [using (6.57)]

$$\mathcal{E}\{\hat{R}'_x[l]\} = \frac{1}{N_s - l} \sum_{n=0}^{N_s-1-l} \mathcal{E}\{x[n+l]x^*[n]\}$$

$$= \frac{1}{N_s - l} \sum_{n=0}^{N_s-1-l} R_x[l] = R_x[l] \tag{6.59}$$

In the case where $x[n]$ is a zero-mean complex Gaussian random process the variance of the sample correlation function can be computed as follows.[12] From (6.57) the second moment of $\hat{R}'_x[l]$ for $l \geq 0$ is

$$\mathcal{E}\{|\hat{R}'_x[l]|^2\} = \frac{1}{[N_s - l]^2} \sum_{n_1=0}^{N_s-1-l} \sum_{n_2=0}^{N_s-1-l} \mathcal{E}\{x[n_1+l]x^*[n_1]x[n_2]x^*[n_2+l]\} \tag{6.60}$$

Since this expression involves fourth moments of the process, its computation would be difficult in general. However, for a *Gaussian* random process all higher-order moments can be expressed in terms of first and second moments. In particular if v_1, v_2, v_3, and v_4 are complex Gaussian random variables, it is known that [15, p. 85]

$$\mathcal{E}\{v_1 v_2^* v_3 v_4^*\} = \mathcal{E}\{v_1 v_2^*\}\,\mathcal{E}\{v_3 v_4^*\} + \mathcal{E}\{v_1 v_4^*\}\,\mathcal{E}\{v_3 v_2^*\}$$

Applying this formula to (6.60) leads to the expression

$$\mathcal{E}\{|\hat{R}'_x[l]|^2\} = \frac{1}{[N_s - l]^2} \sum_{n_1=0}^{N_s-1-l} \sum_{n_2=0}^{N_s-1-l} \{R_x[l]R_x[-l] + R_x[n_1-n_2]R_x[n_2-n_1]\}$$

$$= \frac{1}{[N_s - l]^2} \sum_{n_1=0}^{N_s-1-l} \sum_{n_2=0}^{N_s-1-l} \left(|R_x[l]|^2 + |R_x[n_1-n_2]|^2\right) \tag{6.61}$$

If the mean (6.59) of $\hat{R}'_x$ is now squared and subtracted from this expression, the result is the variance

$$\text{Var}\left[\hat{R}'_x[l]\right] = \frac{1}{[N_s - l]^2} \sum_{n_1=0}^{N_s-1-l} \sum_{n_2=0}^{N_s-1-l} |R_x[n_1-n_2]|^2 \tag{6.62}$$

[12]The variance is defined here in the usual way for complex random variables; however, in this application it is a useful but *limited* description. Since the moments of $\hat{R}'_x[l]$ do not satisfy the required symmetry conditions for complex random variables, nothing can be deduced, for example, about the variances of the real and imaginary parts of $\hat{R}'_x[l]$ except that

$$\text{Var}\left[\hat{R}'_x[l]\right] = \text{Var}\left[\text{Re}\,\hat{R}'_x[l]\right] + \text{Var}\left[\text{Im}\,\hat{R}'_x[l]\right]$$

This formula can be further simplified by letting $k = n_1 - n_2$, collecting terms, and noticing that a similar expression must hold for both positive and negative values of l. This yields the final expression

$$\text{Var}\left[\hat{R}'_x[l]\right] = \frac{1}{[N_s - |l|]^2} \sum_{k=-[N_s-1-|l|]}^{N_s-1-|l|} [N_s - |l| - |k|] |R_x[k]|^2$$

or

$$\text{Var}\left[\hat{R}'_x[l]\right] = \frac{1}{N_s - |l|} \sum_{k=-[N_s-1-|l|]}^{N_s-1-|l|} \left(1 - \frac{|k|}{N_s - |l|}\right) |R_x[k]|^2$$

(complex random process) (6.63)

When $x[n]$ is a *real* random process, a slightly different result obtains (see Problem 6.11), namely

$$\text{Var}\left[\hat{R}'_x[l]\right] = \frac{1}{N_s - |l|} \sum_{k=-[N_s-1-|l|]}^{N_s-1-|l|} \left(1 - \frac{|k|}{N_s - |l|}\right) \left(R_x^2[k] + R_x[k+|l|]R_x[k-|l|]\right)$$

(real random process) (6.64)

For non-Gaussian random processes the expressions (6.63) and (6.64) are a good approximation to the variance for $|l| << N_s$ [16]. If the process has nonzero mean, then another term is added that does not need to be of concern here. Our primary concern is to notice that for a fixed lag l the variance approaches zero as $N_s \to \infty$. Since the estimate is also unbiased, it follows that this estimate of the correlation function is *consistent*.

Although the sample correlation function (6.57) is unbiased and consistant, it has one major problem. This estimate is *not* always positive semidefinite. A simple example involving a three-point sequence is shown in Fig. 6.7. The correlation value $\hat{R}'_x[2]$ is greater than $\hat{R}'_x[0]$. Since any positive definite sequence must have its largest value at the origin, $\hat{R}'_x[l]$ is not positive semidefinite. Of course, there are many cases when this estimate does result in a positive semidefinite correlation estimate; the problem is that the property is *not guaranteed*.

It was stated earlier that there are two ways to define the sample correlation function. The alternative way to define the sample correlation function is

$$\boxed{\begin{aligned} \hat{R}_x[l] &= \frac{1}{N_s} \sum_{n=0}^{N_s-1-l} x[n+l]x^*[n]; \qquad 0 \le l < N_s \\ \hat{R}_x[l] &= \frac{1}{N_s} \sum_{n=0}^{N_s-1-|l|} x[n]x^*[n+|l|]; \quad -N_s < l < 0 \end{aligned}} \qquad (6.65)$$

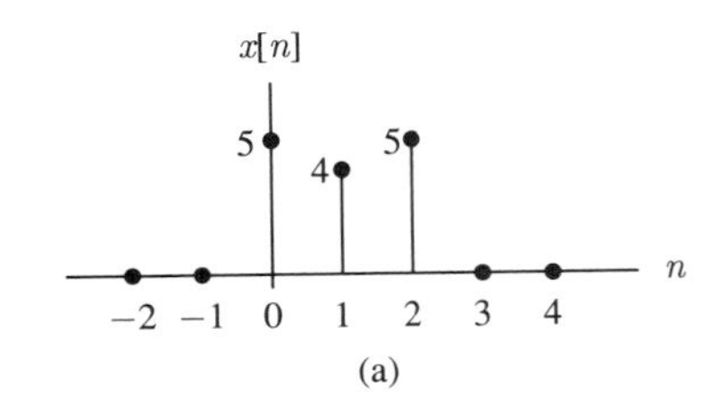

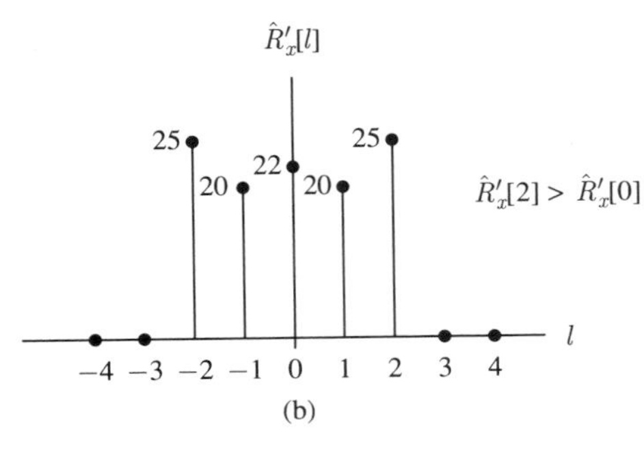

Figure 6.7 Demonstration that the unbiased corrrelation function estimate may not be positive semidefinite. (a) Data sequence. (b) Estimated correlation function.

where the only difference between this definition and (6.57) is the division by N_s instead of $N_s - |l|$. This estimate is *biased.* This follows because

$$\hat{R}_x[l] = \frac{N_s - |l|}{N_s}\hat{R}'_x[l] \tag{6.66}$$

Therefore,

$$\mathcal{E}\{\hat{R}_x[l]\} = \frac{N_s - |l|}{N_s}\mathcal{E}\{\hat{R}'_x[l]\} = \frac{N_s - |l|}{N_s}R_x[l] \tag{6.67}$$

and the bias is

$$\mathcal{E}\{\hat{R}_x[l]\} - R_x[l] = \frac{-|l|}{N_s}R_x[l] \tag{6.68}$$

However, since

$$\lim_{N_s\to\infty} \mathcal{E}\{\hat{R}_x[l]\} = \lim_{N_s\to\infty} \frac{N_s - |l|}{N_s}R_x[l] = R_x[l]$$

this form of the sample correlation function is *asymptotically unbiased.*

The variance can be found by using the relation (6.66) and the previous results (6.63) and (6.64) for the unbiased correlation function. For a complex random process this yields

$$\begin{aligned}
\text{Var}\left[\hat{R}_x[l]\right] &= \left(\frac{N_s - |l|}{N_s}\right)^2 \text{Var}\left[\hat{R}'_x[l]\right] \\
&= \left(\frac{N_s - |l|}{N_s}\right)^2 \frac{1}{N_s - |l|} \sum_{k=-[N_s-1-|l|]}^{N_s-1-|l|} \left(1 - \frac{|k|}{N_s - |l|}\right) |R_x[k]|^2
\end{aligned}$$

or

$$\text{Var}\left[\hat{R}_x[l]\right] = \frac{1}{N_s^2} \sum_{k=-[N_s-1-|l|]}^{N_s-1-|l|} \big(N_s - |l| - |k|\big) \left|R_x[k]\right|^2 \tag{6.69}$$

(complex random process)

and correspondingly, for a real random process

$$\text{Var}\left[\hat{R}_x[l]\right] = \frac{1}{N_s^2} \sum_{k=-[N_s-1-|l|]}^{N_s-1-|l|} \big(N_s - |l| - |k|\big) \big(R_x^2[k] + R_x[k+|l|] \cdot R_x[k-|l|]\big)$$

(real random process) (6.70)

Since both expressions approach zero as $N_s \to \infty$, the estimate is consistent. Further, as shown in the next subsection, *this estimate is guaranteed to be positive semidefinite.*

The sample cross-correlation function can be defined in a method analogous to either (6.57) or (6.65). The latter form is given by

$$\begin{aligned} \hat{R}_{xy}[l] &= \frac{1}{N_s} \sum_{n=0}^{N_s-1-l} x[n+l]y^*[n]; \qquad 0 \le l < N_s \\ \hat{R}_{xy}[l] &= \frac{1}{N_s} \sum_{n=0}^{N_s-1-|l|} x[n]y^*[n+|l|]; \quad -N_s < l < 0 \end{aligned} \tag{6.71}$$

Observe that since the cross-correlation function is not symmetric, *the computations in the second equation are essential and cannot be ignored.* This form of the sample cross-correlation is biased and the behavior of the variance is virtually identical to that for the biased sample autocorrelation function [17]. In particular, the expected value is given by

$$\mathcal{E}\big\{\hat{R}_{xy}[l]\big\} = \frac{N_s - |l|}{N_s} R_{xy}[l] \tag{6.72}$$

and the variance for small values of lag is inversely proportional to N_s.

6.2.2 Estimating the Correlation Matrix

For some problems it is necessary to estimate the entire correlation or covariance matrix for a random process. Only the correlation matrix is discussed here since an estimate for the covariance matrix can be formed by removing the sample mean from the data and then applying any of the methods described below.

One way to estimate the correlation matrix is equivalent to estimating the sample correlation function from (6.65) and forming the corresponding Toeplitz matrix. However, the procedure described here is somewhat more convenient. (It is similar to a general method for random vectors described in Chapter 2.) Again suppose that there are N_s data points $x[0], x[1], \ldots, x[N_s - 1]$. To estimate a $P \times P$ correlation matrix (for $P \leq N_s$) form the matrix

$$\boldsymbol{X} = \begin{bmatrix} x[0] & 0 & \cdots & 0 \\ x[1] & x[0] & \cdots & 0 \\ \vdots & \vdots & \ddots & \vdots \\ x[P-1] & x[P-2] & \cdots & x[0] \\ x[P] & x[P-1] & \cdots & x[1] \\ \vdots & \vdots & & \vdots \\ x[N_s-1] & x[N_s-2] & \cdots & x[N_s-P] \\ 0 & x[N_s-1] & \cdots & \vdots \\ \vdots & \vdots & \ddots & \vdots \\ 0 & 0 & \cdots & x[N_s-1] \end{bmatrix} \tag{6.73}$$

This data matrix $\boldsymbol{X}$ is of size $(N_s + P - 1) \times P$. The columns of $\boldsymbol{X}$ contain the entire data sequence $x[0], \ldots, x[N_s - 1]$ successively displaced and zero-padded; the rows contain the data samples that would appear under a sliding window just P samples long that moves along the data sequence. (These samples are in reverse order.) The correlation matrix estimate is then given by

$$\hat{\mathbf{R}}_x = \frac{1}{N_s} \boldsymbol{X}^{*T} \boldsymbol{X} \tag{6.74}$$

The scale factor can alternatively be taken as $1/(N_s + P - 1)$, where the denominator represents the number of products of row and column elements of $\boldsymbol{X}$ required to compute each term. Frequently, the scale factor is left out entirely since subsequent calculations may be invariant to the scale factor. (Examples of this appear in Chapter 9.)

To illustrate the computations, consider the case for $P = 3$ and $N_s = 50$. The estimate in (6.74) is

$$\frac{1}{50} \begin{bmatrix} x^*[0] & x^*[1] & x^*[2] & \cdots & x^*[49] & 0 & 0 \\ 0 & x^*[0] & x^*[1] & \cdots & x^*[48] & x^*[49] & 0 \\ 0 & 0 & x^*[0] & \cdots & x^*[47] & x^*[48] & x^*[49] \end{bmatrix} \begin{bmatrix} x[0] & 0 & 0 \\ x[1] & x[0] & 0 \\ x[2] & x[1] & x[0] \\ \vdots & \vdots & \vdots \\ x[49] & x[48] & x[47] \\ 0 & x[49] & x[48] \\ 0 & 0 & x[49] \end{bmatrix}$$

It can be seen that the terms in this matrix are exactly those given by (6.65) and that the matrix is Toeplitz. This estimate is also positive definite since if $\mathbf{a}$ is *any* vector of size P,

$$\mathbf{a}^{*T} \hat{\mathbf{R}}_x \mathbf{a} = \frac{1}{N_s} \mathbf{a}^{*T} \boldsymbol{X}^{*T} \boldsymbol{X} \mathbf{a} = \frac{1}{N_s} |\boldsymbol{X}\mathbf{a}|^2 > 0 \tag{6.75}$$

Note here that since the columns of $\boldsymbol{X}$ are independent, the matrix is *strictly* positive definite, not just positive semidefinite. Since the matrix is positive definite for any size P, this also shows that the sample correlation function (6.65) is positive definite.

Although the estimate (6.74) is positive definite and asymptotically unbiased, it can be objected to on the basis that it averages in zeros in the sequence in the computation of its terms. That is, it effectively assumes that the data is zero outside the range $0 \leq n < N_s$, or *windows* the data. Therefore, for any *finite* value of N_s it is *biased*. An alternative method that overcomes this objection is to define the $(N_s - P + 1) \times P$ matrix

$$\boldsymbol{X}'' = \begin{bmatrix} x[P-1] & x[P-2] & \cdots & x[0] \\ x[P] & x[P-1] & \cdots & x[1] \\ \vdots & \vdots & & \vdots \\ x[N_s-1] & x[N_s-2] & \cdots & x[N_s-P] \end{bmatrix} \tag{6.76}$$

[note that this is just the middle section of (6.73)] and use the correlation matrix estimate

$$\hat{\mathbf{R}}''_x = \frac{1}{N_s - P + 1} (\boldsymbol{X}'')^{*T} \boldsymbol{X}'' \tag{6.77}$$

Again, the scale factor is sometimes left out. This estimate is *unbiased* (see Problem 6.28), and since its quadratic form can be expressed similarly to (6.75), it is positive *semi*definite. (Columns in this case could be linearly dependent.) The estimate is *not* Toeplitz, however. Consider again the case of $P = 3$ and $N_s = 50$. The estimate has the form

$$\frac{1}{48} \begin{bmatrix} x^*[2] & x^*[3] & \cdots & x^*[49] \\ x^*[1] & x^*[2] & \cdots & x^*[48] \\ x^*[0] & x^*[1] & \cdots & x^*[47] \end{bmatrix} \begin{bmatrix} x[2] & x[1] & x[0] \\ x[3] & x[2] & x[1] \\ \vdots & \vdots & \vdots \\ x[49] & x[48] & x[47] \end{bmatrix}$$

which does not yield a Toeplitz matrix. (Consider, for example, elements (1,1) and (2,2) of the product matrix. Note that since their computation does not involve the same set of terms, they are in general not equal.) Both of these methods arise naturally in the least squares linear prediction and signal modeling problems discussed in Chapters 8 and 9. Equation 6.74 is referred to as the "autocorrelation method" estimate and (6.77) is called the "covariance method" estimate. These names are really unfortunate since they have nothing to do with the terms "autocorrelation" and "covariance" as they are used in statistics; either method could be used to estimate either a correlation matrix *or* a covariance matrix (if the mean is removed). These names seem to have found a niche in signal processing, however, and are unlikely to change.

Since the autocorrelation method uses the biased estimate of the correlation function, you may wonder if it is possible to construct an estimate for the correlation matrix using the unbiased estimate (6.57). Ignoring the fact that the unbiased estimate may not be positive semidefinite, this is a valid question to ask. If you think about the differences between the two correlation function estimates, it becomes clear that an estimate for the correlation matrix using the unbiased method amounts to weighting the terms in the autocorrelation method estimate by an inverted triangular window [see (6.66)]. For example, consider a 3×3 correlation matrix $\hat{\mathbf{R}}_x$ estimated from (6.74). The matrix corresponding to the unbiased correlation function would be obtained by multiplying $\hat{\mathbf{R}}_x$ element by element by the array

$$\begin{bmatrix} 1 & \frac{3}{2} & 3 \\ \frac{3}{2} & 1 & \frac{3}{2} \\ 3 & \frac{3}{2} & 1 \end{bmatrix} \tag{6.78}$$

It also becomes clear that this method cannot be expressed as the product of two matrices like (6.74), and it is easy to imagine how such inverse weighting can destroy the positive semidefinite nature of the original matrix. Let us pursue this just a bit further in terms of the example in Fig. 6.7. The autocorrelation matrix estimate corresponding to the sequence $\{5\ 4\ 5\}$ is

$$\frac{1}{3}\begin{bmatrix} 5 & 4 & 5 & 0 & 0 \\ 0 & 5 & 4 & 5 & 0 \\ 0 & 0 & 5 & 4 & 5 \end{bmatrix}\begin{bmatrix} 5 & 0 & 0 \\ 4 & 5 & 0 \\ 5 & 4 & 5 \\ 0 & 5 & 4 \\ 0 & 0 & 5 \end{bmatrix} = \begin{bmatrix} 22 & 13.33 & 8.33 \\ 13.33 & 22 & 13.33 \\ 8.33 & 13.33 & 22 \end{bmatrix}$$

This matrix is positive definite; its eigenvalues are found to be $\lambda_1 = 45.5, \lambda_2 = 13.7$, and $\lambda_3 = 6.86$. The matrix corresponding to the unbiased correlation estimate is

$$\begin{bmatrix} 22 & 20 & 25 \\ 20 & 22 & 20 \\ 25 & 20 & 22 \end{bmatrix}$$

Note that the terms are related by the factors in (6.78). This matrix is *not* positive definite. Its eigenvalues are $\lambda_1 = 65.4, \lambda_2 = 3.58$, and $\lambda_3 = -3$. These considerations should indicate why this method of estimating the correlation matrix is not frequently used.

Estimates for the cross-correlation matrix can be obtained by analogous procedures. Let us assume that N_s samples of the sequence $y[n]$ are also available. If data matrices $\boldsymbol{Y}$ and $\boldsymbol{Y}''$ analogous to (6.73) and (6.76) are formed, then estimates of the $P \times P$ square cross-correlation matrix are given by

$$\hat{\mathbf{R}}_{xy} = \frac{1}{N_s}\boldsymbol{Y}^{*T}\boldsymbol{X} \tag{6.79}$$

(autocorrelation method) and

$$\hat{\mathbf{R}}''_{xy} = \frac{1}{N_s - P + 1}(\boldsymbol{Y}'')^{*T}\boldsymbol{X}'' \tag{6.80}$$

(covariance method). Note that although the estimate is for $\mathbf{R}_{xy}$, *the data matrices appear in commuted order*. These estimates, of course, have no particular symmetry and are not necessarily positive semidefinite.

6.3 BAYES ESTIMATION OF RANDOM VARIABLES

In this section we turn to the estimation of random variables. As mentioned earlier, slightly different methods apply here. The topic is developed in the common framework of Bayes estimation, although the two specific methods treated here are distinct. It is sufficient for the topics in this book to treat only the case of a single random random variable and a set of scalar observations. The results could easily be extended to higher-dimensional (vector) quantities however.

Suppose that there exists a set of random variables $x_1, x_2, \ldots, x_N$ and a related random variable y. It is assumed that the x_i can be directly observed but that y cannot. For example, y may be the value of a random signal at some particular time, but the signal is embedded in a hostile environment and cannot be received without significant distortion. The observations may represent various receptions of the signal corrupted by noise, dispersion, multipath, other interfering signals, or whatever. The situation is depicted in Fig. 6.8. The problem to be considered here is the estimation of y from the observations $x_1, x_2, \ldots, x_N$. Mathematically, it is desired to find an estimate which is a function of the observations

$$\hat{y} = \hat{y}(\boldsymbol{x}) = \phi(x_1, x_2, \ldots, x_N) \tag{6.81}$$

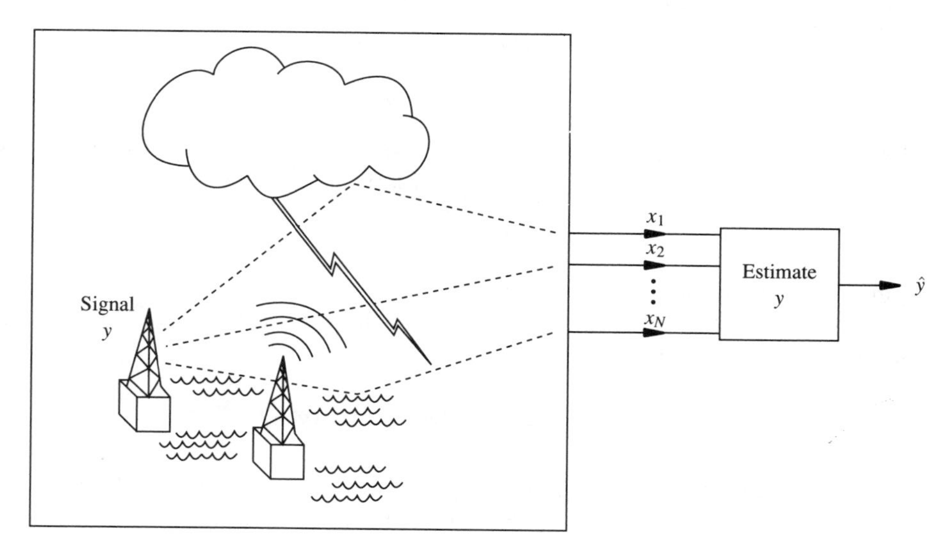

Figure 6.8 Estimation of a random variable y from related observations $x_1, x_2, \ldots, x_N$.

that is optimal in some sense. It is obvious that to solve this problem the observations must be *related* to y, as was stated, and that the relation between them is *known*. Although the specific relation is problem dependent, ultimately the conditional density $f_{y|\boldsymbol{x}}$ is what is needed.

The Bayes procedure for choosing $\hat{y}$ is to define a nonnegative cost function $\mathcal{C}(y, \hat{y})$ which depends on the random variable and its estimate and to minimize the *risk*

$$\mathcal{R} = \mathcal{E}\{\mathcal{C}(y, \hat{y})\} \tag{6.82}$$

Since both y and $\hat{y}$ are random, the risk can be expressed as

$$\begin{aligned}\mathcal{R} &= \int_{-\infty}^{\infty}\int_{-\infty}^{\infty} \mathcal{C}(\mathrm{y}, \hat{y}(\mathbf{x})) f_{y\boldsymbol{x}}(\mathrm{y}, \mathbf{x}) d\mathrm{y} d\mathbf{x} \\ &= \int_{-\infty}^{\infty} \mathcal{I}(\hat{y}) f_{\boldsymbol{x}}(\mathbf{x}) d\mathbf{x}\end{aligned} \tag{6.83}$$

where

$$\mathcal{I}(\hat{y}) = \int_{-\infty}^{\infty} \mathcal{C}(\mathrm{y}, \hat{y}) f_{y|\boldsymbol{x}}(\mathrm{y}|\mathbf{x}) d\mathrm{y} \tag{6.84}$$

Since (6.83) is the expected value of the positive quantity $\mathcal{I}(\hat{y})$, it is necessary to minimize $\mathcal{I}(\hat{y})$ for every $\mathbf{x}$ in order to minimize $\mathcal{R}$. Let us therefore focus attention on the integral (6.84) and its minimization.

Although many different cost functions are possible, only two specific cases are considered here.[13] These cost functions are illustrated in Fig. 6.9.

Case 1. The *mean-square* cost function is given by

$$\mathcal{C}(y, \hat{y}) = |y - \hat{y}|^2 \tag{6.85}$$

[see Fig. 6.9(a)]. For this case large errors have large costs and the incremental cost of an error increases as the error gets larger. The risk, which is to be minimized, is

$$\mathcal{R} = \mathcal{E}\{|y - \hat{y}|^2\} \tag{6.86}$$

This is commonly called the *mean-square error*.

To find the optimal estimation algorithm in this case, substitute (6.85) into (6.84). The integral can then be written as

$$\mathcal{I}(\hat{y}) = \int_{-\infty}^{\infty} |\mathrm{y} - \hat{y}|^2 f_{y|\boldsymbol{x}}(\mathrm{y}|\boldsymbol{x}) d\mathrm{y} \tag{6.87}$$

where the dependence of this quantity on the random vector is indicated by the italic $\boldsymbol{x}$ in the argument of $f_{y|\boldsymbol{x}}$. A necessary condition for the minimum is

$$\nabla_{\hat{y}^*} \mathcal{I}(\hat{y}) = -\int_{-\infty}^{\infty} (\mathrm{y} - \hat{y}) f_{y|\boldsymbol{x}}(\mathrm{y}|\boldsymbol{x}) d\mathrm{y} = 0$$

or

$$\int_{-\infty}^{\infty} \mathrm{y} f_{y|\boldsymbol{x}}(\mathrm{y}|\boldsymbol{x}) d\mathrm{y} - \hat{y} \int_{-\infty}^{\infty} f_{y|\boldsymbol{x}}(\mathrm{y}|\boldsymbol{x}) d\mathrm{y} = 0 \tag{6.88}$$

[13]For a discussion of other cost functions and their corresponding estimation schemes, see [7].

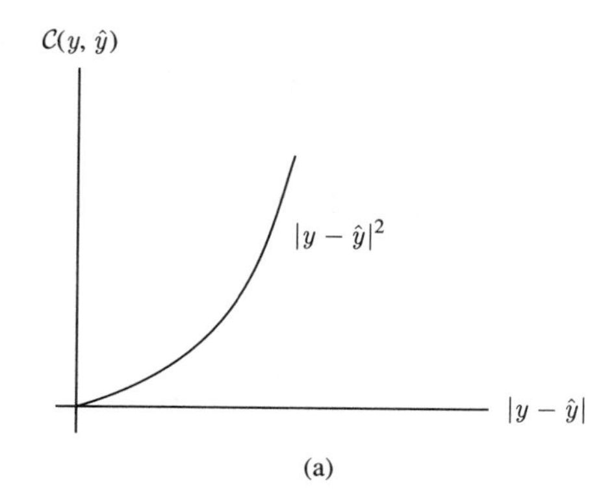

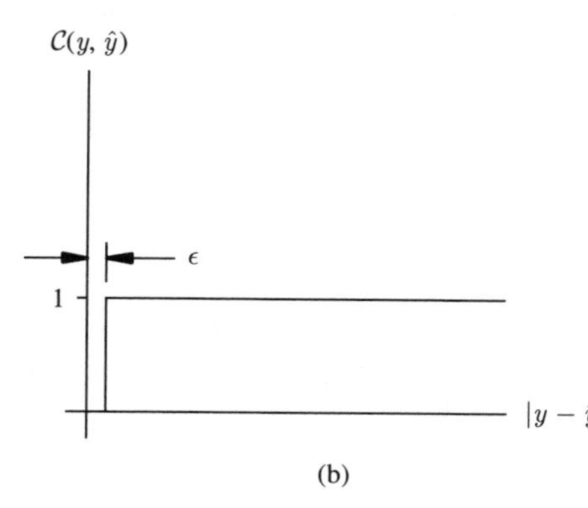

Figure 6.9 Cost functions for Bayes estimation. (a) Mean-square. (b) Uniform.

Since the last integral of the density function is 1, this implies that

$$\boxed{\hat{y}_{ms}(\boldsymbol{x}) = \int_{-\infty}^{\infty} \mathrm{y} f_{y|\boldsymbol{x}}(\mathrm{y}|\boldsymbol{x})dy} \tag{6.89}$$

That is, *the mean-square estimate is the* <u>*mean*</u> *of the conditional density.*

For a real random variable y, the mean-square error of *any* estimate $\hat{y}$ can be shown to satisfy the following inequalities analogous to the Cramér-Rao bound:

$$\mathcal{E}\{(y-\hat{y})^2\} \geq \frac{1}{\mathcal{E}\left\{\left(\frac{\partial \ln f_{y\boldsymbol{x}}(y,\boldsymbol{x})}{\partial y}\right)^2\right\}} \qquad (a)$$

$$\mathcal{E}\{(y-\hat{y})^2\} \geq \frac{1}{-\mathcal{E}\left\{\frac{\partial^2 \ln f_{y\boldsymbol{x}}(y,\boldsymbol{x})}{\partial y^2}\right\}} \qquad (b)$$

$$\tag{6.90}$$

(real random variable y)

where the expectation here is over both $\boldsymbol{x}$ and y. This bound assumes that the corresponding partial derivatives exist and are absolutely integrable with respect to both **x** and y. The bound is met with equality when

$$y - \hat{y}(\boldsymbol{x}) = -K \cdot \frac{\partial \ln f_{y\boldsymbol{x}}(y, \boldsymbol{x})}{\partial y} \tag{6.91}$$

in which case

$$K = \frac{1}{\mathcal{E}\left\{\left(\frac{\partial \ln f_{y\boldsymbol{x}}(y,\boldsymbol{x})}{\partial y}\right)^2\right\}} = \frac{1}{-\mathcal{E}\left\{\frac{\partial^2 \ln f_{y\boldsymbol{x}}(y,\boldsymbol{x})}{\partial y^2}\right\}} \tag{6.92}$$

The proof of this bound is similar to the proof of the Cramér-Rao bound and is explored in Problem 6.18. The mean-square estimate, defined by (6.89), satisfies this bound with equality.

Case 2. The *uniform* cost function shown in Fig. 6.9(b) is defined by

$$\mathcal{C}(y, \hat{y}) = \begin{cases} 0 & |y - \hat{y}| < \epsilon \\ 1 & \text{otherwise} \end{cases} \tag{6.93}$$

The cost of an error with this cost function does not degrade gracefully. This cost function is probably the right one to use for Navy pilots landing their aircraft on a carrier in bad weather. Here the cost of missing the deck by a few yards is as much as missing it by a mile!

For the uniform cost function, (6.84) becomes

$$\begin{aligned} \mathcal{I}(\hat{y}) &= \int_{|y-\hat{y}|\geq\epsilon} f_{y|\boldsymbol{x}}(\mathrm{y}|\boldsymbol{x})dy \\ &= 1 - \int_{|y-\hat{y}|<\epsilon} f_{y|\boldsymbol{x}}(\mathrm{y}|\boldsymbol{x})dy \end{aligned} \tag{6.94}$$

Since the integral is over an arbitrarily small region around $\hat{y}$, the quantity $\mathcal{I}(\hat{y})$ is minimized when $f_{y|\boldsymbol{x}}(\mathrm{y}|\boldsymbol{x})$ takes on its largest value in this region. Therefore, denoting this estimate by $\hat{y}_{MAP}$, we have

$$\boxed{\hat{y}_{MAP}(\boldsymbol{x}) = \underset{\mathrm{y}}{\text{argmax}} \quad f_{y|\boldsymbol{x}}(\mathrm{y}|\boldsymbol{x})} \tag{6.95}$$

That is, *the optimal estimate is that value of* y *that* <u>*maximizes*</u> *the conditional density*. The notation "MAP" derives from the fact that the conditional density function is also referred to as the *posterior* density. (It is the density function for y *after* observing the random vector $\boldsymbol{x}$.) The estimate is thus called the *maximum a posteriori* (MAP) estimate of the random variable.

The calculation of mean-square and MAP estimates is illustrated in the following example.

Example 6.6

The power y in some unobserved signal is at most equal to the square of the magnitude x of a signal that *is* observed. The joint density function for the two random variables x and y is

$$f_{xy}(\mathrm{x}, \mathrm{y}) = \begin{cases} 10\mathrm{y} & 0 \le \mathrm{y} \le \mathrm{x}^2, \quad 0 \le \mathrm{x} \le 1 \\ 0 & \text{otherwise} \end{cases}$$

It is desired to estimate the power y from an observation x using both mean-square and MAP estimation.

The marginal density for x can be computed by integrating over the strip shown in Fig. Ex6.6a.

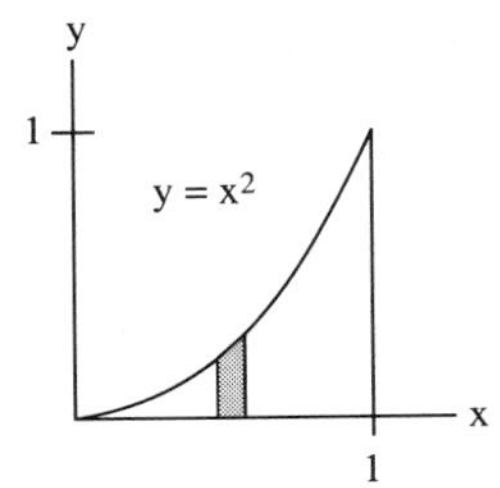

Figure EX6.6a

$$f_x(\mathrm{x}) = \int_0^{\mathrm{x}^2} 10\mathrm{y}d\mathrm{y} = 5\mathrm{y}^2\Big|_0^{\mathrm{x}^2} = 5\mathrm{x}^4$$

The conditional density for y is therefore

$$f_{y|x}(\mathrm{y}|\mathrm{x}) = \frac{10\mathrm{y}}{5\mathrm{x}^4} = \frac{2\mathrm{y}}{\mathrm{x}^4}; \qquad 0 \le \mathrm{y} \le \mathrm{x}^2$$

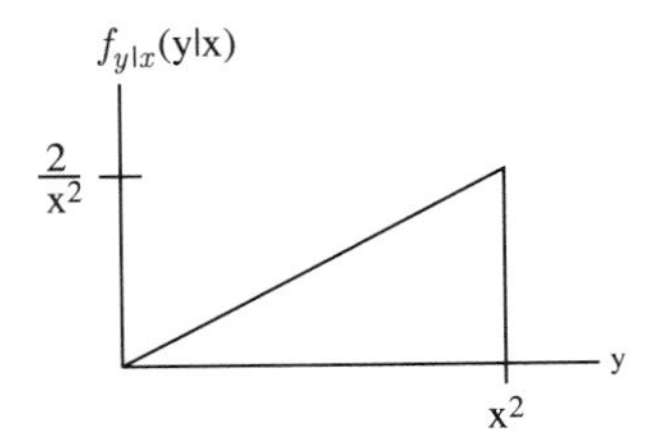

Figure EX6.6b

From Fig. Ex6.6b it is clear that the maximum of the conditional density occurs at $\mathrm{y} = \mathrm{x}^2$. Therefore,

$$\hat{y}_{MAP} = x^2$$

The mean of the conditional density is given by

$$\int_{-\infty}^{\infty} \mathrm{y} f_{y|x}(\mathrm{y}|\mathrm{x})d\mathrm{y} = \int_0^{\mathrm{x}^2} \frac{2\mathrm{y}^2}{\mathrm{x}^4}d\mathrm{y} = \frac{2}{3}\frac{\mathrm{y}^3}{\mathrm{x}^4}\Big|_0^{\mathrm{x}^2} = \frac{2}{3}\mathrm{x}^2$$

Thus

$$\hat{y}_{ms} = \frac{2}{3}x^2$$

The minimum mean-square error is given by

$$\mathcal{E}\{(y - \hat{y}_{ms})^2\} = \int_0^1 \int_0^{x^2} (y - \tfrac{2}{3}x^2)^2 10y\,dy\,dx = \frac{5}{162} = 0.0309$$

To show that this mean-square error is less than, say, the mean-square error for the MAP estimate, compute

$$\mathcal{E}\{(y - \hat{y}_{MAP})^2\} = \int_0^1 \int_0^{x^2} (y - x^2)^2 10y\,dy\,dx = \frac{5}{54} = 0.0926$$

which turns out to be exactly three times the previous mean-square error.

Another case that occurs commonly is the case where the random variables x and y have a joint Gaussian density. Some especially interesting relations result here. This is illustrated in the next example.

Example 6.7

The real random variables x and y have a joint Gaussian density with parameters

$$\mathbf{m}_{xy} = \begin{bmatrix} m_x \\ m_y \end{bmatrix}$$

$$\mathbf{C}_{xy} = \begin{bmatrix} \sigma_x^2 & c_{xy} \\ c_{xy} & \sigma_y^2 \end{bmatrix} = \begin{bmatrix} \sigma_x^2 & \rho_{xy}\sigma_x\sigma_y \\ \rho_{xy}\sigma_x\sigma_y & \sigma_y^2 \end{bmatrix}$$

where ρ_{xy} is the normalized correlation coefficient $c_{xy}/\sigma_x\sigma_y$. The joint density can be written as

$$f_{xy}(\mathrm{x}, \mathrm{y}) = \frac{1}{(2\pi)|\mathbf{C}_{xy}|^{\frac{1}{2}}} e^{-\frac{1}{2}[\mathrm{x}-m_x, \mathrm{y}-m_y]\mathbf{C}_{xy}^{-1}\begin{bmatrix} \mathrm{x}-m_x \\ \mathrm{y}-m_y \end{bmatrix}}$$

$$= \frac{1}{2\pi\sigma_x\sigma_y\sqrt{1-\rho_{xy}^2}} e^{-\frac{1}{2(1-\rho_{xy}^2)}\left\{\frac{(\mathrm{x}-m_x)^2}{\sigma_x^2} - 2\rho_{xy}\frac{(\mathrm{x}-m_x)(\mathrm{y}-m_y)}{\sigma_x\sigma_y} + \frac{(\mathrm{y}-m_y)^2}{\sigma_y^2}\right\}}$$

The marginal density for x can be found by integration to be

$$f_x(\mathrm{x}) = \frac{1}{\sqrt{2\pi}\sigma_x} e^{-\frac{(\mathrm{x}-m_x)^2}{2\sigma_x^2}}$$

Then by forming the ratio

$$f_{y|x}(\mathrm{y}|\mathrm{x}) = \frac{f_{xy}(\mathrm{x}, \mathrm{y})}{f_x(\mathrm{x})}$$

and simplifying, we obtain

$$f_{y|x}(\mathrm{y}|\mathrm{x}) = \frac{1}{\sqrt{2\pi}\sigma} e^{-\frac{(\mathrm{y}-\mu(\mathrm{x}))^2}{2\sigma^2}}$$

where

$$\mu(\mathrm{x}) = m_y + \rho_{xy}\frac{\sigma_y}{\sigma_x}(\mathrm{x} - m_x) = \left(\rho_{xy}\frac{\sigma_y}{\sigma_x}\right)\mathrm{x} + \left(m_y - \rho_{xy}\frac{\sigma_y}{\sigma_x}m_x\right)$$

and

$$\sigma^2 = \sigma_y^2(1 - \rho_{xy}^2)$$

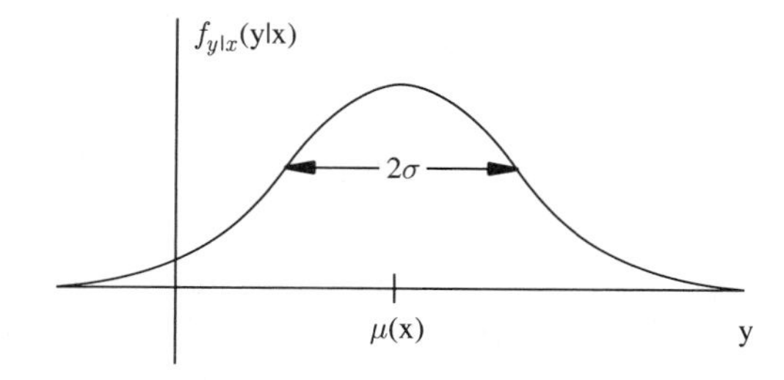

Figure EX6.7

Since the conditional density is Gaussian, *the mean and the maximum occur at the same place;* consequently,

$$\hat{y}_{ms} = \hat{y}_{MAP} = \mu(x) = \left(\rho_{xy}\frac{\sigma_y}{\sigma_x}\right)x + \left(m_y - \rho_{xy}\frac{\sigma_y}{\sigma_x}m_x\right)$$

Note that the function $\mu(x)$ plots as a straight line. It is common in this case to say that the estimate is a *linear* function of x, although strictly speaking, $\mu(x)$ is not a linear function.[14] Let us instead say that $\hat{y}$ has a *linear dependence* on x. A repetition of this analysis for the complex case shows that the last formula also applies when x and y are complex.

The mean-square error corresponding to this estimate can be shown to be equal to the variance σ^2 of the conditional density (see Problem 6.17). This is also the lower bound given by (6.90). To show this it is easiest to use the second form of the bound in (6.90). The logarithm of the joint density is

$$\ln f_{xy}(x,y) = \text{const.} - \frac{1}{2(1-\rho_{xy}^2)}\left\{\frac{(x-m_x)^2}{\sigma_x^2} - 2\rho_{xy}\frac{(x-m_x)(y-m_y)}{\sigma_x\sigma_y} + \frac{(y-m_y)^2}{\sigma_y^2}\right\}$$

Thus

$$\frac{\partial \ln f_{xy}(x,y)}{\partial y} = -\frac{1}{2(1-\rho_{xy}^2)}\left\{-2\rho_{xy}\frac{(x-m_x)}{\sigma_x\sigma_y} + \frac{2(y-m_y)}{\sigma_y^2}\right\}$$

and

$$\frac{\partial^2 \ln f_{xy}(x,y)}{\partial y^2} = -\frac{1}{2(1-\rho_{xy}^2)}\left\{\frac{2}{\sigma_y^2}\right\} = -\frac{1}{\sigma_y^2(1-\rho_{xy}^2)} = -\frac{1}{\sigma^2}$$

Substituting this in (6.90) shows that the mean-square error for *any* estimate $\hat{y}$ is lower bounded by

$$\mathcal{E}\left\{(y-\hat{y})^2\right\} \geq \sigma^2$$

[14] To be linear the function would have to satisfy $\mu(ax_1 + bx_2) = a\mu(x_1) + b\mu(x_2)$, where a and b are arbitrary constants. This is clearly not the case unless $m_y = m_x = 0$.

MAP estimation can also be applied to the estimation of parameters if the parameter is regarded as a *random variable* instead of an entity that is simply "unknown." This can have advantages if there is some prior knowlege about the possible range of values that the parameter can take on. For example, suppose that it is known that the parameter θ of a density function $f_{x;\theta}(\mathbf{x};\theta)$ is likely to be close to some value θ_o but not exactly equal to this value. In this case it may be appropriate to regard θ as a random variable and represent this prior knowledge as (for example) a Gaussian density function $f_\theta(\theta)$ with mean equal to θ_o and variance representative of the spread of possible values that are likely to be expected. Alternatively, if nothing in particular were known about likely values of θ except that θ is restricted to lie within a range $A < \theta < B$, then this restriction could be modeled as a *uniform* density for θ over the interval $A < \theta < B$. In either case the posterior (conditional) density can be computed from Bayes rule as

$$f_{\theta|x}(\theta|\mathbf{x}) = \frac{f_{x|\theta}(\mathbf{x}|\theta)f_\theta(\theta)}{f_x(\mathbf{x})} \tag{6.96}$$

Since θ is now regarded as a random variable, the density $f_{x|\theta}(\mathbf{x}|\theta)$ is equivalent to what was previously written as $f_{x;\theta}(\mathbf{x};\theta)$. An MAP estimate, for instance, could then be found by choosing the value of θ that maximizes $f_{\theta|x}$. Since the denominator does not depend on θ it is sufficient to maximize the numerator of (6.96). This procedure is illustrated in the following example.

Example 6.8

Suppose that the signal in Example 6.1, which was assumed to be constant, is now a random variable with density

$$f_A(\mathrm{A}) = \frac{1}{\pi\sigma_A^2} e^{-\frac{|\mathrm{A}-\mathrm{A_o}|^2}{\sigma_A^2}}$$

It is desired to find the MAP estimate for the signal. In order to maximize $f_{A|x}$ it is sufficient to maximize

$$f_{x|A}(\boldsymbol{x}|\mathrm{A}) \cdot f_A(\mathrm{A})$$

or equivalently, to maximize

$$\begin{aligned} &\ln f_{x|A}(\boldsymbol{x}|\mathrm{A}) + \ln f_A(\mathrm{A}) \\ &= -N_s \ln(\pi\sigma_w^2) - \sum_{n=0}^{N_s-1} \frac{|x[n]-\mathrm{A}|^2}{\sigma_w^2} - \ln(\pi\sigma_A^2) - \frac{|\mathrm{A}-\mathrm{A_o}|^2}{\sigma_A^2} \end{aligned}$$

To maximize this, apply the scalar form of the complex gradient and set it to zero, to obtain (see Table A.3, Appendix A)

$$\sum_{n=0}^{N_s-1} \frac{x[n]-\mathrm{A}}{\sigma_w^2} - \frac{\mathrm{A}-\mathrm{A_o}}{\sigma_A^2} = 0$$

Solving this for A produces the MAP estimate

$$\hat{\mathrm{A}}_{MAP} = \frac{1}{1/\sigma_A^2 + N_s/\sigma_w^2}\left[\frac{1}{\sigma_w^2}\sum_{n=0}^{N_s-1} x[n] + \frac{\mathrm{A_o}}{\sigma_A^2}\right]$$

This result can be further interpreted if we recall that the variance of the sample mean is equal to σ_w^2/N_s and define σ_p^2 as the average reciprocal variance

$$\frac{1}{\sigma_p^2} = \frac{1}{\sigma_A^2} + \frac{1}{\sigma_w^2/N_s}$$

Then the MAP estimate can be put in the form

$$\hat{A}_{MAP} = \sigma_p^2 \left[\frac{1}{\sigma_w^2/N_s} \left(\frac{1}{N_s} \sum_{n=0}^{N_s-1} x[n] \right) + \frac{\mathrm{A_o}}{\sigma_A^2} \right]$$

This is a weighted average of the prior mean $\mathrm{A_o}$ and the sample mean (computed from the observations), with weights that are inversely proportional to their variances. Note that in the limit as $\sigma_A^2 \to \infty$, implying that there is no prior information, the MAP estimate reduces to the maximum likelihood estimate of Example 6.1.

The connection between MAP and maximum likelihood estimates for a parameter demonstrated in the example is in fact a general property that can be seen directly from (6.96). In the limit where the prior density for the parameter has infinite variance or is uniform over all values of θ, the second term in the numerator becomes a constant. In this case maximizing $f_{\theta|x}$ is the same as maximizing $f_{x|\theta}$, so the MAP and the maximum likelihood estimates are identical.

6.4 LINEAR MEAN-SQUARE ESTIMATION

The types of estimates for random variables developed in the preceding section are usually nonlinear functions of the observations. Further, since they involve the conditional density of the random variable given the observations, this density function must be either known or computable from the information in the problem. All of these things frequently make the estimates of the preceding section difficult to compute and to implement. A simpler procedure results when the form of the estimate is constrained to have linear dependence on the observations. Further, if a mean-square cost function is chosen, it turns out that the estimate requires knowledge of only the first and second moments and the solution of *linear* equations. This whole procedure fits in well with estimation problems involving random processes since signal processing operations are usually linear and random processes are often described in terms of their first and second moments.

6.4.1 Estimating a Real Random Variable from a Single Related Observation

In Example 6.7 of the preceding section it was seen that when x and y are jointly Gaussian, the best mean-square estimate of y from x has the form of a straight line

$$\hat{y} = \left(\rho_{xy} \frac{\sigma_y}{\sigma_x} \right) x + \left(m_y - \rho_{xy} \frac{\sigma_y}{\sigma_x} m_x \right) \tag{6.97}$$

where ρ_{xy} is the correlation coefficient

$$\rho_{xy} = \frac{c_{xy}}{\sigma_x \sigma_y} \tag{6.98}$$

Figure 6.10 gives an interpretation of this result for three different cases involving real random variables x and y. In Fig. 6.10(a) the random variables are positively correlated. Recall from Chapter 2 that in this case the ellipse representing the contours of the joint density function has its major axis aligned along a direction with positive slope. The function that specifies the mean-square estimate is a straight line passing through the center of the ellipse (the mean) also with positive slope. The slope of this line is determined by the covariance c_{xy} through the parameter ρ_{xy}. Figure 6.10(a) also depicts some typical points from the distribution (as a scatter plot). Clearly for these points the estimated covariance

$$\hat{c}_{xy} = \frac{1}{N} \sum_{i=1}^{N} (x^{(i)} - m_x)(y^{(i)} - m_y)$$

is positive and the curve that best approximates the value $y^{(i)}$ corresponding to a given value $x^{(i)}$ is the straight line.[15] In the limiting case, when ρ_{xy} approaches 1, the ellipse becomes infinitesimally thin, and the data and its major eigenvector lie along the straight line.

In the case of Fig. 6.10(b), where the correlation is negative, the ellipse is tilted along a line with negative slope, and the best mean-square estimate of y is also a line with negative slope passing through the center. In both cases (a) and (b), the curve $\phi(x)$ that best approximates the data in the sense that the expected squared deviation $\mathcal{E}\{|y - \phi(x)|^2\}$ is minimized, is the straight line.

In Fig. 6.10(c) the random variables are uncorrelated and the ellipse is thus oriented so that its axes are parallel to the x and y axes. Here since c_{xy} and ρ_{xy} are zero, the mean-square estimate reduces to the mean of y [see (6.97)]. This result is intuitively resonable since in this case the value of x tells nothing about the likely value of y (points in the upper half of the ellipse are just as likely to occur as points in the lower half). As a result, the best estimate of y is in fact the mean.

Figure 6.11 shows a case where the joint density for x and y is not Gaussian, and the optimal mean-square estimate $\phi(x)$ that minimizes the expected squared deviation of the points is truly a curve. In this case the correlation is said to be nonlinear. In spite of the nonlinear correlation, we could still try to find the best straight line which when placed through the points would minimize the expected squared deviation. This line is the optimal *linear mean-square* estimate for y. While determining the best nonlinear mean-square estimate $\phi(x)$ requires knowledge of the joint density function f_{xy}, determining the best linear mean-square estimate, as mentioned previously, requires knowledge of only the first and second moments of the distribution. In addition, finding the best linear mean-square estimate involves solution of only linear equations. Linear mean-square estimates are thus easier to compute than general mean-square estimates and consequently are more frequently used.

[15]In statistical data analysis, the line is called the *regression* of the data, and the analysis of data to determine such a straight-line approximation is called regression analysis.

To pose the problem of linear mean-square estimation formally, consider an estimate of the form

$$\hat{y} = ax + b \tag{6.99}$$

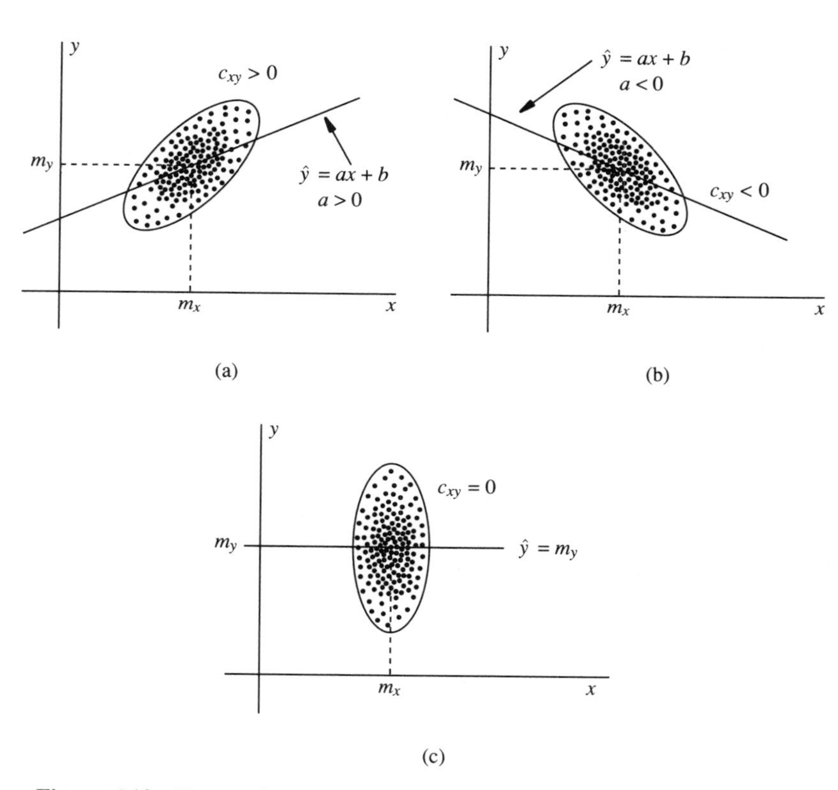

Figure 6.10 Forms of correlation for Gaussian random variables. (a) x and y positively correlated. (b) x and y negatively correlated. (c) x and y uncorrelated.

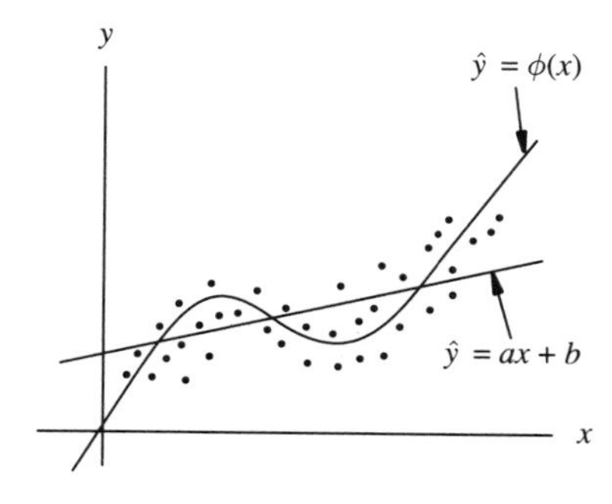

Figure 6.11 Possible form of correlation for non-Gaussian random variables.

where the real constants a and b are to be chosen to minimize the mean-square error

$$\mathcal{E}_{lms} = \mathcal{E}\{(y - \hat{y})^2\} \tag{6.100}$$

To find a and b we could substitute (6.99) into (6.100), take derivatives with respect to a and b, and set them equal to zero. This would yield two linear equations in two unknowns which could be solved for a and b. This procedure is straightforward but a little tedious.

Instead, consider the following approach, which leads more quickly to the desired result, and provides a little more insight for the problem. Define the error

$$\varepsilon = y - \hat{y} \tag{6.101}$$

Then since $\mathcal{E}_{lms}$ is the second moment of ε, it follows that

$$\mathcal{E}_{lms} = \mathcal{E}\{\varepsilon^2\} = m_\varepsilon^2 + \sigma_\varepsilon^2 \tag{6.102}$$

where m_ε and σ_ε^2 are the mean and variance of ε respectively. Clearly, to minimize $\mathcal{E}_{lms}$ the mean m_ε should be set equal to zero. Thus

$$m_\varepsilon = \mathcal{E}\{y - \hat{y}\} = \mathcal{E}\{y - ax - b\} = m_y - am_x - b = 0$$

or

$$b = m_y - am_x \tag{6.103}$$

Therefore, it follows from (6.99) and (6.103) that the optimal estimate is of the form

$$\hat{y} = a(x - m_x) + m_y \tag{6.104}$$

Now it is only necessary to find the parameter a. Substituting (6.104) into (6.100) yields

$$\begin{aligned} \mathcal{E}_{lms} = \mathcal{E}\{(y - \hat{y})^2\} &= \mathcal{E}\{[(y - m_y) - a(x - m_x)]^2\} \\ &= \mathcal{E}\{(y - m_y)^2 - 2a(x - m_x)(y - m_y) + a^2(x - m_x)^2\} \\ &= \sigma_y^2 - 2ac_{xy} + a^2\sigma_x^2 \end{aligned} \tag{6.105}$$

To minimize this quantity, choose a such that

$$\frac{d\mathcal{E}_{lms}}{da} = -2c_{xy} + 2a\sigma_x^2 = 0 \tag{6.106}$$

or

$$a = \frac{c_{xy}}{\sigma_x^2} \tag{6.107}$$

Then by using the definition (6.98), (6.107) becomes

$$a = \rho_{xy}\frac{\sigma_y}{\sigma_x} \tag{6.108}$$

and from (6.104) the optimal linear mean-square estimate is given by

$$\hat{y} = \left(\rho_{xy}\frac{\sigma_y}{\sigma_x}\right)x + \left(m_y - \rho_{xy}\frac{\sigma_y}{\sigma_x}m_x\right)$$

This is the same as (6.97), which is also the expression obtained in Example 6.7.

The minimum mean-square error is obtained by substituting (6.107) into (6.105). This yields

$$\begin{aligned}\mathcal{E}_{lms} &= \sigma_y^2 - 2\frac{c_{xy}^2}{\sigma_x^2} + \frac{c_{xy}^2}{\sigma_x^4}\sigma_x^2 \\ &= \sigma_y^2 - \frac{c_{xy}^2}{\sigma_x^2}\end{aligned} \tag{6.109}$$

Then by virtue of (6.98), the minimum mean-square error can be written neatly as

$$\mathcal{E}_{lms} = \sigma_y^2(1 - \rho_{xy}^2) \tag{6.110}$$

(which incidentally, is the same as the *variance* σ^2 obtained in Example 6.7). Since the correlation coefficient ρ_{xy} is always less than 1 in absolute value, this shows that the minimum mean-square error is always less than the initial variance of y.

To illustrate the application of linear mean-square estimation, let us consider applying it to one of the examples considered earlier.

Example 6.9

The random variables in Example 6.6 have the joint density function

$$f_{xy}(x, y) = \begin{cases} 10y & 0 \le y \le x^2, \quad 0 \le x \le 1 \\ 0 & \text{otherwise} \end{cases}$$

Here the linear mean-square estimate of y given x is computed and compared to the optimal (nonlinear) mean-square estimate computed in that example. The needed first and second moment parameters are computed as follows.

$$m_x = \int_0^1 \int_0^{x^2} x \cdot 10y\,dy\,dx = \frac{5}{6}$$

$$m_y = \int_0^1 \int_0^{x^2} 10y^2\,dy\,dx = \frac{10}{21}$$

$$\mathcal{E}\{x^2\} = \int_0^1 \int_0^{x^2} 10x^2y\,dy\,dx = \frac{5}{7}$$

$$\sigma_x^2 = \mathcal{E}\{x^2\} - m_x^2 = \frac{5}{7} - \left(\frac{5}{6}\right)^2 = \frac{5}{252}$$

$$\mathcal{E}\{y^2\} = \int_0^1 \int_0^{x^2} 10y^3 dy dx = \frac{5}{18}$$

$$\sigma_y^2 = \mathcal{E}\{y^2\} - m_y^2 = \frac{5}{18} - \left(\frac{10}{21}\right)^2 = \frac{5}{98}$$

$$\mathcal{E}\{xy\} = \int_0^1 \int_0^{x^2} 10xy^2 dy dx = \frac{5}{12}$$

$$c_{xy} = \mathcal{E}\{xy\} - m_x m_y = \frac{5}{12} - \left(\frac{5}{6}\right)\left(\frac{10}{21}\right) = \frac{5}{252}$$

$$\rho_{xy} = \frac{c_{xy}}{\sigma_x \sigma_y} = \sqrt{\frac{(5/252)^2}{(5/252)(5/98)}} = \sqrt{\frac{7}{18}}$$

Now from (6.107) and (6.104) we can compute

$$a = \frac{c_{xy}}{\sigma_x^2} = \frac{5/252}{5/252} = 1$$

and

$$b = m_y - a m_x = \frac{10}{21} - \frac{5}{6} = -\frac{5}{14}$$

Therefore, the linear mean-square estimate is

$$\hat{y}_{lms} = ax + b = x - \frac{5}{14}$$

This estimate actually has a problem that will be addressed shortly, but for the moment let us continue. The corresponding mean-square error is

$$\mathcal{E}_{lms} = \sigma_y^2(1 - \rho_{xy}^2) = \frac{5}{98}\left(1 - \frac{7}{18}\right) = \frac{55}{1764} = 0.0312$$

A comparison to the results of Example 6.6 shows that this is only slightly worse than the mean-square error for the optimal (nonlinear) mean-square estimate, which produced $\mathcal{E}_{lms} = 0.0309$.

Now let us return to the problem cited in the example above, which demonstrates that a certain aberration can occur with linear mean-square estimation. Notice that for values of the observations in the range $0 \leq x < 5/14$ the estimate is negative, while the conditions on the density indicate that y can never take on negative values. This can happen in linear mean-square estimation since the estimate uses no information about the density except its moments. Given only the moments, there is no way to know that y has to be greater than zero. The moments could have been derived from a Gaussian distribution in which case negative values for y and $\hat{y}$ would be perfectly valid. Given the constraint $0 \leq y \leq x$ the optimal mean-square estimate is said to be *inadmissible* for some values of x. There are at least two ways to get around this difficulty. One is to constrain the constant b to be equal to zero and solve the mean-square problem. This results in an overall larger mean-square error but an estimate that is admissible for the entire range of values of x. The other is to use the linear mean-square estimate only in the range $5/14 \leq x \leq 1$ and to use some other estimate in the nonadmissible region. In this case the estimate is essentially *nonlinear*.

6.4.2 Estimating a Random Variable from Multiple Related Observations

Consider now the general case of estimating a random variable y from a number of related random variables $x_1, x_2, \ldots, x_N$, which can be considered to be components of a random vector $\boldsymbol{x}$. The random variables may be complex in general, and an estimate is sought in the form

$$\hat{y} = \mathbf{a}^{*T}\boldsymbol{x} + b \tag{6.111}$$

where

$$\mathbf{a} = \begin{bmatrix} a_1 \\ a_2 \\ a_3 \\ \vdots \\ a_N \end{bmatrix} \tag{6.112}$$

and where the a_i and b are in general complex coefficients chosen to minimize the mean-square error

$$\mathcal{E}_{lms} = \mathcal{E}\{|y - \hat{y}|^2\} \tag{6.113}$$

As before, to minimize the mean-square error, we must have

$$\mathcal{E}\{y - \hat{y}\} = 0$$

which implies that

$$b = m_y - \mathbf{a}^{*T}\mathbf{m}_x \tag{6.114}$$

The estimate must therefore be of the form

$$\hat{y} = \mathbf{a}^{*T}(\boldsymbol{x} - \mathbf{m}_x) + m_y \tag{6.115}$$

To find $\mathbf{a}$, first substitute (6.115) in (6.113) to obtain

$$\mathcal{E}_{lms} = \mathcal{E}\{|(y - m_y) - \mathbf{a}^{*T}(x - \mathbf{m}_x)|^2\}$$

and then expand this to find

$$\begin{aligned} \mathcal{E}_{lms} &= \mathcal{E}\{[(y - m_y) - \mathbf{a}^{*T}(\boldsymbol{x} - \mathbf{m}_x)][(y - m_y) - \mathbf{a}^{*T}(\boldsymbol{x} - \mathbf{m}_x)]^*\} \\ &= \mathcal{E}\{|y - m_y|^2 - (y - m_y)(\boldsymbol{x} - \mathbf{m}_x)^{*T}\mathbf{a} \\ &\qquad -\mathbf{a}^{*T}(\boldsymbol{x} - \mathbf{m}_x)(y - m_y)^* + \mathbf{a}^{*T}(\boldsymbol{x} - \mathbf{m}_x)(\boldsymbol{x} - \mathbf{m}_x)^{*T}\mathbf{a}\} \end{aligned} \tag{6.116}$$

Then define the quantities

$$\mathbf{c}_{xy} = \mathcal{E}\{(\boldsymbol{x} - \mathbf{m}_x)(y - m_y)^*\} \tag{6.117}$$

and

$$\mathbf{C}_{\boldsymbol{x}} = \mathcal{E}\{(\boldsymbol{x} - \mathbf{m}_{\boldsymbol{x}})(\boldsymbol{x} - \mathbf{m}_{\boldsymbol{x}})^{*T}\} \tag{6.118}$$

and substitute them into (6.116) to find

$$\mathcal{E}_{lms} = \sigma_y^2 - \mathbf{c}_{\boldsymbol{x}y}^{*T}\mathbf{a} - \mathbf{a}^{*T}\mathbf{c}_{\boldsymbol{x}y} + \mathbf{a}^{*T}\mathbf{C}_{\boldsymbol{x}}\mathbf{a} \tag{6.119}$$

Finally, take the complex gradient of (6.119) using the formulas of Table A.2, Appendix A, and set it equal to $\mathbf{0}$ to find

$$\nabla_{\mathbf{a}^*}\mathcal{E}_{lms} = \mathbf{0} - \mathbf{0} - \mathbf{c}_{\boldsymbol{x}y} + \mathbf{C}_{\boldsymbol{x}}\mathbf{a} = \mathbf{0}$$

or

$$\boxed{\mathbf{C}_{\boldsymbol{x}}\mathbf{a} = \mathbf{c}_{\boldsymbol{x}y}} \tag{6.120}$$

Equation (6.120) represents a set of linear equations that can be solved to find the optimum value of $\mathbf{a}$. A formal representation of the solution (assuming that $\mathbf{C}_{\boldsymbol{x}}$ is of full rank) is

$$\boxed{\mathbf{a} = \mathbf{C}_{\boldsymbol{x}}^{-1}\mathbf{c}_{\boldsymbol{x}y}} \tag{6.121}$$

To find the minimum mean-square error, write (6.119) as

$$\mathcal{E}_{lms} = \sigma_y^2 - \mathbf{c}_{\boldsymbol{x}y}^{*T}\mathbf{a} + \mathbf{a}^{*T}(\mathbf{C}_{\boldsymbol{x}}\mathbf{a} - \mathbf{c}_{\boldsymbol{x}y})$$

Because of (6.120) the last term is equal to $\mathbf{0}$; this equation thus reduces to

$$\mathcal{E}_{lms} = \sigma_y^2 - \mathbf{c}_{\boldsymbol{x}y}^{*T}\mathbf{a} \tag{6.122}$$

or, upon substituting (6.121),

$$\boxed{\mathcal{E}_{lms} = \sigma_y^2 - \mathbf{c}_{\boldsymbol{x}y}^{*T}\mathbf{C}_{\boldsymbol{x}}^{-1}\mathbf{c}_{\boldsymbol{x}y}} \tag{6.123}$$

The following example illustrates the computation of a linear mean-square estimate involving multiple random variables.

Example 6.10

Two real random variables x_1 and x_2 and a related random variable y are jointly distributed. It is known that if $\boldsymbol{v}$ is defined by

$$\boldsymbol{v} = \begin{bmatrix} y \\ x_1 \\ x_2 \end{bmatrix}$$

then the mean vector and covariance matrix of $\boldsymbol{v}$ are

$$\mathbf{m}_{\boldsymbol{v}} = \begin{bmatrix} \frac{1}{4} \\ \frac{1}{2} \\ \frac{1}{2} \end{bmatrix} \qquad \mathbf{C}_{\boldsymbol{v}} = \begin{bmatrix} \frac{7}{10} & \frac{1}{10} & \frac{1}{10} \\ \frac{1}{10} & \frac{3}{10} & -\frac{1}{10} \\ \frac{1}{10} & -\frac{1}{10} & \frac{3}{10} \end{bmatrix}$$

It is desired to find the best linear mean-square estimate of y using x_1 and x_2.

From the given information it follows that

$$\mathbf{m}_x = \begin{bmatrix} \frac{1}{2} \\ \frac{1}{2} \end{bmatrix} \qquad m_y = \tfrac{1}{4}$$

$$\mathbf{C}_x = \begin{bmatrix} \frac{3}{10} & -\frac{1}{10} \\ -\frac{1}{10} & \frac{3}{10} \end{bmatrix} \qquad \mathbf{c}_{xy} = \begin{bmatrix} \frac{1}{10} \\ \frac{1}{10} \end{bmatrix} \qquad \sigma_y^2 = \tfrac{7}{10}$$

Therefore, to find **a** form (6.120):

$$\begin{bmatrix} \frac{3}{10} & -\frac{1}{10} \\ -\frac{1}{10} & \frac{3}{10} \end{bmatrix} \begin{bmatrix} a_1 \\ a_2 \end{bmatrix} = \begin{bmatrix} \frac{1}{10} \\ \frac{1}{10} \end{bmatrix}$$

and solve this to obtain

$$a_1 = a_2 = \tfrac{1}{2}$$

Then from (6.114)

$$b = m_y - \mathbf{a}^{*T}\mathbf{m}_x = \tfrac{1}{4} - \begin{bmatrix} \frac{1}{2} & \frac{1}{2} \end{bmatrix} \cdot \begin{bmatrix} \frac{1}{2} \\ \frac{1}{2} \end{bmatrix} = -\tfrac{1}{4}$$

so the optimal linear mean-square estimate is

$$\hat{y} = \tfrac{1}{2}x_1 + \tfrac{1}{2}x_2 - \tfrac{1}{4}$$

The mean-square error from (6.122) is

$$\mathcal{E}_{lms} = \sigma_y^2 - \mathbf{c}_{xy}^{*T}\mathbf{a} = \tfrac{7}{10} - \begin{bmatrix} \frac{1}{10} & \frac{1}{10} \end{bmatrix} \cdot \begin{bmatrix} \frac{1}{2} \\ \frac{1}{2} \end{bmatrix} = \tfrac{3}{5}$$

The examples in this section illustrate that linear mean-square estimation has some definite advantages over other types of estimation in terms of its simplicity and ease of computation. It is the natural thing to use for estimation problems related to random processes if the solution is to involve a linear filter and the analysis is based on at most second moment properties of the random process. Linear mean-square estimation thus forms the basis for most of the work on optimal filtering discussed in Chapter 7.

6.5 CHAPTER SUMMARY

This chapter deals with the problem of statistical estimation. A fundamental distinction is made between estimation of fixed parameters and estimation of random variables. Maximum likelihood is a form of parameter estimation and requires knowledge of the density function for the observations and its dependence on the parameter. The density, viewed as a function of the parameter, for fixed values of the random vector, is called the likelihood function. The EM algorithm is an iterative technique to compute a maximum likelihood estimate when the observed data can be regarded as "incomplete." If a maximum likelihood estimate exists, it satisfies the Cramér-Rao bound with equality and so is a minimum variance, unbiased estimate. In general, the ideas of bias, consistency, and efficiency apply to any estimate and serve as properties by which different estimates can be compared.

The usual estimate of the mean of a random process is the sample mean. This estimate is unbiased and consistent; that is, its expectation is equal to the true mean of the process and its variance decreases with the number of samples. Two different estimates are discussed in this chapter for the correlation function. The unbiased sample correlation function has a variance that decreases with the number of samples and is therefore consistent. However, this estimate may not be positive semidefinite. The biased sample correlation function is asymptotically unbiased and is also consistent. Further, since this estimate is guarranteed to be positive semidefinite, it is the preferred estimate of the correlation function for many applications. Two methods for estimating the correlation *matrix* are also discussed in this chapter. The procedure called the autocorrelation method is equivalent to estimating the biased correlation function and constructing the corresponding Toeplitz matrix. It is pointed out that this method implicitly "windows" the data and is biased. An alternative related procedure that avoids windowing the data but does not lead to a Toeplitz matrix is called the "covariance method." This estimate, however, is unbiased.

Some general types of estimates for random variables are based on Bayes estimation. Two specific estimates are discussed in this chapter. The mean-square estimate minimizes the expected value of a quadratic cost function. This estimate is the *mean* of the conditional density function. The MAP estimate derives from minimizing the risk for a uniform cost function. This estimate is the value of the random variable that *maximizes* the conditional density function. Both of these estimates are usually nonlinear functions of the observations. For Gaussian random variables, however, the two estimates are identical and depend linearly on the observations.

Linear mean-square estimates are mean-square estimates constrained to have linear dependence on the observations. Their computation involves at most second moments of the random variables and solution of linear equations. Although linear mean-square estimates are suboptimal with respect to general mean-square estimates (except in the Gaussian case), their simple form and dependence on only first and second moment properties makes these types of estimates the ones most frequently used for optimal filtering of random processes.

REFERENCES

1. Athanasios Papoulis. *Probability, Random Variables, and Stochastic Processes*, 3rd ed. McGraw-Hill, New York, 1991.
2. Harold J. Larson and Bruno O. Schubert. *Probabilistic Models in Engineering Sciences*, Volume II. John Wiley & Sons, New York, 1979.
3. A. P. Dempster, N. M. Laird, and D. B. Rubin. Maximum likelihood from incomplete data via the EM algorithm. *Journal of the Royal Statistical Society, Series B*, 39(1):1–38, 1977.

4. R. David Cox and David Oakes. *Analysis of Survival Data*. Chapman and Hall, New York, 1984. (Monograph series on statistics and applied probability.)
5. Meir Feder and Ehud Weinstein. Parameter estimation of superimposed signals using the EM algorithm. *IEEE Transactions on Acoustics, Speech, and Signal Processing*, 36(4):477–489, April 1988.
6. Harold J. Larson and Bruno O. Schubert. *Probabilistic Models in Engineering Sciences*, Volume I. John Wiley & Sons, New York, 1979.
7. Harry L. Van Trees. *Detection, Estimation, and Modulation Theory,* Part I. John Wiley & Sons, New York, 1968.
8. Harald Cramér. *Mathematical Methods of Statistics*. Princeton University Press, Princeton, New Jersey, 1974. (Originally published in 1946.)
9. C. R. Rao. Information and accuracy attainable in the estimation of statistical parameters. *Bull. Calcutta Math. Soc.*, 37:81–91, 1945.
10. R. A. Fisher. On the mathematical foundations of theoretical statistics. *Phil Trans. Roy. Soc. London*, 222:309–368, 1922.
11. Francis B. Hildebrand. *Methods of Applied Mathematics*, 2nd ed. Prentice Hall, Inc., Englewood Cliffs, New Jersey, 1965.
12. Gilbert Strang. *Introduction to Applied Mathematics*. Wellesley Cambridge Press, Wellesley, Massachusetts, 1986.
13. Samuel S. Wilks. *Mathematical Statistics*. John Wiley & Sons, New York, 1962.
14. Louis L. Scharf. *Statistical Signal Processing: Detection, Estimation, and Time Series Analysis*. Addison-Wesley, Reading, Massachusetts, 1991.
15. Kenneth S. Miller. *Complex Stochastic Processes*. Addison-Wesley, Reading, Massachusetts, 1974.
16. M. S. Bartlett. On the theoretical specification and sampling properties of autocorrelated time series. *Journal of the Royal Statistical Society*, B 8(27), 1946.
17. Gwilym M. Jenkins and Donald G. Watts. *Spectral Analysis and Its Applications*. Holden-Day, Oakland, California, 1968.
18. Richard W. Hamming. *Coding and Information Theory*, 2nd ed. Prentice Hall, Inc., Englewood Cliffs, New Jersey, 1986.

PROBLEMS

6.1. A random variable x is described by the Gaussian density function

$$f_x(\mathrm{x}) = \frac{1}{\sqrt{2\pi\sigma^2}} e^{-\frac{\mathrm{x}^2}{2\sigma^2}}$$

(a) Based on a *single* observation of the random variable, compute the maximum likelihood estimate for the variance. (Treat σ^2, not σ, as the parameter to be estimated.)

(b) What is the maximum likelihood estimate of σ^2 if there are N independent observations of the random variable?

6.2. A random variable x has the uniform density

$$f_x(\mathrm{x}) = \begin{cases} 1/a & 0 \leq \mathrm{x} \leq a \\ 0 & \text{otherwise} \end{cases}$$

(a) Determine the likelihood function $f_{x_1 x_2 \ldots x_N;a}$ for $N = 1$ and $N = 2$ and sketch it. Find the maximum likelihood estimate of the parameter a for these two cases.

(b) Determine the maximum likelihood estimate of the parameter a for arbitrary N.

6.3. Show that the maximum likelihood estimates for the parameter P in the multinomial distributions (6.11) and (6.14) are given by the solutions to (6.12) and (6.15) respectively. Verify that with the specific data (6.22) the maximum likelihood solution is given by (6.23).

6.4.[16] Consider the expectation

$$\mathcal{E}\left\{ \ln \frac{f_{x;\theta}(\boldsymbol{x};\boldsymbol{\theta}')}{f_{x;\theta}(\boldsymbol{x};\boldsymbol{\theta})} \middle| \boldsymbol{\theta} \right\}$$

By applying the inequality $\ln y \leq y - 1$ which holds true for all values of y, derive the relation

$$\mathcal{E}\{ \ln f_{x;\theta}(\boldsymbol{x};\boldsymbol{\theta}') | \, \boldsymbol{\theta}\} \leq \mathcal{E}\{ \ln f_{x;\theta}(\boldsymbol{x};\boldsymbol{\theta}) | \, \boldsymbol{\theta}\}$$

Use this fact to establish (6.21). The quantity on the right is known as the *entropy* of the distribution. In most cases it is finite; however, it is possible for a finite distribution to have infinite entropy.

6.5. A set of M observations y_k, $k = 1, 2, \ldots, M$ from the exponential density function

$$f_y(\mathrm{y}) = \begin{cases} \alpha e^{-\alpha \mathrm{y}} & \mathrm{y} \geq 0 \\ 0 & \text{otherwise} \end{cases}$$

are given.

(a) By assuming these observations are "incomplete data" and part of a larger set of observations x_k, $k = 1, 2, \ldots N$ with $N > M$ and $x_k = y_k$ for $k \leq M$, use the EM algorithm to derive the iterative formula

$$\hat{\alpha}^{(i+1)} = \frac{N}{(N-M)/\hat{\alpha}^{(i)} + \sum_{k=1}^{M} y_k}$$

for computing the maximum likelihood estimate of the parameter α from the given data.

(b) Using the data

$$\{y_1, y_2, y_3, y_4\} = \{0.5, 1.4, 3.7, 2.4\}$$

and starting with an initial value $\hat{\alpha}^{(0)} = 1.0$, apply the formula to iteratively compute $\hat{\alpha}$ for $N = 5$ and $N = 8$. Compute the estimate exactly (without iteration) using the result of Example 6.2. How many iterations does it require for each of these values of N to compute the estimate to three significant figures?

6.6. (a) Determine the mean of the exponential density function

$$f_x(\mathrm{x}) = \begin{cases} \alpha e^{-\alpha \mathrm{x}} & \mathrm{x} \geq 0 \\ 0 & \text{otherwise} \end{cases}$$

and express the density in terms of the mean parameter $\mu = \mathcal{E}\{x\}$.

[16] With special thanks to Professor Richard W. Hamming, whose elegant proof of Gibbs inequality [18, p. 117] provided the basis of this result.

(b) Assume that you are given N independent samples $x_1, x_2, \ldots, x_N$ of the random variable x. What is the maximum likelihood estimate for the mean μ?

(c) Is this estimate unbiased?

(d) Is it consistent?

(e) What is the variance of the estimate? Is this a minimum-variance estimate?

6.7. (a) Show that the maximum likelihood estimate for the parameter a in Problem 6.2 is a biased estimate for $N = 1$ and $N = 2$.

(b) Find the expected value of the maximum likelihood estimate of a for arbitrary N and show that the estimate is asymptotically unbiased.

6.8. Show that the mean of the score vector $\boldsymbol{s}(\boldsymbol{x}; \boldsymbol{\theta})$, defined by (6.43), is zero and therefore that $\mathbf{J}(\boldsymbol{\theta})$ of (6.42) is a real covariance matrix.

6.9. By considering the covariance matrix for the vector

$$\begin{bmatrix} \hat{\theta}_i - \theta_i \\ \boldsymbol{s}(\boldsymbol{x}; \boldsymbol{\theta}) \end{bmatrix}$$

and noting that it is positive semidefinite, derive the bounds (6.45) directly. Show that the bound is satisfied with equality if and only if the condition (6.46) holds. Show further that if all of the conditions (6.46) are satisfied for $i = 1, 2, \ldots, N$, i.e., if (6.47) holds, then the matrix $\mathbf{K}(\boldsymbol{\theta})$ is given by

$$\mathbf{K}(\boldsymbol{\theta}) = \mathbf{J}^{-1}(\boldsymbol{\theta})$$

and

$$\mathbf{C}_{\hat{\theta}} = \mathbf{J}^{-1}$$

Hint: To solve the first part of this problem, note that the determinant of the covariance matrix is nonnegative. Expand the determinant using cofactors.

6.10. This problem develops a complex form of the Cramér-Rao bound for a vector parameter. It is useful, however, only if the real and imaginary parts of the estimate and the score possess the appropriate symmetry in their covariance (see Chapter 2). A case is illustrated where the symmetry conditions are in fact satisfied.

(a) Define the complex score vector as

$$\boldsymbol{s}(\boldsymbol{x}; \boldsymbol{\theta}) = \nabla_{\boldsymbol{\theta}^*} \ln f_{\boldsymbol{x};\boldsymbol{\theta}}(\boldsymbol{x}; \boldsymbol{\theta})$$

Show that it satisfies the property

$$\mathcal{E}\left\{\boldsymbol{s} \cdot (\hat{\boldsymbol{\theta}} - \boldsymbol{\theta})^{*T}\right\} = \mathbf{I}$$

(b) By considering the complex covariance matrix of the vector

$$\begin{bmatrix} \hat{\boldsymbol{\theta}}(\boldsymbol{x}) \\ \boldsymbol{s}(\boldsymbol{x}; \boldsymbol{\theta}) \end{bmatrix}$$

derive the bound

$$\mathbf{C}_{\hat{\theta}} \geq \mathbf{J}^{-1}$$

where

$$\mathbf{C}_{\hat{\theta}} = \mathcal{E}\left\{(\hat{\boldsymbol{\theta}} - \boldsymbol{\theta})(\hat{\boldsymbol{\theta}} - \boldsymbol{\theta})^{*T}\right\} \qquad \mathbf{J} = \mathcal{E}\left\{\boldsymbol{s}\boldsymbol{s}^{*T}\right\}$$

(c) As an example of application of this bound consider estimating the mean of a complex Gaussian density function

$$\frac{1}{\pi^d|\mathbf{C}_x|}e^{-(\mathbf{x}-\mathbf{m})^{*T}\mathbf{C}_x^{-1}(\mathbf{x}-\mathbf{m})}$$

from samples $\boldsymbol{x}^{(1)}, \boldsymbol{x}^{(2)}, \ldots, \boldsymbol{x}^{(N)}$. The second moments of the random vector are of course assumed to satisfy the symmetry conditions so that use of the complex Gaussian density is meaningful. Show that $\boldsymbol{s}$ is a *linear* function of the observations, and therefore its complex covariance matrix $\mathbf{J} = \mathcal{E}\{\boldsymbol{s}\boldsymbol{s}^{*T}\}$ provides a complete second moment description of the score.

(d) Now consider the sample mean

$$\hat{\boldsymbol{m}}_N = \frac{1}{N}\sum_{i=1}^{N}\boldsymbol{x}^{(i)}$$

Its complex covariance matrix $\mathbf{C}_{\hat{m}_N} = \mathcal{E}\{(\hat{\boldsymbol{m}}_N - \mathbf{m})(\hat{\boldsymbol{m}}_N - \mathbf{m})^{*T}\}$ provides a complete second moment description of the estimate. Show that this estimate satisfies the bound in part (b) with equality.

6.11. The expectation for the product of four zero-mean *real* Gaussian random variables can be expressed as [1, p. 197]

$$\mathcal{E}\{v_1v_2v_3v_4\} = \mathcal{E}\{v_1v_2\}\,\mathcal{E}\{v_3v_4\} + \mathcal{E}\{v_1v_3\}\,\mathcal{E}\{v_2v_4\} + \mathcal{E}\{v_1v_4\}\,\mathcal{E}\{v_2v_3\}$$

Use this result to derive the expression (6.64) for the variance of the unbiased sample correlation function for a real random process.

6.12. You are given a length of data consisting of the following values:

$$1,\ 2,\ -1,\ -2$$

Assume that the data is from a stationary random process.

(a) Compute three points $R(0), R(1), R(2)$ of the sample correlation function using both the unbiased and biased estimates.

(b) Compute a 3×3 correlation matrix using the *autocorrelation method.*

(c) Compute a 3×3 correlation matrix using the *covariance method.*

6.13. Given the stationary random sequence $x[n] = 3, 1, 2, 1, 3, 1, 3$:

(a) Compute the sample mean for the sequence.

(b) Compute and sketch the sample correlation function of the sequence for lags $l = 0, \pm 1, \pm 2, \pm 3$ using both the biased and the unbiased estimates.

(c) Compute the sample covariance function by subtracting the sample mean from the data and computing the correlation function of the new data. See if these estimates satisfy the relations

$$\hat{C}_x = \hat{R}_x - \hat{m}_x^2$$
$$\hat{C}'_x = \hat{R}'_x - \hat{m}_x^2$$

6.14. A random variable x is given by

$$x = y + \eta$$

where y is uniformly distributed over the interval $-\alpha$ to α and η is uniformly distributed over the interval $-\beta$ to β. Assume that $|\alpha| < |\beta|$.

(a) Compute and sketch the density functions $f_{x|y}$, f_{xy}, f_x and finally, $f_{y|x}$. The latter can be determined by using the density form of Bayes' rule.

(b) The random variable x is observed. Think of η as noise. What is the mean-square estimate of y?

(c) What is the MAP estimate of y? Does a unique estimate exist?

6.15. Random variables x and y are characterized by a two-dimensional uniform density function

$$f_{xy}(\mathrm{x}, \mathrm{y}) = \begin{cases} 1 & \text{for } 0 \leq \mathrm{x} \leq 1; \;\; 0 \leq \mathrm{y} \leq 1 \\ 0 & \text{otherwise} \end{cases}$$

Given that we observe x, what is the mean-square estimate of y?

6.16. The joint probability density function for two random variables x and y is given by

$$f_{xy}(\mathrm{x}, \mathrm{y}) = \begin{cases} 6\mathrm{x} & 0 \leq \mathrm{x} \leq 1; \; 0 \leq \mathrm{y} \leq 1 - \mathrm{x} \\ 0 & \text{otherwise} \end{cases}$$

The area where the density is nonzero is shown shaded in Fig. PR6.16.

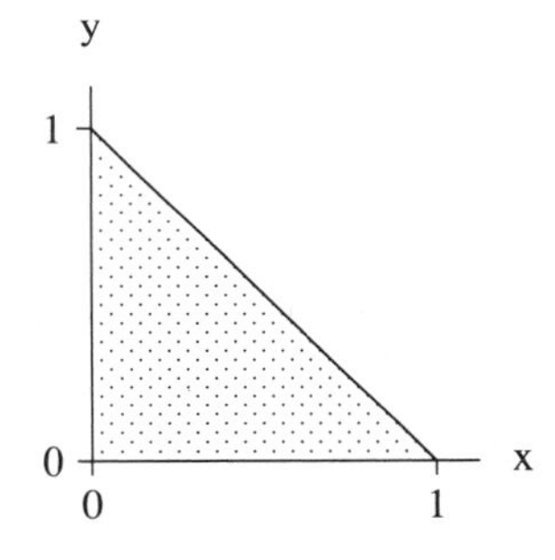

Figure PR6.16

(a) Compute the conditional density function $f_{y|x}(\mathrm{y}|\mathrm{x})$. Sketch it below as a function of y for a typical value of x between 0 and 1.

(b) The random variable x is observed but y is not. What is the minimum mean-square estimate for y?

(c) What is the minimum linear mean-square estimate for y?

(d) What is the MAP estimate for y?

6.17. By directly evaluating the expression $\mathcal{E}\left\{(y - \hat{y}_{ms})^2\right\}$, show that the mean-square error for the estimate in Example 6.7 is equal to $\sigma^2 = \sigma_y^2(1 - \rho_{xy}^2)$; thus it is equal to the variance of the conditional density.

6.18. Consider the expression

$$\mathcal{E}\{y - \hat{y}(\boldsymbol{x})|y\} = \int_{-\infty}^{\infty} (y - \hat{y}(\mathbf{x})) f_{x|y}(\mathbf{x}|y) d\mathbf{x}$$

(a) By first differentiating, then integrating with respect to y and using the assumption

$$\mathcal{E}\{\mathrm{y} - \hat{y}(\boldsymbol{x})|y\} \cdot f_y(\mathrm{y})\big|_{\mathrm{y}=\pm\infty} = 0$$

show that

$$\int_{-\infty}^{\infty} \int_{-\infty}^{\infty} \frac{\partial f_{xy}}{\partial y}(y - \hat{y}(\boldsymbol{x}))\, d\mathbf{x}\, d\mathrm{y} = -1$$

(b) Now by following the steps in the proof of the Cramér-Rao bound given in Section 6.1.4, prove the bound (6.90). Show that the bound is satisfied with equality if and only if (6.91) obtains.

6.19. The joint density function of random variables x and y is given by

$$f_{xy}(\mathrm{x}, \mathrm{y}) = \begin{cases} 8\mathrm{xy} & 0 \leq \mathrm{y} \leq \mathrm{x} \leq 1 \\ 0 & \text{otherwise} \end{cases}$$

(a) Find and sketch the conditional density function $f_{y|x}(\mathrm{y}|\mathrm{x})$.
(b) Determine the MAP estimate of y.
(c) Determine the mean-square estimate of y.
(d) What is the linear mean-square estimate of y?

6.20. Assume that the parameter α of the exponential density in Example 6.2 is a random variable with prior density

$$f_{\alpha}(\alpha) = \begin{cases} \alpha\beta^2 e^{-\alpha\beta} & \alpha \geq 0 \\ 0 & \text{otherwise} \end{cases}$$

(a) What is the conditional density function $f_{\alpha|T_1 T_2 \ldots T_N}$?
(b) What is the MAP estimate of α?
(c) What is the mean-square estimate of α?
Note: $\int_0^\infty y^n e^{-y} dy = n!$

6.21. A two-dimensional random vector $\boldsymbol{x}$ and a random variable y have the joint density function

$$f_{\boldsymbol{x},y}(\mathbf{x}, \mathrm{y}) = \begin{cases} (\mathrm{y} + 3\mathrm{x}_1)\mathrm{x}_2 & 0 \leq \mathrm{x}_1, \mathrm{x}_2, \mathrm{y} \leq 1 \\ 0 & \text{otherwise} \end{cases}$$

Assume that $\boldsymbol{x}$ represents the observations.
(a) Find the mean-square estimate of y.
(b) Find the MAP estimate of y.

6.22. Verify the expressions given for $\mu(\mathrm{x})$ and σ^2 in Example 6.7.

6.23. Find the best *linear* mean-square estimate of y given $\boldsymbol{x}$ for the data in Problem 6.21 and the corresponding mean-square error.

6.24. A zero-mean stationary random signal $s[n]$ has the correlation function $R_s[l] = \rho^{|l|}$. A random process $x[n]$ is related to $s[n]$ by

$$x[n] = s[n] + w[n]$$

where $w[n]$ is a zero-mean white noise process with variance σ_o^2, independent of $s[n]$.
(a) What are the coefficients a_1 and a_2 in

$$\hat{s}[2] = a_1 x[1] + a_2 x[2]$$

such that $\hat{s}[2]$ is the best linear mean-square estimate of $s[2]$? What is the mean-square error?
(b) Suppose now that $x[1]$ is not observed. What is the linear mean-square estimate of $s[2]$ based solely on $x[2]$? What is the mean-square error?

(c) Repeat part (a) when $x[n]$ is given by

$$x[n] = s[n] \cdot w[n]$$

(i.e., when the noise is multiplicative).

6.25. The joint density function for two random variables x and y is

$$f_{xy}(\mathrm{x}, \mathrm{y}) = \begin{cases} 2 & 0 \le \mathrm{y} \le \mathrm{x} \le 1 \\ 0 & \text{otherwise} \end{cases}$$

Assume that x is observed and y is to be estimated.

(a) Find and sketch the conditional density function

$$f_{y|x}(\mathrm{y}|\mathrm{x})$$

(b) What is the mean-square estimate of y?
(c) What is the linear mean-square estimate of y?
(d) What is the MAP estimate of y?

6.26. The correlation matrix for two zero-mean random variables x and y is

$$\mathbf{R}_{\left[\begin{smallmatrix} x \\ y \end{smallmatrix}\right]} = \begin{bmatrix} 2 & 0.2 \\ 0.2 & 1 \end{bmatrix}$$

It is desired to develop a linear mean-square estimate for y in the form

$$\hat{y} = ax$$

(a) What are the values of the parameter a and the corresponding mean-square error?
(b) If x and y are jointly Gaussian, what is the mean-square estimate of y? What is the MAP estimate of y?

6.27. Let the random vector $\boldsymbol{x} = [\ x_1 \quad x_2 \quad x_3\ ]^T$ have a mean vector and covariance matrix

$$\mathbf{m}_x = \begin{bmatrix} 0 \\ 1 \\ 1 \end{bmatrix} \qquad \mathbf{C}_x = \begin{bmatrix} 4 & -2 & -3 \\ -2 & 4 & 0 \\ -3 & 0 & 4 \end{bmatrix}$$

(a) Find the best linear mean-square estimate of x_1 given x_2 and x_3.
(b) What is the mean-square error in the estimate?
(c) Find the best linear mean-square estimate of x_3 given x_1 and x_2.
(d) What is the mean-square error in the estimate?

6.28. Show that the expected value of the covariance method estimate for the correlation function is Toeplitz and equal to the true correlation matrix for the random process. In other words, show that the covariance method estimate is *unbiased*.

6.29. Show that the optimal mean-square estimate when y and $\boldsymbol{x}$ are jointly Gaussian is given by (6.115), where $\mathbf{a}$ is given by (6.121). What parameter of the conditional Gaussian density corresponds to $\mathcal{E}_{lms}$? Show that it is the same as (6.122).
Hint: Use the expressions for the conditional Gaussian density given in Chapter 2(Section 2.3).

COMPUTER ASSIGNMENTS

6.1. Compute and plot the sample correlation function for each of the data sets S00, S01, S02, and S03 using the estimates (6.57) and (6.65). Plot both estimates for $0 \leq l \leq 511$. Also plot the theoretical correlation functions. S00 is a realization of a white noise process with $\sigma_o^2 = 1$. The other data sets have exponential correlation functions of the form

$$R_x[l] = \frac{\sigma_o^2}{1 - \rho^2}\rho^{|l|}$$

with $\sigma_o^2 = 1$ and $\rho = 0.95, 0.70, -0.95$ for S01, S02, S03. Compute and plot the variances of these estimates using (6.64) and (6.70) and the theoretical values of the correlation functions. *Optional*: Repeat the the variance computations using the estimated correlation values in (6.64) and (6.70).

6.2. **(a)** Compute and print an 8×8-dimensional correlation matrix for the data sets S00, S01, S02, and S03 using both the autocorrelation method and the covariance method. Compare the results for

i. $N_s = 32$ (first 32 samples of the data set)
ii. $N_s = 128$
iii. $N_s = 256$
iv. $N_s = 512$ (optional)

(b) Normalize the correlation matrix estimates by pre- and postmultiplying by a diagonal matrix whose elements are the reciprocal square roots of the diagonal elements of the original matrix. Do the normalized estimates appear to be more consistent?

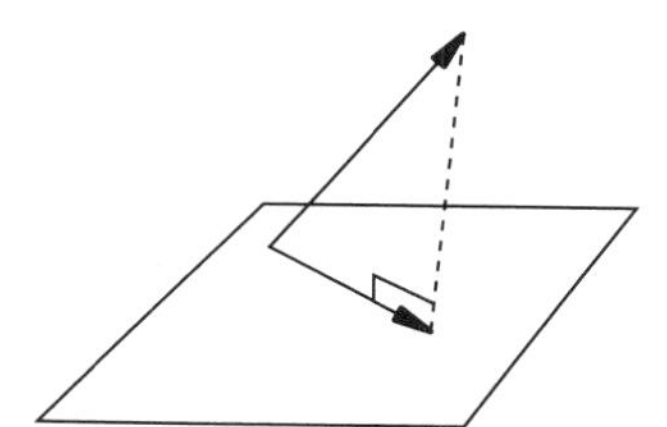

7

Optimal Filtering

Optimal filtering deals with design of filters to process a class of signals with *statistically* similar characteristics. The topic is based on estimation, particularly mean-square estimation, as applied to signal processing. Consequently, this is where the topics of Chapter 6 and those of the earlier chapters come together.

Essentially, all of the topics in linear mean-square estimation and optimal filtering can be developed from a single fact known as the *orthogonality principle*. Since this result is so important and central to all of the remaining topics, it is developed early in the chapter as a theorem, and given a geometric interpretation.

Most of the rest of the chapter concentrates on various forms of optimal filtering. Linear prediction is introduced as a simple kind of optimal filtering problem and used to further strengthen the geometric ideas of the orthogonality principle. The more general Wiener filter is then studied in both the FIR and the IIR forms. A special case study provides an in-depth understanding of several points related to the IIR form. The same case, with a recursive formulation, is the scalar version of the Kalman filter. This recursive form of the filter is also developed.

A final result is presented in this chapter that is more of theoretical than of practical interest. However, it adds to the understanding of random processes and rigorously establishes the representation of the spectrum as a sum of continuous and discrete parts. This result is the Wold decomposition. It shows that any random process has a decomposition where the process corresponding to the continuous part of the spectrum is regular and therefore has an innovations representation, and the process corresponding to the discrete part is completely predictable, that is, it can be predicted by a suitable optimal filter with zero mean-square error.

A particular issue that sometimes confuses people when they first study optimal filtering concerns what is known and what is unknown. In the treatment in this chapter it is assumed that the second moment statistics (i.e., correlation functions, spectral density functions, and/or complex spectral density functions) are *known*. That is, the average second moment behavior of a class of signals is fully characterized. The behavior of any *particular* signal to which the filter may be applied is of course unknown. There would be no need to design a filter to perform prediction of a signal for example, if the signal were already known. In practice, the second moment properties such as the correlation function may have to be estimated from given data. But that is a separate issue. In any event, the data used to *design* the filter would *not* be the same as the data to which it would ultimately be applied (i.e., the filters here are *not* adaptive). With these remarks and precautions in mind, let us proceed to the topic of optimal filtering.

7.1 THE ORTHOGONALITY PRINCIPLE

Optimal filtering as developed in this chapter is based upon linear mean-square estimation. Although linear mean-square estimation is developed in a general way in Chapter 6, it is redeveloped here from the more basic principle of orthogonality. The advantages of this approach are many, but perhaps the greatest advantage is that it permits solving a large number of related linear filtering problems with relatively little effort.

7.1.1 Linear Mean-Square Estimation Revisited

Consider a set of random variables $x_1, x_2, \ldots, x_N$ and a related random variable y. It is desired to form an estimate for y of the form

$$\hat{y} = \mathbf{a}^{*T}\boldsymbol{x} \tag{7.1}$$

where $\boldsymbol{x}$ is the vector of observations $x_1, x_2, \ldots, x_N$ and $\mathbf{a}$ is the vector of weighting coefficients chosen to minimize the mean-square error

$$\mathcal{E}\{|y - \hat{y}|^2\} \tag{7.2}$$

Note that unlike in Chapter 6, there is no constant term b, so the estimate is truly a *linear* function. It will be seen that the optimal estimate consequently depends on correlation terms instead of covariance terms, as before. Alternatively, it can be assumed that all of the random variables are made to have zero mean by subtracting out their mean values. We prefer to take the latter point of view. As a result the constant term b in the estimate for the zero-mean random variables is equal to zero. The $\mathbf{a}$ parameter for the nonzero-mean random variables then is the same as for their zero-mean counterparts and the b parameter can be computed if needed from

$$b = m_y - \mathbf{a}^{*T}\mathbf{m}_x \tag{7.3}$$

Now recall from Chapter 2 that two random variables u and v are said to be *orthogonal* if their correlation $\mathcal{E}\{uv^*\}$ is equal to zero. (The reason for this terminology will become clear in the next subsection.) We have the following important theorem:

Theorem 7.1 (Orthogonality). Let $\varepsilon = y - \hat{y}$ be the error in estimation. Then $\mathbf{a}$ minimizes the mean-square error $\sigma_\varepsilon^2 = \mathcal{E}\{|y - \hat{y}|^2\}$ if $\mathbf{a}$ is chosen such that $\mathcal{E}\{x_i\varepsilon^*\} = \mathcal{E}\{\varepsilon x_i^*\} = 0, \quad i = 1, 2, \ldots, N$, that is, if the error is *orthogonal* to the observations. Further, the minimum mean-square error is given by $\sigma_\varepsilon^2 = \mathcal{E}\{y\varepsilon^*\} = \mathcal{E}\{\varepsilon y^*\}$.

In certain contexts this theorem is also called the Projection Theorem, for reasons that will become clear shortly. It can be proven as follows. Let $\mathbf{a}$ be any weighting vector and $\mathbf{a}^\perp$ be the weighting vector that results in orthogonality. Then

$$\varepsilon = y - \mathbf{a}^{*T}\boldsymbol{x} = y - (\mathbf{a}^\perp)^{*T}\boldsymbol{x} + (\mathbf{a}^\perp - \mathbf{a})^{*T}\boldsymbol{x}$$

or

$$\varepsilon = \varepsilon^\perp + (\mathbf{a}^\perp - \mathbf{a})^{*T}\boldsymbol{x}$$

where $\varepsilon^\perp$ is the error that results from using $\mathbf{a}^\perp$. Now observe that

$$\begin{aligned}
\sigma_\varepsilon^2 = \mathcal{E}\{|\varepsilon|^2\} &= \mathcal{E}\left\{(\varepsilon^\perp + (\mathbf{a}^\perp - \mathbf{a})^{*T}\boldsymbol{x})(\varepsilon^\perp + (\mathbf{a}^\perp - \mathbf{a})^{*T}\boldsymbol{x})^*\right\} \\
&= \mathcal{E}\{|\varepsilon^\perp|^2\} + (\mathbf{a}^\perp - \mathbf{a})^{*T}\mathcal{E}\{\boldsymbol{x}(\varepsilon^\perp)^*\} \\
&\quad + \mathcal{E}\{\varepsilon^\perp\boldsymbol{x}^{*T}\}(\mathbf{a}^\perp - \mathbf{a}) + \mathcal{E}\left\{(|\mathbf{a}^\perp - \mathbf{a})^{*T}\boldsymbol{x}|^2\right\}
\end{aligned}$$

Since the middle terms are zero by assumption, this reduces to

$$\sigma_\varepsilon^2 = \mathcal{E}\{|\varepsilon^\perp|^2\} + \mathcal{E}\left\{|(\mathbf{a}^\perp - \mathbf{a})^{*T}\boldsymbol{x}|^2\right\} \tag{7.4}$$

The expression is therefore minimized when $\mathbf{a} = \mathbf{a}^\perp$ and the minimum value is $\mathcal{E}\{|\varepsilon^\perp|^2\}$. Since the minimum value of σ_ε^2 is equal to $\mathcal{E}\{|\varepsilon^\perp|^2\}$, it follows that

$$(\sigma_\varepsilon^2)_{MIN} = \mathcal{E}\{\varepsilon^\perp(\varepsilon^\perp)^*\} = \mathcal{E}\left\{(y - (\mathbf{a}^\perp)^{*T}\boldsymbol{x})(\varepsilon^\perp)^*\right\} = \mathcal{E}\{y(\varepsilon^\perp)^*\}$$

where the last term again disappears because of the orthogonality condition. This proves both parts of the theorem.

The Orthogonality Theorem can be used to easily derive the previous estimation results. The theorem requires that

$$\mathcal{E}\{\boldsymbol{x}\varepsilon^*\} = \mathcal{E}\{\boldsymbol{x}(y - \mathbf{a}^{*T}\boldsymbol{x})^*\} = \mathcal{E}\{\boldsymbol{x}(y^* - \boldsymbol{x}^{*T}\mathbf{a})\} = \mathbf{0} \tag{7.5}$$

Taking the expectation then yields

$$\mathbf{R}_x \mathbf{a} = \mathbf{r}_{xy} \tag{7.6}$$

where

$$\mathbf{R}_x = \mathcal{E}\{\boldsymbol{x}\boldsymbol{x}^{*T}\} \tag{7.7}$$

and

$$\mathbf{r}_{xy} = \mathcal{E}\{\boldsymbol{x}y^*\} \tag{7.8}$$

The minimum mean-square error follows directly from the second part of the theorem. That is,

$$\sigma_\varepsilon^2 = \mathcal{E}\{y\varepsilon^*\} = \mathcal{E}\{y(y^* - \boldsymbol{x}^{*T}\mathbf{a})\} = \sigma_y^2 - \mathbf{r}_{xy}^{*T}\mathbf{a} \tag{7.9}$$

The results are the same as those in Chapter 6 with the correlation substituted for the covariance.

7.1.2 Vector Space Interpretation

The orthogonality principle and linear mean-square estimation can be interpreted in terms of vectors in an abstract vector space. The "vectors" or elements of the space are not the usual vectors represented by column matrices that were discussed in Chapter 2. Rather, the elements are random variables, as will be seen shortly.

Mathematically, a *vector space* $\mathcal{V}$ is a set of elements $u, v, \ldots$ such that if $u \in \mathcal{V}$ and $v \in \mathcal{V}$, then there is a unique element

$$u + v \ \in \mathcal{V}$$

called the *sum*. Further, if c is an element from an associated field such as the field of real or complex numbers, then the *scalar product*

$$c \cdot u$$

with certain associative and distributive properties, is also an element of $\mathcal{V}$ [1, 2]. A vector space is an inner product space or a *Hilbert space* if an inner product (u, v) between elements is defined.

For purposes of the orthogonality principle a vector space is defined whose elements are the random variables $x_1, x_2, \ldots, x_N$ and y, and linear combinations of these. Addition and the scalar product are defined in the usual way with scalars selected from the field of real or complex numbers. The inner product is defined to be the *expectation* $\mathcal{E}\{uv^*\}$ so that two elements are *orthogonal* if $\mathcal{E}\{uv^*\} = 0$. This explains our previous use of this term.

Figure 7.1 illustrates the principle of orthogonality for $N = 2$. The two observations x_1, x_2 represented by vectors form a two-dimensional subspace (a plane). The vector y is in general not contained in this subspace. However, any estimate $\hat{y}$ which is a linear combination of x_1 and x_2 must be contained in the subspace. Since by definition y is equal to $\hat{y} + \varepsilon$, the vector ε must be drawn as shown. From these geometrical considerations it is clear that the vector ε has minimum length (i.e., the error is minimized) when ε is

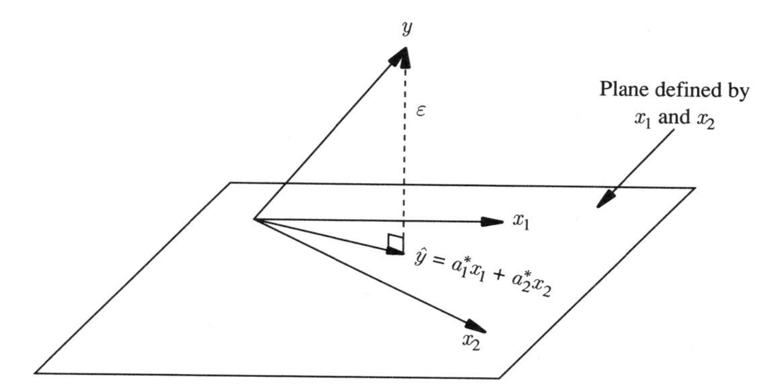

Figure 7.1 Vector space interpretation of linear mean-square estimation.

orthogonal to the subspace defined by x_1 and x_2. Thus it follows that the (minimum) error must be orthogonal to the observations.

To carry the geometrical interpretation further, note that the squared magnitude of the vector is the inner product $(\varepsilon, \varepsilon)$, which is defined as $\mathcal{E}\{\varepsilon\varepsilon^*\} = \sigma_\varepsilon^2$. Since ε and $\hat{y}$ are orthogonal components of y, the inner product of ε with y is the same as the inner product of ε with itself (see Fig. 7.1). Thus the squared magnitude of the error vector is $\sigma_\varepsilon^2 = \mathcal{E}\{y\varepsilon^*\}$, as the Orthogonality Theorem states.

7.2 LINEAR PREDICTIVE FILTERING

Having defined the orthogonality principle, we can now apply it to the estimation of random signals. A particularly convenient place to start is with the problem of linear prediction.

7.2.1 Prediction as an Optimal Filtering Problem

In linear prediction, it is assumed that all the past values of a random sequence x, say up to $x[n-1]$, have been observed, and it is desired to estimate the current value $x[n]$ (which has not yet been observed). There are many possible reasons for wanting to do this. One is to remove redundant information from $x[n]$ as in the coding problem described in Chapter 1. Another is in tracking the position of a target; if $x[n]$ represents the target position, a sensor tracking the target must be able to predict the target's next likely position before it arrives. If it is assumed that only the P previous values of the sequence are used in the estimation, then a correspondence

$$y \leftrightarrow x[n] \qquad x_i \leftrightarrow x[n-i] \tag{7.10}$$

between the variables in the linear prediction problem and those in the general linear mean-square estimation problem can be made (see Fig. 7.2).

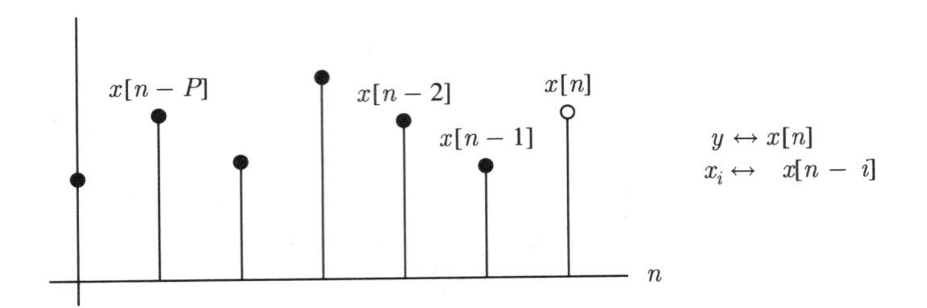

Problem: From P past samples $x[n-1], \bullet\bullet\bullet, x[n-P]$ estimate (predict) $x[n]$.

Figure 7.2 Linear prediction of a random sequence.

The problem can now be stated as follows. Find coefficients $-a_1^*, -a_2^*, \ldots, -a_P^*$ to produce an estimate

$$\hat{x}[n] = -a_1^* x[n-1] - a_2^* x[n-2] - \cdots - a_P^* x[n-P] \tag{7.11}$$

such that if

$$\varepsilon[n] = x[n] - \hat{x}[n] \tag{7.12}$$

then the mean-square error

$$\sigma_\varepsilon^2 = \mathcal{E}\{|\varepsilon[n]|^2\} = \mathcal{E}\{|x[n] - \hat{x}[n]|^2\} \tag{7.13}$$

is minimized. In the problem statement, the coefficients $-a_i^*$ are deliberately defined conjugated and with a minus sign for later convenience. In the context of linear prediction the quantity σ_ε^2 is known as the *prediction error variance* and the parameters $\{a_i\}$ and σ_ε^2 together are called the "linear prediction parameters."

In the usual treatment of linear prediction the random process $x[n]$ is assumed to be wide-sense stationary. The results can easily be generalized to the nonstationary case (see Problem 7.8), but the stationary case is of the most practical interest. In this case the coefficients $\{a_i\}$ are constant (independent of n). Since (7.11) can be written in the equivalent form

$$\hat{x}[n] = \sum_{k=1}^{P} -a_k^* x[n-k] \tag{7.14}$$

it is seen that $\hat{x}[n]$ can be produced as the output of a FIR linear shift-invariant filter with impulse response $h'[k] = -a_k^*,\ k = 1, 2, \ldots, P$. Further, if we *define*

$$a_0 \equiv 1 \tag{7.15}$$

then the error process $\varepsilon[n]$ can be written as

$$\varepsilon[n] = x[n] - \hat{x}[n] = \sum_{k=0}^{P} a_k^* x[n-k] \tag{7.16}$$

which is also the output of an FIR filter with impulse response $h[k] = a_k^*, k = 0, 1, 2, \ldots, P$. This is the the reason for initially defining the coefficients of the estimate in (7.11) with a minus sign. The relation between the two filters, known as the linear predictive filter and the prediction error filter (sometimes abbreviated PEF), is depicted in Fig. 7.3.

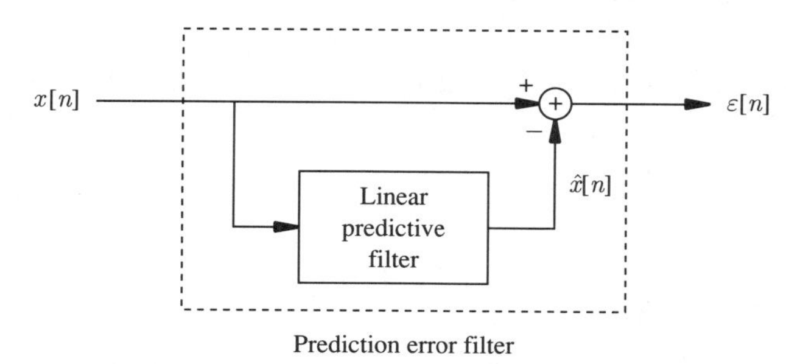

Figure 7.3 Filters involved in linear prediction.

To solve for the linear prediction filter coefficients and the prediction error variance we can invoke the Orthogonality Theorem. For convenience it is restated here in terms of the linear prediction problem.

Theorem 7.2 (Orthogonality: Linear Prediction). Let $\varepsilon[n] = x[n] - \hat{x}[n]$ be the prediction error. Then the prediction error filter with coefficients $1, a_1^*, a_2^*, \ldots, a_P^*$ minimizes the prediction error variance σ_ε^2 if the filter coefficients are chosen such that $\mathcal{E}\{x[n-i]\varepsilon^*[n]\} = 0, \quad i = 1, 2, \ldots, P$, that is, if the error is *orthogonal* to the observations. Further, the (minimum) prediction error variance is given by $\sigma_\varepsilon^2 = \mathcal{E}\{x[n]\varepsilon^*[n]\}$.

The orthogonality principle for linear prediction states that

$$\mathcal{E}\{x[n-i]\varepsilon^*[n]\} = 0; \qquad i = 1, 2, \ldots, P$$

Substituting (7.16) and taking the expectation leads to

$$\sum_{k=0}^{P} a_k \mathcal{E}\{x[n-i]x^*[n-k]\} = 0; \quad i = 1, 2, \ldots, P$$

or

$$\sum_{k=0}^{P} a_k R_x[k-i] = 0; \quad i = 1, 2, \ldots, P \tag{7.17}$$

If the values of the correlation function are known, this represents a system of P linear equations in the P unknowns $a_1, a_2, \ldots, a_P$. For example, for $P = 3$ (7.17) yields the three equations

$$\begin{aligned}
R_x[-1] + a_1 R_x[0] + a_2 R_x[1] + a_3 R_x[2] &= 0 \\
R_x[-2] + a_1 R_x[-1] + a_2 R_x[0] + a_3 R_x[1] &= 0 \\
R_x[-3] + a_1 R_x[-2] + a_2 R_x[-1] + a_3 R_x[0] &= 0
\end{aligned} \tag{7.18}$$

If the terms that do not depend on the a_i are transferred to the right-hand side, then this can be written as a single matrix equation

$$\begin{bmatrix} R_x[0] & R_x[1] & R_x[2] \\ R_x[-1] & R_x[0] & R_x[1] \\ R_x[-2] & R_x[-1] & R_x[0] \end{bmatrix} \begin{bmatrix} a_1 \\ a_2 \\ a_3 \end{bmatrix} = \begin{bmatrix} -R_x[-1] \\ -R_x[-2] \\ -R_x[-3] \end{bmatrix} \tag{7.19}$$

which can be solved formally by inverting the matrix. The general form, which is the matrix equivalent of (7.17), is

$$\begin{bmatrix} R_x[0] & R_x[1] & \cdots & R_x[P-1] \\ R_x[-1] & R_x[0] & \cdots & R_x[P-2] \\ \vdots & \vdots & \cdots & \vdots \\ R_x[-P+1] & R_x[-P+2] & \cdots & R_x[0] \end{bmatrix} \begin{bmatrix} a_1 \\ a_2 \\ \vdots \\ a_P \end{bmatrix} = \begin{bmatrix} -R_x[-1] \\ -R_x[-2] \\ \vdots \\ -R_x[-P] \end{bmatrix} \tag{7.20}$$

and just represents a special case of (7.6) with the correspondences (7.10). The matrix on the left of (7.20) is Toeplitz; but note that it appears to be the transpose of the correlation matrix defined in Chapter 4. In fact, this matrix is the better thought of as the *reversal* rather than the transpose of the other correlation matrix. More is said about this later in this chapter and in Chapter 8.

The prediction error variance can be obtained from the second part of the Orthogonality Theorem, namely

$$\sigma_\varepsilon^2 = \mathcal{E}\{x[n]\varepsilon^*[n]\}$$

Substituting (7.16) in this expression and taking the expectation of the terms under the sum yields

$$\sigma_\varepsilon^2 = \sum_{k=0}^{P} a_k R_x[k] \tag{7.21}$$

which can be easily evaluated once the a_i are known. This expression is a special case of (7.9).

In linear prediction it is convenient to combine the equations for the filter coefficients with the equation for the prediction error variance and write the result as one single matrix equation. To see how this is done, let us write out (7.21) explicitly for $P = 3$. Since $a_0 = 1$ this yields

$$\sigma_\varepsilon^2 = R_x[0] + a_1 R_x[1] + a_2 R_x[2] + a_3 R_x[3]$$

This equation can be combined with the three equations (7.18) to write

$$\begin{bmatrix} R_x[0] & R_x[1] & R_x[2] & R_x[3] \\ R_x[-1] & R_x[0] & R_x[1] & R_x[2] \\ R_x[-2] & R_x[-1] & R_x[0] & R_x[1] \\ R_x[-3] & R_x[-2] & R_x[-1] & R_x[0] \end{bmatrix} \begin{bmatrix} 1 \\ a_1 \\ a_2 \\ a_3 \end{bmatrix} = \begin{bmatrix} \sigma_\varepsilon^2 \\ 0 \\ 0 \\ 0 \end{bmatrix} \tag{7.22}$$

We refer to these as the *Normal equations*[1].

[1]Some authors refer to (7.19) or (7.20) as the Normal equations and to (7.22) as the augmented Normal equations. Since subsequent work with the Normal equations deals primarily with the augmented form, we prefer to drop the term "augmented" and use the present terminology.

From (7.20) and (7.21) it can be seen that the general form of the Normal equations is

$$\begin{bmatrix} R_x[0] & R_x[1] & \cdots & R_x[P] \\ R_x[-1] & R_x[0] & \cdots & R_x[P-1] \\ \vdots & \vdots & \cdots & \vdots \\ R_x[-P] & R_x[-P+1] & \cdots & R_x[0] \end{bmatrix} \begin{bmatrix} 1 \\ a_1 \\ \vdots \\ a_P \end{bmatrix} = \begin{bmatrix} \sigma_\varepsilon^2 \\ 0 \\ \vdots \\ 0 \end{bmatrix} \tag{7.23}$$

This represents a set of $P+1$ equations in $P+1$ unknowns which can either be solved as is, or broken up into the two separate sets of equations for the filter coefficients and the prediction error variance.

Let us conclude this subsection with a simple numerical example that illustrates specifically how the linear prediction parameters are computed from a given correlation function.

Example 7.1

A certain real random process has the exponential correlation function

$$R_x[l] = (0.5)^{|l|}$$

It is desired to compute the coefficients of the second-order linear predictive filter and the corresponding prediction error variance. In this case the required correlation terms are $R_x[0] = 1$, $R_x[1] = R_x[-1] = 0.5$, and $R_x[2] = R_x[-2] = 0.25$. The Normal equations (7.23) have the form

$$\begin{bmatrix} 1 & 0.5 & 0.25 \\ 0.5 & 1 & 0.5 \\ 0.25 & 0.5 & 1 \end{bmatrix} \begin{bmatrix} 1 \\ a_1 \\ a_2 \end{bmatrix} = \begin{bmatrix} \sigma_\varepsilon^2 \\ 0 \\ 0 \end{bmatrix}$$

This can be broken up into a set of equations for the unknown filter coefficients

$$\begin{bmatrix} 1 & 0.5 \\ 0.5 & 1 \end{bmatrix} \begin{bmatrix} a_1 \\ a_2 \end{bmatrix} = \begin{bmatrix} -0.5 \\ -0.25 \end{bmatrix}$$

which has the solution $a_1 = -0.5$ and $a_2 = 0$, a separate equation for the prediction error variance which can then be solved as

$$\begin{aligned} \sigma_\varepsilon^2 &= R_x[0] + a_1 R_x[1] + a_2 R_x[2] \\ &= 1 + (-0.5)(0.5) + (0)(0.25) = 0.75 \end{aligned}$$

The fact that a_2 is zero for this particular correlation function is no coincidence and will be understood clearly when we investigate linear prediction further in Chapter 8.

7.2.2 Whitening Property of the Prediction Error Filter

If the order of the prediction error filter is allowed to increase with each successive observation of the random signal, then the resulting sequence of error terms is orthogonal. To see this, suppose that a Pth-order filter is used to compute $\varepsilon[n]$. Then $\varepsilon[n]$ is orthogonal to $x[n-1], x[n-2], \ldots, x[n-P]$. If now a $(P+1)^{\text{st}}$ order optimal filter is employed to compute $\varepsilon[n+1]$, then $\varepsilon[n+1]$ is orthogonal to the set of terms

$$x[n], x[n-1], x[n-2], \ldots, x[n-P]$$

Since $\varepsilon[n]$ is a linear combination of these exact same terms [see (7.16)], $\varepsilon[n+1]$ is also orthogonal to $\varepsilon[n]$. Thus a growing prediction error filter produces a sequence of orthogonal error terms.

To allow the filter to grow, a *different*-order linear prediction problem has to be solved with each new observation of the random signal. This produces a different set of filter coefficients and a different prediction error variance at each step. Thus the filter that produces the error sequence is no longer shift-invariant and the error sequence (because it has a different variance at each step) is not stationary.

If a fixed-order prediction error filter is used and the order is taken to be reasonably high, then the terms in the error sequence have approximately constant variance and are approximately orthogonal to each other. This approximation can be made as close as desired by chosing a sufficiently high prediction order, and in some cases the approximation is exact. Since the sequence $\varepsilon[n]$ then becomes white noise, the prediction error filter can be thought of as a causal whitening filter. The whitening property is a very important property of the prediction error filter, and a more rigorous discussion of this property is given in Chapter 8.

7.2.3 Vector Space Interpretation of Linear Prediction

Let us consider the random sequence $x[n]$ to represent consecutive samples of some temporal random process. Figure 7.4 gives a vector space interpretation of linear prediction as the random sequence evolves in time. The variables $a_i^{(p)}, \quad p = 1, 2$ are the coefficients for the pth-order linear prediction problem. First a single sample $x[0]$ is observed. This is represented by a single vector drawn in some arbitrary direction. Next a prediction is made and the next sample $x[1]$ is observed. The predicted value $\hat{x}[1]$ is the orthogonal projection of the vector $x[1]$ onto the vector $x[0]$. A linear prediction $\hat{x}[2]$ is then made based on the samples $x[1]$ and $x[0]$. This is the projection of the vector $x[2]$ onto the subspace formed by $x[1]$ and $x[0]$. The foregoing procedure continues as the order of the predictor is allowed to grow with each successive prediction corresponding to a projection of the observed vector on a subspace of reduced dimensionality.

Since each of the error terms is orthogonal to all of the previous observations, each error $\varepsilon[n]$ can be thought of as representing only the "new information" contained in the observation $x[n]$ (i.e., the information that is not already contained in $x[0] \ldots x[n-1]$). As a result, $\varepsilon[n]$ is referred to as the *innovations process* corresponding to the process $x[n]$. It will be seen later that this concept relates to the general innovations representation of random processes introduced in Chapter 5.

One further interpretation of linear prediction is possible. Since the set of error terms produced by increasing orders of linear prediction are mutually orthogonal, they can be used to define a coordinate system for the vector space as shown in Fig. 7.5. Here the orthogonal coordinate system denoted by $\mathrm{u}_0, \mathrm{u}_1, \mathrm{u}_2, \ldots$ is chosen such that $\varepsilon[0]$ $(= x[0])$ lies along the direction of u_0, $\varepsilon[1]$ lies along u_1, and so on. With this geometric interpretation linear prediction can be viewed as a Gram-Schmidt orthogonalization of the observations $x[0], x[1], \ldots$ of the original random signal (see Problem 7.9). This in turn implies that the term $x[n]$ can be formed as a linear combination of the $\varepsilon[i]$. This fact is further explored in Chapter 8 in the study of autoregressive models for random processes.

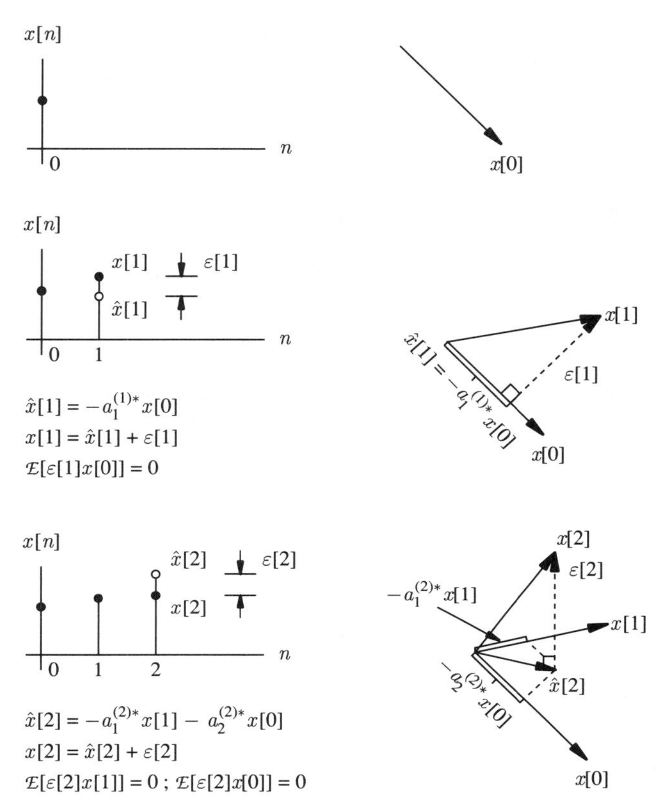

Figure 7.4 Vector space interpretation of linear prediction.

7.3 GENERAL OPTIMAL FILTERING—THE FIR CASE

Linear prediction is just one type of optimal filtering. A more general type of optimal filtering is known as Wiener filtering.[2] The form of Wiener filtering involving a linear FIR filter is considered in this section.

Suppose that a stationary random process $x[n]$ is observed which is somehow related to another jointly stationary random process $d[n]$ that *cannot* be observed directly. The desired random process may, for example, represent a signal which is subject to various forms of distortion and interference. The situation is depicted in Fig. 7.6. The goal of the signal processing is to estimate the sequence $d[n]$ from the observed sequence $x[n]$. Specifically, it is required to use the present value $x[n]$, and the last $P-1$ values, $x[n-1], x[n-2], \ldots, x[n-P+1]$, to estimate the value $d[n]$. This is to be done for all values n by a linear FIR filter designed to minimize the mean-square error.

The sequence $d[n]$ is called the "desired" sequence and can arise in a variety of different contexts. Table 7.1 shows a number of typical problems, the form of the observations, and the definition of the desired sequence. The fourth problem is the linear prediction

[2]After Norbert Wiener (1894–1964), who developed the theory for continuous processes.

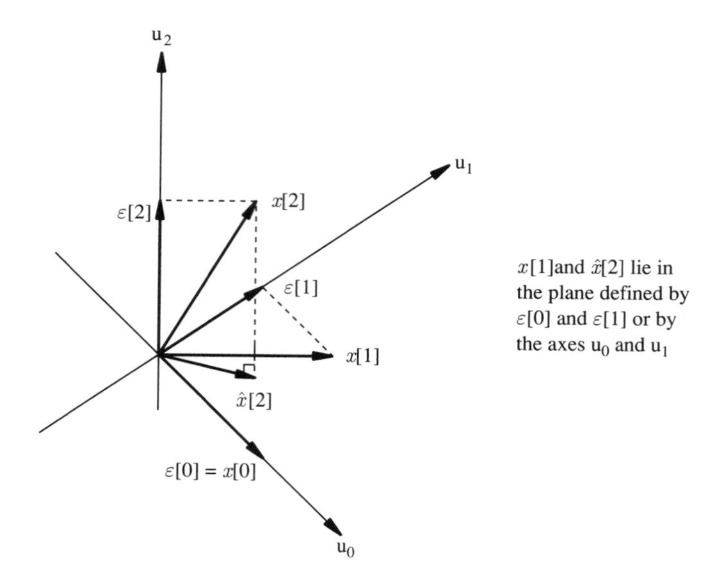

Figure 7.5 Coordinate system defined by the prediction errors.

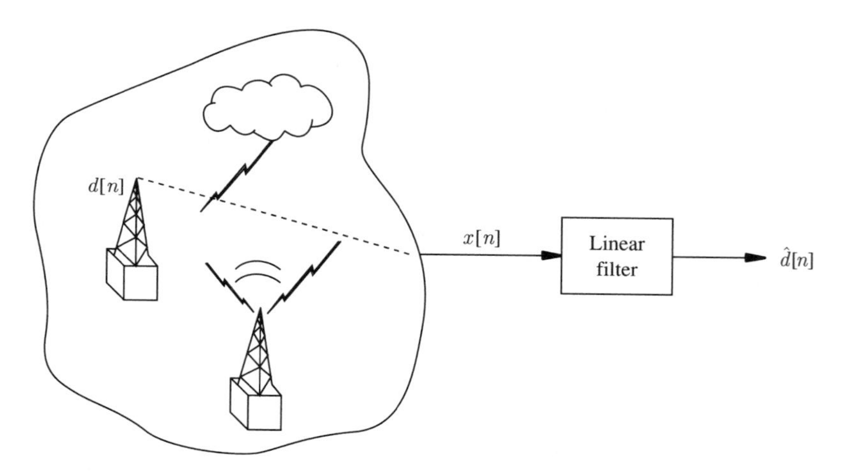

Figure 7.6 General linear signal estimation problem.

problem just treated. The last problem is nonlinear and illustrates that although the *estimate* is linear, the relation between $x[n]$ and $d[n]$ need *not* be linear.

As in all of the previous linear mean-square estimation problems, finding the optimal estimation coefficients involves solving a set of linear equations. The coefficients in this problem correspond to the terms of the impulse response of the optimal filter, and the linear equations are known as Wiener-Hopf equations. These equations are derived in various forms, as follows.

TABLE 7.1 TYPICAL WIENER FILTERING PROBLEMS

Problem	Form of observations	Desired sequence
Filtering of signal in noise	$x[n] = s[n] + \eta[n]$	$d[n] = s[n]$
Prediction of signal in noise	$x[n] = s[n] + \eta[n]$	$d[n] = s[n+p]; \quad p > 0$
Smoothing of signal in noise	$x[n] = s[n] + \eta[n]$	$d[n] = s[n-q]; \quad q > 0$
Linear prediction	$x[n] = s[n-1]$	$d[n] = s[n]$
General nonlinear problem	$x[n] = G(s[n], \eta[n])$	$d[n] = s[n]$

7.3.1 Discrete Wiener-Hopf Equation

To begin, let the estimate $\hat{d}[n]$ be generated by a general linear filter (not necessarily shift-invariant) that operates on $x[n]$ and the $P-1$ past values of the process. The estimate is given by

$$\hat{d}[n] = \sum_{k=n-P+1}^{n} \mathrm{h}[n,k]x[k] = \sum_{l=0}^{P-1} \mathrm{h}[n,n-l]x[n-l] \tag{7.24}$$

where $\mathrm{h}[n,k]$ is the impulse response of the optimal filter, to be determined. Let the error $\varepsilon[n]$ be defined as

$$\varepsilon[n] = d[n] - \hat{d}[n] \tag{7.25}$$

Since this is a problem in linear mean-square estimation, the orthogonality principle applies. Thus if $x[n-i]$ represents any one of the observations, we can write, using (7.24) and (7.25),

$$\mathcal{E}\{x[n-i]\varepsilon^*[n]\} = \mathcal{E}\left\{x[n-i]\left(d^*[n] - \sum_{l=0}^{P-1} \mathrm{h}^*[n,n-l]x^*[n-l]\right)\right\} = 0$$

or

$$\boxed{\sum_{l=0}^{P-1} \mathrm{R}_x[n-i,n-l]\mathrm{h}^*[n,n-l] = \mathrm{R}_{dx}^*[n,n-i]; \quad i = 0,1,\ldots,P-1} \tag{7.26}$$

This is the general discrete Wiener-Hopf equation; it can be solved for the filter coefficients $\{\mathrm{h}[n,k]\}$ by matrix methods. For example, when $P = 3$ the Wiener-Hopf equation can be written in matrix form as

$$\begin{bmatrix} \mathrm{R}_x[n,n] & \mathrm{R}_x[n,n-1] & \mathrm{R}_x[n,n-2] \\ \mathrm{R}_x[n-1,n] & \mathrm{R}_x[n-1,n-1] & \mathrm{R}_x[n-1,n-2] \\ \mathrm{R}_x[n-2,n] & \mathrm{R}_x[n-2,n-1] & \mathrm{R}_x[n-2,n-2] \end{bmatrix} \begin{bmatrix} \mathrm{h}^*[n,n] \\ \mathrm{h}^*[n,n-1] \\ \mathrm{h}^*[n,n-2] \end{bmatrix} = \begin{bmatrix} R_{dx}^*[n,n] \\ R_{dx}^*[n,n-1] \\ R_{dx}^*[n,n-2] \end{bmatrix}$$

This can then be solved for the filter coefficients.

To find the minimum mean-square error, we can apply the second part of the orthogonality theorem to write

$$\sigma_\varepsilon^2[n] = \mathcal{E}\{d[n]\varepsilon^*[n]\} = \mathcal{E}\left\{d[n]\left(d^*[n] - \sum_{l=0}^{P-1} \mathrm{h}^*[n,n-l]x^*[n-l]\right)\right\}$$

or

$$\boxed{\sigma_\varepsilon^2[n] = \mathrm{R}_d[n,n] - \sum_{l=0}^{P-1} \mathrm{h}^*[n,n-l]\mathrm{R}_{dx}[n,n-l]} \tag{7.27}$$

Once (7.26) has been solved for $\{\mathrm{h}[n,k]\}$, $\sigma_\varepsilon^2[n]$ can be computed directly from (7.27).

Since the solution to the general optimal filtering problem involves a filter which is not shift-invariant, (7.26) has to be solved in principle for each value of n. One application of this general case is explored in Section 7.3.3. In practice the case that proves most useful is when the random processes are stationary. Since in this case the statistical properties of the random processes do not change with n, the filter that is required is also shift-invariant and the estimate (7.24) can be written in the form

$$\hat{d}[n] = \sum_{k=n-P+1}^{n} h[n-k]x[k] = \sum_{l=0}^{P-1} h[l]x[n-l] \tag{7.28}$$

where $h[n]$ is the impulse response sequence of the shift-invariant FIR filter, to be determined. The Wiener-Hopf equation (7.26) becomes

$$\boxed{\sum_{l=0}^{P-1} R_x[l-i]h^*[l] = R_{dx}^*[i]; \quad i = 0, 1, \ldots, P-1} \tag{7.29}$$

When these equations are written in matrix form the correlation matrix is Toeplitz. For the example of $P = 3$ considered previously the equations are

$$\begin{bmatrix} R_x[0] & R_x[1] & R_x[2] \\ R_x[-1] & R_x[0] & R_x[1] \\ R_x[-2] & R_x[-1] & R_x[0] \end{bmatrix} \begin{bmatrix} h^*[0] \\ h^*[1] \\ h^*[2] \end{bmatrix} = \begin{bmatrix} R_{dx}^*[0] \\ R_{dx}^*[1] \\ R_{dx}^*[2] \end{bmatrix} \tag{7.30}$$

which by conjugating can be written simply as

$$\begin{bmatrix} R_x[0] & R_x[-1] & R_x[-2] \\ R_x[1] & R_x[0] & R_x[-1] \\ R_x[2] & R_x[1] & R_x[0] \end{bmatrix} \begin{bmatrix} h[0] \\ h[1] \\ h[2] \end{bmatrix} = \begin{bmatrix} R_{dx}[0] \\ R_{dx}[1] \\ R_{dx}[2] \end{bmatrix} \tag{7.31}$$

As long as the correlation function R_x is (strictly) positive definite, the correlation matrix is nonsingular, so the linear equations have a unique solution. The minimum mean-square error (7.27) is constant and takes the form

$$\boxed{\sigma_\varepsilon^2 = R_d[0] - \sum_{l=0}^{P-1} h^*[l]R_{dx}[l]} \tag{7.32}$$

The procedure is illustrated in the following example.

Example 7.2

A real-valued random signal is observed in white noise which is uncorrelated with the signal. The observed sequence is given by

$$x[n] = s[n] + \eta[n]$$

where

$$R_s[l] = 2(0.8)^{|l|}; \qquad R_\eta[l] = 2\delta[l]$$

It is desired to estimate $s[n]$ using the present and previous two observations.

This problem corresponds to the first row in Table 7.1, the filtering of a signal in noise. For this problem $d[n] = s[n]$, consequently,

$$R_d[l] = R_s[l] = 2(0.8)^{|l|}$$

$$\begin{aligned} R_{dx}[l] &= \mathcal{E}\{s[n]x[n-l]\} = \mathcal{E}\{s[n](s[n-l] + \eta[n-l])\} \\ &= R_s[l] = 2(0.8)^{|l|} \end{aligned}$$

and

$$\begin{aligned} R_x[l] &= \mathcal{E}\{(s[n] + \eta[n])(s[n-l] + \eta[n-l])\} \\ &= R_s[l] + R_\eta[l] = 2(0.8)^{|l|} + 2\delta[l] \end{aligned}$$

Then evaluating these expressions for the needed terms yields

$$R_d[0] = 2$$

$$R_x[0] = 4.00 \qquad R_x[1] = R_x[-1] = 1.60 \qquad R_x[2] = R_x[-2] = 1.28$$

$$R_{dx}[0] = 2.00 \qquad R_{dx}[1] = 1.60 \qquad R_{dx}[2] = 1.28$$

The Wiener-Hopf equation (7.29) or (7.30) thus becomes

$$\begin{bmatrix} 4.00 & 1.60 & 1.28 \\ 1.60 & 4.00 & 1.60 \\ 1.28 & 1.60 & 4.00 \end{bmatrix} \begin{bmatrix} h[0] \\ h[1] \\ h[2] \end{bmatrix} = \begin{bmatrix} 2.00 \\ 1.60 \\ 1.28 \end{bmatrix}$$

which can be solved to find

$$h[0] = 0.3824 \qquad h[1] = 0.2000 \qquad h[2] = 0.1176$$

The mean-square error, from (7.32), is

$$\sigma_\varepsilon^2 = 2.000 - \sum_{l=0}^{2} h[l] \cdot 2(0.8)^{|l|} = 0.7647$$

A slight modification of this example illustrates the problem of *prediction* for a random signal in noise.

Example 7.3

Let us consider the problem of estimating the sequence in Example 7.2 two points ahead, using the same three observations. This corresponds to the prediction problem listed as the second line in Table 7.1. Note that it is not the same as the problem of "linear prediction," which is represented in the fourth line of the table. (Be sure you see clearly what the difference is.) For this problem we have $d[n] = s[n+2]$ and

$$\begin{aligned} R_{dx}[l] &= \mathcal{E}\{s[n+2]x[n-l]\} = \mathcal{E}\{s[n+2](s[n-l] + \eta[n-l])\} \\ &= R_s[l+2] = 2(0.8)^{|l+2|} \end{aligned}$$

which yields the values

$$R_{dx}[0] = 1.280 \qquad R_{dx}[1] = 1.024 \qquad R_{dx}[2] = 0.8192$$

$R_d[l]$ and $R_x[l]$ are found to be the same as before. The Wiener-Hopf equation (7.30) becomes

$$\begin{bmatrix} 4.00 & 1.60 & 1.28 \\ 1.60 & 4.00 & 1.60 \\ 1.28 & 1.60 & 4.00 \end{bmatrix} \begin{bmatrix} h[0] \\ h[1] \\ h[2] \end{bmatrix} = \begin{bmatrix} 1.280 \\ 1.024 \\ 0.8192 \end{bmatrix}$$

and solving for the filter terms yields

$$h[0] = 0.2447 \qquad h[1] = 0.1280 \qquad h[2] = 0.07529$$

The mean-square error in this case is

$$\sigma_\varepsilon^2 = 2.000 - \sum_{l=0}^{2} h[l] \cdot 2(0.8)^{|l+2|} = 1.494$$

This is larger than the mean-square error in Example 7.2 by about a factor of 2. This indicates that prediction of the signal two points ahead is considerably less accurate than estimation of its current value.

7.3.2 Matrix Form of the Wiener-Hopf Equation

The Wiener-Hopf equation can be derived directly in matrix form by writing (7.24) as

$$\hat{d}[n] = (\tilde{\boldsymbol{x}}[n])^T \mathbf{h} \tag{7.33}$$

where if

$$\boldsymbol{x}[n] \stackrel{\text{def}}{=} \begin{bmatrix} x[n-P+1] \\ x[n-P] \\ \vdots \\ x[n] \end{bmatrix} \tag{7.34}$$

then

$$\tilde{\boldsymbol{x}}[n] = \begin{bmatrix} x[n] \\ x[n-1] \\ \vdots \\ x[n-P+1] \end{bmatrix}$$

and where $\mathbf{h}$ is defined as

$$\mathbf{h} = \begin{bmatrix} \mathrm{h}[n,n] \\ \mathrm{h}[n,n-1] \\ \vdots \\ \mathrm{h}[n,n-P+1] \end{bmatrix} \tag{7.35}$$

in the general nonstationary case and as

$$\mathbf{h} = \begin{bmatrix} h[0] \\ h[1] \\ \vdots \\ h[P-1] \end{bmatrix} \tag{7.36}$$

in the stationary case. Applying the orthogonality principle yields

$$\mathcal{E}\{\tilde{\boldsymbol{x}}[n]\varepsilon^*[n]\} = \mathcal{E}\{\tilde{\boldsymbol{x}}[n](d^*[n] - (\tilde{\boldsymbol{x}}[n])^{*T}\mathbf{h}^*)\} = \mathbf{0}$$

or

$$\boxed{\tilde{\mathbf{R}}_x \mathbf{h}^* = \tilde{\mathbf{r}}_{dx}^*} \tag{7.37}$$

where

$$\mathbf{R}_x = \mathcal{E}\{\boldsymbol{x}[n]\boldsymbol{x}^{*T}[n]\} \tag{7.38}$$

and

$$\mathbf{r}_{dx} = \mathcal{E}\{d[n](\boldsymbol{x}[n])^*\} \tag{7.39}$$

Equation 7.37 is the matrix form of the Wiener-Hopf equation. In the stationary case $\tilde{\mathbf{R}}_x^* = \mathbf{R}_x$, so the Wiener-Hopf equation can be written as

$$\mathbf{R}_x \mathbf{h} = \tilde{\mathbf{r}}_{dx} \tag{7.40}$$

The minimum mean-square error can be evaluated from the second part of the orthogonality theorem as

$$\sigma_\varepsilon^2 = \mathcal{E}\{d[n]\varepsilon^*[n]\} = \mathcal{E}\{d[n](d^*[n] - (\tilde{\boldsymbol{x}}[n])^{*T}\mathbf{h}^*)\}$$

or

$$\boxed{\sigma_\varepsilon^2 = R_d[0] - \tilde{\mathbf{r}}_{dx}^T \mathbf{h}^* = R_d[0] - \mathbf{h}^{*T}\tilde{\mathbf{r}}_{dx}} \tag{7.41}$$

Equations 7.37 and 7.41 can be combined to form the augmented Wiener-Hopf equation

$$\begin{bmatrix} R_d[0] & | & \tilde{\mathbf{r}}_{dx}^T \\ --- & | & -- \; --- \\ \tilde{\mathbf{r}}_{dx}^* & | & \tilde{\mathbf{R}}_x \end{bmatrix} \begin{bmatrix} 1 \\ -- \\ -\mathbf{h}^* \end{bmatrix} = \begin{bmatrix} \sigma_\varepsilon^2 \\ -- \\ \mathbf{0} \end{bmatrix} \tag{7.42}$$

or

$$\begin{bmatrix} R_d[0] & | & \tilde{\mathbf{r}}_{dx}^{*T} \\ --- & | & -- \; --- \\ \tilde{\mathbf{r}}_{dx} & | & \tilde{\mathbf{R}}_x^* \end{bmatrix} \begin{bmatrix} 1 \\ -- \\ -\mathbf{h} \end{bmatrix} = \begin{bmatrix} \sigma_\varepsilon^2 \\ -- \\ \mathbf{0} \end{bmatrix} \tag{7.43}$$

In this rearrangement the equations are similar in form to the Normal equations of linear prediction.

7.3.3 Filters of Increasing Order

The discussion involving a shift-invariant filter in the previous subsections assumes that a steady-state condition exists for the problem. That is, it assumes that the processes have started in the infinitely distant past and that there are always enough observations to perform the filtering. A slightly different situation exists if the process is observed beginning at some initial time, say $n = 0$, and it is desired to use whatever number of observations are currently available to perform the filtering. In other words, $d[0]$ is estimated using just the observation $x[0]$; $d[1]$ is estimated using the observations $x[0]$ and $x[1]$; and finally, $d[n]$ is estimated using all of the observations $x[0], x[1], \ldots, x[n]$. This problem requires a filter that changes with each new observation, or equivalently, an entire *set* of shift-invariant filters and has a mathematically interesting relation to the triangular decomposition of the correlation matrix for the observations [3]. The errors generated by this procedure are mutually orthogonal and thus form a (nonstationary) white noise process.

Let us now consider this problem in detail. Let $d[0], d[1], \ldots, d[n]$ represent the entire set of values of the desired sequence to be estimated. If a causal filter is used throughout, then from (7.33), the estimate $\hat{d}[k]$ of the desired signal at point k is given by

$$\hat{d}[k] = \begin{bmatrix} x[k] & x[k-1] & \cdots & x[0] \end{bmatrix} \begin{bmatrix} h[k,k] \\ h[k,k-1] \\ \vdots \\ h[k,0] \end{bmatrix}$$

Combining the set of these equations for $0 \leq k \leq n$ produces the single matrix equation

$$\begin{bmatrix} \hat{d}[n] & \hat{d}[n-1] & \cdots & \hat{d}[0] \end{bmatrix}$$

$$= \begin{bmatrix} x[n] & x[n-1] & \cdots & x[0] \end{bmatrix} \begin{bmatrix} h[n,n] & 0 & \cdots & 0 \\ h[n,n-1] & h[n-1,n-1] & \cdots & 0 \\ \vdots & \vdots & & \vdots \\ h[n,0] & h[n-1,0] & \cdots & h[0,0] \end{bmatrix}$$

This last equation can be written as

$$(\tilde{\hat{\mathbf{d}}})^T = (\tilde{\boldsymbol{x}}')^T \mathbf{H} \tag{7.44}$$

where

$$\mathbf{H} = \begin{bmatrix} h[n,n] & 0 & \cdots & 0 \\ h[n,n-1] & h[n-1,n-1] & \cdots & 0 \\ \vdots & \vdots & \vdots & \vdots \\ h[n,0] & h[n-1,0] & \cdots & h[0,0] \end{bmatrix} \tag{7.45}$$

$$\hat{\mathbf{d}} = \begin{bmatrix} \hat{d}[0] \\ \hat{d}[1] \\ \vdots \\ \hat{d}[n] \end{bmatrix} \tag{7.46}$$

and

$$\boldsymbol{x}' = \begin{bmatrix} x[0] \\ x[1] \\ \vdots \\ x[n] \end{bmatrix} \tag{7.47}$$

The product of $\tilde{\boldsymbol{x}}'^T$ with each column of **H** can be thought of as the convolution of the data with the impulse response of the kth-order filter. Now further define

$$\mathbf{d} = \begin{bmatrix} d[0] \\ d[1] \\ \vdots \\ d[n] \end{bmatrix} \tag{7.48}$$

and

$$\boldsymbol{\varepsilon} = \begin{bmatrix} \varepsilon[0] \\ \varepsilon[1] \\ \vdots \\ \varepsilon[n] \end{bmatrix} = \mathbf{d} - \hat{\mathbf{d}} \tag{7.49}$$

and define the cross-correlation matrix

$$\mathbf{R}_{x'\varepsilon} = \mathcal{E}\{\boldsymbol{x}'\boldsymbol{\varepsilon}^{*T}\} \tag{7.50}$$

Since the orthogonality principle states that $\mathcal{E}\{x[i]\varepsilon^*[k]\} = 0$ for $i \leq k$, this matrix is *strictly lower triangular*. (By "strictly" it is meant that all elements *on* the main diagonal are also zero.) Then from (7.44), (7.49), and (7.50) it follows that the *reversed* matrix is

$$\tilde{\mathbf{R}}_{x'\varepsilon} = \mathcal{E}\{\tilde{\boldsymbol{x}}'\tilde{\boldsymbol{\varepsilon}}^{*T}\} = \mathcal{E}\{\tilde{\boldsymbol{x}}'(\tilde{\mathbf{d}}^{*T} - (\tilde{\boldsymbol{x}}')^{*T}\mathbf{H}^*)\} = \tilde{\mathbf{R}}_{x'\mathbf{d}} - \tilde{\mathbf{R}}_{x'}\mathbf{H}^* \tag{7.51}$$

which is strictly *upper* triangular.

Now recall from Chapter 2 (see Table 2.7) that $\tilde{\mathbf{R}}_{x'}$ can be written in upper–lower triangular form as

$$\tilde{\mathbf{R}}_{x'} = \tilde{\mathbf{L}}\tilde{\mathbf{D}}_L\tilde{\mathbf{L}}^{*T} \tag{7.52}$$

where $\tilde{\mathbf{L}}$ is unit *upper* triangular and $\tilde{\mathbf{D}}_L$ is diagonal.[3] Substituting this last equation in (7.51) yields

$$\tilde{\mathbf{R}}_{x'\varepsilon} = \tilde{\mathbf{R}}_{x'\mathbf{d}} - \tilde{\mathbf{L}}\tilde{\mathbf{D}}_L\tilde{\mathbf{L}}^{*T}\mathbf{H}^*$$

and premultiplying by $\tilde{\mathbf{L}}^{-1}$, which is also unit upper triangular, results in

$$\tilde{\mathbf{L}}^{-1}\tilde{\mathbf{R}}_{x'\varepsilon} = \tilde{\mathbf{L}}^{-1}\tilde{\mathbf{R}}_{x'\mathbf{d}} - \tilde{\mathbf{D}}_L\tilde{\mathbf{L}}^{*T}\mathbf{H}^* \tag{7.53}$$

Now since $\tilde{\mathbf{L}}^{-1}$ is upper triangular, both sides of this equation are still strictly upper triangular. It follows that the lower triangular parts (i.e., all terms including those on the diagonal) of both terms on the right-hand side of (7.53) must sum to zero. If the lower triangular part of a matrix $\mathbf{A}$ is denoted by the expression $[\mathbf{A}]_+$, then

$$\left[\tilde{\mathbf{L}}^{-1}\tilde{\mathbf{R}}_{x'\mathbf{d}}\right]_+ - \left[\tilde{\mathbf{D}}_L\tilde{\mathbf{L}}^{*T}\mathbf{H}^*\right]_+ = [\mathbf{0}] \tag{7.54}$$

Finally, observe that since $\mathbf{H}$ was constructed to be lower triangular, the second term in this equation is already lower triangular, and we can remove the $[\ ]_+$. Then solving this equation for $\mathbf{H}$ yields

$$\mathbf{H} = \left((\tilde{\mathbf{L}}^{-1})^{*T}\tilde{\mathbf{D}}_L^{-1}\left[\tilde{\mathbf{L}}^{-1}\tilde{\mathbf{R}}_{x'\mathbf{d}}\right]_+\right)^* \tag{7.55}$$

It is seen that the solution for the causal Wiener filter involves a triangular decomposition of the correlation matrix $\tilde{\mathbf{R}}_{x'}$ into factors representing causal and anticausal transformations and then a construction of the solution from these separate parts. In the next section it is shown that the solution for the causal IIR Wiener filter also involves such a separation, except that the separation is done in the frequency domain. In that way there is a close analogy between the signal-domain and frequency-domain methods.

Before ending this section, let us just examine what happens if the causal constraint is removed from the Wiener filter, that is, if $\mathbf{H}$ is allowed to be a full matrix. Since in this case, the error is orthogonal to *all* the observations, past, present, and future, *all* of the terms in the cross-correlation matrix $\tilde{\mathbf{R}}_{x'\varepsilon}$ are equal to zero. Since $\tilde{\mathbf{R}}_{x'\varepsilon}$ is identically zero, it follows from (7.51) that

$$\mathbf{H} = \left(\tilde{\mathbf{R}}_{x'}^{-1}\tilde{\mathbf{R}}_{x'\mathbf{d}}\right)^* \tag{7.56}$$

Formally, this is the same as what would be obtained by removing the $[\ ]_+$ operation from (7.55).

7.4 GENERAL OPTIMAL FILTERING—THE IIR CASE

In many cases it is more desirable to use an IIR (recursive) filter to perform a given linear operation than to use an FIR filter. The main advantage of IIR filters is that there are fewer parameters in the design; these few parameters specify $h[n]$ for $0 \le n < \infty$. This section develops the procedures for solving the optimal filtering problem with a causal IIR filter.

[3]Recall that since $\tilde{\mathbf{L}}$ is the reversal of a lower triangular matrix $\mathbf{L}$, it is therefore upper triangular.

The preceding section showed how to solve for optimal filters of the general form

$$\hat{d}[n] = \sum_{k=n-P+1}^{n} \mathrm{h}[n,k]x[k]$$

A first thought may be to consider the previous results in the limit as $P \to \infty$ if a causal filter is desired or as both P and the upper limit of the sum approach infinity if a noncausal solution is allowed. However, this procedure can be dismissed quickly since it would lead to an infinite-dimensional matrix equation for the optimal filter. Even if it were possible to solve this infinite system, the result would not be desirable since it would provide the impulse response of the filter as an infinite number of individual terms rather than as a simple closed-form expression. Fortunately, it is possible to solve the problem efficiently by an alternative approach if the filter is restricted to be *shift-invariant*. This section considers the problem and develops solutions for both the causal and noncausal cases.

7.4.1 The Causal Wiener Filter

Let us look at the problem of developing an estimate for $d[n]$ given observations $\{x[k], -\infty < k \leq n\}$ of a stationary random process. If $h[n]$ represents the impulse response of the desired causal filter, then the estimate has the form

$$\hat{d}[n] = \sum_{k=-\infty}^{n} h[n-k]x[k] \tag{7.57}$$

By applying the orthogonality principle and repeating the steps that led to (7.29) and (7.32), it is easy to show that the filter satisfies the (infinite-order) Wiener-Hopf equation

$$\sum_{l=0}^{\infty} R_x[l-i]h^*[l] = R^*_{dx}[i]; \quad 0 \leq i < \infty \tag{7.58}$$

and that the mean-square error is given by

$$\sigma_\varepsilon^2 = R_d[0] - \sum_{l=0}^{\infty} h^*[l]R_{dx}[l] \tag{7.59}$$

The errors in the estimate are a sequence of mutually orthogonal random variables with constant variance and therefore constitute a stationary *white noise process*. The problem is to solve (7.58) by some method that leads to a closed-form expression for $h[n]$. Then with this in hand the mean-square error can be found in principle from (7.59).

Since the approach to this problem is rather lengthy and involves several steps, let us give a brief preview of the solution. First, we observe that h was defined to be a *causal* filter, and it is just that property which makes the problem more difficult. The noncausal problem, considered in the next subsection, can be solved very easily by transform methods. Wiener [4] originally solved the continuous version of this causal problem by spectral factorization methods, and that is exactly what will be done here.[4] The key to the solution is to observe that if $R_x[l] = \sigma_o^2\delta[l]$ (i.e., if x is a white noise process), then substituting this correlation function into (7.58) produces

$$\sum_{l=0}^{\infty} \sigma_o^2 \delta[l-i]h^*[l] = R_{dx}^*[i]; \quad 0 \leq i < \infty$$

and leads to the simple solution

$$h[l] = \begin{cases} \frac{1}{\sigma_o^2} R_{dx}[l] & l \geq 0 \\ 0 & l < 0 \end{cases}$$

The procedure then is to first whiten the random process x and then to solve for the filter as above. This approach is now considered in detail.

Solution of the Wiener-Hopf Equation. Begin by representing the optimal filter as a cascade of two filters g and h' as shown in Fig. 7.7. Since h is to be causal, both g and h' must also be causal. It will be shown (because it is not obvious) that if g is known *and has a causal inverse,* then the problem of finding the optimal overall filter h can be reduced to the problem of finding an optimal filter h'. With this result in hand, we then show how to choose a whitening filter g that has the desired properties, so that finding the optimal h' is almost trivial.

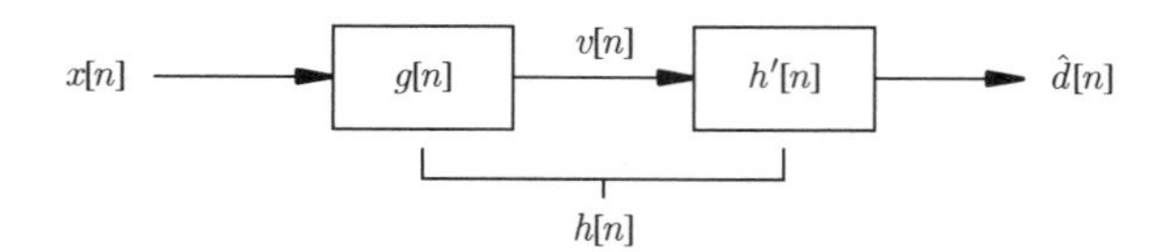

Figure 7.7 Representation of the optimal IIR filter as a cascade of two simpler filters.

For now, assume that the filter g is known and has a causal inverse as stated. We can show that if h' is the optimal filter for estimating $d[n]$ from the intermediate sequence $v[n]$, then h is the optimal filter for estimating $d[n]$ from $x[n]$. To do this, define the estimation error to be

$$\varepsilon[n] = d[n] - \hat{d}[n]$$

as before. Then if h' is the optimal filter, the error satisfies the orthogonality condition

$$\mathcal{E}\{v[n-i]\varepsilon^*[n]\} = 0; \qquad i = 0, 1, 2, \ldots \tag{7.60}$$

Now $x[n]$ can be represented as the convolution of the inverse filter g^{-1} with $v[n]$. Since g^{-1} is assumed to be causal, this convolution takes the form

$$x[n] = \sum_{k=-\infty}^{n} g^{-1}[n-k]v[k] \tag{7.61}$$

[4]Wiener's report, originally bound in a yellow cover with limited distribution, was termed the "yellow peril" by engineers of the day. Little did they know of the perilous work to follow by others working in the fields of statistics, information theory, and modern control theory!

Then from (7.60) and (7.61)

$$\mathcal{E}\{x[n-i]\varepsilon^*[n]\} = \sum_{k=-\infty}^{n-i} g^{-1}[n-i-k]\mathcal{E}\{v[k]\varepsilon^*[n]\} = 0; \qquad i = 0, 1, 2, \ldots \quad (7.62)$$

which follows since all the expectation terms on the right of (7.62) are zero. Since $\varepsilon[n]$ is thus orthogonal to $x[n-i]$, $i = 0, 1, 2, \ldots$, the overall filter $h = g * h'$ is optimal for estimating $d[n]$ from $x[n]$. This proves the original contention.

Now suppose that g is chosen so that $v[n]$ is a white noise process with average power σ_v^2. It follows that

$$R_v[l] = \sigma_v^2[l]$$

and the optimal filter h' satisfies the Wiener-Hopf equation

$$\sum_{k=0}^{\infty} R_v[l-i]h'^*[l] = R_{dv}^*[i]; \quad 0 \leq i < \infty \qquad (7.63)$$

which we previously observed has the solution

$$h'[l] = \begin{cases} \frac{1}{\sigma_v^2} R_{dv}[l] & l \geq 0 \\ 0 & l < 0 \end{cases} \qquad (7.64)$$

Since x and v are related by the filter g, the required cross-correlation function is given by

$$R_{vd}[l] = g[l] * R_{xd}[l]$$

(see Section 5.1.3) or

$$R_{dv}[l] = R_{dx}[l] * g^*[-l] \qquad (7.65)$$

All that is needed now is to identify the causal and causally invertible whitening filter g. This is easily done, however, since it is known from Section 5.5 that the complex spectral density function $S_x(z)$ can be factored as

$$S_x(z) = \mathcal{K}_o H_{ca}(z) H_{ca}^*(1/z^*) \qquad (7.66)$$

and that the filter

$$G(z) = H_{ca}^{-1}(z) \qquad (7.67)$$

has just the required properties. The corresponding white noise variance is given by

$$\sigma_v^2 = \mathcal{K}_o \qquad (7.68)$$

A complete solution to the problem can now be formulated. It follows from (7.65) that

$$S_{dv}(z) = S_{dx}(z) \cdot G^*(1/z^*)$$

or, since $G(z)$ is given by (7.67),

$$S_{dv}(z) = \frac{S_{dx}(z)}{H^*_{ca}(1/z*)} \tag{7.69}$$

Then from (7.64) and (7.68) the filter $H'(z)$ can be expressed as

$$H'(z) = \frac{1}{\mathcal{K}_o}\left[\frac{S_{dx}(z)}{H^*_{ca}(1/z^*)}\right]_+ \tag{7.70}$$

where the notation $[\ ]_+$ means that the $h'[l]$ is computed from this z-transform by taking only the part for $l \geq 0$ as required in (7.64). Then since $H(z) = G(z)H'(z)$ with G and H' given by (7.67) and (7.70), the final solution to the problem is

$$\boxed{H(z) = \frac{1}{\mathcal{K}_o H_{ca}(z)}\left[\frac{S_{dx}(z)}{H^*_{ca}(1/z^*)}\right]_+} \tag{7.71}$$

The solution of the IIR Wiener filtering problem consists of several steps but is actually quite straightforward to apply. The procedure is illustrated in the following example.

Example 7.4

It is again desired to estimate the value of a real signal in white noise which is uncorrelated with the signal. The observed sequence is given by

$$x[n] = s[n] + \eta[n]$$

where

$$R_s[l] = 2(0.8)^{|l|}; \qquad R_\eta[l] = 2\delta[l]$$

and

$$\mathcal{E}\{s[n]\eta^*[n-l]\} = 0$$

for all values of n and l. The needed correlation functions are given, as in Example 7.2, by

$$R_{dx}[l] = R_s[l] = 2(0.8)^{|l|}$$

$$R_x[l] = R_s[l] + R_\eta[l] = 2(0.8)^{|l|} + 2\delta[l]$$

The complex cross-spectral density $S_{dx}(z)$ is the z-transform of $R_{dx}[l]$ and has the form (see Table 4.2)

$$S_{dx}(z) = S_s(z) = \frac{2(1-(0.8)^2)}{(1-0.8z^{-1})(1-0.8z)} = \frac{0.72}{(1-0.8z^{-1})(1-0.8z)}$$

The complex spectral density function for the observation sequence is

$$S_x(z) = S_s(z) + S_\eta(z) = \frac{0.72}{(1-0.8z^{-1})(1-0.8z)} + 2$$

$$= \frac{-1.6z + 4.0 - 1.6z^{-1}}{(1-0.8z^{-1})(1-0.8z)} = 3.2 \cdot \frac{1-0.5z^{-1}}{1-0.8z^{-1}} \cdot \frac{1-0.5z}{1-0.8z}$$

From this it is seen that

$$\mathcal{K}_o = 3.2 \quad \text{and} \quad H_{ca}(z) = \frac{1 - 0.5z^{-1}}{1 - 0.8z^{-1}}$$

Since all of the random processes are real, we form

$$\frac{S_{dx}(z)}{H_{ca}(z^{-1})} = \frac{0.72}{(1 - 0.8z^{-1})(1 - 0.8z)} \cdot \frac{1 - 0.8z}{1 - 0.5z} = \frac{0.72}{(1 - 0.8z^{-1})(1 - 0.5z)}$$

Now express this all in terms of z^{-1}, and expand by partial fractions to obtain

$$\frac{S_{dx}(z)}{H_{ca}(z^{-1})} = \frac{-1.44z^{-1}}{(1 - 0.8z^{-1})(1 - 2z^{-1})} = \frac{1.2}{1 - 0.8z^{-1}} - \frac{1.2}{1 - 2z^{-1}}$$

Since the second term corresponds to the noncausal part of the response, it is dropped to obtain

$$\left[\frac{S_{dx}(z)}{H_{ca}(z^{-1})}\right]_+ = \frac{1.2}{1 - 0.8z^{-1}}$$

Finally, putting this all together produces

$$\begin{aligned} H(z) &= \frac{1}{\mathcal{K}_o H_{ca}(z)} \left[\frac{S_{dx}(z)}{H_{ca}(z^{-1})}\right]_+ \\ &= \frac{1}{3.2} \cdot \frac{1 - 0.8z^{-1}}{1 - 0.5z^{-1}} \cdot \frac{1.2}{1 - 0.8z^{-1}} = \frac{0.375}{1 - 0.5z^{-1}} \end{aligned}$$

Evaluation of the Mean-Square Error. Once the filter impulse response is known, one can in principle find the mean-square error from (7.59). Since this equation may be difficult to evaluate in many cases, however, it is worthwhile to explore other methods for computing the mean-square error in the frequency or transform (z) domain.

The seemingly most straightforward way to evaluate the mean-square error in the frequency domain would be to use the relation

$$\sigma_\varepsilon^2 = R_\varepsilon[0] = \frac{1}{2\pi} \int_{-\infty}^{\infty} S_\varepsilon(e^{\jmath\omega}) d\omega \tag{7.72}$$

but since $S_\varepsilon(z)$ is frequently a ratio of two polynomials, this formula is not easy to use in practice. A better approach is to use contour integration. The z-transform inversion formula states that

$$R_\varepsilon[l] = \frac{1}{2\pi\jmath} \oint S_\varepsilon(z) z^{l-1} dz$$

Therefore, σ_ε^2 can be computed from

$$\sigma_\varepsilon^2 = R_\varepsilon[0] = \frac{1}{2\pi\jmath} \oint S_\varepsilon(z) z^{-1} dz \tag{7.73}$$

The foregoing methods require prior calculation of $S_\varepsilon(z)$. An alternative method, which is less obvious but which often produces a simple and more direct solution, is as follows. Consider the cross-correlation function

$$R_{d\varepsilon}[l] = \mathcal{E}\{d[n]\varepsilon^*[n-l]\} \tag{7.74}$$

and note that by the orthogonality principle

$$\sigma_\varepsilon^2 = \mathcal{E}\{d[n]\varepsilon^*[n]\} = R_{d\varepsilon}[0] \tag{7.75}$$

The cross-correlation function $R_{d\varepsilon}[l]$ can be computed as follows:

$$\mathcal{E}\left\{d[n](d[n-l] - \hat{d}[n-l])^*\right\} = \mathcal{E}\{d[n]d^*[n-l]\} - \mathcal{E}\left\{d[n]\hat{d}^*[n-l]\right\}$$

The first term in this expression is $R_d[l]$; the second term is $R_{d\hat{d}}[l]$, which can be found in a manner identical to the procedure that led to (7.65). In particular, since $\hat{d}[n] = x[n] * h[n]$, the second term in the expression above is equal to $R_{dx}[l] * h^*[-l]$. Thus it follows that

$$R_{d\varepsilon}[l] = R_d[l] - R_{dx}[l] * h^*[-l] \tag{7.76}$$

and correspondingly,

$$S_{d\varepsilon}(z) = S_d(z) - S_{dx}(z)H^*(1/z^*) \tag{7.77}$$

Now since σ_ε^2 is given by (7.75) and

$$R_{d\varepsilon}[l] = \frac{1}{2\pi j}\oint S_{d\varepsilon}(z)z^{l-1}dz$$

it follows that

$$\boxed{\sigma_\varepsilon^2 = R_{d\varepsilon}[0] = \frac{1}{2\pi j}\oint S_{d\varepsilon}(z)z^{-1}dz} \tag{7.78}$$

This last integral can be evaluated by using residues.

To apply the residue theory to (7.78) we first need to examine the region of convergence of $S_{d\varepsilon}(z)$. A typical $S_{d\varepsilon}(z)$ is shown in Fig. 7.8(a). Since the function extends for both positive and negative values of l the region of convergence is an annulus surrounding the origin. Further, since the cross power density spectrum $S_{d\varepsilon}(e^{j\omega})$ is finite, $S_{d\varepsilon}(z)$ must converge on the unit circle. In summary, the region of convergence is an annular region that includes the unit circle, as shown in Fig. 7.8(b).

The foregoing considerations are the key to evaluating the mean-square error. The integral (7.78) is defined over a closed path (contour) in the region of convergence. Since the unit circle is such a closed path it can be taken as the contour of integration. Evaluation of (7.78) therefore amounts to summing the residues corresponding to all poles of the function $S_{d\varepsilon}(z)z^{-1}$ *within the unit circle*.

Before continuing let us just mention in passing that it is sometimes easier to evaluate

$$S_{\varepsilon d}(z) = S_d(z) - S_{dx}^*(1/z^*)H(z) \tag{7.79}$$

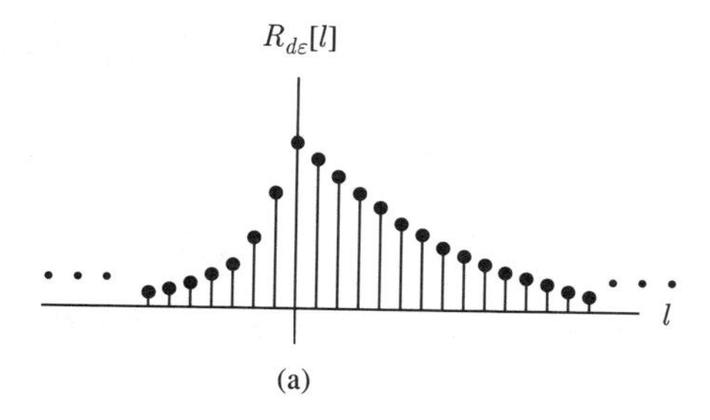

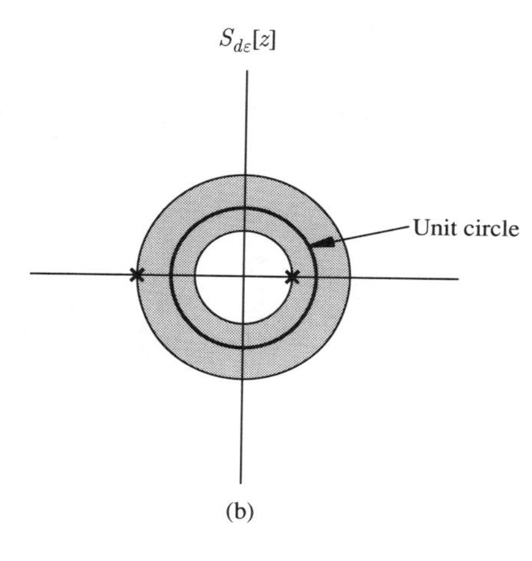

Figure 7.8 Region of convergence of $S_{d\varepsilon}(z)$. (a) Typical cross-correlation function $R_{d\varepsilon}[l]$. (b) Region of convergence of $S_{d\varepsilon}(z)$.

and to find σ_ε^2 from

$$\sigma_\varepsilon^2 = R_{\varepsilon d}[0] = \frac{1}{2\pi \jmath} \oint S_{\varepsilon d}(z) z^{-1} dz \tag{7.80}$$

instead of (7.78). Both methods are equally valid.

The following example illustrates the procedure of finding the mean-square error by integration in the complex plane.

Example 7.5

It is desired to evaluate the mean-square error for the optimal filter of Example 7.4. Note that $R_d[l] = R_s[l]$, and so according to Example 7.4,

$$S_d(z) = S_{dx}(z) = S_s(z) = \frac{0.72}{(1 - 0.8z^{-1})(1 - 0.8z)}$$

and

$$H(z) = \frac{0.375}{1 - 0.5z^{-1}}$$

Then since all of the random processes are real, (7.77) becomes

$$\begin{aligned} S_{d\varepsilon}(z) &= S_d(z) - S_{dx}(z)H(z^{-1}) \\ &= \frac{0.72}{(1 - 0.8z^{-1})(1 - 0.8z)} - \frac{0.72}{(1 - 0.8z^{-1})(1 - 0.8z)} \cdot \frac{0.375}{1 - 0.5z} \end{aligned}$$

$$= \frac{0.72}{(1-0.8z^{-1})(1-0.8z)}\left(1-\frac{0.375}{1-0.5z}\right) = \frac{(0.72)(0.625)(1-0.8z)}{(1-0.8z^{-1})(1-0.8z)(1-0.5z)}$$

Canceling the common term and simplifying then yields

$$S_{d\varepsilon}(z) = \frac{0.45}{(1-0.8z^{-1})(1-0.5z)} = \frac{-0.9z}{(z-0.8)(z-2)}$$

From this it can be seen that $S_{d\varepsilon}(z)z^{-1}$ has only one pole inside the unit circle at $z = 0.8$. Therefore, evaluating its residue, we have

$$\sigma_\varepsilon^2 = \frac{-0.9}{(z-0.8)(z-2)} \cdot (z-0.8)\bigg|_{z=0.8} = 0.75$$

In this particular case it is not very difficult to evaluate σ_ε^2 directly from (7.59) to check the result. Note that

$$h[n] = 0.375(0.5)^n \qquad n \geq 0$$

so (7.59) becomes

$$\sigma_\varepsilon^2 = 2.0 - \sum_{l=0}^{\infty}(0.375)(0.5)^l(2)(0.8)^l = 2\left(1 - 0.375\sum_{l=0}^{\infty}(0.4)^l\right)$$

$$= 2\left(1 - \frac{0.375}{1-0.4}\right) = 0.75$$

7.4.2 The Noncausal Wiener Filter

For certain problems that do not require a real-time solution, a noncausal filter is useful. The study of the noncausal Wiener filter also helps put the causal Wiener filter in perspective. This subsection therefore presents the general noncausal IIR optimal filtering problem and its solution.

For the noncausal filtering problem the estimate of the desired signal is of the form

$$\hat{d}[n] = \sum_{k=-\infty}^{\infty} h[n-k]x[k] \tag{7.81}$$

where this differs from (7.57) in that the upper limit of the sum is $+\infty$ and the impulse response $h[n]$ is allowed to have nonzero values for $n < 0$. Application of the orthogonality principle as in Section 7.3.1 leads to the Wiener-Hopf equation

$$\sum_{l=-\infty}^{\infty} R_x[l-i]h^*[l] = R_{dx}^*[i]; \qquad -\infty < i < \infty \tag{7.82}$$

and the equation for the mean-square error

$$\sigma_\varepsilon^2 = R_d[0] - \sum_{l=-\infty}^{\infty} h^*[l]R_{dx}[l] \tag{7.83}$$

Observe, in particular, that (7.82) must hold for *all* values of i, $-\infty < i < \infty$. If (7.82) is now conjugated and written as

$$\sum_{l=-\infty}^{\infty} h[l]R_x[i-l] = R_{dx}[i]; \quad -\infty < i < \infty \tag{7.84}$$

the left side of (7.84) can be recognized as the *convolution* of h with R_x. Therefore, the z-transform is

$$H(z) \cdot S_x(z) = S_{dx}(z)$$

which yields the solution

$$\boxed{H_{nc}(z) = \frac{S_{dx}(z)}{S_x(z)}} \tag{7.85}$$

where the subscript "*nc*" has been added to emphasize that the filter is noncausal. Note that the solution in this case is straightforward and does not require spectral factorization of $S_x(z)$. A convenient mnemonic device therefore is to remove the "+" subscript in (7.71); then in view of (7.66) the two equations (7.85) and (7.71) become identical.

The mean-square error can be evaluated from (7.77) and (7.78) or (7.79) and (7.80), with $H(z)$ taken to be the optimal noncausal filter. The steps taken to derive these equations do not change when $H(z)$ is noncausal.

To illustrate the procedure, consider the optimal noncausal filter for the problem presented in Example 7.4.

Example 7.6

It is desired to estimate the value of a real signal in white noise using a noncausal filter. The observed sequence is given by

$$x[n] = s[n] + \eta[n]$$

where

$$R_s[l] = 2(0.8)^{|l|}; \quad R_\eta[l] = 2\delta[l]$$

and η is uncorrelated with s. Proceeding as in Example 7.4, we find that

$$S_{dx}(z) = S_d(z) = S_s(z) = \frac{0.72}{(1-0.8z^{-1})(1-0.8z)}$$

and

$$S_x(z) = \frac{3.2(1-0.5z^{-1})(1-0.5z)}{(1-0.8z^{-1})(1-0.8z)}$$

Then, from (7.85),

$$H_{nc}(z) = \frac{0.72}{(1-0.8z^{-1})(1-0.8z)} \cdot \frac{(1-0.8z^{-1})(1-0.8z)}{3.2(1-0.5z^{-1})(1-0.5z)} = \frac{0.225}{(1-0.5z^{-1})(1-0.5z)}$$

To find the mean-square error, first compute

$$
\begin{aligned}
S_{d\varepsilon}(z) &= S_d(z) - S_{dx}(z)H_{nc}(z^{-1}) = S_s(z)\left(1 - H_{nc}(z^{-1})\right) \\
&= \frac{0.72}{(1-0.8z^{-1})(1-0.8z)}\left(1 - \frac{0.225}{(1-0.5z)(1-0.5z^{-1})}\right) \\
&= \frac{0.72(-0.5z + 1.025 - 0.5z^{-1})}{(1-0.8z^{-1})(1-0.8z)(1-0.5z^{-1})(1-0.5z)} \\
&= \frac{0.45(1-0.8z^{-1})(1-0.8z)}{(1-0.8z^{-1})(1-0.8z)(1-0.5z^{-1})(1-0.5z)} \\
&= \frac{0.45}{(1-0.5z^{-1})(1-0.5z)} = \frac{0.45z}{(z-0.5)(1-0.5z)}
\end{aligned}
$$

The function $S_{d\varepsilon}(z)z^{-1}$ thus has a single pole inside the unit circle at $z = 0.5$. Evaluating its residue yields

$$
\sigma_\varepsilon^2 = \frac{0.45}{(z-0.5)(1-0.5z)} \cdot (z-0.5)\bigg|_{z=0.5} = \frac{0.45}{0.75} = 0.60
$$

The performance of the noncausal Wiener filter is thus seen to be better than the performance of the causal Wiener filter. The mean-square error of the causal filter (from Example 7.5) was found to be 0.75.

7.4.3 A Case Study

The previous subsections present the IIR forms of the Wiener filter and illustrate their application through numerical examples. Although the examples sometimes involve a considerable amount of algebraic manipulation, the final expressions are relatively simple. To gain more insight about the Wiener filter, we now pursue the same cases considered in the examples but develop the results symbolically rather than numerically. The random processes are allowed to be complex in general since it involves only slightly more effort. It is seen here how physical quantities like the signal-to-noise ratio affect the bandwidth of the optimal filter and how the three cases of filtering, prediction, and smoothing relate to each other.

The case to be studied consists of an observation sequence

$$
x[n] = s[n] + \eta[n] \tag{7.86}
$$

where $s[n]$ is a random signal with exponential correlation function

$$
R_s[l] = \begin{cases} \sigma_s^2 \alpha^l & l \geq 0 \\ \sigma_s^2 (\alpha^*)^{-l} & l < 0 \end{cases} \tag{7.87}
$$

and $\eta[n]$ is white noise orthogonal to (uncorrelated with) the signal with

$$
R_\eta[l] = \sigma_o^2 \delta[l] \tag{7.88}
$$

Both random processes have zero mean. We first consider the causal Wiener filter and examine the cases of filtering, prediction, and smoothing separately. We then consider the noncausal Wiener filter and compare it to the causal forms developed.

Case (a)—Causal Filtering. For the filtering case the desired signal is defined by

$$d[n] = s[n] \tag{7.89}$$

and the following relations hold:

$$R_{dx}[l] = R_d[l] = R_s[l] \tag{7.90}$$

$$R_x[l] = R_s[l] + R_\eta[l] \tag{7.91}$$

The corresponding z-transforms are

$$S_{dx}(z) = S_d(z) = S_s(z) = \frac{\sigma_s^2(1-|\alpha|^2)}{(1-\alpha z^{-1})(1-\alpha^* z)} \tag{7.92}$$

and

$$\begin{aligned} S_x(z) &= S_s(z) + S_\eta(z) \\ &= \frac{\sigma_s^2(1-|\alpha|^2)}{(1-\alpha z^{-1})(1-\alpha^* z)} + \sigma_o^2 \\ &= \sigma_o^2 \cdot \frac{-\alpha^* z + \left(1+|\alpha|^2 + \frac{\sigma_s^2}{\sigma_o^2}\left(1-|\alpha|^2\right)\right) - \alpha z^{-1}}{-\alpha^* z + (1+|\alpha|^2) - \alpha z^{-1}} \end{aligned} \tag{7.93}$$

For the development that follows it is convenient to define the parameters

$$\overline{\alpha} = \frac{1+|\alpha|^2}{2\alpha^*} = \frac{1}{2}\left(\alpha + \frac{1}{\alpha^*}\right) \tag{7.94}$$

and

$$\gamma = \frac{\sigma_s^2(1-|\alpha|^2)}{\sigma_o^2(1+|\alpha|^2)} \tag{7.95}$$

The parameter γ is proportional to the signal-to-noise ratio σ_s^2/σ_o^2. Equation 7.93 can then be written in the form

$$S_x(z) = \sigma_o^2 \frac{-\alpha^* z + 2\alpha^*\overline{\alpha}(1+\gamma) - \alpha z^{-1}}{-\alpha^* z + 2\alpha^*\overline{\alpha} - \alpha z^{-1}} \tag{7.96}$$

or in the factored form

$$S_x(z) = \sigma_o^2 \frac{\alpha^*}{\beta^*} \frac{(1-\beta z^{-1})(1-\beta^* z)}{(1-\alpha z^{-1})(1-\alpha^* z)} \tag{7.97}$$

where β and $1/\beta^*$ are the roots of the numerator, given by

$$\beta = \overline{\alpha}(1+\gamma) - \sqrt{\overline{\alpha}^2(1+\gamma)^2 - \alpha/\alpha^*} \quad (a)$$
$$1/\beta^* = \overline{\alpha}(1+\gamma) + \sqrt{\overline{\alpha}^2(1+\gamma)^2 - \alpha/\alpha^*} \quad (b) \tag{7.98}$$

Here for $\alpha < 1$, β is defined as the root lying *within* the unit circle.

It is appropriate now to make some observations about these roots that will be important later. Notice from (7.95) and (7.94) that γ is real and that $\overline{\alpha}$ has the same phase as α. Call this phase

$$\phi = \angle\alpha \tag{7.99}$$

Then since each of the terms under the square root in (7.98) has a phase of 2ϕ, β itself has a phase equal to ϕ. Thus β has the form

$$\beta = |\beta|e^{j\phi} \quad (a)$$
$$|\beta| = |\overline{\alpha}|(1+\gamma) - \sqrt{|\overline{\alpha}|^2(1+\gamma)^2 - 1} \quad (b) \tag{7.100}$$

It is not difficult to show that for $\gamma = 0$ the root β is equal to α and otherwise that $|\beta| < |\alpha|$ (see Problem 7.25). Thus the roots β and $1/\beta^*$ of the numerator have the behavior shown in Fig. 7.9. In the real case of course, both α and β lie along the positive or negative real axis.

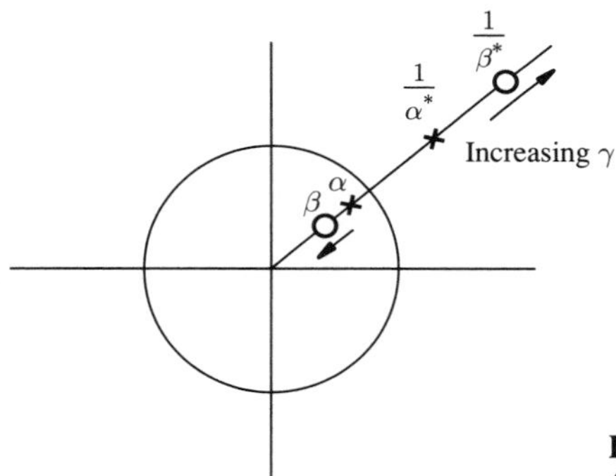

Figure 7.9 Location of zeros of $S_x(z)$ as a function of γ.

Equation 7.97 is the spectral factorization of S_x. From this equation it follows that

$$H_{ca}(z) = \frac{1-\beta z^{-1}}{1-\alpha z^{-1}} \tag{7.101}$$

and

$$\mathcal{K}_o = \sigma_o^2\frac{\alpha}{\beta} \tag{7.102}$$

where we have used the fact that α and β have the same phase to write

$$\frac{\alpha^*}{\beta^*} = \frac{|\alpha|}{|\beta|} = \frac{\alpha}{\beta} \tag{7.103}$$

Now form

$$\frac{1}{\mathcal{K}_o}\frac{S_{dx}(z)}{H^*_{ca}(1/z^*)} = \frac{1}{\mathcal{K}_o}\frac{\sigma_s^2(1-|\alpha|^2)}{(1-\alpha z^{-1})(1-\alpha^* z)}\cdot\frac{1-\alpha^* z}{1-\beta^* z}$$
$$= \frac{-\sigma_s^2(1-|\alpha|^2)(1/\beta\mathcal{K}_o)z^{-1}}{(1-\alpha z^{-1})(1-(1/\beta^*)z^{-1})} \tag{7.104}$$

This can be expanded by partial fractions to obtain

$$\frac{1}{\mathcal{K}_o}\frac{S_{dx}(z)}{H^*_{ca}(1/z^*)} = \frac{C}{1-\alpha z^{-1}} - \frac{C}{1-(1/\beta^*)z^{-1}} \tag{7.105}$$

where the constant C is found to be

$$C = \frac{\sigma_s^2(1-|\alpha|^2)}{\mathcal{K}_o(1-\alpha\beta^*)} = \frac{\sigma_s^2(1-|\alpha|^2)\beta}{\sigma_o^2(1-\alpha\beta^*)\alpha} = \frac{\sigma_s^2}{\sigma_o^2}\frac{\left(\frac{1}{\alpha^*}-\alpha\right)}{\left(\frac{1}{\beta^*}-\alpha\right)} \tag{7.106}$$

and (7.102) has been substituted for $\mathcal{K}_o$ to arrive at the last expression.

The sequence corresponding to (7.105) is given by

$$C\alpha^n u[n] + C\left(\frac{1}{\beta^*}\right)^n u[-n-1] \tag{7.107}$$

where $u[\cdot]$ is the unit step function. This sequence is illustrated in Fig. 7.10(a). The portion of the sequence in 7.10(a) with open circles and dashed lines is set to zero to form the causal filter. From (7.107) and (7.105) the causal portion is given by

$$h'[n] = C\alpha^n u[n] \tag{7.108}$$

and

$$H'(z) = \frac{1}{\mathcal{K}_o}\left[\frac{S_{dx}(z)}{H^*_{ca}(1/z^*)}\right]_+ = \frac{C}{1-\alpha z^{-1}} \tag{7.109}$$

Finally, using (7.101) and (7.109), the optimal filter is

$$H(z) = \frac{1}{H_{ca}(z)}\cdot\frac{1}{\mathcal{K}_o}\left[\frac{S_{dx}(z)}{H_{ca}(1/z^*)}\right]_+$$
$$= \frac{1-\alpha z^{-1}}{1-\beta z^{-1}}\cdot\frac{C}{1-\alpha z^{-1}} = \frac{C}{1-\beta z^{-1}} \tag{7.110}$$

A further simplification comes in the evaluation of the constant C. By using the fact that β is a root of the polynomial

$$z^2 - 2\overline{\alpha}(1+\gamma)z + 1$$

you can show that the expression (7.106) is equivalent to

$$C = 1 - \frac{\beta}{\alpha} \tag{7.111}$$

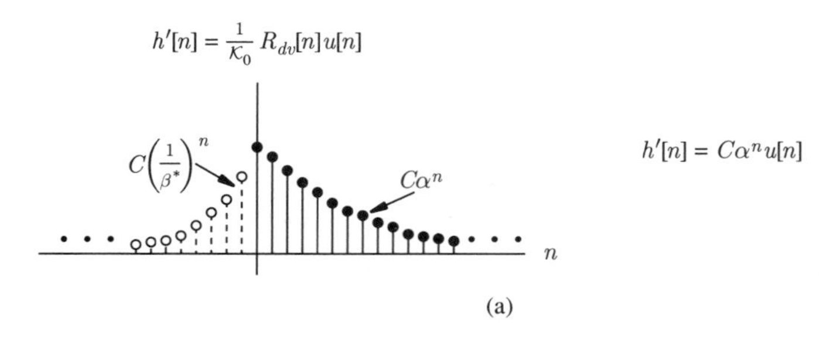

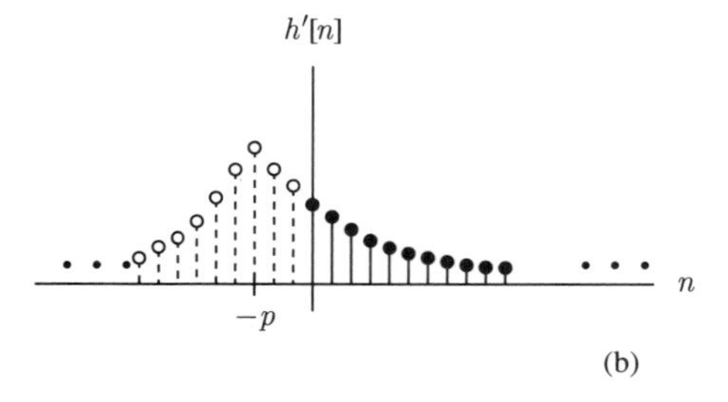

$h'[n] = C\alpha^{n+p}u[n]$

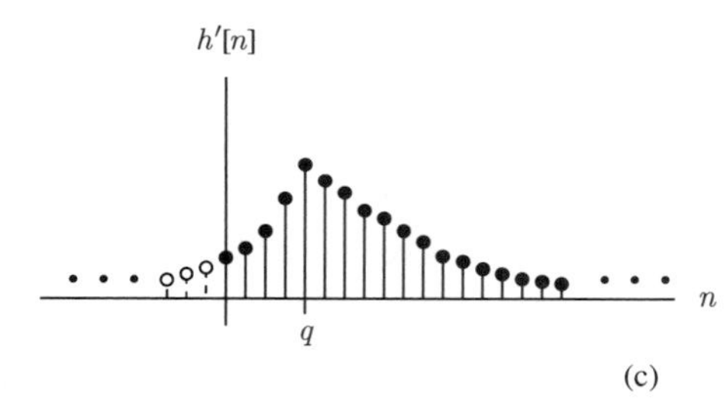

$h'[n] = C\alpha^{n-q}u[n-q] + C\left(\frac{1}{\beta^*}\right)^{n-q}(u[n] - u[n-q-1])$

Figure 7.10 Formation of the causal Wiener filter. (a) Filtering problem. (b) Prediction problem. (c) Smoothing problem.

(see Problem 7.26). Substituting this in (7.110) leads to the final expression

$$\boxed{H(z) = \left(1 - \frac{\beta}{\alpha}\right)\frac{1}{1 - \beta z^{-1}}} \tag{7.112}$$

The corresponding impulse response of the optimal filter is given by

$$h[n] = \left(1 - \frac{\beta}{\alpha}\right)\beta^n u[n] \tag{7.113}$$

The Wiener filter for this signal-in-white noise case is seen to be a single pole filter with pole at location β. This pole lies at the same angular position as the pole α of the signal process and has magnitude less than or equal to α as mentioned before. The three cases of α real and greater than zero, α real and less than zero, and α complex correspond to a low-pass signal, a high-pass signal, and a (one-sided) bandpass signal respectively. The noise, which is white, has power in all portions of the band.

Now consider the case where α is real and positive, so that the signal is low-pass; in this case β is also positive and $H(z)$ is a low-pass filter whose gain and critical frequency both depend on β. From (7.95) and (7.100), β in turn depends on the signal-to-noise ratio $\sigma_s^2/\sigma_\text{o}^2$ through the parameter γ. The location of the two poles at α and β are shown in Fig. 7.11.

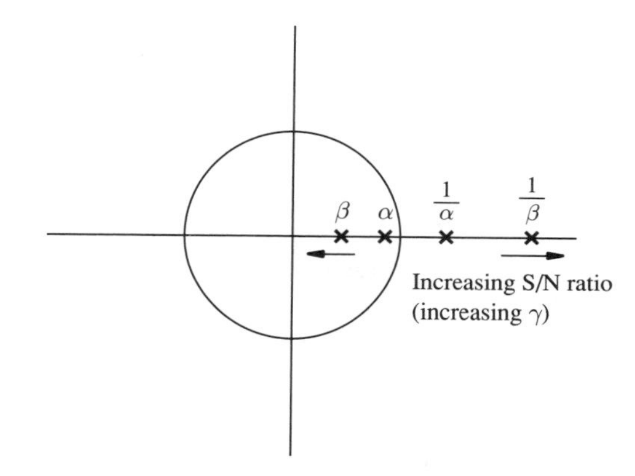

Figure 7.11 Location of poles (α) for a low-pass signal process and those of the corresponding Wiener filter (β).

Let us examine the characteristics of the Wiener filter in two limiting cases. In the case of zero signal-to-noise ratio the signal cannot be observed at all. It can be seen from (7.95) and (7.100) that for $\sigma_s^2/\sigma_\text{o}^2 = 0$ the value of β is equal to α and the gain of the filter in (7.112) is zero. Therefore, the output of the filter is zero, which is the mean of the random signal and thus the intuitively most reasonable estimate. As the signal-to-noise ratio increases, the pole at β moves away from the pole at α and the bandwidth of the filter opens up. The gain of the filter also increases. In the limiting case as $\sigma_s^2/\sigma_\text{o}^2 \to \infty$ the pole moves to the origin and the filter passes *all* frequencies with unit gain. This behavior is also intuitively reasonable since in the case of infinite signal-to-noise ratio (no noise) the filter should pass all of the received signal. You should convince yourself that the behavior of the filter is also intuitively reasonable in the cases of high-pass and bandpass signals.

Let us now proceed to derive an expression for the mean-square error of the optimal filter. Beginning with (7.77) we can write

$$S_{d\varepsilon}(z) = S_d(z) - S_{dx}(z)H^*(1/z^*) = S_s(z)\left[1 - H^*(1/z^*)\right]$$

Then substituting the quantities (7.92) and (7.112) leads to

$$\begin{aligned} S_{d\varepsilon}(z) &= \frac{\sigma_s^2(1-|\alpha|^2)}{(1-\alpha z^{-1})(1-\alpha^* z)}\left(1 - \left(1 - \frac{\beta^*}{\alpha^*}\right)\frac{1}{1-\beta^* z}\right) \\ &= \frac{\sigma_s^2(1-|\alpha|^2)((\beta^*/\alpha^*) - \beta^* z)}{(1-\alpha z^{-1})(1-\alpha^* z)(1-\beta^* z)} \end{aligned}$$

$$= \frac{\sigma_s^2(1-|\alpha|^2)(\beta^*/\alpha^*)(1-\alpha^* z)}{(1-\alpha z^{-1})(1-\alpha^* z)(1-\beta^* z)}$$

$$= \frac{\sigma_s^2(1-|\alpha|^2)(\beta/\alpha)z}{(z-\alpha)(1-\beta^* z)} \tag{7.114}$$

where (7.103) was used in the last step. Since there is only one pole within the unit circle, the mean-square error is obtained by finding the residue at this pole. Thus

$$\sigma_\varepsilon^2 = \left.\frac{\sigma_s^2(1-|\alpha|^2)(\beta/\alpha)}{(z-\alpha)(1-\beta^* z)}\cdot(z-\alpha)\right|_{z=\alpha} = \frac{\sigma_s^2(1-|\alpha|^2)(\beta/\alpha)}{1-\alpha\beta^*}$$

$$= \sigma_s^2\left(\frac{\frac{1}{\alpha^*}-\alpha}{\frac{1}{\beta^*}-\alpha}\right) \tag{7.115}$$

By comparing the last expression with (7.106) and using the alternative form (7.111) for C, the mean-square error can be written in the simpler form

$$\boxed{\sigma_\varepsilon^2 = \sigma_o^2 C = \sigma_o^2\left(1-\frac{\beta}{\alpha}\right)} \tag{7.116}$$

These equations for the mean-square error also satisfy intuition. First observe that since the phase of α and β are identical and $|\beta| \leq |\alpha|$, the mean-square error is less than or equal to either the signal *or* the noise variance. In the limiting cases of zero or infinite signal-to-noise ratios, however, we have to proceed with caution. Either of (7.115) or (7.116) may be indeterminant, depending on how the signal-to-noise ratio approaches its limit. For example, consider the case of zero signal-to-noise ratio. Recall that in this case the estimate is always zero; and since the signal has variance σ_s^2, the variance of the error is also equal to σ_s^2. Let us consider the two ways in which the signal-to-noise ratio can approach zero. Suppose first that $\sigma_s^2 \to 0$ while σ_o^2 remains finite. Then we find that $\sigma_s^2/\sigma_o^2 \to 0$ and $\beta \to \alpha$. In this case both (7.115) and (7.116) give the result $\sigma_\varepsilon^2 = 0$, which is equal to the variance of the signal and is the correct answer. On the other hand, suppose that σ_s^2 remains finite while $\sigma_o^2 \to \infty$, then once again $\sigma_s^2/\sigma_o^2 \to 0$ and $\beta \to \alpha$. However, in this case (7.115) gives the correct result $\sigma_\varepsilon^2 = \sigma_s^2$, while (7.116) is indeterminant.

In the case of infinite signal-to-noise ratio a similar situation occurs. Since in this case the signal dominates the noise, the variance in estimating the signal is zero. You can also verify that when $\sigma_s^2 \to \infty$ while σ_o^2 remains finite, (7.115) is indeterminant while (7.116) gives the correct result $\sigma_\varepsilon^2 = 0$. Alternatively, when σ_s^2 remains finite while $\sigma_o^2 \to 0$, both formulas give the correct result of zero.

Case (b)—Prediction. Here the problem is to estimate

$$d[n] = s[n+p] \tag{7.117}$$

where p is a positive integer. For this case we find

$$R_{dx}[l] = R_s[l+p] \tag{7.118}$$

and therefore

$$S_{dx}(z) = S_s(z)z^p = \frac{\sigma_s^2(1-|\alpha|^2)}{(1-\alpha z^{-1})(1-\alpha^* z)}\, z^p \tag{7.119}$$

The other quantities are the same as for Case (a).

Let us proceed to compute the term

$$\frac{1}{\mathcal{K}_o}\frac{S_{dx}(z)}{H_{ca}^*(1/z^*)} = \frac{-\sigma_s^2(1-|\alpha|^2)(1/\beta\mathcal{K}_o)z^{-1}}{(1-\alpha z^{-1})(1-(1/\beta^*)z^{-1})}\, z^p \tag{7.120}$$

which differs from the corresponding equation (7.104) in the filtering case by only the factor z^p. We could use a formal approach to extract the causal part of this expression; however, it is easier to notice that the sequence corresponding to (7.120) is just the corresponding sequence of Case (a) shifted to the left [see Fig. 7.10(b)]. From the figure the causal part of this sequence is given by

$$h'[n] = C\alpha^{n+p}u[n]$$

and therefore

$$H'(z) = \frac{1}{\mathcal{K}_o}\left[\frac{S_{dx}(z)}{H_{ca}^*(1/z^*)}\right]_+ = \frac{C\alpha^p}{1-\alpha z^{-1}} \tag{7.121}$$

The desired optimal prediction filter is then given by

$$H(z) = \frac{1}{H_{ca}(z)}\cdot\frac{1}{\mathcal{K}_o}\left[\frac{S_{dx}(z)}{H_{ca}^*(1/z^*)}\right]_+ = \frac{C\alpha^p}{1-\beta z^{-1}}$$

and using (7.111) for C yields

$$\boxed{H(z) = \left(1-\frac{\beta}{\alpha}\right)\frac{\alpha^p}{1-\beta z^{-1}}} \tag{7.122}$$

The corresponding impulse response of this filter is

$$h[n] = \left(1-\frac{\beta}{\alpha}\right)\alpha^p\beta^n u[n] \tag{7.123}$$

The impulse response of this prediction filter and that of the filter for estimating the current value of the signal are compared in Fig. 7.12. Note that the output of the filter in part (b) is just equal to the output of the filter in part (a) multiplied by α^p, so

$$\hat{s}[n+p] = \hat{s}[n]\alpha^p$$

Since this signal with exponential correlation function satisfies a first-order difference equation with similar response, this seems like a reasonable estimate. In other words, the prediction filter merely takes the optimal estimate for the current value of the signal and projects it forward according to the signal model. This fact is discussed in some further detail in the next section on recursive filtering.

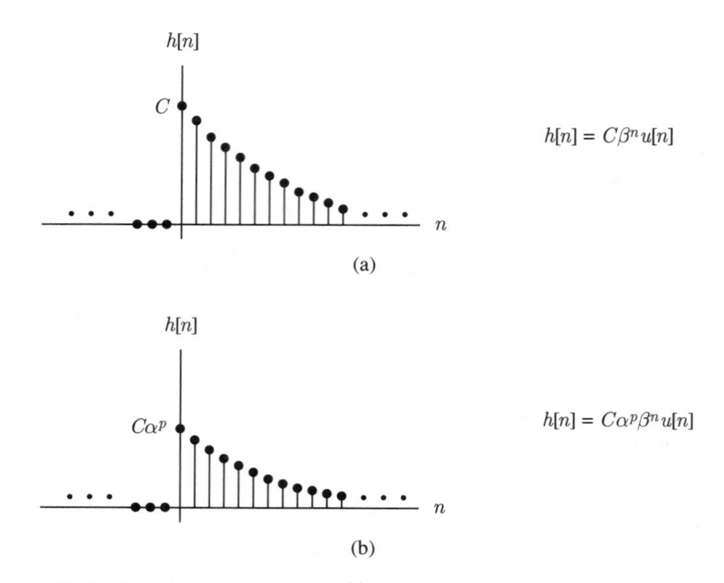

Figure 7.12 Impulse response of the causal Wiener filter. (a) Filtering problem. (b) Prediction problem.

The computation of the mean-square error for a general value of p is rather involved and is not particularly instructive. Let it suffice to say that this mean-square error is larger than that for the filter of Case (a) and so satisfies our intuition that it is more difficult to estimate future values than present values of the signal.

Case (c)—Smoothing. The final use of the causal Wiener filter is for the smoothing problem. For this case the desired signal is defined as

$$d[n] = s[n-q] \tag{7.124}$$

where q is a positive integer. Here observations are used both before and after the point at which the signal is to be estimated. For this case all of the statistical moments are again the same as for Case (a) except that

$$R_{dx}[l] = R_s[l-q] \tag{7.125}$$

Therefore, S_{dx} is given by

$$S_{dx}(z) = S_s(z)z^{-q} = \frac{\sigma_s^2(1-|\alpha|^2)}{(1-\alpha z^{-1})(1-\alpha^* z)} z^{-q} \tag{7.126}$$

Up to this point the problem looks very much like the prediction problem except that p is replaced by $-q$. However, significant differences arise in forming the causal part of the filter. Again, although a formal procedure could be followed, it is easier to use a picture to support our reasoning as before. From the form of (7.126) it can be seen that now the corresponding sequence is shifted to the *right*, as shown in Fig. 7.10(c). The causal part of this sequence is thus given by

$$h'[n] = C\alpha^{n-q}u[n-q] + C\left(\frac{1}{\beta^*}\right)^{n-q}(u[n] - u[n-q-1])$$

This corresponds to the z-transform

$$H'(z) = \frac{1}{\mathcal{K}_o}\left[\frac{S_{dx}(z)}{H^*_{ca}(1/z^*)}\right]_+ = \frac{Cz^{-q}}{1-\alpha z^{-1}} + \frac{C((\beta^*)^q - z^{-q})}{1-(1/\beta^*)z^{-1}} \tag{7.127}$$

where C is the same constant as that used in the previous two cases. The optimal filter is then given by

$$H(z) = \frac{1}{H_{ca}(z)} \cdot \frac{1}{\mathcal{K}_o}\left[\frac{S_{dx}(z)}{H_{ca}(z^{-1})}\right]_+ = C\left(\frac{1-\alpha z^{-1}}{1-\beta z^{-1}}\right)\left(\frac{z^{-q}}{1-\alpha z^{-1}} + \frac{(\beta^*)^q - z^{-q}}{1-(1/\beta^*)z^{-1}}\right)$$

Combining the last expression over a common denominator yields

$$\begin{aligned} H(z) &= C\left(\frac{1-\alpha z^{-1}}{1-\beta z^{-1}}\right)\left(\frac{z^{-q}(1-(1/\beta^*)z^{-1}) + ((\beta^*)^q - z^{-q})(1-\alpha z^{-1})}{(1-\alpha z^{-1})(1-(1/\beta^*)z^{-1})}\right) \\ &= C\left(\frac{(\beta^*)^q(1-\alpha z^{-1}) + (\alpha - 1/\beta^*)z^{-(1+q)}}{(1-\beta z^{-1})(1-(1/\beta^*)z^{-1})}\right) \end{aligned} \tag{7.128}$$

or, using (7.111),

$$\boxed{H(z) = \left(1 - \frac{\beta}{\alpha}\right)\frac{(\beta^*)^q(1-\alpha z^{-1}) + (\alpha - 1/\beta^*)z^{-(1+q)}}{(1-\beta z^{-1})(1-(1/\beta^*)z^{-1})}} \tag{7.129}$$

which is the final expression for the optimal filter.

The impulse response of the smoothing filter can be obtained by inverting (7.129). We will not do this completely since the expression is complicated for $0 \leq n < q$ and does not provide any particular enlightenment. For $n \geq q$, however, the behavior depends on β^{n-q} and thus is similar to the other two filters (see Fig. 7.13). The peak value of the impulse response occurs at $n = q$. This value can be found from (7.129) by using residues and turns out to be

$$M_o = \left(1 - \frac{\beta}{\alpha}\right)\frac{(1-\alpha\beta^*) + |\beta|^{2q}(\alpha\beta^* - |\beta|^2)}{1-|\beta|^2} \tag{7.130}$$

Noting that $\left(1 - \frac{\beta}{\alpha}\right) = C$, we can check this for $q = 0$ to find

$$M_o = C\frac{(1-\alpha\beta^*) + (\alpha\beta^* - |\beta|^2)}{1-|\beta|^2} = C\frac{1-|\beta|^2}{1-|\beta|^2} = C$$

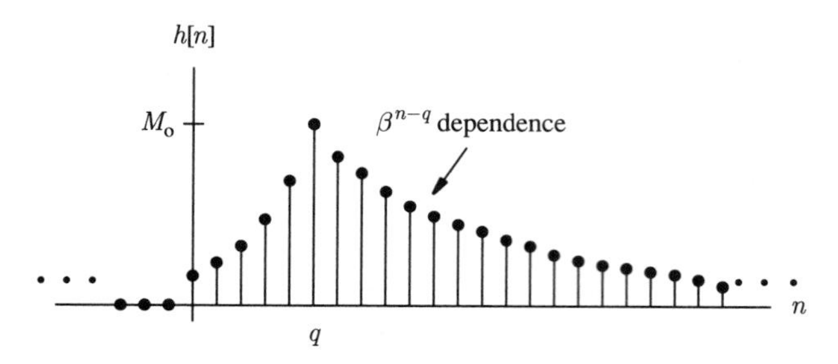

Figure 7.13 Impulse response of the causal Wiener filter for the smoothing problem.

which corresponds to the peak value of the impulse response in Case (a). The impulse response of this smoothing filter is discussed further after development of results for the noncausal Wiener filter.

Noncausal Filtering. Let us now develop the noncausal Wiener filter for the problem in Case (a) and find an expression for the mean-square error. From (7.85) the optimal noncausal filter is just the ratio of $S_{dx}(z)$ to $S_x(z)$. Thus from (7.85), (7.92), and (7.97) we have directly

$$H_{nc}(z) = \frac{S_{dx}(z)}{S_x(z)} = \frac{\sigma_s^2(1-|\alpha|^2)\beta}{\sigma_o^2\alpha} \cdot \frac{1}{(1-\beta z^{-1})(1-\beta^* z)}$$

If the constant C' is now defined as

$$C' = \frac{\sigma_s^2(1-|\alpha|^2)\beta}{\sigma_o^2(1-|\beta|^2)\alpha} = \frac{\sigma_s^2}{\sigma_o^2}\frac{\left(\frac{1}{\alpha^*}-\alpha\right)}{\left(\frac{1}{\beta^*}-\beta\right)} \tag{7.131}$$

then the filter can be written as

$$\boxed{H_{nc}(z) = \frac{C'(1-|\beta|^2)}{(1-\beta z^{-1})(1-\beta^* z)}} \tag{7.132}$$

This filter has the bilateral impulse response

$$h_{nc}[n] = \begin{cases} C'\beta^n & n \geq 0 \\ C'(\beta^*)^{-n} & n < 0 \end{cases} \tag{7.133}$$

The impulse response is sketched in Fig. 7.14.

The mean-square error for the noncausal filter can be evaluated quite easily by residue methods. To do so it is necessary first to compute the term $S_{d\varepsilon}(z)$. It will be shown that (for this case *only*) $S_{d\varepsilon}$ has a particularly simple relation to the optimal filter. To show this we use the fact that $S_{dx}(z) = S_d(z)$ and that

$$S_x(z) = S_d(z) + S_\eta(z)$$

which holds because the signal and noise are uncorrelated. Further, observe from (7.132) that

$$H_{nc}^*(1/z^*) = H_{nc}(z)$$

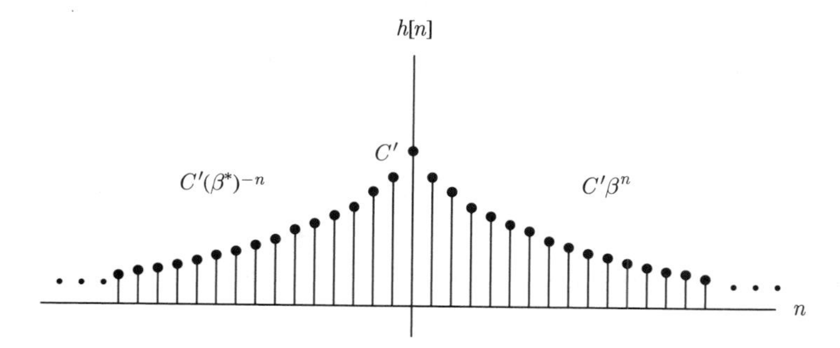

Figure 7.14 Impulse response of the noncausal Wiener filter.

which arises because both $S_{dx}(z)$ and $S_x(z)$ satisfy a similar condition. Thus it follows from (7.77), (7.85), and the observations above that

$$\begin{aligned}
S_{d\varepsilon}(z) &= S_d(z) - S_{dx}(z)H_{nc}(z) = S_{dx}(z)\left(1 - \frac{S_{dx}(z)}{S_x(z)}\right) \\
&= S_{dx}(z)\left(\frac{S_x(z) - S_{dx}(z)}{S_x(z)}\right) \\
&= S_{dx}(z)\left(\frac{S_d(z) + S_\eta(z) - S_{dx}(z)}{S_x(z)}\right) \\
&= \frac{S_{dx}(z)S_\eta(z)}{S_x(z)} = H_{nc}(z)S_\eta(z)
\end{aligned} \tag{7.134}$$

Finally, using the fact $S_\eta(z) = \sigma_o^2$ and the filter expression (7.132) produces

$$S_{d\varepsilon}(z) = \frac{\sigma_o^2 C'(1 - |\beta|^2)}{(1 - \beta z^{-1})(1 - \beta^* z)} = \frac{\sigma_o^2 C'(1 - |\beta|^2)z}{(z - \beta)(1 - \beta^* z)} \tag{7.135}$$

Finding σ_ε^2 requires finding the residues of poles of $S_{d\varepsilon}(z)z^{-1}$ inside the unit circle. Since this term has only a single pole inside the unit circle at $z = \beta$, σ_ε^2 is given by

$$\begin{aligned}
\sigma_\varepsilon^2 &= \left.\frac{\sigma_o^2 C'(1 - |\beta|^2)}{(z - \beta)(1 - \beta^* z)}(z - \beta)\right|_{z=\beta} \\
&= \frac{\sigma_o^2 C'(1 - |\beta|^2)}{1 - \beta^*\beta} = \sigma_o^2 C'
\end{aligned} \tag{7.136}$$

or, using (7.131),

$$\sigma_\varepsilon^2 = \sigma_s^2 \frac{\left(\frac{1}{\alpha^*} - \alpha\right)}{\left(\frac{1}{\beta^*} - \beta\right)} \tag{7.137}$$

It is of interest to compare the mean-square error for the noncausal filter to that of the causal filter. Using (7.136) and (7.116) followed by (7.137) and (7.115) yields

$$\frac{\sigma^2_{\varepsilon(nc)}}{\sigma^2_{\varepsilon(causal)}} = \frac{C'}{C} = \frac{\left(\frac{1}{\beta^*} - \alpha\right)}{\left(\frac{1}{\beta^*} - \beta\right)} = \frac{1 - \alpha\beta^*}{1 - |\beta|^2} \tag{7.138}$$

Since $|\beta| < |\alpha|$ for any nonzero signal-to-noise ratio, it follows that this ratio is correspondingly less than 1.

Comparison of the Noncausal Filter to the Causal Smoothing Filter. A final topic of interest is to compare the causal smoothing filter in the limit as the delay q gets very large to the noncausal filter. We may expect that the causal filter should approximate the optimal noncausal filter (with delay of q). Begin with the expression for the causal smoothing filter (7.128) and let $q \rightarrow \infty$. Note that since $|\beta| < 1$ the first term in the numerator goes to zero and leaves

$$\begin{aligned}\lim_{q\rightarrow\infty} H(z) &= C\frac{(\alpha - 1/\beta^*)z^{-(1+q)}}{(1 - \beta z^{-1})(1 - (1/\beta^*)z^{-1})} \\ &= C\frac{(1 - \alpha\beta^*)z^{-q}}{(1 - \beta z^{-1})(1 - \beta^* z)}\end{aligned} \tag{7.139}$$

Then substituting the ratio C/C' from (7.138) produces

$$\begin{aligned}\lim_{q\rightarrow\infty} H(z) &= C'\frac{1 - |\beta|^2}{1 - \alpha\beta^*}\frac{(1 - \alpha\beta^*)z^{-q}}{(1 - \beta z^{-1})(1 - \beta^* z)} \\ &= C'\frac{(1 - |\beta|^2)}{(1 - \beta z^{-1})(1 - \beta^* z)}z^{-q}\end{aligned} \tag{7.140}$$

Comparing this with (7.132) shows that

$$\lim_{q\rightarrow\infty} H(z) \rightarrow H_{nc}(z)z^{-q}$$

We can also check the peak value of the impulse response in Fig. 7.13. Noting that $\left(1 - \frac{\beta}{\alpha}\right) = C$, we have from (7.130)

$$\lim_{q\rightarrow\infty} M_o = C\frac{1 - \alpha\beta^*}{1 - |\beta|^2} = C' \tag{7.141}$$

where the last equality follows from (7.138). This analysis shows that the impulse response in Fig. 7.13 approaches a delayed version of the impulse response in Fig. 7.14, which agrees with intuition.

7.5 RECURSIVE FILTERING

The signal in additive noise problem studied in the preceding section provides the setting for another important technique: that of recursive filtering. The general vector form of this procedure is commonly referred to as Kalman filtering.[5] In this treatment, however, attention is restricted to the scalar form of the procedure and the special problems to which it applies. Although to maintain generality, time is not referred to specifically in the description, it is nevertheless natural and convenient to think of the random process as evolving in time and to think of the parameter n for the random process as representing the time index.

7.5.1 The Recursive Filtering Problem

This section examines the problem of estimating a random process at a particular point in its evolution using observations from some original point of reference, say $n = 0$, to the present. The filter must naturally change with each new observation; however, we want to generate the estimate and to update the filter parameters recursively.

Estimation of a Dynamic Random Signal in Noise. Let us again consider a very specific form of estimation problem consisting of a signal in additive white noise. The observations are of the form (7.86) where $s[n]$ is a random signal with correlation function $R_s[l]$ and $\eta[n]$ is a white noise process with variance σ_o^2, uncorrelated with s. It is desired to find the best linear mean-square estimate of $s[n]$ from the observations

$$x[k] = s[k] + \eta[k]; \quad k = 0, 1, \ldots, n \tag{7.142}$$

From the results of the previous sections it is clear that an optimal linear mean-square estimate can be found that is of the form[6]

$$\hat{s}[n] = c_0 x[0] + c_1 x[1] + \cdots + c_n x[n] \tag{7.143}$$

where the c_i are constant coefficients. However, it is desired to express the estimate recursively, that is, to put it in the form

$$\hat{s}[n] = B_n \hat{s}[n-1] + K_n x[n] \tag{7.144}$$

where $\hat{s}[n-1]$ is the estimate of $s[n-1]$ at the previous point and $x[n]$ is the present observation, and to find the dynamically changing coefficients B_n and K_n. Note that it is not *always* possible to generate the estimate recursively according to (7.144); however, we shall see under what conditions this can be done.

To proceed with this recursive estimation problem, define the error in the estimate as

$$\varepsilon[n] = s[n] - \hat{s}[n] \tag{7.145}$$

[5] After Rudolf E. Kalman, who published an early paper on the method and brought it to the fore [5]. R. S. Bucy also developed the method and later published a joint paper with Kalman [6] that treated the continuous-time problem. As a result, the method is sometimes referred to as Kalman-Bucy filtering.

[6] In recursive estimation, which deals specifically with observations of the form (7.142), it is conventional to drop the notation $d[n]$ and $\hat{d}[n]$ and use a more explicit reference to the signal s and its estimate.

and the error variance

$$\sigma^2_{\varepsilon[n]} = \text{Var}\,[\varepsilon[n]] \tag{7.146}$$

Also note, in preparation for the analysis to follow, that since the noise is white and uncorrelated with the signal, we have, as before [see (7.90) and (7.91)],

$$\mathcal{E}\{s[n]x^*[n-l]\} = R_s[l] \tag{7.147}$$

$$\mathcal{E}\{x[n]x^*[n-l]\} = R_s[l] + \sigma_o^2\delta[l] \tag{7.148}$$

and

$$\mathcal{E}\{x[n]\eta^*[n]\} = \mathcal{E}\{(s[n] + \eta[n])\eta^*[n]\} = \sigma_o^2 \tag{7.149}$$

Now to find the coefficients B_n and K_n, apply the orthogonality principle in the form

$$\mathcal{E}\{\varepsilon[n]x^*[n-l]\} = 0; \quad l = 0, 1, \ldots, n \tag{7.150}$$

and consider the implications. For the specific case $l = 0$ substituting (7.142) for $x[n]$ yields

$$\mathcal{E}\{\varepsilon[n](s^*[n] + \eta^*[n])\} = \mathcal{E}\{\varepsilon[n]s^*[n]\} + \mathcal{E}\{\varepsilon[n]\eta^*[n]\} = 0 \tag{7.151}$$

Note that by the second part of the Orthogonality Theorem the first term on the right of (7.151) is equal to $\sigma^2_{\varepsilon[n]}$. Then use this fact and substitute (7.145) and (7.144) in (7.151) to obtain

$$\begin{aligned}
&\sigma^2_{\varepsilon[n]} + \mathcal{E}\{(s[n] - B_n\hat{s}[n-1] - K_n x[n])\eta^*[n]\} \\
&\quad = \sigma^2_{\varepsilon[n]} + \mathcal{E}\{s[n]\eta^*[n]\} - B_n\mathcal{E}\{\hat{s}[n-1]\eta^*[n]\} - K_n\mathcal{E}\{x[n]\eta^*[n]\} = 0
\end{aligned} \tag{7.152}$$

Now observe that since the signal is orthogonal to the noise, the second term on the right is zero. In addition, since the noise is white, and $\hat{s}[n-1]$ is a function of only signal values and noise values up to $n-1$, the third term on the right is zero. Making these observations and substituting (7.149) in the last term of (7.152) yields

$$\sigma^2_{\varepsilon[n]} - K_n\sigma_o^2 = 0$$

or

$$K_n = \frac{\sigma^2_{\varepsilon[n]}}{\sigma_o^2} \tag{7.153}$$

Note that K_n is *real* and *positive*. This is the implication of (7.150) for $l = 0$.

For $l > 0$ (7.150) can be written as

$$\begin{aligned}
\mathcal{E}\{\varepsilon[n]x^*[n-l]\} &= \mathcal{E}\{(s[n] - B_n\hat{s}[n-1] - K_n x[n])x^*[n-l]\} \\
&= \mathcal{E}\{s[n]x^*[n-l]\} - B_n\mathcal{E}\{\hat{s}[n-1]x^*[n-l]\} \\
&\quad - K_n\mathcal{E}\{x[n]x^*[n-l]\} = 0
\end{aligned} \tag{7.154}$$

Now note from (7.147) and (7.148) with $l > 0$ that the expectations in both the first and the last terms on the right of (7.154) are equal to $R_s[l]$. Using this fact and the relation (7.145), we can write (7.154) as

$$(1 - K_n)R_s[l] - B_n \mathcal{E}\{(s[n-1] - \varepsilon[n-1])x^*[n-l]\} = 0 \qquad (7.155)$$

Then since $\varepsilon[n-1]$ is orthogonal to $x[n-l]$ for l in the present range of consideration, this equation can be written using (7.147) as

$$R_s[l] - \frac{B_n}{1 - K_n} R_s[l-1] = 0; \qquad l > 0 \qquad (7.156)$$

Recall that at the start of this section we sought to find an optimal estimate in the recursive form (7.144). It was desired to see under what conditions this could be done. The analysis so far has shown that if an estimate of the form (7.144) is possible, then the correlation function for the signal must satisfy (7.156). This implies that $R_s[l]$ is of the form

$$R_s[l] = R_s[0]\alpha^l; \qquad l > 0 \qquad (7.157)$$

where

$$\alpha = \frac{B_n}{1 - K_n} \qquad (7.158)$$

Now recall from Chapter 5 that a signal that has an exponential correlation function such as (7.157) can be generated by a model of the form[7]

$$s[n] = \alpha s[n-1] + w[n] \qquad (7.159)$$

where w is a white noise process, *different* from η. Specifically, the correlation function of this signal is

$$R_s[l] = \sigma_s^2 \alpha^l; \qquad l \geq 0 \qquad (7.160)$$

where

$$\sigma_s^2 = R_s[0] = \frac{\sigma_w^2}{1 - |\alpha|^2} \qquad (7.161)$$

and where σ_w^2 is the variance of $w[n]$. Thus it is seen that with the signal generated by (7.159), an estimate in the form of (7.144) satisfies the orthogonality principle and is therefore the optimal estimate. Now solving (7.158) for B_n and substituting the result in (7.144) produces the explicit form

$$\hat{s}[n] = \alpha(1 - K_n)\hat{s}[n-1] + K_n x[n] \qquad (7.162)$$

[7]In the literature on Kalman filtering the signal is usually assumed to be generated according to a model

$$s[n+1] = \alpha s[n] + w[n]$$

which has the same correlation function. We follow a convention for the signal model that is more standard in signal processing.

or

$$\boxed{\hat{s}[n] = \alpha\hat{s}[n-1] + K_n\left(x[n] - \alpha\hat{s}[n-1]\right)} \qquad (7.163)$$

where K_n is given by (7.153).

Although (7.163) is the desired result, it is also common to write it as a combination of two equations as follows:[8]

$$\boxed{\begin{aligned} \hat{s}[n|n-1] &= \alpha\hat{s}[n-1] \\ \hat{s}[n] &= \hat{s}[n|n-1] + K_n\left(x[n] - \hat{s}[n|n-1]\right) \end{aligned}} \qquad (7.164)$$

These equations show that when the signal model is (7.159) and the observations are of the form (7.142), the estimate is the sum of two terms. The term $\hat{s}[n|n-1]$, defined by the first equation, is a simple forward prediction of the previous estimate according to the signal model. In other words, if the signal at $n-1$ were truly equal to the estimate $\hat{s}[n-1]$, then our best guess for the signal at n prior to any new observations would be simply $\alpha\hat{s}[n-1]$. The second equation shows that the final estimate is this simple forward prediction plus a second term which is a "correction" of that simple prediction. In making this correction the difference between the new observation $x[n]$ and the forward predicted estimate is used to form an error or "residual":

$$r[n] = x[n] - \hat{s}[n|n-1]$$

This residual is a white noise process which is nonstationary (because the variance depends on n) and represents the "new information" in the observations. In the "steady state," as n becomes large, the variance approaches a constant and the residual becomes the *innovations process* for the observations. That is, $r[n]$ becomes a stationary white noise process which when applied to a suitable linear filter would generate the observation process $x[n]$. This fact and several other important related ideas are explored in Problem 7.33.

Figure 7.15 shows block diagrams for the signal and observation model [(a)] and the optimal linear mean-square estimator [(b)]. Note that the loop representing the signal model occurs also in the block diagram for the estimator. Figure 7.15 also shows the fundamentally different roles of the white noise processes w and η. The process w is a driving term used to *produce* the signal. If w were zero, the signal s would be zero. The process η, on the other hand, is observation noise. It is noise added to the signal, for example by the receiver. This noise is not desirable, but it is unavoidably present and gives rise to the estimation problem. If the observation noise were not present, there would be no need to estimate the signal, since it could be observed directly.

One final ingredient is required to make the estimator complete, namely an expression for the error variance $\sigma^2_{\varepsilon[n]}$. This in turn defines the gain K_n through (7.153). It turns out that the error variance and therefore the gain can also be computed recursively, although it is not always necessary to do so.

The analysis proceeds as follows. Using the second part of the orthogonality theorem, substituting (7.145) and (7.161), and taking note of (7.147), (7.159), and the orthogonality

[8] In the literature on Kalman filtering the estimate $\hat{s}[n]$ is sometimes also written as $\hat{s}[n|n]$.

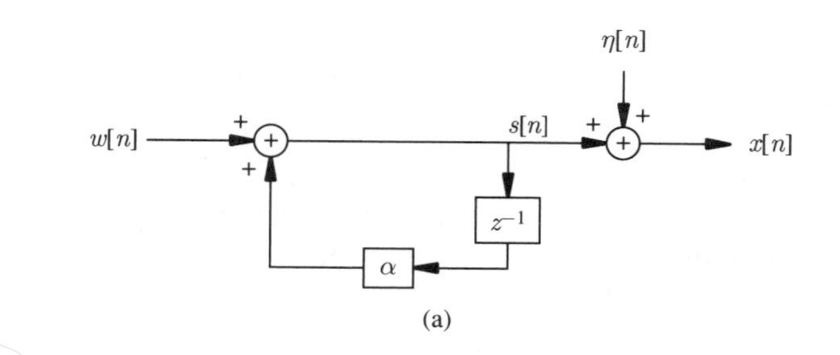

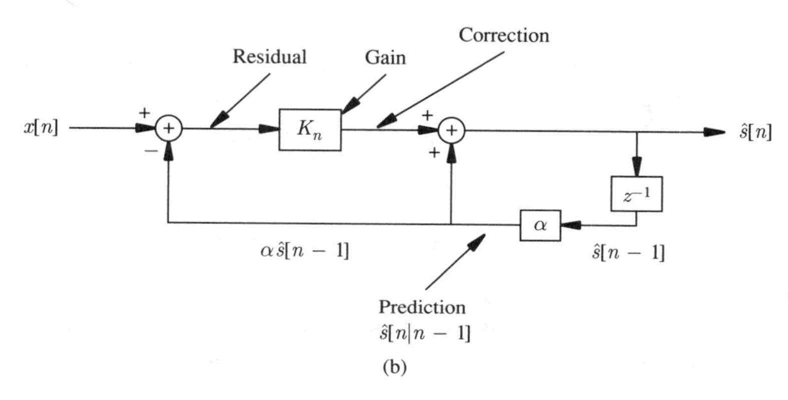

Figure 7.15 Systems involved in recursive filtering. (a) Signal and observation model. (b) Recursive filter.

properties of the white noise, we can write

$$\begin{aligned}
\sigma^2_{\varepsilon[n]} &= \mathcal{E}\{s[n]\varepsilon^*[n]\} = \mathcal{E}\{s[n][s^*[n] - \alpha^*(1-K_n)\hat{s}^*[n-1] - K_n x^*[n]]\} \\
&= \sigma_s^2(1-K_n) - \alpha^*(1-K_n)\mathcal{E}\{s[n]\hat{s}^*[n-1]\} \\
&= (1-K_n)\left[\sigma_s^2 - \alpha^*\mathcal{E}\{s[n]\hat{s}^*[n-1]\}\right] \\
&= (1-K_n)\left[\sigma_s^2 - \alpha^*\mathcal{E}\{(\alpha s[n-1] + w[n])\hat{s}^*[n-1]\}\right] \\
&= (1-K_n)\left[\sigma_s^2 - |\alpha|^2\mathcal{E}\{s[n-1]\hat{s}^*[n-1]\}\right]
\end{aligned} \tag{7.165}$$

For $n = 0$, one of two approaches is generally taken. In the first, it is assumed that an observation is made at $n = -1$ and $\hat{s}[-1]$ is taken to be $x[-1]$. In this case the last term becomes

$$\mathcal{E}\{s[-1]\hat{s}^*[-1]\} = R_{sx}[0] = \sigma_s^2$$

so (7.165) can be written as

$$\sigma^2_{\varepsilon[0]} = (1-K_0)[\sigma_s^2 - |\alpha|^2\sigma_s^2] \tag{7.166}$$

After substituting (7.153) and rearranging terms, this becomes

$$\sigma^2_{\varepsilon[0]}\left(1 + \frac{\sigma_s^2}{\sigma_o^2}(1-|\alpha|^2)\right) = \sigma_s^2(1-|\alpha|^2)$$

or

$$\sigma^2_{\varepsilon[0]} = \frac{\sigma_s^2(1-|\alpha|^2)}{1+\sigma_s^2(1-|\alpha|^2)/\sigma_o^2} \tag{7.167}$$

In the second approach no prior observations are assumed to be made, so $\hat{s}[-1]$ is taken to be zero. This makes the last term in (7.165) also equal to zero. Then performing the same operations that led to (7.167) results in

$$\sigma^2_{\varepsilon[0]} = \frac{\sigma_s^2}{1+\sigma_s^2/\sigma_o^2} \tag{7.168}$$

By using the relation $\sigma_w^2 = (1-|\alpha|^2)\sigma_s^2$ implied by (7.161) we can write (7.167) and (7.168) in the alternative forms

$$\sigma^2_{\varepsilon[0]} = \frac{\sigma_w^2}{1+\sigma_w^2/\sigma_o^2} \tag{7.169}$$

and

$$\sigma^2_{\varepsilon[0]} = \frac{\sigma_w^2\sigma_o^2}{\sigma_w^2+\sigma_o^2(1-|\alpha|^2)} \tag{7.170}$$

which can be used if σ_w^2 rather than σ_s^2 is given.

To find $\sigma^2_{\varepsilon[n]}$ for $n>0$ the last term in (7.165) needs to be evaluated further. To do so, notice that the error variance at $n-1$ can be written as

$$\begin{aligned}\sigma^2_{\varepsilon[n-1]} &= \mathcal{E}\{s[n-1]\varepsilon^*[n-1]\} = \mathcal{E}\{s[n-1]\left(s^*[n-1]-\hat{s}^*[n-1]\right)\}\\ &= \sigma_s^2 - \mathcal{E}\{s[n-1]\hat{s}^*[n-1]\}\end{aligned} \tag{7.171}$$

If (7.171) is solved for $\mathcal{E}\{s[n-1]\hat{s}^*[n-1]\}$ and the result is substituted in (7.165), then this equation becomes

$$\begin{aligned}\sigma^2_{\varepsilon[n]} &= (1-K_n)\left[\sigma_s^2+|\alpha|^2(\sigma^2_{\varepsilon[n-1]}-\sigma_s^2)\right]\\ &= (1-K_n)[\sigma_s^2(1-|\alpha|^2)+|\alpha|^2\sigma^2_{\varepsilon[n-1]}]\\ &= (1-K_n)[\sigma_w^2+|\alpha|^2\sigma^2_{\varepsilon[n-1]}]\end{aligned} \tag{7.172}$$

where (7.160) was used in the last step. Substituting (7.153) for K_n and solving for $\sigma^2_{\varepsilon[n]}$ in terms of $\sigma^2_{\varepsilon[n-1]}$ then yields

$$\sigma^2_{\varepsilon[n]} = \frac{\sigma_w^2+|\alpha|^2\sigma^2_{\varepsilon[n-1]}}{\sigma_o^2+\sigma_w^2+|\alpha|^2\sigma^2_{\varepsilon[n-1]}}\cdot\sigma_o^2; \qquad n>0 \tag{7.173}$$

This is the desired recursive expression for the variance.

An alternative equivalent expression for the variance (see Problem 7.30) is

$$\sigma^2_{\varepsilon[n]} = \left(\frac{1}{\sigma_o^2}+\frac{1}{|\alpha|^2\sigma^2_{\varepsilon[n-1]}+\sigma_w^2}\right)^{-1}; \qquad n>0 \tag{7.174}$$

From this, note that when σ_w^2 is very large compared to σ_o^2, that is, the signal power is much larger than the noise power, the value of $\sigma_{\varepsilon[n]}^2$ becomes approximately equal to σ_o^2. It follows from (7.153) that the gain K_n approaches a value of 1. In this case the estimate (7.161) or (7.164) for the signal becomes equal to just the observation $x[n]$. This corresponds to what we would intuitively expect in a high signal-to-noise ratio environment.

Before considering an example, let us point out that although the error variance and the filter gain K_n are computed recursively, *neither of these terms depends on the data* $x[n]$. Therefore, if desired, both quantities can be computed "off-line" prior to any filtering and stored for later use. This is an advantage in real-time filtering problems since only the estimate $\hat{s}[n]$ needs to be computed in real time. The following example now illustrates the computation of the parameters for the recursive filter.

Example 7.7

The signal of Example 7.2 is to be estimated using a recursive filter. The signal has a correlation function

$$R_s[l] = 2(0.8)^{|l|}$$

and the noise has a variance of $\sigma_o^2 = 2.0$. The parameters of the recursive filter are to be computed for $n = 0, 1, 2$. If it is assumed that $s[-1] = 0$, then the parameters can be computed using (7.153), (7.168), and (7.174).

First, observe that this is a suitable problem for recursive filtering. Since the correlation function of the signal is exponential, the signal can be generated by a model of the required form

$$s[n] = 0.8s[n-1] + w[n]$$

where from (7.161) the noise variance is given by

$$\sigma_w^2 = \sigma_s^2(1 - |\alpha|^2) = 2(1 - (0.8)^2) = 0.72$$

The solution begins by computing the initial variance from (7.168):

$$\sigma_{\varepsilon[0]}^2 = \frac{2.0}{1 + 2.0/2.0} = 1.0$$

Then the first and second order error variance can be computed by applying (7.174) recursively:

$$\sigma_{\varepsilon[1]}^2 = \left(\frac{1}{2.0} + \frac{1}{(0.8)^2(1.0) + 0.72}\right)^{-1} = 0.8095$$

$$\sigma_{\varepsilon[2]}^2 = \left(\frac{1}{2.0} + \frac{1}{(0.8)^2(0.8095) + 0.72}\right)^{-1} = 0.7647$$

(Observe that this last value is exactly the same as the error variance computed in Example 7.2 using three observations.) Once the variances are known, the filter gains can then be computed from (7.153) as

$$K_0 = 1.0/2.0 = 0.5 \qquad K_1 = 0.8095/2.0 = 0.4048 \qquad K_2 = 0.7647/2.0 = 0.3824$$

This completes the solution.

Estimation of a Constant Signal in Noise. In the special case where the random signal is equal to a constant value the recursive estimation procedure takes on an especially simple form. This case can be handled formally by taking $\alpha = 1$ and $w[n] = 0$ in (7.159). This yields

$$s[n] = s[n-1] = s \quad \text{(a constant)} \tag{7.175}$$

For this case the estimate (7.164) simplifies to

$$\hat{s}[n] = \hat{s}[n-1] + K_n(x[n] - \hat{s}[n-1]) \tag{7.176}$$

Since $w[n] = 0$ implies that $\sigma_w^2 = 0$, the expression (7.173) for the variance simplifies to

$$\sigma_{\varepsilon[n]}^2 = \frac{\sigma_{\varepsilon[n-1]}^2}{1 + \sigma_{\varepsilon[n-1]}^2 / \sigma_o^2} \tag{7.177}$$

Note that since the term in the denominator is always greater than 1, this implies that the error variance decreases with increasing n.

Equation (7.176) shows that the estimate for the signal is obtained by taking the previous estimate and adding to it a fraction of the residual. Since the normalized error variance (7.177) is continually decreasing, the estimate will converge to the true signal value. The following example illustrates this application.

Example 7.8

Here the problem is one of estimating a constant random signal in noise. The observation sequence is

$$x[n] = s + \eta[n]$$

where s has mean zero and variance $\sigma_s^2 = 2.0$ and η is white noise with variance $\sigma_o^2 = 2.0$. If it is again assumed that $s[-1] = 0$, the equations for computing the error variance in this case are (7.168) and (7.177). From (7.168) we can compute

$$\sigma_{\varepsilon[0]}^2 = \frac{2.0}{1 + 2.0/2.0} = 1.0$$

Then from (7.177) we have

$$\sigma_{\varepsilon[1]}^2 = \frac{1.0}{1 + 1.0/2.0} = \frac{2}{3}$$

and

$$\sigma_{\varepsilon[2]}^2 = \frac{2/3}{1 + (2/3)(1/2.0)} = \frac{1}{2}$$

Notice that the error variances after $\sigma_{\varepsilon[0]}^2$ are lower than those in Example 7.7. This makes intuitive sense since the signal is not changing. The filter gains, from (7.153), are then

$$K_0 = 1.0/2.0 = 0.5 \qquad K_1 = 0.6667/2.0 = 0.3333 \qquad K_2 = 0.50/2.0 = 0.25$$

7.5.2 Recursive Prediction

The preceding subsection dealt with estimating the *present* value of a signal in noise, given observations up to the present. This subsection deals with estimating the *next* value (i.e., the value of the signal at point $n+1$ given values of the observations only up to point n). It will be shown that this can also be done recursively and that the result is related to the previously derived estimate for the signal in a very simple way. In particular, the estimate is just the predicted estimate in the first of the equations (7.164). That is,

$$\hat{s}[n+1|n] = \alpha\hat{s}[n] \tag{7.178}$$

This estimate is clearly recursive since it is derived from an estimate $\hat{s}[n]$ that is computed recursively. To show that it is the *optimal* mean-square estimate of $s[n+1]$, it is only necessary to show that it satisfies the orthogonality principle.

Let us begin by defining the error in the estimate as

$$\varepsilon[n+1|n] = s[n+1] - \hat{s}[n+1|n] \tag{7.179}$$

Then using (7.178), (7.179), (7.159), (7.145), (7.150), and the white noise property of w, we can show that the expectation of the observations times the conjugated error is given by

$$\begin{aligned}
\mathcal{E}\{x[n-l]\varepsilon^*[n+1|n]\} &= \mathcal{E}\{x[n-l]\left(s^*[n+1] - \hat{s}^*[n+1|n]\right)\} \\
&= \mathcal{E}\{x[n-l]\left(\alpha^* s^*[n] + w^*[n+1] - \alpha^*\hat{s}^*[n]\right)\} \\
&= \alpha^*\mathcal{E}\{x[n-l]\left(s^*[n] - \hat{s}^*[n]\right)\} \\
&= \alpha^*\mathcal{E}\{x[n-l]\varepsilon^*[n]\} = 0; \qquad l = 0, 1, 2, \ldots, n
\end{aligned} \tag{7.180}$$

where the last expression is equal to zero because of (7.150). Thus the orthogonality condition is satisfied and $\hat{s}[n+1|n]$ as given by (7.178) must be the optimal estimate.

By the second part of the Orthogonality Theorem the prediction error variance is given by

$$\begin{aligned}
\sigma^2_{\varepsilon[n+1|n]} &= \mathcal{E}\{s[n+1]\varepsilon^*[n+1|n]\} \\
&= \mathcal{E}\{s[n+1]\left(s^*[n+1] - \hat{s}^*[n+1|n]\right)\} \\
&= \mathcal{E}\{s[n+1]\left(\alpha^* s^*[n] + w^*[n+1] - \alpha^*\hat{s}^*[n]\right)\} \\
&= \mathcal{E}\{\left(\alpha\ s[n] + w[n+1]\right)\left(\alpha^*\varepsilon^*[n] + w^*[n+1]\right)\} \\
&= |\alpha|^2\mathcal{E}\{s[n]\varepsilon^*[n]\} + \mathcal{E}\{w[n+1]w^*[n+1]\} \\
&= |\alpha|^2\sigma^2_{\varepsilon[n]} + \sigma^2_w
\end{aligned} \tag{7.181}$$

where the same conditions and equations used to derive (7.180) have been used here.

Since the optimal predicted estimate for the signal is given by (7.178), the predicted estimate can be generated by the system shown in Fig. 7.15 if the delay and the gain α are simply interchanged. The result is shown in Fig. 7.16.

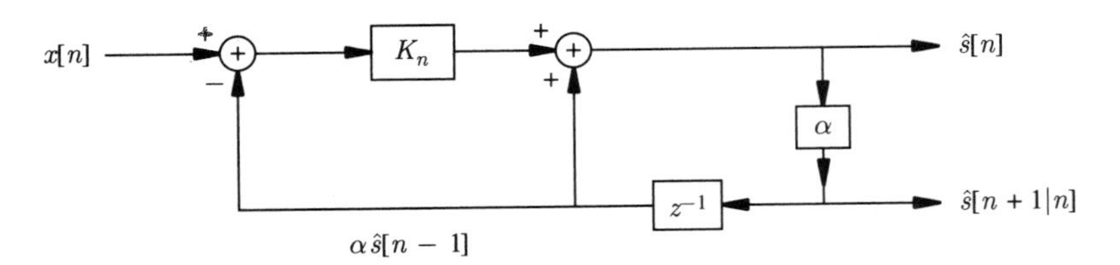

Figure 7.16 System to perform simultaneous filtering and prediction.

The recursive nature of the predicted estimate can be expressed more explicitly by starting with (7.164) and multiplying both sides of the equation by α.

$$\alpha\hat{s}[n] = \alpha(\alpha\hat{s}[n-1]) + \alpha K_n\,(x[n] - \alpha\hat{s}[n-1])$$

Then by virtue of the relation (7.178) this last equation can be written as

$$\hat{s}[n+1|n] = \alpha\hat{s}[n|n-1] + \alpha K_n(x[n] - \hat{s}[n|n-1]) \qquad (7.182)$$

This equation, which is depicted as a block diagram in Fig. 7.17, gives an explicit recursive relation for the estimate $\hat{s}[n+1|n]$. It is also possible to show the additional relations

$$\alpha K_n = \frac{\sigma^2_{\varepsilon[n+1|n]} - \sigma^2_w}{\alpha^*\sigma^2_o} \qquad (7.183)$$

and

$$\sigma^2_{\varepsilon[n+1|n]} = \sigma^2_w + \frac{|\alpha|^2\sigma^2_o\sigma^2_{\varepsilon[n|n-1]}}{\sigma^2_o + \sigma^2_{\varepsilon[n|n-1]}} \qquad (7.184)$$

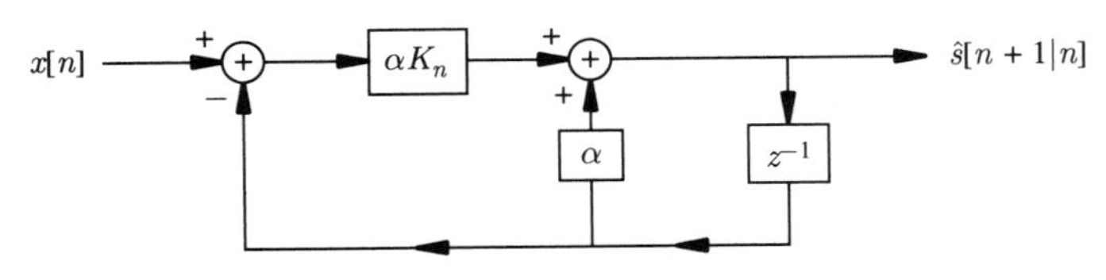

Figure 7.17 Direct realization of the recursive predictor.

which provide explicit recursions for the filter gain and the prediction error variance. The derivation of these relations is left as a problem (Problem 7.31).

The following example illustrates use of the recursive filter in prediction.

Example 7.9

Here we compute the error variances for the *prediction* of the signal specified in Example 7.7. The filter gain parameter K_n is of course the same as in that example. Using (7.181) and the results from Example 7.7 yields

$$\sigma^2_{\varepsilon[1|0]} = (0.8)^2(1.0) + 0.72 = 1.360$$

$$\sigma^2_{\varepsilon[2|1]} = (0.8)^2(0.8095) + 0.72 = 1.238$$

$$\sigma^2_{\varepsilon[3|2]} = (0.8)^2(0.7647) + 0.72 = 1.209$$

Notice that the error variance is worse (i.e., higher) than it is for the filtering problem and that it does not rapidly improve with increasing numbers of observations.

As a final topic we shall show that when the observation noise is zero, the recursive predictor is the same as the linear predictive filter of Section 7.2. In particular, in the limit as σ_o^2 approaches zero in (7.174) or (7.173), the mean-square error $\sigma^2_{\varepsilon[n]}$ becomes equal to σ_o^2 and thus from (7.153) the gain is given by $K_n = 1$. As a result, (7.182) becomes

$$\hat{s}[n+1|n] = \alpha x[n] \tag{7.185}$$

Since with no observation noise $x[n]$ is equal to $s[n]$, it is seen that (7.185) is the same as the linear predictive filter for the signal.

7.5.3 Steady-state Characterization

In this final subsection let us consider the behavior of the filter in Example 7.7 for some larger values of n. Table 7.2 lists the values of K_n and $\sigma^2_{\varepsilon[n]}$ for values of n from 0 to 8. Note that the filter gain and the mean-square error appear to converge to constant steady-state values. For large values of n we would expect the recursive filter to approach the causal Wiener filter and the steady-state mean-square error of the two filters to be identical. In fact, the mean-square error in Table 7.2 apparently converges to 0.75. This is the same value that was computed for the causal Wiener filter in Example 7.5. The gain also appears to converge to the constant value 0.375.

TABLE 7.2 GAIN AND MEAN-SQUARE ERROR OF A TYPICAL RECURSIVE FILTER

n	Gain K_n	Mean-square error $\sigma^2_{\varepsilon[n]}$
0	0.5000	1.0000
1	0.4048	0.8095
2	0.3824	0.7647
3	0.3768	0.7537
4	0.3755	0.7509
5	0.3751	0.7502
6	0.3750	0.7501
7	0.3750	0.7500
8	0.3750	0.7500

To investigate the steady-state behavior in the general case let

$$\sigma^2_{\varepsilon[n]} \rightarrow \sigma^2_\varepsilon = \text{const.}$$

and let us show that σ^2_ε is the same value of mean-square error obtained for the causal Wiener filter in the case study. Substituting σ^2_ε for both $\sigma^2_{\varepsilon[n]}$ and $\sigma^2_{\varepsilon[n-1]}$ in (7.173) and cross-multiplying yields

$$\sigma^2_\varepsilon(\sigma^2_o + \sigma^2_w) + |\alpha|^2\sigma^4_\varepsilon = \sigma^2_w\sigma^2_o + |\alpha|^2\sigma^2_\varepsilon\sigma^2_o$$

which after rearranging results in

$$|\alpha|^2\sigma^4_\varepsilon + \left(\sigma^2_o(1 - |\alpha|^2) + \sigma^2_w\right)\sigma^2_\varepsilon - \sigma^2_w\sigma^2_o = 0 \tag{7.186}$$

The steady-state mean-square error must satisfy this quadratic equation. Direct solution for the roots does not immediately lead to a simple expression for σ^2_ε. However, it can be shown by direct substitution that the expression

$$\sigma^2_\varepsilon = \sigma^2_o\left(1 - \frac{\beta}{\alpha}\right)$$

from (7.116) satisfies the equation and so must be the steady-state mean-square error. This is explored in Problem 7.32. From (7.153) and (7.116) the steady-state gain K is then given by

$$K = \lim_{n\rightarrow\infty} K_n = \frac{\sigma^2_\varepsilon}{\sigma^2_o} = 1 - \frac{\beta}{\alpha} \tag{7.187}$$

You can easily check this result for the case given in Example 7.7 and Table 7.2.

In summary, the steady-state recursive filter behaves like a simple first-order filter with impulse response given by (7.113). The steady-state gain is a simple function of the ratio β/α.

7.6 WOLD DECOMPOSITION

This section turns to a basic result about random processes that it is now possible to address. In Chapter 4 the complex spectral density function for a random process is introduced and written as

$$S_x(z) = S'_x(z) + \sum_i 2\pi \mathsf{P}_i\delta_c(z - e^{\jmath\omega_i}) \tag{7.188}$$

where the first term represents the continuous part and the second term represents the discrete part (the lines) of the spectrum. In Chapter 5 it is shown that if S'_x is *regular* (i.e., if this term can be factored), then the corresponding part of the process can be modeled as a causal and causally invertible filter driven by white noise (the innovations representation). A necessary and sufficient condition for the regularity is the Paley-Wiener condition, which was assumed to hold for S'_x. This section presents the final part of the story, which shows that (7.188) is a completely general representation for a stationary random process and that the process corresponding to S'_x *is* in fact regular. Further, it gives an interpretation of the representation in terms of linear prediction, which is known as the Wold decomposition [7]. The Wold decomposition theorem states that any random process can be written as the sum of two mutually orthogonal random processes, one of which is regular and one that can be predicted by a suitable linear predictive filter with *zero* mean-square error. The regular component corresponds to S'_x while the latter component, which is sometimes referred to as the *singular* component, corresponds to the discrete terms in the spectrum.

The approach here follows the development by Papoulis [8, 9]. We begin by defining the concept of a predictable random process and show that it corresponds to the discrete terms of the spectrum. We then formally state and prove the Wold decomposition theorem.

7.6.1 Predictable Random Processes

Begin by considering the linear prediction problem for a general random process, where the linear predictive filter is possibly of infinite order.

$$\hat{x}[n] = \sum_{k=1}^{\infty} -a_k^* x[n-k] \tag{7.189}$$

The prediction error is

$$\varepsilon[n] = x[n] - \hat{x}[n] = \sum_{k=0}^{\infty} a_k^* x[n-k] \tag{7.190}$$

where $a_0 = 1$. The process is said to be *predictable* if the $\{a_k\}$ can be chosen such that

$$\sigma_\varepsilon^2 = \mathcal{E}\{|\varepsilon[n]|^2\} = 0 \tag{7.191}$$

Otherwise, the process is not predictable.

The phrase "not predictable" is somewhat misleading. Linear prediction can be applied to *any* process, whether predictable or not, with satisfactory results. If a process is not predictable, it just means that the prediction error variance is not zero. It will be shown in fact that *if a random process is regular, then it is not predictable.* As a simple example consider the random process defined by

$$x[n] = \rho x[n-1] + w[n]$$

where $|\rho| < 1$ and w is a white noise process with variance σ_w^2. The optimal set of prediction coefficients is

$$\{a_k\} = \{-\rho^*, 0, 0, \ldots\}$$

and the corresponding prediction error variance is

$$\sigma_\varepsilon^2 = \sigma_w^2 \neq 0$$

Thus, according to the definition, the process is not predictable.

When a process is predictable, there is a deterministic relationship between its samples, although its parameters may be random. Recall the simple random process introduced in Chapter 3,

$$x[n] = A\cos(\omega_o n + \phi) \tag{7.192}$$

where A and ϕ are both real random variables. Two samples of this process determine the values of the random parameters and thus permit this process to be predicted with no error. To view this another way, the sequence satisfies a second-order homogeneous difference equation which forms the basis of a linear predictive filter that can predict the process with zero mean-square error. This is illustrated in the example at the end of this subsection.

Now consider the most general form for a predictable random process. It will be shown that a random process is predictable *if and only if* the complex spectral density function has the form

$$S_x(z) = \sum_{i=1}^{L} 2\pi \mathsf{P}_i \delta_c(z - e^{\jmath\omega_i}) \tag{7.193}$$

or

$$S_x(e^{\jmath\omega}) = \sum_{i=1}^{L} 2\pi \mathsf{P}_i \delta_c(e^{\jmath\omega} - e^{\jmath\omega_i}) \tag{7.194}$$

(where L may be infinite), that is, if its spectrum consists only of lines. Since all periodic and almost periodic processes satisfy this condition, they are therefore predictable.

Before considering the proof of the foregoing statements, let us just mention an interesting extension. The process defined by (7.193) or (7.194) is a very special type of bandlimited process. It can be shown [8, 9] that *any* process that is bandlimited [i.e., has one or more regions where $S_x(e^{\jmath\omega})$ is zero] can be predicted with a prediction error variance that is arbitrarily small. Such processes have therefore been termed *weakly predictable.*

It will now be shown that (7.193) implies that the corresponding random process is predictable, and vice versa. We first prove sufficiency of the condition. Suppose that the spectrum is in the form (7.193) or (7.194). Then form the filter

$$G'(z) = \prod_{i=1}^{L} (1 - z^{-1} e^{\jmath\omega_i}) \tag{7.195}$$

and observe that since $G'(e^{\jmath\omega})$ is zero at each ω_i, it follows that

$$S_x(e^{\jmath\omega})|G'(e^{\jmath\omega})|^2 = \sum_{i=1}^{L} 2\pi \mathsf{P}_i |G'(e^{\jmath\omega})|^2 \delta_c(e^{\jmath\omega} - e^{\jmath\omega_i}) = 0 \tag{7.196}$$

Now the filter $G'(z)$ in (7.195) can be written as

$$G'(z) = 1 + \sum_{i=1}^{L} g'_i z^{-i} \tag{7.197}$$

which has the form of a prediction error filter; this filter is depicted in Fig. 7.18. Then it follows from (7.196) that

$$\sigma_\varepsilon^2 = \mathcal{E}\{|\varepsilon[n]|^2\} = \frac{1}{2\pi}\int_{-\pi}^{\pi} S_x(e^{\jmath\omega})|G'(e^{\jmath\omega})|^2 d\omega = 0 \tag{7.198}$$

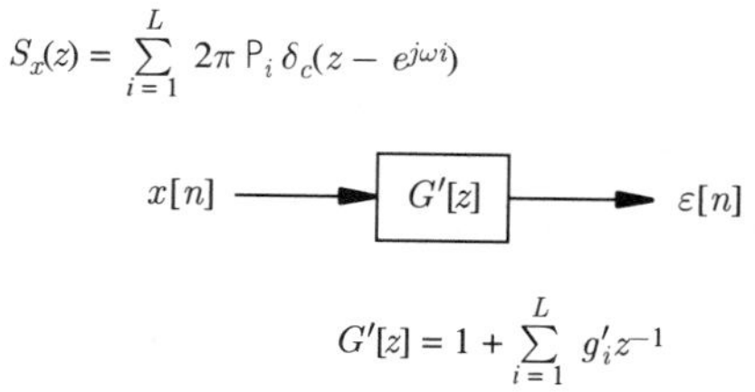

$$S_x(z) = \sum_{i=1}^{L} 2\pi \, \mathsf{P}_i \, \delta_c(z - e^{j\omega_i})$$

$x[n] \longrightarrow \boxed{G'[z]} \longrightarrow \varepsilon[n]$

$$G'[z] = 1 + \sum_{i=1}^{L} g_i' z^{-1}$$

Figure 7.18 Prediction error filter for a process with line spectrum.

This shows that the condition is sufficient.

To show that (7.193) or (7.194) is necessary, note from (7.198) that if σ_ε^2 is zero, then since $S_x(e^{j\omega})|G'(e^{j\omega})|^2$ cannot be negative, this expression must be zero (except at possibly a countable discrete set of points). Now define the function

$$S(z) = G'(z)G'^*(1/z^*)$$

and observe that since it is factorable, $S(e^{j\omega})$ corresponds to the spectrum of a regular process. Consequently, $S(e^{j\omega})$ satisfies the Paley-Wiener condition

$$\int_{-\pi}^{\pi} |\ln S(e^{j\omega})| d\omega < \infty$$

which in turn implies that $S(e^{j\omega}) = |G'(e^{j\omega})|^2$ *cannot* be zero except at a countable set of points. Since the integrand in (7.198) has to be zero and the term $|G'(e^{j\omega})|^2$ cannot be zero, this implies that $S_x(e^{j\omega})$ has to be zero everywhere except at a countable set of points; that is, $S_x(e^{j\omega})$ has the form (7.194). This proves the necessity of the condition.

Let us conclude this subsection with an example to illustrate the construction of the prediction filter.

Example 7.10

A random process has the form

$$x[n] = A\cos(\pi n/4 + \phi)$$

where A and ϕ are real random variables with ϕ uniformly distributed between 0 and 2π. This process has a power spectral density function consisting of two impulses

$$S_x(e^{j\omega}) = 2\pi \mathsf{P}_o \delta_c(e^{j\omega} - e^{j\pi/4}) + 2\pi \mathsf{P}_o \delta_c(e^{j\omega} - e^{-j\pi/4})$$

where $\mathsf{P}_o = \mathcal{E}\left\{A^2\right\}/2$. The prediction filter is constructed from (7.195) as

$$\begin{aligned} G'(z) &= \left(1 - z^{-1}e^{j\pi/4}\right)\left(1 - z^{-1}e^{-j\pi/4}\right) \\ &= 1 - 2\cos\frac{\pi}{4}z^{-1} + z^{-2} \\ &= 1 - \sqrt{2}z^{-1} + z^{-2} \end{aligned}$$

Since any realization of the random process satisfies the difference equation

$$x[n] - \sqrt{2}x[n-1] + x[n-2] = 0$$

the random process is *predictable*.

7.6.2 The Wold Decomposition Theorem

Having examined the concept of a predictable random process, we can now present the main result. The Wold decomposition theorem can be stated as

Theorem 7.3 (Wold Decomposition). A general random process can be written as a sum

$$x[n] = x_r[n] + x_p[n]$$

where $x_r[n]$ is a regular process and $x_p[n]$ is a predictable process, and $x_r[n]$ is orthogonal to $x_p[n]$, that is,

$$\mathcal{E}\{x_r[n_0]x_p^*[n_1]\} = 0; \qquad \forall n_0, n_1$$

Before proceeding with the proof of this theorem, let us just note its implications. Since x_r and x_p are orthogonal, the correlation function of the process is just the sum

$$R_x[l] = R_{x_r}[l] + R_{x_p}[l]$$

Therefore, the complex spectral density function is of the form (7.188) where S'_x is the complex spectral density of the regular process and the sum of impulses is the complex spectral density of the predictable process. Since S'_x is regular, an innovations representation always exists for this part of the process. Further implications of the decomposition can be found in [10].

The proof of this theorem is somewhat lengthy, but the gist of it is as follows. Let $\varepsilon[n]$ be the random process that results from linear prediction of $x[n]$ using a possibly infinite-order predictor. Write

$$\varepsilon[n] = x[n] - \hat{x}[n] = \sum_{k=0}^{\infty} a_k^* x[n-k] \tag{7.199}$$

where $a_0 = 1$ as always. Now consider a different estimation problem, namely that of estimating $x[n]$ *from the error process* $\varepsilon[n]$ with a causal linear filter. Denote this estimate by $\check{x}[n]$ and *note that it is different from the linear predictive estimate* $\hat{x}[n]$. The estimate $\check{x}[n]$, which uses both present and past values of ε, is given by

$$\check{x}[n] = \sum_{k=0}^{\infty} c_k \varepsilon[n-k] \tag{7.200}$$

The relations between the random processes are depicted in Fig. 7.19(a). The estimation error is

$$e[n] = x[n] - \check{x}[n] \tag{7.201}$$

and the coefficients $\{c_k\}$ are chosen to minimize $\mathcal{E}\{|e[n]|^2\}$. It will be shown that $\check{x}[n]$ is the regular process x_r, and that $e[n]$ is the predictable process x_p of the theorem.

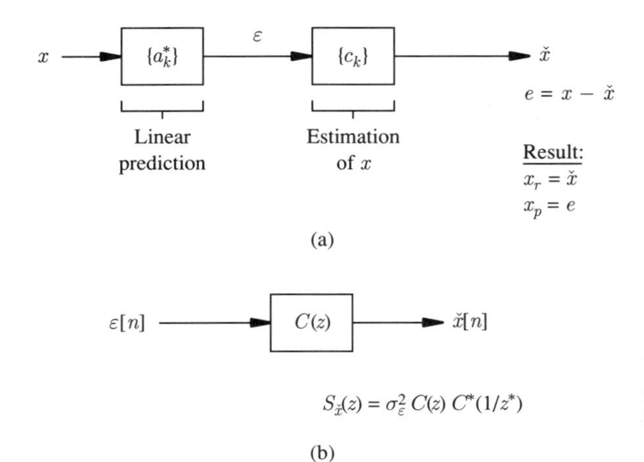

Figure 7.19 Random processes in the Wold decomposition. (a) Relations between $x[n]$, $\varepsilon[n]$, and $\check{x}[n]$. (b) Innovations representation of $\check{x}[n]$.

To begin the proof, let us first show that the two processes $e[n]$ and $\check{x}[n]$ are orthogonal. The following three statements can be made. First,

$$\mathcal{E}\{\varepsilon[n_1]e^*[n_0]\} = 0; \qquad n_1 \leq n_0 \tag{7.202}$$

which follows from the orthogonality principle because e is the error and ε are the observations in the estimation problem (7.200). Second,

$$\mathcal{E}\{x[k]\varepsilon^*[n]\} = 0; \qquad k < n \tag{7.203}$$

This again follows from the orthogonality principle because *in the linear prediction problem* $\varepsilon[n]$ is the error and $\{x[k],\ k < n\}$ are the observations. Finally,

$$\mathcal{E}\{\check{x}[k]\varepsilon^*[n]\} = 0; \qquad k < n \tag{7.204}$$

This is because $\check{x}[k]$ is a linear function of the terms $\varepsilon[l]$ for $l \leq k$ and $\mathcal{E}\{\varepsilon[l]\varepsilon^*[n]\} = 0$ for $l < n$ since ε is a white noise sequence (see Section 7.2.2). Now we can further write

$$\mathcal{E}\{\varepsilon[n_1]e^*[n_0]\} = \mathcal{E}\{\varepsilon[n_1](x^*[n_0] - \check{x}^*[n_0])\} = 0; \qquad n_1 > n_0 \tag{7.205}$$

which follows because from (7.203) and (7.204) each term on the right is zero. The combination of (7.202) and (7.205) yields

$$\mathcal{E}\{\varepsilon[n_1]e^*[n_0]\} = 0; \qquad \forall n_1, n_0 \tag{7.206}$$

Since $\check{x}[n]$ is a linear function of $\varepsilon[n]$ through (7.200), it also follows that

$$\mathcal{E}\{\check{x}[n_1]e^*[n_0]\} = 0; \qquad \forall n_1, n_0 \tag{7.207}$$

In other words, the two random processes e and $\check{x}$ are orthogonal.

Now it will be shown that $\check{x}[n]$ is regular. From (7.201) and the orthogonality (7.207) it follows that

$$\mathcal{E}\{|x[n]|^2\} = \mathcal{E}\{|\check{x}[n]|^2\} + \mathcal{E}\{|e[n]|^2\} \tag{7.208}$$

Then since $\varepsilon[n]$ is a white noise process, (7.200) and (7.208) imply that

$$\mathcal{E}\{|\check{x}[n]|^2\} = \sigma_\varepsilon^2 \sum_{k=0}^{\infty} |c_k|^2 \leq \mathcal{E}\{|x[n]|^2\} = R_x[0] \tag{7.209}$$

where σ_ε^2 is the variance of $\varepsilon[n]$. This bound on the sum implies that the z-transform

$$C(z) = \sum_{k=0}^{\infty} c_k z^{-k} \tag{7.210}$$

exists and converges for $|z| \geq 1$.[9] Thus the operation defined by (7.200) can be implemented by a *causal linear system* with system function $C(z)$ as shown in Fig. 7.19(b). As a result, $S_{\check{x}}(z)$ has a representation

$$S_{\check{x}}(z) = \sigma_\varepsilon^2 C(z) C^*(1/z^*) \tag{7.211}$$

or

$$S_{\check{x}}(e^{j\omega}) = \sigma_\varepsilon^2 |C(e^{j\omega})|^2 \tag{7.212}$$

In other words, the process $\check{x}[n]$ is regular.

The final step in the proof of the theorem is to show that $e[n]$ is predictable. For this, return to the linear prediction problem and the prediction error filter (7.199). The filter is depicted in Fig. 7.20. When the input $x[n]$ is applied to the filter, the error $\varepsilon[n]$ can be represented as the sum of two terms: a component $\varepsilon_r[n]$ corresponding to the term $\check{x}[n]$, and a component $\varepsilon_p[n]$ corresponding to the term $e[n]$ [see Fig. 7.20(a)]. It will be shown that the filter $A(z)$ is also an optimal PEF for $e[n]$ and in particular that

$$\mathcal{E}\{|\varepsilon_p[n]|^2\} = 0 \tag{7.213}$$

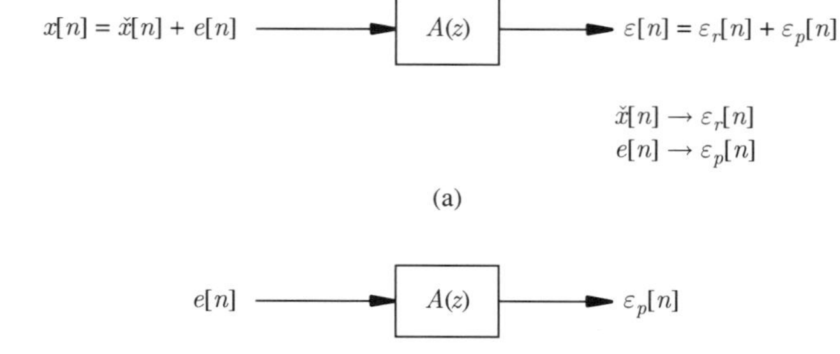

Figure 7.20 Prediction error filter applied to a random process. (a) Response due to entire random process. (b) Response due to predictable component only.

so that $e[n]$ is predictable.

To show the result first write ε_p as

$$\varepsilon_p[n] = \varepsilon[n] - \varepsilon_r[n] \tag{7.214}$$

[9] See [11] for a discussion of this point.

Now notice that since $\check{x}[n]$ depends only on $\varepsilon[n]$ and its past, $\varepsilon_r[n]$ and therefore $\varepsilon_p[n]$ depends only on $\varepsilon[n]$ and its past. This is also obvious from Fig. 7.21, which shows the relations between all the quantities in block diagram form. Therefore, the orthogonality statement (7.206) also implies that

$$\mathcal{E}\{\varepsilon_p[n_1]e^*[n_0]\} = 0; \qquad \forall n_1, n_0 \tag{7.215}$$

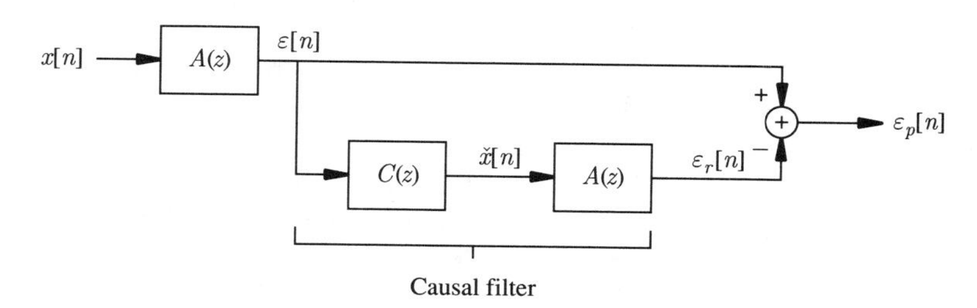

Figure 7.21 Processing involved in proof of Wold decomposition theorem.

In particular, this implies the two conditions

$$\mathcal{E}\{\varepsilon_p[n]e^*[n-k]\} = 0; \qquad k > 0 \tag{7.216}$$

and

$$\mathcal{E}\{\varepsilon_p[n]e^*[n]\} = 0 \tag{7.217}$$

which can be interpreted in the context of Fig. 7.20(b). (Here $e[n]$ is the process to be predicted and $\varepsilon_p[n]$ is the prediction error.) Equation 7.216 is a statement of the orthogonality principle and shows that $A(z)$ is the optimal PEF for $e[n]$. Equation 7.217 is just the second part of the Orthogonality Theorem and states that the prediction error variance is zero.[10] This completes the proof.

7.7 CHAPTER SUMMARY

This chapter deals with the problem of optimal filtering. The optimal filters described are based upon linear mean-square estimation and the associated orthogonality principle. This principle, which is presented as a theorem, states that the optimal linear mean-square estimate of a random variable is found when the error is made *orthogonal* to the observations. The minimum mean-square error is then the expected value of the product of the error and the quantity being estimated. All of this has a convenient interpretation in a vector space where the elements are random variables and the inner product is expectation.

Linear prediction is described here as a first example of optimal filtering. According to the orthogonality principle, the optimal filter for predicting the present value of a random process from some number of past values is the filter whose error process is orthogonal to

[10] An alternative argument is the following. Since $\varepsilon_p[n]$ is derived from $e[n]$ [see Fig. 7.20(b)] the only way for the expression (7.215) to hold is if $\varepsilon_p[n]$ is zero almost everywhere. Therefore, the variance of $\varepsilon_p[n]$ is zero.

the past observations. This is called the linear predictive filter. The filter that generates the error sequence from the input process is called the prediction error filter (PEF). The output power of the PEF is the minimum mean-square error or the prediction error variance. The filter coefficients and the prediction error variance are found by solving a set of linear equations known as the Normal equations. By choosing a filter of sufficiently high order the error process can be made white or as close to a white noise process as desired.

The most general form of optimal mean-square filtering is called Wiener filtering. A Wiener filter can be constructed to estimate any signal, whether linearly or nonlinearly generated or linearly or nonlinearly related to the observations. The estimate is linear, however, and construction of the filter requires knowledge of the autocorrelation of the observation process and the cross-correlation between the observations and the estimated or "desired" random process. The equation that specifies the optimal filter is called the Wiener-Hopf equation. For the case of an FIR filter the Wiener-Hopf equation can be written as a set of linear equations similar to the Normal equations of linear prediction. For the case of an IIR filter the Wiener-Hopf equation has to be solved in the transform domain. For a causal IIR filter the procedure is based on spectral factorization and requires several steps. If the IIR filter is allowed to be noncausal, then the filter can be found with simple algebraic techniques. The case of a signal with exponential correlation function in additive white noise can be analyzed algebraically and leads to a deeper understanding of the Wiener filter. This is presented in the chapter as a case study.

The special problem analyzed in the case study can be formulated as an optimal recursive filtering problem. The generalization of this procedure to vector-valued random processes (not treated here) is known as Kalman filtering. The recursive filter is a special form of the Wiener filter that is causal but not shift-invariant. It begins by processing a single observation and continues to use more samples of the observation process to estimate the desired signal as those samples become available. Moreover, the filter has an intuitively pleasing interpretation in its use of past and present information. The filter parameters are also computed recursively, but since they do not depend on the observed data, they can be computed "off-line." After a sufficient number of observations the filter approaches a steady-state condition and becomes equivalent in its action and performance to the IIR Wiener filter.

Finally, the Wold decomposition provides a rigorous fundamental characterization of random processes that is based upon prediction and estimation. The Wold decomposition shows that any random process can be represented as the sum of two orthogonal processes corresponding to the continuous and discrete (line) components of the spectrum. The first is a regular process and thus has an innovations representation as shown in Chapter 5. The second, although random, is predictable; a suitable linear predictive filter can be found that predicts this component with zero mean-square error.

REFERENCES

1. Garrett Birkhoff and Saunders MacLane. *A Survey of Modern Algebra*, 4th ed. Macmillan, New York, 1977.
2. Paul R. Halmos. *Finite-Dimensional Vector Spaces*. Springer-Verlag, New York, 1974.

3. Sophocles J. Orfanidis. *Optimum Signal Processing: An Introduction*, 2nd ed. Macmillan, New York, 1988.
4. Norbert Wiener. *Extrapolation, Interpolation, and Smoothing of Stationary Time Series*. MIT Press (formerly Technology Press), Cambridge, Massachusetts, 1949.
5. R. E. Kalman. A new approach to linear filtering and prediction problems. *Transactions of the ASME, Series D, Journal of Basic Engineering*, 82:34–45, 1960.
6. R. E. Kalman and R. S. Bucy. New results in linear filtering and prediction theory. *Transactions of the ASME, Series D, Journal of Basic Engineering*, 83:95–107, 1961.
7. H. Wold. *The Analysis of Stationary Time Series*, 2nd ed. Almquist and Wicksell, Uppsala, Sweden, 1954. (Originally published in 1938 as *A Study in the Analysis of Stationary Time Series*.)
8. A. Papoulis. Predictable processes and Wold's decomposition: a review. *IEEE Transactions on Acoustics, Speech, and Signal Processing*, ASSP-33(4):933–938, August 1985.
9. A. Papoulis. Levinson's algorithm, Wold's decomposition, and spectral estimation. *SIAM Review*, 27(3):405–441, September 1985.
10. Peter Whittle. *Prediction and Regulation by Linear Least-Squares Methods*, 2nd ed. University of Minnesota Press, Minneapolis, Minnesota, 1983.
11. Alan V. Oppenheim and Ronald W. Schafer. *Discrete-Time Signal Processing*. Prentice Hall, Inc., Englewood Cliffs, New Jersey, 1989.

PROBLEMS

7.1. A random variable x is described by

$$x = A\cos\theta + b$$

Another random variable y is described by

$$y = A\sin\theta + c$$

where A, b, c, and θ are all random variables. Assume that x and y are observed. It is desired to estimate θ as a *linear* combination of x and y. State the *orthogonality principle* involved in this estimation problem. State it in words using the variables in this problem. Be sure to define what you mean by all of the terms that you use.

7.2. A constant random signal s is observed in white noise $w[n]$ independent of s. Two observations are taken:

$$x[0] = s + w[0]$$

$$x[1] = s + w[1]$$

The signal has mean zero and variance σ_s^2. The noise has mean zero and variance σ_w^2.

(a) What is the optimal linear mean-square estimate of s given only the single observation $x[0]$? What is the corresponding error variance?

(b) What is the optimal linear mean-square estimate of s given both $x[0]$ and $x[1]$? What is the corresponding error variance?

7.3. The Orthogonality Theorem of linear mean-square estimation states that the error must be orthogonal to the observations. A more general result for (nonlinear) mean-square estimation is that the error must be orthogonal to *any function* of the observations. In other words, if $g(\boldsymbol{x})$ is some function of the observations $x_1, x_2, \ldots, x_N$, and $\varepsilon = y - \hat{y}_{ms}$ where $\hat{y}_{ms}$ is the mean-square estimate of y (not necessarily linear), then

$$\mathcal{E}\{\varepsilon g^*(\boldsymbol{x})\} = 0$$

Prove this more general result.

7.4. In Section 7.2 it was shown that the Normal equations of linear prediction could be arrived at by invoking the orthogonality theorem. The Normal equations can also be derived straightforwardly by minimizing the mean-square error with respect to the filter coefficients. Assume that the random process is real.

(a) Write an expression for the mean-square error of linear prediction in terms of the signal values and the filter coefficients $a_1, a_2, \ldots a_P$.

(b) By taking derivatives of the expression with respect to a_i for $i = 1, 2, \ldots P$, derive the Normal equations.

7.5. Find the first- and second-order prediction error filters and the prediction error variance for random processes with the following correlation functions:

(a)

$$R_x[0] = 3\,; \quad R_x[\pm 1] = 2\,; \quad R_x[\pm 2] = 1$$

$$R_x[l] = 0\,; \quad |l| > 2$$

(b)

$$R_x[l] = 2(0.8)^{|l|}$$

7.6. Find the coefficients of the first-, second-, third- and fourth-order prediction error filters and the corresponding prediction error variances for a random process with the correlation function

$$R_x[0] = 2\,; \quad R_x[\pm 1] = 1\,; \quad R_x[l] = 0\,;\ |l| > 2$$

7.7. A first-order real random process is defined by the difference equation

$$x[n] = \rho x[n-1] + w[n]$$

where $w[n]$ is a white noise process with variance σ_w^2. The correlation matrix for any order P is Toeplitz and has the form

$$\mathbf{R}_x = \frac{\sigma_w^2}{1-\rho^2}\begin{bmatrix} 1 & \rho & \rho^2 & \cdots & \rho^P \\ \rho & 1 & \rho & \cdots & \rho^{P-1} \\ \vdots & & & & \vdots \\ \rho^P & & & \cdots & 1 \end{bmatrix}$$

Show that the linear predictive filter parameters for such a random process are given by

$$\{a_1, a_2, \ldots, a_P\} = \{-\rho, 0, 0, \ldots, 0\}$$

for any order $P \geq 1$ and that the prediction error variance is equal to σ_w^2. That is, show that this solution satisfies the Normal equations.

7.8. A nonstationary random process with autocorrelation function $R_x[n_1, n_0]$ is to be predicted by a general linear filter of the form

$$\hat{x}[n] = \sum_{k=1}^{P} -a_k^*[n]x[n-k]$$

Derive the Normal equations for this nonstationary case.

7.9. Given a set of linearly independent vectors $v_0, v_1, \ldots, v_{N-1}$, it is possible to construct a set of mutually orthogonal unit vectors $u_0', u_1', \ldots, u_{N-1}'$ through the Gram-Schmidt orthogonalization procedure:

$$u_0' = v_0/|v_0|$$

$$u_1 = v_1 - (v_1, u_0')u_0'$$

$$u_1' = u_1/|u_1|$$

$$u_2 = v_2 - (v_2, u_1')u_1' - (v_2, u_0')u_0'$$

$$u_2' = u_2/|u_2|$$

$$\vdots$$

$$u_{N-1} = v_{N-1} - \sum_{i=0}^{N-2}(v_{N-1}, u_i')u_i'$$

$$u_{N-1}' = u_{N-1}/|u_{N-1}|$$

where $|u|$ denotes the vector "magnitude"

$$|u|^2 = (u, u)$$

Identify the vectors $v_0, v_1, \ldots, v_{N-1}$ with the observations $x[0], x[1], \ldots, x[N-1]$ of a random process. Further, identify the vectors $u_0, u_1, \ldots, u_{N-1}$ with the errors $\varepsilon[0], \varepsilon[1], \ldots, \varepsilon[N-1]$ in successive orders of linear prediction of the random process. The purpose of this problem is to show that generation of the prediction errors is mathematically equivalent to performing the Gram-Schmidt orthogonalization of the vectors.

(a) Start with $\varepsilon[0] = x[0]$. What is the normalized unit vector corresponding to u_0'? Call this $\varepsilon'[0]$.

(b) Now write

$$\varepsilon[1] = x[1] + a_1^{(1)*}x[0]$$

By applying the orthogonality principle, solving for $a_1^{(1)*}$, and substituting back in this equation, show that this corresponds to the next step in the Gram-Schmidt procedure. Also write an expression for $\varepsilon'[1]$, the normalized version of $\varepsilon[1]$.

(c) The expression for $\varepsilon[2]$ is

$$\varepsilon[2] = x[2] + a_1^{(2)*}x[1] + a_2^{(2)*}x[0]$$

Since $\varepsilon[1]$ and $\varepsilon[0]$ are linear combinations of $x[1]$ and $x[0]$, observe that $\varepsilon[2]$ can be written as

$$\varepsilon[2] = x[2] - c_1\varepsilon[1] - c_0\varepsilon[0]$$

for some suitable choice of the constants c_1 and c_0. Now think of $\varepsilon[1]$ and $\varepsilon[0]$ as the observations. By applying the orthogonality principle to this last equation (in terms of the new "observations") solve for c_1 and c_0 and show that the last equation corresponds to the next step in the Gram-Schmidt procedure. Also write an expression for $\varepsilon'[2]$, the normalized error.

(d) Generalize step (c) of this problem to show that linear prediction involving the k observations $x[0], x[1], \ldots, x[k-1]$ corresponds to the kth step in the Gram-Schmidt procedure.

7.10. In the following, the sequence $\{x[n]; -\infty < n < \infty\}$ is composed of independent identically-distributed random variables. The density function for any one of the $x[n]$ is uniform:

$$f_{x[n]}(x) = \begin{cases} \frac{1}{2} & -1 \leq x \leq 1 \\ 0 & \text{otherwise} \end{cases}$$

A random process $y[n]$ is defined by applying $x[n]$ to a causal linear shift-invariant system. The system is described by the difference equation

$$y[n] - \tfrac{1}{3}y[n-1] = x[n]$$

State if the following statements are correct or incorrect and why:

(a) The best linear mean-square estimate of $y[n]$ given $y[n-1]$ and $y[n-2]$ is

$$\hat{y}[n] = \tfrac{1}{2}\left(y[n-1] + y[n-2]\right)$$

(b) The input power density spectrum is given by

$$S_x(e^{\jmath\omega}) = \tfrac{1}{3} \qquad (\text{all } \omega)$$

(c) As long as the input sequence has the same mean and variance, the power spectrum of the output is independent of the form of the probability density $f_{x[n]}$.

(d) The probability density function of the output has the same form as the probability density function of the input but may have different parameters.

(e) The present value of the output is uncorrelated with future values of the output.

(f) The values $R_y[0]$ and $R_y[1]$ determine the values of the output correlation function $R_y[l]$ for all l.

7.11. A real random process has the exponential correlation function

$$R_x[l] = 2^{-|l|}$$

It is desired to estimate $x[n+2]$ using an FIR linear shift-invariant filter of the form

$$\hat{x}[n+2] = \sum_{k=0}^{P-1} h[k]x[n-k]$$

(a) Write the equations to find the filter coefficients $h[k]$ where $P = 2$. You do not need to solve these equations for $h[0]$ and $h[1]$.

(b) Write the equations for the mean-square error.

$$\mathcal{E}\left\{(x[n+2] - \hat{x}[n+2])^2\right\}$$

Write the answer in terms of $h[0]$ and $h[1]$. You do not need to compute the value of the mean-square error for this problem.

(c) Show that the minimizing values of $h[0]$ and $h[1]$ are given by the solution to the equations in part (a).

7.12. It is desired to estimate a random signal $s[n]$ using an FIR Wiener filter of length $P = 2$. The observations are of the form

$$x[n] = s[n] - s[n-1] + \eta[n]$$

where $\eta[n]$ is white noise with variance $\sigma_o^2 = 10$, independent of the signal. The correlation function for the signal is

$$R_s[l] = 6(0.8)^{|l|}$$

(a) What is the impulse response of the optimal filter?
(b) What is the corresponding mean-square error?

7.13. Consider a real random signal $s[n]$ with correlation function

$$R_s[l] = \frac{\sigma_w^2}{1-\rho^2}\rho^{|l|}$$

(a) Suppose at first that the signal can be observed directly, that is, the observations $x[n]$ are given by

$$x[n] = s[n]$$

Find the coefficient a_1 of the first-order linear predictive filter and the corresponding prediction error variance σ_ϵ^2.

(b) Now suppose that the signal is observed in noise. That is,

$$x[n] = s[n] + \eta[n]$$

where $\eta[n]$ is a zero-mean white noise process with variance σ_o^2 and uncorrelated with $s[n]$. What is the value of the filter coefficient $h[0]$ and the prediction error variance in the first-order Wiener filter? The estimate is defined by

$$\hat{s}[n+1] = h[0]x[n]$$

7.14.[11] Given an observation sequence of the form

$$x[n] = s[n] + \eta[n]$$

where s and η are orthogonal and where

$$R_s[l] = 2(0.8)^{|l|}; \qquad R_\eta[l] = 2(0.5)^{|l|}$$

find the FIR Wiener filter and the corresponding mean-square error for estimating

$$d[n] = s[n]$$

(a) Use a filter with length $P = 2$.
(b) Use a filter with length $P = 3$.

[11]Problems 7.14 to 7.22 and 7.24 are intended to be solved "from scratch," not simply by using the formulas derived in the Case Study.

7.15. Repeat Problem 7.14 for $d[n] = s[n+1]$.

7.16. Repeat Problem 7.14 for $d[n] = s[n-1]$.

7.17. A random process is observed which is of the form

$$x[n] = s[n] + \eta[n]$$

where $s[n]$ is a random signal with complex spectral density function

$$S_s(z) = \frac{0.82}{(1-0.6z^{-1})(1-0.6z)}$$

and $\eta[n]$ is white noise with variance $\sigma_o^2 = 1.0$ uncorrelated with the signal. Find the causal IIR Wiener filter for estimating the signal and the minimum mean-square error.

7.18. Find the optimal IIR Wiener filter and the corresponding mean-square error for the processes of Example 7.4 with
(a) $d[n] = s[n+2]$.
(b) $d[n] = s[n-2]$.

7.19. Find the optimal IIR Wiener filter and the corresponding mean-square error for the processes of Problem 7.14. Assume that $d[n] = s[n]$.

7.20. Repeat Problem 7.19 for
(a) $d[n] = s[n+2]$.
(b) $d[n] = s[n-2]$.

7.21. Find the *noncausal* Wiener filter for the signal $s[n]$ in Problem 7.17 and the corresponding mean-square error.

7.22. Find the *noncausal* IIR Wiener filter for the random processes of Problem 7.14 with $d[n] = s[n]$ and evaluate the mean-square error.

7.23. In traveling through the ocean, a sonar signal is subject to dispersion as well as additive noise. A simple model for the received signal is

$$x[n] = s[n] * b[n] + \eta[n]$$

where $b[n]$ is the impulse response of a linear system representing the ocean medium. Find the noncausal Wiener filter to estimate $s[n]$ when s is a real random signal with correlation function

$$R_s[l] = 6(0.8)^{|l|}$$

$\eta[n]$ is white noise with variance $\sigma_o^2 = 5$, independent of the signal, and the impulse response $b[n]$ is given by

$$b[n] = (0.9)^n u[n]$$

where $u[n]$ is the unit step function. Also find the corresponding mean-square error.

7.24. Consider a random process

$$x[n] = s[n] + \eta[n]$$

where

$$s[n] = \left(\frac{1}{\sqrt{2}}\right)^n u[n]$$

and $\eta[n]$ is white noise with unit variance uncorrelated with the signal. The signal complex spectral density function is

$$S_s(z) = \frac{1/2}{(1-(1/\sqrt{2})z^{-1})(1-(1/\sqrt{2})z)}$$

The causal Wiener filter is known to have the form

$$H(z) = \frac{C}{1-\beta z^{-1}}$$

Without developing a complete solution for the filter, find the pole location β. [Work out the steps in deriving the Wiener filter as far as necessary. Do not simply use the formula (7.98).]

7.25. Show that the quantities α and β defined in Section 7.4.3 satisfy the condition $|\beta| \leq |\alpha|$. *Hint*: Notice that when $\gamma = 0$, $\beta = \alpha$.

7.26. By equating the two quantities and showing that β is a root of the resulting equation, show that the two expressions (7.106) and (7.111) are equivalent.

7.27. Check that the expressions (7.98), (7.102), (7.112), (7.115), (7.116), (7.131), (7.132), and (7.137) are correct by evaluating them for conditions of Examples 7.4, 7.5, and 7.6.

7.28. A constant random signal with a uniform distribution

$$f_s(s) = \begin{cases} 1/2 & 0 \leq s \leq 2 \\ 0 & \text{otherwise} \end{cases}$$

is observed in zero mean white Gaussian noise $\eta[n]$ with variance $\sigma_o^2 = 1$. The observation sequence is

$$x[n] = s + \eta[n]$$

The values of $x[n]$ given in the table below are observed. Compute the optimal linear least mean-square estimate of the signal corresponding to each of the observations and the variance of the error.

n	$x[n]$	$\hat{s}[n]$	Error variance
0	–	1	1/3
1	1.7		
2	1.3		
3	0.9		
4	1.5		

7.29. In estimation of a constant but unknown random signal in noise, we use the recursive filter

$$\hat{s}[n] = \hat{s}[n-1] + K_n\Big(x[n] - \hat{s}[n-1]\Big)$$

where from (7.153) and (7.177) it is seen that the gain K_n satisfies the recursion

$$K_n = \frac{K_{n-1}}{1+K_{n-1}}$$

In this problem we want to study the response of the filter when the "constant" signal undergoes a change. We will assume that the observation $x[n]$ is equal to the signal itself (i.e., there is no noise in the observations).

(a) Beginning with the value $K_0 = 1$, compute and tabulate the gain K_n versus n for $n = 0, 1, 2, 3, 4$.

(b) Suppose that the signal is constant but changes to another value at the $n = 2$. Specifically, assume that the observation sequence $x[n]$ is

$$1 \quad 1 \quad 2 \quad 2 \quad 2 \quad \cdots$$

Assume that $\hat{s}[-1] = 0$. Compute the estimated signal $\hat{s}[n]$ for $n = 0, 1, 2, 3, 4$. It will be easiest to do this if you organize your work into the table

n	$x[n]$	$\hat{s}[n-1]$	$x[n] - \hat{s}[n-1]$	K_n	$\hat{s}[n]$
0	1	0			
1	1				
2	2				
3	2				
4	2				

Will the estimate converge to the new value of the signal?

(c) Suppose that the signal had changed from 1 to 2 at a much later point, for example at $n = 10$. What can you say about the response of the recursive filter? How would the history of the estimate compare to that for part (b)?

7.30. Show that the two recursions for the error variance (7.173) and (7.174) are equivalent.

7.31. Derive the recursive relations (7.183) and (7.184) for the recursive predictor. You can begin with the relations that were derived for the filter in Section 7.5.1.

7.32. Show that the expression

$$\sigma_\varepsilon^2 = \sigma_\mathrm{o}^2 \left(1 - \frac{\beta}{\alpha}\right)$$

of (7.116) satisfies the quadratic equation (7.186) that determines the steady-state mean-square error for the recursive filter. Do this by substituting this expression in (7.186) and noting that β is a root of the resulting quadratic equation.

7.33. **(a)** Show that the residual

$$\begin{aligned} r[n] &= x[n] - \alpha\hat{s}[n-1] \\ &= x[n] - \hat{s}[n|n-1] \end{aligned}$$

in the recursive filter of Section 7.5 is a white noise process. Determine its variance.

(b) Show in the steady state ($K_n \to K$ and $\sigma_{\varepsilon[n]}^2 \to \sigma_\varepsilon^2$) that $r[n]$ is the innovations process for the observation process $x[n]$. In other words, show that $x[n]$ can be derived by applying $r[n]$ to a suitable linear shift-invariant filter. Determine this filter.

(c) Show that $r[n]$ is orthogonal to the past values of the observations, $x[n-1], x[n-2], \ldots$, and therefore we can define

$$\hat{x}[n] = \alpha\hat{s}[n-1] = \hat{s}[n|n-1]$$

as the optimal estimate of $x[n]$ given the past observations.

(d) Now derive the filter for the residual $r[n]$ expressing $r[n]$ in terms of present and past values of the observations and past values of the residual. Show that this filter is the inverse of the filter found in part (b).

(e) Using the expression (7.187) for the steady-state gain, express the filters in parts (b) and (d) in terms of the parameters α and β.

COMPUTER ASSIGNMENTS

7.1. In this computer assignment you will write a program to do linear predictive filtering of the data sets S00, S01, S02, and S03.

(a) Using the autocorrelation method of Chapter 6, estimate a 3×3 Toeplitz correlation matrix for the signal data. (This is the same as using the *biased* estimate for the correlation function and constructing the Toeplitz matrix.) Print this matrix.

(b) Solve a set of Normal equations involving the correlation matrix that you just generated to obtain the second-order linear predictive filter parameters and the prediction error variance.

(c) Apply the filter to the original data set and generate the prediction error signal. Plot this signal and compute its variance. Does it compare well with the theoretical prediction error variance you obtained by solving the Normal equations?

(d) Take the upper 2×2 block of the correlation matrix you generated in part (a) and solve for the coefficients and prediction error variance of a first-order linear predictive filter. Do this using pencil and paper, not using the computer. How does this first-order prediction filter compare to the second-order filter you generated on the computer in part (b)?

7.2. The sequence S00.DAT is a realization of a white noise process with variance 1. Assume that the sequence S01.DAT is observed in additive white noise uncorrelated with the signal. That is, the observed sequence is given by

$$x[n] = s[n] + \sigma_o w[n]$$

where $s[n]$ is S01.DAT and $w[n]$ is S00.DAT and s and w are assumed to be statistically independent.

The variance parameter σ_o^2 is determined so that the signal-to-noise ratio, defined as

$$\text{SNR} = 10 \log_{10} \frac{\text{Var}\,[s[n]]}{\sigma_o^2} = 10 \log_{10} \frac{R_s[0]}{\sigma_o^2}$$

is equal to 0 dB.

(a) Let $d[n] = s[n]$. Estimate the sample correlation functions $R_x[l]$ and $R_{dx}[l]$ from the given data. Use the relation

$$R_x[l] = R_s[l] + \sigma_o^2 R_w[l]$$

to save computation time when using different values of SNR.

(b) Using the estimated correlation functions, compute the FIR Wiener filter for

1. $P = 4$
2. $P = 8$
3. $P = 16$

For each case plot the sequence $s[n]$, the estimate $\hat{s}[n]$, and the error $\varepsilon[n]$. Compare the theoretical mean-square error to the signal average-squared error $\langle \varepsilon^2[n] \rangle$.

(c) Repeat steps (a) and (b) with σ_o^2 chosen to yield a signal-to-noise ratio of -10 dB.

(d) *(optional)* Repeat steps (a) through (c) using S02.DAT and S03.DAT for $s[n]$.

7.3. Repeat Computer Assignment 7.2 with $d[n] = s[n+3]$.

7.4. Repeat Computer Assignment 7.2 with $d[n] = s[n-3]$.

7.5. Consider the situation in Computer Assignment 7.2. Assume that the sequences S01.DAT, S02.DAT, and S03.DAT satisfy the conditions of the case study. You can further assume that $\alpha = 0.95, 0.70, \text{and} - 0.95$ for S01, S02, and S03 respectively and use the values of σ_s^2 computed in Computer Assignment 7.2.

(a) Find the optimal IIR filter for estimating $s[n]$ and filter the data using the recursive form of the difference equation for the filter. Also compute the theoretical mean-square error.

(b) Plot the estimation error $\varepsilon[n]$ and compute $\langle \varepsilon^2[n] \rangle$. Compare the results to those of Computer Assignment 7.2. Some values of the theoretical mean-square error may be higher than those in Computer Assignment 7.2. Can you explain why?

7.6. Consider the situation in Computer Assignment 7.2.

(a) Find the optimum recursive filter for estimating $s[n]$ and filter the data.

(b) Plot the estimation error and compute $\langle \varepsilon^2[n] \rangle$. Compare the results to those of Computer Assignments 7.2 and 7.5.

8

Linear Prediction

In Chapter 7 linear prediction was introduced as a special type of optimal filtering. Although linear prediction is indeed an optimal filtering problem, it is much more than just that. Methods of linear prediction have been used in numerous problems relating to signal processing [1, 2]. Digital processing of speech in particular uses methods of linear prediction for speech synthesis, recognition, coding, and many other applications [3, 4]. Mathematically, it also is a very rich area with a number of important connections to other topics.

This chapter takes a comprehensive look at linear prediction and the sister topic of autoregressive modeling. Although linear prediction seeks to estimate a random process while autoregression seeks to model it, the two topics are like two sides of the same coin. Thus it is most appropriate to treat them together. In the discussion of linear prediction it is found that that an algorithm known as the Levinson recursion, originally developed as a fast method to solve the Normal equations, brings a wealth of new insight to the problem and leads to a number of important new developments. In fact, linear prediction is central to a large number of topics in signal processing and its role does not end in this chapter. A number of additional connections of linear prediction to other topics are found in the subsequent chapters that discuss linear modeling and spectral estimation.

8.1 ANOTHER LOOK AT LINEAR PREDICTION

To begin this chapter, let us review the linear prediction problem introduced in Chapter 7. Recall that the problem is to find an estimate of the current value of a random process $x[n]$ from P previous values. If the estimate is written as

$$\hat{x}[n] = -a_1^* x[n-1] - a_2^* x[n-2] - \cdots - a_P^* x[n-P] \tag{8.1}$$

then the error in the estimate is given by the difference

$$\varepsilon[n] = x[n] - \hat{x}[n] \tag{8.2}$$

or

$$\varepsilon[n] = x[n] + a_1^* x[n-1] + a_2^* x[n-2] + \cdots + a_P^* x[n-P] \tag{8.3}$$

The Normal equations, whose solution provides the optimal coefficients, can be derived directly in matrix form by first defining the vectors

$$\boldsymbol{x}[n] \stackrel{\text{def}}{=} \begin{bmatrix} x[n-P] \\ x[n-P+1] \\ \vdots \\ x[n] \end{bmatrix} \tag{8.4}$$

and

$$\boldsymbol{a} \stackrel{\text{def}}{=} \begin{bmatrix} 1 \\ a_1 \\ \vdots \\ a_P \end{bmatrix} \tag{8.5}$$

The vector $\boldsymbol{a}$ is the vector of linear prediction coefficients. This vector and the prediction error variance $\sigma_\varepsilon^2 = \mathcal{E}\{|\varepsilon[n]|^2\}$ constitute the *linear prediction parameters*. With the definitions (8.4) and (8.5) the equation (8.3) for the prediction error can be written as

$$\varepsilon[n] = \boldsymbol{a}^{*T} \tilde{\boldsymbol{x}}[n] \tag{8.6}$$

where $\tilde{\boldsymbol{x}}[n]$ is the reversal of $\boldsymbol{x}[n]$, that is,

$$\tilde{\boldsymbol{x}}[n] = \begin{bmatrix} x[n] \\ x[n-1] \\ \vdots \\ x[n-P] \end{bmatrix}$$

To find the optimal filter coefficients apply the Orthogonality Theorem which states that $\mathcal{E}\{x[n-i]\varepsilon^*[n]\} = 0, \quad i = 1, 2, \ldots, P$ and that $\sigma_\varepsilon^2 = \mathcal{E}\{x[n]\varepsilon^*[n]\}$. These conditions can be stated very compactly in matrix notation as

$$\mathcal{E}\{\tilde{\boldsymbol{x}}[n]\varepsilon^*[n]\} = \begin{bmatrix} \sigma_\varepsilon^2 \\ 0 \\ \vdots \\ 0 \end{bmatrix} = \sigma_\varepsilon^2 \boldsymbol{\iota} \tag{8.7}$$

where the vector $\boldsymbol{\iota}$ is defined by

$$\boldsymbol{\iota} \stackrel{\text{def}}{=} \begin{bmatrix} 1 \\ 0 \\ \vdots \\ 0 \end{bmatrix} \tag{8.8}$$

Now dropping the argument $[n]$ for simplicity and substituting (8.6) into (8.7) yields

$$\mathcal{E}\{\tilde{\boldsymbol{x}}(\boldsymbol{a}^{*T}\tilde{\boldsymbol{x}})^*\} = \mathcal{E}\{\tilde{\boldsymbol{x}}(\tilde{\boldsymbol{x}}^{*T}\boldsymbol{a})\} = \mathcal{E}\{\tilde{\boldsymbol{x}}\tilde{\boldsymbol{x}}^{*T}\}\,\boldsymbol{a} = \sigma_\varepsilon^2 \boldsymbol{\iota}$$

Then since $\mathcal{E}\{\tilde{\boldsymbol{x}}\tilde{\boldsymbol{x}}^{*T}\} = \tilde{\mathbf{R}}_x$, it follows that the linear prediction parameters satisfy

$$\boxed{\tilde{\mathbf{R}}_x \boldsymbol{a} = \sigma_\varepsilon^2 \boldsymbol{\iota}} \tag{8.9}$$

This represents the Normal equations written in concise matrix notation.[1] By substituting the explicit form of the matrices, the equations become

$$\begin{bmatrix} R_x[0] & R_x[1] & \cdots & R_x[P] \\ R_x[-1] & R_x[0] & \cdots & R_x[P-1] \\ \vdots & \vdots & \vdots & \vdots \\ R_x[-P] & R_x[-P+1] & \cdots & R_x[0] \end{bmatrix} \begin{bmatrix} 1 \\ a_1 \\ \vdots \\ a_P \end{bmatrix} = \begin{bmatrix} \sigma_\varepsilon^2 \\ 0 \\ \vdots \\ 0 \end{bmatrix} \tag{8.10}$$

as before.

8.2 THE AUTOREGRESSIVE (AR) MODEL

It was seen earlier that any regular stationary random process can be represented as the output of a linear shift-invariant filter driven by white noise. This is the innovations representation. One common method for modeling random signals is to represent them as the output of an *all-pole* linear filter driven by white noise. Although this may not be the most efficient representation of the random process (in terms of the number of parameters required) it

[1]This derivation clearly shows that the correlation matrix that appears in the Normal equations is the *reversed* correlation matrix as noted in Chapter 7.

nevertheless leads to a simple modeling procedure. Since the power spectrum of the filter output is given by the constant noise spectrum (σ_w^2) multiplied by the squared magnitude of the filter, random signals with desired spectral characteristics can be produced by choosing a filter with an appropriate denominator polynomial.

The linear prediction problem leads to a method for performing this modeling. Recall that the prediction error filter that produces $\varepsilon[n]$ from $x[n]$ is an FIR filter whose output is given by (8.3). This filter, which is depicted in Fig. 8.1(a), has a transfer function given by

$$A(z) = 1 + a_1^* z^{-1} + a_2^* z^{-2} + \cdots + a_P^* z^{-P} \tag{8.11}$$

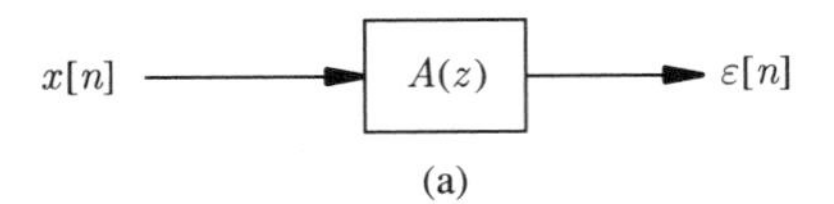

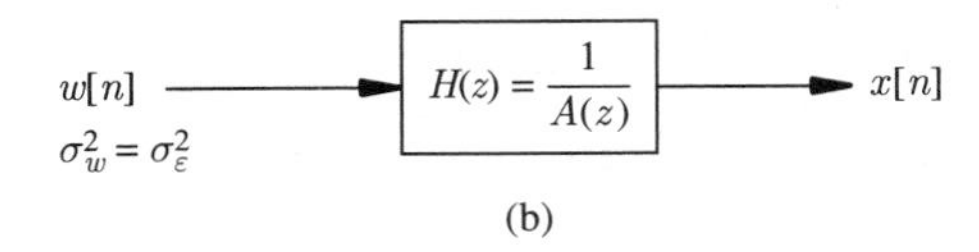

Figure 8.1 Comparison of linear prediction and autoregressive modeling. (a) Prediction error filter. (b) AR model.

Further recall that if the prediction order P of the filter is sufficiently large, then the output is (approximately) a white noise process with variance σ_ε^2.

Suppose now that this procedure is turned around. That is, the filter is inverted and driven with a white noise sequence of variance $\sigma_w^2 = \sigma_\varepsilon^2$ as shown in Fig. 8.1(b). This system will then produce a random sequence with the same statistical characteristics as those of the original sequence $x[n]$. Thus it represents a model for the process $x[n]$. Let us thus consider a system of the form

$$x[n] = -a_1 x[n-1] - a_2 x[n-2] - \cdots - a_P x[n-P] + w[n] \tag{8.12}$$

This equation is equivalent to (8.3) except the coefficients a_i^* have been replaced by the coefficients a_i and $\varepsilon[n]$ has been replaced by $w[n]$. The equation has also been rewritten in recursive form. As such, it is in the form of a statistical *regression*, where the "dependent" variable $x[n]$ is represented as a linear combination of the "independent" variables $x[n-1]$ through $x[n-P]$. Since both independent and dependent variables belong to the same random process, $x[n]$ is called an *autoregressive* or AR process; the process is seen to be "regressed upon itself."

The filter in the AR model is an IIR filter with transfer function

$$H(z) = \frac{1}{A(z)} \tag{8.13}$$

where

$$A(z) = 1 + a_1 z^{-1} + a_2 z^{-2} + \cdots + a_P z^{-P} \tag{8.14}$$

Since $A(z)$ involves only negative powers of z, the filter has a Pth-order zero at the origin but otherwise has only poles. The AR model is therefore said to be an "all-pole" model.

Moreover, it is later shown that as long as the denominator coefficients a_i are found by solving Normal equations with a positive definite correlation function, the poles all lie within the unit circle. Since $H(z)$ corresponds to a *causal* filter, this implies that it is *minimum-phase* (i.e., it is a causal stable filter with a causal stable inverse). Direct form realizations for the prediction error filter and the corresponding AR model filter are shown in Fig. 8.2. Other more convenient forms for both filters are derived later in this chapter.

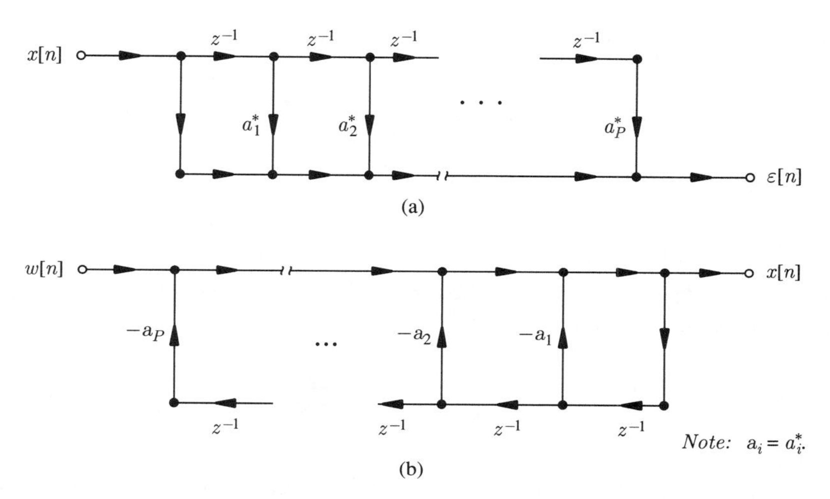

Figure 8.2 Direct form realizations. (a) Prediction error filter. (b) AR model (all-pole) filter.

The simplest AR model is one specified by the first-order difference equation

$$x[n] = \rho x[n-1] + w[n] \tag{8.15}$$

where $w[n]$ is a white noise sequence with mean zero and variance σ_w^2. The correlation function of the resulting random process is

$$R_x[l] = \begin{cases} \frac{\sigma_w^2}{1-|\rho|^2}\rho^l; & l \geq 0 \\ \\ \frac{\sigma_w^2}{1-|\rho|^2}(\rho^*)^{-l}; & l < 0 \end{cases} \tag{8.16}$$

(see Example 5.2).

Since the AR model is obtained by inverting the prediction error filter, it should be possible to show directly that the parameters $a_1, a_2, \ldots, a_P$ and σ_w^2 of any AR model can be found by solving Normal equations. This can be done as follows. *Begin* with a postulated model for a random process of the form (8.12) and write this as

$$x[n] + a_1 x[n-1] + \cdots + a_P x[n-P] = w[n] \tag{8.17}$$

Then it follows from the results of Chapter 5 that the correlation function satisfies the difference equation

$$R_x[l] + \mathrm{a}_1 R_x[l-1] + \cdots + \mathrm{a}_P R_x[l-P] = R_{wx}[l] \tag{8.18}$$

Now further recall that if $h[n]$ is the impulse response of (8.17), then R_{xw} is given by

$$R_{xw}[l] = h[l] * R_w[l] = h[l] * \sigma_w^2 \delta[l] = \sigma_w^2 h[l]$$

so

$$R_{wx}[l] = \sigma_w^2 h^*[-l] \tag{8.19}$$

Substituting this into (8.18) produces

$$R_x[l] + \mathrm{a}_1 R_x[l-1] + \cdots + \mathrm{a}_P R_x[l-P] = \sigma_w^2 h^*[-l] \tag{8.20}$$

Finally, since $h[n]$ is the impulse response of a causal filter, $h[n]$ is equal to 0 for $n < 0$, and from the Initial Value Theorem[2]

$$h[0] = \lim_{z\to\infty} H(z) = \lim_{z\to\infty} \frac{1}{1 + \mathrm{a}_1 z^{-1} + \cdots + \mathrm{a}_P z^{-P}} = 1 \tag{8.21}$$

With these considerations (8.20) can be evaluated for $l = 0, 1, 2, \ldots, P$. For $l = 0$ the right side of (8.20) will be equal to σ_w^2, while for $l > 0$ the right side will be equal to 0. If this set of equations is written in matrix form the result is

$$\begin{bmatrix} R_x[0] & R_x[-1] & \cdots & R_x[-P] \\ R_x[1] & R_x[0] & \cdots & R_x[-P+1] \\ \vdots & \vdots & \vdots & \vdots \\ R_x[P] & R_x[P-1] & \cdots & R_x[0] \end{bmatrix} \begin{bmatrix} 1 \\ \mathrm{a}_1 \\ \vdots \\ \mathrm{a}_P \end{bmatrix} = \begin{bmatrix} \sigma_w^2 \\ 0 \\ \vdots \\ 0 \end{bmatrix} \tag{8.22}$$

We refer to these equations for the AR model as the *Yule–Walker equations* [6]. By conjugating both sides and noting that $R_x^*[l] = R_x[-l]$ the equations can be written as

$$\begin{bmatrix} R_x[0] & R_x[1] & \cdots & R_x[P] \\ R_x[-1] & R_x[0] & \cdots & R_x[P-1] \\ \vdots & \vdots & \vdots & \vdots \\ R_x[-P] & R_x[-P+1] & \cdots & R_x[0] \end{bmatrix} \begin{bmatrix} 1 \\ \mathrm{a}_1^* \\ \vdots \\ \mathrm{a}_P^* \end{bmatrix} = \begin{bmatrix} \sigma_w^2 \\ 0 \\ \vdots \\ 0 \end{bmatrix} \tag{8.23}$$

which is identical in form to the Normal equations of linear prediction (8.10). The essential difference between the Normal equations (8.10) and the Yule–Walker equations (8.22) is that the former involve the linear prediction parameters and the *reversed* correlation matrix, while the latter involve the parameters of the AR model and the *unreversed* correlation

[2] See, e.g., [5], p. 179.

matrix. For the real case, the matrix is symmetric rather than Hermitian symmetric and both sets of equations are identical.[3]

Note that in the derivation of the Yule–Walker equations the correlation function R_x *is assumed to be that of the AR model.* The correlation function of the process produced by the model indeed matches the correlation function in the Yule–Walker equations for all lags up to the given order P. Beyond that, the correlation function is one that satisfies (8.20). In particular, for $l > P$ the term $h^*[-l]$ in the equation remains zero, so additional values of the correlation function can be computed from previous values according to

$$R_x[l] = -\mathrm{a}_1 R_x[l-1] - \cdots - \mathrm{a}_P R_x[l-P]; \quad l = P+1, P+2, \ldots \tag{8.24}$$

This is known as the *correlation extension property* of the AR model. More is said about this property in the section on maximum entropy spectral estimation in Chapter 10. It turns out that the extension provided by the AR model belongs to a process which is in a sense the "most random" of all possible processes that match the first $P+1$ values of the correlation function. Problem 8.5 demonstrates how the AR model extends the correlation function in a simple case.

The fact that the AR model parameters satisfy Yule–Walker equations provides a practical method for signal modeling. That is, suppose it is desired to represent some given random signal $x[n]$ by an AR model. The correlation function $R_x[l]$ can be estimated and used in (8.22) to solve for the model parameters. Evidently the procedure is identical to that involved in solving for the filter coefficients in a linear prediction problem. In the next section we look at this relation more closely.

8.3 LINEAR PREDICTION FOR AR PROCESSES

The last two sections show that both the linear prediction problem and the AR modeling problem lead to the same set of linear equations. This section further cements the relation between the two procedures by considering what happens when linear prediction is applied to a sequence that is truly generated by an AR model. The situation is depicted in Fig. 8.3. It turns out that if the order P' of linear prediction is at least as high as the order P of the AR model, then the prediction error filter coefficients correspond to the coefficients in the difference equation for the AR model, and higher orders of linear prediction do not result in any change in the prediction error filter.

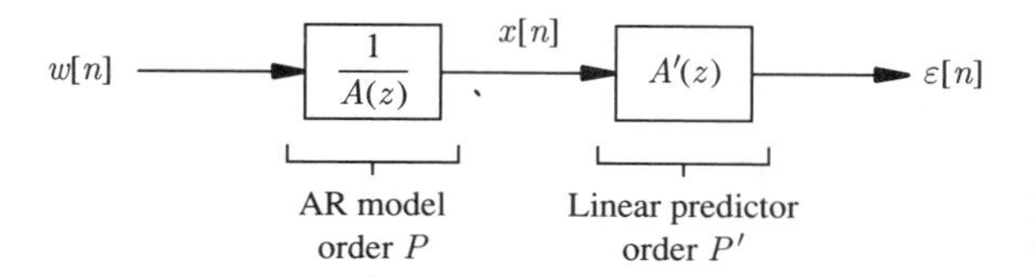

Figure 8.3 Linear prediction applied to an autoregressive process.

[3] You may notice that it is possible to write both sets of equations in terms of the unreversed correlation matrix if we had initially defined the linear prediction coefficients in (8.1) without the complex conjugate. To do so, however, is to hide some of the essential features of the problem.

Begin by considering an AR model of some given order P. The random process produced by the model satisfies the difference equation (8.17) and the model parameters satisfy the Yule–Walker equations (8.22). Now consider applying linear prediction to this P^{th}-order AR process using a prediction order P'. The prediction error filter coefficients will be denoted by $a_i^{(P')}$ for $i = 1, 2, \ldots, P'$ and the prediction error variance will be denoted by $\sigma^2_{\varepsilon_{P'}}$. Three cases will be considered, namely $P' = P$, $P' > P$, and $P' < P$.

Case 1 **($P' = P$).** For this case the Normal equations of linear prediction are of the form

$$\begin{bmatrix} R_x[0] & R_x[1] & \cdots & R_x[P] \\ R_x[-1] & R_x[0] & \cdots & R_x[P-1] \\ \vdots & \vdots & \vdots & \vdots \\ R_x[-P] & R_x[-P+1] & \cdots & R_x[0] \end{bmatrix} \begin{bmatrix} 1 \\ a_1^{(P)} \\ \vdots \\ a_P^{(P)} \end{bmatrix} = \begin{bmatrix} \sigma^2_{\varepsilon_P} \\ 0 \\ \vdots \\ 0 \end{bmatrix}$$

Since these are identical in form to (8.23), and since the equations have a unique solution, it follows that

$$a_i^{(P)} = \mathrm{a}_i^*; \quad i = 1, 2, \ldots, P$$

and that

$$\sigma^2_{\varepsilon_P} = \sigma^2_w$$

That is, the linear prediction parameters of order P correspond to the AR model parameters.

Case 2 **($P' > P$).** For this case the Normal equations of linear prediction have the form

$$\begin{bmatrix} R_x[0] & R_x[1] & \cdots & R_x[P] & \cdots & R_x[P'] \\ R_x[-1] & R_x[0] & \cdots & R_x[P-1] & \cdots & R_x[P'-1] \\ \vdots & \vdots & \vdots & \vdots & \vdots & \vdots \\ R_x[-P] & R_x[-P+1] & \cdots & R_x[0] & \cdots & R_x[P'-P] \\ \vdots & \vdots & \vdots & \vdots & \vdots & \vdots \\ R_x[-P'] & R_x[-P'+1] & \cdots & R_x[P-P'] & \cdots & R_x[0] \end{bmatrix} \begin{bmatrix} 1 \\ a_1^{(P')} \\ \vdots \\ a_P^{(P')} \\ \vdots \\ a_{P'}^{(P')} \end{bmatrix} = \begin{bmatrix} \sigma^2_{\varepsilon_{P'}} \\ 0 \\ \vdots \\ 0 \\ \vdots \\ 0 \end{bmatrix}$$

Again, since the correlation matrix is nonsingular, the Normal equations have a unique solution. Further it is contended that the following *postulated solution* satisfies the Normal equations and thus *is* the unique solution.

$$\begin{bmatrix} R_x[0] & R_x[1] & \cdots & R_x[P] & \cdots & R_x[P'] \\ R_x[-1] & R_x[0] & \cdots & R_x[P-1] & \cdots & R_x[P'-1] \\ \vdots & \vdots & \vdots & \vdots & \vdots & \vdots \\ R_x[-P] & R_x[-P+1] & \cdots & R_x[0] & \cdots & R_x[P'-P] \\ \vdots & \vdots & \vdots & \vdots & \vdots & \vdots \\ R_x[-P'] & R_x[-P'+1] & \cdots & R_x[-P'+P] & \cdots & R_x[0] \end{bmatrix} \begin{bmatrix} 1 \\ \mathrm{a}_1^* \\ \vdots \\ \mathrm{a}_P^* \\ \vdots \\ 0 \end{bmatrix} = \begin{bmatrix} \sigma^2_w \\ 0 \\ \vdots \\ 0 \\ \vdots \\ 0 \end{bmatrix}$$

It is actually quite easy to see that this is true if this equation is conjugated and written as

$$\begin{bmatrix} R_x[0] & R_x[-1] & \cdots & R_x[-P] & \cdots & R_x[-P'] \\ R_x[1] & R_x[0] & \cdots & R_x[-P+1] & \cdots & R_x[-P'+1] \\ \vdots & \vdots & \vdots & \vdots & \vdots & \vdots \\ R_x[P] & R_x[P-1] & \cdots & R_x[0] & \cdots & R_x[-P'+P] \\ \vdots & \vdots & \vdots & \vdots & \vdots & \vdots \\ R_x[P'] & R_x[P'-1] & \cdots & R_x[P'-P] & \cdots & R_x[0] \end{bmatrix} \begin{bmatrix} 1 \\ \mathrm{a}_1 \\ \vdots \\ \mathrm{a}_P \\ \vdots \\ 0 \end{bmatrix} = \begin{bmatrix} \sigma_w^2 \\ 0 \\ \vdots \\ 0 \\ \vdots \\ 0 \end{bmatrix}$$

Since the last $P' - P$ terms in the filter coefficient vector are zero, this matrix equation just represents the difference equation (8.20) evaluated for $l = 0, 1, 2, \ldots, P'$. Thus the postulated solution does indeed satisfy the Normal equations. The linear prediction parameters for $P' > P$ are therefore given by

$$a_i^{(P')} = \begin{cases} \mathrm{a}_i^*; & 0 \leq i \leq P \\ 0; & P < i \leq P' \end{cases}$$

and

$$\sigma_{\varepsilon_{P'}}^2 = \sigma_w^2$$

Thus using a higher-order prediction filter for the AR process can do no better than using the Pth-order filter; the prediction error variance remains the same.[4]

Case 3 ($P' < P$)**.** This case leads to a smaller set of Normal equations

$$\begin{bmatrix} R_x[0] & R_x[1] & \cdots & R_x[P'] \\ R_x[-1] & R_x[0] & \cdots & R_x[P'-1] \\ \vdots & \vdots & \vdots & \vdots \\ R_x[-P'] & R_x[-P'+1] & \cdots & R_x[0] \end{bmatrix} \begin{bmatrix} 1 \\ a_1^{(P')} \\ \vdots \\ a_{P'}^{(P')} \end{bmatrix} = \begin{bmatrix} \sigma_{\varepsilon_{P'}}^2 \\ 0 \\ \vdots \\ 0 \end{bmatrix}$$

with solution $a_i^{(P')} \neq \mathrm{a}_i^*$, $i = 1, 2, \ldots, P'$ and $\sigma_{\varepsilon_{P'}}^2 \neq \sigma_w^2$. These linear prediction parameters correspond to an AR process of lower order; however, they are still useful. In particular, suppose that a random process with given correlation function $R_x[l]$ is being modeled by an AR process of order P. In producing a random signal from the model, say beginning at $n = 0$, the samples $x[0], x[1], \ldots, x[P-1]$ have to be computed initially. If these samples are computed directly from (8.12), using zeros for the samples of the process where $n - i < 0$, the samples so generated will not have the required statistical properties. That is, the zero initial conditions and constant variance of the input generate a transient

[4]At this point you may want to refer back to Example 7.1, where the linear prediction parameters for a random process with correlation function

$$R_x[l] = (0.5)^{|l|}$$

are computed. This correlation function is of the form (8.16) and so corresponds to a first-order AR process of the form (8.15). The reason that the linear prediction parameter a_2 in that example turned out to be zero should now be clear.

condition and the resulting random process will not have the correct correlation function R_x until the transient condition has died out. This effect is illustrated for a first order-system in Example 3.8. The transient condition can be eliminated, however, by generating the first P values of the random process using AR models of orders beginning at 0 and increasing to the final value P. That is, $x[0]$ is taken to be white noise with variance $R_x[0]$; $x[1]$ is generated by a first-order model whose parameters are determined by solving first-order Normal equations; $x[2]$ is generated similarly from a second-order model; and so on. (Note that the variance of the white noise input at each step changes as well as the filter parameters.) If this procedure is followed, then the resulting random process by design will match the desired correlation function over the initial interval $0 \leq n \leq P-1$ as well as over the subsequent interval $n \geq P$. Thus the lower-order model parameters provide a smooth transition and eliminate the transient and the associated nonstationarity in the random process when it is started up initially.

The computation of parameters for all of the reduced-order AR models in the latter problem is not as burdensome as it might seem. In fact, the model parameters for order $p+1$ can conveniently be expressed in terms of the parameters for order p through an efficient procedure known as the Levinson recursion [7] and developed later in this chapter. In addition, the Levinson recursion leads to a tremendous amount of new insight and structure to problems of linear prediction and AR modeling. This includes an efficient lattice structure for the filters, simple tests for stability, and further efficient algorithms.

8.4 BACKWARD LINEAR PREDICTION AND THE ANTICAUSAL AR MODEL

In order to fully develop the Levinson recursion and the topics that follow it is necessary to consider a problem which at first seems rather artificial, namely the topic of backward linear prediction. However, it turns out that the backward problem is intimately related to the forward problem already considered and it is the combination of the two that provides the needed mathematical structure for the topics to follow. The following subsections deal with the problem of backward linear prediction, the associated anticausal AR model, and finally the connection between forward and backward linear prediction.

8.4.1 Backward Linear Prediction

To define the problem of backward linear prediction, consider the random sequence and the data points $x[n-P]$ through $x[n]$ shown in Fig. 8.4. The forward linear prediction problem deals with predicting the point $x[n]$ from the P previous points $x[n-1], \ldots, x[n-P]$. The backward problem instead deals with predicting the point $x[n-P]$ from the points $x[n-P+1], \ldots, x[n]$. If n represents a time index, then it is as if the random process were evolving backward in time and it is desired to predict the next *earlier* value. The estimate is of the form

$$\hat{x}[n-P] = -b_1^* x[n-P+1] - b_2^* x[n-P+2] - \cdots - b_P^* x[n] \tag{8.25}$$

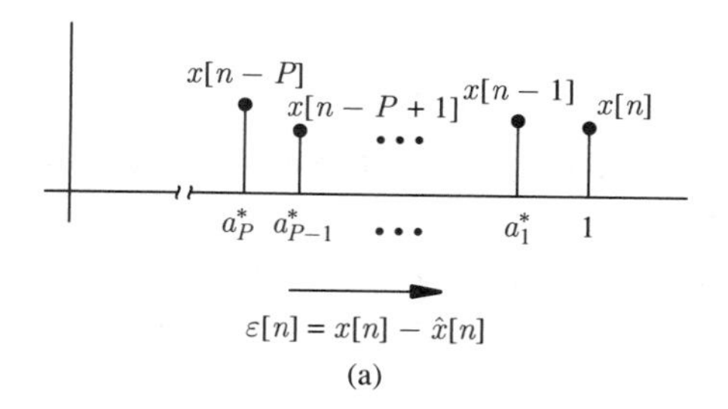

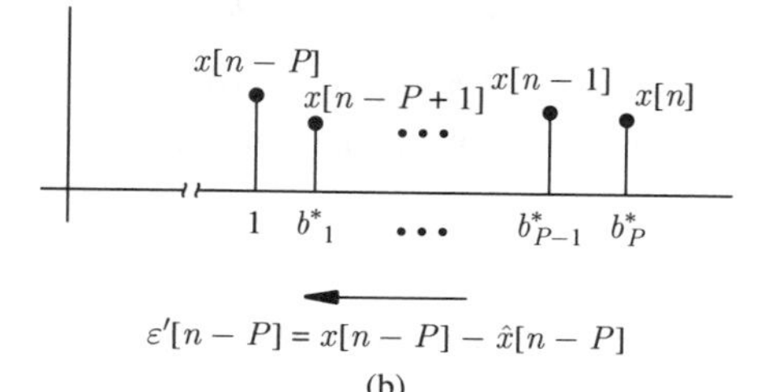

Figure 8.4 Linear prediction of a random process. (a) Forward prediction. (b) Backward prediction.

where the b_i's are the coefficients of the backward linear predictive filter, to be determined.

The error in the estimate is defined as the difference

$$\varepsilon'[n-P] = x[n-P] - \hat{x}[n-P] \tag{8.26}$$

or

$$\varepsilon'[n-P] = x[n-P] + b_1^* x[n-P+1] + \cdots + b_P^* x[n] \tag{8.27}$$

If the backward filter coefficient vector is defined as

$$\boldsymbol{b} \stackrel{\text{def}}{=} \begin{bmatrix} 1 \\ b_1 \\ \vdots \\ b_P \end{bmatrix} \tag{8.28}$$

then the error can be written as

$$\varepsilon'[n-P] = \boldsymbol{b}^{*T} \boldsymbol{x}[n] \tag{8.29}$$

where $\boldsymbol{x}[n]$ is given by (8.4), and the orthogonality principle takes the form

$$\mathcal{E}\{\boldsymbol{x}[n]\varepsilon'^*[n-P]\} = \begin{bmatrix} \sigma^2_{\varepsilon'} \\ 0 \\ \vdots \\ 0 \end{bmatrix} = \sigma^2_{\varepsilon'} \boldsymbol{\iota} \tag{8.30}$$

where $\sigma^2_{\varepsilon'}$ is the backward prediction error variance. Substituting (8.29) into (8.30) and again dropping the argument $[n]$ for simplicity yields

$$\mathcal{E}\{\boldsymbol{x}(\boldsymbol{b}^{*T}\boldsymbol{x})^*\} = \mathcal{E}\{\boldsymbol{x}\boldsymbol{x}^{*T}\}\,\boldsymbol{b} = \sigma^2_{\varepsilon'} \boldsymbol{\iota}$$

or

$$\mathbf{R}_x \boldsymbol{b} = \sigma^2_{\varepsilon'} \boldsymbol{\iota} \tag{8.31}$$

These are the Normal equations for the backward linear prediction problem; they can be written in more explicit form as

$$\begin{bmatrix} R_x[0] & R_x[-1] & \cdots & R_x[-P] \\ R_x[1] & R_x[0] & \cdots & R_x[-P+1] \\ \vdots & \vdots & \vdots & \vdots \\ R_x[P] & R_x[P-1] & \cdots & R_x[0] \end{bmatrix} \begin{bmatrix} 1 \\ b_1 \\ \vdots \\ b_P \end{bmatrix} = \begin{bmatrix} \sigma^2_{\varepsilon'} \\ 0 \\ \vdots \\ 0 \end{bmatrix} \tag{8.32}$$

Observe by comparing (8.31) with (8.9) or (8.32) with (8.10) that the backward Normal equations involve the *unreversed* correlation matrix, and that this is the only difference.

8.4.2 The Anticausal AR Model

The backward linear prediction problem can be associated with an anticausal AR model that has the difference equation

$$x[n] = -b_1 x[n+1] - b_2 x[n+2] - \cdots - b_P x[n+P] + w'[n] \tag{8.33}$$

where $w'[n]$ is a white noise process. This AR model can be obtained formally from (8.27) by replacing $n-P$ by n, substituting b_i for b_i^* and $w'[n]$ for $\varepsilon'[n]$, and rearranging terms. The difference equation is evaluated recursively by *decrementing* the index n; that is, the process is generated in a *backward* direction.

To continue the discussion let us write (8.33) in the form

$$x[n] + b_1 x[n+1] + \cdots + b_P x[n+P] = w'[n] \tag{8.34}$$

and observe that the correlation function for the random process satisfies the corresponding difference equation

$$R_x[l] + b_1 R_x[l+1] + \cdots + b_P R_x[l+P] = R_{w'x}[l] \tag{8.35}$$

Further if $h'[n]$ is the impulse response of the system represented by (8.33), then

$$R_{xw'}[l] = h'[l] * R_{w'}[l] = \sigma^2_{w'} h'[l]$$

or

$$R_{w'x}[l] = \sigma^2_{w'} h'^*[-l] \tag{8.36}$$

If this is substituted in (8.35), the result is

$$R_x[l] + b_1 R_x[l+1] + \cdots + b_P R_x[l+P] = \sigma^2_{w'} h'^*[-l] \tag{8.37}$$

Now observe that the backward AR model involves a filter with transfer function

$$H'(z) = \frac{1}{B(z)} \tag{8.38}$$

where

$$B(z) = 1 + b_1 z + b_2 z^2 + \cdots + b_P z^P \tag{8.39}$$

The value $h'[0]$ can be obtained from an "initial value theorem" for anticausal systems of the form[5]

$$h'[0] = \lim_{z \to 0} H'(z) = \lim_{z \to 0} \frac{1}{1 + b_1 z + \cdots + b_P z^P} = 1 \tag{8.40}$$

In addition, since the system is anticausal, $h'[n] = 0$ for $n > 0$. Therefore if (8.37) is evaluated for $l = 0, -1, -2, \ldots, -P$ the result is the Yule–Walker equations for the backward model:

$$\begin{bmatrix} R_x[0] & R_x[1] & \cdots & R_x[P] \\ R_x[-1] & R_x[0] & \cdots & R_x[P-1] \\ \vdots & \vdots & \vdots & \vdots \\ R_x[-P] & R_x[-P+1] & \cdots & R_x[0] \end{bmatrix} \begin{bmatrix} 1 \\ b_1 \\ \vdots \\ b_P \end{bmatrix} = \begin{bmatrix} \sigma_{w'}^2 \\ 0 \\ \vdots \\ 0 \end{bmatrix} \tag{8.41}$$

If these equations are conjugated, they are identical in form to the Normal equations (8.32) obtained from the viewpoint of backward linear prediction.

8.4.3 Relation Between Forward and Backward Problems

The four sets of linear equations involved in linear prediction and AR modeling are shown in Table 8.1, where $\boldsymbol{a}$ and $\boldsymbol{b}$ are the vectors of linear prediction coefficients and $\mathbf{a}$ and $\mathbf{b}$ are the vectors of AR model coefficients. The relation between the linear prediction parameters and the AR model parameters has already been discussed. Let us now turn to the relation between the forward and backward linear prediction parameters. Since

$$\tilde{\mathbf{R}}_x = \mathbf{R}_x^*$$

TABLE 8.1 LINEAR EQUATIONS INVOLVED IN LINEAR PREDICTION AND AR MODELING

	Forward	Backward
Normal equations (linear prediction)	$\tilde{\mathbf{R}}_x \boldsymbol{a} = \sigma_\varepsilon^2 \boldsymbol{\iota}$	$\mathbf{R}_x \boldsymbol{b} = \sigma_{\varepsilon'}^2 \boldsymbol{\iota}$
Yule–Walker equations (AR model)	$\mathbf{R}_x \mathbf{a} = \sigma_w^2 \boldsymbol{\iota}$	$\tilde{\mathbf{R}}_x \mathbf{b} = \sigma_{w'}^2 \boldsymbol{\iota}$

the items in the first row of the table (equations 8.9 and 8.31) imply that

$$\boxed{\boldsymbol{b} = \boldsymbol{a}^*} \tag{8.42}$$

[5]You may want to verify this result by writing down at the definition of the z-transform for the impulse response of an anticausal system and taking the limit as $z \to 0$.

and

$$\boxed{\sigma_{\varepsilon}^{2} = \sigma_{\varepsilon'}^{2}} \tag{8.43}$$

Thus solving any one set of the equations in Table 8.1 solves them all. For purposes of this chapter, we focus on the forward Normal equations (8.9).

For *real* random processes, the considerations of this section show that the forward and backward AR models are statistically identical, and that the forward prediction error filter is identical to the backward prediction error filter. *This is a remarkable property.* The equivalence of forward and backward modeling and prediction is unique to Gaussian random processes, or equivalently to the second moment characterization of general random processes. It can be demonstrated that if a random process is not Gaussian and the characterization is carried out using more than just the second moment quantities, the symmetry in the forward and backward characterization of these random processes disappears [8, 9].

8.5 THE LEVINSON RECURSION

In the days when Wiener was developing the theory of statistical filtering, Norman Levinson, who was working with Wiener, described an efficient recursive method for solving a discrete version of the Wiener-Hopf equations [7][10, Appendix B]. In later work Durbin [11] applied this method to the more specialized problem of solving Normal equations in AR modeling and as a result the Levinson recursion is sometimes referred to as the "Levinson–Durbin" recursion.

The Levinson recursion provides a fast method for solving the Normal equations by beginning with a filter of order 0 and recursively generating filters of order 1, 2, 3, and so on, up to the desired order P. At each stage the filter parameters of order p are computed by simple and elegant operations on the filter parameters of order $p-1$. The total computational effort grows as P^2 and thus the method is more computationally efficient than solving the Normal equations by matrix inversion, where the computational effort grows as P^3.

Although the original motivation for the Levinson recursion was efficient solution of the Normal equations, the insights brought on by the recursive formulation are more far-reaching than anyone would at first suspect. They lead, among other things, to an efficient lattice structure for the filter that has several advantages over the common direct form or "tapped delay line" implementation given in Fig. 8.2. This lattice structure is discussed in the next section.

The Levinson recursion can be derived by considering the (forward) Normal equations of order p, which are written here as

$$\tilde{\mathbf{R}}_{\boldsymbol{x}}^{(p)} \boldsymbol{a}_p = \begin{bmatrix} \sigma_p^2 \\ 0 \\ \vdots \\ 0 \end{bmatrix} \tag{8.44}$$

where the notation $\sigma^2_{\varepsilon_p}$ has been dropped and replaced with σ^2_p for convenience, and where

$$\tilde{\mathbf{R}}_x^{(p)} = \begin{bmatrix} R_x[0] & R_x[1] & \cdots & R_x[p] \\ R_x[-1] & R_x[0] & \cdots & R_x[p-1] \\ \vdots & \vdots & \vdots & \vdots \\ R_x[-p] & R_x[-p+1] & \cdots & R_x[0] \end{bmatrix} \tag{8.45}$$

and

$$\boldsymbol{a}_p = \begin{bmatrix} 1 \\ a_1^{(p)} \\ \vdots \\ a_p^{(p)} \end{bmatrix} \tag{8.46}$$

The backward Normal equations of order p likewise have the form

$$\mathbf{R}_x^{(p)} \boldsymbol{b}_p = \begin{bmatrix} \sigma'^2_p \\ 0 \\ \vdots \\ 0 \end{bmatrix} \tag{8.47}$$

where σ'^2_p replaces $\sigma^2_{\varepsilon'_p}$ and where

$$\mathbf{R}_x^{(p)} = \begin{bmatrix} R_x[0] & R_x[-1] & \cdots & R_x[-p] \\ R_x[1] & R_x[0] & \cdots & R_x[1-p] \\ \vdots & \vdots & \vdots & \vdots \\ R_x[p] & R_x[p-1] & \cdots & R_x[0] \end{bmatrix} \tag{8.48}$$

and

$$\boldsymbol{b}_p = \begin{bmatrix} 1 \\ b_1^{(p)} \\ \vdots \\ b_p^{(p)} \end{bmatrix} \tag{8.49}$$

In considering the backward problem, we deliberately use the variables $\boldsymbol{b}_p$ and σ'^2_p and temporarily ignore the relations (8.42) and (8.43) that exist between the forward and backward parameters. This is done to more clearly show the form of the recursion. At the end, the relations between the forward and backward parameters are taken into account and used to develop the most efficient computational method.[6]

[6]It is possible, at least for the cases considered here, to derive the Levinson recursion without ever explicitly considering the backward problem. In fact, the derivation is a little shorter when it is done this way. However, the forward and backward problems really form the essence of the method. They are fundamental in the lattice realization of the prediction error filter and are also the key to extending it to more advanced topics such as multichannel filtering.

Now return to (8.45) and (8.48) and note that if the term $\mathbf{r}_p$ is defined as

$$\mathbf{r}_p = \begin{bmatrix} R_x[1] \\ R_x[2] \\ \vdots \\ R_x[p+1] \end{bmatrix} \tag{8.50}$$

then the reversed correlation matrix of order p can be partitioned as

$$\tilde{\mathbf{R}}_{\boldsymbol{x}}^{(p)} = \left[\begin{array}{c|c} \tilde{\mathbf{R}}_{\boldsymbol{x}}^{(p-1)} & \tilde{\mathbf{r}}_{p-1} \\ \hline \tilde{\mathbf{r}}_{p-1}^{*T} & R_x[0] \end{array}\right] \tag{8.51}$$

while the (unreversed) correlation matrix can be partitioned as

$$\mathbf{R}_{\boldsymbol{x}}^{(p)} = \left[\begin{array}{c|c} \mathbf{R}_{\boldsymbol{x}}^{(p-1)} & \tilde{\mathbf{r}}_{p-1}^{*} \\ \hline \tilde{\mathbf{r}}_{p-1}^{T} & R_x[0] \end{array}\right] \tag{8.52}$$

Now assume that the linear prediction parameters of order $p-1$ are known. These satisfy (8.44) with p replaced by $p-1$. Then consider an augmented set of Normal equations for the forward problem of the form

$$\tilde{\mathbf{R}}_{\boldsymbol{x}}^{(p)} \left[\begin{array}{c} \boldsymbol{a}_{p-1} \\ \hline 0 \end{array}\right] = \left[\begin{array}{c|c} \tilde{\mathbf{R}}_{\boldsymbol{x}}^{(p-1)} & \tilde{\mathbf{r}}_{p-1} \\ \hline \tilde{\mathbf{r}}_{p-1}^{*T} & R_x[0] \end{array}\right] \left[\begin{array}{c} \boldsymbol{a}_{p-1} \\ \hline 0 \end{array}\right] = \begin{bmatrix} \sigma_{p-1}^2 \\ 0 \\ \vdots \\ \Delta_p \end{bmatrix} \tag{8.53}$$

where the term Δ_p is seen to be

$$\Delta_p = \tilde{\mathbf{r}}_{p-1}^{*T} \boldsymbol{a}_{p-1} = \mathbf{r}_{p-1}^{*T} \tilde{\boldsymbol{a}}_{p-1} \tag{8.54}$$

Also consider a corresponding augmented set of Normal equations for the backward problem

$$\mathbf{R}_{\boldsymbol{x}}^{(p)} \left[\begin{array}{c} \boldsymbol{b}_{p-1} \\ \hline 0 \end{array}\right] = \left[\begin{array}{c|c} \mathbf{R}_{\boldsymbol{x}}^{(p-1)} & \tilde{\mathbf{r}}_{p-1}^{*} \\ \hline \tilde{\mathbf{r}}_{p-1}^{T} & R_x[0] \end{array}\right] \left[\begin{array}{c} \boldsymbol{b}_{p-1} \\ \hline 0 \end{array}\right] = \begin{bmatrix} \sigma_{p-1}'^2 \\ 0 \\ \vdots \\ \Delta_p' \end{bmatrix} \tag{8.55}$$

where

$$\Delta_p' = \tilde{\mathbf{r}}_{p-1}^{T} \boldsymbol{b}_{p-1} = \mathbf{r}_{p-1}^{T} \tilde{\boldsymbol{b}}_{p-1} \tag{8.56}$$

Suppose now that all of the terms in (8.55) are reversed. The result is

$$\tilde{\mathbf{R}}_{\boldsymbol{x}}^{(p)} \begin{bmatrix} 0 \\ -- \\ \\ \tilde{\boldsymbol{b}}_{p-1} \end{bmatrix} = \begin{bmatrix} \Delta'_p \\ 0 \\ \vdots \\ \sigma'^2_{p-1} \end{bmatrix} \tag{8.57}$$

If this equation is multiplied by a constant c_1 and added to (8.53), the combined result is

$$\tilde{\mathbf{R}}_{\boldsymbol{x}}^{(p)} \left[\begin{bmatrix} \boldsymbol{a}_{p-1} \\ \\ -- \\ 0 \end{bmatrix} + c_1 \begin{bmatrix} 0 \\ -- \\ \\ \tilde{\boldsymbol{b}}_{p-1} \end{bmatrix} \right] = \begin{bmatrix} \sigma^2_{p-1} \\ 0 \\ \vdots \\ \Delta_p \end{bmatrix} + c_1 \begin{bmatrix} \Delta'_p \\ 0 \\ \vdots \\ \sigma'^2_{p-1} \end{bmatrix} \tag{8.58}$$

Now compare this result to (8.44), the Normal equations of order p. Since the solution to (8.44) is unique, if c_1 is chosen so that

$$\begin{bmatrix} \sigma^2_{p-1} \\ 0 \\ \vdots \\ \Delta_p \end{bmatrix} + c_1 \begin{bmatrix} \Delta'_p \\ 0 \\ \vdots \\ \sigma'^2_{p-1} \end{bmatrix} = \begin{bmatrix} \sigma^2_p \\ 0 \\ \vdots \\ 0 \end{bmatrix} \tag{8.59}$$

then it must follow that

$$\begin{bmatrix} \boldsymbol{a}_{p-1} \\ \\ -- \\ 0 \end{bmatrix} + c_1 \begin{bmatrix} 0 \\ -- \\ \\ \tilde{\boldsymbol{b}}_{p-1} \end{bmatrix} = \boldsymbol{a}_p \tag{8.60}$$

Equation (8.59) requires that

$$\sigma^2_{p-1} + c_1 \Delta'_p = \sigma^2_p \tag{8.61}$$

and

$$\Delta_p + c_1 \sigma'^2_{p-1} = 0 \tag{8.62}$$

All of these relations turn out to be necessary in the recursion.

Now let us repeat this procedure for the backward system of equations. First reverse all of the terms in (8.53) to obtain

$$\mathbf{R}_{\boldsymbol{x}}^{(p)} \begin{bmatrix} 0 \\ -- \\ \\ \tilde{\boldsymbol{a}}_{p-1} \end{bmatrix} = \begin{bmatrix} \Delta_p \\ 0 \\ \vdots \\ \sigma^2_{p-1} \end{bmatrix} \tag{8.63}$$

Then multiply this equation by a constant c_2 and add it to (8.55):

$$\mathbf{R}_{\boldsymbol{x}}^{(p)} \left[\begin{bmatrix} \boldsymbol{b}_{p-1} \\ \\ -- \\ 0 \end{bmatrix} + c_2 \begin{bmatrix} 0 \\ -- \\ \\ \tilde{\boldsymbol{a}}_{p-1} \end{bmatrix} \right] = \begin{bmatrix} \sigma'^2_{p-1} \\ 0 \\ \vdots \\ \Delta'_p \end{bmatrix} + c_2 \begin{bmatrix} \Delta_p \\ 0 \\ \vdots \\ \sigma^2_{p-1} \end{bmatrix} \tag{8.64}$$

Finally, compare this to the backward Normal equations (8.47). If it is required that

$$\sigma'^2_{p-1} + c_2\Delta_p = \sigma'^2_p \tag{8.65}$$

and

$$\Delta'_p + c_2\sigma^2_{p-1} = 0 \tag{8.66}$$

then it follows that

$$\begin{bmatrix} \boldsymbol{b}_{p-1} \\ \\ -- \\ 0 \end{bmatrix} + c_2 \begin{bmatrix} 0 \\ -- \\ \\ \tilde{\boldsymbol{a}}_{p-1} \end{bmatrix} = \boldsymbol{b}_p \tag{8.67}$$

The equations (8.54), (8.56), (8.60), (8.61), (8.65), and (8.67), form the main part of the recursion. To complete the procedure, the needed constants c_1 and c_2 can be found from (8.62) and (8.66). In particular, solving (8.62) for c_1 yields

$$c_1 = -\Delta_p/\sigma'^2_{p-1} \tag{8.68}$$

and solving (8.66) for c_2 produces

$$c_2 = -\Delta'_p/\sigma^2_{p-1} \tag{8.69}$$

This completes the process. Since c_1, c_2, Δ_p, and Δ'_p are defined in terms of the correlation function and the parameters of order $p-1$, these quantities can be computed immediately. Once these are determined, the backward and forward filter parameters and the prediction error variances can be computed from (8.60), (8.61), (8.65), and (8.67).

To summarize the results, let $\gamma_p = -c_1$ and $\gamma'_p = -c_2$. These parameters are known as the forward and backward *reflection coefficients* because of their analogy with similar quantities that appear in the analysis of propagating waves (see, e.g., [12], also [3]). They are also called partial correlation or PARCOR coefficients because of their statistical interpretation, which is explored in Section 8.7. With the identification of these parameters, the Levinson recursion can be restated as shown in Table 8.2.

The recursion is initialized with $\mathbf{r}_0 = R_x[1]$, $\boldsymbol{a}_0 = \boldsymbol{b}_0 = 1$, and $\sigma^2_0 = \sigma'^2_0 = R_x[0]$. An alternative formula for step IV is given in the table. This is easily obtained by solving (8.66) for Δ'_p and substituting the result in (8.61) as follows:

$$\sigma^2_p = (1 - c_1c_2)\sigma^2_{p-1} = (1 - \gamma_p\gamma'_p)\sigma^2_{p-1} \tag{8.70}$$

Similarly solving (8.62) for Δ_p and substituting in (8.65) yields

$$\sigma'^2_p = (1 - c_2c_1)\sigma'^2_{p-1} = (1 - \gamma'_p\gamma_p)\sigma'^2_{p-1} \tag{8.71}$$

These equations are generally more convenient than the original forms.

TABLE 8.2 GENERAL FORM OF THE LEVINSON RECURSION, SHOWING THE ROLES OF THE FORWARD AND BACKWARD VARIABLES

Initialization: $\mathbf{r}_0 = R_x[1], \quad \boldsymbol{a}_0 = \boldsymbol{b}_0 = 1, \quad \sigma_0^2 = \sigma_0'^2 = R_x[0]$		
I	$\Delta_p = \mathbf{r}_{p-1}^{*T}\tilde{\boldsymbol{a}}_{p-1}$	$\Delta_p' = \mathbf{r}_{p-1}^{T}\tilde{\boldsymbol{b}}_{p-1}$
II	$\gamma_p = \dfrac{\Delta_p}{\sigma_{p-1}'^2}$	$\gamma_p' = \dfrac{\Delta_p'}{\sigma_{p-1}^2}$
III	$\boldsymbol{a}_p = \begin{bmatrix} \boldsymbol{a}_{p-1} \\ -- \\ 0 \end{bmatrix} - \gamma_p \begin{bmatrix} 0 \\ -- \\ \tilde{\boldsymbol{b}}_{p-1} \end{bmatrix}$	$\boldsymbol{b}_p = \begin{bmatrix} \boldsymbol{b}_{p-1} \\ -- \\ 0 \end{bmatrix} - \gamma_p' \begin{bmatrix} 0 \\ -- \\ \tilde{\boldsymbol{a}}_{p-1} \end{bmatrix}$
IV	$\sigma_p^2 = \sigma_{p-1}^2 - \gamma_p\Delta_p'$ or $\sigma_p^2 = (1-\gamma_p\gamma_p')\sigma_{p-1}^2$	$\sigma_p'^2 = \sigma_{p-1}'^2 - \gamma_p'\Delta_p$ or $\sigma_p'^2 = (1-\gamma_p'\gamma_p)\sigma_{p-1}'^2$

Although the equations of Table 8.2 clearly show the interaction between the forward and the backward models, they take no advantage of the relations between the models that were shown in Section 8.4. Specifically, we know that $\boldsymbol{b}_{p-1} = \boldsymbol{a}_{p-1}^*$ and $\sigma_{p-1}'^2 = \sigma_{p-1}^2$. Therefore, it follows from the first step in Table 8.2 that

$$\Delta_p' = \mathbf{r}_{p-1}^T\tilde{\boldsymbol{b}}_{p-1} = \mathbf{r}_{p-1}^T\tilde{\boldsymbol{a}}_{p-1}^* = \Delta_p^* \tag{8.72}$$

and application of this result to the second step in the table produces

$$\gamma_p' = \frac{\Delta_p'}{\sigma_{p-1}^2} = \frac{\Delta_p^*}{\sigma_{p-1}'^2} = \gamma_p^* \tag{8.73}$$

By using these additional relations and combining steps I and II, the equations in Table 8.2 can be written as a single-thread recursion, which is all that is necessary for purposes of computation. The equations are

$$\begin{aligned} \gamma_p &= \frac{\mathbf{r}^{*T}_{p-1}\tilde{\boldsymbol{a}}_{p-1}}{\sigma^2_{p-1}} \quad (a) \\ \boldsymbol{a}_p &= \begin{bmatrix} \boldsymbol{a}_{p-1} \\ -- \\ 0 \end{bmatrix} - \gamma_p \begin{bmatrix} 0 \\ -- \\ \tilde{\boldsymbol{a}}^*_{p-1} \end{bmatrix} \quad (b) \\ \sigma^2_p &= (1 - |\gamma_p|^2)\sigma^2_{p-1} \quad (c) \end{aligned} \tag{8.74}$$

which are applied for $p = 1, 2, \ldots, P$ with initial conditions

$$\boldsymbol{a}_0 = [1]\ ; \quad \mathbf{r}_0 = R_x[1]; \quad \sigma^2_0 = R_x[0]. \tag{8.75}$$

Note that since the last element of the vector $\tilde{\boldsymbol{a}}^*_{p-1}$ is equal to 1, it follows from (8.74)(b) that

$$a^{(p)}_p = -\gamma_p \tag{8.76}$$

This is a useful fact to remember.

It was mentioned before that the Levinson recursion provides new insight to the problem of linear prediction that may not otherwise be apparent. A first example of this is as follows. Look at (8.74)(c) and note that since both $\sigma^2_p \geq 0$ and $\sigma^2_{p-1} \geq 0$, it follows that

$$|\gamma_p| \leq 1 \tag{8.77}$$

This in turn implies that

$$\sigma^2_p \leq \sigma^2_{p-1} \tag{8.78}$$

The interpretation of this in terms of linear prediction is eminently reasonable. Equation 8.78 states that the pth-order predictor always has a prediction error variance (mean-square error) less than or equal to that of the $(p-1)^{\text{th}}$-order predictor.

The condition (8.77) on the reflection coefficients is extremely important in itself. This condition follows from the property $\sigma^2_p \geq 0$ of the prediction error variances, which is a consequence of the positive semidefinite property of the correlation matrix. (The last statement will become clear from the material in Section 8.11.) Observe further that for any *regular* process the quantities σ^2_p and σ^2_{p+1} are both *strictly* positive; therefore, the reflection coefficients must satisfy the condition

$$0 \leq |\gamma_p| < 1 \tag{8.79}$$

where the upper inequality is *strict.* It will be seen later in this chapter that this condition guarantees the stability of the inverse filter and the strict positive definiteness of the correlation matrix. Equation 8.78 can still be satisfied with equality since it derives from (8.74)(c) and it is possible for a regular process to have $\gamma_p = 0$.

The following is a simple example of applying the Levinson recursion to compute the linear prediction parameters.

Example 8.1

Consider the linear prediction of a real random process with the correlation function shown in Fig. EX8.1.[7]

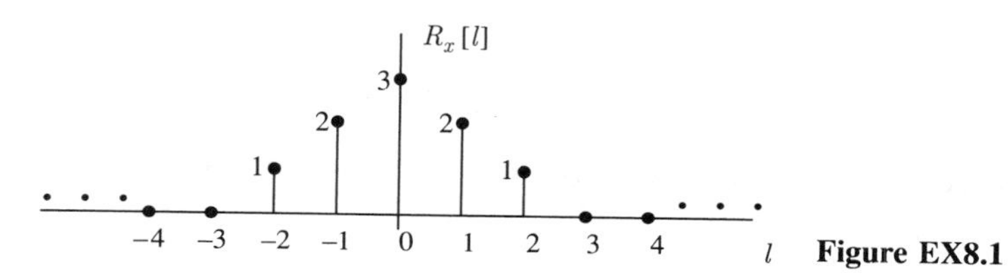

Figure EX8.1

It is desired to find the second-order linear prediction parameters for the random process $x[n]$ and the values of the reflection coefficients. The first three values of the correlation function are

$$R_x[0] = 3; \quad R_x[1] = R_x[-1] = 2; \quad R_x[2] = R_x[-2] = 1$$

Therefore, the Normal equations for this problem are

$$\begin{bmatrix} 3 & 2 & 1 \\ 2 & 3 & 2 \\ 1 & 2 & 3 \end{bmatrix} \begin{bmatrix} 1 \\ a_1 \\ a_2 \end{bmatrix} = \begin{bmatrix} \sigma_2^2 \\ 0 \\ 0 \end{bmatrix}$$

The equations are solved by applying the Levinson recursion in the form (8.74)(a)–(c).

Begin with $\mathbf{r}_0 = R_x[1] = 2$, $\boldsymbol{a}_0 = 1$, and $\sigma_0^2 = R_x[0] = 3$ and proceed through the first stage of the recursion (for $p = 1$):

$$\gamma_1 = \frac{\mathbf{r}_0 \boldsymbol{a}_0}{\sigma_0^2} = \frac{2 \cdot 1}{3} = \tfrac{2}{3}$$

$$\boldsymbol{a}_1 = \begin{bmatrix} \boldsymbol{a}_0 \\ -- \\ 0 \end{bmatrix} - \gamma_1 \begin{bmatrix} 0 \\ -- \\ \tilde{\boldsymbol{a}}_0^* \end{bmatrix} = \begin{bmatrix} 1 \\ 0 \end{bmatrix} - \tfrac{2}{3} \begin{bmatrix} 0 \\ 1 \end{bmatrix} = \begin{bmatrix} 1 \\ -\tfrac{2}{3} \end{bmatrix}$$

$$\sigma_1^2 = (1 - |\gamma_1|^2)\sigma_0^2 = (1 - (\tfrac{2}{3})^2) \cdot 3 = \tfrac{5}{3}$$

[7] This process is actually somewhat degenerate since $S_x(z)$ has (double) roots on the unit circle at $z = e^{\jmath 2\pi/3}$ and $z = e^{-\jmath 2\pi/3}$. The correlation function has simple integer values, however, so the computations are convenient and easy to follow.

This determines the first-order parameters, including the reflection coefficient γ_1. The next stage of recursion proceeds in a similar fashion (for $p = 2$):

$$\gamma_2 = \frac{\mathbf{r}_1^{*T}\tilde{\boldsymbol{a}}_1}{\sigma_1^2} = \frac{1}{5/3}\begin{bmatrix} 2 & 1 \end{bmatrix}\begin{bmatrix} -\frac{2}{3} \\ 1 \end{bmatrix} = -\frac{1}{5} = -0.2$$

$$\boldsymbol{a}_2 = \begin{bmatrix} \boldsymbol{a}_1 \\ -- \\ 0 \end{bmatrix} - \gamma_2 \begin{bmatrix} 0 \\ -- \\ \tilde{\boldsymbol{a}}_1^* \end{bmatrix} = \begin{bmatrix} 1 \\ -\frac{2}{3} \\ 0 \end{bmatrix} - (-\tfrac{1}{5})\begin{bmatrix} 0 \\ -\frac{2}{3} \\ 1 \end{bmatrix} = \begin{bmatrix} 0 \\ -\frac{4}{5} \\ \frac{1}{5} \end{bmatrix} = \begin{bmatrix} 1 \\ -0.8 \\ 0.2 \end{bmatrix}$$

$$\sigma_2^2 = (1 - |\gamma_2|^2)\sigma_1^2 = (1 - |-\tfrac{1}{5}|^2)\cdot\tfrac{5}{3} = \tfrac{8}{5} = 1.6$$

This completes the solution.

8.6 LATTICE REPRESENTATION FOR THE PREDICTION ERROR FILTER

Some of the previous sections showed that there is a very close relation between linear prediction and AR modeling. In particular, whenever a random process $x[n]$ is suitably represented by an AR model, the parameters of the AR model can be generated by considering the linear prediction problem for the random process. The coefficients of the AR model will then correspond to those of the linear predictive filter. It was further noted that the backward linear prediction problem and the corresponding anticausal AR model play an essential role in the development of the Levinson recursion. This section further extends those relations and shows that the Levinson recursion and the idea of forward and backward filtering leads to a very convenient lattice representation for both the prediction error filter and the inverse filter that appears in the AR model. The reflection coefficients of all orders, $\gamma_p,\ p = 0, 1, 2, \ldots, P$ turn out to be the only parameters needed in the lattice representation. These will naturally be available if the Levinson recursion is used to solve the Normal equations. However, if some other procedure, such as direct matrix inversion, is used to solve for the linear prediction parameters, the complete set of reflection coefficients will not be known. A procedure is therefore discussed for obtaining the reflection coefficients from the linear prediction parameters. This procedure is known as the reverse-order Levinson recursion.

8.6.1 Derivation of the Lattice Representation

To develop the lattice representation for the prediction error filter let us begin by defining the vector

$$\boldsymbol{x}_p[n] = \begin{bmatrix} x[n-p] \\ x[n-p+1] \\ \vdots \\ x[n-1] \\ x[n] \end{bmatrix} \tag{8.80}$$

(When $p = P$ this vector consists of the points shown in Fig. 8.4.) Then the prediction error for the pth-order forward predictor is given by

$$\varepsilon_p[n] = \sum_{i=0}^{p} a_i^{(p)*} x[n-i] = \boldsymbol{a}_p^{*T} \tilde{\boldsymbol{x}}_p[n] = \tilde{\boldsymbol{a}}_p^{*T} \boldsymbol{x}_p[n] \tag{8.81}$$

It is convenient here to define the *backward prediction error* $\varepsilon_p^b[n]$ as

$$\varepsilon_p^b[n] = \varepsilon_p'[n-p] \tag{8.82}$$

Note the important distinction between $\varepsilon_p'[n]$ and $\varepsilon_p^b[n]$; namely, $\varepsilon_p'[n]$ requires *future* values for its computation (see Fig. 8.4) and thus requires noncausal filtering while $\varepsilon_p^b[n]$ is a delayed version of $\varepsilon_p'[n]$ and can be computed by a *causal* filter. Although it introduces a certain amount of asymmetry to the problem, it is conventional to formulate the results from now on in terms of $\varepsilon_p^b[n]$. The backward prediction error [see (8.27)] is then given by

$$\varepsilon_p^b[n] = \sum_{i=0}^{p} b_i^{(p)*} x[n-p+i] = \boldsymbol{b}_p^{*T} \boldsymbol{x}_p[n] = \tilde{\boldsymbol{b}}_p^{*T} \tilde{\boldsymbol{x}}_p[n] \tag{8.83}$$

Now observe that the vector $\boldsymbol{x}_p$ can be partitioned in two ways, namely

$$\boldsymbol{x}_p[n] = \begin{bmatrix} x[n-p] \\ ---- \\ \boldsymbol{x}_{p-1}[n] \end{bmatrix} = \begin{bmatrix} \boldsymbol{x}_{p-1}[n-1] \\ ---- \\ x[n] \end{bmatrix} \tag{8.84}$$

where

$$\boldsymbol{x}_{p-1}[n] = \begin{bmatrix} x[n-p+1] \\ \vdots \\ x[n-1] \\ x[n] \end{bmatrix}$$

and correspondingly, the reversed vector can be written as

$$\tilde{\boldsymbol{x}}_p[n] = \begin{bmatrix} \tilde{\boldsymbol{x}}_{p-1}[n] \\ ---- \\ x[n-p] \end{bmatrix} = \begin{bmatrix} x[n] \\ ---- \\ \tilde{\boldsymbol{x}}_{p-1}[n-1] \end{bmatrix} \tag{8.85}$$

A recursion for the forward and backward prediction errors can now be developed.

The recursion for the forward filter parameters (see step III of Table 8.2) and (8.85) can be used to write

$$\boldsymbol{a}_p^{*T}\tilde{\boldsymbol{x}}_p[n] = \left[\ \boldsymbol{a}_{p-1}^{*T} \quad \middle|\ 0\ \right] \begin{bmatrix} \tilde{\boldsymbol{x}}_{p-1}[n] \\ ---- \\ x[n-p] \end{bmatrix} - \gamma_p^* \left[\ 0 \ \middle| \quad \tilde{\boldsymbol{b}}_{p-1}^{*T}\ \right] \begin{bmatrix} x[n] \\ ---- \\ \tilde{\boldsymbol{x}}_{p-1}[n-1] \end{bmatrix}$$

or [by (8.81) and (8.83)]

$$\varepsilon_p[n] = \varepsilon_{p-1}[n] - \gamma_p^* \varepsilon_{p-1}^b[n-1] \tag{8.86}$$

Likewise, the recursion for the backward filter and (8.84) imply that

$$\boldsymbol{b}_p^{*T}\boldsymbol{x}_p[n] = \left[\ \boldsymbol{b}_{p-1}^{*T} \quad \middle|\ 0\ \right] \begin{bmatrix} \boldsymbol{x}_{p-1}[n-1] \\ ---- \\ x[n] \end{bmatrix}$$

$$- \gamma_p'^* \left[\ 0 \ \middle| \quad \tilde{\boldsymbol{a}}_{p-1}^{*T}\ \right] \begin{bmatrix} x[n-p] \\ ---- \\ \boldsymbol{x}_{p-1}[n] \end{bmatrix}$$

or

$$\varepsilon_p^b[n] = \varepsilon_{p-1}^b[n-1] - \gamma_p'^* \varepsilon_{p-1}[n] \tag{8.87}$$

These last two numbered equations can be combined to write

$$\boxed{\begin{bmatrix} \varepsilon_p[n] \\ \varepsilon_p^b[n] \end{bmatrix} = \begin{bmatrix} 1 & -\gamma_p^* \\ -\gamma_p'^* & 1 \end{bmatrix} \begin{bmatrix} \varepsilon_{p-1}[n] \\ \varepsilon_{p-1}^b[n-1] \end{bmatrix}} \tag{8.88}$$

Equation 8.88 shows that if it were somehow possible to compute the $(p-1)^{\text{th}}$ order-forward and backward prediction errors, then those for the pth-order filter could be obtained in a simple way. This is illustrated in Fig. 8.5, where it is seen that a pth-order filter can be formed from the $(p-1)^{\text{th}}$-order filter by adding a simple lattice section. Evidently the entire prediction error filter can be formed by cascading lattice sections, each characterized by its reflection coefficients. A general P^{th}-order filter is shown in Fig. 8.6.

The inverse filter needed in the AR model can be formed by inverting the structure of Fig. 8.6 to produce the signal $x[n]$ from a white noise source. To do this requires inverting each of the lattice sections so that $\varepsilon_{p-1}[n]$ is formed from $\varepsilon_p[n]$ and $\varepsilon_{p-1}^b[n]$ as shown in Fig. 8.7. The desired inverse lattice section can be obtained by writing (8.86) as

$$\varepsilon_{p-1}[n] = \varepsilon_p[n] + \gamma_p^* \varepsilon_{p-1}^b[n-1] \tag{8.89}$$

and using this in conjunction with (8.87). Note that the section of Fig. 8.7(a) implements (8.86) and (8.87) while the section of Fig. 8.7(b) implements (8.87) and (8.89). A complete inverse P^{th}-order filter is shown in Fig. 8.8.

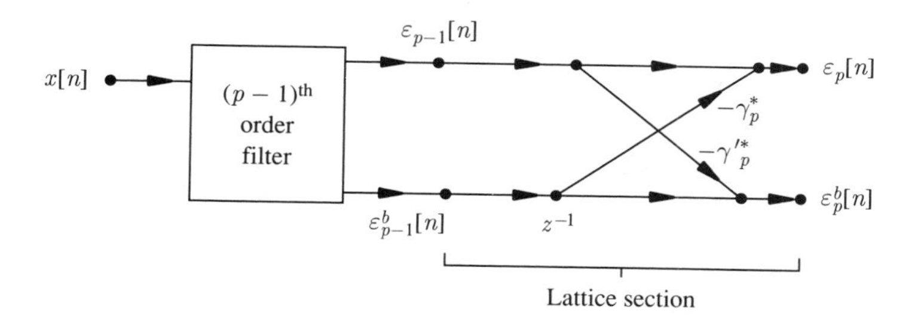

Note: $\gamma'^*_p = \gamma_p$.

Figure 8.5 Formation of a pth-order prediction error filter by adding a lattice section.

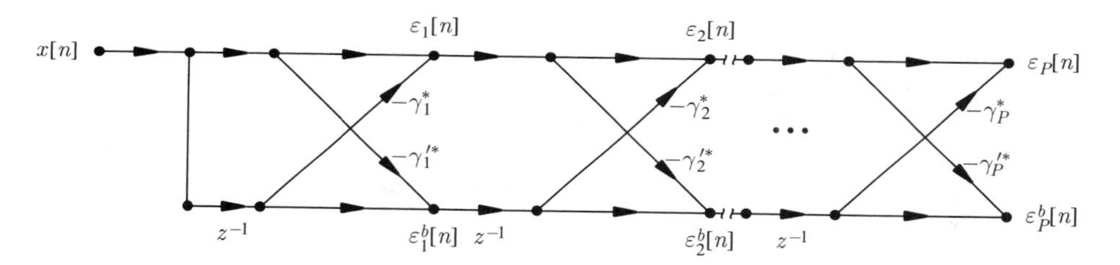

Note: $\gamma'^*_p = \gamma_p$.

Figure 8.6 Lattice realization of a prediction error filter.

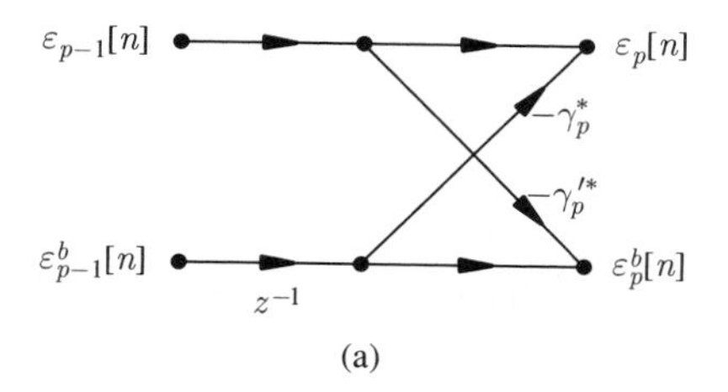

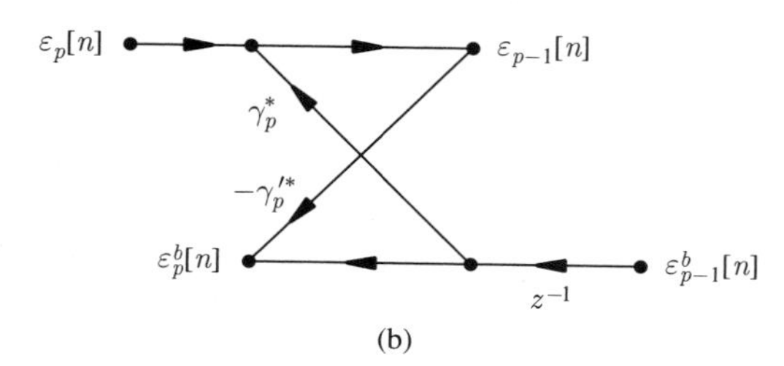

Note: $\gamma'^*_p = \gamma_p$.

Figure 8.7 Forward and inverse lattice sections. (a) Forward section. (b) Inverse section.

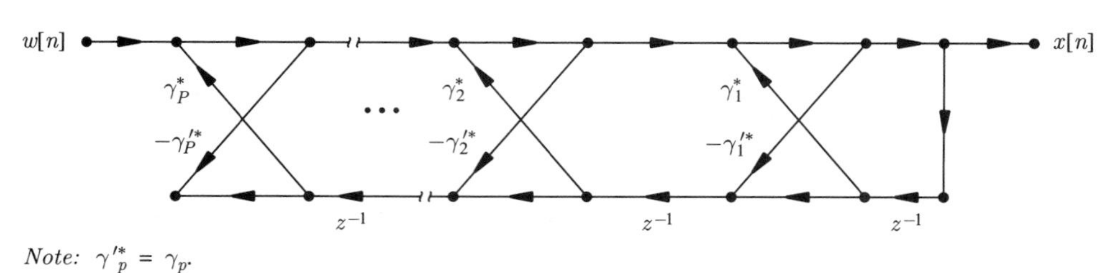

Figure 8.8 Lattice realization of the inverse (all-pole) filter.

8.6.2 Reversing the Levinson Recursion

It is obvious that from the complete set of reflection coefficients it is always possible to compute the filter coefficient vector $\boldsymbol{a}_P$ by recursively applying step III in Table 8.2 or the simpler relation (8.74)(b). It turns out that it is also possible to do the reverse; that is, given the coefficients of some P^{th}-order prediction error filter, to go back and compute all of the reflection coefficents. In so doing the linear prediction parameters $a_k^{(p)}$ of all orders up to order P are also computed. The procedure is thus known as the *reverse-order* or "backward" Levinson recursion (not to be confused with backward linear prediction).

To derive this recursion, begin with the filter parameter update equations (step III) of Table 8.2:

$$\boldsymbol{a}_p = \begin{bmatrix} \boldsymbol{a}_{p-1} \\ -- \\ 0 \end{bmatrix} - \gamma_p \begin{bmatrix} 0 \\ -- \\ \tilde{\boldsymbol{b}}_{p-1} \end{bmatrix} \tag{8.90}$$

and

$$\boldsymbol{b}_p = \begin{bmatrix} \boldsymbol{b}_{p-1} \\ -- \\ 0 \end{bmatrix} - \gamma'_p \begin{bmatrix} 0 \\ -- \\ \tilde{\boldsymbol{a}}_{p-1} \end{bmatrix} \tag{8.91}$$

If the vectors in (8.91) are reversed and the equation is rearranged, we have

$$\begin{bmatrix} 0 \\ -- \\ \tilde{\boldsymbol{b}}_{p-1} \end{bmatrix} = \boldsymbol{b}_p + \gamma'_p \begin{bmatrix} \boldsymbol{a}_{p-1} \\ -- \\ 0 \end{bmatrix}$$

Substitution of this in (8.90) then yields

$$\boldsymbol{a}_p = \begin{bmatrix} \boldsymbol{a}_{p-1} \\ -- \\ 0 \end{bmatrix} - \gamma_p \left(\boldsymbol{b}_p + \gamma'_p \begin{bmatrix} \boldsymbol{a}_{p-1} \\ -- \\ 0 \end{bmatrix} \right)$$

or

$$\begin{bmatrix} \boldsymbol{a}_{p-1} \\ -- \\ 0 \end{bmatrix} = \frac{1}{1-\gamma_p \gamma_p'} \left[\boldsymbol{a}_p + \gamma_p \tilde{\boldsymbol{b}}_p \right] \tag{8.92}$$

A similar relation can be derived for $\boldsymbol{b}_{p-1}$ but it is redundant. Substituting the relations $\boldsymbol{b}_p = \boldsymbol{a}_p^*$ and $\gamma_p' = \gamma_p^*$ in (8.92) produces the simple recursion

$$\boxed{\begin{bmatrix} \boldsymbol{a}_{p-1} \\ -- \\ 0 \end{bmatrix} = \frac{1}{1-|\gamma_p|^2} \left[\boldsymbol{a}_p + \gamma_p \tilde{\boldsymbol{a}}_p^* \right]} \tag{8.93}$$

By using the recursion above and noting that $\gamma_p = -a_p^{(p)}$ [see (8.76)] it is possible to compute all of the lower-order filter parameters and reflection coefficients. The prediction error variances can also be computed by solving (8.74)(c) to obtain

$$\sigma_{p-1}^2 = \frac{\sigma_p^2}{1-|\gamma_p|^2} \tag{8.94}$$

To illustrate the use of these equations let us compute the reflection coefficients and prediction error variances for the second-order filter found in Example 8.1.

Example 8.2

It is desired to find the reflection coefficients in the lattice representation of the prediction error filter with coefficients

$$\boldsymbol{a}_2 = \begin{bmatrix} 1 \\ -0.8 \\ 0.2 \end{bmatrix}$$

Begin by observing that

$$\gamma_2 = -a_2^{(2)} = -0.2$$

Then from (8.93)

$$\begin{bmatrix} \boldsymbol{a}_1 \\ -- \\ 0 \end{bmatrix} = \frac{1}{1-|\gamma_2|^2} \left[\boldsymbol{a}_2 + \gamma_2 \tilde{\boldsymbol{a}}_2^* \right]$$

$$= \frac{1}{1-|-0.2|^2} \left[\begin{bmatrix} 1 \\ -0.8 \\ 0.2 \end{bmatrix} + (-0.2) \begin{bmatrix} 0.2 \\ -0.8 \\ 1 \end{bmatrix} \right] = \begin{bmatrix} 1 \\ -0.666\ldots \\ 0 \end{bmatrix}$$

Thus

$$\gamma_1 = +0.666\ldots$$

These correspond to the reflection coefficients found in Example 8.1. Obviously, if the original filter order had been higher, the process could have been continued until all the reflection coefficients were calculated. The prediction error variance of the first-order filter in this example is obtained from (8.94) as

$$\sigma_1^2 = \frac{\sigma_2^2}{1 - |\gamma_2|^2} = \frac{1.6}{1 - |-0.2|^2} = 1.666\ldots$$

which again agrees with the result of Example 8.1.

8.7 PARTIAL CORRELATION INTERPRETATION OF THE REFLECTION COEFFICIENTS

The reflection coefficients γ_p and γ_p' are seen to play an ever increasingly important role in linear prediction and AR modeling. It has already been seen that

$$a_p^{(p)} = -\gamma_p$$

Since $a_0^{(p)} = 1$ it also follows from (8.91) that

$$b_p^{(p)} = -\gamma_p' \tag{8.95}$$

In other words, *γ_p and γ_p' are the negative last elements of $\mathbf{a}_p$ and $\mathbf{b}_p$*. It was also noted that the reflection coefficients have direct analogs in the theory of propagating waves. In this section they are given a statistical interpretation.

Let us begin by considering a set of random variables $\{u, w_1, w_2, \ldots, w_L, v\}$. (The w_i are not to be confused with white noise samples.) It could happen that u and v are both highly correlated with the w's and as a result that u is highly correlated with v. It may be desirable, however, to know whether u has any *direct* influence on v, that is, if u remains correlated with v once the effect of the intermediate variables is removed. This direct type of correlation is known as *partial correlation*.[8]

[8] As an illustration of this idea, suppose there are three random variables $u, v,$ and w with u a function of w and w a function of v. We write (using slightly abusive notation)

$$u = u(w)$$

$$w = w(v)$$

It is obvious that u and v are correlated in general, but if the dependence of u on w were removed, for example by letting v vary in such a way that w remains fixed, then v would have no effect on u. On the other hand, if u depends explicitly on both w and v, that is,

$$u = u(w, v)$$

then even if the dependence on w is removed, v has a direct effect on u. This explicit dependence of u on v produces partial correlation.

In order to develop a geometric picture of partial correlation, let us use the vector space ideas introduced in Chapter 7. Recall that the inner product (u, v) in this space is synonymous with the correlation $\mathcal{E}\{uv^*\}$ of the two random variables. Further, when $\mathcal{E}\{uv^*\} = 0$ the two random variables are orthogonal, while when u and v are perfectly correlated they lie along the same line.

To remove the influence of the intermediate variables w_i on u and v we project u and v on the subspace $\mathcal{W}$ defined by $w_1, w_2, \ldots, w_L$ and deal only with the residuals. This is illustrated in Fig. 8.9(a) for a one-dimensional subspace ($L = 1$). The projections on the subspace are simply the best linear mean-square estimates of u and v using the elements of $\mathcal{W}$. If these estimates are denoted by $\hat{u}$ and $\hat{v}$, then the residuals are the estimation errors

$$\epsilon_u = u - \hat{u} \tag{8.96}$$

and

$$\epsilon_v = v - \hat{v} \tag{8.97}$$

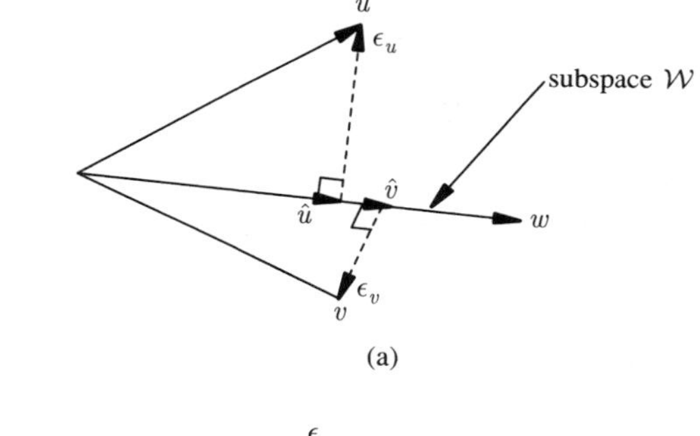

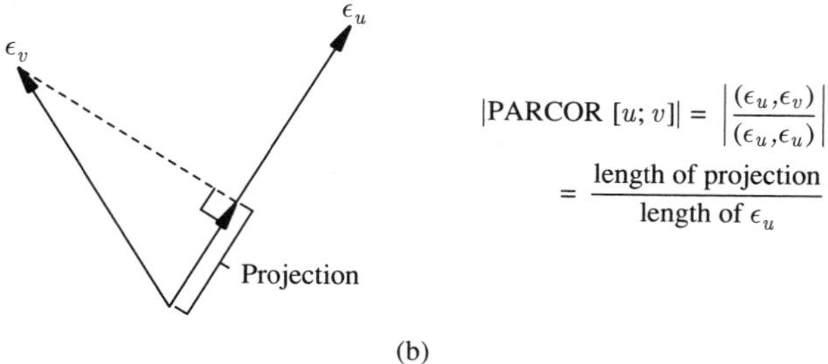

Figure 8.9 Geometric interpretation of partial correlation. (a) Projection of random variables u and v on subspace $\mathcal{W}$ and and definition of errors. (b) Magnitude of partial correlation in terms of errors.

The correlation between the two random vectors can now be expressed as

$$\mathcal{E}\{uv^*\} = \mathcal{E}\{(\hat{u} + \epsilon_u)(\hat{v} + \epsilon_v)^*\} = \mathcal{E}\{\hat{u}\hat{v}^*\} + \mathcal{E}\{\epsilon_u \epsilon_v^*\} \tag{8.98}$$

where the cross-terms are zero because of orthogonality [$\hat{u}$ and $\hat{v}$ both lie in the same subspace and ϵ_u and ϵ_v are both orthogonal to that subspace; see Fig. 8.9(a)]. The first term on the right of (8.98) represents the indirect correlation due to the presence of the random variables w_i while the second term is the *partial* (i.e., direct) correlation. It is the correlation of the errors. Partial correlation is usually measured as a normalized quantity called the PARCOR coefficient and defined as

$$\text{PARCOR}\,[u;v] \stackrel{\text{def}}{=} \frac{\mathcal{E}\{\epsilon_u \epsilon_v^*\}}{\mathcal{E}\{|\epsilon_u|^2\}} \tag{8.99}$$

Note that according to the definition of the inner product for this vector space, PARCOR $[u;v]$ is the inner product of ϵ_u and ϵ_v normalized by the inner product of ϵ_u with itself. The magnitude is therefore the ratio of the length of the projection of ϵ_v on ϵ_u to the length of ϵ_u [see Fig. 8.9(b)]. When the errors are orthogonal the partial correlation is zero. Also, because of the normalization, $|\text{PARCOR}\,[u;v]| \neq |\text{PARCOR}\,[v;u]|$ for general random varables u and v.

The partial correlation interpretation is easy to apply to the reflection coefficients. Consider the $p+1$ data points shown in Fig. 8.10 and the associated $(p-1)^{\text{th}}$-order forward and backward linear prediction problems. Let us identify $x[n-p]$ with u and $x[n]$ with v and the points in between with the w_i. Note that both the forward prediction of $x[n]$ and the backward prediction of $x[n-p]$ use this common set C of intermediate points. The error residuals corresponding to $x[n]$ and $x[n-p]$ are $\varepsilon_{p-1}[n]$ and $\varepsilon'_{p-1}[n-p] = \varepsilon^b_{p-1}[n-1]$. Thus the partial correlation between $x[n-p]$ and $x[n]$ is given by

$$\text{PARCOR}\,[x[n-p];x[n]] = \frac{\mathcal{E}\{\varepsilon^b_{p-1}[n-1]\varepsilon^*_{p-1}[n]\}}{\mathcal{E}\left\{|\varepsilon^b_{p-1}[n-1]|^2\right\}} \tag{8.100}$$

It will be shown that this quantity is just equal to γ_p. The proof is rather short and easy.

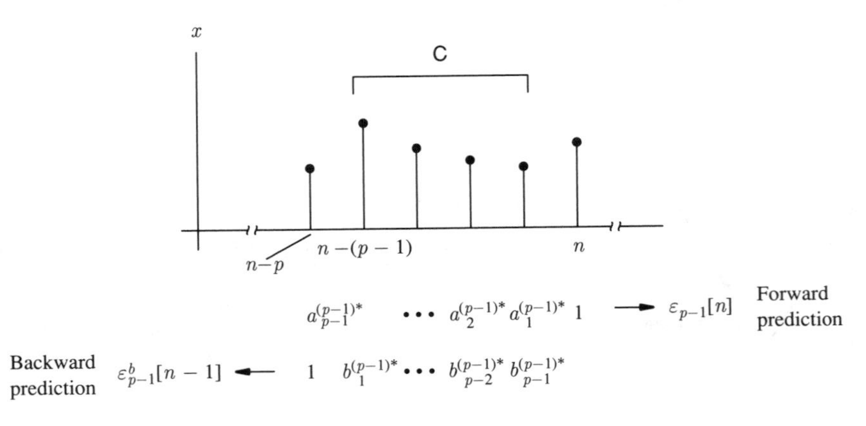

Figure 8.10 Points used in interpretation of partial correlation for linear prediction.

To begin, look at the full set of points $x[n-p], \ldots, x[n]$, which is redrawn in Fig. 8.11(a). Observe that the backward error $\varepsilon^b_{p-1}[n-1]$ is a linear combination of the points in the set A while the forward error $\varepsilon_p[n]$ is *orthogonal* to the points in this set. Therefore, it follows that

$$\mathcal{E}\{\varepsilon^b_{p-1}[n-1]\varepsilon^*_p[n]\} = 0 \tag{8.101}$$

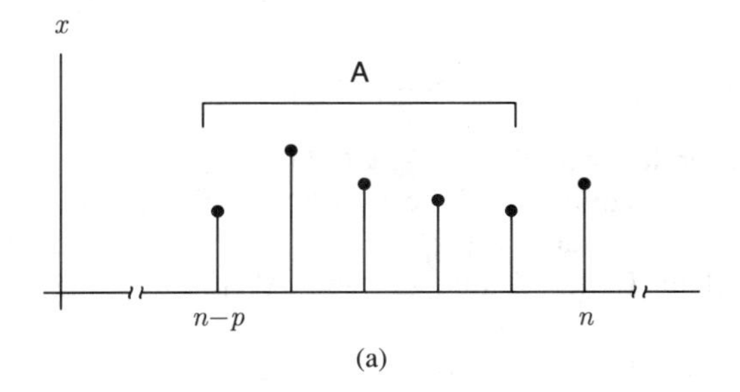

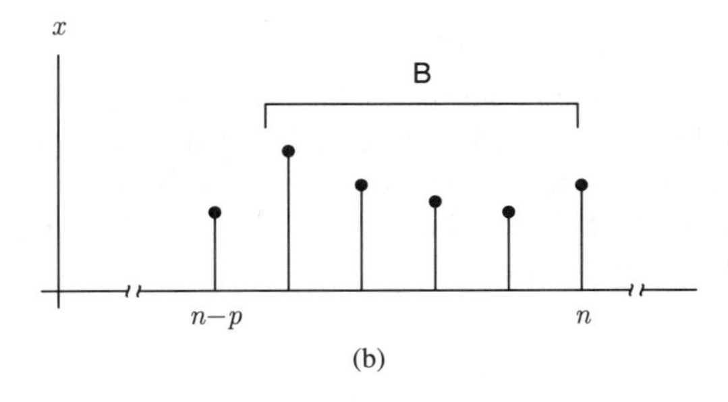

Figure 8.11 Points used in linear prediction. (a) Points used to show that $\mathcal{E}\{\varepsilon^b_{p-1}[n-1]\varepsilon^*_p[n]\} = 0$. (b) Points used to show that $\mathcal{E}\{\varepsilon_{p-1}[n]\varepsilon^{b*}_p[n]\} = 0$.

Substituting (8.86) for $\varepsilon_p[n]$ then leads to

$$\mathcal{E}\left\{\varepsilon^b_{p-1}[n-1]\left(\varepsilon_{p-1}[n] - \gamma^*_p\varepsilon^b_{p-1}[n-1]\right)^*\right\} = 0$$

or

$$\gamma_p = \frac{\mathcal{E}\{\varepsilon^b_{p-1}[n-1]\varepsilon^*_{p-1}[n]\}}{\mathcal{E}\left\{|\varepsilon^b_{p-1}[n-1]|^2\right\}} \tag{8.102}$$

which proves the result.

It can be shown by an almost identical argument that PARCOR $[x[n]; x[n-p]]$ is equal to γ'_p. Look at the grouping of points shown in Fig. 8.11(b). Since $\varepsilon_{p-1}[n]$ is a linear combination of only the data in set B while $\varepsilon^b_p[n]$ is orthogonal to this same data, it follows that

$$\mathcal{E}\{\varepsilon_{p-1}[n]\varepsilon^{b*}_p[n]\} = 0 \tag{8.103}$$

Substitution of (8.87) then yields

$$\mathcal{E}\left\{\varepsilon_{p-1}[n]\left(\varepsilon^b_{p-1}[n-1] - \gamma'^*_p\varepsilon_{p-1}[n]\right)^*\right\} = 0$$

or

$$\gamma'_p = \frac{\mathcal{E}\{\varepsilon_{p-1}[n]\varepsilon^{b*}_{p-1}[n-1]\}}{\mathcal{E}\{|\varepsilon_{p-1}[n]|^2\}} \tag{8.104}$$

In summary, it has been shown that

$$\gamma_p = \text{PARCOR}\,[x[n-p]; x[n]] \tag{8.105}$$

while

$$\gamma'_p = \text{PARCOR}\,[x[n]; x[n-p]] \tag{8.106}$$

Note that although for general random variables $|\text{PARCOR}\,[u;v]\,| \neq |\text{PARCOR}\,[v;u]\,|$, for the case of linear prediction, the magnitudes of the two PARCOR coefficients *are* equal. Further, since $\mathcal{E}\{|\varepsilon_{p-1}[n]|^2\} = \sigma^2_{p-1}$ and $\mathcal{E}\{|\varepsilon^b_{p-1}[n-1]|^2\} = \sigma'^2_{p-1}$, it follows from (8.102) and (8.104) and step II of Table 8.2 that Δ_p and Δ'_p must be the *unnormalized* partial correlations

$$\begin{aligned} \Delta_p &= \mathcal{E}\{\varepsilon^b_{p-1}[n-1]\varepsilon^*_{p-1}[n]\} \quad (a) \\ \Delta'_p &= \mathcal{E}\{\varepsilon_{p-1}[n]\varepsilon^{b*}_{p-1}[n-1]\} \quad (b) \end{aligned} \tag{8.107}$$

Let us see how the partial correlation concept applies to a first-order AR process. In particular consider the process defined by (8.15) with correlation function given by (8.16). A typical correlation function (for real ρ) is depicted in Fig. 8.12. It is obvious that $x[n-p]$ and $x[n]$ are correlated for any value of p and the degree of correlation is represented by the value of the correlation function at $l = p$.

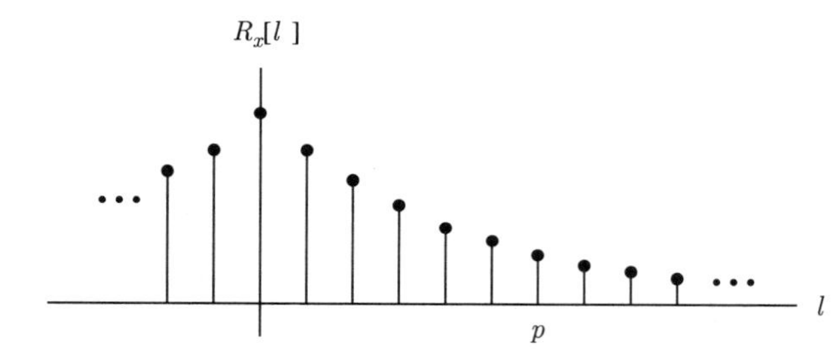

Figure 8.12 Correlation function for a first-order AR process.

Now consider the *partial* correlation, which represents the *direct* influence of $x[n-p]$ on $x[n]$; this is given by γ_p. The filter coefficient vector for this first-order AR process is

$$\boldsymbol{a}_p = \begin{bmatrix} 1 \\ -\rho \\ 0 \\ \vdots \\ 0 \end{bmatrix} \tag{8.108}$$

and so from (8.76)

$$\gamma_p = -a_p^{(p)} = \begin{cases} \rho; & p = 1 \\ 0; & p > 1 \end{cases} \tag{8.109}$$

In other words, the partial correlation of $x[n]$ and $x[n-1]$ is equal to ρ, but the partial correlation of $x[n]$ and any earlier points is *zero*. This is not a surprising conclusion given the first-order form (8.15) for the process. Any correlation between nonadjacent points in the sequence is indirect and due to the presence of the mutually correlated intervening points.

8.8 MINIMUM-PHASE PROPERTY OF THE PREDICTION ERROR FILTER

In this section it is shown that the condition

$$|\gamma_p| = |\gamma'_p| < 1 \tag{8.110}$$

satisfied by the reflection coefficients of a regular random process guarantees the stability of the inverse filter. In other words, the prediction error filter is minimum-phase.

Begin by defining the polynomials

$$A_p(z) = \sum_{k=0}^{p} a_k^{(p)*} z^{-k} \tag{8.111}$$

and

$$B_p(z) = \sum_{k=0}^{p} b_k^{(p)*} z^{k} \tag{8.112}$$

The polynomial $A_p(z)$ is just the transfer function of the pth-order forward prediction error filter, while $B_p(z)$ is the transfer function of the pth-order backward filter that computes ε'_p from x [see (8.27)]. Note that this filter is *noncausal*, since it requires values to the right of $\varepsilon'_p[n]$ to compute $\varepsilon'_p[n]$. Since $b_k^{(p)} = a_k^{(p)*}$ it follows that

$$B_p(z) = A_p^*(1/z^*) \tag{8.113}$$

It will be shown that $A_p(z)$ is a minimum-phase polynomial and therefore that $B_p(z)$ is just the corresponding maximum-phase polynomial. Their interpretation in terms of AR models is illustrated in Fig. 8.13. $A_p(z)$ is the denominator polynomial in the (causal) p^{th}-order AR model for a random process; $B_p(z)$ is the denominator polynomial of the anticausal model. Because of the relation (8.113) *both models produce random processes with identical second moment statistics.*

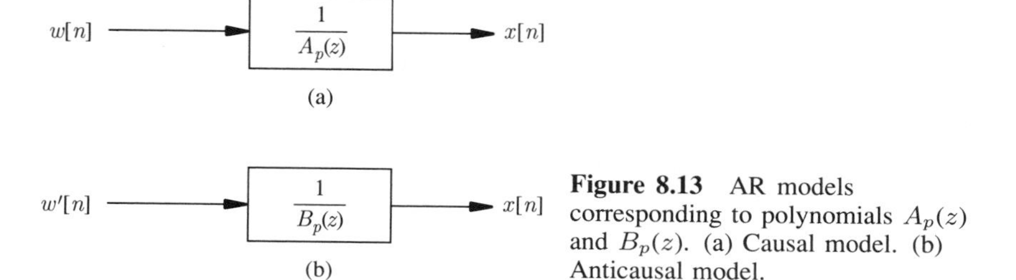

Figure 8.13 AR models corresponding to polynomials $A_p(z)$ and $B_p(z)$. (a) Causal model. (b) Anticausal model.

By equating similar powers of z and using the relations (8.90) and (8.91) it follows that the polynomials satisfy the recursions

$$A_p(z) = A_{p-1}(z) - \gamma_p^* z^{-p} B_{p-1}(z) \tag{8.114}$$

and

$$B_p(z) = B_{p-1}(z) - \gamma_p'^{*} z^p A_{p-1}(z) \tag{8.115}$$

These equations can also be used to derive the reverse-order Levinson recursion (see Problem 8.15).

An alternative form of these relations can be expressed in terms of the backward *prediction error filter* that computes ε_p^b from x. Since $\varepsilon_p^b[n] = \varepsilon'[n-p]$ this filter has a transfer function

$$B_p^b(z) = z^{-p} B_p(z) \tag{8.116}$$

Note that $B_p^b(z)$ is causal while $B_p(z)$ is not [see discussion following (8.82)]. The relation between the two filters is depicted in Fig. 8.14. Substituting (8.116) in (8.114) and (8.115) and simplifying yields

$$A_p(z) = A_{p-1}(z) - \gamma_p^* z^{-1} B_{p-1}^b(z) \tag{8.117}$$

and

$$B_p^b(z) = z^{-1} B_{p-1}^b(z) - \gamma_p'^{*} A_{p-1}(z) \tag{8.118}$$

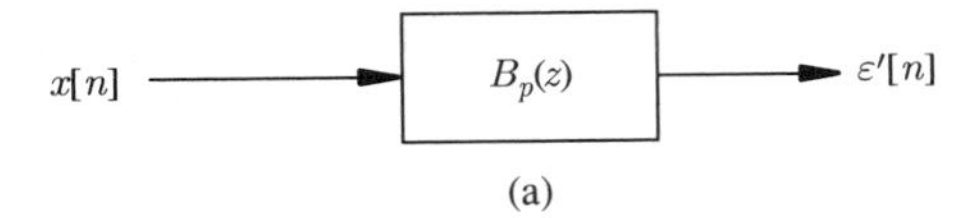

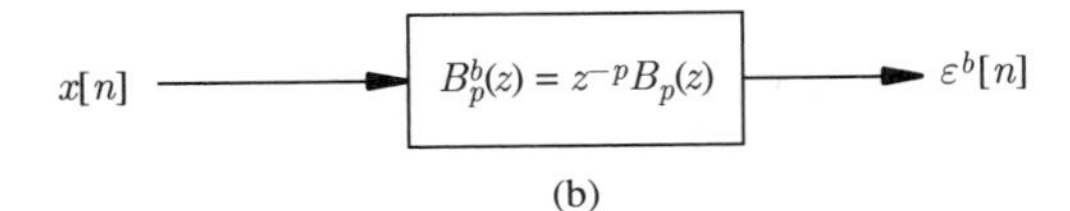

Figure 8.14 Backward prediction error filters and the error processes produced. (a) Noncausal filter $B_p(z)$. (b) Causal filter $B_p^b(z)$.

These results can also be derived by taking z-transforms of (8.86) and (8.87) (see Problem 8.16).

With the foregoing relations in place, it is now possible to prove the minimum-phase property of $A_p(z)$. Substituting (8.113) into (8.114) produces

$$A_p(z) = A_{p-1}(z) - \gamma_p^* z^{-p} A_{p-1}^*(1/z^*) \tag{8.119}$$

This is a simple order recursion for the forward filter which is equivalent to (8.74)(b). Let us define $D_p(z)$ to be the second term in this recursion, i.e.,

$$D_p(z) \stackrel{\text{def}}{=} -\gamma_p^* z^{-p} A_{p-1}^*(1/z^*) \tag{8.120}$$

and examine the frequency response of this term. Observe that it satisfies the inequality

$$|D_p(z)|_{z=e^{\jmath\omega}} = \left|\gamma_p^* e^{-\jmath\omega p} A_{p-1}^*(e^{\jmath\omega})\right| = |\gamma_p| \cdot |A_{p-1}(e^{\jmath\omega})| < |A_{p-1}(e^{\jmath\omega})| \tag{8.121}$$

which follows since $|\gamma_p| < 1$.

Now consider the change in phase undergone by $A_p(z)$ as the point z moves counterclockwise around the unit circle (see Fig. 8.15). The net change of phase of $A_p(z)$ is equal to

$$2\pi\left[\#\text{ poles } A_p(z) - \#\text{ zeros } A_p(z)\right] \tag{8.122}$$

where in this expression only poles and zeros *within the unit circle* are considered. Poles and zeros outside the unit circle contribute no net change of phase (see Fig. 8.15). Now look at the form of $A_p(z)$ [see (8.111)]. It can be written as

$$A_p(z) = \frac{z^p + a_1^{(p)*} z^{p-1} + \cdots + a_p^{(p)*}}{z^p} \tag{8.123}$$

All of the p poles of $A_p(z)$ are at the origin and therefore inside the unit circle. By definition, $A_p(z)$ is minimum-phase, if and only if all of its zeros are also inside the unit circle. Thus, according to (8.122), *$A_p(z)$ is minimum-phase, if and only if the net change of phase of the vector $A(e^{\jmath\omega})$ as $e^{\jmath\omega}$ moves around the unit circle is* <u>*zero*</u>.

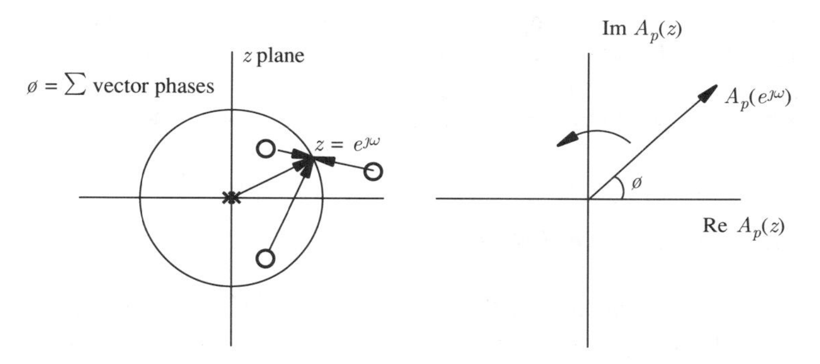

Figure 8.15 Change of phase of polynomial $A_p(z)$ as z moves around the unit circle.

Now consider that according to (8.119) and (8.120), $A_p(e^{\jmath\omega})$ is the sum of the complex vectors (phasors) $A_{p-1}(z)$ and $D_p(z)$ for $z = e^{\jmath\omega}$. This is shown in Fig. 8.16. The minimum-phase property can now be shown by induction. Suppose that $A_{p-1}(z)$ is minimum-phase. Then according to the argument above, $A_{p-1}(z)$ has all of its zeros inside the unit circle and $A_{p-1}(e^{\jmath\omega})$ undergoes no net change of phase as z traverses the unit circle. In other words, the vector $A_{p-1}(e^{\jmath\omega})$ in Fig. 8.16 may move about back and forth but all in all does not encircle the origin. The vector $D_p(e^{\jmath\omega})$ may encircle the tip of the larger vector any number of times, but since $|D_p(e^{\jmath\omega})| < |A_{p-1}(e^{\jmath\omega})|$ [see (8.121)] the net change in phase of $A_p(e^{\jmath\omega})$ is also zero. This implies that $A_p(z)$ also has all of its zeros inside the unit circle and so it too is minimum-phase. Since the zero-order polynomial $A_0(z) = 1$ is trivially minimum-phase, and the magnitudes of all the reflection coefficients are less than 1, (8.121) is satisfied for any value p and consequently the prediction error filter of *any* order is minimum-phase.

The minimum-phase property shown here depends on the theoretical fact that the reflection coefficients have magnitude less than 1. In practice it is sometimes useful to

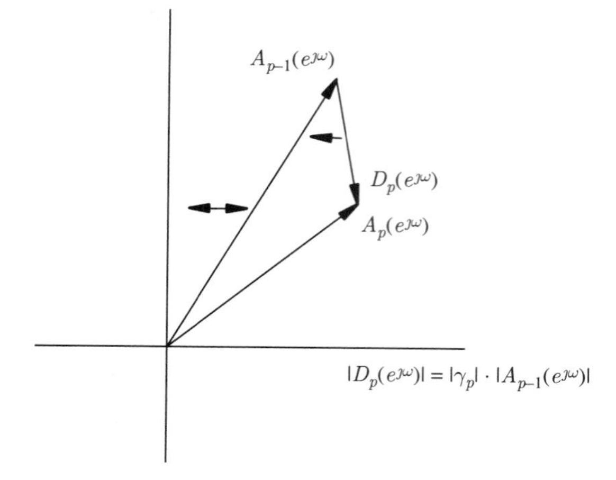

Figure 8.16 Relation between $A_p(z)$ and $A_{p-1}(z)$ as z moves around the unit circle.

construct the prediction error filter by estimating the reflection coefficients directly. Thus it is useful to know that *a necessary and sufficient condition for the lattice filter to be minimum phase is that the reflection coefficients have magnitude less than 1*. The proof of this fact is similar to the above and is left as a problem (Problem 8.17).

8.9 THE SCHUR ALGORITHM

With the development of the Levinson recursion and the related concepts of the reflection coefficients we have been able to take a much broader perspective of linear prediction than would have at first been suspected. Observe, among other things, that if a given random process is approximated by an autoregressive model of order P, then there are three equivalent sets of parameters that can be used to describe that model. These are the $P+1$ terms of the correlation function $R_x[0], R_x[1], \ldots, R_x[P]$; the linear prediction parameters $a_1^{(P)}, \ldots, a_P^{(P)}$ and σ_P^2 (or the AR model parameters $\mathrm{a}_1^{(P)}, \ldots, \mathrm{a}_P^{(P)}$ and σ_P^2); and the reflection coefficients and prediction error variance $\gamma_1, \ldots, \gamma_P$ and σ_P^2. The equivalence of these three sets of parameters is depicted in Fig. 8.17 and one set can be obtained from another by using the Levinson recursion, the AR model and lattice equations, and the reverse-order Levinson recursion.

It is unfortunate that the Levinson recursion, whose original purpose was efficient computation of the linear prediction parameters, has a serious drawback with respect to computation on highly parallel pipelined computer architectures which have become of increasing interest in the implementation of signal processing algorithms. The drawback is the need to compute an inner product in one of the steps (step I of Table 8.2), which destroys any hope of "parallelizing" the algorithm. This limitation has led to the search for other efficient computational algorithms and in particular has brought interest in the Schur algorithm [13], which had its origins in the analysis of certain functions bounded within the unit circle [14, 15]. The functions involved in Schur's original problem have a close

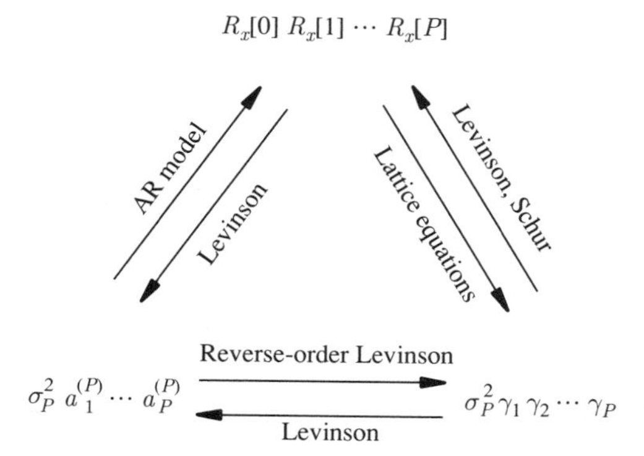

Figure 8.17 Equivalent parameters in the description of a Pth-order autoregressive process.

relation to positive-real functions and scattering parameter representations in network theory (see [13]). Schur's algorithm, when applied to the linear prediction problem, completely eliminates the need for the computation of an inner product, which is the computational bottleneck in any attemped parallel implementation of the Levinson algorithm.

8.9.1 Derivation of the Schur Algorithm

The Schur algorithm can be derived rather simply by considering the set of points shown in Fig. 8.11(a) and noting that if the p points in set A are used to predict $x[n]$, then by the orthogonality principle

$$\mathcal{E}\{\varepsilon_p[n]x^*[n-l]\} = 0; \qquad l = 1, 2, \ldots, p \tag{8.124}$$

Substituting (8.86) yields

$$\mathcal{E}\left\{\left(\varepsilon_{p-1}[n] - \gamma_p^* \varepsilon_{p-1}^b[n-1]\right) x^*[n-l]\right\} = 0$$

or

$$R_{\varepsilon_{p-1}x}[l] - \gamma_p^* R_{\varepsilon_{p-1}^b x}[l-1] = 0; \qquad l = 1, 2, \ldots, p \tag{8.125}$$

Before proceeding further let us say a few words about the cross-correlation functions $R_{\varepsilon_p x}$ and $R_{\varepsilon_p^b x}$. Since the conjugates of these functions are the primary variables in the Schur algorithm it is convenient to use the simpler notation

$$\begin{aligned} g_p[l] &\stackrel{\text{def}}{=} R_{\varepsilon_p x}^*[l] = \mathcal{E}\left\{\varepsilon_p^*[n]x[n-l]\right\} \ (a) \\ g_p^b[l] &\stackrel{\text{def}}{=} R_{\varepsilon_p^b x}^*[l] = \mathcal{E}\left\{\varepsilon_p^{b*}[n]x[n-l]\right\} \ (b) \end{aligned} \tag{8.126}$$

but not to lose sight of the fact that these are actually cross-correlation functions. Now notice that because of the orthogonality principle the functions $g_p[l]$ and $g_p^b[l-1]$ have the properties

$$g_p[l] = 0; \qquad 1 \leq l \leq p \; (a)$$
$$g_p^b[l] = 0; \qquad 0 \leq l \leq p-1 \; (b) \tag{8.127}$$

Because of these regions of zeros, these functions have been called the forward and backward "gapped functions" [12]. In the context of the Schur algorithm they are sometimes also referred to in the literature as "Schur variables." It will also be shown later that g_p and g_p^b are related as

$$g_p^b[l] = g_p^*[p-l] \tag{8.128}$$

so these functions are not really distinct. Typical plots of $g_p[l]$ are shown in Fig. 8.18.

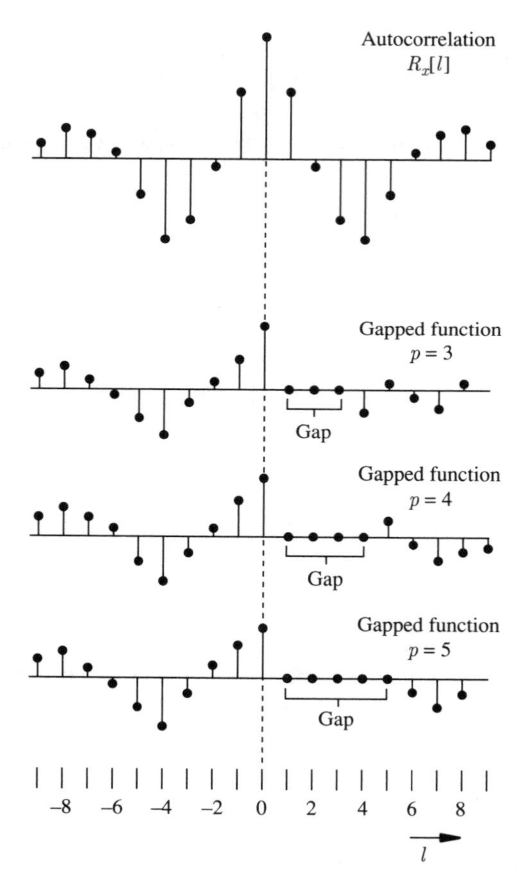

Figure 8.18 Typical plots of the functions $g_p[l] = R^*_{\varepsilon_p x}[l]$ for $p = 3, 4, 5$. (From E. A. Robinson and S. Trietel, Maximum entropy and the relationship of the partial autocorrelation to the reflection coefficients of a layered system, *IEEE Transactions on Acoustics, Speech, and Signal Processing*, April 1980, ©IEEE 1980, reproduced by permission.)

Since $R_{\varepsilon_{p-1}x}[l]$ and $R_{\varepsilon^b_{p-1}x}[l-1]$ are both zero in the range $1 \leq l \leq p-1$, (8.125) for l in this range is a trivial identity. For $l = p$ the functions are *not* equal to zero, however, so (8.125) implies that

$$\gamma_p = \frac{R^*_{\varepsilon_{p-1}x}[p]}{R^*_{\varepsilon^b_{p-1}x}[p-1]} = \frac{g_{p-1}[p]}{g^b_{p-1}[p-1]} \tag{8.129}$$

which is a simple way to compute the reflection coefficients. In fact, (8.129) is not a new relation. It can be seen from (8.126) and the second part of the Orthogonality Theorem that

$$\begin{aligned} g_p[0] &= R^*_{\varepsilon_p x}[0] = \sigma_p^2 \quad (a) \\ g^b_p[p] &= R^*_{\varepsilon^b_p x}[p] = \sigma_p'^2 \quad (b) \end{aligned} \tag{8.130}$$

Thus $g^b_{p-1}[p-1] = \sigma'^2_{p-1}$, so by comparing (8.129) to step I in Table 8.2 it follows that

$$g_{p-1}[p] = \Delta_p \tag{8.131}$$

What *is* new is the recursive procedure that will be developed for these functions that permits computation of the reflection coefficients from (8.129).

The recursion follows directly from the lattice network which is depicted in Fig. 8.19. Each lattice section is represented by a two-input two-output system with input/output relation (8.88). For later convenience define the transfer matrix

$$\mathbf{H}_p = \begin{bmatrix} 1 & -\gamma_p^* \\ -\gamma_p'^* & 1 \end{bmatrix} \tag{8.132}$$

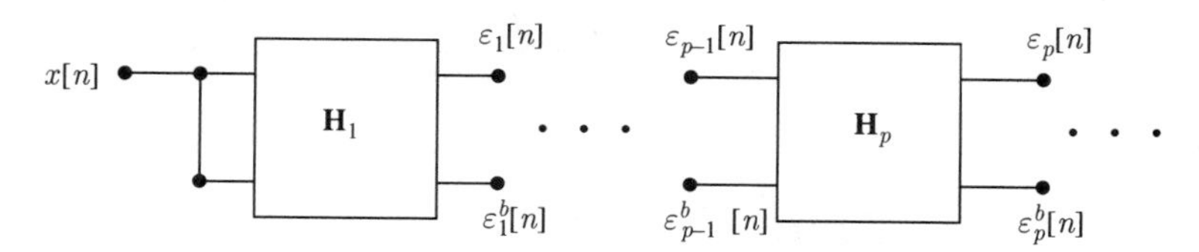

Figure 8.19 Representation of a lattice network as a cascade of two-input two-output systems.

that characterizes each lattice section. To develop the recursion, multiply both sides of (8.88) by $x[n-l]$ and take the expectation to find

$$\begin{bmatrix} R_{\varepsilon_p x}[l] \\ R_{\varepsilon^b_p x}[l] \end{bmatrix} = \begin{bmatrix} 1 & -\gamma_p^* \\ -\gamma_p'^* & 1 \end{bmatrix} \begin{bmatrix} R_{\varepsilon_{p-1}x}[l] \\ R_{\varepsilon^b_{p-1}x}[l-1] \end{bmatrix}$$

or

$$\begin{bmatrix} g_p[l] \\ g^b_p[l] \end{bmatrix} = \begin{bmatrix} 1 & -\gamma_p \\ -\gamma_p' & 1 \end{bmatrix} \begin{bmatrix} g_{p-1}[l] \\ g^b_{p-1}[l-1] \end{bmatrix} \tag{8.133}$$

which is the desired relation.

The complete Schur algorithm, now to be described, is essentially a recursive application of (8.129), (8.130), and (8.133). The initial conditions are given by

$$\boxed{\begin{aligned} g_0[l] &= R^*_{\varepsilon_0 x}[l] = R^*_x[l] \ (a) \\ g^b_0[l] &= R^*_{\varepsilon^b_0 x}[l] = R^*_x[l] \ (b) \end{aligned}} \tag{8.134}$$

A convenient way to carry out the algorithm is as follows.

Initialization. Begin with the "generator matrix"

$$\mathbf{G}_0 \stackrel{\text{def}}{=} \begin{bmatrix} g_0[0] & g_0[1] & g_0[2] & \cdots & g_0[P] \\ g^b_0[0] & g^b_0[1] & g^b_0[2] & \cdots & g^b_0[P] \end{bmatrix}$$
$$= \begin{bmatrix} \times & R^*_x[1] & R^*_x[2] & \cdots & R^*_x[P] \\ R_x[0] & R^*_x[1] & R^*_x[2] & \cdots & R^*_x[P] \end{bmatrix}$$

The symbol $\times$ indicates that the value of the element is not used in any further computation. It may be set to zero or just ignored. In the following part of the description the superscript "$(shift)$" applied to the generator matrix denotes the operation of shifting the bottom row one place to the right.

Step 1. $(p = 1)$

(a) Define

$$\gamma_1 = \frac{g_0[1]}{g^b_0[0]} = \frac{R^*_x[1]}{R^*_x[0]} \quad \text{and} \quad \gamma'_1 = \gamma^*_1$$

(b) Form

$$\mathbf{G}_1 = \mathbf{H}^*_1 \mathbf{G}_0^{(shift)}$$
$$= \begin{bmatrix} 1 & -\gamma_1 \\ -\gamma'_1 & 1 \end{bmatrix} \begin{bmatrix} \times & R^*_x[1] & R^*_x[2] & \cdots & R^*_x[P] \\ \times & R_x[0] & R^*_x[1] & \cdots & R^*_x[P-1] \end{bmatrix}$$
$$= \begin{bmatrix} \times & 0 & g_1[2] & \cdots & g_1[P] \\ \times & g^b_1[1] & g^b_1[2] & \cdots & g^b_1[P] \end{bmatrix}$$

(c) Set

$$\sigma^2_1 = \sigma'^2_1 = g^b_1[1] \quad \text{(which is real)}$$

Step 2. $(p = 2)$

(a) Define

$$\gamma_2 = \frac{g_1[2]}{g^b_1[1]} \quad \text{and} \quad \gamma'_2 = \gamma^*_2$$

(b) Form

$$\mathbf{G}_2 = \mathbf{H}_2^*\mathbf{G}_1^{(shift)}$$

$$= \begin{bmatrix} 1 & -\gamma_2 \\ -\gamma_2' & 1 \end{bmatrix} \begin{bmatrix} \times & 0 & g_1[2] & \cdots & g_1[P] \\ \times & \times & g_1^b[1] & \cdots & g_1^b[P-1] \end{bmatrix}$$

$$= \begin{bmatrix} \times & \times & 0 & g_2[3] & \cdots & g_2[P] \\ \times & \times & g_2^b[2] & g_2^b[3] & \cdots & g_2^b[P] \end{bmatrix}$$

(c) Set

$$\sigma_2^2 = \sigma_2'^2 = g_2^b[2]$$

In general, the generator matrix of order $p-1$ is

$$\mathbf{G}_{p-1} = \begin{bmatrix} \times & \cdots & \times & 0 & g_{p-1}[p] & \cdots & g_{p-1}[P] \\ \times & \cdots & \times & g_{p-1}^b[p-1] & g_{p-1}^b[p] & \cdots & g_{p-1}^b[P] \end{bmatrix} \tag{8.135}$$

the shifted generator matrix is

$$\mathbf{G}_{p-1}^{(shift)} \begin{bmatrix} \times & \cdots & \times & 0 & g_{p-1}[p] & \cdots & g_{p-1}[P] \\ \times & \cdots & \times & \times & g_{p-1}^b[p-1] & \cdots & g_{p-1}^b[P-1] \end{bmatrix} \tag{8.136}$$

and the recursion is

$$\mathbf{G}_p = \mathbf{H}_p^*\mathbf{G}_{p-1}^{(shift)} \tag{8.137}$$

In view of (8.128) it may appear that the algorithm performs excess computation in updating both the forward and backward functions. This is not so, however, because the nonzero values of these functions are defined over disjoint intervals. To see this, use (8.128) to write (8.135) as

$$\mathbf{G}_{p-1} = \begin{bmatrix} \times & \cdots & \times & 0 & g_{p-1}[p] & \cdots & g_{p-1}[P] \\ \times & \cdots & \times & g_{p-1}[0] & g_{p-1}[-1] & \cdots & g_{p-1}[p-(P+1)] \end{bmatrix}$$

from which it is clear that the terms in the top row are in general not the same as the terms in the bottom row.

To more specifically illustrate the computations in the Schur algorithm let us consider its use for the problem of Example 8.1.

Example 8.3

It is desired to compute the reflection coefficients and prediction error variances for the prediction error filter of Example 8.1 by the Schur algorithm. The algorithm is initialized with the real correlation function

$$\mathbf{G}_0 = \begin{bmatrix} \times & R_x^*[1] & R_x^*[2] \\ R_x[0] & R_x^*[1] & R_x^*[2] \end{bmatrix} = \begin{bmatrix} \times & 2 & 1 \\ 3 & 2 & 1 \end{bmatrix}$$

In the first step ($p = 1$) γ_1, $\mathbf{G}_1$, and σ_1^2 are computed:

$$\gamma_1 = \frac{g_0[1]}{g_0^b[0]} = \frac{2}{3} = \tfrac{2}{3}$$

$$\mathbf{G}_1 = \mathbf{H}_1^* \mathbf{G}_0^{(shift)}$$

$$= \begin{bmatrix} 1 & -\frac{2}{3} \\ -\frac{2}{3} & 1 \end{bmatrix} \begin{bmatrix} \times & 2 & 1 \\ \times & 3 & 2 \end{bmatrix} = \begin{bmatrix} \times & 0 & -\frac{1}{3} \\ \times & \frac{5}{3} & \frac{4}{3} \end{bmatrix}$$

$$\sigma_1^2 = g_1^b[1] = \tfrac{5}{3}$$

In the second step ($p = 2$) the remaining terms are computed similarly:

$$\gamma_2 = \frac{g_1[2]}{g_1^b[1]} = \frac{-1/3}{5/3} = -\tfrac{1}{5}$$

$$\mathbf{G}_2 = \mathbf{H}_2^* \mathbf{G}_1^{(shift)}$$

$$= \begin{bmatrix} 1 & \frac{1}{5} \\ \frac{1}{5} & 1 \end{bmatrix} \begin{bmatrix} \times & 0 & -\frac{1}{3} \\ \times & \times & \frac{5}{3} \end{bmatrix} = \begin{bmatrix} \times & \times & 0 \\ \times & \times & \frac{8}{5} \end{bmatrix}$$

$$\sigma_2^2 = g_2^b[2] = \tfrac{8}{5}$$

The results are seen to agree with those of Example 8.1.

The Schur algorithm is especially well-suited for computation on machines involving systolic arrays or wavefront processors. These machines have a highly parallel pipelined structure which can take maximum advantage of the steps in the algorithm. A typical systolic architecture [16] is shown in Fig. 8.20. This machine is synchronous and performs similar operations on each clock cycle. The square boxes represent multiply/accumulate (MAC) cells which perform the simple operation

$$C \leftarrow C + AB$$

where C is the contents of the cell and A and B are the two inputs. A typical cell in the top row performs the computation

$$g_p[l] = g_{p-1}[l] - \gamma_p g_{p-1}^b[l-1]$$

while one on the bottom row performs

$$g_p^b[l] = g_{p-1}^b[l-1] - \gamma_p' g_{p-1}[l]$$

The round cell is a divider and computes γ_p from (8.129).

The system begins with the initial configuration of Fig. 8.20(a), where the labels on the boxes indicate the initial contents of the cell. On the first clock cycle the divider cell computes γ_1 (and $\gamma_1' = \gamma_1^*$), which is then broadcast (i.e., made available as an input to all cells). The cells then compute $g_1[l]$ and $g_1^b[l]$ as above and the top cells (only) shift their

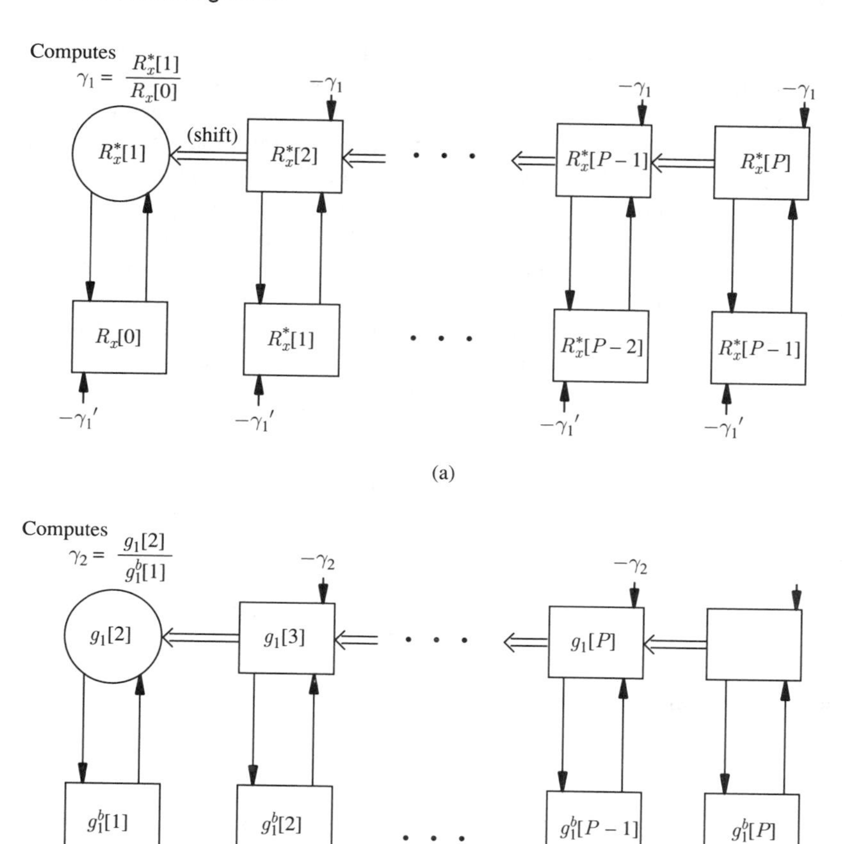

Figure 8.20 Systolic implementation of the Schur algorithm. (a) First clock cycle. (b) Second clock cycle.

contents left to the next cell. The state at the beginning of the next clock cycle is shown in Fig. 8.20(b). On this next clock cycle γ_2, $g_2[l]$, and $g_2^b[l]$ are computed and the process continues in this manner, computing one reflection coefficient at each clock cycle. The prediction error variances are also available because of the relation (8.130b). These appear in the bottom left cell at each clock cycle.

8.9.2 An Alternative Formulation of the Schur Algorithm

In the preceding subsection the Schur algorithm was presented in its more-or-less traditional form. Although this form is completely adequate for implementing the algorithm, it is not in the vector form (8.74) in which the Levinson recursion was presented. Therefore, direct comparison of the two algorithms is somewhat difficult. This subsection presents an alternative derivation of the Schur algorithm that results in a vector-oriented form directly comparable to (8.74). You may also prefer this form for implementation. Direct comparison of the two algorithms, however, is only part of our motivation for this section. In the process of the derivation, the relation between the functions g_p and g_p^b of the Schur algorithm and the filter parameter vectors $\boldsymbol{a}_p$ and $\boldsymbol{b}_p$ of the Levinson recursion are brought out and given further interpretation. This interpretation leads to a deeper understanding of the entire linear prediction problem.

Let us begin by expressing the function g_p in terms of the filter parameters and the autocorrelation function. From the definitions it follows that

$$\begin{aligned} g_p^*[l] &= R_{\varepsilon_p x}[l] = \mathcal{E}\{\varepsilon_p[n]x^*[n-l]\} \\ &= \mathcal{E}\left\{\sum_{k=0}^{p} a_k^{(p)*} x[n-k]x^*[n-l]\right\} = \sum_{k=0}^{p} a_k^{(p)*} R_x[l-k] \end{aligned}$$

or

$$\boxed{g_p[l] = \sum_{k=0}^{p} R_x[k-l]a_k^{(p)}} \tag{8.138}$$

In a similar manner it can be seen that

$$\begin{aligned} g_p^{b*}[l] &= R_{\varepsilon_p^b x}[l] = \mathcal{E}\{\varepsilon_p^b[n]x^*[n-l]\} \\ &= \mathcal{E}\left\{\sum_{k=0}^{p} b_k^{(p)*} x[n-p+k]x^*[n-l]\right\} = \sum_{k=0}^{p} b_k^{(p)*} R_x[l-p+k] \end{aligned}$$

or

$$\boxed{g_p^b[l] = \sum_{k=0}^{p} R_x[p-k-l]b_k^{(p)} = \sum_{k'=0}^{p} R_x[k'-l]b_{p-k'}^{(p)}} \tag{8.139}$$

where in the last step the variable k' was defined as $k' = p - k$. It follows from (8.138) and (8.139) that

$$\begin{aligned} g_p^b[l] &= \left(\sum_{k=0}^{p} R_x^*[p-k-l]b_k^{(p)*}\right)^* \\ &= \left(\sum_{k=0}^{p} R_x[k-(p-l)]a_k^{(p)}\right)^* = g_p^*[p-l] \end{aligned}$$

which is the relation (8.128) previously stated without proof.

Equations 8.138 and 8.139 are very important relations. Two interpretations of the former relation are given in Figs. 8.21 and 8.22. In Fig. 8.21, Equation 8.138 is regarded

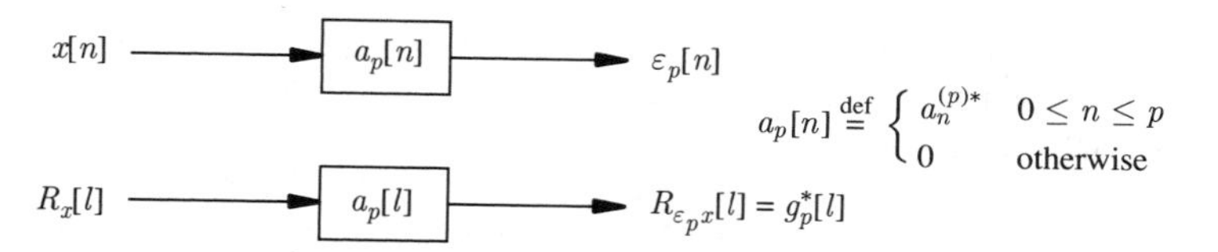

Figure 8.21 Interpretation of Equation 8.138 as a transformation of the correlation function by the prediction error filter.

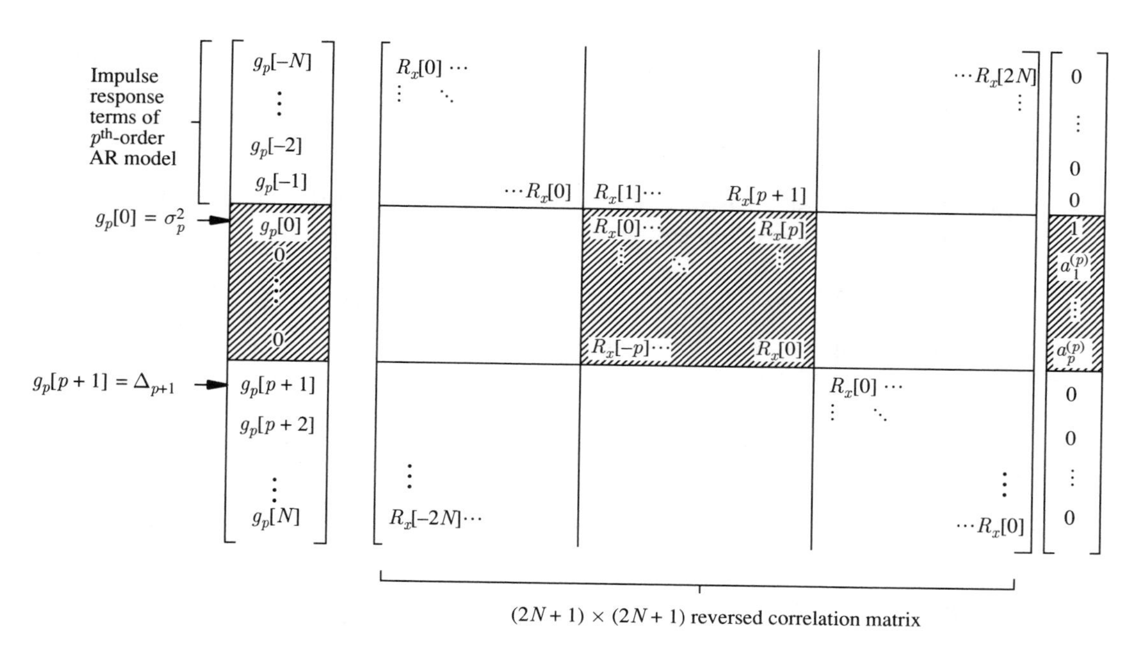

Figure 8.22 Interpretation of Equation 8.138 as a transformation of the linear prediction filter coefficients by the correlation matrix.

as a transformation applied to the correlation function by a linear system (the prediction error filter). This shows that (8.138) is equivalent to the familiar relation for transforming a random process derived in Chapter 5. A similar interpretation can be given to (8.139). In Fig. 8.22 the roles of the $a_i^{(p)}$ and the correlation are interchanged. That is, *the correlation matrix provides the transformation of the filter coefficients.* This provides the all-important link between the $a_i^{(p)}$ and the $g_p[l]$. Notice that the terms in the left-hand vector are known constants (zero) at the positions where the terms in the right hand vector are variables, and vice versa. The shaded block from this larger set of equations represents the Normal equations. The terms extend both upward and downward, however, and the equations in *any* block selected (not just the Normal equations) are satisfied. That property will be exploited shortly. One other property is worth mentioning, however. If the correlation function in the matrix belongs to that of a pth-order AR model, then the upper terms $g_p[0], g_p[-1], g_p[-2], \ldots$ are proportional to the impulse response terms $h[0], h[1], h[2], \ldots$

of the AR model [see (8.20)] and the lower terms $g_p[p+1], g_p[p+2], \ldots$ are all zero. It should be emphasized that this occurs *only when the correlation function R_x is that of a pth-order AR model.*

Let us now turn to the problem of rederiving the Schur algorithm. If l is allowed to take on values from p to P in (8.138), then the result can be written in matrix form as

$$\begin{bmatrix} 0 \\ g_p[p+1] \\ \vdots \\ g_p[P-1] \\ g_p[P] \end{bmatrix} = \begin{bmatrix} R_x[-p] & R_x[-p+1] & \cdots & R_x[0] \\ R_x[-p-1] & R_x[-p] & \cdots & R_x[-1] \\ \vdots & \vdots & \vdots & \vdots \\ R_x[-P] & R_x[-P+1] & \cdots & R_x[-P+p] \end{bmatrix} \begin{bmatrix} 1 \\ a_1^{(p)} \\ \vdots \\ a_{p-1}^{(p)} \\ a_p^{(p)} \end{bmatrix} \tag{8.140}$$

This is of course just a subset of the equations in Fig. 8.22. If (8.139) is similarly written for l in the same range of values, the result is

$$\begin{bmatrix} g_p^b[p] \\ g_p^b[p+1] \\ \vdots \\ g_p^b[P-1] \\ g_p^b[P] \end{bmatrix} = \begin{bmatrix} R_x[-p] & R_x[-p+1] & \cdots & R_x[0] \\ R_x[-p-1] & R_x[-p] & \cdots & R_x[-1] \\ \vdots & \vdots & \vdots & \vdots \\ R_x[-P] & R_x[-P+1] & \cdots & R_x[-P+p] \end{bmatrix} \begin{bmatrix} b_p^{(p)} \\ b_{p-1}^{(p)} \\ \vdots \\ b_1^{(p)} \\ 1 \end{bmatrix} \tag{8.141}$$

Two more set of equations can be written from (8.138) and (8.139) with p replaced by $p+1$. These are

$$\begin{bmatrix} g_{p-1}[p] \\ g_{p-1}[p+1] \\ \vdots \\ g_{p-1}[P] \end{bmatrix} = \begin{bmatrix} R_x[-p] & R_x[-p+1] & \cdots & R_x[0] \\ R_x[-p-1] & R_x[-p] & \cdots & R_x[-1] \\ \vdots & \vdots & \vdots & \vdots \\ R_x[-P] & R_x[-P+1] & \cdots & R_x[-P+p] \end{bmatrix} \begin{bmatrix} 1 \\ a_1^{(p-1)} \\ \vdots \\ a_{p-1}^{(p-1)} \\ 0 \end{bmatrix} \tag{8.142}$$

and

$$\begin{bmatrix} g_{p-1}^b[p-1] \\ g_{p-1}^b[p] \\ \vdots \\ g_{p-1}^b[P-1] \end{bmatrix} = \begin{bmatrix} R_x[-p] & R_x[-p+1] & \cdots & R_x[0] \\ R_x[-p-1] & R_x[-p] & \cdots & R_x[-1] \\ \vdots & \vdots & \vdots & \vdots \\ R_x[-P] & R_x[-P+1] & \cdots & R_x[-P+p] \end{bmatrix} \begin{bmatrix} 0 \\ b_{p-1}^{(p-1)} \\ \vdots \\ b_1^{(p-1)} \\ 1 \end{bmatrix} \tag{8.143}$$

where the vectors on the right have been extended with an appropriately placed 0 so that the *same* correlation matrix appears in all of the last four equations.

Now define the vectors

$$\boldsymbol{g}_p^{\dagger} \stackrel{\text{def}}{=} \begin{bmatrix} g_p[p+1] \\ \vdots \\ g_p[P-1] \\ g_p[P] \end{bmatrix} \tag{8.144}$$

$$\boldsymbol{g}_p^{b(s)} \stackrel{\text{def}}{=} \begin{bmatrix} g_p^b[p] \\ g_p^b[p+1] \\ \vdots \\ g_p^b[P-1] \end{bmatrix} \tag{8.145}$$

and

$$\boldsymbol{g}_p^{b} = \begin{bmatrix} g_p^b[p] \\ g_p^b[p+1] \\ \vdots \\ g_p^b[P-1] \\ g_p^b[P] \end{bmatrix} = \begin{bmatrix} \boldsymbol{g}_p^{b(s)} \\ ----- \\ g_p^b[P] \end{bmatrix} \tag{8.146}$$

Then (8.140) through (8.143) can be written as

$$\begin{bmatrix} 0 \\ \boldsymbol{g}_p^{\dagger} \end{bmatrix} = \mathbf{R}\boldsymbol{a}_p \tag{8.147}$$

$$\boldsymbol{g}_p^{b} = \begin{bmatrix} \boldsymbol{g}_p^{b(s)} \\ g_p^b[P] \end{bmatrix} = \mathbf{R}\tilde{\boldsymbol{b}}_p \tag{8.148}$$

$$\boldsymbol{g}_{p-1}^{\dagger} = \mathbf{R}\begin{bmatrix} \boldsymbol{a}_{p-1} \\ 0 \end{bmatrix} \tag{8.149}$$

and

$$\boldsymbol{g}_{p-1}^{b(s)} = \mathbf{R}\begin{bmatrix} 0 \\ \tilde{\boldsymbol{b}}_{p-1} \end{bmatrix} \tag{8.150}$$

where $\mathbf{R}$ is the rectangular correlation matrix that appears in (8.140) through (8.143). Finally, multiplying both sides of (8.90) and the reversal of (8.91) by $\mathbf{R}$ and using (8.147) through (8.150) results in the recursions

$$\begin{bmatrix} 0 \\ \boldsymbol{g}_p^{\dagger} \end{bmatrix} = \boldsymbol{g}_{p-1}^{\dagger} - \gamma_p \boldsymbol{g}_{p-1}^{b(s)} \tag{8.151}$$

and

$$\boldsymbol{g}_p^b = \begin{bmatrix} \boldsymbol{g}_p^{b(s)} \\ \\ g_p^b[P] \end{bmatrix} = \boldsymbol{g}_{p-1}^{b(s)} - \gamma_p' \boldsymbol{g}_{p-1}^{\dagger} \tag{8.152}$$

The last two equations are the essence of the Schur algorithm. As a final step let us combine these equations with (8.129) and (8.130) to write the Schur algorithm as a single procedure. To simplify the indexing the notation "top(**v**)" is used to represent the first element of any vector **v**. Thus from (8.144),

$$g_{p-1}[p] = \text{top}(\boldsymbol{g}_{p-1}^{\dagger})$$

and from (8.130)(a) and (8.146),

$$\sigma_p^2 = g_p^b[p] = \text{top}(\boldsymbol{g}_p^b)$$

With this notation, the complete algorithm becomes

$$\begin{aligned}
&\gamma_p = \frac{\text{top}(\boldsymbol{g}_{p-1}^{\dagger})}{\sigma_{p-1}^2} && (a) \\
&\left.\begin{aligned}
&\begin{bmatrix} 0 \\ \boldsymbol{g}_p^{\dagger} \end{bmatrix} = \boldsymbol{g}_{p-1}^{\dagger} - \gamma_p \boldsymbol{g}_{p-1}^{b(s)} \\
&\boldsymbol{g}_p^b \left\{ \begin{bmatrix} \boldsymbol{g}_p^{b(s)} \\ \times \end{bmatrix} \right. = \boldsymbol{g}_{p-1}^{b(s)} - \gamma_p^* \boldsymbol{g}_{p-1}^{\dagger}
\end{aligned}\right\} && (b) \\
&\sigma_p^2 = \text{top}(\boldsymbol{g}_p^b) && (c)
\end{aligned} \tag{8.153}$$

which is applied for $p = 1, 2, \ldots, P$ with initial conditions

$$\sigma_0^2 = R_x[0]; \qquad \boldsymbol{g}_0^{\dagger} = \begin{bmatrix} R_x^*[1] \\ R_x^*[2] \\ \vdots \\ R_x^*[P] \end{bmatrix}; \qquad \boldsymbol{g}_0^{b(s)} = \begin{bmatrix} R_x[0] \\ R_x^*[1] \\ \vdots \\ R_x^*[P-1] \end{bmatrix} \tag{8.154}$$

The symbol $\times$ in (8.153)(b) indicates that this element (actually, $g_p^b[P]$) does not need to be computed since it is discarded at each step until the last, where it is found to be the variance σ_P^2.

In comparing the Schur algorithm (8.153) with the Levinson recursion (8.74) several observations can be made. First, of course, there is no inner product in step (a) of the Schur algorithm. This permits the completely parallel implementation. Step (b) of the Schur requires about twice as much computation as step (b) of the Levinson algorithm. However, since the Levinson algorithm requires computation of an inner product at each stage of the recursion, the total computational complexity of both algorithms is about the same. Finally, step (c) of the Schur algorithm is trivial (i.e., it requires no computation).

A basic difference in the two algorithms is that in the Levinson algorithm the vector $\boldsymbol{a}_p$ starts with a length of 1 and grows to a full length of $P+1$. In the Schur algorithm the vectors $\boldsymbol{g}_p^\dagger$ and $\boldsymbol{g}_p^{b(s)}$ start at size P and *shrink* in dimension with each stage of the recursion. This is clear from (8.153)(b), where it is seen that the partition of the vector that propagates to the next stage of the recursion is reduced in size by one.

The preceding example (Example 8.3) is not repeated here for the vector form of the Schur algorithm since the computations are identical. However, you may want to repeat that example yourself to be sure that the formulas in (8.153) are clear.

Implementation of the Schur algorithm in an array-oriented language such as MATLAB is particularly simple. The MATLAB code for the main loop is shown in the box below. (The initialization of variables is not shown.)

MAIN LOOP OF FUNCTION
(Schur algorithm):

```
L=length(g);
for p=2:P+1;
    gamma(p) = top(g)/sigma(p-1);
    g1=g - gamma(p)*gbs;
    gb= gbs - conj(gamma(p))*g;
    g=g1(2:L);
    gbs=gb(1:L-1);
    sigma(p) = top(gb);
    L=L-1;
end;
```

FUNCTION top:

```
function y=top(v)
% gets first element of a vector.
y= v(1);
```

Note that since MATLAB (unfortunately!) does not allow array indices to begin at 0, `sigma(1)` through `sigma(P+1)` correspond to $\sigma_0^2, \ldots, \sigma_P^2$. Likewise, `gamma(2)` through `gamma(P+1)` correspond to $\gamma_1, \ldots, \gamma_P$.

8.10 "SPLIT" ALGORITHMS[9]

It is tempting to think, as many people did, that the Levinson and Schur algorithms provide the most efficient possible way to solve Normal equations involving a Toeplitz matrix. It turns out, however, that an additional factor of 2 can be wrung out of the computations. More specifically, it was discovered by careful reformulation of the problem that the number of multiplications in the Levinson and Schur algorithms can be reduced by approximately 50%, while the number of additions remains approximately the same. The key to this discovery lies in the fact that by forming a vector of variables that has symmetry, and using it to replace those in the Levinson and Schur algorithms, a version of these algorithms can be generated where only half of the terms needs to be computed. All of the variables in these new algorithms are analogous to those in the Levinson and Schur algorithms but they do not have the same natural relation to the basic linear prediction problem. Consequently, from a practical point of view, these algorithms are of interest mainly because of their reduced computation.

The algorithms are called "split" algorithms because some of the variables are split into a symmetric and an antisymmetric part, either of which can be used to develop a procedure with reduced computational requirements. For example, in the original split Levinson algorithm [17] new variables are defined by

$$\boldsymbol{s}_p = K_p\left(\boldsymbol{a}_p + \tilde{\boldsymbol{b}}_p\right)$$

and

$$\boldsymbol{s}'_p = K_p\left(\boldsymbol{a}_p - \tilde{\boldsymbol{b}}_p\right)$$

where K_p is a suitably defined constant, and which, assuming that $\boldsymbol{a}_p$ and $\boldsymbol{b}_p$ are both real, satisfy the conditions $\boldsymbol{s}_p = \tilde{\boldsymbol{s}}_p$ and $\boldsymbol{s}'_p = -\tilde{\boldsymbol{s}}'_p$. Either one of these can be propagated in equations analogous to (8.74) but involving three terms in the vector update step instead of two. Corresponding definitions for the other variables involved in linear prediction lead to a split lattice structure and a split Schur algorithm [18].

Unfortunately, these original algorithms achieved the computational reductions only for the case of real sequences. Bistritz, whose work in stability theory initially inspired the split algorithms, and others subsequently extended the work to the complex case [19–21]. Bistritz et al. in particular pointed out that the original split algorithms were only one group of a family of algorithms with similar properties [19, 22, 23]. They showed that while the original algorithms did not extend in any computationally efficient way, another group in the family could indeed be extended to the complex case in a way that would retain the computational advantage. This later work was quite general and in fact applied not only to stationary random processes but also to nonstationary processes whose correlation matrix had a certain "quasi-Toeplitz" form [21, 23].

All of the possible generalizations of the split algorithms are not of interest here. Algorithms that apply to the complex as well as the real case are of interest, however. Further, the original algorithms of Desalte and Genin are also of interest, although they

[9]This section can be skipped without disturbing the continuity of the chapter.

apply only to the real case, because they are particularly efficient and convenient to use when the problem does not involve complex data. We therefore approach the derivation in a way that is general enough to apply to either the original or the complex version and later specialize the results to generate one or the other of these computationally efficient forms. Although essentially all of the original work on split algorithms was developed in terms of polynomials in z, the topic is presented here with a vector-oriented approach that is more consistent with previous developments in this chapter and leads more directly to the desired final results.

8.10.1 Split Levinson Algorithms

Development of the split Levinson algorithms is the most lengthy part of this section on split algorithms. The other results follow more easily once the split Levinson algorithms are in place. To keep the steps in better perspective, the derivation is divided among subsections that culminate in the two split Levinson algorithms. The first subsection develops the basic recursion; the second discusses the special form of Normal equations and their implications. The third subsection develops the general procedure for recovering the linear prediction parameters from the "split" variables, and finally, the last subsection presents the two split algorithms. The algorithms will be called the "original split Levinson algorithm" and the "complex split Levinson algorithm" although the last name is a little misleading, since this algorithm can be applied to the case of real data as well.

Development of the Three-Term Recursion. As mentioned above, the key to developing a Levinson-type algorithm with reduced computation is to design an algorithm around a variable that has proper symmetry. Since the primary variables in the Levinson recursion are the filter coefficient vectors, which satisfy $\boldsymbol{a}_p = \tilde{\boldsymbol{b}}_p^*$, a fairly general way to define the new symmetric variable is

$$\boxed{\boldsymbol{s}_p = K_p \boldsymbol{a}_p + K_p^* \tilde{\boldsymbol{b}}_p} \tag{8.155}$$

where K_p is a generally complex constant which depends on the order and which will be specified later. Since the new vector $\boldsymbol{s}_p$ satisfies the symmetry condition

$$\tilde{\boldsymbol{s}}_p^* = \boldsymbol{s}_p \tag{8.156}$$

only half of its components need to be computed. Note also that it is possible to define a vector $\boldsymbol{s}'_p$ involving the *difference* instead of the sum of the two quantities above, which satisfies the antisymmetric condition $\tilde{\boldsymbol{s}}'^*_p = -\boldsymbol{s}'_p$. This results in a slightly different set of algorithms with identical computational properties [17, 18], which is not pursued here. Substituting (8.90) and (8.91) in (8.155) yields

$$\boldsymbol{s}_p = K_p \left(\begin{bmatrix} \boldsymbol{a}_{p-1} \\ 0 \end{bmatrix} - \gamma_p \begin{bmatrix} 0 \\ \tilde{\boldsymbol{b}}_{p-1} \end{bmatrix} \right) + K_p^* \left(\begin{bmatrix} 0 \\ \tilde{\boldsymbol{b}}_{p-1} \end{bmatrix} - \gamma'_p \begin{bmatrix} \boldsymbol{a}_{p-1} \\ 0 \end{bmatrix} \right)$$

or, noting that $\gamma'_p = \gamma^*_p$,

$$\boldsymbol{s}_p = C_p \begin{bmatrix} \boldsymbol{a}_{p-1} \\ 0 \end{bmatrix} + C^*_p \begin{bmatrix} 0 \\ \tilde{\boldsymbol{b}}_{p-1} \end{bmatrix} \tag{8.157}$$

where

$$C_p \stackrel{\text{def}}{=} K_p - \gamma^*_p K^*_p \tag{8.158}$$

is another complex constant that will be seen to play a role of equal importance to that of K_p.

The two different expressions (8.155) and (8.157) are now used to derive the recursion for $\boldsymbol{s}_p$. To begin, replace p by $p-1$ in (8.155) and solve that equation for $\tilde{\boldsymbol{b}}_p$, to obtain

$$\tilde{\boldsymbol{b}}_{p-1} = \frac{1}{K^*_{p-1}} \left(\boldsymbol{s}_{p-1} - K_{p-1} \boldsymbol{a}_{p-1} \right)$$

Substituting this result in (8.157) and rearranging then produces

$$\begin{bmatrix} \boldsymbol{a}_{p-1} \\ 0 \end{bmatrix} = \frac{1}{C_p} \boldsymbol{s}_p - \frac{C^*_p}{C_p K^*_{p-1}} \left(\begin{bmatrix} 0 \\ \boldsymbol{s}_{p-1} \end{bmatrix} - K_{p-1} \begin{bmatrix} 0 \\ \boldsymbol{a}_{p-1} \end{bmatrix} \right) \tag{8.159}$$

It will later be shown that this relation can be used to determine $\boldsymbol{a}_p$ when $\boldsymbol{s}_p$ and $\boldsymbol{s}_{p+1}$ are both known. For the moment, let us continue to develop a similar relation for $\tilde{\boldsymbol{b}}_{p-1}$. Again replacing p by $p-1$ in (8.155), solving for $\boldsymbol{a}_{p-1}$, and substituting the result in (8.157) yields

$$\boldsymbol{s}_p = \frac{C_p}{K_{p-1}} \left(\begin{bmatrix} \boldsymbol{s}_{p-1} \\ 0 \end{bmatrix} - K^*_{p-1} \begin{bmatrix} \tilde{\boldsymbol{b}}_{p-1} \\ 0 \end{bmatrix} \right) + C^*_p \begin{bmatrix} 0 \\ \tilde{\boldsymbol{b}}_{p-1} \end{bmatrix}$$

Then rearranging this produces

$$\begin{bmatrix} \tilde{\boldsymbol{b}}_{p-1} \\ 0 \end{bmatrix} = -\frac{K_{p-1}}{C_p K^*_{p-1}} \boldsymbol{s}_p + \frac{1}{K^*_{p-1}} \begin{bmatrix} \boldsymbol{s}_{p-1} \\ 0 \end{bmatrix} + \frac{C^*_p K_{p-1}}{C_p K^*_{p-1}} \begin{bmatrix} 0 \\ \tilde{\boldsymbol{b}}_{p-1} \end{bmatrix} \tag{8.160}$$

Now append a zero to all of the vectors in (8.90) and write

$$\begin{bmatrix} \boldsymbol{a}_p \\ \\ 0 \end{bmatrix} = \begin{bmatrix} \boldsymbol{a}_{p-1} \\ 0 \\ 0 \end{bmatrix} - \gamma_p \begin{bmatrix} 0 \\ \tilde{\boldsymbol{b}}_{p-1} \\ 0 \end{bmatrix}$$

Then substitute (8.159) (twice) and (8.160) in this equation to obtain

$$\frac{1}{C_{p+1}}\boldsymbol{s}_{p+1} - \frac{C_{p+1}^*}{C_{p+1}K_p^*}\left(\begin{bmatrix} 0 \\ \\ \boldsymbol{s}_p \end{bmatrix} - K_p \begin{bmatrix} 0 \\ \\ \boldsymbol{a}_p \end{bmatrix}\right) \tag{8.161}$$
$$= \left\{\frac{1}{C_p}\begin{bmatrix} \boldsymbol{s}_p \\ \\ 0 \end{bmatrix} - \frac{C_p^*}{C_pK_{p-1}^*}\left(\begin{bmatrix} 0 \\ \boldsymbol{s}_{p-1} \\ 0 \end{bmatrix} - K_{p-1}\begin{bmatrix} 0 \\ \boldsymbol{a}_{p-1} \\ 0 \end{bmatrix}\right)\right\}$$
$$-\gamma_p\left\{-\frac{-K_{p-1}}{C_pK_{p-1}^*}\begin{bmatrix} 0 \\ \\ \boldsymbol{s}_p \end{bmatrix} + \frac{1}{K_{p-1}^*}\begin{bmatrix} 0 \\ \boldsymbol{s}_{p-1} \\ 0 \end{bmatrix} + \frac{C_p^*K_{p-1}}{C_pK_{p-1}^*}\begin{bmatrix} 0 \\ 0 \\ \tilde{\boldsymbol{b}}_{p-1} \end{bmatrix}\right\}$$

Now observe that a simple recursion would result if the terms involving $\boldsymbol{a}_p$, $\boldsymbol{a}_{p-1}$, and $\tilde{\boldsymbol{b}}_{p-1}$ were to cancel. This can happen if the coefficient multiplying each of the vectors is the same. Therefore, let us force this situation by requiring that

$$\boxed{\frac{C_p^*K_{p-1}}{C_pK_{p-1}^*} = 1} \tag{8.162}$$

Although this condition is well motivated, it may still seem rather arbitrary or it may even appear that we are being somewhat naive to assume that it could hold. In fact, it is the *key* condition that permits development of the split algorithms. It will be used not just here, but several times again in the development that follows. Condition (8.162) can be stated equivalently as

$$\mu_p \stackrel{\text{def}}{=} \frac{K_{p-1}}{C_p} = \frac{K_{p-1}^*}{C_p^*} \tag{8.163}$$

where the variable μ_p is therefore seen to be real.

With the condition (8.162) and the relation (8.90), the final terms in each of the three lines of (8.161) cancel. Now collect the remaining terms and using (8.162) to obtain term A below, rewrite this as

$$\frac{1}{C_{p+1}}\boldsymbol{s}_{p+1} = \frac{1}{C_p}\begin{bmatrix} \boldsymbol{s}_p \\ \\ 0 \end{bmatrix} + \underbrace{\left(\frac{1}{K_p} + \gamma_p\frac{1}{C_p^*}\right)}_{\text{A}}\begin{bmatrix} 0 \\ \\ \boldsymbol{s}_p \end{bmatrix}$$
$$- \frac{1}{K_{p-1}^*}\underbrace{\left(\frac{C_p^*}{C_p} + \gamma_p\right)}_{\text{B}}\begin{bmatrix} 0 \\ \boldsymbol{s}_{p-1} \\ 0 \end{bmatrix}$$

Now some further simplification needs to be done. By using the definition (8.158) for C_p, term A can be reduced to

$$\frac{1}{K_p} + \gamma_p \frac{1}{C_p^*} = \frac{C_p^* + \gamma_p K_p}{K_p C_p^*} = \frac{K_p^* - \gamma_p K_p + \gamma_p K_p}{K_p C_p^*} = \frac{K_p^*}{K_p C_p^*}$$

while term B can be written as

$$\frac{C_p^*}{C_p} + \gamma_p = \frac{K_p^* - \gamma_p K_p + \gamma_p(K_p - \gamma_p^* K_p^*)}{C_p} = \frac{K_p^*(1 - |\gamma_p|^2)}{C_p}$$

Thus applying these results and multiplying through by C_{p+1} produces

$$\boldsymbol{s}_{p+1} = \frac{C_{p+1}}{C_p} \begin{bmatrix} \boldsymbol{s}_p \\ \\ 0 \end{bmatrix} + \frac{K_p^* C_{p+1}}{K_p C_p^*} \begin{bmatrix} 0 \\ \\ \boldsymbol{s}_p \end{bmatrix} - \frac{C_{p+1} K_p^*}{C_p K_{p-1}^*}(1 - |\gamma_p|^2) \begin{bmatrix} 0 \\ \boldsymbol{s}_{p-1} \\ 0 \end{bmatrix}$$

Finally, note that the coefficient $\frac{K_p^* C_{p+1}}{K_p C_p^*}$ on the right-hand side can be simplified using (8.162) to $\frac{C_{p+1}^*}{C_p^*}$. Therefore, this last equation can finally be written as

$$\boxed{\boldsymbol{s}_{p+1} = \beta_p \begin{bmatrix} \boldsymbol{s}_p \\ \\ 0 \end{bmatrix} + \beta_p^* \begin{bmatrix} 0 \\ \\ \boldsymbol{s}_p \end{bmatrix} - \beta_p \alpha_p \begin{bmatrix} 0 \\ \boldsymbol{s}_{p-1} \\ 0 \end{bmatrix}} \tag{8.164}$$

where

$$\boxed{\alpha_p \stackrel{\text{def}}{=} \left(\frac{K_p}{K_{p-1}}\right)^* (1 - |\gamma_p|^2)} \tag{8.165}$$

and

$$\boxed{\beta_p \stackrel{\text{def}}{=} \frac{C_{p+1}}{C_p}} \tag{8.166}$$

This is the desired three-term recursion for $\boldsymbol{s}_p$.

Split Normal Equations and Other Relations. Although the three-term recursion for the filter parameter vector is the cornerstone of the split Levinson algorithm, the complete algorithm depends on a number of other relations. Some of these are derived now.

To begin, observe from (8.155), (8.44), and (8.47) that

$$\tilde{\mathbf{R}}_{\boldsymbol{x}}^{(p)} \boldsymbol{s}_p = \tilde{\mathbf{R}}_{\boldsymbol{x}}^{(p)}(K_p \boldsymbol{a}_p + K_p^* \tilde{\boldsymbol{b}}_p) = K_p \begin{bmatrix} \sigma_p^2 \\ 0 \\ \vdots \\ 0 \end{bmatrix} + K_p^* \begin{bmatrix} 0 \\ \vdots \\ 0 \\ \sigma_p'^2 \end{bmatrix}$$

where $\tilde{\mathbf{R}}_{\boldsymbol{x}}^{(p)}$ is the reversed correlation matrix (8.45). Therefore, $\boldsymbol{s}_p$ is seen to satisfy the "split" Normal equations

$$\tilde{\mathbf{R}}_{\boldsymbol{x}}^{(p)} \boldsymbol{s}_p = \begin{bmatrix} \tau_p \\ 0 \\ \vdots \\ 0 \\ \tau_p' \end{bmatrix} \tag{8.167}$$

where

$$\begin{aligned} \tau_p &= K_p \sigma_p^2 \ (a) \\ \tau_p' &= K_p^* \sigma_p'^2 \ (b) \end{aligned} \tag{8.168}$$

Since $\sigma_p^2 = \sigma_p'^2$ it follows that

$$\tau_p' = \tau_p^* \tag{8.169}$$

Equation 8.167 states that τ_p is given by the inner product of the top row of $\tilde{\mathbf{R}}_{\boldsymbol{x}}^{(p)}$ with $\boldsymbol{s}_p$:

$$\tau_p = \left[\ R_x[0] \quad R_x[1] \quad \cdots \quad R_x[p] \ \right] \boldsymbol{s}_p = \sum_{k=0}^{p} R_x[k] s_k^{(p)} \tag{8.170}$$

where the $s_k^{(p)}$ are the components of the vector $\boldsymbol{s}_p$. Now the parameter α_p that appears in the three-term recursion can be expressed in terms of the parameters τ_p and τ_{p-1}. To show this note that since

$$\sigma_p^2 = \sigma_{p-1}^2 (1 - |\gamma_p|^2)$$

it follows from (8.168) that

$$\frac{\tau_p}{K_p} = \frac{\tau_{p-1}}{K_{p-1}} (1 - |\gamma_p|^2)$$

Rearranging this equation and using (8.165) then yields

$$\frac{\tau_p}{\tau_{p-1}} = \frac{K_p}{K_{p-1}} (1 - |\gamma_p|^2) = \alpha_p^*$$

or

$$\boxed{\alpha_p = \frac{\tau_p^*}{\tau_{p-1}^*} = \frac{\tau_p'}{\tau_{p-1}'}} \tag{8.171}$$

In the algorithms to be presented the parameters α_p are computed from τ_p and τ_{p-1} using (8.171) and β_p is computed by a similar procedure. The computation of τ_p is analogous to step I in Table 8.2 [or (8.74)(a)] of the Levinson recursion in that it represents an inner product of two vectors. The computational savings of the split Levinson algorithm are due in part to the fact that τ_p can be computed with essentially half the number of multiplications as its counterpart in the Levinson recursion. For the real case the components of $\boldsymbol{s}_p$ satisfy $s_k^{(p)} = s_{p-k}^{(p)}$, so (8.170) can be rewritten as

$$\tau_p = \begin{cases} \sum_{k=0}^{(p-1)/2} (R_x[k] + R_x[p-k])\, s_k^{(p)} & \text{for } p \text{ odd} \\ R_x[p/2] s_{p/2}^{(p)} + \sum_{k=0}^{(p/2)-1} (R_x[k] + R_x[p-k])\, s_k^{(p)} & \text{for } p \text{ even} \end{cases} \tag{8.172}$$

This reduces the required multiplications by essentially a factor of 2.

For the complex case the procedure is less straightforward, but the computational savings still obtain. The trick is to represent the complex variables in terms of real and imaginary parts, and then group the terms so that the number of *real* multiplications is reduced. Specifically, let

$$R_x[k] = R_\mathrm{r}[k] + \jmath R_\mathrm{i}[k] \tag{8.173}$$

and

$$s_k^{(p)} = s_{\mathrm{r}k} + \jmath s_{\mathrm{i}k} \tag{8.174}$$

where because of (8.156)

$$s_{p-k}^{(p)} = (s_k^{(p)})^* = s_{\mathrm{r}k} - \jmath s_{\mathrm{i}k} \tag{8.175}$$

Then, assuming for the moment that p is odd, apply these results to write (8.170) as

$$\tau_p = \sum_{k=0}^{(p-1)/2} (R_\mathrm{r}[k] + \jmath R_\mathrm{i}[k])(s_{\mathrm{r}k} + \jmath s_{\mathrm{i}k}) + (R_\mathrm{r}[p-k] + \jmath R_\mathrm{i}[p-k])(s_{\mathrm{r}k} - \jmath s_{\mathrm{i}k})$$

or

$$\begin{aligned} \tau_p = \sum_{k=0}^{(p-1)/2} & \{(R_\mathrm{r}[k] + R_\mathrm{r}[p-k])\, s_{\mathrm{r}k} - (R_\mathrm{i}[k] - R_\mathrm{i}[p-k])\, s_{\mathrm{i}k}\} \\ & + \jmath \{(R_\mathrm{r}[k] - R_\mathrm{r}[p-k])\, s_{\mathrm{i}k} + (R_\mathrm{i}[k] + R_\mathrm{i}[p-k])\, s_{\mathrm{r}k}\} \end{aligned}$$

$$(p \text{ odd}) \tag{8.176}$$

In this arrangement there are four *real* multiplications for each term in the sum [the same as for (8.170)] but only half the number of terms in the sum. When p is even, the corresponding result is

$$\begin{aligned} \tau_p = & \left\{R_\mathrm{r}[p/2] s_{\mathrm{r}p/2} - R_\mathrm{i}[p/2] s_{\mathrm{i}p/2}\right\} + \jmath \left\{R_\mathrm{r}[p/2] s_{\mathrm{i}p/2} + R_\mathrm{i}[p/2] s_{\mathrm{r}p/2}\right\} \\ & + \sum_{k=0}^{(p/2)-1} \{(R_\mathrm{r}[k] + R_\mathrm{r}[p-k])\, s_{\mathrm{r}k} - (R_\mathrm{i}[k] - R_\mathrm{i}[p-k])\, s_{\mathrm{i}k}\} \\ & \qquad + \jmath \{(R_\mathrm{r}[k] - R_\mathrm{r}[p-k])\, s_{\mathrm{i}k} + (R_\mathrm{i}[k] + R_\mathrm{i}[p-k])\, s_{\mathrm{r}k}\} \end{aligned}$$

$$(p \text{ even}) \tag{8.177}$$

To avoid writing these complicated formulas again later let us simply *define the shorthand notation*

$$\tau_p = \sum_{k=0}^{p} R_x[k] s_k^{(p)} \qquad \text{(compute half)} \tag{8.178}$$

to mean (8.172) in the real case and (8.176) or (8.177) in the complex case.

Recovering the Linear Prediction Parameters. The relations developed so far provide a neat recursive way of computing the symmetric filter vector $\boldsymbol{s}_p$ and other associated parameters. These parameters are of interest, however, only if they can be used to find the linear prediction parameters $\boldsymbol{a}_P$ and σ_P^2. Fortunately, this is possible. To obtain an equation for $\boldsymbol{a}_P$ it is only necessary to let p be equal to $P+1$ in (8.159). If (8.162) is then used to simplify the constant, the result is

$$\boxed{\begin{bmatrix} \boldsymbol{a}_P \\ 0 \end{bmatrix} = \frac{1}{C_{P+1}} \boldsymbol{s}_{P+1} - \frac{1}{K_P} \begin{bmatrix} 0 \\ \boldsymbol{s}_P \end{bmatrix} + \begin{bmatrix} 0 \\ \boldsymbol{a}_P \end{bmatrix}} \tag{8.179}$$

Now provided that the multipliers $\frac{1}{C_{P+1}}$ and $\frac{1}{K_P}$ and the last two vectors $\boldsymbol{s}_P$ and $\boldsymbol{s}_{P+1}$ are known, this equation can be solved recursively for the components of $\boldsymbol{a}_P$. That is, beginning with the known value $a_0^{(P)} = 1$ on the right side of the equation, we can use (8.179) to find $a_1^{(P)}$ on the left. Then, with $a_1^{(P)}$ in hand, we can use the equation to find $a_2^{(P)}$; and so on. This technique is demonstrated in Example 8.4 at the end of this section.

The constants K_P and C_{P+1} can be computed recursively as shown later; however, there is a simple nonrecursive procedure as well. Recall the notation "top(**v**)" introduced previously to represent the first element of any vector **v**. Then because $\text{top}(\boldsymbol{a}_P) = 1$, it follows from (8.179) that C_{P+1} *must* be equal to $\text{top}(\boldsymbol{s}_{P+1})$. Since this last relation is true for any value of P, it must be true in general that

$$C_p = \text{top}(\boldsymbol{s}_p) \tag{8.180}$$

This can be used specifically to determine C_{P+1} from $\boldsymbol{s}_{P+1}$.

The expression for K_P is also rather simple, but less straightforward to derive. Again let **v** be any vector and denote by $\boldsymbol{\Sigma}(\mathbf{v})$ the *sum* of the elements of **v**. Then define ξ_p as

$$\xi_p \stackrel{\text{def}}{=} \boldsymbol{\Sigma}(\boldsymbol{s}_p) = \sum_{k=0}^{p} s_k^{(p)} \tag{8.181}$$

and note that since $\boldsymbol{s}_p = \tilde{\boldsymbol{s}}_p^*$, ξ_p is *real*. Now from (8.155) and the fact that $\boldsymbol{a}_p = \boldsymbol{b}_p^*$ it follows that

$$\begin{aligned} \xi_p = \boldsymbol{\Sigma}(\boldsymbol{s}_p) &= K_p \boldsymbol{\Sigma}(\boldsymbol{a}_p) + K_p^* \boldsymbol{\Sigma}(\tilde{\boldsymbol{b}}_p) \\ &= K_p \boldsymbol{\Sigma}(\boldsymbol{a}_p) + K_p^* \boldsymbol{\Sigma}(\tilde{\boldsymbol{a}}_p^*) = 2\text{Re}\left(K_p \boldsymbol{\Sigma}(\boldsymbol{a}_p)\right) \end{aligned}$$

But since ξ_p is real, the last equation implies that

$$\xi_p = 2K_p\mathbf{\Sigma}(\boldsymbol{a}_p) \tag{8.182}$$

Now look at the alternative expression (8.157) with p replaced by $p+1$. From this expression it follows that

$$\begin{aligned}\xi_{p+1} = \mathbf{\Sigma}(\boldsymbol{s}_{p+1}) &= C_p\mathbf{\Sigma}(\boldsymbol{a}_p) + C_p^*\mathbf{\Sigma}(\tilde{\boldsymbol{b}}_p) \\ &= C_p\mathbf{\Sigma}(\boldsymbol{a}_p) + C_p^*\mathbf{\Sigma}(\tilde{\boldsymbol{a}}_p^*) = 2\text{Re}\left(C_p\mathbf{\Sigma}(\boldsymbol{a}_p)\right)\end{aligned}$$

or

$$\xi_{p+1} = 2C_p\mathbf{\Sigma}(\boldsymbol{a}_p) \tag{8.183}$$

Dividing (8.182) by (8.183) yields

$$\frac{\xi_p}{\xi_{p+1}} = \frac{K_p}{C_{p+1}}$$

This, incidentally, is the quantity μ_{p+1} defined in (8.163). Solving for K_p then yields

$$\boxed{K_p = \frac{\xi_p}{\xi_{p+1}}C_{p+1} = \frac{\mathbf{\Sigma}(\boldsymbol{s}_p)}{\mathbf{\Sigma}(\boldsymbol{s}_{p+1})}C_{p+1}} \tag{8.184}$$

Thus knowledge of C_{P+1}, $\boldsymbol{s}_P$, and $\boldsymbol{s}_{P+1}$ permits computation of K_P.

Once K_P is known, the prediction error variance follows directly from (8.168)(a). In particular, with $p = P$, the prediction error variance is

$$\boxed{\sigma_P^2 = \frac{1}{K_P}\tau_P} \tag{8.185}$$

The Split Levinson Algorithms. All of the relations are now in place that are necessary for stating the split Levinson algorithms. The two forms of the algorithm considered are obtained by forcing conditions on the parameters in (8.164) that result in minimal computation.

To begin, consider the requirement $\beta_p = 1$ for all values of p. With this requirement the number of multiplications involved in the three-term recursion are minimized since only one vector is multiplied by a nontrivial scale factor. From (8.166) and (8.158) the requirement can be stated as

$$\beta_p = \frac{C_{p+1}}{C_p} = \frac{K_{p+1} - \gamma_{p+1}^*K_{p+1}^*}{K_p - \gamma_p^*K_p^*} = 1 \tag{8.186}$$

Unfortunately, this requirement is inconsistent with the key condition (8.162) in the general complex case (see Problem 8.27). *This is the fundamental reason why the original split*

Levinson algorithm [17] *does not extend to the complex case.* If all of the parameters are real, however, (8.162) and (8.186) can both be satisfied by choosing

$$K_p = \frac{1}{1-\gamma_p} \tag{8.187}$$

Note from the definition (8.158) that (8.187) implies that $C_p = 1$ for *all* values of p. To develop the initial conditions we can start with $\gamma_0 = 0$ and apply (8.187) to obtain $K_0 = 1$. This in turn implies through (8.155), (8.157), (8.168) and the fact $\boldsymbol{a}_0 = \boldsymbol{b}_0 = [1]$ that $\boldsymbol{s}_0 = [2]$, $\boldsymbol{s}_1 = \begin{bmatrix} 1 \\ 1 \end{bmatrix}$, and $\tau_0 = R_x[0]$. The resulting *original split Levinson algorithm* is summarized below.

$$\begin{array}{ll}
\tau_p = \sum_{k=0}^{p} R_x[k] s_k^{(p)} & \text{(compute half)} \ \ \text{[see (8.172)]} \ (a) \\
\alpha_p = \dfrac{\tau_p}{\tau_{p-1}} & (b) \\
\boldsymbol{s}_{p+1} = \begin{bmatrix} \boldsymbol{s}_p \\ \\ 0 \end{bmatrix} + \begin{bmatrix} 0 \\ \\ \boldsymbol{s}_p \end{bmatrix} - \alpha_p \begin{bmatrix} 0 \\ \boldsymbol{s}_{p-1} \\ 0 \end{bmatrix} & \text{(compute half)} \ (c)
\end{array} \tag{8.188}$$

with initial conditions

$$\tau_0 = R_x[0]; \quad \boldsymbol{s}_0 = [2]\,; \quad \boldsymbol{s}_1 = \begin{bmatrix} 1 \\ 1 \end{bmatrix} \tag{8.189}$$

The final relations for $\boldsymbol{a}_P$ and γ_P are given by (8.179) and (8.185) with K_P defined by (8.184) and $C_{P+1} = 1$:

$$\begin{array}{ll}
\begin{bmatrix} \boldsymbol{a}_P \\ \\ 0 \end{bmatrix} = \boldsymbol{s}_{P+1} - \dfrac{1}{K_P} \begin{bmatrix} 0 \\ \\ \boldsymbol{s}_P \end{bmatrix} + \begin{bmatrix} 0 \\ \\ \boldsymbol{a}_P \end{bmatrix} & (a) \\
\sigma_P^2 = \dfrac{1}{K_P} \tau_P & (b) \\
\text{where } \dfrac{1}{K_P} = \dfrac{\Sigma(\boldsymbol{s}_{P+1})}{\Sigma(\boldsymbol{s}_P)} & (c)
\end{array} \tag{8.190}$$

The complex split Levinson algorithm is obtained by applying a different constraint on the parameters α_p and β_p in (8.164), namely

$$\beta_p = \frac{1}{\alpha_p} \tag{8.191}$$

This forces the last coefficient in the three-term recursion to be trivial and therefore reduces the number of multiplications accordingly. In the case where the quantities are all real the first two vectors on the right of (8.164) can first be added before multiplying by the common scale factor β_p. Thus the number of multiplications is the same as in the original split Levinson algorithm. In the general complex case, the fact that the coefficients of the vectors are complex conjugates permits the computation to be carried out with an approximate 50% reduction in the number of multiplications. The procedure is similar to that used to compute (8.178) in the complex case (see Problem 8.28). The initial conditions are the same as for the original split Levinson algorithm.[10] In view of (8.191) and (8.171),β_p has the simple form

$$\beta_p = \frac{\tau^*_{p-1}}{\tau^*_p} = \frac{\tau'_{p-1}}{\tau'_p} \tag{8.192}$$

The *complex split Levinson algorithm* can be summarized as

$$\begin{aligned}
\tau_p &= \sum_{k=0}^{p} R_x[k] s_k^{(p)} \qquad \text{(compute half)[see (8.176),(8.177)]} \ (a) \\
\beta_p^* &= \frac{\tau_{p-1}}{\tau_p} \qquad (b) \\
\boldsymbol{s}_{p+1} &= \beta_p \begin{bmatrix} \boldsymbol{s}_p \\ \\ 0 \end{bmatrix} + \beta_p^* \begin{bmatrix} 0 \\ \\ \boldsymbol{s}_p \end{bmatrix} - \begin{bmatrix} 0 \\ \boldsymbol{s}_{p-1} \\ 0 \end{bmatrix} \qquad \text{(compute half)} \ (c)
\end{aligned} \tag{8.193}$$

with initial conditions

$$\tau_0 = R_x[0]\,; \quad \boldsymbol{s}_0 = [2]\ ; \quad \boldsymbol{s}_1 = \begin{bmatrix} 1 \\ 1 \end{bmatrix} \tag{8.194}$$

[10]These initial conditions are not unique, however. An alternative set is $\gamma_0 = -1$, $K_0 = \frac{1}{2}$, $\boldsymbol{s}_0 = [1]$, $\boldsymbol{s}_1 = \left[\frac{1}{2}, \frac{1}{2}\right]^T$, and $\tau_0 = \frac{1}{2} R_x[0]$ [21].

The final relations for $\boldsymbol{a}_P$ and γ_P are obtained from (8.179), (8.180), (8.184), and (8.185):

$$
\begin{array}{ll}
\begin{bmatrix} \boldsymbol{a}_P \\ 0 \end{bmatrix} = \dfrac{1}{C_{P+1}} \boldsymbol{s}_{P+1} - \dfrac{1}{K_P} \begin{bmatrix} 0 \\ \boldsymbol{s}_P \end{bmatrix} + \begin{bmatrix} 0 \\ \boldsymbol{a}_P \end{bmatrix} & (a) \\
\sigma_P^2 = \dfrac{1}{K_P} \tau_P & (b) \\
\text{where } \dfrac{1}{C_{P+1}} = \dfrac{1}{\text{top}(\boldsymbol{s}_{P+1})}; \quad \dfrac{1}{K_P} = \dfrac{\Sigma(\boldsymbol{s}_{P+1})}{\Sigma(\boldsymbol{s}_P)} \dfrac{1}{C_{P+1}} & (c)
\end{array}
\tag{8.195}
$$

If you are extremely alert, you may have noticed that there is one other constraint that could be applied to (8.164) that would also lead to reduced computation (at least in the real case). This is to require that

$$\alpha_p = 1$$

In this case the multipliers β_p can be factored from all three terms and the three vectors can be summed before performing the multiplication. It turns out that while this procedure works well for the real case and leads to the same reduction in complexity, its computational advantages do not extend to the complex case. For the real case, however, it is another alternative.

To illustrate the application of the split Levinson algorithm, let us compute the linear prediction parameters for Example 8.1 by this method. Since the problem involves a real random process, either the original or the complex split Levinson algorithm can be used. We use the complex form to illustrate how it applies in the real case.

Example 8.4

It is desired to compute the second-order linear prediction parameters for the random process whose correlation function is given in Example 8.1 (Recall that $R_x[0] = 3$, $R_x[1] = 2$, and $R_x[2] = 1$.) The computation is carried out using the complex split Levinson algorithm. The algorithm is initialized according to (8.194) with

$$\tau_0 = R_x[0] = 3; \quad \boldsymbol{s}_0 = 2; \quad \boldsymbol{s}_1 = \begin{bmatrix} 1 \\ 1 \end{bmatrix}$$

In the first step of the recursion ($p = 1$) τ_1, β_1, and $\boldsymbol{s}_2$ are computed.

$$
\begin{aligned}
\tau_1 &= \sum_{k=0}^{0} (R_x[k] + R_x[1-k])\, s_k^{(1)} \\
&= (R_x[0] + R_x[1])\, s_0^{(1)} = (3 + 2) \cdot 1 = 5
\end{aligned}
$$

$$\beta_1 = \frac{\tau_0}{\tau_1} = \frac{3}{5} = \tfrac{3}{5}$$

$$\boldsymbol{s}_2 = \beta_1 \left(\begin{bmatrix} \boldsymbol{s}_1 \\ \\ 0 \end{bmatrix} + \begin{bmatrix} 0 \\ \\ \boldsymbol{s}_1 \end{bmatrix} \right) - \begin{bmatrix} 0 \\ \boldsymbol{s}_0 \\ 0 \end{bmatrix}$$

$$= \tfrac{3}{5} \left(\begin{bmatrix} 1 \\ 1 \\ 0 \end{bmatrix} + \begin{bmatrix} 0 \\ 1 \\ 1 \end{bmatrix} \right) - \begin{bmatrix} 0 \\ 2 \\ 0 \end{bmatrix} = \begin{bmatrix} \tfrac{3}{5} \\ -\tfrac{4}{5} \\ \times \end{bmatrix}$$

(The symbol $\times$ here indicates that this element does not have to be computed due to the symmetry.)

In the next step of the recursion ($p = 2$) τ_2, β_2, and $\boldsymbol{s}_3$ are computed.

$$\begin{aligned} \tau_2 &= R_x[1]s_1^{(2)} + \sum_{k=0}^{0} (R_x[k] + R_x[2-k])\, s_k^{(1)} \\ &= R_x[1]s_1^{(2)} + (R_x[0] + R_x[2])\, s_0^{(2)} \\ &= 2\left(-\tfrac{4}{5}\right) + (3+1)\tfrac{3}{5} = \tfrac{4}{5} \end{aligned}$$

$$\beta_2 = \frac{\tau_1}{\tau_2} = \frac{5}{4/5} = \tfrac{25}{4}$$

$$\boldsymbol{s}_3 = \beta_2 \left(\begin{bmatrix} \boldsymbol{s}_2 \\ \\ 0 \end{bmatrix} + \begin{bmatrix} 0 \\ \\ \boldsymbol{s}_2 \end{bmatrix} \right) - \begin{bmatrix} 0 \\ \boldsymbol{s}_1 \\ \\ 0 \end{bmatrix}$$

$$= \tfrac{25}{4} \left(\begin{bmatrix} \tfrac{3}{5} \\ -\tfrac{4}{5} \\ \tfrac{3}{5} \\ 0 \end{bmatrix} + \begin{bmatrix} 0 \\ \tfrac{3}{5} \\ -\tfrac{4}{5} \\ \tfrac{3}{5} \end{bmatrix} \right) - \begin{bmatrix} 0 \\ 1 \\ 1 \\ 0 \end{bmatrix} = \begin{bmatrix} \tfrac{15}{4} \\ -\tfrac{9}{4} \\ \times \\ \times \end{bmatrix}$$

The filter coefficients are then computed from (8.195)(a):

$$\begin{bmatrix} \boldsymbol{a}_2 \\ \\ 0 \end{bmatrix} = \frac{1}{C_3}\boldsymbol{s}_3 - \frac{1}{K_2} \begin{bmatrix} 0 \\ \\ \boldsymbol{s}_2 \end{bmatrix} + \begin{bmatrix} 0 \\ \\ \boldsymbol{a}_2 \end{bmatrix}$$

where
$$\frac{1}{C_3} = \frac{1}{\text{top}(\boldsymbol{s}_3)} = \tfrac{4}{15}$$

and
$$\frac{1}{K_2} = \frac{\boldsymbol{\Sigma}(\boldsymbol{s}_3)}{\boldsymbol{\Sigma}(\boldsymbol{s}_2)} \frac{1}{C_3} = \frac{3}{2/5} \cdot \tfrac{4}{15} = 2$$

Writing this in component form we have

$$\begin{bmatrix} 1 \\ a_1^{(2)} \\ a_2^{(2)} \\ 0 \end{bmatrix} = \tfrac{4}{15} \begin{bmatrix} \tfrac{15}{4} \\ -\tfrac{9}{4} \\ -\tfrac{9}{4} \\ \tfrac{15}{4} \end{bmatrix} - 2 \begin{bmatrix} 0 \\ \tfrac{3}{5} \\ -\tfrac{4}{5} \\ \tfrac{3}{5} \end{bmatrix} + \begin{bmatrix} 0 \\ 1 \\ a_1^{(2)} \\ a_2^{(2)} \end{bmatrix}$$

The components are then computed one at a time.

$$a_1^{(2)} = \tfrac{4}{15}\left(-\tfrac{9}{4}\right) - 2\left(\tfrac{3}{5}\right) + 1 = -\tfrac{4}{5}$$

$$\begin{aligned} a_2^{(2)} &= \tfrac{4}{15}\left(-\tfrac{9}{4}\right) - 2\left(-\tfrac{4}{5}\right) + a_1^{(2)} \\ &= -\tfrac{3}{5} + \tfrac{8}{5} - \tfrac{4}{5} = \tfrac{1}{5} \end{aligned}$$

The coefficient vector is therefore

$$\boldsymbol{a}_2 = \begin{bmatrix} 1 \\ -\frac{4}{5} \\ \frac{1}{5} \end{bmatrix}$$

Finally, σ_2^2 is computed from (8.195)(b).

$$\sigma_2^2 = \frac{1}{K_2}\tau_2 = 2\left(\tfrac{4}{5}\right) = \tfrac{8}{5}$$

These are seen to check with the results of Example 8.1.

8.10.2 Split Lattice Representations

The split Levinson algorithm implies the existence of a split lattice structure which provides yet another realization of the prediction error filter and its inverse. To derive this structure, begin by defining the split error $e_p[n]$ in analogy to (8.81) and (8.83) as:

$$e_p[n] \stackrel{\text{def}}{=} \boldsymbol{s}_p^{*T}\tilde{\boldsymbol{x}}_p[n] \tag{8.196}$$

where $\boldsymbol{x}_p[n]$ is defined by (8.80). Substitution of (8.155) yields

$$e_p[n] = K_p^*\boldsymbol{a}_p^{*T}\tilde{\boldsymbol{x}}_p[n] + K_p\tilde{\boldsymbol{b}}^{*T}\tilde{\boldsymbol{x}}_p[n]$$

or [from (8.81) and (8.83)]

$$e_p[n] = K_p^*\varepsilon[n] + K_p\varepsilon^b[n] \tag{8.197}$$

This relates the split error to the forward and backward prediction errors. Additionally, substitution of (8.157) and (8.85) in (8.196) yields

$$e_p[n] = C_p^*\left[\begin{array}{cc} \boldsymbol{a}_{p-1}^{*T} & 0 \end{array}\right]\begin{bmatrix} \tilde{\boldsymbol{x}}_{p-1}[n] \\ x[n-p] \end{bmatrix} + C_p\left[\begin{array}{cc} 0 & \tilde{\boldsymbol{b}}_{p-1}^{*T} \end{array}\right]\begin{bmatrix} x[n] \\ \tilde{\boldsymbol{x}}_{p-1}[n-1] \end{bmatrix}$$

or [again by (8.81) and (8.83)]

$$e_p[n] = C_p^*\varepsilon_{p-1}[n] + C_p\varepsilon_{p-1}^b[n-1] \tag{8.198}$$

which provides an alternative relation to (8.197).

To obtain a recursion for $e_p[n]$ first use (8.164) and (8.85) to write

$$\mathbf{s}_{p+1}^{*T}\tilde{\mathbf{x}}_{p+1}[n] = \beta_p^* \begin{bmatrix} \mathbf{s}_p^{*T} & 0 \end{bmatrix} \begin{bmatrix} \tilde{\mathbf{x}}_p[n] \\ x[n-(p+1)] \end{bmatrix}$$

$$+ \beta_p \begin{bmatrix} 0 & \mathbf{s}_p^{*T} \end{bmatrix} \begin{bmatrix} x[n] \\ \tilde{\mathbf{x}}_p[n-1] \end{bmatrix} - \beta_p^* \alpha_p^* \begin{bmatrix} 0 & \mathbf{s}_p^T & 0 \end{bmatrix} \begin{bmatrix} x[n] \\ \tilde{\mathbf{x}}_{p-1}[n-1] \\ x[n-(p+1)] \end{bmatrix}$$

Then apply (8.196) to obtain

$$\boxed{e_{p+1}[n] = \beta_p^* e_p[n] + \beta_p e_p[n-1] - \beta_p^* \alpha_p^* e_{p-1}[n-1]} \tag{8.199}$$

This is the relation that defines one section in the split lattice structure. The lattice section flow graph is shown in Fig. 8.23(a) and (b) for odd and even values of p, respectively. The "odd" and "even" sections are mirror images of each other and are simply defined here, so they fit together more easily in the complete lattice.

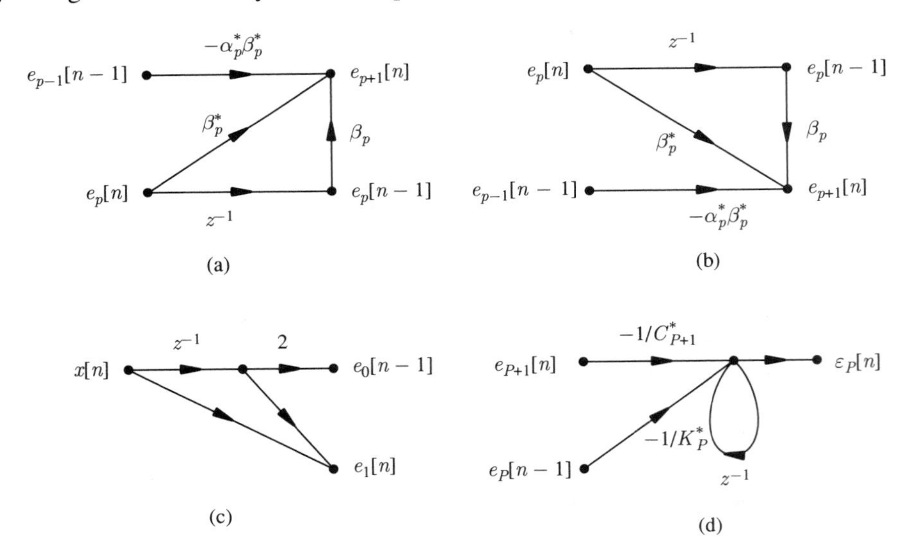

Figure 8.23 Sections of a split lattice prediction error filter. (a) Odd section. (b) Even section. (c) Initial section. (d) Final section.

To implement the entire prediction error filter an initial section is needed that transforms $x[n]$ to $e_0[n]$ and $e_1[n]$, and a final section is needed that converts $e_P[n]$ and $e_{P+1}[n]$ to $\varepsilon_P[n]$. The initial section is based on the relations

$$e_0[n] = \mathbf{s}_0^{*T}\,[x[n]] = 2x[n] \tag{8.200}$$

and

$$e_1[n] = \boldsymbol{s}_1^{*T}\tilde{\boldsymbol{x}}_1[n] = \begin{bmatrix} 1 & 1 \end{bmatrix} \begin{bmatrix} x[n] \\ x[n-1] \end{bmatrix} = x[n] + x[n-1] \tag{8.201}$$

which are both obtained from the initial conditions (8.189) or (8.194) on $\boldsymbol{s}_p$. The corresponding signal flow graph is shown in Fig. 8.23(c).

To obtain the equation for the final section, combine (8.179) and (8.85) to write

$$\begin{bmatrix} \boldsymbol{a}_P^{*T} & 0 \end{bmatrix} \begin{bmatrix} \tilde{\boldsymbol{x}}_P[n] \\ x[n-(P+1)] \end{bmatrix} = \frac{1}{C_{P+1}^*} \boldsymbol{s}_{P+1}^{*T} \tilde{\boldsymbol{x}}_{P+1}[n]$$

$$- \frac{1}{K_P^*} \begin{bmatrix} 0 & \boldsymbol{s}_P^{*T} \end{bmatrix} \begin{bmatrix} x[n] \\ \tilde{\boldsymbol{x}}_P[n-1] \end{bmatrix} + \begin{bmatrix} 0 & \boldsymbol{a}_P^{*T} \end{bmatrix} \begin{bmatrix} x[n] \\ \tilde{\boldsymbol{x}}_P[n-1] \end{bmatrix}$$

or

$$\boxed{\varepsilon_P[n] = \frac{1}{C_{P+1}^*} e_{P+1}[n] - \frac{1}{K_P^*} e_P[n-1] + \varepsilon_P[n-1]} \tag{8.202}$$

The signal flow graph is depicted in Fig. 8.23(d). When constructing the entire lattice structure, the sections are specialized for the original algorithm (where $\beta_p = 1$ and $C_{P+1} = 1$) or the complex algorithm (where $\alpha_p \beta_p = 1$). Lattices for each of these possibilities (also constructed with different numbers of sections) are shown in Fig. 8.24(a) and (b).

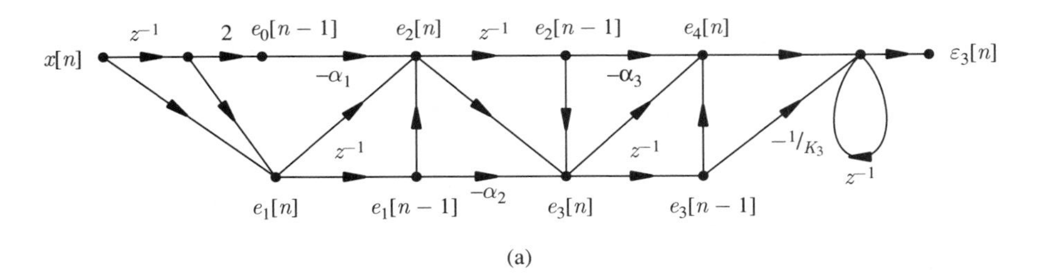

(a)

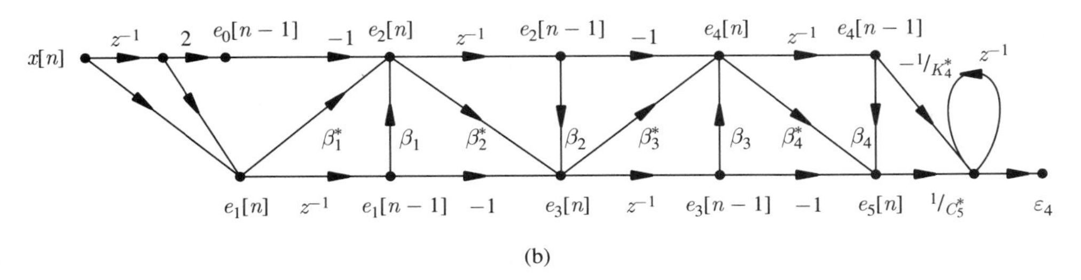

(b)

Figure 8.24 Split lattice realizations for a prediction error filter. (a) Original algorithm form for $P = 3$. (b) Complex algorithm form for $P = 4$.

8.10.3 Split Schur Algorithms

The split Levinson algorithms and their associated lattice structure suggest the existence of split Schur algorithms. These algorithms also involve a three-term recursion and have computational advantages over the ordinary Schur algorithm. To continue the general approach here, the algorithm is derived in terms of both the α and the β parameters, and then specialized to obtain the original and the complex versions at the end. Since the development is rather lengthy, this section is broken down into smaller subsections, following the procedure used in development of the split Levinson algorithm. The first of these derives the basic three-term recursion; the next derives additional necessary relations among the variables; finally, the last subsection presents and summarizes the original and complex split Schur algorithms.

Basic Recursion. Let us begin as in Section 8.9.1, by defining the function

$$v_p[l] \stackrel{\text{def}}{=} R^*_{e_p x}[l] = \left(\mathcal{E}\{e_p[n]x^*[n-l]\}\right)^* \tag{8.203}$$

Note from (8.197) that

$$R_{e_p x}[l] = \mathcal{E}\left\{\left(K^*_p \varepsilon_p[n] + K_p \varepsilon^b_p[n]\right) x^*[n-l]\right\} = K^*_p R_{\varepsilon_p x}[l] + K_p R_{\varepsilon^b_p x}[l]$$

so by (8.203) and (8.126),

$$v_p[l] = K_p g_p[l] + K^*_p g^b_p[l] \tag{8.204}$$

Alternatively, from (8.198),

$$R_{e_p x}[l] = \mathcal{E}\left\{\left(C^*_p \varepsilon_{p-1}[n] + C_p \varepsilon^b_{p-1}[n]\right) x^*[n-l]\right\} = C^*_p R_{\varepsilon_{p-1} x}[l] + C_p R_{\varepsilon^b_{p-1} x}[l]$$

which results in the other relation,

$$v_p[l] = C_p g_{p-1}[l] + C^*_p g^b_{p-1}[l] \tag{8.205}$$

These equations define the functions $v_p[l]$ used in the split Schur algorithms in terms of the forward and backward functions $g_p[l]$ and $g^b_p[l]$ that are used in the ordinary Schur algorithm. Now it follows from (8.199) that

$$\begin{aligned}\mathcal{E}\{e_{p+1}[n]x^*[n-l]\} = \beta^*_p \mathcal{E}\{e_p[n]x^*[n-l]\} + \beta_p \mathcal{E}\{e_p[n-1]x^*[n-l]\} \\ -\beta^*_p \alpha^*_p \mathcal{E}\{e_{p-1}[n-1]x^*[n-l]\}\end{aligned}$$

or, using (8.203),

$$\boxed{v_{p+1}[l] = \beta_p v_p[l] + \beta^*_p v_p[l-1] - \beta_p \alpha_p v_{p-1}[l-1]} \tag{8.206}$$

This recursion is the basis for the split Schur algorithms.

The initial conditions for the recursion can be derived from the initial conditions of the ordinary Schur algorithm with the help of (8.204) and (8.205). First note according to (8.180), (8.184), and the initial conditions (8.189) or (8.194) on the split Levinson algorithms that

$$C_1 = \text{top}(\boldsymbol{s}_1) = 1 \tag{8.207}$$

and

$$K_0 = \frac{\boldsymbol{\Sigma}(\boldsymbol{s}_0)}{\boldsymbol{\Sigma}(\boldsymbol{s}_1)} C_1 = \frac{2}{2} \cdot 1 = 1 \tag{8.208}$$

Then by using (8.204) and (8.205) with (8.207), (8.208), and (8.134), the initial conditions on the split Schur algorithm are specified as

$$\begin{aligned} v_0[l] &= K_0 g_0[l] + K_0^* g_0^b[l] \\ &= R_x^*[l] + R_x^*[l] = 2R_x^*[l] \text{ for } 0 \leq l \leq P-1 \end{aligned} \tag{8.209}$$

and

$$\begin{aligned} v_1[l] &= C_1 g_0[l] + C_1^* g_0^b[l-1] \\ &= R_x^*[l] + R_x^*[l-1] \quad \text{for } 1 \leq l \leq P \end{aligned} \tag{8.210}$$

This completes the basic recursion.

Additional Relations. Although (8.206) and the initial conditions form the basis of the split Schur algorithm, there are more details that need to be filled in before the algorithm is complete. In particular, recursions are needed for the parameters α_p and β_p; and all of these quantities need to be related to the linear prediction lattice parameters. (Recall that in the Schur algorithm, γ_p and σ_p^2 are the parameters computed.) First note, using (8.127)(a), (8.130)(b), and (8.168), that for $p > 0$

$$v_p[p] = K_p \underbrace{g_p[p]}_{0} + K_p^* \underbrace{g_p^b[p]}_{\sigma_p'^2} = K_p^* \sigma_p'^2 = \tau_p' = \tau_p^* \tag{8.211}$$

or

$$\boxed{\tau_p = v_p^*[p]} \tag{8.212}$$

Then it follows from (8.171) that

$$\boxed{\alpha_p = \frac{\tau_p^*}{\tau_{p-1}^*} = \frac{v_p[p]}{v_{p-1}[p-1]}} \tag{8.213}$$

Although this provides the needed recursion for α_p, there is a slight problem with the last two equations arising from the fact that $v_0[0]$ is equal to $2\tau_0$ instead of simply τ_0. [Note that we were careful to say that (8.211) holds for $p > 0$ since for $p = 0$, $g_0[0]$ is equal to

σ_0^2, not 0.] The easiest way to solve this problem is to change the initial conditions (8.209) slightly, replacing them by

$$v_0[l] = \begin{cases} R_x[0] & \text{for } l = 0 \\ 2R_x^*[l] & \text{for } 1 \leq l \leq P-1 \end{cases} \tag{8.214}$$

Then (8.212) and (8.213) can be used for all required values of p.

To find a recursive solution for γ_p, first conjugate (8.158) and solve for γ_p to obtain

$$\boxed{\gamma_p = \frac{1}{K_p}\left(K_p^* - C_p^*\right)} \tag{8.215}$$

Unfortunately, C_p cannot be obtained from (8.180) since $\boldsymbol{s}_p$ is not being computed in the split Schur algorithm. C_p can be computed recursively from the definition (8.166), however. That is, if β_p and C_p are known, C_{p+1} is given by

$$\boxed{C_{p+1} = \beta_p C_p} \tag{8.216}$$

Finally, K_p can be computed from (8.184) if C_{p+1}, ξ_p, and ξ_{p+1} are known. Fortunately, it is possible to compute $\xi_p = \boldsymbol{\Sigma}(\boldsymbol{s}_p)$ without having to compute $\boldsymbol{s}_p$ itself. Specifically, it follows from the recursion (8.164) that

$$\boldsymbol{\Sigma}(\boldsymbol{s}_{p+1}) = \beta_p \boldsymbol{\Sigma}(\boldsymbol{s}_p) + \beta_p^* \boldsymbol{\Sigma}(\boldsymbol{s}_p) - \beta_p \alpha_p \boldsymbol{\Sigma}(\boldsymbol{s}_{p-1})$$

or

$$\boxed{\xi_{p+1} = (\beta_p + \beta_p^*)\xi_p - \beta_p \alpha_p \xi_{p-1}} \tag{8.217}$$

Therefore, ξ_p can be computed recursively beginning with the initial conditions

$$\xi_0 = \boldsymbol{\Sigma}(\boldsymbol{s}_0) = 2\,; \qquad \xi_1 = \boldsymbol{\Sigma}(\boldsymbol{s}_1) = 2 \tag{8.218}$$

The combination of (8.215) through (8.217) and (8.184) permits the recursive computation of γ_p.

As a final step, note that the prediction error variance σ_p^2 can be found by combining (8.168)(a) and (8.212) to obtain

$$\boxed{\sigma_p^2 = \frac{\tau_p}{K_p} = \frac{v_p^*[p]}{K_p}} \tag{8.219}$$

This completes the development of all of the needed relations.

The Split Schur Algorithms. The two forms of the split Schur algorithm can now be presented. The algorithms are presented in vector form because this provides a simple computer implementation. A procedure can also be developed in terms of the generator matrix formulation (see Problem 8.29).

Begin by writing (8.206) for $l = p, p+1, \ldots, P$ in the form

$$\begin{bmatrix} v_{p+1}[p] \\ v_{p+1}[p+1] \\ v_{p+1}[p+2] \\ \vdots \\ v_{p+1}[P] \end{bmatrix} = \beta_p \begin{bmatrix} v_p[p] \\ v_p[p+1] \\ v_p[p+2] \\ \vdots \\ v_p[P] \end{bmatrix} + \beta_p^* \begin{bmatrix} v_p[p-1] \\ v_p[p] \\ v_p[p+1] \\ \vdots \\ v_p[P-1] \end{bmatrix} - \beta_p\alpha_p \begin{bmatrix} v_{p-1}[p-1] \\ v_{p-1}[p] \\ v_{p-1}[p+1] \\ \vdots \\ v_{p-1}[P-1] \end{bmatrix} \tag{8.220}$$

Notice from (8.205) and (8.127) that for any $p > 0$,

$$v_{p+1}[p] = C_{p+1} \underbrace{g_p[p]}_{0} + C_{p+1}^* \underbrace{g_p^b[p-1]}_{0} = 0$$

so that the top elements in the first and third vectors above are zero. Now define the vectors

$$\boldsymbol{v}_p^{(s)} \stackrel{\text{def}}{=} \begin{bmatrix} v_p[p] \\ v_p[p+1] \\ \vdots \\ v_p[P-1] \end{bmatrix} \tag{8.221}$$

and

$$\boldsymbol{v}_p \stackrel{\text{def}}{=} \begin{bmatrix} v_p[p] \\ v_p[p+1] \\ \vdots \\ v_p[P-1] \\ v_p[P] \end{bmatrix} = \begin{bmatrix} \boldsymbol{v}_p^{(s)} \\ \\ v_p[P] \end{bmatrix} \tag{8.222}$$

The recursion (8.220) can then be written as

$$\boxed{\begin{bmatrix} 0 \\ \boldsymbol{v}_{p+1}^{(s)} \\ \\ v_{p+1}[P] \end{bmatrix} = \beta_p \begin{bmatrix} \boldsymbol{v}_p^{(s)} \\ \\ \\ v_p[P] \end{bmatrix} + \beta_p^* \begin{bmatrix} 0 \\ \\ \\ \boldsymbol{v}_p^{(s)} \end{bmatrix} - \beta_p\alpha_p\boldsymbol{v}_{p-1}^{(s)}} \tag{8.223}$$

Notice from (8.221) and (8.223) that the size of the vector $\boldsymbol{v}_p^{(s)}$ *decreases* with each step in the recursion.

Let us now consider the specific conditions on α_p and β_p and develop the two corresponding specific forms of the split Schur algorithm. First consider the case of real variables with the constraint $\beta_p = 1$. This will be called the "original" split Schur algorithm [18]. Although (8.215) can be used to compute γ_p, in this case, it is possible to develop a simpler direct expression. In particular, the original definition (8.165) for α_p combined with the special expression (8.187) for K_p that holds in this case produces

$$\alpha_p = \frac{K_p}{K_{p-1}}(1-\gamma_p^2) = \frac{1-\gamma_{p-1}}{1-\gamma_p}(1-\gamma_p)(1+\gamma_p)$$
$$= (1-\gamma_{p-1})(1+\gamma_p)$$

This can be solved for γ_p to obtain

$$\gamma_p = \frac{\alpha_p}{1-\gamma_{p-1}} - 1 \tag{8.224}$$

A simple expression for σ_p^2 is likewise obtained in this case by substituting (8.187) into (8.219) to yield

$$\sigma_p^2 = (1-\gamma_p)v_p[p] \tag{8.225}$$

Finally, (8.213) and (8.221) through (8.225) can be combined into a single equation which represents the *original split Schur algorithm.*

$$\begin{aligned}
\alpha_p &= \frac{\text{top}(\boldsymbol{v}_p)}{\text{top}(\boldsymbol{v}_{p-1}^{(s)})} && (a)\\
\gamma_p &= \frac{\alpha_p}{1-\gamma_{p-1}} - 1 && (b)\\
\sigma_p^2 &= (1-\gamma_p)\text{top}(\boldsymbol{v}_p) && (c)\\
\boldsymbol{v}_{p+1}\left\{\begin{bmatrix} 0 \\ \boldsymbol{v}_{p+1}^{(s)} \\ \\ v_{p+1}[P] \end{bmatrix}\right. &= \begin{bmatrix} \boldsymbol{v}_p^{(s)} \\ \\ \\ v_p[P] \end{bmatrix} + \begin{bmatrix} 0 \\ \\ \\ \boldsymbol{v}_p^{(s)} \end{bmatrix} - \alpha_p \boldsymbol{v}_{p-1}^{(s)} && (d)
\end{aligned} \tag{8.226}$$

The initial conditions [see (8.214) and (8.210)] are given by

$$\boldsymbol{v}_0^{(s)} = \begin{bmatrix} R_x[0] \\ 2R_x[1] \\ \vdots \\ 2R_x[P-1] \end{bmatrix}; \quad \boldsymbol{v}_1 = \begin{bmatrix} \boldsymbol{v}_1^{(s)} \\ \\ v_1[P] \end{bmatrix} = \begin{bmatrix} R_x[1]+R_x[0] \\ R_x[2]+R_x[1] \\ \vdots \\ \hline R_x[P]+R_x[P-1] \end{bmatrix};$$
$$\gamma_0 = 0 \tag{8.227}$$

Steps (a) through (d) of (8.226) are computed for $p = 1$ through $p = P$ except that step (d) is skipped for $p = P$ since the vectors are then null and the step is not needed. Some MATLAB code for this algorithm is shown in the box below.

MAIN LOOP OF FUNCTION
(Split Schur algorithm):

```
gamma(1)=0
L=length(v);
for p=2:P+1;
    alpha = top(v)/top(vs_1);
    gamma(p) = alpha/(1-gamma(p-1)) -1;
    sigma(p) = (1-gamma(p))*top(v);
    if p < P+1;
        vs = v(1:L-1);
        vP = v(L);
        vector = [vs,vP] + [0,vs] - alpha*vs_1;
        v = vector(2:L);
        L=L-1;
        vs_1 = vs;
    end;
end;
```

Again since MATLAB does not allow arrays to begin at 0, `sigma(1)` through `sigma(P+1)` correspond to $\sigma_0^2, \ldots, \sigma_P^2$ and `gamma(2)` through `gamma(P+1)` correspond to $\gamma_1, \ldots, \gamma_P$.

For the complex split Schur algorithm it is required that $\beta_p \alpha_p = 1$, which with (8.213) implies that

$$\beta_p = \frac{1}{\alpha_p} = \frac{v_{p-1}[p-1]}{v_p[p]} \tag{8.228}$$

Since in this case there are no special simplifications for computing γ_p and σ_p^2, there are a few more steps to be performed. Fortunately, these steps involve mostly scalar computations, so the overall computational requirements are not high. To summarize, combine (8.215) through (8.217), (8.219), (8.223), and (8.228) into a single equation which represents the *complex split Schur algorithm:*

$$
\begin{aligned}
\beta_p &= \frac{\text{top}(\boldsymbol{v}_{p-1}^{(s)})}{\text{top}(\boldsymbol{v}_p)} && (a)\\
\xi_{p+1} &= (2\text{Re}\beta_p)\xi_p - \xi_{p-1} && (b)\\
C_{p+1} &= \beta_p C_p && (c)\\
K_p &= C_{p+1}\frac{\xi_p}{\xi_{p+1}} && (d)\\
\gamma_p &= \frac{1}{K_p}(K_p^* - C_p^*) && (e)\\
\sigma_p^2 &= \frac{1}{K_p}(\text{top}(\boldsymbol{v}_p))^* && (f)
\end{aligned}
\tag{8.229}
$$

$$
\boldsymbol{v}_{p+1}\left\{\begin{bmatrix} 0 \\ \boldsymbol{v}_{p+1}^{(s)} \\ \\ v_{p+1}[P] \end{bmatrix}\right. = \beta_p \begin{bmatrix} \boldsymbol{v}_p^{(s)} \\ \\ \\ v_p[P] \end{bmatrix} + \beta_p^* \begin{bmatrix} 0 \\ \\ \\ \boldsymbol{v}_p^{(s)} \end{bmatrix} - \boldsymbol{v}_{p-1}^{(s)} \quad (g)
$$

with initial conditions

$$
\boldsymbol{v}_0^{(s)} = \begin{bmatrix} R_x[0] \\ 2R_x^*[1] \\ \vdots \\ 2R_x^*[P-1] \end{bmatrix}; \quad \boldsymbol{v}_1 = \begin{bmatrix} \boldsymbol{v}_1^{(s)} \\ \\ v_1[P] \end{bmatrix} = \begin{bmatrix} R_x^*[1] + R_x[0] \\ R_x^*[2] + R_x^*[1] \\ \vdots \\ ----------- \\ R_x^*[P] + R_x^*[P-1] \end{bmatrix};
$$

$$
C_1 = 1; \qquad \xi_0 = \xi_1 = 2 \tag{8.230}
$$

The computational saving arises from the fact that the only step that involves other than scalar multiplications is step (g) and the number of real multiplications in this step can be reduced to about half by using techniques similar to those used to develop (8.176) and (8.177) (see Problem 8.28). Once again the vector update [step (g)] is skipped the last time through the recursion ($p = P$).

To close this section, let us compute the lattice parameters for our by now standard example which has appeared throughout this chapter.

Example 8.5

It is desired to compute the reflection coefficients and prediction error variances up to second order for the random process in Example 8.1 with

$$
R_x[0] = 3; \qquad R_x[1] = 2; \qquad R_x[2] = 1
$$

This is done using the complex split Schur algorithm. The algorithm is initialized according to (8.230) with

$$\boldsymbol{v}_0^{(s)} = \begin{bmatrix} 3 \\ 4 \end{bmatrix}; \qquad \boldsymbol{v}_1 = \begin{bmatrix} \boldsymbol{v}_1^{(s)} \\ \\ v_1[2] \end{bmatrix} = \begin{bmatrix} 5 \\ 3 \end{bmatrix}; \qquad C_1 = 1; \qquad \xi_1 = \xi_0 = 2$$

The first step ($p = 1$) consists of the following computations:

$$\beta_1 = \frac{\text{top}(\boldsymbol{v}_0)}{\text{top}(\boldsymbol{v}_1)} = \frac{3}{5} = \tfrac{3}{5}$$

$$\xi_2 = 2\beta_1\xi_1 - \xi_0 = \tfrac{6}{5}\cdot 2 - 2 = \tfrac{2}{5}$$

$$C_2 = \beta_1 C_1 = \tfrac{3}{5}\cdot 1 = \tfrac{3}{5}$$

$$K_1 = C_2\frac{\xi_1}{\xi_2} = \tfrac{3}{5}\frac{2}{2/5} = 3$$

$$\gamma_1 = \frac{1}{K_1}(K_1 - C_1) = \frac{1}{3}(3-1) = \tfrac{2}{3}$$

$$\sigma_1^2 = \frac{\text{top}(\boldsymbol{v}_1)}{K_1} = \tfrac{5}{3}$$

and finally,

$$\begin{bmatrix} 0 \\ \boldsymbol{v}_2 \end{bmatrix} = \begin{bmatrix} 0 \\ v_2[2] \end{bmatrix} = \beta_1\left(\begin{bmatrix} \boldsymbol{v}_1^{(s)} \\ \\ v_1[2] \end{bmatrix} + \begin{bmatrix} 0 \\ \\ \boldsymbol{v}_1^{(s)} \end{bmatrix}\right) - \boldsymbol{v}_0^{(s)}$$

$$= \tfrac{3}{5}\left(\begin{bmatrix} 5 \\ 3 \end{bmatrix} + \begin{bmatrix} 0 \\ 5 \end{bmatrix}\right) - \begin{bmatrix} 3 \\ 4 \end{bmatrix} = \begin{bmatrix} 0 \\ \tfrac{4}{5} \end{bmatrix}$$

This produces the first-order parameters γ_1, σ_1^2 and the vector $\boldsymbol{v}_2$ needed for the next time through the recursion.

At the next and final step ($p = 2$) the process is repeated.

$$\beta_2 = \frac{\text{top}(\boldsymbol{v}_1)}{\text{top}(\boldsymbol{v}_2)} = \frac{5}{4/5} = \tfrac{25}{4}$$

$$\xi_3 = 2\beta_2\xi_2 - \xi_1 = \tfrac{25}{2}\cdot\tfrac{2}{5} - 2 = 3$$

$$C_3 = \beta_2 C_2 = \tfrac{25}{4}\cdot\tfrac{3}{5} = \tfrac{15}{4}$$

$$K_2 = C_3\frac{\xi_2}{\xi_3} = \tfrac{15}{4}\frac{2/5}{3} = \tfrac{1}{2}$$

$$\gamma_2 = \frac{1}{K_2}(K_2 - C_2) = 2\left(\tfrac{1}{2} - \tfrac{3}{5}\right) = -\tfrac{1}{5}$$

$$\sigma_2^2 = \frac{1}{K_2}\text{top}(\boldsymbol{v}_2) = 2\cdot\tfrac{4}{5} = \tfrac{8}{5}$$

This completes the computation of the parameters γ_p and σ_p^2 up to order 2. The computation of $\boldsymbol{s}_3$ is skipped since all of the desired parameters have been found and the vector on the left of (8.229)(g) has been reduced as far as possible.

The results check with those obtained in the earlier examples.

8.11 RELATIONS TO TRIANGULAR DECOMPOSITION

So far a number of interesting topics have been explored that relate to the problem of linear prediction. This section explores a further set of relations that exist with some purely linear algebraic operations discussed in Chapter 2. Recall that one way to transform a random vector to one with orthogonal components is through triangular decomposition of the correlation matrix. For random vectors whose components represent samples of a random process the *lower–upper* triangular decomposition results in a *causal* linear transformation. Since the prediction error filter for a random process is also such a causal transformation, there must be a relation. In fact, as we show here, they are the same thing!

The backward linear prediction problem relates in a similar manner to the *upper–lower* decomposition of the correlation matrix. Recall from Chapter 2 that the (causal and anticausal) transformations are given by the inverse matrices $\mathbf{L}^{-1}$ and $\mathbf{U}^{-1}$. It will be seen that these matrices can be expressed explicitly in terms of the linear prediction parameters. It also turns out that the *uninverted* matrices $\mathbf{L}$ and $\mathbf{U}$ can be expressed in terms of the functions g_p and g_p^b involved in the Schur algorithm. Thus everything fits nicely into place.

The first subsection below discusses the relation of the linear prediction parameters to triangular decomposition. The next subsection shows the role of the Schur algorithm in the decomposition. The final subsection mentions some additional relations to polynomials orthogonal on the unit circle and states some final conclusions.

8.11.1 Linear Prediction as Triangular Decomposition

Let us approach this topic by first recalling the triangular decomposition of the correlation matrix discussed in Chapter 2. Begin with the random vector

$$\boldsymbol{x} = \begin{bmatrix} x[0] \\ x[1] \\ \vdots \\ x[P] \end{bmatrix} \tag{8.231}$$

and the corresponding correlation matrix

$$\mathbf{R}_x = \mathcal{E}\{\boldsymbol{x}\boldsymbol{x}^{*T}\} = \begin{bmatrix} R_x[0] & R_x[-1] & \cdots & R_x[-P] \\ R_x[1] & R_x[0] & \cdots & R_x[1-P] \\ \vdots & \vdots & \vdots & \vdots \\ R_x[P] & R_x[P-1] & \cdots & R_x[0] \end{bmatrix} \tag{8.232}$$

For simplicity $\boldsymbol{x}$ is defined as starting with $x[0]$, although this is by no means necessary. Then recall that the correlation matrix can be written as

$$\mathbf{R}_x = \mathbf{L}\mathbf{D}_L\mathbf{L}^{*T} \tag{8.233}$$

where $\mathbf{L}$ is a unit lower triangular matrix and $\mathbf{D}_L$ is a diagonal matrix with positive terms. This is the lower–upper triangular decomposition of the correlation matrix. Since

$$\mathbf{L}^{-1}\mathbf{R}_x(\mathbf{L}^{-1})^{*T} = \mathbf{D}_L \tag{8.234}$$

the matrix $\mathbf{D}_L$ is the correlation matrix of a random vector defined by the transformation

$$\boldsymbol{x}'' = \mathbf{L}^{-1}\boldsymbol{x} \tag{8.235}$$

and because $\mathbf{D}_L$ is diagonal, the vector $\boldsymbol{x}''$ has orthogonal components. Since $\mathbf{L}^{-1}$ is also a unit lower trangular matrix, $\mathbf{L}^{-1}$ represents a *causal* transformation of the sequence $x[0], x[1], \ldots, x[P]$. It will be shown that this transformation is precisely the transformation provided by the (forward) prediction error filters of orders 0 through P.

To show this, begin with the Normal equations of any order p where $0 < p \leq P$:

$$\begin{bmatrix} R_x[0] & R_x[1] & \cdots & R_x[p] \\ R_x[-1] & R_x[0] & \cdots & R_x[p-1] \\ \vdots & \vdots & \vdots & \vdots \\ R_x[-p] & R_x[-p+1] & \cdots & R_x[0] \end{bmatrix} \begin{bmatrix} 1 \\ a_1^{(p)} \\ \vdots \\ a_p^{(p)} \end{bmatrix} = \begin{bmatrix} \sigma_p^2 \\ 0 \\ \vdots \\ 0 \end{bmatrix} \tag{8.236}$$

If the matrix and vectors are reversed in this equation, then the equation can be written as

$$\begin{bmatrix} R_x[0] & R_x[-1] & \cdots & R_x[-p] \\ R_x[1] & R_x[0] & \cdots & R_x[-p+1] \\ \vdots & \vdots & \vdots & \vdots \\ R_x[p] & R_x[p-1] & \cdots & R_x[0] \end{bmatrix} \begin{bmatrix} a_p^{(p)} \\ \vdots \\ a_1^{(p)} \\ 1 \end{bmatrix} = \begin{bmatrix} 0 \\ \vdots \\ 0 \\ \sigma_p^2 \end{bmatrix} \tag{8.237}$$

Now augment these equations to the full-size P and write

$$\begin{bmatrix} R_x[0] & R_x[-1] & \cdots & R_x[-p] & \cdots & R_x[-P] \\ R_x[1] & R_x[0] & \cdots & R_x[-p+1] & \cdots & r_x[-P+1] \\ \vdots & \vdots & \ddots & \vdots & & \vdots \\ R_x[p] & R_x[p-1] & \cdots & R_x[0] & \cdots & R_x[-P+p] \\ \vdots & \vdots & \vdots & \vdots & \ddots & \vdots \\ R_x[P] & R_x[P-1] & \cdots & R_x[P-p] & \cdots & R_x[0] \end{bmatrix} \begin{bmatrix} a_p^{(p)} \\ \vdots \\ a_1^{(p)} \\ 1 \\ 0 \\ \vdots \\ 0 \end{bmatrix} = \begin{bmatrix} 0 \\ \vdots \\ 0 \\ \sigma_p^2 \\ \times \\ \vdots \\ \times \end{bmatrix} \tag{8.238}$$

where the vector on the left has been augmented with zeros, and the vector on the right has been filled in with the appropriate nonzero values ($\times$).[11]

[11]These values are actually the extended values of the correlation function, but it is sufficient here to note that they are nonzero in general.

Now form the quadratic product with the filter of order q and apply (8.238).

$$\begin{bmatrix} a_q^{(q)*} & \cdots & a_1^{(q)*} & 1 & 0 & \cdots & 0 \end{bmatrix} \mathbf{R}_x \begin{bmatrix} a_p^{(p)} \\ \vdots \\ a_1^{(p)} \\ 1 \\ 0 \\ \vdots \\ 0 \end{bmatrix}$$

$$= \begin{bmatrix} a_q^{(q)*} & \cdots & a_1^{(q)*} & 1 & 0 & \cdots & 0 \end{bmatrix} \begin{bmatrix} 0 \\ \vdots \\ 0 \\ \sigma_p^2 \\ \times \\ \vdots \\ \times \end{bmatrix}$$

Notice that this product is equal to zero for $q < p$ and equal to σ_p^2 for $q = p$. For $q > p$ observe that the same quadratic product can be written as

$$\begin{bmatrix} a_q^{(q)*} & \cdots & a_1^{(q)*} & 1 & 0 & \cdots & 0 \end{bmatrix} \mathbf{R}_x \begin{bmatrix} a_p^{(p)} \\ \vdots \\ a_1^{(p)} \\ 1 \\ 0 \\ \vdots \\ 0 \end{bmatrix}$$

$$= \begin{bmatrix} 0 & \cdots & 0 & \sigma_q^2 & \times & \cdots & \times \end{bmatrix} \begin{bmatrix} a_p^{(p)} \\ \vdots \\ a_1^{(p)} \\ 1 \\ 0 \\ \vdots \\ 0 \end{bmatrix}$$

which is also zero.[12] Therefore, all of these results can be combined to write

$$\begin{bmatrix} 1 & 0 & \cdots & 0 \\ a_1^{(1)*} & 1 & \cdots & 0 \\ \vdots & \vdots & \ddots & \vdots \\ a_P^{(P)*} & a_{P-1}^{(P)*} & \cdots & 1 \end{bmatrix} \underbrace{\begin{bmatrix} R_x[0] & \cdots & R_x[-P] \\ R_x[1] & \cdots & R_x[1-P] \\ \vdots & \vdots & \vdots \\ R_x[P] & \cdots & R_x[0] \end{bmatrix}}_{\mathbf{R}_x} \begin{bmatrix} 1 & a_1^{(1)} & \cdots & a_P^{(P)} \\ 0 & 1 & \cdots & a_{P-1}^{(P)} \\ \vdots & \vdots & \ddots & \vdots \\ 0 & 0 & \cdots & 1 \end{bmatrix}$$

$$= \begin{bmatrix} \sigma_0^2 & 0 & \cdots & 0 \\ 0 & \sigma_1^2 & \cdots & 0 \\ \vdots & \vdots & \ddots & \vdots \\ 0 & 0 & \cdots & \sigma_P^2 \end{bmatrix} \tag{8.239}$$

Since this is of the form (8.234), and since the triangular decomposition is unique, it follows that

$$\boxed{\mathbf{L}^{-1} = \begin{bmatrix} 1 & 0 & & \cdots & 0 \\ \tilde{\boldsymbol{a}}_1^{*T} & \rightarrow & & \cdots & 0 \\ & \vdots & & \ddots & \vdots \\ \leftarrow & \tilde{\boldsymbol{a}}_{P-1}^{*T} & & \rightarrow & 0 \\ \leftarrow & -- & \tilde{\boldsymbol{a}}_P^{*T} & & \rightarrow \end{bmatrix}} \tag{8.240}$$

and

$$\boxed{\mathbf{D}_L = \begin{bmatrix} \sigma_0^2 & 0 & \cdots & 0 \\ 0 & \sigma_1^2 & \cdots & 0 \\ \vdots & \vdots & \ddots & \vdots \\ 0 & 0 & \cdots & \sigma_P^2 \end{bmatrix}} \tag{8.241}$$

Note also that the first column of $\mathbf{L}^{-1}$ contains the conjugated reflection coefficients $\gamma_p^* = a_p^{(p)*}$. From (8.231), (8.235), and (8.240) the components of the vector $\boldsymbol{x}''$ are seen to be the successive terms of the forward prediction error.

$$\boldsymbol{x}'' = \begin{bmatrix} \varepsilon_0[0] \\ \varepsilon_1[1] \\ \vdots \\ \varepsilon_P[P] \end{bmatrix} \tag{8.242}$$

It follows that solving the Normal equations is mathematically equivalent to finding the factors in the lower–upper triangular decomposition of the (inverse) correlation matrix.

It can similarly be shown that backward linear prediction is related to the upper–lower decomposition of the correlation matrix. To show this, first write $\mathbf{R}_x$ in the form

$$\mathbf{R}_x = \mathbf{U}_1 \mathbf{D}_U \mathbf{U}_1^{*T} \tag{8.243}$$

[12]The vectors comprised of the filter coefficients are said to be orthogonal relative to the matrix $\mathbf{R}_x$.

where $\mathbf{U}_1$ is unit upper triangular and $\mathbf{D}_U$ is diagonal. Since

$$\mathbf{U}_1^{-1}\mathbf{R}_x(\mathbf{U}_1^{-1})^{*T} = \mathbf{D}_U \tag{8.244}$$

the transformation

$$\boldsymbol{x}''' = \mathbf{U}_1^{-1}\boldsymbol{x} \tag{8.245}$$

is an *anticausal* transformation of the random vector that results in a diagonal correlation matrix $\mathbf{D}_U$. Now from arguments analogous to those given for the forward case it can be shown that the quadratic product

$$\begin{bmatrix} 0 & \cdots & 1 & b_1^{(q)*} & \cdots & b_q^{(q)*} \end{bmatrix} \mathbf{R}_x \begin{bmatrix} 0 \\ \vdots \\ 1 \\ b_1^{(p)} \\ \vdots \\ b_p^{(p)} \end{bmatrix}$$

is equal to $\sigma_p'^2$ when $q = p$ and zero when $q \neq p$. Therefore, it follows that

$$\begin{bmatrix} 1 & b_1^{(P)*} & \cdots & b_P^{(P)*} \\ 0 & 1 & \cdots & b_{P-1}^{(P-1)*} \\ \vdots & \vdots & \ddots & \vdots \\ 0 & 0 & \cdots & 1 \end{bmatrix} \underbrace{\begin{bmatrix} R_x[0] & \cdots & R_x[-P] \\ R_x[1] & \cdots & R_x[1-P] \\ \vdots & \vdots & \vdots \\ R_x[P] & \cdots & R_x[0] \end{bmatrix}}_{\mathbf{R}_x} \begin{bmatrix} 1 & 0 & \cdots & 0 \\ b_1^{(P)} & 1 & \cdots & 0 \\ \vdots & \vdots & \ddots & \vdots \\ b_P^{(P)} & b_{P-1}^{(P-1)} & \cdots & 1 \end{bmatrix}$$

$$= \begin{bmatrix} \sigma_P'^2 & 0 & \cdots & 0 \\ 0 & \sigma_{P-1}'^2 & \cdots & 0 \\ \vdots & \vdots & \ddots & \vdots \\ 0 & 0 & \cdots & \sigma_0'^2 \end{bmatrix} \tag{8.246}$$

By comparing this equation to (8.244) it can be seen that the backward prediction parameters relate to the *upper–lower* decomposition of the correlation matrix. Specifically, the matrices in (8.244) must be

$$\boxed{\mathbf{U}_1^{-1} = \begin{bmatrix} \leftarrow & \boldsymbol{b}_P^{*T} & -- & \rightarrow \\ 0 & \leftarrow & \boldsymbol{b}_{P-1}^{*T} & \rightarrow \\ \vdots & \vdots & \ddots & \vdots \\ 0 & \cdots & \leftarrow & \boldsymbol{b}_1^{*T} \\ 0 & 0 & \cdots & 1 \end{bmatrix}} \tag{8.247}$$

and

$$\mathbf{D}_U = \begin{bmatrix} \sigma_P'^2 & 0 & \cdots & 0 \\ 0 & \sigma_{P-1}'^2 & \cdots & 0 \\ \vdots & \vdots & \ddots & \vdots \\ 0 & 0 & \cdots & \sigma_0'^2 \end{bmatrix} \tag{8.248}$$

From (8.231), (8.245), and (8.247) it follows that the vector $\boldsymbol{x}'''$ is comprised of the backward prediction error terms.

$$\boldsymbol{x}''' = \begin{bmatrix} \varepsilon'_P[0] \\ \varepsilon'_1[1] \\ \vdots \\ \varepsilon'_0[P] \end{bmatrix} = \begin{bmatrix} \varepsilon^b_P[P] \\ \varepsilon^b_{P-1}[P] \\ \vdots \\ \varepsilon^b_0[P] \end{bmatrix} \tag{8.249}$$

Since the b parameters are the complex conjugates of the a parameters and $\sigma_p^2 = \sigma_p'^2$, a comparison of (8.240) with (8.247) and (8.241) with (8.248) also shows that

$$\mathbf{U}_1 = \tilde{\mathbf{L}}^* \tag{8.250}$$

and

$$\mathbf{D}_U = \tilde{\mathbf{D}}_L \tag{8.251}$$

These facts are also pointed out in Chapter 2.

8.11.2 The Schur Algorithm as Triangular Decomposition

The triangular decompositions of both the correlation matrix and its inverse are shown in Table 8.3. Since the prediction error filter coefficients determine the elements of the matrices $\mathbf{L}^{-1}$ and $\mathbf{U}_1^{-1}$ it is probably most accurate to say that the linear prediction parameters determine the triangular decomposition of the *inverse* correlation matrix $\mathbf{R}_x^{-1}$. If this is the case, then it is reasonable to ask if the triangular decomposition of $\mathbf{R}_x$ itself is related to anything that has been studied in this chapter. The answer to this question is "yes"; it turns out that the factors $\mathbf{L}$ and $\mathbf{U}_1$ in the triangular decomposition of $\mathbf{R}_x$ can be expressed in terms of the quantities involved in the Schur algorithm.

TABLE 8.3 TRIANGULAR DECOMPOSITION RELATIONS FOR THE CORRELATION MATRIX AND ITS INVERSE

Matrix	Lower–upper decomposition	Upper–lower decomposition
$\mathbf{R}_x$	$\mathbf{R}_x = \mathbf{L}\mathbf{D}_L\mathbf{L}^{*T}$	$\mathbf{R}_x = \mathbf{U}_1\mathbf{D}_U\mathbf{U}_1^{*T}$
$\mathbf{R}_x^{-1}$	$\mathbf{R}_x^{-1} = (\mathbf{U}_1^{-1})^{*T}\mathbf{D}_U^{-1}\mathbf{U}_1^{-1}$	$\mathbf{R}_x^{-1} = (\mathbf{L}^{-1})^{*T}\mathbf{D}_L^{-1}\mathbf{L}^{-1}$

To derive the relation, begin with the linear prediction parameters of any order p $(0 < p \leq P)$ and their relation to the Schur variables as specified by (8.138). Writing this relation out explicitly for $p - P \leq l \leq p$ results in the matrix equation

$$\begin{bmatrix} g_p[p-P] \\ g_p[p-P+1] \\ \vdots \\ g_p[0] \\ 0 \\ \vdots \\ 0 \end{bmatrix} = \begin{bmatrix} R_x[0] & \cdots & R_x[P-p] & \cdots & R_x[P] \\ \vdots & \ddots & \vdots & & \vdots \\ R_x[p-P] & \cdots & R_x[0] & \cdots & R_x[p] \\ \vdots & & & \ddots & \vdots \\ R_x[-P] & \cdots & R_x[-p] & \cdots & R_x[0] \end{bmatrix} \begin{bmatrix} 0 \\ 0 \\ \vdots \\ 1 \\ a_1^{(p)} \\ \vdots \\ a_p^{(p)} \end{bmatrix}$$

where the zeros on the left arise because of (8.127)(a) (also see Fig. 8.22). If a new vector $\boldsymbol{g}_p$ is defined by

$$\boldsymbol{g}_p \stackrel{\text{def}}{=} \begin{bmatrix} g_p[p-P] \\ \vdots \\ g_p[-1] \\ g_p[0] \end{bmatrix} = \tilde{\boldsymbol{g}}_p^{b*} \tag{8.252}$$

then the previous matrix equation can be written as

$$\begin{bmatrix} \boldsymbol{g}_p \\ \\ \mathbf{0} \end{bmatrix} = \tilde{\mathbf{R}}_x \begin{bmatrix} \mathbf{0} \\ \\ \boldsymbol{a}_p \end{bmatrix}$$

or, after reversing all quantities,

$$\begin{bmatrix} \mathbf{0} \\ \\ \tilde{\boldsymbol{g}}_p \end{bmatrix} = \mathbf{R}_x \begin{bmatrix} \tilde{\boldsymbol{a}}_p \\ \\ \mathbf{0} \end{bmatrix} \tag{8.253}$$

Now combine the set of equations (8.253) for $p = 0, 1, \ldots, P$ and write the result as a single matrix equation:

$$\mathbf{R}_x \begin{bmatrix} 1 & \tilde{\boldsymbol{a}}_1 & & \uparrow \\ 0 & \downarrow & \cdots & \tilde{\boldsymbol{a}}_P \\ \vdots & \vdots & \ddots & | \\ 0 & 0 & \cdots & \downarrow \end{bmatrix} = \begin{bmatrix} \uparrow & 0 & \cdots & 0 \\ \tilde{\boldsymbol{g}}_0 & \uparrow & & \vdots \\ | & \tilde{\boldsymbol{g}}_1 & \ddots & 0 \\ \downarrow & \downarrow & \cdots & g_P[0] \end{bmatrix}$$

$$= \begin{bmatrix} \uparrow & 0 & \cdots & 0 \\ \tilde{\boldsymbol{g}}_0/\sigma_0^2 & \uparrow & & \vdots \\ | & \tilde{\boldsymbol{g}}_1/\sigma_1^2 & \ddots & 0 \\ \downarrow & \downarrow & \cdots & 1 \end{bmatrix} \begin{bmatrix} \sigma_0^2 & 0 & \cdots & 0 \\ 0 & \sigma_1^2 & \cdots & 0 \\ \vdots & \vdots & \ddots & \vdots \\ 0 & 0 & \cdots & \sigma_P^2 \end{bmatrix}$$

where we note that $g_P[0]/\sigma_P^2 = 1$ (see (8.130)(a)).

This last equation can be written as the lower–upper triangular decomposition

$$\mathbf{R}_x = \begin{bmatrix} \uparrow & 0 & \cdots & 0 \\ \tilde{g}_0/\sigma_0^2 & \uparrow & & \vdots \\ | & \tilde{g}_1/\sigma_1^2 & \ddots & 0 \\ \downarrow & \downarrow & \cdots & 1 \end{bmatrix} \begin{bmatrix} \sigma_0^2 & 0 & \cdots & 0 \\ 0 & \sigma_1^2 & \cdots & 0 \\ \vdots & \vdots & \ddots & \vdots \\ 0 & 0 & \cdots & \sigma_P^2 \end{bmatrix} \begin{bmatrix} 1 & \tilde{\boldsymbol{a}}_1 & & \uparrow \\ 0 & \downarrow & \cdots & \tilde{\boldsymbol{a}}_P \\ \vdots & \vdots & \ddots & | \\ 0 & 0 & \cdots & \downarrow \end{bmatrix}^{-1} \tag{8.254}$$

It is already known that the middle factor is the matrix $\mathbf{D}_L$ and the last factor is $\mathbf{L}^{*T}$. Since the decomposition is unique, it must also follow that

$$\boxed{\mathbf{L} = \begin{bmatrix} \uparrow & 0 & \cdots & 0 \\ \tilde{g}_0/\sigma_0^2 & \uparrow & & \vdots \\ | & \tilde{g}_1/\sigma_1^2 & \ddots & 0 \\ \downarrow & \downarrow & \cdots & 1 \end{bmatrix} = \begin{bmatrix} 1 & 0 & \cdots & 0 \\ \tilde{\boldsymbol{a}}_1^{*T} & \rightarrow & \cdots & 0 \\ & \vdots & \ddots & \vdots \\ \leftarrow & \tilde{\boldsymbol{a}}_{P-1}^{*T} & \rightarrow & 0 \\ \leftarrow & -- & \tilde{\boldsymbol{a}}_P^{*T} & \rightarrow \end{bmatrix}^{-1}} \tag{8.255}$$

This provides the explicit relation of the quantities in the Schur algorithm to the lower–upper triangular decomposition.

The upper–lower decomposition can be related to the *backward* quantities. Since the steps are similar, we go over them just quickly. First the relation (8.139) is written in the matrix form

$$\begin{bmatrix} 0 \\ \vdots \\ 0 \\ g_p^b[p] \\ g_p^b[p+1] \\ \vdots \\ g_p^b[P] \end{bmatrix} = \begin{bmatrix} R_x[0] & \cdots & R_x[P-p] & \cdots & R_x[P] \\ \vdots & \ddots & \vdots & & \vdots \\ R_x[p-P] & \cdots & R_x[0] & \cdots & R_x[p] \\ \vdots & & & \ddots & \vdots \\ R_x[-P] & \cdots & R_x[-p] & \cdots & R_x[0] \end{bmatrix} \begin{bmatrix} b_p^{(p)} \\ \vdots \\ b_1^{(p)} \\ 1 \\ 0 \\ \vdots \\ 0 \end{bmatrix}$$

where the zeros on the left arise because of (8.127)(b). Reversing all of the quantities then produces

$$\begin{bmatrix} \tilde{\boldsymbol{g}}_p^b \\ \mathbf{0} \end{bmatrix} = \mathbf{R}_x \begin{bmatrix} \mathbf{0} \\ \tilde{\boldsymbol{b}}_p \end{bmatrix} \tag{8.256}$$

where $\boldsymbol{g}_p^b$ and $\boldsymbol{b}_p$ are defined by (8.146) and (8.49), respectively. Now writing this as a single matrix equation for $p = 0, 1, \ldots, P$ and rearranging produces

$$\mathbf{R}_x = \begin{bmatrix} 1 & \uparrow & \uparrow & \\ 0 & \cdots & \tilde{g}_1^b/\sigma_1'^2 & | \\ \vdots & \ddots & \downarrow & \tilde{g}_0^b/\sigma_0'^2 \\ 0 & \cdots & 0 & \downarrow \end{bmatrix} \begin{bmatrix} \sigma_P'^2 & 0 & \cdots & 0 \\ 0 & \ddots & \cdots & 0 \\ \vdots & \vdots & \sigma_1'^2 & \vdots \\ 0 & 0 & \cdots & \sigma_0'^2 \end{bmatrix} \begin{bmatrix} \uparrow & 0 & & \cdots & 0 & 0 \\ \boldsymbol{b}_P & \uparrow & & & \vdots & \vdots \\ | & \boldsymbol{b}_{P-1} & & \ddots & \uparrow & 0 \\ \downarrow & \downarrow & & \cdots & \boldsymbol{b}_1 & 1 \end{bmatrix}^{-1} \tag{8.257}$$

Finally, by comparing this equation to (8.243) it follows that

$$\mathbf{U}_1 = \begin{bmatrix} 1 & \uparrow & \uparrow & \\ 0 & \cdots & \tilde{g}_1^b/\sigma_1'^2 & | \\ \vdots & \ddots & \downarrow & \tilde{g}_0^b/\sigma_0'^2 \\ 0 & \cdots & 0 & \downarrow \end{bmatrix} = \begin{bmatrix} \leftarrow & \boldsymbol{b}_P^{*T} & -- & \rightarrow \\ 0 & \leftarrow & \boldsymbol{b}_{P-1}^{*T} & \rightarrow \\ \vdots & \vdots & \ddots & \vdots \\ 0 & \cdots & \leftarrow & \boldsymbol{b}_1^{*T} \\ 0 & 0 & \cdots & 1 \end{bmatrix}^{-1} \tag{8.258}$$

This provides an explicit relation for the quantities in the Schur algorithm to the upper–lower triangular decomposition.

8.11.3 Other Relations and Observations

The equations (8.239) and (8.246) are actually very closely related to results of another purely mathematical topic, namely the theory of polynomials orthogonal on the unit circle. These are called Szegö polynomials [24]. If $Q_i(z)$ and $Q_j(z)$ are such polynomials, then a complex spectral density function such as $S_x(z)$ can be used to define a weighted inner product on the unit circle

$$\int_{-\pi}^{\pi} Q_i(e^{\jmath\omega})Q_j(e^{\jmath\omega})S_x(e^{\jmath\omega})d\omega$$

which is zero when $i \neq j$ and nonzero when $i = j$. Polynomials defined by the prediction error filters of orders $0, 1, \ldots, P$ turn out to be orthogonal with respect to this inner product. In fact, the inner product and the orthogonality condition can be shown to be the same as the conditions expressed by (8.239) and (8.246). Recursive relations for Szegö polynomials of the general form (8.114) and (8.115) have been known by mathematicians for some time. Moreover, the relation between these polynomials and the inverse of an associated Toeplitz matrix has been explored [25, 26] and bears essentially the same relation to linear prediction as the triangular factorization of the correlation matrix explored here.

To conclude this section we point out one other important fact. From (8.239) and (8.246) it is clear that the matrix $\mathbf{R}_x$ is positive definite if and only if all of the prediction error variances σ_p^2 (or $\sigma_p'^2$) are strictly positive. Because

$$\sigma_p^2 = (1 - |\gamma_p|^2)\sigma_{p-1}^2$$

these variances are positive if and only if the magnitudes of all the reflection coefficients are strictly less than 1. Thus the strict inequality (8.110) satisfied by the reflection coefficients of a regular process guarantees the positive definiteness of the correlation matrix $\mathbf{R}_x$. This in turn guarantees the positive definiteness of the correlation function and therefore the strict positivity of the power spectrum. In summary, *the correlation function or correlation matrix of a regular process is always strictly positive definite and the power spectral density is always strictly greater than zero.* For a predictable process, the prediction error variance at and beyond some order L is zero. Therefore, the correlation matrix for orders greater than or equal to L (and therefore the correlation function) is only positive semidefinite.

8.12 LATTICE REPRESENTATION FOR THE FIR WIENER FILTER

The relations established in the preceding section permit extending the lattice structure for the prediction error filter to a lattice realization for the (shift-invariant) FIR Wiener filter. This is a very powerful result since it permits developing the optimal filter for the estimation of *any* random process in lattice form. (The procedure is sometimes called "joint process estimation.")

Begin by recalling that the estimate produced by the shift-invariant Wiener filter can be expressed as

$$\hat{d}[n] = (\tilde{\boldsymbol{x}}[n])^T \mathbf{h} = \tilde{\mathbf{h}}^T \boldsymbol{x}[n] \tag{8.259}$$

where $\boldsymbol{x}[n]$ is defined as the vector of observations

$$\boldsymbol{x}[n] \stackrel{\text{def}}{=} \boldsymbol{x}_{P-1}[n] = \begin{bmatrix} x[n-P+1] \\ \vdots \\ x[n-1] \\ x[n] \end{bmatrix} \tag{8.260}$$

and $\mathbf{h}$ is the vector of filter coefficients

$$\mathbf{h} = \begin{bmatrix} h[0] \\ h[1] \\ \vdots \\ h[P-1] \end{bmatrix} \tag{8.261}$$

The filter coefficient vector satisfies the Wiener–Hopf equation

$$\tilde{\mathbf{R}}_x \mathbf{h}^* = \tilde{\mathbf{r}}_{dx}^*$$

or

$$\mathbf{R}_x \tilde{\mathbf{h}}^* = \mathbf{r}_{dx}^* \tag{8.262}$$

Now writing $\mathbf{R}_x$ in upper–lower triangular form yields

$$(\mathbf{U}_1 \mathbf{D}_U \mathbf{U}_1^{*T}) \tilde{\mathbf{h}}^* = \mathbf{r}_{dx}^*$$

which implies that

$$\tilde{\mathbf{h}}^* = (\mathbf{U}_1^{-1})^{*T} \mathbf{D}_U^{-1} \mathbf{U}_1^{-1} \mathbf{r}_{dx}^*$$

Then taking the conjugate transpose and substituting the resulting expression in (8.259) produces

$$\hat{d}[n] = \mathbf{r}_{dx}^T(\mathbf{U}_1^{-1})^{*T}\mathbf{D}_U^{-1}\mathbf{U}_1^{-1}\boldsymbol{x}[n] = \tilde{\mathbf{w}}^{*T}\boldsymbol{\varepsilon}^b[n] \tag{8.263}$$

where the vector $\boldsymbol{\varepsilon}^b[n]$ is defined as

$$\boldsymbol{\varepsilon}^b[n] \stackrel{\text{def}}{=} \mathbf{U}_1^{-1}\boldsymbol{x}[n] = \begin{bmatrix} 1 & b_1^{(P-1)*} & \cdots & b_{P-1}^{(P-1)*} \\ 0 & 1 & \cdots & b_{P-2}^{(P-2)*} \\ \vdots & \vdots & \ddots & \vdots \\ 0 & 0 & \cdots & 1 \end{bmatrix} \begin{bmatrix} x[n-P+1] \\ \vdots \\ x[n-1] \\ x[n] \end{bmatrix} = \begin{bmatrix} \varepsilon_{P-1}^b[n] \\ \varepsilon_{P-2}^b[n] \\ \vdots \\ \varepsilon_0^b[n] \end{bmatrix} \tag{8.264}$$

and the weight vector

$$\mathbf{w} = \begin{bmatrix} w_0 \\ w_1 \\ \vdots \\ w_{P-1} \end{bmatrix} \tag{8.265}$$

is given in its reversed form by

$$\tilde{\mathbf{w}} = \mathbf{D}_U^{-1}\mathbf{U}_1^{-1}\mathbf{r}_{dx}^* = \begin{bmatrix} 1/\sigma_{P-1}'^2 & 0 & \cdots & 0 \\ 0 & 1/\sigma_{P-2}'^2 & \cdots & 0 \\ \vdots & \vdots & \ddots & \vdots \\ 0 & 0 & \cdots & 1/\sigma_0'^2 \end{bmatrix} \begin{bmatrix} 1 & b_1^{(P-1)*} & \cdots & b_{P-1}^{(P-1)*} \\ 0 & 1 & \cdots & b_{P-2}^{(P-2)*} \\ \vdots & \vdots & \ddots & \vdots \\ 0 & 0 & \cdots & 1 \end{bmatrix} \mathbf{r}_{dx}^* \tag{8.266}$$

This shows that the estimate can be formed as a linear combination of the backward error terms in the lattice filter, so the Wiener filter has the realization shown in Fig. 8.25(a). The weights w_i are essentially the normalized cross-correlations between the backward errors and the desired sequence d. Since all of the quantities in (8.266) are known, the weights can be computed directly from that equation. However, one additional interpretation is possible. By comparing (8.266) to (8.264) and observing that $\mathbf{D}_U^{-1}$ is a diagonal matrix it can be seen that $\tilde{\mathbf{w}}$ is just a normalized version of the lattice network response that would result if the sequence corresponding to $\mathbf{r}_{dx}^*$ were applied as an input [see Fig. 8.25(b)]. The cross-correlation sequence $R_{dx}^*[l]$ is fed in and when the sample $R_{dx}^*[P-1]$ appears at the input, the weights w_i appear in the places shown. This provides a convenient alternative method to compute the weights in terms of the reflection coefficients and eliminates the need to determine the $b_i^{(p)}$ explicitly in order to evaluate (8.266).

In summary, the developments of this section show that the lattice structure is useful not only for the prediction error filter and its inverse, but also for the implementation of *any* FIR Wiener filter. The implementation is order-recursive, so that filters of any order can be constructed by adding additional lattice sections. In so doing the previously computed parameters of the (lower order) filter remain unchanged and do not have to be recomputed.

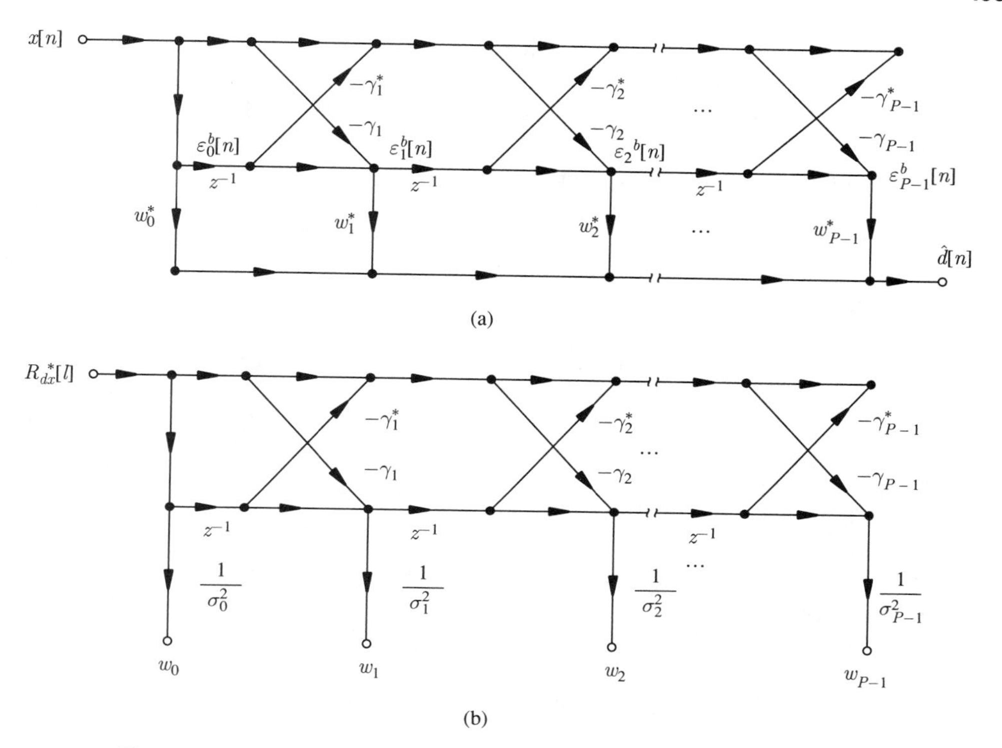

Figure 8.25 Lattice realization for the FIR Wiener filter. (a) Wiener filter. (b) Procedure for obtaining weights.

8.13 CHAPTER SUMMARY

This chapter deals with linear prediction and the closely related problem of autoregressive modeling. In linear prediction an FIR filter is applied to a general random process to estimate its next value. The prediction error so generated, given a sufficiently long prediction filter, is either exactly or approximately white noise. In AR modeling white noise is applied to the inverse of the prediction error filter to regenerate the random process. The AR model is sometimes called an all-pole model because the IIR inverse filter has only a nontrivial denominator polynomial. All zeros are at the origin.

The concept of backward linear prediction, or predicting backward along a random process, plays a major role in the theory of linear prediction. The corresponding AR model is anticausal. The backward prediction error filter coefficients of any order are complex conjugates of the forward prediction error filter coefficients and the prediction error variances for the backward and the forward problems are identical. The simultaneous consideration of forward and backward linear prediction problems leads to an efficient procedure known as the Levinson recursion for computing the linear prediction parameters. For purposes of the computation, the Levinson recursion can be expressed in terms of only the forward parameters.

The Levinson recursion leads among other things to representations of both the prediction error filter and the inverse filter as lattice structures. The lattice structures are completely specified by a set of reflection coefficients whose magnitudes for any stationary regular random process are less than 1. The reflection coefficients have the statistical interpretation of partial correlation. Specifically, γ_p represents the normalized cross-correlation between $x[n]$ and $x[n-p]$ when the effect of correlation through the intermediate points is removed. One major advantage of the lattice structure is that the realization is modular. That is, when the prediction order of the filter is increased, all that is necessary is to add on an additional lattice section. The other lattice sections and the values of their reflection coefficients do not change.

The prediction error filter corresponding to a regular random process is a minimum-phase filter. This means that all poles and zeros lie strictly within the unit circle; the filter is causal and stable and has a causal stable inverse. The minimum-phase property is a direct consequence of the fact that the reflection coefficients of any regular process all have magnitude less than 1.

The Schur algorithm is another efficient algorithm for computing the parameters of the lattice structure for a random process. The Schur algorithm computes the complete set of reflection coefficients and prediction error variances. The Schur algorithm is based on a recursion of certain functions, initially defined as cross-correlations between the input and prediction errors and sometimes known as "gapped functions." The Schur algorithm is unique in that it requires no inner product computations. Thus it is ideal for computation on machines with highly parallel pipelined architecture.

"Split algorithms" is the name given to a class of algorithms based on a three-term recursion that achieves a somewhat greater reduction in computation than the standard Levinson and Schur algorithms. These algorithms require about the same number of additions as the standard algorithms but require a number of multiplications that is up to 50% less. The split Levinson algorithm begins with a splitting of a weighted filter coefficient vector into symmetric and antisymmetric parts, either of which can be used to develop a recursive algorithm. It is the symmetry of these new parameters that is primarily responsible for the computational reduction. Once a split Levinson algorithm is realized, a split lattice structure and a split Schur algorithm can also be developed. Since the variables propagated in these algorithms are not the standard linear prediction parameters, an extra step needs to be added at the end to recover the linear prediction parameters.

The causal whitening transformation produced by linear prediction is closely related to the linear algebraic operation of triangular decomposition first discussed in Chapter 2. In particular, the linear prediction parameters appear directly as elements of the matrices when the *inverse* correlation matrix is factored. The forward parameters relate to the lower–upper decomposition, while the backward parameters relate to the upper–lower decomposition. The functions appearing in the Schur algorithm relate to the triangular decomposition of the correlation matrix itself (rather than its inverse). The forward quantities appear directly in the lower–upper decomposition, while the backward quantities appear in the upper–lower decomposition.

Finally, the lattice structure can be applied to more than just linear prediction. In particular, any FIR Wiener filter can be implemented in a form that has the linear prediction lattice structure as its principal component. The resulting implementation is therefore modular and has most of the other important features associated with the lattice structure for the prediction error filter.

REFERENCES

1. John Makhoul. Linear prediction: a tutorial review. *Proceedings of the IEEE*, 63:561–580, April 1975.
2. Simon Haykin. Radar signal processing. *IEEE ASSP Magazine*, 2(2):2–18, April 1985.
3. Lawrence R. Rabiner and Ronald W. Schafer. *Digital Processing of Speech Signals*. Prentice Hall, Inc., Englewood Cliffs, New Jersey, 1978.
4. John D. Markel and Augustine H. Gray, Jr. *Linear Prediction of Speech*. Springer-Verlag, New York, 1976.
5. Alan V. Oppenheim and Ronald W. Schafer. *Discrete-Time Signal Processing*. Prentice Hall, Inc., Englewood Cliffs, New Jersey, 1989.
6. George E. P. Box and Gwilym M. Jenkins. *Time Series Analysis: Forecasting and Control*, rev. ed. Holden-Day, Oakland, California, 1976.
7. Norman Levinson. The Wiener RMS (root mean square) error criterion in filter design and prediction. *Journal of Mathematics and Physics*, 25(4):261–278, January 1947. (See also Appendix B in [10].)
8. Gideon Weiss. Time-reversibility of linear stochastic processes. *Journal of Applied Probability*, 12:831–836, 1975.
9. A. J. Lawrance. Directionality and reversibility in time series. *International Statistical Review*, 59(1):67–79, 1991.
10. Norbert Wiener. *Extrapolation, Interpolation, and Smoothing of Stationary Time Series*. The MIT Press (formerly Technology Press), Cambridge, Massachusetts, 1949.
11. J. Durbin. The fitting of time-series models. *Revue de l'Institute International de Statistique*, 28(3):233–243, 1960.
12. Enders A. Robinson and Sven Treitel. Maximum entropy and the relationship of the partial autocorrelation to the reflection coefficients of a layered system. *IEEE Transactions on Acoustics, Speech, and Signal Processing*, ASSP-28(2):224–235, April 1980.
13. Thomas Kailath. A theorem of I. Schur and its impact on modern signal processing. In I. Gohberg, editor, *I. Schur Methods in Operator Theory and Signal Processing*, pages 9–30. Birkhäuser Verlag, Basel, 1986.
14. I. Schur. On power series which are bounded in the interior of the unit circle I. In I. Gohberg, editor, *I. Schur Methods in Operator Theory and Signal Processing*, pages 31–59. Birkhäuser Verlag, Basel, 1986. First published in German in *Journal für die Reine und Angewandte Mathematik*, 147:205–232, 1917.
15. I. Schur. On power series which are bounded in the interior of the unit circle II. In I. Gohberg, editor, *I. Schur Methods in Operator Theory and Signal Processing*, pages 61–88. Birkhäuser Verlag, Basel, 1986. First published in German in *Journal für die Reine und Angewandte Mathematik*, 148: 122–145, 1918.

16. Sun-Yuan Kung and Yu Hen Hu. A highly concurrent algorithm and pipelined architecture for solving Toeplitz systems. *IEEE Transactions on Acoustics, Speech, and Signal Processing*, ASSP-31(1):66–76, February 1983.
17. Philippe Delsarte and Yves V. Genin. The split Levinson algorithm. *IEEE Transactions on Acoustics, Speech, and Signal Processing*, ASSP-34(3):470–478, June 1986.
18. Philippe Delsarte and Yves V. Genin. On the splitting of classical algorithms in linear prediction theory. *IEEE Transactions on Acoustics, Speech, and Signal Processing*, ASSP-35(5):645–653, May 1987.
19. Yuval Bistritz, Hanoch Lev-Ari, and Thomas Kailath. Complexity reduced lattice filters for digital speech processing. In *Proceedings of the IEEE International Conference on Acoustics, Speech, and Signal Processing*, pages 21–24, IEEE ASSP, April 1987 (Dallas, Texas).
20. Hari Krishna and Salvatore D. Morgera. The Levinson recurrence and fast algorithms for solving Toeplitz systems of linear equations. *IEEE Transactions on Acoustics, Speech, and Signal Processing*, ASSP-35(6):839–848, June 1987.
21. Yuval Bistritz, Hanoch Lev-Ari, and Thomas Kailath. Immittance-type three-term Schur and Levinson recursions for quasi-Toeplitz and complex Hermitian matrices. *SIAM Journal on Matrix Analysis and Applications*, 12(3): 497–520, July 1991.
22. Yuval Bistritz, Hanoch Lev-Ari, and Thomas Kailath. Immittance-domain Levinson algorithms. In *Proceedings of the IEEE International Conference on Acoustics, Speech, and Signal Processing*, pages 253–256, IEEE ASSP, April 1986 (Tokyo).
23. Yuval Bistritz, Hanoch Lev-Ari, and Thomas Kailath. Immittance-domain Levinson algorithms. *IEEE Transactions on Information Theory*, 35(3):675–682, May 1989.
24. Gabor Szegö. *Orthogonal Polynomials*, 6th ed. American Mathematics Society, Providence, Rhode Island, 1975.
25. Ulf Grenander and Gabor Szegö. *Toeplitz Forms and Their Applications*. Chelsea Publishing Company, New York, second edition, 1984.
26. James H. Justice. The Szegö recursion relation and inverses of positive definite Toeplitz matrices. *SIAM Journal of Mathematical Analysis*, 5(3):503–508, May 1974.
27. I. C. Gohberg and I. A. Feĺdman. *Convolution Equations and Projection Methods for their Solution*. American Mathmatical Society, Providence, RI, 1974. Volume 41, Translations of Mathematical Monographs.
28. Thomas Kailath, A. Viera, and M. Morf. Inverses of Toeplitz operators, innovations, and orthogonal polynomials. *SIAM Review*, 20(1):106–119, January 1978.

PROBLEMS

8.1. A certain random process has an autocorrelation function given by

$$R_x[l] = 2^{-|l|}$$

(a) Compute the first-order linear prediction parameters for this random process.
(b) Compute the second-order linear prediction parameters for this process.

Suppose now that

$$R_x[l] = 2^{-|l|} + \delta[l]$$

(c) Compute the first-order linear prediction parameters for this new random process.
(d) Compute the second-order linear prediction parameters for this new process.

8.2. (a) The following numbers are sixteen points of a zero-mean Gaussian white noise process $w[n]$ with variance $\sigma_w^2 = 1$.

$$0.226,\ 0.403,\ 0.559,\ 0.266,\ -0.724,\ -0.526,\ 0.176,\ -0.169,$$
$$-0.557,\ 0.625,\ -1.521,\ 0.033,\ -0.349,\ 0.520,\ 1.325,\ 0.355$$

Compute the first 16 points of a first-order autoregressive process with parameter $\rho = 0.95$. Use the given values of the white noise and the initial condition $x[0] = w[0]$. Plot the result.

(b) Repeat part (a) for $\rho = -0.95$. Observe the difference in the results.

8.3. You are given three values of an autocorrelation function—$R[0] = 2.280$, $R[1] = 0.768$, and $R[2] = 0.4608$ and are told that these values correspond to a random process which is to be modeled as shown in Fig. PR8.3. The system is described by the difference equation

$$x[n] + a_1 x[n-1] + a_2 x[n-2] = w[n]$$

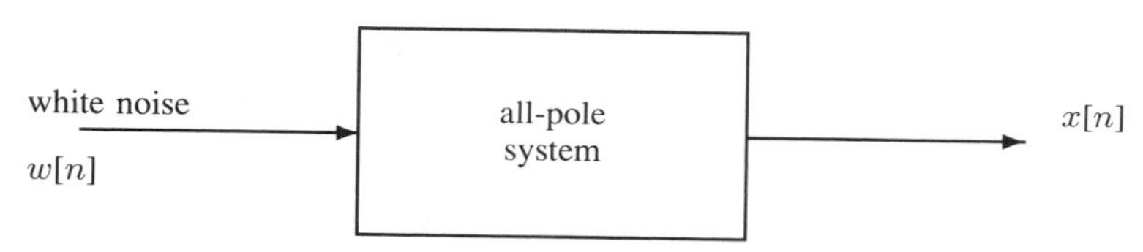

Figure PR8.3

(a) Find the model parameters a_1, a_2, and σ_w^2 from the data given.
(b) Derive an expression for extrapolation of the correlation function and use it to find the values $R[3]$ and $R[4]$ of the modeled random process.

8.4. Let the values of the correlation function in Problem 8.3 be represented simply by $R[0]$, $R[1]$, and $R[2]$.

(a) Find an expression for the model parameters a_1, a_2, and σ_w^2 in terms of arbitrary values $R[0]$, $R[1]$, and $R[2]$ of the autocorrelation function.
(b) Derive an explicit formula for extrapolation of the model correlation function in terms of the known but arbitrary values $R[0]$, $R[1]$, and $R[2]$ of the given correlation function.

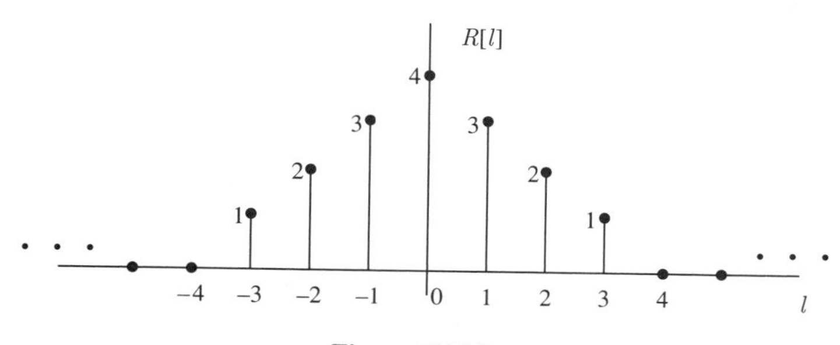

Figure PR8.5

8.5. Given the triangular autocorrelation function shown in Fig. PR8.5.

(a) Find the coefficients of the second-order linear predictive filter ($P = 2$) by solving the Normal equations.

(b) Find the values of the correlation function $R_x[l]$ produced by the AR model for $l = 3, 4, 5$ using (8.24). Note that they are not equal to the values of the original correlation function.

8.6. The correlation function for a particular random sequence is given by

$$R_x[l] = 2^{-l^2}$$

(a) Write the Yule–Walker equations for a first-order AR model that would approximate the given random process.

(b) Solve the equations for the filter parameters and white noise variance.

(c) The correlation function values $R[0]$ and $R[1]$ of the AR model match the given correlation function exactly. What values for $R[2]$ and $R[3]$ would the AR model produce? How do they differ from the given correlation function? Plot the two autocorrelation functions for lag values between 0 and 10.

(d) Repeat parts (a), (b), and (c) for a second-order AR model.

8.7. Define the *length* of an autocorrelation function to be the lag value at which the magnitude of $R[l]$ drops to less than 63% of its initial value $R[0]$. Define the *prediction order* of a random sequence to be the largest number of nonzero coefficients needed to optimally predict the sequence from past values. What does the length of the autocorrelation function for a sequence have to do with the prediction order? The answer to this question consists of a single word. (*Hint*: Consider a first-order AR process with parameter ρ. How is the length of the correlation function and the prediction order determined by the value of ρ?)

8.8. Solve the Normal equations to find the second-order forward and backward linear prediction parameters for a complex random process with correlation function whose first three values are

$$R_x[0] = 1.707\,; \quad R_x[1] = 0.5 + 0.5\jmath\,; \quad R_x[2] = 0.707\jmath$$

8.9. Use the Levinson recursion to solve for the linear prediction parameters and the reflection coefficients up to order 4 for the correlation function of Problem 8.5.

8.10. Use the Levinson recursion to solve the Normal equations of Problem 8.8. Draw the lattice realization for the prediction error filter and the inverse (all-pole) filter.

8.11. A real-valued random process is defined by

$$x[n] = y[n] + \eta[n]$$

where $y[n]$ is a signal with correlation function

$$R_y[l] = 1.28(0.6)^{|l|}$$

and $\eta[n]$ is white noise with variance $\sigma_\eta^2 = 1.0$.

(a) Write the Normal equations for a second-order predictor ($P = 2$) for the process $x[n]$.

(b) Find the reflection coefficients γ_1 and γ_2 and draw the lattice network for the prediction error filter.

(c) Suppose now that $x[n] = y[n]$ (i.e., there is no noise). What are the reflection coefficients γ_1 and γ_2 now?

8.12. Find the reflection coefficients in the lattice representation of the complex prediction error filter with coefficients

$$a_1 = -0.2071 + 0.2071\jmath$$
$$a_2 = 0.2929\jmath$$

8.13. Find the reflection coefficients corresponding to the prediction error filter with coefficients $a_1 = 0.35$, $a_2 = -0.65$, and $a_3 = -\frac{2}{3}$.

8.14. Consider a two-stage lattice filter with reflection coefficients $\gamma_1 = 0.9$ and $\gamma_2 = -1.1$.

(a) Find the filter coefficients a_0, a_1, a_2 and the corresponding transfer function $A(z)$ of the prediction error filter.

(b) By finding the roots of $A(z)$, demonstrate that the filter is not minimum-phase.

8.15. By writing out the relations (8.114) and (8.115) explicitly in powers of z, show that these relations are equivalent to those of step III in Table 8.2 (equations 8.90 and 8.91). Then solve these equations for $A_{p-1}(z)$ and $B_{p-1}(z)$ in terms of $A_p(z)$ and $B_p(z)$ and apply (8.113) to develop a recursion for the filter that proceeds backward in order. Finally, express this result in terms of the filter parameters to obtain the reverse-order Levinson recursion.

8.16. Define $A_p(z)$ to be the transfer function relating the z-transform of $x[n]$ to the z-transform of $\varepsilon_p[n]$. Define $B_p^b(z)$ to be the transfer function between the z-transforms of $x[n]$ and $\varepsilon_p^b[n]$. Then using (8.86) and (8.87) derive (8.117) and (8.118) directly.

8.17. By considering geometric arguments similar to the one presented in Section 8.8, show that if the magnitude of the reflection coefficient γ_p is less than 1, then the resulting filter is minimum-phase while if the magnitude of γ_p is greater than 1 $A_p(z)$ is guaranteed to have a zero outside the unit circle. Thus prove that the condition $\gamma_p < 1$ for all p is both necessary and sufficient for the prediction error filter to be minimum-phase.

8.18. Use the Schur algorithm to compute the reflection coefficients up to order 4 for the data in Problem 8.5. Also find the prediction error variance for each of the various order linear predictors. Check your results with those of Problem 8.9.

8.19. Use the Schur algorithm to find the reflection coefficients and prediction error variances for the correlation matrix in Problem 8.8. Check your results with those of Problem 8.10.

8.20. Find the linear prediction parameters for the problem solved in Example 8.4 using the original split Levinson algorithm.

8.21. Apply the original split Levinson algorithm to compute the linear prediction parameters for Problem 8.5. Repeat using the complex split Levinson algorithm.

8.22. Apply the complex split Levinson algorithm to compute the linear prediction parameters for Problem 8.8.

8.23. Repeat Example 8.5 using the original split Schur algorithm.

8.24. Compute the lattice parameters γ_p and σ_p^2 for Problem 8.5 using the original split Schur algorithm. Repeat using the complex split Schur algorithm.

8.25. Apply the complex split Schur algorithm to compute the standard lattice parameters γ_p and σ_p^2 for Problem 8.8.

8.26. Starting with the condition $0 \leq |\gamma_p| < 1$, show that the parameter α_p in the original split Levinson algorithm satisfies the condition

$$0 < \alpha_p < 4$$

8.27. Show that the requirement (8.186) is inconsistent with the condition (8.162) unless all variables are real.

8.28. By following the development culminating in (8.176) and (8.177), show explicitly how the computations in the three-term recursion of the complex split Levinson algorithm can be grouped to reduce the number of real multiplications. Show also how the same technique can be applied to the complex split Schur algorithm in (8.229)(g).

8.29. By following the description in Section 8.9.1, develop a canonical implementation of the split Schur algorithm based on the idea of a "generator" matrix. The generator matrix should have three rows, corresponding to the three terms in the basic recursion. Demonstrate your algorithm by using it to solve the problem of Example 8.5.

8.30. In Problem 2.39 you were asked to perform the triangular decomposition of the matrix

$$\mathbf{R}_x = \begin{bmatrix} 3 & 2 & 1 \\ 2 & 3 & 2 \\ 1 & 2 & 3 \end{bmatrix}$$

Note that this is the 3×3 correlation matrix for the random process introduced in Example 8.1 and verify that the filter coefficients do in fact correspond to the rows of the matrix $\mathbf{L}^{-1}$ and that the prediction error variances are the diagonal elements of the matrix $\mathbf{D}_L$. You will probably want to use a programmable calculator or computer to do the computations in the rest of this problem.

(a) Form a 5×5 correlation matrix for the random process of Example 8.1 and note that it is banded. Compute the linear prediction parameters of all orders up to $P = 4$. By forming the appropriate $\mathbf{L}^{-1}$ and $\mathbf{D}_L$ matrices and writing the inverse correlation matrix in the factored form

$$\mathbf{R}^{-1} = \mathbf{L}^{T^{-1}} \mathbf{D}_L^{-1} \mathbf{L}^{-1}$$

compute the 5×5 inverse correlation matrix. Verify that your inverse matrix is correct by forming the product with the 5×5 banded correlation matrix and checking that the result is equal to the identity matrix.

(b) Suppose now that the random process is approximated by a second-order AR model. By forming $\mathbf{R}^{-1}$ as in part (a) and inverting it directly, compute the 5×5 correlation matrix corresponding to the AR model. Note that the matrix $\mathbf{R}^{-1}$ is banded but the matrix $\mathbf{R}$ for the AR process is *not* banded. The correlation matrix is Toeplitz, however, and has extrapolated values for the correlation function $R[3]$ and $R[4]$. The values $R[0]$, $R[1]$, and $R[2]$ match the original correlation function.

(c) Show that the extrapolated values of the correlation function obtained in part (b) can also be computed recursively from the difference equation

$$R[l] + a_1 R[l-1] + a_2 R[l-2] = 0$$

where a_1 and a_2 are the coefficients of the second order AR model.

8.31. The Golberg-Semencul formula [27, 28] expresses the inverse of a Toeplitz correlation matrix in terms of the full order forward and backward linear prediction parameters. If $\boldsymbol{a}$ and $\boldsymbol{b}$ as given by (8.5) and (8.28) are the solutions to (8.9) and (8.31) respectively, then the formula states that

$$\mathbf{R}_x^{-1} = \frac{1}{\sigma_\varepsilon^2}\left\{\begin{bmatrix} 1 & 0 & 0 & \cdots & 0 \\ b_1 & 1 & 0 & & 0 \\ b_2 & b_1 & 1 & & 0 \\ \vdots & \vdots & \vdots & \ddots & \vdots \\ b_P & b_{P-1} & b_{P-2} & \cdots & 1 \end{bmatrix}\begin{bmatrix} 1 & a_1 & a_2 & \cdots & a_P \\ 0 & 1 & a_1 & \cdots & a_{P-1} \\ 0 & 0 & 1 & \cdots & a_{P-2} \\ \vdots & & & \vdots & \vdots \\ 0 & 0 & 0 & \cdots & 1 \end{bmatrix}\right.$$

$$\left. - \begin{bmatrix} 0 & 0 & 0 & \cdots & 0 \\ a_P & 0 & 0 & & 0 \\ a_{P-1} & a_P & 0 & & 0 \\ \vdots & \vdots & \vdots & \ddots & \vdots \\ a_1 & a_2 & a_3 & \cdots & 0 \end{bmatrix}\begin{bmatrix} 0 & b_P & b_{P-1} & \cdots & b_1 \\ 0 & 0 & b_P & \cdots & b_2 \\ 0 & 0 & 0 & \cdots & b_3 \\ \vdots & & & \vdots & \vdots \\ 0 & 0 & 0 & \cdots & 0 \end{bmatrix}\right\}$$

Use the filter coefficients found in Problem 8.8 to evaluate this formula. Multiply the result by the correlation matrix and thus verify by direct computation that the formula is correct.

8.32. Find the parameters and draw the lattice representation for the FIR Wiener filter of Problem 7.14.

8.33. Find the parameters and draw the lattice representation for the FIR Wiener filter of Problem 7.15.

COMPUTER ASSIGNMENTS

8.1. **(a)** Write a computer algorithm to implement the Levinson recursion. Test the algorithm by computing the linear prediction parameters for a real correlation function whose first five values are

$$2.0 \quad 0.95 \quad 0.9025 \quad 0.8573 \quad 0.8145$$

Print out the parameters for all orders up to the maximum order (4). Repeat for a correlation function whose first five values are

$$2.25 \quad 1.4125 \quad 1.6181 \quad 0.9860 \quad 1.2582$$

Check your results by forming a set of Normal equations for the problem and solving for the fourth-order linear prediction parameters by any standard method for solving linear equations.

(b) Test your algorithm for complex data by using it to solve Problem 8.8. Check your results with those of Prob. 8.10.

8.2. **(a)** Write a computer algorithm to solve for the lattice parameters γ_p and σ_p^2 by the Schur algorithm. Solve for the lattice parameters up to fourth order corresponding to the data in Computer Assignment 8.1(a). Check your results with those of Computer Assignment 8.1.

(b) Test your algorithm for complex data by using it to solve Problem 8.8. Again, check your results.

8.3. Write a computer algorithm to implement the complex split Levinson algorithm. Follow the procedures for testing in Computer Assignment 8.1.

8.4. Write a computer algorithm to implement the complex split Schur algorithm. Follow the procedures for testing in Computer Assignment 8.2.

8.5. Write a computer algorithm to implement the original split Levinson algorithm. Follow the procedures for testing in Computer Assignment 8.1 [part (a) only].

8.6. Write a computer algorithm to implement the original split Schur algorithm. Follow the procedures for testing in Computer Assignment 8.2 [part (a) only].

B/A

9

Linear Models

This chapter deals with modeling an observed data sequence as the output of a linear filter. This is referred to as a *linear model*. Usually, the sequence is considered to be a realization of a stationary random process and the input to the filter is taken to be white noise. In the initial approach to this topic we presume that the second moment statistics of the random process are known and develop the equations whose solution provides the model parameters. This provides the basic framework for the topic. In most practical applications of the theory, however, the fixed quantities appearing in the equations, namely the values of the correlation function, are *not* known a priori but need to be estimated from data. This introduces errors and various other complications and brings up the whole issue of estimation of the model parameters. Maximum likelihood estimation is explored briefly as a method of estimating the model parameters. This method is based directly on the data rather than on an estimated correlation function, but several considerations generally deter its use.

Other methods based directly on the observed data are then discussed. The key to developing a data-oriented approach lies in the choice of the criterion that measures how well the model fits the data. Specifically, the data-oriented methods seek to minimize a sum of squares of the error data rather than the theoretical mean-square error. Thus the procedures are referred to as "least squares" methods.

Least squares methods are practical and widely employed in modern signal processing since they lead to algorithms that can be used directly on the available data and may have better computational properties than methods that begin with an estimated correlation function or other second moment quantity. Unfortunately, all of the least squares methods do not have the same starting point. Some emulate statistical procedures, while others are better thought of as purely deterministic. Still others begin with statistical procedures and try to extend them in some way. The collection is really a "mixed bag." This chapter attempts to introduce some of the more widely used methods and the principles upon which they are based.

9.1 LINEAR MODELING OF RANDOM PROCESSES

It was seen earlier that when white noise with power σ_w^2 is passed through a stable linear filter with system function $H(z)$ the power spectral density function of the resulting process is given by

$$\sigma_w^2 |H(e^{\jmath\omega})|^2$$

Since the shape of the power spectrum depends only upon the magnitude of the filter frequency response, it seems that a random process with almost any desired second moment characteristics can be produced by applying white noise to an appropriate linear filter. In fact, the innovations representation is proof that any regular random process can be generated by a model of exactly this form. The design of filters to represent random processes that have some desired power density spectrum or correlation function is referred to as signal *modeling*. If the procedure is actually used to *produce* the random process (as in some speech applications), then it is known as signal *synthesis*. In either case the same set of techniques apply.

9.1.1 Types of Models

We restrict attention to the case where $H(z)$ is a rational polynomial and look at the three basic types of linear models depicted in Fig. 9.1. It is conventional in the literature on modeling to assume that the white noise sequence has unit variance ($\sigma_w^2 = 1$) and that the filter provides the appropriate gain. This is different from the convention used in spectral factorization, where it was assumed that $H(z)$ is a ratio of comonic polynomials. The first type of model, shown in Fig. 9.1(a), is the "all-pole" or autoregressive model encountered in Chapter 8. The filter in the AR model is recursive (IIR) and has only a nontrivial denominator polynomial (i.e., all zeros are at the origin). Of the three, this model is the one most frequently used since it is relatively easy to design (the equations that determine its parameters are linear) and there is a great deal of theory available to support its analysis. Although the AR model has already been studied in Chapter 8, more specific methods for its design are developed later in this chapter.

The second type of model, shown in Fig. 9.1(b), involves an FIR filter, so the corresponding transfer function has only a (nontrivial) numerator polynomial. This model is known as an all-zero or *moving average* (MA) model. Although its relatively simple form could lead you to suspect that this model would be easy to develop, the estimation of parameters is actually much more difficult for the MA than for the AR model.

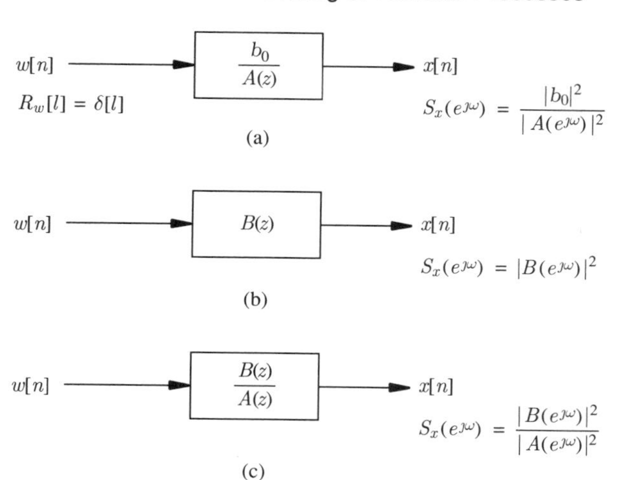

Figure 9.1 Types of linear models. (a) AR. (b) MA. (c) ARMA.

The last type of model, which is depicted in Fig. 9.1(c), involves a filter with nontrivial numerator *and* denominator polynomials and is therefore called a pole-zero or *autoregressive moving average* (ARMA) model. This model is the most flexible and frequently requires considerably fewer parameters to match a desired power spectral density function than an equivalent AR or MA model. This form of linear model is used in a number of practical applications, but its design and analysis is more difficult than that of either the AR or the MA model.

The terms "autoregressive" and "moving average" were originally coined by statisticians in the analysis of time series (see [1]) and were carried over to the emerging field of statistical digital signal processing during the late 1960s and 70s. Recall that "autoregressive" refers to the fact that the difference equation resembles a statistical regression. In fact, the output sequence $x[n]$ can be thought of as a regression upon itself (see Chapter 8). The name "moving average" also derives from the difference equation for the model which is of the form

$$x[n] = b_0 w[n] + b_1 w[n-1] + \cdots + b_Q w[n-Q]$$

The output here is seen to be a weighted average of the input that moves along the process with n. The difference equation for the ARMA model of course has aspects of both the AR and the MA models.

9.1.2 Linear Models and the Yule–Walker Equations

Let us now consider the equations that govern the parameters of a general ARMA model. Begin with a model characterized by the difference equation

$$x[n] + a_1 x[n-1] + \cdots + a_P x[n-P] \\ = b_0 w[n] + b_1 w[n-1] + \cdots + b_Q w[n-Q] \tag{9.1}$$

where without any loss of generality the coefficient of $x[n]$ is set equal to 1 and $w[n]$ is assumed to be a white noise process with unit variance. The $\{a_k\}$ are referred to as the "AR parameters" while the $\{b_k\}$ are called the "MA parameters." For later convenience it is worthwhile to define the corresponding vectors

$$\mathbf{a} = \begin{bmatrix} 1 & a_1 & \cdots & a_P \end{bmatrix}^T \tag{9.2}$$

and

$$\mathbf{b} = \begin{bmatrix} b_0 & b_1 & \cdots & b_Q \end{bmatrix}^T \tag{9.3}$$

Now recall from Chapter 5 that the correlation and cross-correlation functions satisfy the same difference equation as the random process, namely

$$R_x[l] + a_1 R_x[l-1] + \cdots + a_P R_x[l-P] = b_0 R_{wx}[l] + b_1 R_{wx}[l-1] + \cdots + b_Q R_{wx}[l-Q] \tag{9.4}$$

Further recall that if $h[n]$ represents the impulse response of the system (9.1), then the cross-correlation function is given by

$$R_{xw}[l] = h[l] * R_w[l] = h[l] * \delta[l] = h[l]$$

Therefore, R_{wx} has the form

$$R_{wx}[l] = h^*[-l] \tag{9.5}$$

and substituting this in (9.4) yields

$$\boxed{\begin{aligned} &R_x[l] + a_1 R_x[l-1] + \cdots + a_P R_x[l-P] \\ &\qquad = b_0 h^*[-l] + b_1 h^*[1-l] + \cdots + b_Q h^*[Q-l] \end{aligned}} \tag{9.6}$$

Evaluating this equation for a set of consecutive values of l leads to a set of simultaneous equations that can be solved for the filter coefficients. These are the Yule–Walker equations for the ARMA model (also called modified Yule–Walker equations). Notice that the equations are nonlinear, since the system impulse response h depends in a complicated way upon the coefficients a_k and b_k, which also appear explicitly in the equation. Note also that (9.6) is very general; it holds for any value of l and applies whether or not the system is causal. Very few assumptions have been made at this point. In the next section it is shown that when $h[n]$ corresponds to a causal system, (9.6) leads to a set of equations whose solution provides both the a_k (AR) and the b_k (MA) parameters of the model.

9.1.3 Solution of Yule–Walker Equations

Let us now consider some specific procedures for determining the model parameters from (9.6). To make the solution easier, first rewrite the equations by defining the deterministic sequence

$$b[l] \stackrel{\text{def}}{=} \begin{cases} b_l; & 0 \leq l \leq Q \\ 0; & \text{otherwise} \end{cases} \tag{9.7}$$

Then the right-hand side of (9.6) can be expressed as

$$c[l] = \sum_{k=0}^{Q} b_k h^*[k-l] = \sum_{k=-\infty}^{\infty} b[k]h^*[k-l]$$

which is the convolution of the sequence $b[l]$ with the sequence $h^*[-l]$. With this result (9.6) can be rewritten as

$$\boxed{R_x[l] + a_1 R_x[l-1] + \cdots + a_P R_x[l-P] = c[l]} \tag{9.8}$$

where

$$\boxed{c[l] = b[l] * h^*[-l]} \tag{9.9}$$

Let us now constrain the impulse response $h[n]$ to be causal. Then the convolution can be expressed explicitly [see Fig. 9.2(a)] as

$$c[l] = \sum_{k=l}^{Q} b[k]h^*[k-l]; \qquad l \le Q \tag{9.10}$$

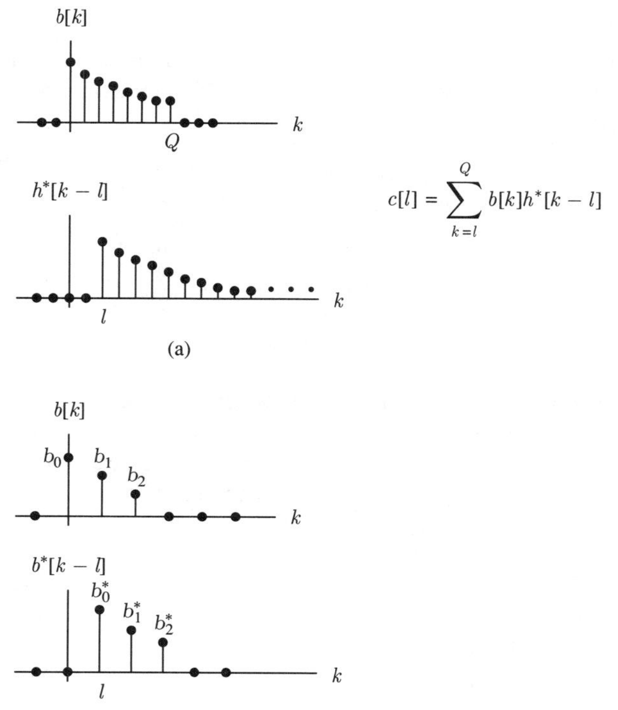

Figure 9.2 Terms under the sum involved in computing the convolution $c[l] = b[l] * h^*[-l]$. (a) General case. (b) FIR case with $Q = 2$.

and

$$c[l] = 0; \qquad l > Q \tag{9.11}$$

Further, since h is causal, $h[0]$ can be obtained from the initial value theorem as

$$h[0] = \lim_{z \to \infty} H(z) = \lim_{z \to \infty} \frac{b_0 + b_1 z^{-1} + \cdots + b_Q z^{-Q}}{1 + a_1 z^{-1} + \cdots + a_P z^{-P}} = b_0$$

Therefore, it follows from (9.10) and (9.7) that

$$c[Q] = b[Q]h^*[0] = b_Q b_0^* \tag{9.12}$$

Finally, if (9.8) is evaluated for $l = 0, 1, 2, \ldots, Q + P$ and written as a matrix equation, the result is

$$\begin{bmatrix} R_x[0] & R_x[-1] & \cdots & R_x[-P] \\ R_x[1] & R_x[0] & \cdots & R_x[1-P] \\ \vdots & \vdots & & \vdots \\ R_x[Q] & R_x[Q-1] & \cdots & R_x[Q-P] \\ R_x[Q+1] & R_x[Q] & \cdots & R_x[Q-P+1] \\ \vdots & \vdots & \ddots & \vdots \\ R_x[Q+P] & \cdots & \cdots & R_x[Q] \end{bmatrix} \begin{bmatrix} 1 \\ a_1 \\ \vdots \\ a_P \end{bmatrix} = \begin{bmatrix} c[0] \\ c[1] \\ \vdots \\ c[Q] \\ 0 \\ \vdots \\ 0 \end{bmatrix} \tag{9.13}$$

These are the Yule–Walker equations. Let us now consider the explicit solution of these equations. We first consider two special cases, the AR and the MA models, and then consider the general ARMA model.

The AR Model ($Q = 0$). For the special case where the order Q of the *numerator* polynomial is zero, then from (9.12) $c[0] = |b_0|^2$, so the Yule–Walker equations become

$$\begin{bmatrix} R_x[0] & R_x[-1] & \cdots & R_x[-P] \\ R_x[1] & R_x[0] & \cdots & R_x[-P+1] \\ \vdots & \vdots & & \vdots \\ R_x[P] & R_x[P-1] & \cdots & R_x[0] \end{bmatrix} \begin{bmatrix} 1 \\ a_1 \\ \vdots \\ a_P \end{bmatrix} = \begin{bmatrix} |b_0|^2 \\ 0 \\ \vdots \\ 0 \end{bmatrix} \tag{9.14}$$

As noted in Chapter 8, these are just the conjugated Normal equations. They are linear in the parameters a_i and several different efficient methods have already been presented for solving them. Since the white noise variance was assumed to be 1, the squared gain $|b_0|^2$ appears in its place. This unknown term is also found by any of the methods in Chapter 8.

The MA Model ($P = 0$). For this case the order of the *denominator* polynomial is zero and the Yule–Walker equations take the form

$$\begin{bmatrix} R_x[0] \\ R_x[1] \\ \vdots \\ R_x[Q] \end{bmatrix} = \begin{bmatrix} c[0] \\ c[1] \\ \vdots \\ c[Q] \end{bmatrix} \tag{9.15}$$

Since the system is an FIR filter, $h[n]$ is equal to $b[n]$. Therefore, it follows from (9.9) that

$$c[l] = b[l] * b^*[-l] \tag{9.16}$$

Since solving for the $\{b_i\}$ essentially involves undoing a convolution, the overall equations are *nonlinear*. To illustrate this, consider the specific form of the equations for $Q = 2$. The terms in the convolution are depicted in Fig. 9.2 (b). The three individual equations represented by (9.15) and (9.16) in this case are

$$\begin{aligned} |b_0|^2 + |b_1|^2 + |b_2|^2 &= R_x[0] \\ b_1 b_0^* + b_2 b_1^* &= R_x[1] \\ b_2 b_0^* &= R_x[2] \end{aligned} \tag{9.17}$$

The easiest way to solve these equations is to divide them all by the term $b_0 b_0^*$. Then if the values of R_x are given, these equations can be solved from the "bottom up" for the normalized variables b_k/b_0 and rescaled at the end. This method works in general because starting at the bottom, each equation brings in at most one new variable. Since the equations are nonlinear, however, the solution of even a single equation may not be simple and in general multiple solutions exist.

An alternative way to solve for the unknowns is by spectral factorization. When $P = 0$, (9.8) states that

$$R_x[l] = c[l]$$

[Note that because of (9.11), $R_x[l]$ is forced to be zero for $l > Q$.] Combining this with (9.16) and taking the z-transform yields

$$S_x(z) = B(z)B^*(1/z^*) \tag{9.18}$$

$S_x(z)$ is therefore a known finite-order polynomial whose coefficients are determined by $R_x[l]$ for $-Q \leq l \leq Q$. Since $S_x(z)$ has roots in conjugate reciprocal locations inside and outside the unit circle, there are usually several possible factorizations of (9.18), as discussed in Chapter 5. This also accounts for the multiple solutions of (9.16). Since we normally want the model to coincide with the innovations representation of the random process whenever possible, the factorization is performed so that $B(z)$ has no zeros outside the unit circle. Note that zeros occurring *on* the unit circle are legitimate as long as they occur with even multiplicities. The example below illustrates the procedure.

Example 9.1

The real random process with correlation function shown in Figure EX9.1 is to be represented by an MA model. If the order is taken to be $Q = 2$, the Yule–Walker equations have the form (9.17), namely

$$\begin{aligned} b_0^2 + b_1^2 + b_2^2 &= 3 \\ b_1 b_0 + b_2 b_1 &= 2 \\ b_2 b_0 &= 1 \end{aligned}$$

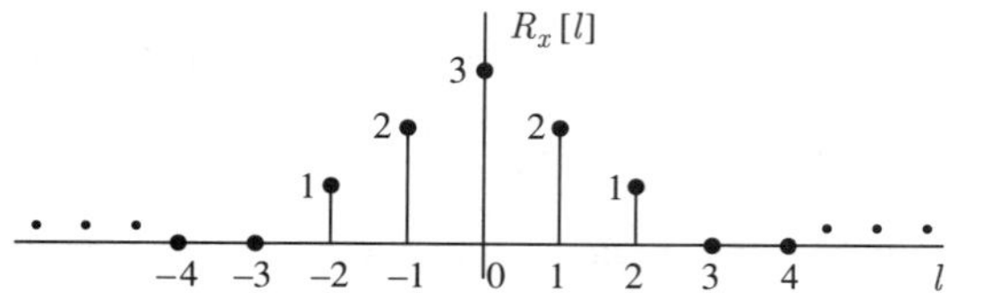

Figure EX9.1

These equations will not be solved here explicitly; however, it is easy to verify that $b_0 = b_1 = b_2 = 1$ is a solution.

The alternative method to find the model parameters is by spectral factorization. The complex spectral density function is easily obtained as

$$S_x(z) = \sum_{l=-Q}^{Q} R_x[l]z^{-l} = z^2 + 2z + 3 + 2z^{-1} + z^{-2}$$

Notice that this can be factored in the form of (9.18) as

$$S_x(z) = \underbrace{(1 + z^{-1} + z^{-2})}_{B(z)}(1 + z + z^2)$$

which again shows that $b_0 = b_1 = b_2 = 1$. In general, the factorization of $S_x(z)$ is not obvious and the polynomial needs to be factored by finding its roots. This particular example is somewhat tricky because there are *double roots* at each location on the unit circle. Specifically, the factors are

$$S_x(z) = (z^{-1} - e^{\jmath 2\pi/3})(z^{-1} - e^{-\jmath 2\pi/3})(z - e^{\jmath 2\pi/3})(z - e^{-\jmath 2\pi/3})$$

The product of the first two factors yields $B(z)$.

The selection of the order for the model can be something of a problem in general. The order $Q = 2$ was just what was needed in this example, but a larger order would lead to the same solution. Since the correlation function goes to zero for values of lag $|l| > 2$, $S_x(z)$ never contains any terms of order higher than $z^{\pm 2}$. This implies that the coefficients b_k for $k > 2$ in $B(z)$ are all zero. An order smaller than $Q = 2$ leads to problems here, however. For $Q = 1$ the polynomial $S_x(z)$ is

$$S_x(z) = 2z + 3 + 2z^{-1}$$

This turns out to have two zeros at conjugate locations on the unit circle and therefore is not a proper complex spectral density function. (The two zeros would need to occur at the *same* location on the unit circle if this second-order polynomial were to qualify as a complex spectral density function.) Thus there is *no solution* for $Q = 1$. This is also evident from another point of view. By choosing $Q = 1$ we are truncating the correlation function. It can be determined (e.g., by examining eigenvalues) that the resulting "correlation sequence"

$$\ldots 0\ 0\ 2\ 3\ 2\ 0\ 0\ \ldots$$

is not positive semidefinite. This rather simple example illustrates that MA modeling can be hazardous. The MA model, unlike the AR model, does *not* extend the correlation function. In fact, it *truncates* it explicitly to zero. The discussion here shows that this truncation can destroy the required positive semidefinite property and therefore fail to generate an appropriate model.

The ARMA Model. In the general case where both Q and P are nonzero, elements of both of the former procedures need to be applied to find the model parameters. Begin by writing (9.13) as

$$\boxed{\begin{bmatrix} \mathbf{R}_B \\ \mathbf{R}_A \end{bmatrix} \mathbf{a} = \begin{bmatrix} \mathbf{c} \\ \mathbf{0} \end{bmatrix}} \tag{9.19}$$

where $\mathbf{R}_B$ is the $(Q+1) \times (P+1)$ matrix

$$\mathbf{R}_B = \begin{bmatrix} R_x[0] & R_x[-1] & \cdots & R_x[-P] \\ R_x[1] & R_x[0] & \cdots & R_x[1-P] \\ \vdots & \vdots & & \vdots \\ R_x[Q] & R_x[Q-1] & \cdots & R_x[Q-P] \end{bmatrix} \tag{9.20}$$

$\mathbf{R}_A$ is the $P \times (P+1)$ matrix

$$\mathbf{R}_A = \begin{bmatrix} R_x[Q+1] & R_x[Q] & \cdots & R_x[Q-P+1] \\ \vdots & \vdots & \ddots & \vdots \\ R_x[Q+P] & \cdots & \cdots & R_x[Q] \end{bmatrix} \tag{9.21}$$

$\mathbf{c}$ is the vector

$$\mathbf{c} = \begin{bmatrix} c[0] \\ c[1] \\ \vdots \\ c[Q] \end{bmatrix} \tag{9.22}$$

and $\mathbf{a}$ is defined by (9.2). Now the equations represented by the bottom partitions of (9.19), namely

$$\boxed{\mathbf{R}_A \mathbf{a} = \mathbf{0}} \tag{9.23}$$

are the higher-order Yule–Walker equations. These are *linear* and can be solved for $\mathbf{a}$. Once $\mathbf{a}$ is known, the equations represented by the top partitions

$$\boxed{\mathbf{c} = \mathbf{R}_B \mathbf{a}} \tag{9.24}$$

determine $c[k]$ for $k = 0, 1, \ldots, Q$. Since the coefficients b_k are related to the $c[k]$ through (9.9), however, this part of the problem is nonlinear.

Fortunately, because it is possible to separately find the $\{a_k\}$ from (9.23), it is possible to find the $\{b_k\}$ implicit in (9.24) through spectral factorization. To do this, first define the finite-length sequence $a[l]$ as

$$a[l] \stackrel{\text{def}}{=} \begin{cases} a_l; & 0 \leq l \leq P \quad (a_0 = 1) \\ 0; & \text{otherwise} \end{cases} \tag{9.25}$$

The system (9.1) then has a transfer function

$$H(z) = \frac{B(z)}{A(z)} \tag{9.26}$$

where $A(z)$ and $B(z)$ are the z-transforms of $a[l]$ and $b[l]$, respectively. Equation 9.8 can now be written using (9.9) and (9.25) as

$$a[l] * R_x[l] = c[l] = b[l] * h^*[-l] \tag{9.27}$$

Taking the z-transform of this produces

$$A(z) \cdot S_x(z) = C(z) = B(z) \cdot \frac{B^*(1/z^*)}{A^*(1/z^*)}$$

where $C(z)$ is the z-transform of $c[l]$. This last equation can be written as

$$S_{x'}(z) = B(z)B^*(1/z^*) \tag{9.28}$$

where

$$S_{x'}(z) \stackrel{\text{def}}{=} A(z)S_x(z)A^*(1/z^*) = C(z)A^*(1/z^*) \tag{9.29}$$

Once $S_{x'}(z)$ is known, $B(z)$ can be found exactly as in the MA case [compare (9.28) and (9.18)].

The concepts involved in the modeling, however, are depicted in Fig. 9.3. The original process $x[n]$ is the input to the system $A(z)$ shown in Fig. 9.3(a), whose parameters are determined from solving the higher-order Yule–Walker equations (9.23). The output $y[n]$ is called the *residual process*; it is the result of applying a Pth-order prediction error filter to the data. An MA model for the residual process is then generated by spectral factorization. This is shown in Fig. 9.3(b). The model for $x[n]$ is a cascade of the all-zero system $B(z)$ and the all-pole system $1/A(z)$. This is shown in Fig. 9.3(c).

Let us now consider the details of forming the complex spectral density function $S_{x'}(z)$. Since $c[l]$ is zero for $l > Q$ [recall (9.11)], the right-hand side of (9.29) has the form

$$C(z)A^*(1/z^*) = (\cdots + c[0] + c[1]z^{-1} + \cdots + c[Q]z^{-Q})\underbrace{(1 + a_1^* z + \cdots + a_P^* z^P)}_{A^*(1/z^*)}$$

where the terms in $C(z)$ before $c[0]$ correspond to *positive* powers of z and are not known. Observe, however, that all the terms corresponding to negative powers of z in the final

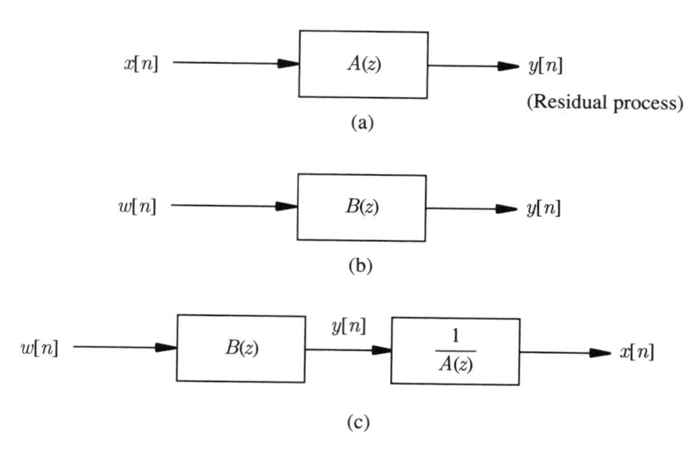

Figure 9.3 Concepts involved in ARMA modeling. (a) Relation between the original and the residual processes. (b) MA model for the residual process. (c) Complete model for the original process.

product can be computed from the known values $c[0], \ldots, c[Q]$ and $a_1, \ldots, a_P$. [Any of the unknown terms when multiplied by the $A^*(1/z^*)$ polynomial can only produce positive powers of z.] Thus the terms in $S_{x'}(z)$ from z^0 to the highest negative power z^{-Q} can be computed from the equation above. The remaining terms involving positive powers of z can then be determined from the symmetry property $S_{x'}(z) = S^*_{x'}(1/z^*)$. The entire procedure for finding the ARMA model parameters is demonstated in the example below.

Example 9.2

A first-order ARMA model of the form

$$H(z) = \frac{b_0 + b_1 z^{-1}}{1 + a_1 z^{-1}}$$

is to be fit to the data with correlation function given in Example 9.1. The Yule–Walker equations (9.13) have the form

$$\begin{bmatrix} 3 & 2 \\ 2 & 3 \\ 1 & 2 \end{bmatrix} \begin{bmatrix} 1 \\ a_1 \end{bmatrix} = \begin{bmatrix} c[0] \\ c[1] \\ 0 \end{bmatrix}$$

The lower partition for the AR parameters consists of the single equation

$$1 + 2a_1 = 0$$

which has the solution $a_1 = -\frac{1}{2}$. Substituting this in the upper partition yields

$$\begin{bmatrix} c[0] \\ c[1] \end{bmatrix} = \begin{bmatrix} 3 & 2 \\ 2 & 3 \end{bmatrix} \begin{bmatrix} 1 \\ -\frac{1}{2} \end{bmatrix} = \begin{bmatrix} 2 \\ \frac{1}{2} \end{bmatrix}$$

The negative power terms in $S_{x'}(z)$ are computed from

$$S_{x'}(z) = C(z)A^*(1/z^*) = (\cdots + 2 + \tfrac{1}{2}z^{-1})(1 - \tfrac{1}{2}z)$$
$$= \cdots + \tfrac{7}{4} + \tfrac{1}{2}z^{-1}$$

Then from the symmetry, the entire function $S_{x'}(z)$ must have the form

$$S_{x'}(z) = \tfrac{1}{2}z + \tfrac{7}{4} + \tfrac{1}{2}z^{-1}$$

Factoring this produces

$$S_{x'}(z) = (0.396 + 1.262z^{-1})(0.396 + 1.262z)$$

which yields $b_0 = 0.396$ and $b_1 = 1.262$. The desired transfer function for the ARMA model is therefore

$$H(z) = \frac{0.396 + 1.262z^{-1}}{1 - 0.5z^{-1}}$$

9.2 ESTIMATION OF MODEL PARAMETERS FROM DATA

Although the approach taken in Section 9.1 to find a linear model for a random process is a valid one, there are two implicit assumptions that limit its use for practical problems. These assumptions are:

1. That the order of the model is known.
2. That the correlation function is known.

If these two conditions are met, the procedure described in Section 9.1 will find the model parameters exactly. This statement can be made with no qualifications. Unfortunately, in most practical problems *neither* of these conditions is met.

From a *theoretical* point of view the first assumption, knowledge of the model order, is less of a problem than the second assumption. If an order larger than the true order is chosen, the excess parameters will theoretically turn out to be zero. In practice choosing the model order is not that simple, for a number of reasons that are discussed later. The second assumption leads to *both* theoretical and practical problems. If the correlation function is not known, it must be estimated from data. This brings up a number of questions, such as "If we estimate the correlation function, how good is the resulting estimate for the model parameters—in a statistical sense?", "Why estimate the correlation function at all when the model parameters are what needs to be estimated?", and "What is the *best* procedure?" Some of these questions are addressed in this section.

9.2.1 Back to Basics: Maximum Likelihood

Let us assume that the observed data for a random process corresponds to one of the models shown in Fig. 9.1. Let us further assume that somehow, the model order is known, and the form of the distribution of the white noise is also known. Then $x[n]$ is a random process whose probability density function can be computed (at least in principle) if the model parameters are given. If the model parameters are not known, then the model parameters must be estimated from a given data sequence x$[n]$. That is the problem addressed in this chapter.

Because of its desirable properties, we are led to consider the maximum likelihood estimate. Note that in order to develop a maximum likelihood estimate for any set of parameters, it is necessary to know the form of the density function. In the estimation of parameters for random processes it is almost always assumed that the random process is Gaussian. That assumption is also made here. Note, however, that *if the Gaussian assumption is not met, then this estimate is not a maximum likelihood estimate and the minimum variance property does not hold.* Frequently, people tend to forget or ignore this fact and refer to the procedure developed in the Gaussian case as *the* maximum likelihood estimate, without acknowledging that a Gaussian random process is assumed. True maximum likelihood estimates can also be developed if the white noise density is not Gaussian, but the procedure is more complicated and is rarely applied. The discussion here follows the treatment in [1, Chapter VII].

Conditional Log Likelihood Function. To develop the maximum likelihood estimate, consider the general ARMA process described by the difference equation (9.1), where $w[n]$ is a Gaussian white noise process. This equation can be written in the inverted form

$$w[n] = \\ (1/\mathrm{b}_0)x[n]+\mathrm{a}_1'x[n-1]+\cdots+\mathrm{a}_P'x[n-P]-b_1'w[n-1]-\cdots-b_Q'w[n-Q] \tag{9.30}$$

where the coefficients a_k' and b_k' represent the normalized quantities

$$\mathrm{a}_k' = \frac{\mathrm{a}_k}{\mathrm{b}_0} \quad k = 1, 2, \ldots, P$$

$$b_k' = \frac{\mathrm{b}_k}{\mathrm{b}_0} \quad k = 1, 2, \ldots, Q \tag{9.31}$$

This permits samples of the white noise to be computed from samples of the observed data. To begin the development, assume that a set of initial values

$$\mathbf{w}_- = \left[\ \mathrm{w}[-Q] \ \cdots \ \mathrm{w}[-2] \ \ \mathrm{w}[-1]\ \right]^T$$

and

$$\mathbf{x}_- = \left[\ \mathrm{x}[-P] \ \cdots \ \mathrm{x}[-2] \ \ \mathrm{x}[-1]\ \right]^T$$

are given. Now just consider evaluating (9.30) for $n = 0, 1, \ldots, N_s - 1$. For $n = 0$, $w[0]$ is expressed in terms of $x[0]$ and the elements of $\mathbf{w}_-$ and $\mathbf{x}_-$. To illustrate, let $P = Q = 2$; then the equation is

$$w[0] = (1/\mathrm{b}_0)x[0] + \underbrace{\mathrm{a}_1'x[-1] + \mathrm{a}_2'x[-2] - b_1'w[-1] - b_2'w[-2]}_{\text{initial conditions}}$$

For $n = 1$, $w[1]$ is expressed in terms of $x[1]$, $x[0]$, $w[0]$ and the elements of $\mathbf{w}_-$ and $\mathbf{x}_-$. To continue the example, this is

$$w[1] = (1/b_0)x[1] + a_1'x[0] + \underbrace{a_2'x[-1]} - b_1'w[0] - \underbrace{b_2'w[-1]}$$

where the brackets again denote the initial conditions. Since $w[0]$ is a function of just $x[0]$ and the initial conditions, however, the first equation can be substituted into the second to express $w[1]$ in terms of just $x[1]$, $x[0]$, and the initial conditions. By continuing in this way, any sample $w[n]$ can be expressed in terms of previous values of x and the initial conditions. Therefore, let $\boldsymbol{x}$ and $\boldsymbol{w}$ represent the set of samples $x[0], x[1], \ldots, x[N_s - 1]$ and $w[0], w[1], \ldots, w[N_s - 1]$, respectively. Then $\boldsymbol{w}$ can be written as

$$\boldsymbol{w} = \mathbf{L}\boldsymbol{x} + \mathbf{M_x}\mathbf{x}_- + \mathbf{M_w}\mathbf{w}_- \tag{9.32}$$

where $\mathbf{M_x}$ and $\mathbf{M_w}$ are some suitable rectangular matrices and the matrix $\mathbf{L}$ is *lower triangular* with terms $1/b_0$ on the main diagonal. Because of the lower triangular form of $\mathbf{L}$, the Jacobian $|\mathbf{L}\mathbf{L}^{*T}|$ of the transformation is $1/|b_0|^{2N_s}$. Thus, *for any fixed values of the initial conditions,* the complex density function for $\boldsymbol{x}$ can be expressed in terms of the density for $\boldsymbol{w}$ as

$$f_{\boldsymbol{x};\mathbf{a},\mathbf{b},\mathbf{x}_-,\mathbf{w}_-}(\mathbf{x}; \mathbf{a}, \mathbf{b}, \mathbf{x}_-, \mathbf{w}_-) = |b_0|^{-2N_s} f_{\boldsymbol{w}}(\mathbf{w}) \tag{9.33}$$

where $\boldsymbol{w}$ is given by (9.32). This is actually the likelihood function for the vector parameters $\mathbf{a}$ and $\mathbf{b}$ given the initial conditions. Since the $w[n]$ are independent Gaussian random variables the *conditional log likelihood function* has the form

$$\ln f_{\boldsymbol{x};\mathbf{a},\mathbf{b},\mathbf{x}_-,\mathbf{w}_-}(\boldsymbol{x}; \mathbf{a}, \mathbf{b}, \mathbf{x}_-, \mathbf{w}_-) = -2N_s \ln|b_0| - \sum_{n=0}^{N_s-1} |w[n]|^2 + \text{const} \tag{9.34}$$

where $w[n]$ is computed from (9.30) and where the *complex* form of the Gaussian density was used in the last equation. (For the real case there would be a factor of $\frac{1}{2}$ multiplying all of the terms.) Notice the sum of squares that appears in the expression. This sum of squares form is common to many of the methods that are described in this chapter; for large values of N_s this term tends to dominate.

The conditional log likelihood function is usually a satisfactory approximation to the true log likelihood function when $N_s >> \max(P, Q)$ and appropriate values are used for the initial conditions. One procedure is to set the initial conditions equal to zero (their unconditional expected value). This may cause problems, however, if some of the poles of the system are close to the unit circle. In this case the actual initial value of x could be considerably different from the expected value of zero and the effect of the initial conditions does not quickly decay. A better procedure may be to modify (9.34) so that the sum starts at $n = P$ and set earlier values of $w[n]$ to 0. With this approach only actual *observed* values of x are used in computing the log likelihood function.

True Log Likelihood Function. The true log likelihood function is a little more difficult to derive, and it is not necessary to develop all of the details here. A complete derivation for the real case can be found in Box and Jenkins [1, Appendix A7.4]. Our purpose is to describe just enough of the procedure to convey what the calculation of the maximum likelihood estimate involves. Since the method does not lead to any closed-form solution and is computationally intensive, it is not discussed any further after this brief introduction.

By beginning with the MA case and writing $w[n]$ as in (9.30) (with $P = 0$), it is possible to show that the (unconditional) log likelihood function is given, within a constant, by

$$g_{MA}(\mathbf{b}) - 2N_s \ln |b_0| - \sum_{n=-Q}^{N_s-1} |\overline{w}[n]|^2 \tag{9.35}$$

where g_{MA} is a function depending only on $\mathbf{b}$ that can be explicitly computed in particular cases, and $\overline{w}[n]$ is the expectation of $w[n]$ *conditioned on* $\mathbf{b}$ and $\boldsymbol{x}$. For any reasonably large value of N, this sum of terms tends to dominate the overall expression. It turns out that the values of $\overline{w}[n]$ are given by

$$\overline{w}[n] = -b_1'\overline{w}[n-1] - \cdots - b_Q'\overline{w}[n-Q] + (1/b_0)x[n] \tag{9.36}$$

for $n = 0, 1, \ldots, N_s - 1$, and by a least squares minimization problem for $-Q \leq n < 0$ [1]. Note that for the MA case only a finite number of values of $\overline{w}[n]$ are needed (going back only to $n - Q$). The values of $\overline{w}[n]$ for $n < 0$ can also be found through a technique called back-prediction,[1] where given values of the sequence are used to estimate earlier unspecified values. This technique involves the following steps. For any given values of the model parameters, a *backward* model is set up that has the same parameters. This backward model has the same second moment statistics as the forward model. This model is then used to estimate values of $x[n]$ for $n < 0$. These can in turn be used to compute values of $\overline{w}[n]$ for $n < 0$. Since a true maximum likelihood estimate involves searching over a range of values of the parameters to maximize the likelihood function, this procedure has to be repeated for every possible set of parameter values for which the log likelihood function is to be evaluated. Needless to say, the procedure is very computationally burdensome.

For the case of an AR or ARMA model the terms of the impulse response can be used (at least conceptually) to form a high-order MA process which approaches the given random process in the limit as the order L of the approximation approaches infinity.

$$x[n] = h[0]w[n] + h[1]w[n-1] + h[2]w[n-2] + \cdots + h[L]w[n-L] \tag{9.37}$$

The results for MA processes thus apply and the likelihood function has the form (9.35) with the lower limit on the sum set to $-\infty$. That is, the true likelihood function for the AR or ARMA case depends on values of $\overline{w}[n]$ going back to $-\infty$. In practice, it is only necessary to use an order L such that the impulse response decays approximately to zero. In this case the log likelihood function becomes

$$g(\mathbf{a}, \mathbf{b}) - 2N_s \ln |b_0| - \sum_{n=-L}^{N_s-1} |\overline{w}[n]|^2 \tag{9.38}$$

[1] Box and Jenkins [1] call the technique back-forecasting.

The terms $w[n]$ for $0 \leq n \leq N_s - 1$ are computed from

$$\overline{w}[n] = \frac{1}{b_0} x[n] + a_1 x[n-1] + \cdots + a_P x[n-P] - b_1 \overline{w}[n-1] - \cdots - b_Q \overline{w}[n-Q] \tag{9.39}$$

using a combination of the observed and the back-predicted data.

It is not necessary to describe any further details of the maximum likelihood method here since maximizing the likelihood function (9.39) [or the conditional likelihood function (9.34)] is not simple. For the case of an AR model, certain simplifying assumptions can be made that lead to linear equations for the model parameters. The method is then essentially identical to the "covariance method" described in Section 9.4.1. In other cases the equations are nonlinear and approximate techniques or gradient search techniques have to be used to attempt to find the maximum. In summary, there has been no easy solution to the maximum likelihood problem. Instead, several other formulations of the modeling problem have been proposed whose performance is acknowledged to be poorer than maximum likelihood but which are much easier to apply. These other methods have been found to give reasonably good results in practice and do not depend on knowing the density function of the white noise source.

9.2.2 Summary of Other Approaches

Because of the difficulty involved in seeking a maximum likelihood estimate (or even an approximate maximum likelihood estimate) a number of other approaches to estimating the parameters of general linear models have evolved. Most of the other approaches developed in this chapter begin directly with the data (rather than the correlation function) and are based on methods of least squares where a sum of squared terms such as that appearing in (9.34) and (9.38) is minimized. Since AR modeling is the inverse problem to linear prediction, AR modeling procedures can be developed by considering the least squares form of linear predictive filtering. The procedure is analogous to what was studied in earlier chapters, with the data-oriented least squares criterion replacing the statistical expectation. Other methods seek to model the given data sequence as the impulse response of a linear system and forgo the statistical approach altogether. Nevertheless, the optimization criterion is one of least squares and some of the results correspond directly to those arising from a statistical formulation of the problem. Finally, let us point out that least squares methods and methods based on the estimated correlation function are not mutually exclusive. In particular it will be seen that the modeling method described in Sections 9.1.2 and 9.1.3 can be made more robust and more likely to succeed if an overdetermined set of Yule–Walker equations is used and a least squares solution to the larger set of equations is found.

9.3 PRINCIPLES OF LEAST SQUARES

This section begins the discussion of least squares methods. All of the methods developed in this and the next two sections begin with measured data rather than with known or estimated statistical quantities. In many cases the approach follows along lines parallel to those of the methods developed previously, but the concept of mean square error is replaced by signal domain averages of the squared error terms. The section begins by revisiting the problem of optimal (Wiener) filtering and developing a least squares orthogonality principle that deals directly with the observed data.

9.3.1 Least Squares Optimal Filtering

This section considers the optimal filtering problem from a slightly different point of view than was taken in Chapter 7. As before, the problem is to estimate a desired sequence $d[n]$ from a related sequence $x[n]$ using an optimal linear shift-invariant filter. In Chapter 7 the sequences d and x were regarded as random processes with known (or previously estimated) second moment statistics. Here there is no presumed knowledge of the statistical properties beforehand. Instead, it is assumed that a typical data sequence for *both* $d[n]$ and $x[n]$ has been measured and recorded and that these sequences can be used to design the filter. You may wonder here about what the point is in designing a filter to estimate $d[n]$ if the sequence is *known* to begin with. There are several answers to this question. For example, the sequences x and d may represent the input and output of a linear system, respectively. Building a filter to estimate d from x then amounts to designing a filter that behaves like the linear system. This is the *system identification* problem. Alternatively, the purpose of the filter may really be to estimate an unknown signal from a known signal (e.g., to extract a clean signal from a noisy one). In this case if the recorded sequences are representative in some sense of a larger class of sequences to which the filter will eventually be applied, then the filter designed where both signals are known will hopefully be useful in applications where $d[n]$ is *not* known. In other words, the recorded sequences can be thought of as *training data* used principally to design the filter.

Although it is possible to use the given data to estimate the second moment parameters (correlation and cross-correlation functions) and apply previous results, a somewhat different approach is taken here. To emphasize that the following development focuses on the given data sequences, and not the random processes, we use a roman font for the variables thoughout this discussion. We therefore represent the given sequences as $\mathrm{x}[0], \ldots, \mathrm{x}[N_s-1]$ and $\mathrm{d}[0], \ldots, \mathrm{d}[N_s-1]$ (see Fig. 9.4). If a causal FIR filter of length P is used, then the estimate for the given data sequence is

$$\hat{\mathrm{d}}[n] = \sum_{k=0}^{P-1} h[k]\mathrm{x}[n-k] \tag{9.40}$$

and the error can be defined as

$$\epsilon[n] = \mathrm{d}[n] - \hat{\mathrm{d}}[n] \tag{9.41}$$

The approach here is to design the filter to minimize the sum of squared errors

$$\mathcal{S} \stackrel{\text{def}}{=} \sum_{n=n_I}^{n_F} |\epsilon[n]|^2 \tag{9.42}$$

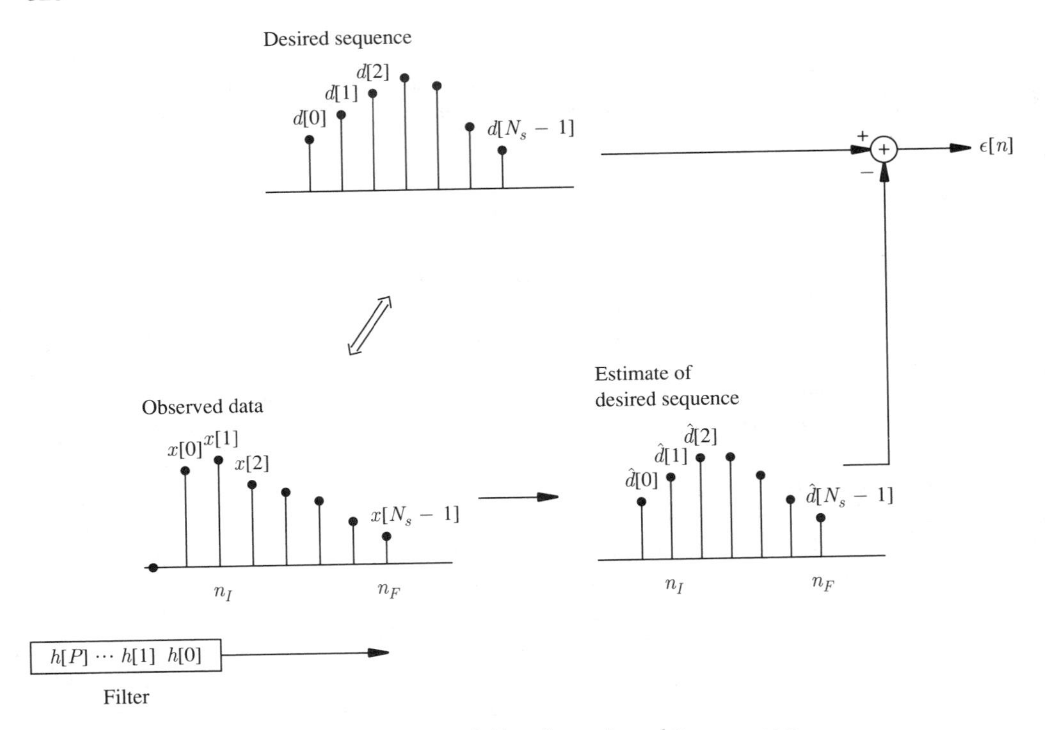

Figure 9.4 Design of filter from given data sequences.

where n_I and n_F are some initial and final values of n that define the interval over which to perform the minimization.

Note that no probabilistic statements have been made in defining this problem. The criterion (9.42) is called a *least squares* criterion and could be regarded as purely deterministic. Because it is desired to minimize the sum of squared-error terms, the data-oriented methods here are called "least squares methods" [2, 3].

In order to solve the least squares problem, imagine running the filter over the given data from the starting point $n = n_I$ to the end point $n = n_F$, as shown in Fig. 9.4. The expression for the estimates $\hat{\mathrm{d}}[n]$ can be written as a matrix equation. If there were $N_s = 50$ data samples and the filter had length $P = 3$ and we took $n_I = P - 1 = 2$ and $n_F = N_s - 1 = 49$, this equation would be

$$\begin{bmatrix} \hat{\mathrm{d}}[2] \\ \hat{\mathrm{d}}[3] \\ \vdots \\ \\ \vdots \\ \hat{\mathrm{d}}[49] \end{bmatrix} = \begin{bmatrix} \mathrm{x}[2] & \mathrm{x}[1] & \mathrm{x}[0] \\ \mathrm{x}[3] & \mathrm{x}[2] & \mathrm{x}[1] \\ \vdots & \vdots & \vdots \\ \\ \vdots & \vdots & \vdots \\ \mathrm{x}[49] & \mathrm{x}[48] & \mathrm{x}[47] \end{bmatrix} \begin{bmatrix} h[0] \\ h[1] \\ h[2] \end{bmatrix}$$

The equation can be written in the more general form

$$\hat{\mathbf{d}} = \mathbf{X}\mathbf{h} \tag{9.43}$$

where

$$\hat{\mathbf{d}} = \begin{bmatrix} \hat{\mathrm{d}}[n_I] \\ \hat{\mathrm{d}}[n_I + 1] \\ \vdots \\ \\ \vdots \\ \hat{\mathrm{d}}[n_F] \end{bmatrix} \tag{9.44}$$

$$\mathbf{X} = \begin{bmatrix} \mathrm{x}[n_I] & \mathrm{x}[n_I - 1] & \cdots & \mathrm{x}[n_I - P + 1] \\ \mathrm{x}[n_I + 1] & \mathrm{x}[n_I] & \cdots & \mathrm{x}[n_I - P + 2] \\ \vdots & \vdots & & \vdots \\ \\ \vdots & \vdots & & \vdots \\ \mathrm{x}[n_F] & \mathrm{x}[n_F - 1] & \cdots & \mathrm{x}[n_F - P + 1] \end{bmatrix} \tag{9.45}$$

and

$$\mathbf{h} = \begin{bmatrix} h[0] \\ h[1] \\ \vdots \\ h[P-1] \end{bmatrix} \tag{9.46}$$

The matrix $\mathbf{X}$ is called the *data matrix* and has dimension $K \times P$ where $K = (n_F - n_I + 1)$. It will be assumed for now that $K > P$. In fact, usually K is *considerably* larger than P, so $\mathbf{X}$ is tall and thin. If $\mathbf{d}$ and $\boldsymbol{\epsilon}$ are the K-dimensional vectors formed from the corresponding samples of $\mathrm{d}[n]$ and $\epsilon[n]$, respectively, then the error vector can be defined as

$$\boldsymbol{\epsilon} = \mathbf{d} - \hat{\mathbf{d}} \tag{9.47}$$

and the problem is to minimize

$$\mathcal{S} = \|\boldsymbol{\epsilon}\|^2 = \boldsymbol{\epsilon}^{*T}\boldsymbol{\epsilon} \tag{9.48}$$

A direct approach to this problem would be as follows. Substitute (9.43) and (9.47) into (9.48) and expand the result to obtain

$$\begin{aligned} \mathcal{S} &= (\mathbf{d} - \mathbf{X}\mathbf{h})^{*T}(\mathbf{d} - \mathbf{X}\mathbf{h}) \qquad (9.49) \\ &= \mathbf{d}^{*T}\mathbf{d} - \mathbf{h}^{*T}\mathbf{X}^{*T}\mathbf{d} - \mathbf{d}^{*T}\mathbf{X}\mathbf{h} + \mathbf{h}^{*T}\mathbf{X}^{*T}\mathbf{X}\mathbf{h} \end{aligned}$$

Then by formal methods of differentiation (using the results of Appendix A), a necessary condition for the minimum can be found to be

$$\boxed{(\mathbf{X}^{*T}\mathbf{X})\mathbf{h} = \mathbf{X}^{*T}\mathbf{d}} \tag{9.50}$$

This is the *least squares Wiener–Hopf equation*. Note that the term $\mathbf{X}^{*T}\mathbf{X}$ has the form of an (unnormalized) estimated correlation matrix[2] and $\mathbf{X}^{*T}\mathbf{d}$ is a similarly computed cross-correlation vector. The correlation matrix is *not always Toeplitz*, however. If $\mathbf{X}$ has independent columns (i.e., if it is of full rank), then $\mathbf{X}^{*T}\mathbf{X}$ is also of full rank and (9.50) has the solution

$$\mathbf{h} = (\mathbf{X}^{*T}\mathbf{X})^{-1}\mathbf{X}^{*T}\mathbf{d} \tag{9.51}$$

If this equation is now written as

$$\boxed{\mathbf{h} = \mathbf{X}^{+}\mathbf{d}} \tag{9.52}$$

where

$$\mathbf{X}^{+} = (\mathbf{X}^{*T}\mathbf{X})^{-1}\mathbf{X}^{*T} \tag{9.53}$$

then (9.51) can also be thought of as representing the "best" solution to the overdetermined set of equations

$$\boxed{\mathbf{X}\mathbf{h} \stackrel{\text{ls}}{=} \mathbf{d}} \tag{9.54}$$

This last equation is interpreted to mean: Find $\mathbf{h}$ to solve

$$\mathbf{X}\mathbf{h} = \mathbf{d} - \epsilon \tag{9.55}$$

where ϵ in this context is called the *equation error* and $\|\epsilon\|^2$ is to be minimized. Equation 9.54 or 9.55 is a statement of the "least squares problem" in mathematics. Both forms are equivalent to (9.50). The matrix $\mathbf{X}^{+}$ that provides the solution to this problem is known as the *Moore–Penrose pseudoinverse* . An alternative expression for the pseudoinverse, which holds even when $\mathbf{X}$ does not satisfy the rank and dimensional conditions assumed here and has much better computational properties, is given in Section 9.3.4.

The sum of the squared errors for the optimal filter can be found by returning to (9.49) and writing

$$\begin{aligned}\mathcal{S} &= (\mathbf{d} - \mathbf{X}\mathbf{h})^{*T}(\mathbf{d} - \mathbf{X}\mathbf{h}) \\ &= \mathbf{d}^{*T}(\mathbf{d} - \mathbf{X}\mathbf{h}) - (\mathbf{X}\mathbf{h})^{*T}(\mathbf{d} - \mathbf{X}\mathbf{h}) \\ &= \mathbf{d}^{*T}\mathbf{d} - \mathbf{d}^{*T}\mathbf{X}\mathbf{h} - \mathbf{h}^{*T}\left(\mathbf{X}^{*T}\mathbf{d} - \mathbf{X}^{*T}\mathbf{X}\mathbf{h}\right)\end{aligned}$$

Since the optimal filter satisfies (9.50), the term in parentheses is zero. This leaves the result

$$\boxed{\mathcal{S} = \mathbf{d}^{*T}\mathbf{d} - \mathbf{d}^{*T}\mathbf{X}\mathbf{h}} \tag{9.56}$$

The following simple example illustrates the procedures in computing the least squares optimal (Wiener) filter.

[2] See Chapter 6.

Example 9.3

The sequence $\{\mathrm{d}[n]\,; 0 \leq n \leq 4\} = \{1, -1,\ 1, -1, 1\}$ is to be estimated from the observation sequence $\{\mathrm{x}[n]\} = \{1, -2,\ 3, -4,\ 5\}$ using an FIR filter of length $P = 2$. Choosing $n_I = 1$ and $n_F = 4$ so that the filter remains within the data leads to the following least squares problem:

$$\underbrace{\begin{bmatrix} -2 & 1 \\ 3 & -2 \\ -4 & 3 \\ 5 & -4 \end{bmatrix}}_{\mathbf{X}} \underbrace{\begin{bmatrix} h[0] \\ h[1] \end{bmatrix}}_{\mathbf{h}} \stackrel{\text{ls}}{=} \underbrace{\begin{bmatrix} -1 \\ 1 \\ -1 \\ 1 \end{bmatrix}}_{\mathbf{d}} \qquad \text{(A)}$$

(You may find it helpful to refer to Fig. 9.4 in examining this equation.) Note that for this choice of n_I and n_F the point d[0] is not used. The pseudoinverse of the data matrix is [from (9.53)]

$$\mathbf{X}^+ = \left(\begin{bmatrix} -2 & 3 & -4 & 5 \\ 1 & -2 & 3 & -4 \end{bmatrix} \begin{bmatrix} -2 & 1 \\ 3 & -2 \\ -4 & 3 \\ 5 & -4 \end{bmatrix} \right)^{-1} \begin{bmatrix} -2 & 3 & -4 & 5 \\ 1 & -2 & 3 & -4 \end{bmatrix}$$
$$= \begin{bmatrix} -1 & 0.5 & 0 & -0.5 \\ -1.3 & 0.6 & 0.1 & -0.8 \end{bmatrix}$$

The filter is then given by

$$\mathbf{h} = \mathbf{X}^+\mathbf{d} = \begin{bmatrix} -1 & 0.5 & 0 & -0.5 \\ -1.3 & 0.6 & 0.1 & -0.8 \end{bmatrix} \begin{bmatrix} -1 \\ 1 \\ -1 \\ 1 \end{bmatrix} = \begin{bmatrix} 1 \\ 1 \end{bmatrix}$$

From (9.56), the minimum sum of squared errors is

$$\mathcal{S} = \begin{bmatrix} -1 & 1 & -1 & 1 \end{bmatrix} \begin{bmatrix} -1 \\ 1 \\ -1 \\ 1 \end{bmatrix} - \begin{bmatrix} -1 & 1 & -1 & 1 \end{bmatrix} \begin{bmatrix} -2 & 1 \\ 3 & -2 \\ -4 & 3 \\ 5 & -4 \end{bmatrix} \begin{bmatrix} 1 \\ 1 \end{bmatrix}$$
$$= 0$$

In this example the prediction is perfect. Checking out the original least squares problem [equation (A)] with $h[0] = h[1] = 1$ shows that indeed every point in the sequence is predicted with zero error.

9.3.2 A Least Squares Orthogonality Principle

In the preceding subsection we indicated how the least squares optimal filtering problem could be solved by formal methods of differentiation. An alternative approach is pursued here that is more advantageous in the long run. Specifically, a least squares version of the orthogonality principle is developed and applied to develop the least squares Wiener–Hopf equation. Consider the following theorem (sometimes called the Projection Theorem):

Theorem 9.1 (Least Squares Orthogonality). Let $\hat{\mathbf{d}} = \mathbf{X}\mathbf{h}$ and $\epsilon = \mathbf{d} - \hat{\mathbf{d}}$. *Then* $\mathbf{h}$ *minimizes the sum of squared errors* $\mathcal{S} = \|\epsilon\|^2$, *if* $\mathbf{h}$ *is chosen such that*

$$\mathbf{X}^{*T}\epsilon = \mathbf{0}$$

Further, the sum of squared errors is given by

$$\mathcal{S} = \mathbf{d}^{*T}\epsilon$$

The proof of this theorem is similar to the proof of the Orthogonality Theorem in Chapter 7. Let $\mathbf{h}$ be any vector of filter coefficients and $\mathbf{h}_\perp$ be the vector that results in orthogonality. Further, let ϵ be the error vector corresponding to $\mathbf{h}$ and $\epsilon_\perp$ be the error vector corresponding to $\mathbf{h}_\perp$. Then it follows that

$$\epsilon = \mathbf{d} - \mathbf{X}\mathbf{h} = (\mathbf{d} - \mathbf{X}\mathbf{h}_\perp) + \mathbf{X}(\mathbf{h}_\perp - \mathbf{h}) = \epsilon_\perp - \mathbf{X}(\mathbf{h}_\perp - \mathbf{h})$$

so that

$$\begin{aligned}\epsilon^{*T}\epsilon &= [\epsilon_\perp - \mathbf{X}(\mathbf{h}_\perp - \mathbf{h})]^{*T}\,[\epsilon_\perp - \mathbf{X}(\mathbf{h}_\perp - \mathbf{h})] \\ &= \epsilon_\perp^{*T}\epsilon_\perp - (\mathbf{h}_\perp - \mathbf{h})^{*T}\mathbf{X}^{*T}\epsilon_\perp - \epsilon_\perp^{*T}\mathbf{X}(\mathbf{h}_\perp - \mathbf{h}) + (\mathbf{h}_\perp - \mathbf{h})^{*T}\mathbf{X}^{*T}\mathbf{X}(\mathbf{h}_\perp - \mathbf{h})\end{aligned}$$

Since $\mathbf{X}^{*T}\epsilon_\perp$ and $\epsilon_\perp^{*T}\mathbf{X}$ are both zero by assumption, this leaves

$$\epsilon^{*T}\epsilon = \epsilon_\perp^{*T}\epsilon_\perp + (\mathbf{h}_\perp - \mathbf{h})^{*T}\mathbf{X}^{*T}\mathbf{X}(\mathbf{h}_\perp - \mathbf{h})$$

which is clearly mimimized when $\mathbf{h} = \mathbf{h}_\perp$. The mimimum sum of squared errors is then

$$\epsilon_\perp^{*T}\epsilon_\perp = (\mathbf{d} - \mathbf{X}\mathbf{h}_\perp)^{*T}\epsilon_\perp = \mathbf{d}^{*T}\epsilon_\perp$$

where the last step again follows because $\mathbf{X}^{*T}\epsilon_\perp = \mathbf{0}$. This proves the theorem.

The results of this theorem can now be applied to solve the optimal least squares filtering problem. The theorem requires that

$$\mathbf{X}^{*T}\epsilon = \mathbf{X}^{*T}(\mathbf{d} - \mathbf{X}\mathbf{h}) = \mathbf{0} \tag{9.57}$$

which leads directly to the Wiener-Hopf equation (9.50) and the solution (9.52). Further, from the theorem, the minimum sum of squared errors is given by

$$\begin{aligned}\mathcal{S} = \mathbf{d}^{*T}\epsilon &= \mathbf{d}^{*T}(\mathbf{d} - \mathbf{X}\mathbf{h}) \\ &= \mathbf{d}^{*T}\mathbf{d} - \mathbf{d}^{*T}\mathbf{X}\mathbf{h}\end{aligned}$$

as before [see (9.56)].

9.3.3 Geometric Interpretation

The least squares orthogonality principle can now be given a geometric interpretation. The interpretation is similar to that of Chapter 7 but it is less abstract and the concepts will be found to be quite familiar. The elements of the vector space are the usual column matrices used to represent data vectors, and the inner product is the usual row-column product of these data vectors.

$$(\mathbf{x}, \mathbf{y}) \stackrel{\text{def}}{=} \mathbf{x}^{*T}\mathbf{y} \tag{9.58}$$

To begin, consider the matrix $\mathbf{X}$ and the vector $\mathbf{d}$ associated with the least squares problem of (9.54). The matrix $\mathbf{X}$ can be written as

$$\mathbf{X} = \begin{bmatrix} | & | & & | \\ \mathbf{x}_0 & \mathbf{x}_1 & \cdots & \mathbf{x}_{P-1} \\ | & | & & | \end{bmatrix} \tag{9.59}$$

where $\mathbf{x}_0, \mathbf{x}_1, \ldots, \mathbf{x}_{P-1}$ are the columns of $\mathbf{X}$ as defined by (9.45). These columns represent a set of K-dimensional vectors ($K = n_F - n_I + 1$). Assume that $P < K$ and that the columns are linearly independent. Then these columns span a P-dimensional subspace of the K-dimensional vector space. The vector $\mathbf{d}$ is generally not a linear combination of the columns of $\mathbf{X}$ and so does not lie in the P-dimensional subspace. A typical situation (for $P = 2$) is depicted in Fig. 9.5(a). It can further be seen from (9.43), (9.46), and (9.59) that

$$\hat{\mathbf{d}} = h[0]\mathbf{x}_0 + h[1]\mathbf{x}_1 + \cdots + h[P-1]\mathbf{x}_{P-1} \tag{9.60}$$

Since $\hat{\mathbf{d}}$ is a linear combination of $\mathbf{x}_0, \ldots, \mathbf{x}_{P-1}$ it lies in the subspace defined by these vectors [see Fig. 9.5(b)]. The orthogonality condition $\mathbf{X}^{*T}\boldsymbol{\epsilon} = \mathbf{0}$ with $\mathbf{X}$ in the form (9.59) means that $\boldsymbol{\epsilon}$ is orthogonal to each of the vectors $\mathbf{x}_0, \ldots, \mathbf{x}_{P-1}$ and so is orthogonal to the subspace defined by them. Clearly, this condition minimizes the squared magnitude of $\boldsymbol{\epsilon}$. Thus the optimal estimate $\hat{\mathbf{d}}$ is the orthogonal projection of $\mathbf{d}$ onto the subspace spanned by the $\mathbf{x}_i$.

It is possible to express this projection idea formally and algebraically as follows. Substituting the solution (9.51) into (9.43) yields

$$\hat{\mathbf{d}} = \mathbf{X}\mathbf{h} = \mathbf{X}(\mathbf{X}^{*T}\mathbf{X})^{-1}\mathbf{X}^{*T}\mathbf{d}$$

which can be written as

$$\boxed{\hat{\mathbf{d}} = \boldsymbol{P}_{\mathbf{X}}\mathbf{d}} \tag{9.61}$$

where

$$\boldsymbol{P}_{\mathbf{X}} = \mathbf{X}(\mathbf{X}^{*T}\mathbf{X})^{-1}\mathbf{X}^{*T} \tag{9.62}$$

The matrix $\boldsymbol{P}_{\mathbf{X}}$ is called an orthogonal projection operator or *projection matrix* because it formally projects *any* vector into the subspace formed by the columns of $\mathbf{X}$. The ideas of projection and the projection matrix can be generalized when the matrix does not satisfy the condition $K > P$ and/or the columns of $\mathbf{X}$ are not linearly independent (see Section 9.3.4).

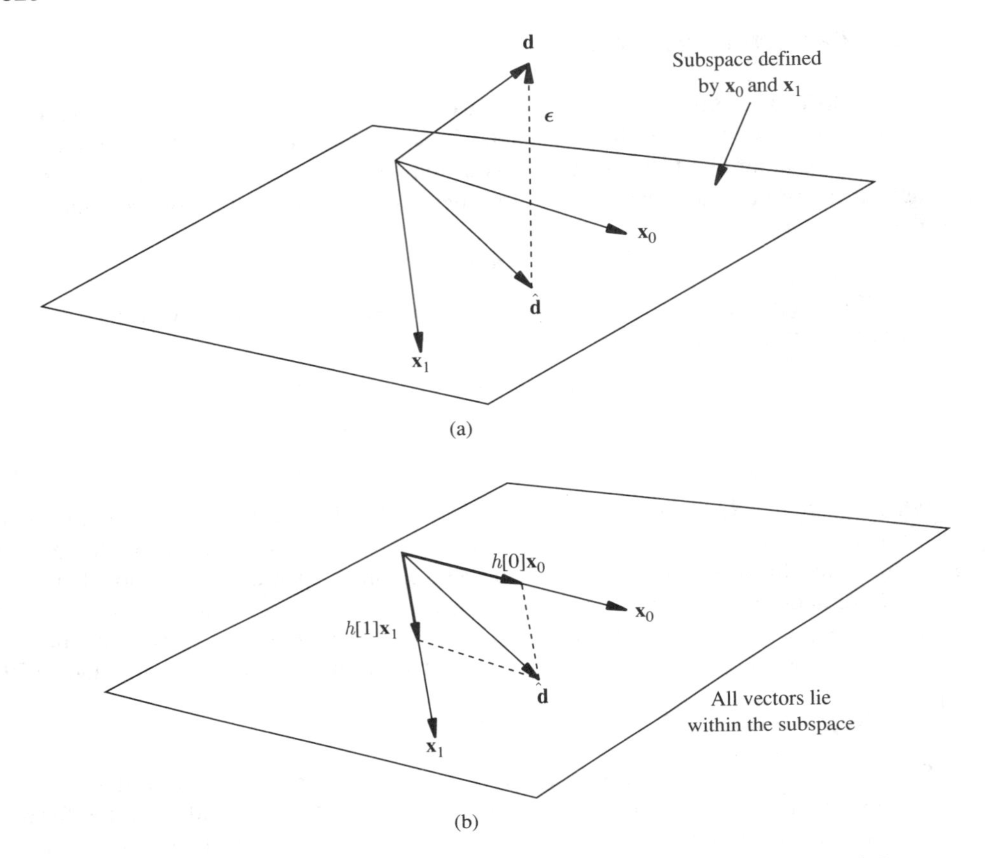

Figure 9.5 Geometric representation of data vectors. (a) Projection of $\mathbf{d}$ into subspace spanned by columns of $\mathbf{X}$. (b) Representation of $\hat{\mathbf{d}}$ as linear combination of column vectors.

A projection matrix has the following two basic properties:

$$\boldsymbol{P}_{\mathbf{X}}\boldsymbol{P}_{\mathbf{X}} = \boldsymbol{P}_{\mathbf{X}} \tag{9.63}$$

which states that it is *idempotent*, and

$$\boldsymbol{P}_{\mathbf{X}}^{*T} = \boldsymbol{P}_{\mathbf{X}} \tag{9.64}$$

which implies that it is an *orthogonal* projection.[3] These properties can easily be verified from (9.62). A projection matrix is obviously in one-to-one correspondence with the subspace that it represents; the subspace can be referred to via the projection matrix, and

[3]It is also possible to form the image of a vector in a subspace by projecting it in a direction that is not necessarily orthogonal to the subspace. The projection matrix corresponding to this type of projection does not satisfy (9.64). Orthogonal projections are the most common and are the only type that are of concern here.

vice versa. Projection matrices are generally (in fact, almost by definition) *not* of full rank. When a projection matrix *is* of full rank it represents the entire vector space and so does not change a vector in any way whatsoever.

The projection matrix associated with the complementary (orthogonal) subspace is

$$\boldsymbol{P}_{\mathbf{X}}^{\perp} = \mathbf{I} - \boldsymbol{P}_{\mathbf{X}} \tag{9.65}$$

and it clearly satisfies[4]

$$\boldsymbol{P}_{\mathbf{X}}\boldsymbol{P}_{\mathbf{X}}^{\perp} = \boldsymbol{P}_{\mathbf{X}}^{\perp}\boldsymbol{P}_{\mathbf{X}} = \mathbf{0} \tag{9.66}$$

The error vector ϵ can thus be written using the projection matrix for the orthogonal subspace as

$$\epsilon = \boldsymbol{P}_{\mathbf{X}}^{\perp}\mathbf{d} = \left(\mathbf{I} - \mathbf{X}(\mathbf{X}^{*T}\mathbf{X})^{-1}\mathbf{X}^{*T}\right)\mathbf{d} \tag{9.67}$$

The projection matrix is a powerful tool in the development and understanding of computationally fast adaptive least squares algorithms [4, 5] and related least squares methods. Additional uses for it are found later in this chapter and in some of the topics in Chapter 10. Let us end this subsection with a simple example of its computation.

Example 9.4

The projection matrix corresponding to the data matrix in Example 9.3 is

$$\begin{aligned}\boldsymbol{P}_{\mathbf{X}} &= \mathbf{X}(\mathbf{X}^{*T}\mathbf{X})^{-1}\mathbf{X}^{*T}\\ &= \begin{bmatrix} -2 & 1 \\ 3 & -2 \\ -4 & 3 \\ 5 & -4 \end{bmatrix} \left(\begin{bmatrix} -2 & 3 & -4 & 5 \\ 1 & -2 & 3 & -4 \end{bmatrix} \begin{bmatrix} -2 & 1 \\ 3 & -2 \\ -4 & 3 \\ 5 & -4 \end{bmatrix} \right)^{-1} \begin{bmatrix} -2 & 3 & -4 & 5 \\ 1 & -2 & 3 & -4 \end{bmatrix}\\ &= \begin{bmatrix} 0.7 & -0.4 & 0.1 & 0.2 \\ -0.4 & 0.3 & -0.2 & 0.1 \\ 0.1 & -0.2 & 0.3 & -0.4 \\ 0.2 & 0.1 & -0.4 & 0.7 \end{bmatrix}\end{aligned}$$

The projection matrix for the orthogonal subspace is given by

$$\boldsymbol{P}_{\mathbf{X}}^{\perp} = \mathbf{I} - \boldsymbol{P}_{\mathbf{X}} = \begin{bmatrix} 0.3 & 0.4 & -0.1 & -0.2 \\ 0.4 & 0.7 & 0.2 & -0.1 \\ -0.1 & 0.2 & 0.7 & 0.4 \\ -0.2 & -0.1 & 0.4 & 0.3 \end{bmatrix}$$

The optimal estimate of the vector $\mathbf{d}$ is given by

$$\hat{\mathbf{d}} = \boldsymbol{P}_{\mathbf{X}}\mathbf{d} = \begin{bmatrix} 0.7 & -0.4 & 0.1 & 0.2 \\ -0.4 & 0.3 & -0.2 & 0.1 \\ 0.1 & -0.2 & 0.3 & -0.4 \\ 0.2 & 0.1 & -0.4 & 0.7 \end{bmatrix} \begin{bmatrix} -1 \\ 1 \\ -1 \\ 1 \end{bmatrix} = \begin{bmatrix} -1 \\ 1 \\ -1 \\ 1 \end{bmatrix}$$

[4]When two or more subspaces are orthogonal and their union is the complete vector space, as is the case here, the vector space is said to be the *direct sum* of the subspaces. In more mathematically oriented treatments the operator $\oplus$ is used to represent the direct sum of subspaces.

The error vector is given by

$$\epsilon = P_{\mathbf{X}}^{\perp}\mathbf{d} = \begin{bmatrix} 0.3 & 0.4 & -0.1 & -0.2 \\ 0.4 & 0.7 & 0.2 & -0.1 \\ -0.1 & 0.2 & 0.7 & 0.4 \\ -0.2 & -0.1 & 0.4 & 0.3 \end{bmatrix} \begin{bmatrix} -1 \\ 1 \\ -1 \\ 1 \end{bmatrix} = \begin{bmatrix} 0 \\ 0 \\ 0 \\ 0 \end{bmatrix}$$

These results confirm those of Example 9.3.

9.3.4 The Role of Singular Value Decomposition

Singular value decomposition (SVD) plays an extremely important role in least squares problems from both a theoretical and a practical point of view. From a theoretical viewpoint it provides a unified setting for the solution of (9.54) when the matrix $\mathbf{X}$ is of full rank, as well as when it is rank deficient. From a practical viewpoint, it provides the best numerical method to compute the pseudoinverse and the projection matrix. Efficient and stable algorithms exist for computing the SVD [6] and research to enhance computation of the SVD is still ongoing. Use of the SVD as it applies to problems of least squares is discussed here only briefly. More information about the topic can be found in several places [2, 6–8].

Recall from Chapter 2 (Section 2.6.2) that the SVD factors the $K \times P$ matrix $\mathbf{X}$ into a product of three terms:

$$\mathbf{X} = \mathbf{U}\mathbf{\Sigma}\mathbf{V}^{*T} \tag{9.68}$$

where $\mathbf{U}$ is the $K \times K$ unitary matrix of left singular vectors

$$\mathbf{U} = \begin{bmatrix} | & | & & | \\ \mathbf{u}_1 & \mathbf{u}_2 & \cdots & \mathbf{u}_K \\ | & | & & | \end{bmatrix} \tag{9.69}$$

and $\mathbf{V}$ is the $P \times P$ unitary matrix of right singular vectors

$$\mathbf{V} = \begin{bmatrix} | & | & & | \\ \mathbf{v}_1 & \mathbf{v}_2 & \cdots & \mathbf{v}_P \\ | & | & & | \end{bmatrix} \tag{9.70}$$

The matrix $\mathbf{\Sigma}$ is the $K \times P$ matrix of nonnegative real singular values, which is written here in block partitioned form as

$$\mathbf{\Sigma} = \begin{bmatrix} \mathbf{S}_1 & \mathbf{0} \\ \mathbf{0} & \mathbf{0} \end{bmatrix} \tag{9.71}$$

where $\mathbf{S}_1$ is a diagonal matrix of the nonzero singular values

$$\mathbf{S}_1 = \begin{bmatrix} \sigma_1 & 0 & \cdots & 0 \\ 0 & \sigma_2 & \cdots & 0 \\ \vdots & \vdots & \ddots & \vdots \\ 0 & 0 & \cdots & \sigma_r \end{bmatrix} \tag{9.72}$$

and r is the rank of $\mathbf{X}$. The Moore–Penrose pseudoinverse is *defined* as

$$\boxed{\mathbf{X}^+ \stackrel{\text{def}}{=} \mathbf{V}\boldsymbol{\Sigma}^+\mathbf{U}^{*T}} \tag{9.73}$$

where you should note that the positions of $\mathbf{U}$ and $\mathbf{V}$ have been interchanged and where $\boldsymbol{\Sigma}^+$, the pseudoinverse of $\boldsymbol{\Sigma}$, is the $P \times K$ matrix

$$\boldsymbol{\Sigma}^+ = \begin{bmatrix} \mathbf{S}_1^{-1} & \mathbf{0} \\ \mathbf{0} & \mathbf{0} \end{bmatrix} \tag{9.74}$$

Since $\mathbf{S}$ is diagonal, its inverse is just a diagonal matrix of the terms $1/\sigma_i$. It is shown later that in the usual case where $K \geq P$ and the matrix $\mathbf{X}$ has full rank P, this definition is equivalent to the expression for the pseudoinverse given previously. *Notice that while* $\mathbf{X}$ *is a* $K \times P$ *matrix, its pseudoinverse* $\mathbf{X}^+$ *is a* $P \times K$ *matrix.* If $\mathbf{X}$ is tall and thin, then $\mathbf{X}^+$ is short and wide. (The same is true for $\boldsymbol{\Sigma}$ and $\boldsymbol{\Sigma}^+$.) If the matrix $\mathbf{X}$ happens to be square and of full rank, then this definition produces $\mathbf{X}^+ = \mathbf{X}^{-1}$. It can also be seen from the equations presented so far that the matrix $\mathbf{X}$ and its pseudoinverse can be written in the alternative forms

$$\mathbf{X} = \sum_{k=1}^{r} \sigma_k \mathbf{u}_k \mathbf{v}_k^{*T} \tag{9.75}$$

and

$$\mathbf{X}^+ = \sum_{k=1}^{r} \frac{1}{\sigma_k} \mathbf{v}_k \mathbf{u}_k^{*T} \tag{9.76}$$

which are frequently the easiest to use.

The product of a square matrix and its inverse is the identity matrix. Let us consider the interpretation of the products $\mathbf{X}^+\mathbf{X}$ and $\mathbf{X}\mathbf{X}^+$. Since $\mathbf{U}$ is unitary, it follows that

$$\begin{aligned} \mathbf{X}^+\mathbf{X} &= \left(\mathbf{V}\boldsymbol{\Sigma}^+\mathbf{U}^{*T}\right)\left(\mathbf{U}\boldsymbol{\Sigma}\mathbf{V}^{*T}\right) = \mathbf{V}\boldsymbol{\Sigma}^+\boldsymbol{\Sigma}\mathbf{V}^{*T} \\ &= \mathbf{V}\begin{bmatrix} \mathbf{I}_{r\times r} & \mathbf{0} \\ \mathbf{0} & \mathbf{0} \end{bmatrix}\mathbf{V}^{*T} \end{aligned} \tag{9.77}$$

where $\mathbf{I}_{r\times r}$ denotes the $r \times r$ identity matrix. Therefore, if $K > P$ and $\mathbf{X}$ has the full rank $r = P$ (which is the usual case for the problems considered in this chapter), (9.77) yields

$$\mathbf{X}^+\mathbf{X} = \mathbf{I}_{P\times P} \tag{9.78}$$

On the other hand, it follows by a procedure identical to the above that

$$\mathbf{X}\mathbf{X}^+ = \mathbf{U}\begin{bmatrix} \mathbf{I}_{r\times r} & \mathbf{0} \\ \mathbf{0} & \mathbf{0} \end{bmatrix}\mathbf{U}^{*T} \tag{9.79}$$

where here the three matrices on the right are all of size $K \times K$. Again, if $K > P$, this matrix is rank deficient. A possible mnemonic for remembering the difference between the two forms of product in the usual case where $K > P$ is that $\mathbf{X}^{+}\mathbf{X}$ (with the "+" on the inside) is a *compression* to a smaller dimension and will usually have full rank whereas $\mathbf{X}\mathbf{X}^{+}$ (with the "+" on the outside) is an *expansion* to the larger dimension and is therefore rank deficient. Equation 9.79 can be written as

$$\mathbf{X}\mathbf{X}^{+} = \sum_{k=1}^{r} \mathbf{u}_k \mathbf{u}_k^{*T} \tag{9.80}$$

and represents the projection matrix $\boldsymbol{P}_{\mathbf{X}}$ defined earlier. To see this, note that if $\mathbf{e}$ is any arbitrary vector, then the product $\mathbf{X}\mathbf{X}^{+}\mathbf{e}$ is

$$\mathbf{X}\mathbf{X}^{+}\mathbf{e} = (\mathbf{u}_1^{*T}\mathbf{e})\mathbf{u}_1 + (\mathbf{u}_2^{*T}\mathbf{e})\mathbf{u}_2 + \cdots + (\mathbf{u}_K^{*T}\mathbf{e})\mathbf{u}_K$$

This is the sum of the components of the vector $\mathbf{e}$ with respect to the basis $\{\mathbf{u}_k\}$ that lie in the subspace spanned by the columns of $\mathbf{X}$. Therefore, the projection matrix can be defined in general by

$$\boxed{\boldsymbol{P}_{\mathbf{X}} \stackrel{\text{def}}{=} \mathbf{X}\mathbf{X}^{+}} \tag{9.81}$$

without regard to the rank r. The special expression (9.62) applies only when $\mathbf{X}$ has full rank.

Let us now see how the SVD applies to the solution of the least squares problem (9.54). Section 9.3.1 shows that when $\mathbf{X}$ has full rank P the solution is given by (9.52) and (9.53). It can be shown that in this case the expressions (9.53) and (9.73) are equivalent. Although this can be done by directly substituting (9.68) into (9.53), in reality it is a little easier to use the alternative expressions (9.75) and (9.76). First observe from (9.75) that since the $\{\mathbf{u}_k\}$ form an orthonormal set

$$\mathbf{X}^{*T}\mathbf{X} = \sum_{k=1}^{P} \sigma_k \mathbf{v}_k \mathbf{u}_k^{*T} \cdot \sum_{j=1}^{P} \sigma_j \mathbf{u}_j \mathbf{v}_j^{*T} = \sum_{k=1}^{P} \sigma_k^2 \mathbf{v}_k \mathbf{v}_k^{*T}$$

In fact, this is just what is known in linear algebra as the "spectral representation" of the correlation matrix $\mathbf{X}^{*T}\mathbf{X}$ in terms of its eigenvalues σ_k^2 and eigenvectors $\mathbf{v}_k$. Since the matrix is of full rank the inverse is

$$(\mathbf{X}^{*T}\mathbf{X})^{-1} = \sum_{k=1}^{P} \frac{1}{\sigma_k^2} \mathbf{v}_k \mathbf{v}_k^{*T}$$

Then it follows from (9.75) again that

$$(\mathbf{X}^{*T}\mathbf{X})^{-1}\mathbf{X}^{*T} = \sum_{k=1}^{P} \frac{1}{\sigma_k^2} \mathbf{v}_k \mathbf{v}_k^{*T} \cdot \sum_{j=1}^{P} \sigma_j \mathbf{v}_j \mathbf{u}_j^{*T} = \sum_{k=1}^{P} \frac{1}{\sigma_k} \mathbf{v}_k \mathbf{u}_k^{*T} = \mathbf{X}^{+}$$

where the last equality follows from (9.76). This shows the equivalence of the two expressions for the pseudoinverse. As mentioned before, the SVD has better numerical properties, so it may be preferable to use the expression (9.73) instead of (9.53) to compute the pseudoinverse even when **X** is of full rank. In special cases such as when **X** is Toeplitz, however, other algorithms that take advantage of the special structure may be preferred to solve the least squares filtering problem.

Consider now the case when the conditions $K \geq P$ and Rank $(\mathbf{X}) = P$ are not satisfied. There are three possibilities:

1. $K \geq P$ but Rank $(\mathbf{X}) < P$.
2. $K < P$ and Rank $(\mathbf{X}) = K$.
3. $K < P$ and Rank $(\mathbf{X}) < K$.

In all of these cases there are many possible solutions for **h** (actually, an *infinite* number). This can be seen, for example, from Fig. 9.5. The solution to the least squares problem is always a projection of **d** into the subspace spanned by the columns of **X**. Suppose, however, that the columns of **X** are not linearly independent, as the foregoing three possibilities imply. To cite a specific case, suppose there are three column vectors $\mathbf{x}_i$ in Fig. 9.5 all lying within the same plane. With two vectors there is only one possible linear combination to form $\hat{\mathbf{d}}$; but with three there is an infinite number of linear combinations that will do it! This is the essence of the problem.

The way out of this dilemma is to put an additional constraint on **h**. It turns out that if the constraint is to require the vector **h** to have the smallest possible magnitude, the unique solution is again given by the Moore–Penrose pseudoinverse. Since the magnitude of the vector is its Euclidean norm, this solution is called the *minimum norm solution.* In other words, the expression (9.52) where $\mathbf{X}^+$ is defined by (9.73) is the unique *minimum norm least squares* solution in *all* cases. From (9.52) and (9.76) the general solution to the least squares Wiener–Hopf equation can then be written as

$$\boxed{\mathbf{h} = \sum_{k=1}^{r} \frac{1}{\sigma_k} (\mathbf{u}_k^{*T} \mathbf{d}) \mathbf{v}_k} \tag{9.82}$$

where r is the rank of **X**. As we have already indicated, this form is better from a computational viewpoint than (9.51).

To see how the minimum norm condition leads to the unique solution given by the pseudoinverse, observe that since the right singular vectors $\{\mathbf{v}_k\}$ form a basis for P–dimensional space, any proposed filter vector can be written as a linear combination of these vectors:

$$\mathbf{h} = \sum_{k=1}^{P} h_k \mathbf{v}_k \tag{9.83}$$

The optimal filter must satisfy the orthogonality condition (9.57). Substituting (9.75) and (9.83) yields

$$\sum_{k=1}^{r} \sigma_k \mathbf{v}_k \mathbf{u}_k^{*T} \left(\mathbf{d} - \sum_{j=1}^{r} \sigma_j \mathbf{u}_j \mathbf{v}_j^{*T} \sum_{i=1}^{P} h_i \mathbf{v}_i \right) = \mathbf{0}$$

Since the $\{\mathbf{u}_k\}$ and $\{\mathbf{v}_k\}$ form orthonormal sets, this reduces to

$$\sum_{k=1}^{r} \sigma_k(\mathbf{u}_k^{*T}\mathbf{d})\mathbf{v}_k - \sum_{k=1}^{r} \sigma_k^2 h_k \mathbf{v}_k = \mathbf{0}$$

Again, since the $\mathbf{v}_k$ are orthonormal vectors, the last equation implies that

$$\sigma_k(\mathbf{u}_k^{*T}\mathbf{d}) = \sigma_k^2 h_k$$

or

$$h_k = \frac{1}{\sigma_k}(\mathbf{u}_k^{*T}\mathbf{d}); \qquad k = 1, 2, \ldots, r$$

Observe that the orthogonality condition places no constraint on h_k for $k = r+1, \ldots, P$. Thus, substituting this expression in (9.83) produces

$$\mathbf{h} = \sum_{k=1}^{r} \frac{1}{\sigma_k}(\mathbf{u}_k^{*T}\mathbf{d})\mathbf{v}_k + \sum_{k=r+1}^{P} h_k \mathbf{v}_k$$

Any of these solutions for *any* choice of the unspecified h_k satisfies the least squares problem. The choice that minimizes $\|\mathbf{h}\|$, however, is clearly $h_{r+1} = \cdots = h_P = 0$. This yields the solution (9.82) produced by the pseudoinverse.

To solidify the concepts of this subsection, let us see how the results of the last two examples can be obtained using the SVD.

Example 9.5

The SVD of the data matrix in Example 9.3 is found to be[5]

$$\underbrace{\begin{bmatrix} -2 & 1 \\ 3 & -2 \\ -4 & 3 \\ 5 & -4 \end{bmatrix}}_{\mathbf{X}} = \underbrace{\begin{bmatrix} -0.24 & 0.80 & -0.41 & 0.37 \\ 0.39 & -0.38 & -0.27 & 0.79 \\ -0.55 & -0.039 & 0.68 & 0.48 \\ 0.70 & 0.46 & 0.54 & 0.059 \end{bmatrix}}_{\mathbf{U}} \underbrace{\begin{bmatrix} 9.2 & 0 \\ 0 & 0.49 \\ 0 & 0 \\ 0 & 0 \end{bmatrix}}_{\mathbf{\Sigma}} \underbrace{\begin{bmatrix} 0.80 & -0.60 \\ -0.60 & -0.80 \end{bmatrix}}_{\mathbf{V}^T}$$

The singular values are thus $\sigma_1 = 9.2$ and $\sigma_2 = 0.49$. The pseudoinverse is then given by

$$\mathbf{X}^+ = \mathbf{V}\mathbf{\Sigma}^+\mathbf{U}^{*T}$$

$$= \begin{bmatrix} 0.80 & -0.60 \\ -0.60 & -0.80 \end{bmatrix} \begin{bmatrix} 1/9.2 & 0 & 0 & 0 \\ 0 & 1/0.49 & 0 & 0 \end{bmatrix} \begin{bmatrix} -0.24 & 0.39 & -0.55 & 0.70 \\ 0.80 & -0.38 & -0.039 & 0.46 \\ -0.41 & -0.27 & 0.68 & 0.54 \\ 0.37 & 0.79 & 0.48 & 0.059 \end{bmatrix}$$

$$= \begin{bmatrix} -1 & 0.5 & 0 & -0.5 \\ -1.3 & 0.6 & 0.1 & -0.8 \end{bmatrix}$$

which checks with the result of Example 9.3.

[5] Numerical values are rounded to two significant figures here. Results can be checked with more precision by using an interactive programming environment such as MATLAB or APL.

The projection matrix is given by

$$P_X = \mathbf{X}\mathbf{X}^+ = \begin{bmatrix} -2 & 1 \\ 3 & -2 \\ -4 & 3 \\ 5 & -4 \end{bmatrix} \begin{bmatrix} -1 & 0.5 & 0 & -0.5 \\ -1.3 & 0.6 & 0.1 & -0.8 \end{bmatrix}$$

$$= \begin{bmatrix} 0.7 & -0.4 & 0.1 & 0.2 \\ -0.4 & 0.3 & -0.2 & 0.1 \\ 0.1 & -0.2 & 0.3 & -0.4 \\ 0.2 & 0.1 & -0.4 & 0.7 \end{bmatrix}$$

which checks with the result obtained in Example 9.4. Finally, the filter coefficient vector can be computed using (9.82):

$$\mathbf{h} = \frac{1}{9.2} \begin{bmatrix} -0.24 & 0.39 & -0.55 & 0.70 \end{bmatrix} \begin{bmatrix} -1 \\ 1 \\ -1 \\ 1 \end{bmatrix} \begin{bmatrix} 0.80 \\ -0.60 \end{bmatrix}$$

$$+ \frac{1}{.49} \begin{bmatrix} 0.80 & -0.38 & -0.039 & 0.46 \end{bmatrix} \begin{bmatrix} -1 \\ 1 \\ -1 \\ 1 \end{bmatrix} \begin{bmatrix} -0.60 \\ -0.80 \end{bmatrix}$$

$$= 0.20 \begin{bmatrix} 0.80 \\ -0.60 \end{bmatrix} - 1.4 \begin{bmatrix} -0.60 \\ -0.80 \end{bmatrix} = \begin{bmatrix} 1 \\ 1 \end{bmatrix}$$

which checks with the result obtained in Example 9.3.

9.3.5 Total Least Squares

In recent years a new type of least squares technique has evolved that can have some significant advantages over traditional least squares methods. The method is known as total least squares (TLS) [6, 9]. In traditional least squares problems an equation of the form

$$\mathbf{X}\mathbf{h} = \mathbf{d} - \epsilon$$

[(9.55)] is solved to minimize the squared norm of the error $\|\epsilon\|^2$. This implicitly assumes that the error is associated with the vector **d**. In total least squares an error is associated with with both **d** and **X**. This is sound logic since both **d** and **X** represent measured data. The total least squares equation is written as

$$(\mathbf{X} - \Delta)\mathbf{h} = \mathbf{d} - \epsilon \tag{9.84}$$

where $\boldsymbol{\Delta}$ is the matrix of errors associated with **X**. If (9.84) is rewritten as

$$(\overline{\mathbf{X}} - \overline{\boldsymbol{\Delta}}) \begin{bmatrix} \mathbf{h} \\ -1 \end{bmatrix} = \mathbf{0} \tag{9.85}$$

where

$$\overline{\mathbf{X}} \stackrel{\text{def}}{=} \left[\, \mathbf{X} \mid \mathbf{d} \,\right] \tag{9.86}$$

and

$$\overline{\boldsymbol{\Delta}} \stackrel{\text{def}}{=} \left[\, \boldsymbol{\Delta} \mid \boldsymbol{\epsilon} \,\right] \tag{9.87}$$

then the total least squares problem seeks a solution for $\mathbf{h}$ that minimizes the *Frobenius norm* of the total error $\left\|\overline{\boldsymbol{\Delta}}\right\|_F$ (hence the term "total least squares").

The solution to this problem involves the SVD of $\overline{\mathbf{X}}$ and the important fact that the squared Frobenius norm of a matrix is equal to the sum of its squared singular values (see Problem 9.11). Let us assume that $K \geq P+1$ and that $\overline{\mathbf{X}}$ is of full rank. Then $\overline{\mathbf{X}}$ can be written as

$$\overline{\mathbf{X}} = \sum_{k=1}^{P+1} \overline{\sigma}_k \overline{\mathbf{u}}_k \overline{\mathbf{v}}_k^{*T} \tag{9.88}$$

where $\overline{\sigma}_k$, $\overline{\mathbf{u}}_k$, and $\overline{\mathbf{v}}_k$ are the terms in its SVD, and its squared Frobenius norm is given by

$$\left\|\overline{\mathbf{X}}\right\|_F^2 = \sum_{k=1}^{P+1} \overline{\sigma}_k^2$$

Now (9.85) states that the columns of the matrix $\overline{\mathbf{X}} - \overline{\boldsymbol{\Delta}}$ are linearly dependent; in other words, the matrix $\overline{\mathbf{X}} - \overline{\boldsymbol{\Delta}}$ is rank deficient. Thus the total least squares problem seeks *the matrix* $\overline{\boldsymbol{\Delta}}$ *of smallest squared Frobenius norm that makes* $\overline{\mathbf{X}} - \overline{\boldsymbol{\Delta}}$ *rank deficient.* A little thought suggests that this must be the matrix

$$\overline{\boldsymbol{\Delta}} = \overline{\sigma}_{P+1} \overline{\mathbf{u}}_{P+1} \overline{\mathbf{v}}_{P+1}^{*T} \tag{9.89}$$

This matrix has a Frobenius norm of $\overline{\sigma}_{P+1}$ (the smallest singular value) and reduces the rank of $\overline{\mathbf{X}}$ by one. [The rigorous proof that (9.89) is actually the desired matrix is left as a problem; see Problem 9.12.]

In order to find the corresponding solution $\mathbf{h}$ use (9.88) and (9.89) to write (9.85) as

$$(\overline{\mathbf{X}} - \overline{\boldsymbol{\Delta}}) \begin{bmatrix} \mathbf{h} \\ -1 \end{bmatrix} = \sum_{k=1}^{P} \overline{\sigma}_k \overline{\mathbf{u}}_k \overline{\mathbf{v}}_k^{*T} \begin{bmatrix} \mathbf{h} \\ -1 \end{bmatrix} = \mathbf{0} \tag{9.90}$$

Clearly, to satisfy this condition the vector $\begin{bmatrix} \mathbf{h} \\ -1 \end{bmatrix}$ must be proportional to $\overline{\mathbf{v}}_{P+1}$. (This is the only vector orthogonal to all the rest of the $\overline{\mathbf{v}}_k$.) Thus

$$\begin{bmatrix} \mathbf{h} \\ -1 \end{bmatrix} = c\overline{\mathbf{v}}_{P+1} \tag{9.91}$$

where c is a constant. If the vector $\overline{\mathbf{v}}_{P+1}$ is partitioned and written as

$$\overline{\mathbf{v}}_{P+1} = \begin{bmatrix} \mathbf{v}' \\ v_1 \end{bmatrix} \tag{9.92}$$

then to satisfy the condition on the last component in (9.91) we must take $c = -1/v_1$. Therefore, the solution to the total least squares problem is given by

$$\boxed{\mathbf{h} = -\frac{1}{v_1}\mathbf{v}'} \tag{9.93}$$

Two possible degeneracies can arise in the total least squares problem. First if the matrix $\overline{\mathbf{X}}$ is not of full rank, then the problem may have no solution at all. Second, if the smallest singular value has multiplicity $M > 1$, there are M possible distinct solutions. Golub and Van Loan suggest that in this case the solution with smallest (Euclidean) norm can be chosen [6].

9.4 AR MODELING VIA LINEAR PREDICTION

With the development of the least squares form of optimal filtering we now turn to the topic of autoregressive modeling and the closely related topic of linear prediction and approach these from a least squares point of view. In Chapter 8 the correlation function describing the data is assumed to be known. Here typical *data* sequences are given instead. The methods of this chapter effectively estimate the required correlation matrices from the data, although the problem is not formulated in this way. In this section two common data-oriented methods, known as the "autocorrelation" and the "covariance" methods, are presented first. The role of backward linear prediction is then described in the context of least squares and two methods for estimating the model parameters based on minimizing both forward and backward prediction errors are presented. One of these is a modified version of the covariance method; the other is a clever method based on the Levinson recursion and the lattice structure. This latter technique is known as Burg's method [10].

9.4.1 Autocorrelation and Covariance Methods

Linear prediction is a special case of optimal filtering where $\mathrm{d}[n] = \mathrm{x}[n]$ and the observations are $\mathrm{x}[n-1], \mathrm{x}[n-2], \ldots, \mathrm{x}[n-P]$. Since the goal in this chapter is principally to describe the AR model, let us change the notation for the linear prediction problem slightly. Instead of our usual notation a_k^* to represent the filter coefficients, let us use the same variables a_k as those describing the AR model and write the estimate as

$$\hat{\mathrm{x}}[n] = -\mathrm{a}_1\mathrm{x}[n-1] - \mathrm{a}_2\mathrm{x}[n-2] - \cdots - \mathrm{a}_P\mathrm{x}[n-P] \tag{9.94}$$

The error is then

$$\epsilon[n] = \mathrm{x}[n] - \hat{\mathrm{x}}[n] = \sum_{k=0}^{P} \mathrm{a}_k\mathrm{x}[n-k] \tag{9.95}$$

where as usual

$$\mathrm{a}_0 \stackrel{\text{def}}{=} 1 \tag{9.96}$$

To formulate the least squares version of this problem, let us use (9.94) to predict the given data from $n = n_I$ to $n = n_F$ and write the equations in matrix form as

$$\begin{bmatrix} \hat{x}[n_I] \\ \hat{x}[n_I+1] \\ \vdots \\ \vdots \\ \hat{x}[n_F] \end{bmatrix} = -\begin{bmatrix} x[n_I-1] & x[n_I-2] & \cdots & x[n_I-P] \\ x[n_I] & x[n_I-1] & \cdots & x[n_I-P+1] \\ \vdots & \vdots & & \vdots \\ \vdots & \vdots & & \vdots \\ x[n_F-1] & x[n_F-2] & \cdots & x[n_F-P] \end{bmatrix} \begin{bmatrix} a_1 \\ a_2 \\ \vdots \\ a_P \end{bmatrix} \tag{9.97}$$

This can be recognized as a standard least squares Wiener filtering problem if the vectors $\mathbf{x}_i$ are defined as

$$\mathbf{x}_i = \begin{bmatrix} x[n_I-i] \\ x[n_I+1-i] \\ \vdots \\ \vdots \\ x[n_F-i] \end{bmatrix} \qquad i = 0, 1, \ldots P \tag{9.98}$$

and $\mathbf{d}$ is identified with $\mathbf{x}_0$. Equation 9.97 can then be written as

$$\hat{\mathbf{x}}_0 = -\mathbf{X}_1\mathbf{a}' \tag{9.99}$$

where

$$\mathbf{X}_1 = \begin{bmatrix} | & | & & | \\ \mathbf{x}_1 & \mathbf{x}_2 & \cdots & \mathbf{x}_P \\ | & | & & | \end{bmatrix} \tag{9.100}$$

and

$$\mathbf{a}' = \begin{bmatrix} a_1 \\ a_2 \\ \vdots \\ a_P \end{bmatrix} \tag{9.101}$$

Since this is in the form of the Wiener filtering problem, the results of the preceding section apply directly. The least squares problem here can be written as

$$-\mathbf{X}_1\mathbf{a}' \stackrel{\text{ls}}{=} \mathbf{x}_0 \tag{9.102}$$

which has the solution

$$\mathbf{a}' = -\mathbf{X}_1^+\mathbf{x}_0 \tag{9.103}$$

It is useful to express the problem in a slightly different way, however, and develop a set of equations in more standard form.

If the error vector for this problem is defined as

$$\boldsymbol{\epsilon} = \mathbf{x}_0 - \hat{\mathbf{x}}_0 \tag{9.104}$$

then it follows that

$$\begin{bmatrix} | & \\ \mathbf{x}_0 & \mathbf{X}_1 \\ | & \end{bmatrix} \begin{bmatrix} 1 \\ \mathbf{a}' \end{bmatrix} = \mathbf{X}\mathbf{a} = \boldsymbol{\epsilon} \tag{9.105}$$

where $\mathbf{X}$ is now defined as[6]

$$\mathbf{X} = \begin{bmatrix} | & \\ \mathbf{x}_0 & \mathbf{X}_1 \\ | & \end{bmatrix} = \begin{bmatrix} | & | & & | \\ \mathbf{x}_0 & \mathbf{x}_1 & \cdots & \mathbf{x}_P \\ | & | & & | \end{bmatrix} \tag{9.106}$$

and

$$\mathbf{a} = \begin{bmatrix} 1 \\ \mathbf{a}' \end{bmatrix} = \begin{bmatrix} 1 \\ a_1 \\ \vdots \\ a_P \end{bmatrix} \tag{9.107}$$

Now if $\mathcal{S} = \boldsymbol{\epsilon}^{*T}\boldsymbol{\epsilon}$ is the minimum sum of squared errors, then the Orthogonality Theorem can be written as a single equation, namely

$$\mathbf{X}^{*T}\boldsymbol{\epsilon} = \begin{bmatrix} -- & \mathbf{x}_0^{*T} & -- \\ & \mathbf{X}_1^{*T} & \end{bmatrix} \boldsymbol{\epsilon} = \begin{bmatrix} \mathcal{S} \\ \mathbf{0} \end{bmatrix} \tag{9.108}$$

Combining (9.105) and (9.108) then yields

$$\boxed{(\mathbf{X}^{*T}\mathbf{X})\mathbf{a} = \begin{bmatrix} \mathcal{S} \\ \mathbf{0} \end{bmatrix}} \tag{9.109}$$

This is the least squares form of the Yule–Walker equations.[7] Since $\mathcal{S}$ is the sum of the squared errors, an estimate for the prediction error variance can be computed from

$$\sigma_{\varepsilon_P}^2 = \frac{1}{n_F - n_I + 1}\mathcal{S} \tag{9.110}$$

This is the same as the term $|b_0|^2$ in the AR model.

To develop two well-known methods of modeling, assume that the prediction error filter again moves over the data as shown in Fig. 9.6. Then (9.105) has the form

$$\begin{bmatrix} x[n_I] & x[n_I-1] & \cdots & x[n_I-P] \\ x[n_I+1] & x[n_I] & \cdots & x[n_I-P+1] \\ \vdots & \vdots & & \vdots \\ \\ \vdots & \vdots & & \vdots \\ x[n_F] & x[n_F-1] & \cdots & x[n_F-P] \end{bmatrix} \begin{bmatrix} 1 \\ a_1 \\ \vdots \\ a_P \end{bmatrix} = \begin{bmatrix} \epsilon[n_I] \\ \epsilon[n_I+1] \\ \vdots \\ \\ \vdots \\ \epsilon[n_F] \end{bmatrix} \tag{9.111}$$

[6]The matrix $\mathbf{X}$ is actually the same as in (9.45) and (9.59) except that the column order is one larger.

[7]Although (9.109) was derived from the linear prediction problem, the "correlation matrix" $\mathbf{X}^{*T}\mathbf{X}$ that appears there is the *unreversed* correlation matrix that normally appears in the Yule–Walker equations for the AR model. This is a consequence of the way in which the prediction coefficients were defined and the way in which the prediction problem was formulated.

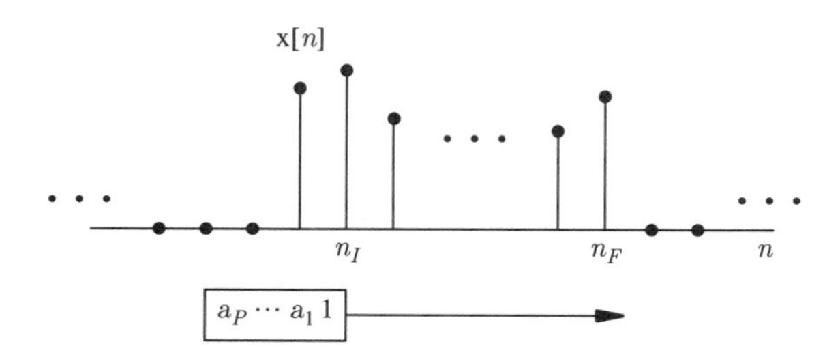

Figure 9.6 Generation of linear prediction error sequence.

The two different methods result from different choices of n_I and n_F. As mentioned in Chapter 6, the names "autocorrelation" and "covariance" (especially the latter) bear no relation to the statistical meaning of these terms and so should not be confused with the statistical definitions. The names for these methods are therefore unfortunate, but they have stuck and are unlikely to be changed.

Autocorrelation Method. In this method the end points are chosen as $n_I = 0$ and $n_F = N_s + P - 1$. Then the filter in Fig. 9.6 runs over the entire length of the data, predicting some of the early points from zeros, and predicting P additional zero values at the end. Since this method uses zeros for data outside the given interval, it can be thought of as applying a rectangular window to the data. For this method the data matrix has the specific structure

$$\mathbf{X} = \begin{bmatrix} \mathrm{x}[0] & 0 & \cdots & 0 \\ \mathrm{x}[1] & \mathrm{x}[0] & \cdots & 0 \\ \vdots & \vdots & \ddots & \vdots \\ \mathrm{x}[P] & \mathrm{x}[P-1] & \cdots & \mathrm{x}[0] \\ \vdots & \vdots & & \vdots \\ \vdots & \vdots & & \vdots \\ \mathrm{x}[N_s-1] & \mathrm{x}[N_s-2] & \cdots & \mathrm{x}[N_s-P-1] \\ 0 & \mathrm{x}[N_s-1] & \cdots & \mathrm{x}[N_s-P] \\ \vdots & \vdots & \ddots & \vdots \\ 0 & 0 & \cdots & \mathrm{x}[N_s-1] \end{bmatrix} \tag{9.112}$$

When formed into the product $\mathbf{X}^{*T}\mathbf{X}$ this data matrix produces a Toeplitz correlation matrix; consequently, the Levinson recursion can be applied to solve the equations. Since the correlation matrix is strictly positive definite [the data matrix (9.112) has full rank], the reflection coefficients are strictly less than one and the prediction error filter is minimum-phase.

Covariance Method. An alternative method is to choose $n_I = P$ and $n_F = N_s - 1$. With this method no zeros are either predicted or used in the prediction. In other words, the limits are chosen so that the filter of Fig. 9.6 always remains within the measured data; no window is applied. For this method the data matrix has the specific form

$$\mathbf{X} = \begin{bmatrix} x[P] & x[P-1] & \cdots & x[0] \\ \vdots & \vdots & \cdots & \vdots \\ \\ \vdots & \vdots & \cdots & \vdots \\ x[N_s-1] & x[N_s-2] & \cdots & x[N_s-P-1] \end{bmatrix} \tag{9.113}$$

A variation of this method called "prewindowed" chooses $n_I = 0$ and $n_F = N_s - 1$ and results in a data matrix that consists of the first N_s rows of (9.112). In the autocorrelation method the data is said to be both prewindowed and "postwindowed."

With the covariance method (or the prewindowed covariance method) the resulting correlation matrix is positive semidefinite but it is *not* Toeplitz. The Levinson recursion and the lattice structure for the filter therefore do not apply and in fact the prediction error filter is *not guaranteed to be minimum-phase*. To state it another way, the filter in the AR model *may be unstable*. Nevertheless, unstable cases seem to occur rarely in practice and the covariance method is often preferred because it makes use of only the measured data. This avoids the bias of the filter coefficients that is caused in the autocorrelation method when the filter runs outside the window. In addition, with some mild conditions, the method can be shown to be equivalent to maximum likelihood [1]. "Fast" algorithms for solving the covariance form of the Yule–Walker equations are available (see, e.g., [11]) but they are considerably more complicated than the Levinson recursion. These algorithms, like the Levinson recursion, are capable of solving the Yule–Walker equations in $\mathcal{O}(P^2)$ operations.

The following example demonstrates use of the autocorrelation and the covariance methods and illustrates some typical differences in the techniques.

Example 9.6

It is desired to estimate the parameters of a second-order AR model for the sequence $\{x[n];\ 0 \leq n \leq 4\} = \{1, -2,\ 3, -4,\ 5\}$ by applying linear prediction to the observed sequence.

Begin with the autocorrelation method. The data matrix is generated conceptually by running the filter over the entire data sequence. This leads to the equation

$$\begin{bmatrix} \varepsilon[0] \\ \varepsilon[1] \\ \varepsilon[2] \\ \varepsilon[3] \\ \varepsilon[4] \\ \varepsilon[5] \\ \varepsilon[6] \end{bmatrix} = \underbrace{\begin{bmatrix} 1 & 0 & 0 \\ -2 & 1 & 0 \\ 3 & -2 & 1 \\ -4 & 3 & -2 \\ 5 & -4 & 3 \\ 0 & 5 & -4 \\ 0 & 0 & 5 \end{bmatrix}}_{\mathbf{X}} \begin{bmatrix} 1 \\ a_1 \\ a_2 \end{bmatrix}$$

The correlation matrix that appears in the Yule–Walker equations is then computed from

$$\mathbf{X}^{*T}\mathbf{X} = \begin{bmatrix} 1 & -2 & 3 & -4 & 5 & 0 & 0 \\ 0 & 1 & -2 & 3 & -4 & 5 & 0 \\ 0 & 0 & 1 & -2 & 3 & -4 & 5 \end{bmatrix} \begin{bmatrix} 1 & 0 & 0 \\ -2 & 1 & 0 \\ 3 & -2 & 1 \\ -4 & 3 & -2 \\ 5 & -4 & 3 \\ 0 & 5 & -4 \\ 0 & 0 & 5 \end{bmatrix}$$

$$= \begin{bmatrix} 55 & -40 & 26 \\ -40 & 55 & -40 \\ 26 & -40 & 55 \end{bmatrix}$$

Notice that the matrix is Toeplitz. The least squares Yule–Walker equations are then

$$\begin{bmatrix} 55 & -40 & 26 \\ -40 & 55 & -40 \\ 26 & -40 & 55 \end{bmatrix} \begin{bmatrix} 1 \\ a_1 \\ a_2 \end{bmatrix} = \begin{bmatrix} \mathcal{S} \\ 0 \\ 0 \end{bmatrix}$$

with solution

$$a_1 = 0.814\,, \quad a_2 = 0.119\,, \quad \text{and} \quad \mathcal{S} = 25.5$$

Since the matrix is Toeplitz, the filter is guaranteed to be minimum-phase. (The roots of the polynomial are at $z = -0.62$ and $z = -0.19$.) The prediction error variance is estimated at

$$\sigma_\varepsilon^2 = \frac{25.5}{7} = 3.64$$

Next apply the covariance method to the same problem. Since the filter stays entirely within the data, the error is evaluated from $n = 2$ to $n = 4$. The data matrix is therefore

$$\mathbf{X} = \begin{bmatrix} 3 & -2 & 1 \\ -4 & 3 & -2 \\ 5 & -4 & 3 \end{bmatrix}$$

and the correlation matrix is

$$\mathbf{X}^{*T}\mathbf{X} = \begin{bmatrix} 3 & -4 & 5 \\ -2 & 3 & -4 \\ 1 & -2 & 3 \end{bmatrix} \begin{bmatrix} 3 & -2 & 1 \\ -4 & 3 & -2 \\ 5 & -4 & 3 \end{bmatrix}$$

$$= \begin{bmatrix} 50 & -38 & 26 \\ -38 & 29 & -20 \\ 26 & -20 & 14 \end{bmatrix}$$

This matrix is clearly *not* Toeplitz. The Yule–Walker equations are

$$\begin{bmatrix} 50 & -38 & 26 \\ -38 & 29 & -20 \\ 26 & -20 & 14 \end{bmatrix} \begin{bmatrix} 1 \\ a_1 \\ a_2 \end{bmatrix} = \begin{bmatrix} \mathcal{S} \\ 0 \\ 0 \end{bmatrix}$$

which have the solution

$$a_1 = 2\,, \quad a_2 = 1\,, \quad \text{and} \quad \mathcal{S} = 0$$

The prediction error variance is correspondingly

$$\sigma_\varepsilon^2 = \frac{0}{3} = 0$$

Evidently, this filter predicts the data perfectly. Indeed, if $\epsilon[n]$ is computed over the chosen range $n = 2$ to $n = 4$, it is found to be zero at every point. The price to be paid for this perfect prediction, however, is an unstable AR model. The prediction error filter

$$1 + 2z^{-1} + z^{-2} = (1 + z^{-1})^2$$

has a double zero at $z = -1$; therefore, the AR model has a double pole at this location. A bounded input into this filter can potentially produce an unbounded output. Further, any errors in computation of the model coefficients can easily put a pole outside the unit circle.

9.4.2 Backward Least Squares Linear Prediction

Since backward prediction plays such an important role in linear prediction, it is worthwhile now to formulate the backward least squares linear prediction problem and examine its solution. Recall that in the forward prediction problem of Fig. 9.6 the equation for the error vector has the form (9.111). Suppose now that a backward prediction error filter is run over the same data, as shown in Fig. 9.7, and the squared error is summed over the same interval. Since in Chapter 8 the backward filter coefficients were found to be the conjugates of the forward filter coefficients, let us designate them here by a_k^*. The vector of backward prediction errors can then be written as

$$\begin{bmatrix} \epsilon^b[n_F] \\ \epsilon^b[n_F-1] \\ \vdots \\ \\ \vdots \\ \epsilon^b[n_I] \end{bmatrix} = \begin{bmatrix} \mathrm{x}[n_F-P] & \cdots & \mathrm{x}[n_F-1] & \mathrm{x}[n_F] \\ \mathrm{x}[n_F-P-1] & \cdots & \mathrm{x}[n_F-2] & \mathrm{x}[n_F-1] \\ \vdots & \cdots & \vdots & \vdots \\ \\ \vdots & \cdots & \vdots & \vdots \\ \mathrm{x}[n_I-P] & \cdots & \mathrm{x}[n_I-1] & \mathrm{x}[n_I] \end{bmatrix} \begin{bmatrix} 1 \\ a_1^* \\ \vdots \\ a_P^* \end{bmatrix} \tag{9.114}$$

From the forms of (9.111) and (9.114) it can be seen that the data matrices for the forward and backward problems are the reversals of each other and the last equation can be written as

$$\tilde{\boldsymbol{\epsilon}}^b = \tilde{\mathbf{X}}\mathbf{a}^* \tag{9.115}$$

where

$$\boldsymbol{\epsilon}^b = \begin{bmatrix} \epsilon^b[n_I] \\ \vdots \\ \\ \vdots \\ \epsilon^b[n_F-1] \\ \epsilon^b[n_F] \end{bmatrix} \tag{9.116}$$

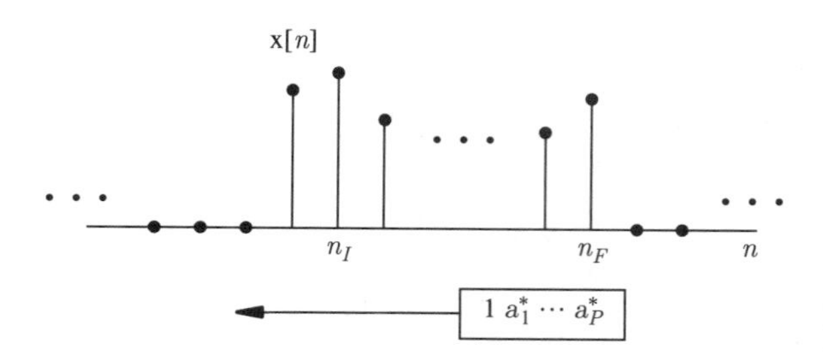

Figure 9.7 Generation of backward prediction error sequence.

It follows that the least squares Yule–Walker equations for the backward problem have the form

$$(\tilde{\mathbf{X}}^{*T}\tilde{\mathbf{X}})\mathbf{a}^* = \begin{bmatrix} \mathcal{S}^b \\ \mathbf{0} \end{bmatrix} \tag{9.117}$$

where $\mathcal{S}^b$ denotes the backward sum of squared errors. Thus the "correlation matrix" in this equation is the reversal of the correlation matrix in the forward least squares Yule–Walker equations.

In this least squares formulation of the linear prediction problem the question naturally arises of whether the sum of squared errors $\mathcal{S}^b$ in this last equation is equal to the sum of squared errors $\mathcal{S}$ in the forward Yule–Walker equations (9.109) and if the vectors $\mathbf{a}$ resulting from solving these two set of equations are identical. The answer depends on whether the reversed and unreversed correlation matrices bear any simple relation to other. If the correlation matrix is Toeplitz, as in the autocorrelation method, then $\tilde{\mathbf{X}}^{*T}\tilde{\mathbf{X}}$ is the complex conjugate of $\mathbf{X}^{*T}\mathbf{X}$. As a result, the solutions to (9.109) and (9.117) are identical and $\mathcal{S}^b = \mathcal{S}$. If the correlation matrix is *not* Toeplitz, as in the covariance method, then no such simple relation exists and the two sets of parameters will be different (see Problem 9.13). This suggests the modified covariance method described in the next subsection.

9.4.3 Modified Covariance Method

The fact that the forward and backward covariance methods give different estimates for the filter parameters is from one point of view troublesome since the theoretical forward and backward filters (i.e., those derived from a known correlation function) are related by conjugation. On the other hand, it is not surprising since the data read forward and backward are actually *different* and theory states only that the *statistical* properties are related. Since there is no compelling reason to prefer forward to backward prediction (or vice versa) for estimating the model parameters, it seems appropriate to consider minimizing the *sum* of the squared forward and backward error terms. This procedure has two advantages. First it eliminates the dilemma of choosing between two different but equally valid estimates for the parameters. Second, it effectively provides twice as much data upon which to base the estimates and so reduces the variance. This procedure is called the forward-backward method [12] or the *modified covariance method.*

The modified covariance method seeks to minimize the combined criteria

$$\mathcal{S}^{fb} = \sum_{n=P}^{N_s-1} \left(|\epsilon[n]|^2 + |\epsilon^b[n]|^2\right) = \|\boldsymbol{\epsilon}\|^2 + \|\boldsymbol{\epsilon}^b\|^2 \tag{9.118}$$

The error terms can be written using (9.105) and (9.115) as

$$\|\boldsymbol{\epsilon}\|^2 = \boldsymbol{\epsilon}^{*T}\boldsymbol{\epsilon} = \mathbf{a}^{*T}\mathbf{X}^{*T}\mathbf{X}\mathbf{a}$$

and

$$\|\boldsymbol{\epsilon}^b\|^2 = \|\tilde{\boldsymbol{\epsilon}}^b\|^2 = (\tilde{\boldsymbol{\epsilon}}^b)^{*T}\tilde{\boldsymbol{\epsilon}}^b = \mathbf{a}^T\tilde{\mathbf{X}}^{*T}\tilde{\mathbf{X}}\mathbf{a}^* = \mathbf{a}^{*T}\tilde{\mathbf{X}}^T\tilde{\mathbf{X}}^*\mathbf{a}$$

where the last expression follows because the quantity is *real*. Substituting these expressions in (9.118) produces

$$\mathcal{S}^{fb} = \mathbf{a}^{*T}\left(\mathbf{X}^{*T}\mathbf{X} + \tilde{\mathbf{X}}^T\tilde{\mathbf{X}}^*\right)\mathbf{a} \tag{9.119}$$

This quantity is to be minimized subject to the constraint that the first component of $\mathbf{a}$ is 1, i.e.,

$$\mathbf{a}^{*T}\boldsymbol{\iota} = 1 \tag{9.120}$$

where $\boldsymbol{\iota}$ is the vector $[\,1 \;\; 0 \;\; \cdots \;\; 0\,]^T$. The problem can be solved formally using a single Lagrange multiplier to incorporate the real constraint. This leads to the condition (see Appendix A)

$$\begin{aligned}\nabla_{\mathbf{a}^*}&\left[\mathbf{a}^{*T}\left(\mathbf{X}^{*T}\mathbf{X} + \tilde{\mathbf{X}}^T\tilde{\mathbf{X}}^*\right)\mathbf{a} + \lambda(1 - \mathbf{a}^{*T}\boldsymbol{\iota})\right] \\ &= \left(\mathbf{X}^{*T}\mathbf{X} + \tilde{\mathbf{X}}^T\tilde{\mathbf{X}}^*\right)\mathbf{a} - \lambda\boldsymbol{\iota} = \mathbf{0}\end{aligned} \tag{9.121}$$

where λ is the Lagrange multiplier. Multiplying through by $\mathbf{a}^{*T}$ yields

$$\mathbf{a}^{*T}\left(\mathbf{X}^{*T}\mathbf{X} + \tilde{\mathbf{X}}^T\tilde{\mathbf{X}}^*\right)\mathbf{a} - \lambda = 0$$

and comparison with (9.119) shows that

$$\lambda = \mathcal{S}^{fb}$$

Finally, substituting this result in (9.121) and rearranging yields the modified covariance Yule–Walker equations

$$\boxed{\left(\mathbf{X}^{*T}\mathbf{X} + \tilde{\mathbf{X}}^T\tilde{\mathbf{X}}^*\right)\mathbf{a} = \begin{bmatrix}\mathcal{S}^{fb} \\ \mathbf{0}\end{bmatrix}} \tag{9.122}$$

Solution of these produces the filter coefficients and the combined least squares error. Generation of the "correlation matrix" in this equation is not as difficult as it may seem. Note that if $\mathbf{R} = \mathbf{X}^{*T}\mathbf{X}$, then the second term $(\tilde{\mathbf{X}}^T\tilde{\mathbf{X}}^*)$ is just $\tilde{\mathbf{R}}^*$. The resulting "correlation matrix" is not Toeplitz but does have a special structure. Marple has developed a fast way to solve these equations that exploits that structure [13].

The following example demonstrates both backward linear prediction and the modified covariance method.

Example 9.7

To apply backward linear prediction to the data in Example 9.6, imagine generating the backward error sequence according to (9.114) (see Fig. 9.7). If the covariance method is used, this equation becomes

$$\begin{bmatrix} \epsilon^b[4] \\ \epsilon^b[3] \\ \epsilon^b[2] \end{bmatrix} = \underbrace{\begin{bmatrix} 3 & -4 & 5 \\ -2 & 3 & -4 \\ 1 & -2 & 3 \end{bmatrix}}_{\tilde{\mathbf{X}}} \begin{bmatrix} 1 \\ a_1 \\ a_2 \end{bmatrix}$$

(The coefficients in this equation are not conjugated because all parameters are real.) The correlation matrix for the Yule–Walker equations is thus

$$\tilde{\mathbf{X}}^T\tilde{\mathbf{X}} = \begin{bmatrix} 3 & -2 & 1 \\ -4 & 3 & -2 \\ 5 & -4 & 3 \end{bmatrix} \begin{bmatrix} 3 & -4 & 5 \\ -2 & 3 & -4 \\ 1 & -2 & 3 \end{bmatrix} = \begin{bmatrix} 14 & -20 & 26 \\ -20 & 29 & -38 \\ 26 & -38 & 50 \end{bmatrix}$$

This correlation matrix is just the *reversal* of the correlation matrix in Example 9.6. The (backward) Yule–Walker equations are

$$\begin{bmatrix} 14 & -20 & 26 \\ -20 & 29 & -38 \\ 26 & -38 & 50 \end{bmatrix} \begin{bmatrix} 1 \\ a_1 \\ a_2 \end{bmatrix} = \begin{bmatrix} \mathcal{S}^b \\ 0 \\ 0 \end{bmatrix}$$

In this problem the solution turns out to be the same as for the forward problem, namely

$$a_1 = 2\,, \quad a_2 = 1\,, \quad \text{and} \quad \mathcal{S}^b = 0$$

This is not normally the case for the covariance method.

For the modified covariance method, the correlation matrix that appears in the Yule–Walker equations is the sum of the correlation matrices that appear in the forward and backward problems:

$$\begin{bmatrix} 50 & -38 & 26 \\ -38 & 29 & -20 \\ 26 & -20 & 14 \end{bmatrix} + \begin{bmatrix} 14 & -20 & 26 \\ -20 & 29 & -38 \\ 26 & -38 & 50 \end{bmatrix} = \begin{bmatrix} 64 & -58 & 52 \\ -58 & 58 & -58 \\ 52 & -58 & 64 \end{bmatrix}$$

Notice that although this matrix has symmetry about both the forward and the reverse diagonals, it is *not* Toeplitz. (The elements on the main diagonal are not all equal.) The Yule–Walker equations involving this matrix have the same solution as that of the forward and backward problems. Again this is not typical. The solution demonstrates, however, that even the modified covariance method is not guaranteed to produce a stable filter.

9.4.4 Burg's Method

In the early 1960s a geophysicist named John Parker Burg developed a method for spectral estimation based on AR modeling that he called the "maximum entropy method" (the method is discussed in detail in Chapter 10). As a part of this method Burg developed an approach to estimate the AR model parameters "directly from the data" without the intermediate step of computing a correlation matrix and solving Yule–Walker equations. Burg's method is based on the Levinson recursion and the lattice structure for the model and has proven to be effective in a number of practical applications.

The Burg procedure is a recursive procedure where at each step in the recursion a single reflection coefficient is estimated. The pth reflection coefficient is chosen to minimize the sum of the forward *and backward* squared prediction errors.

$$\mathcal{S}_p^{fb} = \sum_{n=p}^{N_s-1} \left(|\epsilon_p[n]|^2 + |\epsilon_p^b[n]|^2\right) \tag{9.123}$$

Burg noted, as we did above, that the forward and backward prediction problems are statistically identical and that there is no reason to favor one over the other in estimating the prediction error. Therefore, both ϵ_p and ϵ_p^b should be included in the minimization criterion.

Burg's method can be derived as follows. Consider running the pth order forward and backward filters over the available data to generate the error terms needed in (9.123). According to the lattice equation [see (8.88)] these terms satisfy the order recursions

$$\begin{bmatrix} \epsilon_p[p] \\ \epsilon_p[p+1] \\ \vdots \\ \epsilon_p[N_s-1] \end{bmatrix} = \begin{bmatrix} \epsilon_{p-1}[p] \\ \epsilon_{p-1}[p+1] \\ \vdots \\ \epsilon_{p-1}[N_s-1] \end{bmatrix} - \gamma_p^* \begin{bmatrix} \epsilon_{p-1}^b[p-1] \\ \epsilon_{p-1}^b[p] \\ \vdots \\ \epsilon_{p-1}^b[N_s-2] \end{bmatrix} \tag{9.124}$$

and

$$\begin{bmatrix} \epsilon_p^b[p] \\ \epsilon_p^b[p+1] \\ \vdots \\ \epsilon_p^b[N_s-1] \end{bmatrix} = \begin{bmatrix} \epsilon_{p-1}^b[p-1] \\ \epsilon_{p-1}^b[p] \\ \vdots \\ \epsilon_{p-1}^b[N_s-2] \end{bmatrix} - \gamma_p'^* \begin{bmatrix} \epsilon_{p-1}[p] \\ \epsilon_{p-1}[p+1] \\ \vdots \\ \epsilon_{p-1}[N_s-1] \end{bmatrix} \tag{9.125}$$

(Note that the last equation is written with the terms ordered from $\epsilon_p^b[p]$ to $\epsilon_p^b[N_s-1]$, although in our concept of running the backward filter over the data the terms would be generated in the reverse order, $\epsilon_p^b[N_s-1], \ldots, \epsilon_p^b[p]$.) These equations can be written quite concisely by defining the vectors

$$\boldsymbol{\epsilon}_p = \begin{bmatrix} \epsilon_p[p] \\ -- \\ \\ \mathbf{e}_p^f \end{bmatrix}; \qquad \boldsymbol{\epsilon}_p^b = \begin{bmatrix} \mathbf{e}_p^b \\ \\ ---- \\ \epsilon_p^b[N_s-1] \end{bmatrix} \tag{9.126}$$

where

$$\mathbf{e}_p^f = \begin{bmatrix} \epsilon_p[p+1] \\ \epsilon_p[p+2] \\ \vdots \\ \epsilon_p[N_s-1] \end{bmatrix}; \qquad \mathbf{e}_p^b = \begin{bmatrix} \epsilon_p^b[p] \\ \epsilon_p^b[p+1] \\ \vdots \\ \epsilon_p^b[N_s-2] \end{bmatrix} \tag{9.127}$$

and noting that since $\gamma_p' = \gamma_p^*$, the lattice relations (9.124) and (9.125) can be written as

$$\begin{aligned} \boldsymbol{\epsilon}_p &= \mathbf{e}_{p-1}^f - \gamma_p^* \mathbf{e}_{p-1}^b \\ \boldsymbol{\epsilon}_p^b &= \mathbf{e}_{p-1}^b - \gamma_p \mathbf{e}_{p-1}^f \end{aligned} \tag{9.128}$$

With these definitions and relations (9.123) can be expressed as

$$\begin{aligned} \mathcal{S}_p^{fb} &= \|\boldsymbol{\epsilon}_p\|^2 + \|\boldsymbol{\epsilon}_p^b\|^2 \qquad (9.129) \\ &= (\mathbf{e}_{p-1}^f - \gamma_p^* \mathbf{e}_{p-1}^b)^{*T} (\mathbf{e}_{p-1}^f - \gamma_p^* \mathbf{e}_{p-1}^b) + (\mathbf{e}_{p-1}^b - \gamma_p \mathbf{e}_{p-1}^f)^{*T} (\mathbf{e}_{p-1}^b - \gamma_p \mathbf{e}_{p-1}^f) \\ &= (1 + \gamma_p \gamma_p^*)(\|\mathbf{e}_{p-1}^f\|^2 + \|\mathbf{e}_{p-1}^b\|^2) - 2\gamma_p^* (\mathbf{e}_{p-1}^f)^{*T} \mathbf{e}_{p-1}^b - 2\gamma_p (\mathbf{e}_{p-1}^b)^{*T} \mathbf{e}_{p-1}^f \end{aligned}$$

Then using the expressions for the scalar form of the complex gradient (see Table A.3 in Appendix A) provides the necessary condition

$$\nabla_{\gamma_p^*} \mathcal{S}_p^{fb} = \gamma_p (\|\mathbf{e}_{p-1}^f\|^2 + \|\mathbf{e}_{p-1}^b\|^2) - 2(\mathbf{e}_{p-1}^f)^{*T} \mathbf{e}_{p-1}^b = 0$$

or

$$\gamma_p = \frac{2(\mathbf{e}_{p-1}^f)^{*T} \mathbf{e}_{p-1}^b}{\|\mathbf{e}_{p-1}^f\|^2 + \|\mathbf{e}_{p-1}^b\|^2} \tag{9.130}$$

The form of this equation guarantees that the estimated reflection coefficient satisfies the condition $|\gamma_p| \leq 1$ (see Problem 9.17). Therefore, (unless it turns out that $|\gamma_p| = 1$) *the filter computed by the Burg algorithm is minimum-phase.*

Equations 9.126, 9.128, and 9.130 are summarized below:

$$\gamma_p = \frac{2(\mathbf{e}_{p-1}^f)^{*T} \mathbf{e}_{p-1}^b}{\|\mathbf{e}_{p-1}^f\|^2 + \|\mathbf{e}_{p-1}^b\|^2} \quad (a)$$

$$\begin{bmatrix} \times \\ -- \\ \\ \mathbf{e}_p^f \end{bmatrix} = \mathbf{e}_{p-1}^f - \gamma_p^* \mathbf{e}_{p-1}^b \quad (b)$$

$$\begin{bmatrix} \mathbf{e}_p^b \\ \\ -- \\ \times \end{bmatrix} = \mathbf{e}_{p-1}^b - \gamma_p \mathbf{e}_{p-1}^f \quad (c) \tag{9.131}$$

These are iterated for $p = 1, 2, \ldots, P$ starting from the initial conditions

$$\boxed{\begin{bmatrix} \times \\ -- \\ \\ \mathbf{e}_0^f \end{bmatrix} = \begin{bmatrix} \mathbf{e}_0^b \\ \\ -- \\ \times \end{bmatrix} = \begin{bmatrix} \mathrm{x}[0] \\ \mathrm{x}[1] \\ \vdots \\ \mathrm{x}[N_s - 1] \end{bmatrix}} \tag{9.132}$$

The algorithm is very simple in this form. At each stage of iteration the reflection coefficient is formed from (9.131)(a). The vectors are then combined as in (b) and (c), the elements denoted by $\times$ are dropped, and the steps are repeated. Thus the vectors $\mathbf{e}_p^f$ and $\mathbf{e}_p^b$ decrease in size at each iteration. This makes sense because as the size of the prediction error filter grows and the original data remains fixed in size, there are correspondingly fewer error terms that can be generated. An estimate for the prediction error variance (which is equal to the constant $|b_0|^2$ in the AR model) can be computed by the additional recursion

$$\sigma_{\varepsilon_p}^2 = (1 - |\gamma_p|^2)\sigma_{\varepsilon_{p-1}}^2$$

Finally, if the filter is to be realized in direct form rather than lattice form, the filter parameters can be computed from the recursion

$$\mathbf{a}_p = \begin{bmatrix} \mathbf{a}_{p-1} \\ \\ -- \\ 0 \end{bmatrix} - \gamma_p^* \begin{bmatrix} 0 \\ -- \\ \\ \tilde{\mathbf{a}}_{p-1}^* \end{bmatrix}$$

These two steps can be inserted directly in the loop that implements (9.131) if desired. Some MATLAB code for the algorithm (without these additional computations) is shown in the box below. The test `p < P`

MAIN PART OF FUNCTION
(Burg algorithm):

```
ef = x(2:N);
eb = x(1:N-1);
L=length(ef);
for p=1:P;
    gamma(p) = (2*ef'*eb)/(ef'*ef + eb'*eb)
    if p < P
        tmp1 = ef - conj(gamma(p))*eb;
        tmp2 = eb - gamma(p)*ef;
        ef = tmp1(2:L);
        eb = tmp2(1:L-1);
    end;
    L=L-1;
end;
```

is really unnecessary but so are the computations for `p = P`. An alternative way to organize the code is to compute `ef` and `eb` first, before computing `gamma(p)`.

Burg also suggested a weighted form of this algorithm using a criterion of the form

$$\mathcal{S}_p^{fb'} = \sum_{n=p}^{N_s-1} \mathrm{w}_p[n] \left(|\epsilon_p[n]|^2 + |\epsilon_p^b[n]|^2\right) \tag{9.133}$$

where $\{\mathrm{w}_p[n]\}$ is a positive weighting sequence from some suitable window function (triangular, Hamming, etc.). Such windowing has been found to be effective in the application of the method to spectral analysis of sinusoids in noise. The weighted criterion can be expressed in matrix form as

$$\mathcal{S}_p^{fb'} = \boldsymbol{\epsilon}_p^{*T}\mathbf{W}_p\boldsymbol{\epsilon}_p + (\boldsymbol{\epsilon}_p^b)^{*T}\mathbf{W}_p\boldsymbol{\epsilon}_p^b \tag{9.134}$$

where $\mathbf{W}_p$ is a diagonal matrix with elements $\mathrm{w}_p[n]$. The estimate for γ_p then takes the form

$$\gamma_p = \frac{2(\mathbf{e}_{p-1}^f)^{*T}\mathbf{W}_p\mathbf{e}_{p-1}^b}{(\mathbf{e}_{p-1}^f)^{*T}\mathbf{W}_p\mathbf{e}_{p-1}^f + (\mathbf{e}_{p-1}^b)^{*T}\mathbf{W}_p\mathbf{e}_{p-1}^b} \tag{9.135}$$

This result can be further generalized by letting $\mathbf{W}_p$ be any positive definite matrix. However, this further generalization does not appear to have been studied in any practical applications.

9.4.5 Order Selection

Selecting the order is one of the most difficult problems in developing a linear model for data. If the data is truly described by a finite-order AR model, then the theoretical prediction error variance becomes constant once the model order is reached. Therefore, a procedure that suggests itself is to monitor the prediction error variance as the model order increases and choose the smallest order where the prediction error variance and the model coefficients seem to stabilize. This procedure may not work well in practice for a variety of reasons; the estimated quantities may not seem to converge at all, and if they do it may be difficult to judge when this first happens. As a result, a number of other quantities have been proposed to estimate the model order. The four quantities most well known are listed in Table 9.1. They are Akaike's information-theoretic criteria (AIC) [14], Parzen's criterion autoregressive transfer (CAT) [15, 16], Akaike's final prediction error (FPE) [17], and Schwartz and Rissanen's minimum description length (MDL) [18–20]. Each of these has a distinct minimum at the optimal model order (the AIC, FPE, and MDL each contains a monitonically increasing term that counters the monotonically decreasing prediction error variance). The AIC, MDL, and FPE are all asymptotically equivalent in the sense that

$$\lim_{N_s\to\infty} \frac{\mathrm{AIC}(P)}{N_s} = \lim_{N_s\to\infty} \frac{\mathrm{MDL}(P)}{N_s} = \lim_{N_s\to\infty} \ln \mathrm{FPE}(P)$$

TABLE 9.1 CRITERIA FOR AR MODEL ORDER SELECTION

AIC	$\text{AIC}(P) = N_s \ln \sigma^2_{\varepsilon_P} + 2P$
CAT	$\text{CAT}(P) = \left(\frac{1}{N_s} \sum_{p=1}^{P} \frac{N_s - p}{N_s \sigma^2_{\varepsilon_p}} \right) - \frac{N_s - P}{N_s \sigma^2_{\varepsilon_P}}$
FPE	$\text{FPE}(P) = \sigma^2_{\varepsilon_P} \left(\frac{N_s + P + 1}{N_s - P - 1} \right)$
MDL	$\text{MDL}(P) = N_s \ln \sigma^2_{\varepsilon_P} + P \ln N_s$

Discussion of the statistical properties of these criteria is beyond the scope of this text; moreover, while they seem to work well when the data is truly derived from a finite-order AR model, none of the procedures has been found to be extremely useful for fitting an AR model to non-AR data. This considerably limits their use in practice.

Another procedure that can be applied is as follows. The data matrix $\mathbf{X}$ of any order usually turns out to be of full rank. Its *effective rank*, however, may be somewhat less than full. Let us explain what is meant by "effective rank." Let the rank of $\mathbf{X}$ be denoted by r and consider a rank $r' < r$ approximation to $\mathbf{X}$ in the form

$$\mathbf{X}_{r'} = \sum_{k=1}^{r'} \sigma_k \mathbf{u}_k \mathbf{v}_k^{*T}$$

where σ_k are the singular values, $\mathbf{u}_k$ are the left singular vectors, and $\mathbf{v}_k$ are the right singular vectors of $\mathbf{X}$. If the singular values obey the standard convention $\sigma_1 \geq \sigma_2 \geq \cdots \geq \sigma_{r'}$, then it can be shown that $\mathbf{X}_{r'}$ is an optimal approximation to $\mathbf{X}$ in the sense that the Euclidean norm

$$\|\mathbf{X} - \mathbf{X}_{r'}\|$$

or the Frobenius norm

$$\|\mathbf{X} - \mathbf{X}_{r'}\|_F$$

of the difference matrix is minimal over all possible approximations to $\mathbf{X}$ that have rank r'. This property is analogous to the optimal representation property of the DKLT for random processes with a reduced set of basis functions. A procedure for selecting the order based on the Frobenius norm is as follows. Let the dimension of the data matrix be chosen to be somewhat larger that the suspected order of the model. Next find the singular values of $\mathbf{X}$ and define the reduced rank r' such that

$$\left(\frac{\sigma_1^2 + \sigma_2^2 + \cdots + \sigma_{r'}^2}{\sigma_1^2 + \sigma_2^2 + \cdots + \sigma_r^2} \right)^{\frac{1}{2}} \approx 1$$

This quantity on the left is the ratio of the Frobenius norm of the approximation to the Frobenius norm of the original matrix. The smallest value of r' that satisfies this will be called the "effective rank." Choose the order P of the model to be equal to $r' - 1$.

There is, of course, some subjectivity in defining how close to one the ratio of the singular values needs to be. Therefore, criteria similar to those in Table 9.1 are sometimes applied to the singular values to determine the effective rank.

9.5 ARMA MODELING: A DETERMINISTIC APPROACH

A number of methods for ARMA modeling begin with a deterministic point of view. Although a deterministic approach would seem to have no place in a book on *statistical* signal processing, it is included for at least two reasons. First it leads to a least squares solution and thus ties in well with most of the other topics in this chapter. Second, it provides another practical approach to signal modeling which is frequently used in applications such as speech modeling and spectral estimation. Further, it turns out that although we start from a deterministic viewpoint, the desire to match all of the data with a model of finite limited order leads to terms in the equations that are identical to estimates of the correlation function. Some parts of the problem thus have solutions that are identical to those that have already been developed from other points of view, so the distinction between statistical modeling and deterministic modeling is somewhat blurred.

This section addresses the general ARMA modeling problem since it is the most difficult and the most comprehensive. The specialization of the methods to AR and MA cases are pointed out in the development. There is no attempt to thoroughly cover the collection of possible methods; that would take a whole volume of its own. Even in books devoted more directly to the modeling problem there is some diversity in the methods treated. A good sampling can be found in the texts by Marple [13], Kay [21], and Jackson [22]. Our approach is similar to that of McClellan [11].

9.5.1 Padé Approximation and Prony's Method

Most of the deterministic methods to signal modeling have the goal of representing a given sequence $\mathrm{x}[n]$ as the impulse response of a rational linear system. The approximation problem can be expressed in the transform (z) domain as

$$\mathrm{X}(z) \approx \frac{B(z)}{A(z)} \tag{9.136}$$

where the sense of the approximation is discussed below. What is known as the *direct method* for the approximation problem is shown in Fig. 9.8. The response of the system, denoted by $\check{\mathrm{x}}[n]$, is compared to the desired signal and an error term $\mathrm{e}[n]$ is formed. The linear system is chosen to minimize the sum of the squared errors

$$\mathcal{S}_D = \sum_n |\mathrm{e}[n]|^2$$

where

$$\mathrm{e}[n] = \mathrm{x}[n] - \check{\mathrm{x}}[n]$$

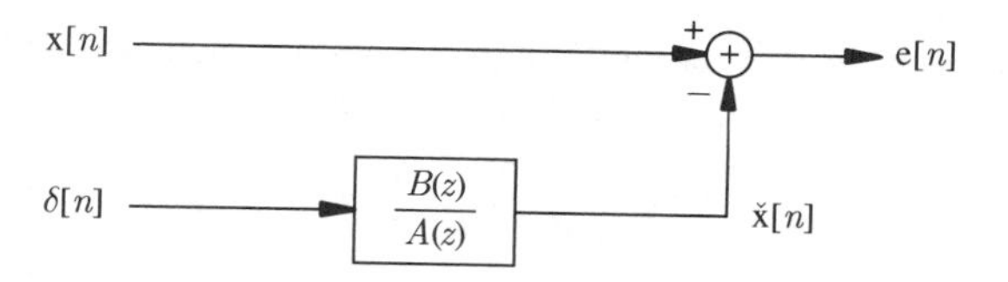

Minimize: $\mathcal{S}_D = \sum_n |\mathrm{e}[n]|^2$

Figure 9.8 Direct method for signal modeling.

Although the problem statement sounds simple and reasonable, the solution is extremely difficult. To demonstrate this first write the error in the transform domain as

$$\begin{aligned}\mathrm{E}(z) &= \mathrm{X}(z) - \check{\mathrm{X}}(z) \\ &= \mathrm{X}(z) - \frac{B(z)}{A(z)}\end{aligned}$$

Then, using Parseval's theorem, $\mathcal{S}_D$ can be expressed in the frequency domain as

$$\mathcal{S}_D = \frac{1}{2\pi}\int_{-\pi}^{\pi}\left|\mathrm{X}(e^{\jmath\omega}) - \frac{B(e^{\jmath\omega})}{A(e^{\jmath\omega})}\right|^2 d\omega$$

Assume for the moment that the filter parameters $\{a_k\}$ and $\{b_k\}$ are real. Then taking partial derivatives with respect to the filter parameters and setting these to zero yields

$$\frac{\partial \mathcal{S}_D}{\partial a_k} = \frac{1}{2\pi}\int_{-\pi}^{\pi} 2\mathrm{Re}\left\{\frac{-B(e^{\jmath\omega})e^{-\jmath\omega k}}{A^2(e^{\jmath\omega})}\left[\mathrm{X}^*(e^{\jmath\omega}) - \frac{B^*(e^{\jmath\omega})}{A^*(e^{\jmath\omega})}\right]\right\} d\omega = 0$$
$$k = 1, 2, \ldots, P$$

and

$$\frac{\partial \mathcal{S}_D}{\partial b_k} = \frac{1}{2\pi}\int_{-\pi}^{\pi} 2\mathrm{Re}\left\{\frac{e^{-\jmath\omega k}}{A(e^{\jmath\omega})}\left[\mathrm{X}^*(e^{\jmath\omega}) - \frac{B^*(e^{\jmath\omega})}{A^*(e^{\jmath\omega})}\right]\right\} d\omega = 0$$
$$k = 0, 1, 2, \ldots, Q$$

The problem thus leads to nonlinear equations that must be solved by iteration or some other numerical technique. Computational requirements are high and existence and uniqueness of the solution are nearly impossible to determine. If we wanted a difficult method, we should have chosen maximum likelihood!

With this brief consideration of the direct method, it seems wise to look for another approach. The input and output of the system satisfies the difference equation

$$\check{\mathrm{x}}[n] + a_1\check{\mathrm{x}}[n-1] + \cdots + a_P\check{\mathrm{x}}[n-P] = b_0\delta[n] + b_1\delta[n-1] + \cdots + b_Q\delta[n-Q] \quad (9.137)$$

Let us try to require that

$$\check{\mathrm{x}}[n] = \mathrm{x}[n]; \qquad n = 0, 1, \ldots N_s - 1$$

where N_s is the length of the given data. Then applying this "requirement" to (9.137) and evaluating that equation for $n = 0, 1, \cdots, N_s - 1$ leads to the matrix equation

$$\begin{bmatrix} x[0] & 0 & 0 & \cdots & 0 \\ x[1] & x[0] & 0 & \cdots & 0 \\ x[2] & x[1] & x[0] & \cdots & 0 \\ \vdots & \vdots & \vdots & \cdots & \vdots \\ \vdots & \vdots & \vdots & & \vdots \\ x[Q] & x[Q-1] & & \cdots & x[Q-P] \\ x[Q+1] & x[Q] & & \cdots & x[Q-P+1] \\ \vdots & \vdots & \vdots & & \vdots \\ \vdots & \vdots & \vdots & \cdots & \vdots \\ x[N_s-1] & x[N_s-2] & & \cdots & x[N_s-P-1] \end{bmatrix} \begin{bmatrix} 1 \\ a_1 \\ a_2 \\ \vdots \\ a_P \end{bmatrix} = \begin{bmatrix} b_0 \\ b_1 \\ b_2 \\ \vdots \\ \vdots \\ b_Q \\ 0 \\ \vdots \\ \vdots \\ 0 \end{bmatrix}$$

This can be written as

$$\boxed{\begin{bmatrix} \mathbf{X}_B \\ \mathbf{X}_A \end{bmatrix} \mathbf{a} = \begin{bmatrix} \mathbf{b} \\ \mathbf{0} \end{bmatrix}} \tag{9.138}$$

where

$$\mathbf{X}_B = \begin{bmatrix} x[0] & 0 & 0 & \cdots & 0 \\ x[1] & x[0] & 0 & \cdots & 0 \\ x[2] & x[1] & x[0] & \cdots & 0 \\ \vdots & \vdots & & \cdots & \vdots \\ \vdots & \vdots & & \cdots & \vdots \\ x[Q] & x[Q-1] & & \cdots & x[Q-P] \end{bmatrix} \tag{9.139}$$

and

$$\mathbf{X}_A = \begin{bmatrix} x[Q+1] & x[Q] & \cdots & x[Q-P+1] \\ \vdots & \vdots & \cdots & \vdots \\ \vdots & \vdots & \cdots & \vdots \\ x[N_s-1] & x[N_s-2] & \cdots & x[N_s-P-1] \end{bmatrix} \tag{9.140}$$

and **a** and **b** are the vectors of filter coefficients [see (9.2) and (9.3)]. This is reminiscent of the Yule–Walker equations in Section 9.1. Suppose first that $N_s = P + Q + 1$. Then the lower set of partitions in (9.138), namely

$$\boxed{\mathbf{X}_A \mathbf{a} = \mathbf{0}} \tag{9.141}$$

represents a set of P linear equations in P unknowns. Assume that these equations are not degenerate and can be solved for **a**. The vector **a** is then known and can be substituted in the upper partitions of (9.138),

$$\boxed{\mathbf{b} = \mathbf{X}_B \mathbf{a}} \tag{9.142}$$

to determine **b**. This exact matching procedure that occurs when $N_s = P+Q+1$ is known as Padé approximation [23]. It derives from the French mathematician Padé's original work on approximating functions by the fitting of rational polynomials.

The Padé approximation is not of much practical value in this application since it is seldom that the number of data points N_s would be chosen to be exactly $P+Q+1$. Usually, N_s is much greater than this number and if we were to match only the first $P+Q+1$ data points exactly the model may not fit the remaining points at all. It does suggest another procedure that is of more practical value. In particular, when the number of points N_s is *greater* than $P+Q+1$, (9.141) is an overdetermined set of equations which can be solved by least squares. Specifically, writing

$$\mathbf{X}_A \mathbf{a} = \mathbf{e}_A$$

where $\mathbf{e}_A$ is the equation error, and finding $\mathbf{a}$ to minimize

$$\mathcal{S}_A = \|\mathbf{e}_A\|^2$$

as in Section 9.4.1, leads to the Yule–Walker equations for the AR model

$$(\mathbf{X}_A^{*T}\mathbf{X}_A)\mathbf{a} = \begin{bmatrix} \mathcal{S}_A \\ \mathbf{0} \end{bmatrix} \tag{9.143}$$

which can be solved for **a** and $\mathcal{S}_A$. Notice from the form of $\mathbf{X}_A$ in (9.140) that (9.143) represents the Yule–Walker equations correponding to the *covariance* method. Once (9.143) is solved, the solution for **a** can then be substituted in (9.142) to find **b**.

The procedure just described is called Prony's method (also the modern Prony method or the extended Prony method) [24]. It has an alternative formulation described in a later section which is closer to Prony's original work and seeks to represent the data in terms of a set of damped exponentials. Prony did not consider the overdetermined case in his original work, so his method was actually closer to the Padé approximation procedure presented above. However, Prony showed how the two sets of variables in the problem could be segregated and determined separately. That is why the modern procedure bears his name.

The Prony method can be interpreted in terms of the indirect modeling procedure depicted in Fig. 9.9. Instead of minimizing the squared magnitude of the error

$$\mathrm{E}(e^{\jmath\omega}) = \mathrm{X}(e^{\jmath\omega}) - \frac{B(e^{\jmath\omega})}{A(e^{\jmath\omega})}$$

this procedure minimizes the squared magnitude of the error

$$\mathrm{E}_A(e^{\jmath\omega}) = \mathrm{X}(e^{\jmath\omega}) \cdot A(e^{\jmath\omega}) - B(e^{\jmath\omega})$$

The expression to minimize in the time domain is

$$\mathcal{S}_A = \sum_{n=Q+1}^{N_s-1} |\mathrm{e}_A[n]|^2$$

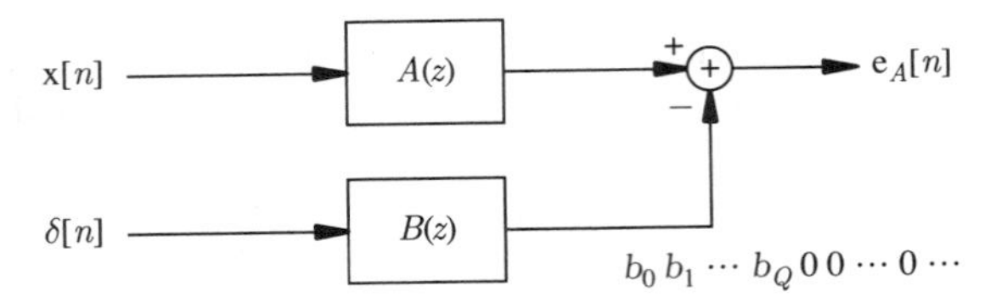

Minimize: $\mathcal{S}_A = \sum_{n=Q+1}^{N_s-1} |e_A[n]|^2$

Figure 9.9 Indirect approach to signal modeling (Prony method).

where $e_A[n]$ is defined by

$$e_A[n] = x[n] * a[n] - b[n]$$

(see Fig. 9.9). By using the lower limit $Q+1$ on the sum, this problem is decoupled from the problem of finding the nonzero terms of $b[n]$ and the equations become linear.

Let us illustrate the methods described in this subsection with a simple example.

Example 9.8

The same sequence $\{x[n]\} = \{1, -2, 3, -4, 5\}$ used in the previous examples is to be modeled as the impulse response of a linear system. If the system is chosen to have order $(P, Q) = (2, 2)$, the transfer function has the form

$$H(z) = \frac{b_0 + b_1 z^{-1} + b_2 z^{-2}}{1 + a_1 z^{-1} + a_2 z^{-2}}$$

For this problem (9.138) becomes

$$\begin{bmatrix} 1 & 0 & 0 \\ -2 & 1 & 0 \\ 3 & -2 & 1 \\ -4 & 3 & -2 \\ 5 & -4 & 3 \end{bmatrix} \begin{bmatrix} 1 \\ a_1 \\ a_2 \end{bmatrix} = \begin{bmatrix} b_0 \\ b_1 \\ b_2 \\ 0 \\ 0 \end{bmatrix}$$

The lower portion of this equation is

$$\begin{bmatrix} -4 & 3 & -2 \\ 5 & -4 & 3 \end{bmatrix} \begin{bmatrix} 1 \\ a_1 \\ a_2 \end{bmatrix} = \begin{bmatrix} 0 \\ 0 \end{bmatrix}$$

which can be solved to find

$$a_1 = 2, \quad a_2 = 1$$

The upper portion of the equation then becomes

$$\begin{bmatrix} b_0 \\ b_1 \\ b_2 \end{bmatrix} = \begin{bmatrix} 1 & 0 & 0 \\ -2 & 1 & 0 \\ 3 & -2 & 1 \end{bmatrix} \begin{bmatrix} 1 \\ a_1 \\ a_2 \end{bmatrix}$$

$$= \begin{bmatrix} 1 & 0 & 0 \\ -2 & 1 & 0 \\ 3 & -2 & 1 \end{bmatrix} \begin{bmatrix} 1 \\ 2 \\ 1 \end{bmatrix} = \begin{bmatrix} 1 \\ 0 \\ 0 \end{bmatrix}$$

Thus the system turns out to be an all-pole model

$$H(z) = \frac{1}{1 + 2z^{-1} + z^{-2}}$$

In fact, this is the same model that was found in Example 9.6 through a different procedure. As observed there, the system is unstable; but no claims were made that this procedure would lead to a stable system.

Since the number of unknown parameters in this example is exactly equal to the number of data points, the method applied here is actually the Padé procedure. If more data were available and the order of the model were kept the same, the bottom set of equations would be larger and would need to be solved by least squares methods. That is the only difference between the Padé and the basic Prony procedures.

9.5.2 Improvements to the Basic Prony Method

Although the procedure described in the preceding subsection can be used as a practical algorithm for data modeling, it has one main drawback. In solving for the MA parameters (b_k), only a portion of the information is used. In particular, the least squares problem addressed by the Prony method is

$$\begin{bmatrix} \mathbf{X}_B \\ \mathbf{X}_A \end{bmatrix} \mathbf{a} = \begin{bmatrix} \mathbf{b} \\ \mathbf{e}_A \end{bmatrix} \tag{9.144}$$

where $\mathbf{a}$ is determined so the squared norm of $\mathbf{e}_A$ is minimized. There is no corresponding error term associated with the top partition. Now given $\mathbf{a}$, the left-hand side of the equation is *fixed* and it remains to find $\mathbf{b}$ so that the entire equation is satisfied in some sense. In determining $\mathbf{a}$, the error in matching $\mathbf{b}$ was deliberately ignored. Now it is well to let that error exist. Forcing $\mathbf{b}$ to match the left side of the equations (as in the basic Prony method) is an unnecessary constraint and may give overall poor results. Some alternative procedures are therefore described for finding the $\mathbf{b}$ coefficients.

Durbin's Method. The idea of Durbin's method [25] is to turn an MA modeling problem into a set of two AR modeling problems each of which can be solved rather easily. Durbin's method applies to *any* problem where it is desired to fit an MA model to data; here is a quick overview. Let $\mathrm{y}[n]$ represent the original data after it is processed by the filter $A(z)$ in Fig. 9.9. (This is what was called the residual process in Section 9.1.3.) The goal is to fit a Qth-order MA model to this data. Instead of fitting an MA model directly, first fit a high-order (say, an Lth-order) AR model to this data. Let this model be represented by $1/G^{(L)}(z)$. Then the approximation in the transform domain is

$$Y(z) \approx \frac{1}{G^{(L)}(z)} \tag{9.145}$$

Now consider the data sequence $g[0], g[1], \ldots, g[L]$, which is the impulse response of the inverse filter $G^{(L)}(z)$. This data sequence is just the set of coefficients of the polynomial $G^{(L)}(z)$. A Qth-order AR model is now matched to this new data sequence. If the denominator polynomial is denoted by $B(z)$, then the approximation is

$$G^{(L)}(z) \approx \frac{1}{B(z)} \tag{9.146}$$

These last two equations imply that $B(z)$ is a Qth-order MA approximation to $Y(z)$.

Let us summarize the method in detail.

1. Denote the data to be approximated by $\mathrm{y}[0], \mathrm{y}[1], \ldots, \mathrm{y}[N_s - 1]$. Find a first-stage AR model for this data by one of the methods in Section 9.4.1. Choose the order L of the model to be significantly higher than the eventual order of the MA system. (L should typically be at least four or five times the order Q of the desired MA model.) Note that in representing the data in the form (9.145) the coefficients resulting from solution of the Yule–Walker equations for the AR model need to be divided by the gain term that would normally appear in the numerator.
2. Let the coefficients of the first stage AR model be denoted by $g[n]$. Find a second stage Q^{th}-order AR model in the form (9.146) for the "data" $g[0], g[1], \ldots, g[L]$, again by one of the procedures in Section 9.4.1. Remember that the coefficients $b_0, b_1, \ldots, b_Q$ of the denominator polynomial are obtained by dividing the coefficients from solution of the Yule–Walker equations for the AR model by the corresponding gain term. The b_k are the desired MA parameters.

If an MA model is truly appropriate to the data and the order L of the first-stage AR model is large enough, then the "data" $g[n]$ will tend to zero for larger values of n. Thus the autocorrelation method can be used in the second step to obtain a minimum-phase solution without the deleterious effects usually caused by windowing. A short example helps to illustrate the steps itemized above.

Example 9.9

Suppose that the data to be modeled is

$$\{\mathrm{y}[n]\} = \{0.9,\ 1.1,\ 0.1, -0.1, -0.1,\ 0.1\}$$

A first-order MA model is to be fit to this data. Note that if this data represented the left-hand side of (9.144) after solving for the AR parameters, then the first-order MA model chosen in the basic Prony method would have $b_0 = 0.9$ and $b_1 = 1.1$. The rest of the data would be disregarded. We expect the model chosen to have both of its coefficients close to 1, but the model that results from merely matching the first two data points and ignoring the rest may not be the best possible choice.

An alternative procedure is to apply Durbin's method. The results arrived at in this example may not necessarily be realistic because the length of the original data sequence $\mathrm{y}[n]$ is very short and the length of the first-stage AR model is made correspondingly short to keep the presentation manageable. Nevertheless, the detailed steps can be illustrated.

If an order $L = 5$ is chosen for the first-stage AR model and the autocorrelation method is used to ensure a stable model, then the data matrix is

$$\mathbf{X} = \begin{bmatrix} 0.9 & 0 & 0 & 0 & 0 & 0 \\ 1.1 & 0.9 & 0 & 0 & 0 & \\ 0.1 & 1.1 & 0.9 & 0 & 0 & \\ -0.1 & 0.1 & 1.1 & 0.9 & 0 & \\ -0.1 & -0.1 & 0.1 & 1.1 & 0.9 & \\ 0.1 & -0.1 & -0.1 & 0.1 & 1.1 & 0.9 \\ 0 & 0.1 & -0.1 & -0.1 & 0.1 & 1.1 \\ 0 & 0 & 0.1 & -0.1 & -0.1 & 0.1 \\ 0 & 0 & 0 & 0.1 & -0.1 & -0.1 \\ 0 & 0 & 0 & 0 & 0.1 & -0.1 \\ 0 & 0 & 0 & 0 & 0 & 0.1 \end{bmatrix}$$

The correlation matrix is given by $\frac{1}{6}\mathbf{X}^T\mathbf{X}$ and results in the AR Yule–Walker equations

$$\frac{1}{6}\begin{bmatrix} 2.06 & 1.09 & -0.04 & -0.19 & 0.02 & 0.09 \\ 1.09 & 2.06 & 1.09 & -0.04 & -0.19 & 0.02 \\ -0.04 & 1.09 & 2.06 & 1.09 & -0.04 & -0.19 \\ -0.19 & -0.04 & 1.09 & 2.06 & 1.09 & -0.04 \\ 0.02 & -0.19 & -0.04 & 1.09 & 2.06 & 1.09 \\ 0.09 & 0.02 & -0.19 & -0.04 & 1.09 & 2.06 \end{bmatrix} \begin{bmatrix} g[0] \\ g[1] \\ g[2] \\ g[3] \\ g[4] \\ g[5] \end{bmatrix} = \begin{bmatrix} 1 \\ 0 \\ 0 \\ 0 \\ 0 \\ 0 \end{bmatrix}$$

where the gain has been normalized to 1 for convenience. Solution of these equations yields (approximately)

$$\{g[0], \ldots, g[5]\} = \{0.87, -0.76,\ 0.58, -0.30,\ 0.12, -0.045\} \cdot 6$$

and the first-stage AR model is

$$G^{(5)}(z) = \frac{\sqrt{0.87/6}}{0.87 - 0.76z^{-1} + 0.58z^{-2} - 0.30z^{-3} + 0.12z^{-4} - 0.045z^{-5}}$$

A "data" matrix is now formed of the model parameters. Call this matrix $\mathbf{X}_G$. It is given by

$$\mathbf{X}_G = \begin{bmatrix} 0.87 & 0 \\ -0.76 & 0.87 \\ 0.58 & -0.76 \\ -0.30 & 0.58 \\ 0.12 & -0.30 \\ -0.045 & 0.12 \\ 0 & -0.045 \end{bmatrix} \cdot \sqrt{\frac{6}{0.87}}$$

The correlation matrix for the second-stage AR model (of order $Q = 1$) is $\frac{1}{6}\mathbf{X}_G^T\mathbf{X}_G$. The resulting Yule–Walker equations are

$$\frac{1}{6}\left(\frac{6}{0.87}\right)\begin{bmatrix} 1.79 & -1.33 \\ -1.33 & 1.79 \end{bmatrix}\begin{bmatrix} b_0 \\ b_1 \end{bmatrix} = \begin{bmatrix} 1 \\ 0 \end{bmatrix}$$

The resulting MA model is $1.04 + 0.77z^{-1}$.

Reverse-Order Levinson Recursion. Another method to obtain the MA parameters in (9.144) is to use the reverse-order Levinson recursion (Chapter 8). It is difficult to justify this method without resorting to statistical arguments, but we can try. The goal is to represent the data sequence $\mathrm{y}[0], \mathrm{y}[1], \ldots, \mathrm{y}[N_s - 1]$ by the shorter sequence $b[0], b[1], \ldots, b[Q]$. The $\mathrm{y}[n]$ can be thought of as the coefficients of an AR model for the "inverse" sequence as discussed in the previous subsection on Durbin's method. This model would be of order N_s. The most consistent model of lower-order $Q < N_s$ in some sense is the one obtained from "stepping down" the higher-order model using the reverse-order Levinson recursion.

The procedure can be summarized as follows. Normalize the data sequence so that the first term is 1. Therefore, let $\mathrm{y}[n]/\mathrm{y}[0]$ play the role of $a_n^{(N_s-1)}$ in the $(N_s - 1)^{\text{th}}$-order AR model and let $1/\mathrm{y}[0]$ play the role of $\sigma^2_{\varepsilon_{N_s-1}}$. Apply the reverse-order Levinson recursion to find the Qth-order parameters. The estimates for b_k are given by

$$b_k = a_k^{(Q)}/\sigma^2_{\varepsilon_Q}; \qquad k = 0, 1, 2, \ldots, Q$$

The example below shows how this procedure is applied to the data in Example 9.9.

Example 9.10

To apply the reverse-order Levinson recursion to the data of Example 9.9, the data is first put into the standard form required by the algorithm. Dividing the data sequence by its first value (0.9) gives the normalized sequence $\{1,\ 1.222,\ 0.111, -0.111, -0.111,\ 0.111\}$. The "prediction error variance" is $1/0.9 = 1.111$. The parameters of the successively reduced-order filters are given below.

Order	Filter coefficients	Error variance
5	$1,\ 1.222,\ 0.111, -0.111, -0.111,\ 0.111$	1.111
4	$1,\ 1.25,\ 0.125, -0.125, -0.25$	1.125
3	$1,\ 1.3,\ 0.17,\ 0.2$	1.20
2	$1,\ 1.32, -0.097$	1.25
1	$1,\ 1.46$	1.26

The final coefficients of the filter are obtained by dividing the filter coefficients from the reverse-order Levinson algorithm by the error variance. This yields $b_0 = 1/1.26 = 0.8$ and $b_1 = 1.46/1.26 = 1.2$.

Shanks' Method. This next procedure is due to Shanks [26] and the combination of this with the least squares procedure for finding the AR parameters is sometimes referred to as Shanks' method. The goal is to represent the given data as in (9.136), where $A(z)$ has already been determined. Define

$$H_A(z) = \frac{1}{A(z)} \tag{9.147}$$

and write (9.136) as

$$X(z) \approx B(z)H_A(z) \tag{9.148}$$

Let $h_A[n]$ be the sequence corresponding to $H_A(z)$. This sequence is actually the impulse response corresponding to the partial system $1/A(z)$. The approximation problem can now be thought of as shown in Fig. 9.10. (Compare this to Fig. 9.9.) The error in the signal domain is given by

$$e_B[n] = x[n] - h_A[n] * b[n] \tag{9.149}$$

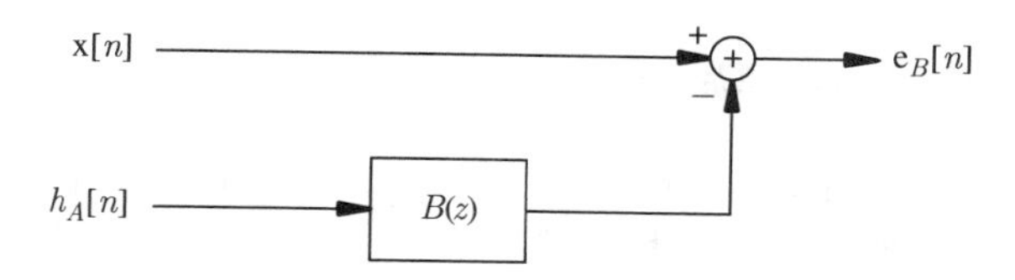

Minimize: $\mathcal{S}_B = \sum_n |e_B[n]|^2$

Figure 9.10 Procedure for estimating the MA parameters in an ARMA model.

If the filter $B(z)$ is chosen to minimize the sum of squared errors

$$\mathcal{S}_B = \sum_{n=0}^{N_s-1} |e_B[n]|^2 \tag{9.150}$$

then this is exactly the least squares Wiener filtering problem. The problem can be written as

$$\mathbf{H}_A\mathbf{b} \overset{\text{ls}}{=} \mathbf{x} \tag{9.151}$$

where

$$\mathbf{H}_A = \begin{bmatrix} h_A[0] & 0 & \cdots & 0 \\ h_A[1] & h_A[0] & \cdots & 0 \\ \vdots & \vdots & \cdots & \vdots \\ h_A[Q] & h_A[Q-1] & \cdots & h_A[0] \\ \vdots & \vdots & \cdots & \vdots \\ h_A[N_s-1] & h_A[N_s-2] & \cdots & h_A[N_s-Q-1] \end{bmatrix} \tag{9.152}$$

and

$$\mathbf{x} = \begin{bmatrix} x[0] \\ x[1] \\ \vdots \\ x[Q] \\ \vdots \\ x[N_s-1] \end{bmatrix} \tag{9.153}$$

and $\mathbf{b}$ is given by (9.3). As an alternative the lower limit in (9.150) can be set to Q, in which case (9.152) and (9.153) will be missing the first Q rows. The solution is

$$\mathbf{b} = \mathbf{H}_A^+ \mathbf{x} \tag{9.154}$$

The needed terms of the sequence $h_A[n]$ can be obtained from the recursive difference equation

$$h_A[n] = -a_1 h_A[n-1] - \cdots - a_P h_A[n-P] + \delta[n] \tag{9.155}$$

Alternatively, it may be feasible to find the roots of $A(z)$ and develop an analytic expression for $h_A[n]$ (at least in simple cases).

9.5.3 Signal Domain Form of Prony's Method

This section describes an alternative form of Prony's method which is actually much closer to Prony's original formulation of the problem. The problem (9.136) can be stated in the signal domain as representing $\mathrm{x}[n]$ by a weighted sum of complex exponentials

$$\mathrm{x}[n] \approx c_1 r_1^n + c_2 r_2^n + \ldots + c_P r_P^n \tag{9.156}$$

where the r_k are the roots of the denominator polynomial $A(z)$ and the c_k are the complex coefficients required for the expansion.

The problem can first be formulated as in Section 9.5.1. In particular we can set up the equations (9.138), ignore the top partition, and solve the bottom partitions (9.141) in a least squares sense to find the coefficients of $A(z)$. The roots r_k of $A(z)$ are then found by a numerical procedure and used in the approximation (9.156). To find the coefficients c_k simply evaluate (9.156) for $n = 0, 1, \ldots, N_s$. This yields the set of equations

$$\begin{bmatrix} 1 & 1 & \cdots & 1 \\ r_1 & r_2 & \cdots & r_P \\ r_1^2 & r_2^2 & \cdots & r_P^2 \\ \vdots & \vdots & \cdots & \vdots \\ r_1^{N_s-1} & r_2^{N_s-1} & \cdots & r_P^{N_s-1} \end{bmatrix} \begin{bmatrix} c_1 \\ c_2 \\ \vdots \\ c_P \end{bmatrix} \stackrel{\text{ls}}{=} \begin{bmatrix} \mathrm{x}[0] \\ x[1] \\ \mathrm{x}[2] \\ \vdots \\ \mathrm{x}[N_s-1] \end{bmatrix} \tag{9.157}$$

which can be solved in a least squares sense to find the coefficients.

If there are multiple roots at the same location a similar procedure applies. For example, if the first root is a double root, the form of the approximation is

$$\mathrm{x}[n] \approx c_1 r_1^n + c_2\, n r_1^n + c_3 r_3^n + \ldots + c_P r_P^n$$

The equations for the coeffients are then

$$\begin{bmatrix} 1 & 0 & 1 & \cdots & 1 \\ r_1 & r_1 & r_3 & \cdots & r_P \\ r_1^2 & 2r_1^2 & r_3^2 & \cdots & r_P^2 \\ \vdots & \vdots & \vdots & \cdots & \vdots \\ r_1^{N_s-1} & (N_s-1)r_1^{N_s-1} & r_3^{N_s-1} & \cdots & r_P^{N_s-1} \end{bmatrix} \begin{bmatrix} c_1 \\ c_2 \\ c_3 \\ \vdots \\ c_P \end{bmatrix} \stackrel{\text{ls}}{=} \begin{bmatrix} \mathrm{x}[0] \\ x[1] \\ \mathrm{x}[2] \\ \vdots \\ \mathrm{x}[N_s-1] \end{bmatrix}$$

The following example illustrates the procedure.

Example 9.11

It is desired to find a signal domain model for the data of Example 9.8. The procedure in that example is followed to determine the AR parameters. This yields the denominator polynomial, which is factored to find the roots of the system:

$$1 + 2z^{-1} + z^{-2} = (1 + z^{-1})^2 = 0$$

The system is seen to have a double root at $z = -1$. Therefore, the approximation is of the form

$$\mathrm{x}[n] \approx c_1(-1)^n + c_2\, n(-1)^n$$

The coefficients can be found from the equations

$$\begin{bmatrix} 1 & 0 \\ -1 & -1 \\ (-1)^2 & 2(-1)^2 \\ (-1)^3 & 3(-1)^3 \\ (-1)^4 & 3(-1)^4 \end{bmatrix} \begin{bmatrix} c_1 \\ c_2 \end{bmatrix} \stackrel{\text{ls}}{=} \begin{bmatrix} 1 \\ -2 \\ 3 \\ -4 \\ 5 \end{bmatrix}$$

In this case the equations can be solved exactly yielding $c_1 = 1$ and $c_2 = 1$. The signal model is therefore

$$\begin{aligned} x[n] &= (-1)^n + n(-1)^n \\ &= (-1)^n(1+n); \qquad 0 \le n \le 4 \end{aligned}$$

The form of Prony's method discussed in this subsection is like a more general DFT expansion with two major differences. First the basis functions are usually complex exponentials with a damped (or growing) envelope instead of the undamped exponentials $e^{\jmath\omega n}$. Second, the basis functions are not selected at the outset to be harmonics, but rather are determined from the data. In this way the basis functions are like those of the DKLT. They are unlike those of the DKLT, however, in that they are in general *not* orthonormal.

9.5.4 Iterative Prefiltering

The last of the deterministic methods to be discussed is the method of iterative prefiltering. This procedure is due to Steiglitz and McBride [27, 28] and is also referred to in the literature as the Steiglitz–McBride method. Iterative prefiltering attempts to provide an efficient iterative approach to the direct modeling problem outlined in the beginning of this section. The block diagram for the direct modeling problem given earlier is repeated in Fig. 9.11(a). This time, however, the numerator and denominator of the model have been represented as separate blocks and the response $h_A[n]$ of the all-pole portion has been indicated in the diagram. The error function for the direct problem can be written in the transform domain as

$$\mathrm{E}(z) = \mathrm{X}(z) - \frac{B(z)}{A(z)} = \frac{\mathrm{X}(z)A(z) - B(z)}{A(z)} \tag{9.158}$$

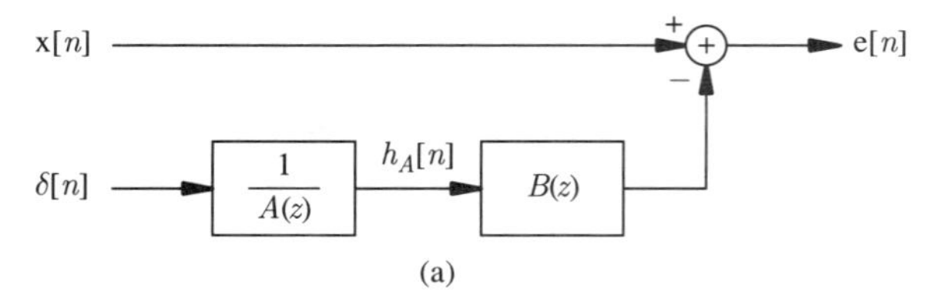

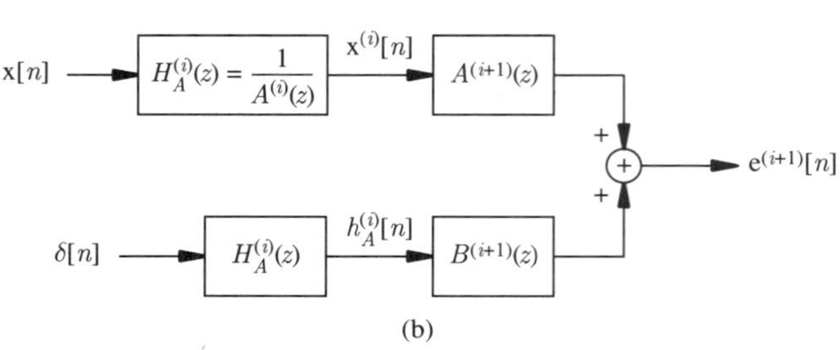

Figure 9.11 Procedure for iterative prefiltering. (a) Block diagram for the direct method. (b) Block diagram for iterative prefiltering.

The iterative prefiltering method replaces this with the iterative error function

$$\mathrm{E}^{(i+1)}(z) = \frac{\mathrm{X}(z)A^{(i+1)}(z) - B^{(i+1)}(z)}{A^{(i)}(z)} \tag{9.159}$$

where the superscripts (i) and $(i+1)$ designate the values of the functions at the ith and $(i+1)^{\text{th}}$ iterations. The terms $A^{(i+1)}$ and $B^{(i+1)}$ are chosen to minimize the corresponding sum of squared errors at each iteration. No proof of convergence has been given for this method; however, if the iteration converges, the solution is the same as that of the direct method. This is easy to see because at convergence $A^{(i)}$ and $A^{(i+1)}$ have the same value, so (9.159) becomes identical to (9.158).

To formulate the method explicitly, let $h_A^{(i)}[n]$ denote the impulse response corresponding to $1/A^{(i)}(z)$. Then (9.159) can be written in the signal domain as

$$\mathrm{e}^{(i+1)}[n] = \mathrm{x}^{(i)}[n] * a^{(i+1)}[n] - b^{(i+1)}[n] * h_A^{(i)}[n] \tag{9.160}$$

where $\mathrm{x}^{(i)}$ is defined as

$$\mathrm{x}^{(i)}[n] \stackrel{\text{def}}{=} \mathrm{x}[n] * h_A^{(i)}[n] \tag{9.161}$$

The situation is depicted in Fig. 9.11(b). For each new iteration, the sequences $\mathrm{x}^{(i)}$ and $h_A^{(i)}$ are known inputs and $A^{(i+1)}$ and $B^{(i+1)}$ are chosen to minimize

$$\mathcal{S}^{(i+1)} = \sum_{n=P}^{N_s-1} |\mathrm{e}^{(i+1)}[n]|^2 \tag{9.162}$$

The sequence $h_A^{(i)}[n]$ can be evaluated as discussed in Section 9.5.2 [see (9.155) and the following].

To make the entire procedure clear, consider the particular case of $P = 2$ and $Q = 1$. The data sequence $\mathrm{x}[n]$ is then modeled as the impulse response of a system with transfer function

$$\frac{B(z)}{A(z)} = \frac{b_0 + b_1 z^{-1}}{1 + a_1 z^{-1} + a_2 z^{-2}}$$

Although the procedure could start with any initial values for the coefficients, it is important in practice that a reasonably good set of initial values is chosen. A suitable procedure is to use Prony's method or one of its variations to compute these initial values. The sequences $h_A^{(i)}[n]$ and $x^{(i)}[n]$ for $n = 0, 1, \ldots, N_s - 1$ are then computed from the recursive difference equations

$$h_A^{(i)}[n] = -a_1^{(i)} h_A^{(i)}[n-1] - a_2^{(i)} h_A^{(i)}[n-2] + \delta[n]$$

and

$$x^{(i)}[n] = -a_1^{(i)} x^{(i)}[n-1] - a_2^{(i)} x^{(i)}[n-2] + x[n]$$

Here the difference equation is used to compute $x^{(i)}[n]$ instead of the convolution (9.161). The computation is possible because the coefficients $a_1^{(i)}$ and $a_2^{(i)}$ are assumed to have already been determined. For any given new estimates $a_1^{(i+1)}$, $a_2^{(i+1)}$, $b_0^{(i+1)}$, $b_1^{(i+1)}$ of the coefficients, the error (9.160) is given by

$$e^{(i+1)}[n] = x^{(i)}[n] + a_1^{(i+1)} x^{(i)}[n-1] + a_2^{(i+1)} x^{(i)}[n-2] - b_0^{(i+1)} h_A^{(i)}[n] - b_1^{(i+1)} h_A^{(i)}[n-1]$$

In order to find the new estimates for the coefficients, the error is written in matrix form for $n = 0, 1, \ldots, N_s - 1$ as

$$\begin{bmatrix} x^{(i)}[2] & x^{(i)}[1] & x^{(i)}[0] & h_A^{(i)}[2] & h_A^{(i)}[1] \\ x^{(i)}[3] & x^{(i)}[2] & x^{(i)}[1] & h_A^{(i)}[3] & h_A^{(i)}[2] \\ \vdots & \vdots & \vdots & \vdots & \vdots \\ x^{(i)}[N_s-1] & x^{(i)}[N_s-2] & x^{(i)}[N_s-3] & h_A^{(i)}[N_s-1] & h_A^{(i)}[N_s-2] \end{bmatrix} \begin{bmatrix} 1 \\ a_1^{(i+1)} \\ a_2^{(i+1)} \\ -b_0^{(i+1)} \\ -b_1^{(i+1)} \end{bmatrix}$$

$$= \begin{bmatrix} e^{(i+1)}[2] \\ e^{(i+1)}[3] \\ \vdots \\ e^{(i+1)}[N_s-1] \end{bmatrix}$$

This equation has the form of the least squares problem

$$\begin{bmatrix} \mathbf{X}^{(i)} & \mathbf{H}_A^{(i)} \end{bmatrix} \begin{bmatrix} \mathbf{a}^{(i+1)} \\ -\mathbf{b}^{(i+1)} \end{bmatrix} = \mathbf{e}^{(i+1)} \tag{9.163}$$

where

$$\mathbf{X}^{(i)} = \begin{bmatrix} x^{(i)}[P] & x^{(i)}[P-1] & \cdots & x^{(i)}[0] \\ x^{(i)}[P+1] & x^{(i)}[P] & \cdots & x^{(i)}[1] \\ \vdots & \vdots & \vdots & \vdots \\ x^{(i)}[N_s-1] & x^{(i)}[N_s-2] & \cdots & x^{(i)}[N_s-1-P] \end{bmatrix} \tag{9.164}$$

and

$$\mathbf{H}_A^{(i)} = \begin{bmatrix} h_A^{(i)}[P] & h_A^{(i)}[P-1] & \cdots & h_A^{(i)}[P-Q] \\ h_A^{(i)}[P+1] & h_A^{(i)}[P] & \cdots & h_A^{(i)}[P-Q+1] \\ \vdots & \vdots & \vdots & \vdots \\ h_A^{(i)}[N_s-1] & h_A^{(i)}[N_s-2] & \cdots & h_A^{(i)}[N_s-1-Q] \end{bmatrix} \tag{9.165}$$

where the first element of $\mathbf{a}^{(i+1)}$ is 1. Since this is exactly in the form of the least squares linear prediction problem (9.105), it leads to the Yule–Walker-type equations

$$\boxed{\left(\begin{bmatrix} \mathbf{X}^{(i)} & \mathbf{H}_A^{(i)} \end{bmatrix}^{*T} \begin{bmatrix} \mathbf{X}^{(i)} & \mathbf{H}_A^{(i)} \end{bmatrix} \right) \begin{bmatrix} \mathbf{a}^{(i+1)} \\ -\mathbf{b}^{(i+1)} \end{bmatrix} = \begin{bmatrix} \mathcal{S}^{(i+1)} \\ 0 \\ \vdots \\ 0 \end{bmatrix}} \tag{9.166}$$

which can be solved for the vector of filter parameters. The steps in the iteration are then repeated.

9.5.5 Order Selection

Order selection for an ARMA model is even more difficult than for an AR model since both a numerator and a denominator order must be chosen. In a number of applications (but not all) the denominator order is more critical than the numerator order because it determines the number of "modes" or natural frequencies present in the signal. In the Prony method it is possible to choose this order separately with the help of some of the methods discussed in the preceding section. The numerator may then be chosen to be of similar order. Alternatively, some of the AR criteria can be adapted to the ARMA case. Although it is derived from stochastic considerations, one commonly used method is the AIC, which for the ARMA case has the form

$$\text{AIC}(P, Q) = N_s \ln \sigma^2_{\varepsilon_{P,Q}} + 2(P+Q) \tag{9.167}$$

In this formula, $\sigma^2_{\varepsilon_{P,Q}}$ represents the prediction error variance for the ARMA model or the would–be power of the white noise driving source. Since in most modeling procedures the white noise is assumed to have unit variance, this term can be replaced by the squared gain of the filter $|b_0|^2$.

9.6 LEAST SQUARES METHODS AND THE YULE–WALKER EQUATIONS

The discussion of the modeling problem began with a statistical formulation in Section 9.1, proceeded through the presentation of maximum likelihood, and continued with several methods based on least squares. Let us now return to the statistical formulation of the problem and see how the techniques afforded by least squares methods can be applied to it. The specific methods presented here are due to Cadzow [29, 30] but others had also suggested the basic idea. The procedure has been called the "least squares modified Yule–Walker" method [21].

Consider the basic equation (9.8) satisfied by the ARMA model. In Section 9.1 this equation was evaluated for $n = 0$ through $n = P + Q$ in order to derive the Yule–Walker equations (9.19). Let us now continue to evaluate (9.8) up to some larger value $n = L > P + Q$ to find an *extended* set of Yule–Walker equations in the form

$$\begin{bmatrix} \mathbf{R}_B \\ \mathbf{R}_E \end{bmatrix} \mathbf{a} = \begin{bmatrix} \mathbf{c} \\ \mathbf{0} \end{bmatrix} \tag{9.168}$$

where the extended correlation matrix $\mathbf{R}_E$ is given by

$$\mathbf{R}_E = \begin{bmatrix} R_x[Q+1] & R_x[Q] & \cdots & R_x[Q-P+1] \\ \vdots & \vdots & \ddots & \vdots \\ R_x[L] & R_x[L-1] & \cdots & R_x[L-P] \end{bmatrix} \tag{9.169}$$

If the model is truly of order P, Q and R_x is the exact correlation function, then the equations of (9.19) are sufficient to solve the problem. The extra equations provided in (9.168) are redundant. They are automatically satisfied and neither help nor hinder in solving for the model parameters. In practical applications of these equations to signal modeling, however, R_x is an estimated, not a *known* correlation function. In this case the extra equations in (9.168) are usually *not* redundant and provide information that can lead to a better estimate of the model parameters. The lower partition represents an overdetermined system of equations that can be solved in a *least squares* sense. Specifically, let ϵ represent the error that results in the lower partition for a given choice of the **a** vector. That is,

$$\mathbf{R}_E\,\mathbf{a} = \epsilon \tag{9.170}$$

Since this equation is exactly in the form of (9.105), choosing **a** to minimize the square magnitude of the error vector in (9.170) leads to an equation analogous to the Yule–Walker equations:

$$\boxed{(\mathbf{R}_E^{*T}\mathbf{R}_E)\mathbf{a} = \begin{bmatrix} \mathcal{S} \\ \mathbf{0} \end{bmatrix}} \tag{9.171}$$

The "correlation matrix" that appears here is a correlation matrix of correlation terms. These equations can be solved for **a** (and $\mathcal{S}$) and **b** can be found from the top partition of (9.168) as before.[8]

In a typical application of this procedure the order P, Q is unknown. Cadzow suggests beginning with an overly high order P', Q' with $P' > P$ and $Q' > Q$ and choosing P as

[8]Actually, if the correlation function has been estimated from a given data sequence, it may be better once the AR parameters are known to use that data sequence and one of the methods discussed in Section 9.5 to estimate the MA parameters.

the effective rank of $\mathbf{R}_E$ by the SVD procedure decribed in Section 9.4.5. Usually, it is possible to make an educated guess so that the values of P' and Q' will exceed those of the actual model order P, Q. Then one of the following two methods can be used.

Method 1: Rank P, Order (P', Q') Model. Let P be chosen to be the effective rank of $\mathbf{R}_E$. The singular values (of $\mathbf{R}_E$) then satisfy

$$\left(\frac{\sigma_1^2 + \sigma_2^2 + \cdots + \sigma_P^2}{\sigma_1^2 + \sigma_2^2 + \cdots + \sigma_{P'}^2}\right)^{\frac{1}{2}} \approx 1$$

A rank P but order P' solution for the AR parameters is found by using the pseudoinverse. Specifically, let the extended correlation matrix $\mathbf{R}_E$ be partitioned as

$$\mathbf{R}_E = \begin{bmatrix} | & \\ \mathbf{r}_\text{o} & \mathbf{R}'_E \\ | & \end{bmatrix} \tag{9.172}$$

The vector $\mathbf{a}'$ of coefficients $a_1, a_2, \ldots, a_{P'}$ is then found from

$$\boxed{\mathbf{a}' = -\mathbf{R}'^{+}_E\,\mathbf{r}_\text{o}} \tag{9.173}$$

where in computing $\mathbf{R}'^{+}_E$, the terms $1/\sigma_k$ for $k > P$ are set to zero. The MA parameters are found in the usual way.

Method 2: Order (P, Q) Model. This method involves submatrices derived from the extended correlation matrix of order (P', Q'). Let matrices $\mathbf{R}_{(k)}$ be defined by selecting $P+1$ contiguous columns of the matrix $\mathbf{R}_E$. Beginning at the first column it is possible to generate $P' - P + 1$ such matrices, which have the form

$$\mathbf{R}_{(k)} = \begin{bmatrix} R_x[Q'+1-k] & \cdots & R_x[Q'-P+1-k] \\ \vdots & \ddots & \vdots \\ R_x[L-k] & \cdots & R_x[L-P-k] \end{bmatrix}; \quad k = 0, 1, \ldots, P'-P \tag{9.174}$$

As a simple example, suppose that $P' = Q' = 2$, $L = 6$, and $P = 1$. The extended correlation matrix (9.169) has the form

$$\mathbf{R}_E = \begin{bmatrix} R_x[3] & R_x[2] & R_x[1] \\ R_x[4] & R_x[3] & R_x[2] \\ R_x[5] & R_x[4] & R_x[3] \\ R_x[6] & R_x[5] & R_x[4] \end{bmatrix}$$

The submatrices are then

$$\mathbf{R}_{(0)} = \begin{bmatrix} R_x[3] & R_x[2] \\ R_x[4] & R_x[3] \\ R_x[5] & R_x[4] \\ R_x[6] & R_x[5] \end{bmatrix} \qquad \mathbf{R}_{(1)} = \begin{bmatrix} R_x[2] & R_x[1] \\ R_x[3] & R_x[2] \\ R_x[4] & R_x[3] \\ R_x[5] & R_x[4] \end{bmatrix}$$

Notice that any one of the submatrices could be used to generate a set of Pth-order equations analogous to the Yule–Walker equations

$$\mathbf{R}_{(k)}^{*T}\mathbf{R}_{(k)}\mathbf{a} = \begin{bmatrix} \mathcal{S} \\ \mathbf{0} \end{bmatrix}$$

This method effectively averages the matrices from all of these individual sets of equations and computes $\mathbf{a}$ from

$$\boxed{\left(\sum_{k=0}^{P'-P} \mathbf{R}_{(k)}^{*T}\mathbf{R}_{(k)}\right)\mathbf{a} = \begin{bmatrix} \mathcal{S} \\ \mathbf{0} \end{bmatrix}} \tag{9.175}$$

The procedure has been found to be effective in some practical applications.

9.7 EXAMPLES AND COMPARISONS

This section demonstrates the application of some of the methods in this chapter. The purpose is not to compare different methods critically, because frequently the best method depends on the problem at hand. The purpose is, rather, to demonstrate that several of the methods can produce satisfactory results and to illustrate their performance in the presence of noise. Discussions here will be qualitative rather than quantitative, and in some cases procedures for finding poles and zeros of the model will be "mixed and matched." That is, we may for instance combine the least squares modified Yule–Walker method for finding the poles with Durbin's method for finding the zeros. This illustrates that there is some inherent flexibility in the procedures described in this chapter.

The examples are based on two order (4,2) ARMA systems selected from Kay [21] and a segment of recorded acoustic data. The Kay models, denoted as ARMA1 and ARMA3, have the pole-zero plots shown in Fig. 9.12. ARMA1 has its poles considerably removed from the unit circle and thus exhibits a significantly damped impulse response. ARMA3 has poles very close to the unit circle and thus has a highly resonant behavior. Parameters for these systems are given in Table 9.2. The acoustic data corresponds to the sound of a ruler being struck on a book. The signal is shown in Fig. 9.13.

TABLE 9.2 COEFFICIENTS FOR ARMA TEST SYSTEMS

	a_1	a_2	a_3	a_4	b_0	b_1	b_2
ARMA1	−1.352	1.338	−0.662	0.240	1.000	−0.200	0.040
ARMA3	−2.760	3.809	−2.654	0.924	1.000	−0.900	0.810

Source: S. M. Kay, *Modern Spectral Estimation*, Prentice-Hall, Englewood Cliffs, New Jersey, 1988.

Both the pole-zero plots and the impulse response of the model are compared in the examples. These performance criteria are more critical than the autocorrelation function

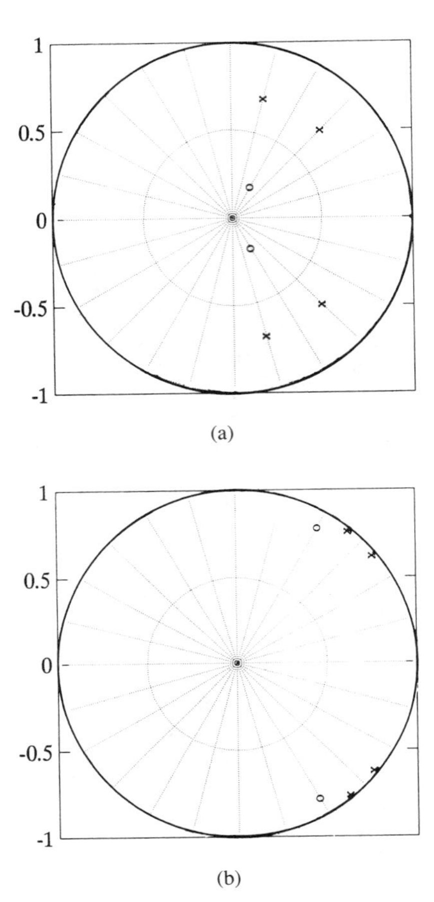

Figure 9.12 Pole-zero plots for ARMA test systems. (a) ARMA1 (damped). (b) ARMA3 (resonant).

that might be compared in a purely stochastic representation of the data, because the autocorrelation function is insensitive to phase and even nonminimum-phase systems can match the correlation function.

Let us begin with the ARMA test cases first. When data is generated from the model and there is no added noise, essentially all of the methods discussed in this chapter can find the exact parameters of the model. This includes the ordinary Yule–Walker method (if a correlation function is first generated from the data) and the Padé approximation. These methods cannot be expected to work well when the data is not exact.

The more realistic and challenging case arises when there is imprecise measurement of the signal. This can be simulated by adding white noise to the data. In this situation none of the methods estimate the model parameters precisely and it is interesting to compare how well the impulse reponse is represented and how accurately the placement of poles and zeros occurs.

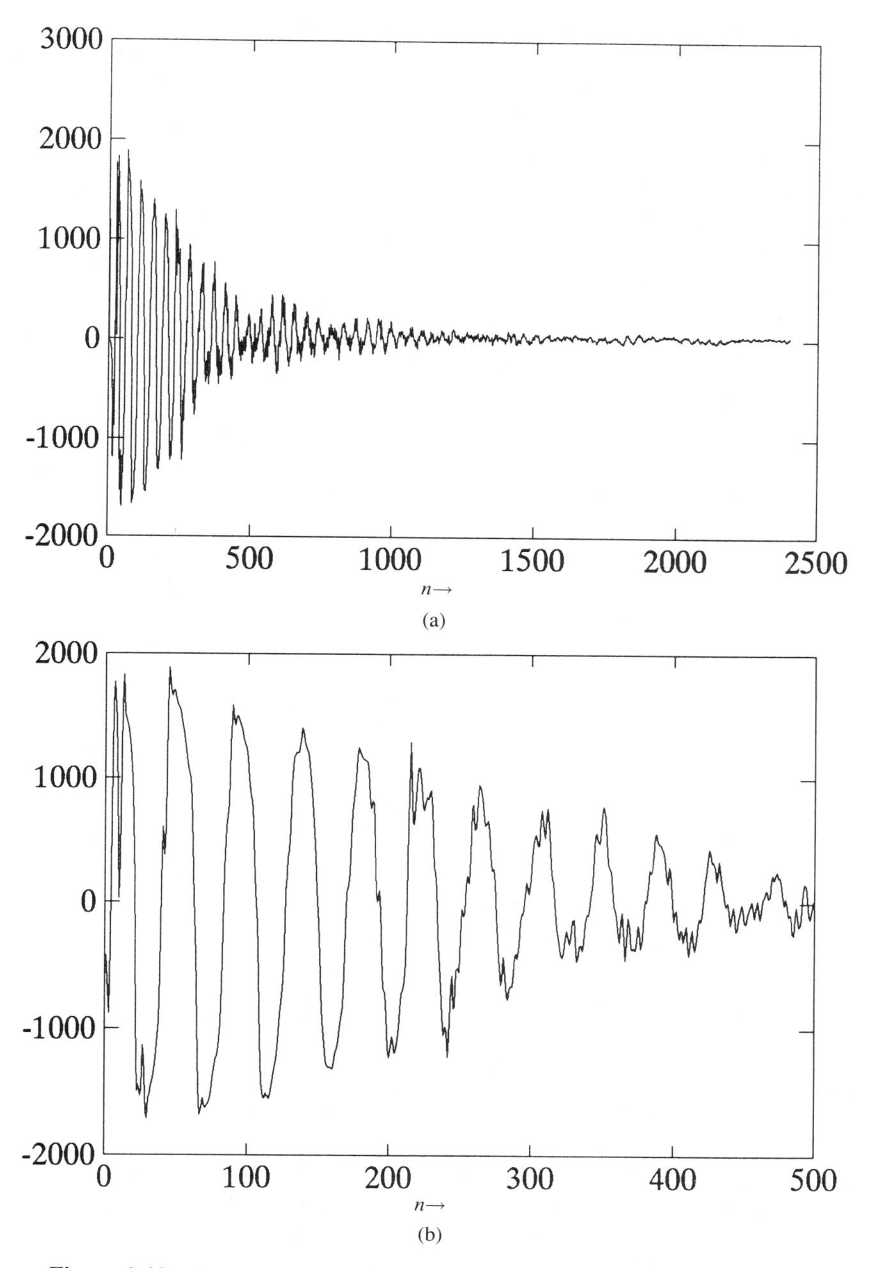

Figure 9.13 Acoustic signal corresponding to ruler struck on a book (sampled at approximately 12.5 kHz). (a) Complete signal. (b) Segment used for modeling.

The iterative prefiltering method was applied to the ARMA1 data with added white noise, resulting in a 20-dB signal-to-noise ratio. The order of the model was chosen to be the same as that of the system which produced the data [i.e., (4,2)]. The results are shown in Fig. 9.14. Although the impulse response of the model appears to match the original data fairly closely, the poles and zeros of the model are not very close to those of the original system. In particular, one pole pair occurs close to the correct distance from the origin (about 0.7) but placed approximately halfway between the actual poles of the system. The other poles occur on the real line. Zeros also occur on the real line far from the zeros of the original system, which are closer to the origin. For this example and this method a reduced-order (2,2) model performs about as well or better. The results are shown in Fig. 9.15. The impulse response looks about the same and the most important pole is in about the same location as it is for the higher-order model. Zeros have moved closer to the origin, nearer to those of the original system.

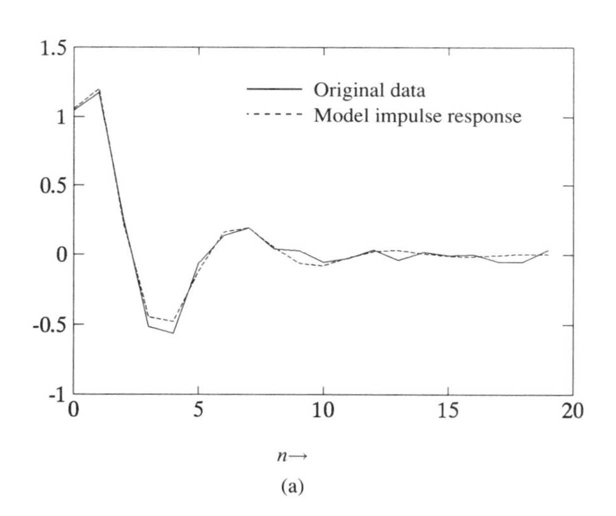

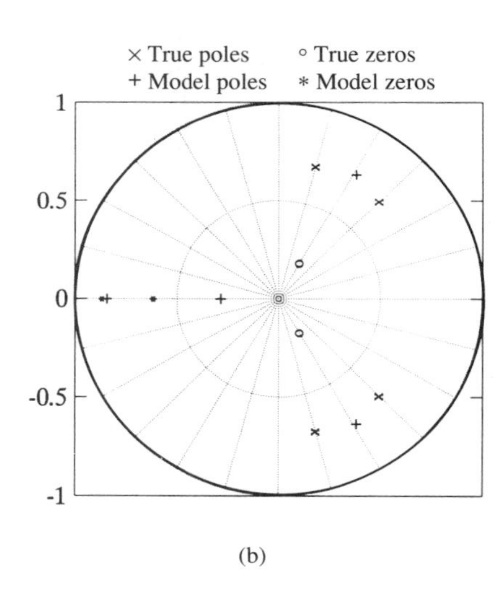

Figure 9.14 Results of iterative prefiltering applied to ARMA1 data: model order (4,2). (a) System impulse response. (b) Pole-zero plot.

In fact, the iterative prefiltering method is much more effective than it would appear from these examples. In tests on more highly resonant systems, iterative prefiltering resulted in an almost exact match of the impulse response and nearly exact placement of the poles and zeros. The present results illustrate that for a system that is not highly resonant, placement of the poles and zeros does not have to be very accurate to provide a reasonable match to the impulse response of the system.

Two other techniques, namely Prony's method and the least squares modified Yule–Walker method, were applied to the ARMA3 data with a 25-dB signal-to-noise ratio. Frequently, when there are multiple poles very close to the unit circle, it is necessary to choose

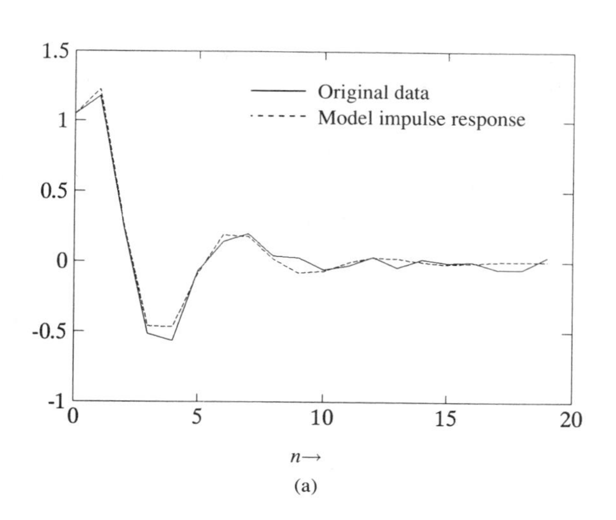

(a)

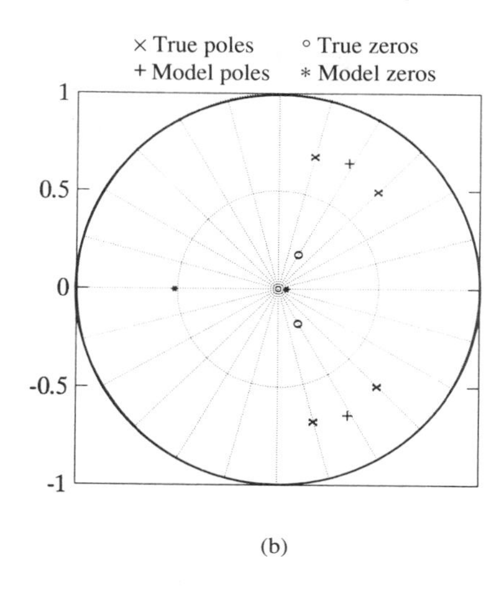

(b)

Figure 9.15 Results of reduced-order iterative prefiltering applied to ARMA1 data: model order (2,2). (a) System impulse response. (b) Pole-zero plot.

a larger model order to ensure that these poles are represented in the model.[9] The model for these examples was chosen to be of order (6,6). Figure 9.16 shows the results of application of the basic Prony method. One pole is very close to the pole of the original system, but the other is a little farther away. This causes the impulse response of the model to decay somewhat more quickly than the original data. A pair of zeros is close but does not precisely coincide with the pair of zeros from the original system. The extra poles and zeros are all in the left half plane and far enough from the unit circle to have little effect on the accuracy of the impulse response of the model. Figure 9.17 shows results of the same example but with zeros computed by Durbin's method. Zeros are slightly closer to those of the original data but the difference in the impulse response is almost imperceptible. Figure 9.18 shows the results of the least squares modified Yule–Walker method applied to the same ARMA3 data. In this case there was no attempt made to reduce the order made by either of the procedures described in Section 9.6 and zeros were determined by the spectral factorization technique. The critical poles are much closer to the correct locations and the impulse response is represented quite accurately throughout the entire range of 200 data points. If anything, the model tends to overshoot in the peaks slightly. Zeros in the vicinity of the unit circle are represented in the model as well; their locations are close but not exact. Extra poles and zeros are far from the critical singularities and well out of the way. Finally, Fig. 9.19 shows the results of the same basic technique but with Durbin's method used to find the zeros. The results are almost the same as those of the previous case, but the two zeros near the unit circle coincide almost exactly with those of the original data.

[9]The larger model order can also result in extra ("spurious") poles close to the unit circle which have an *undesirable* effect on the transient or the frequency response. Thus selection of a higher-order model has to be made with some caution.

The two zeros near the origin have also shifted position, but those zeros are spurious to begin with. The impulse response of the model tracks the original data very closely. This time the peaks in the response are a little less pronounced than those of the original data.

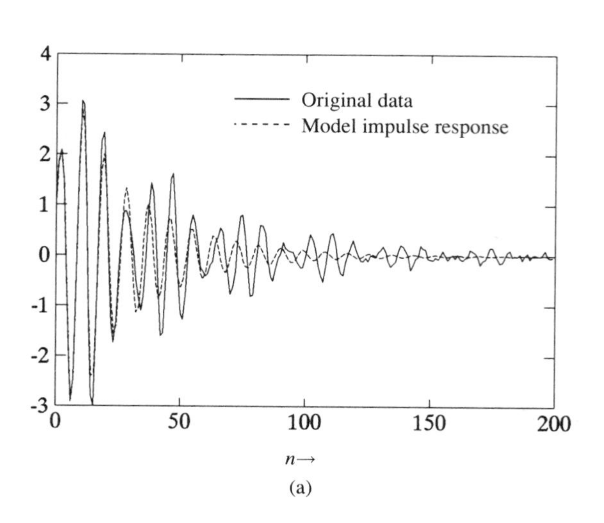

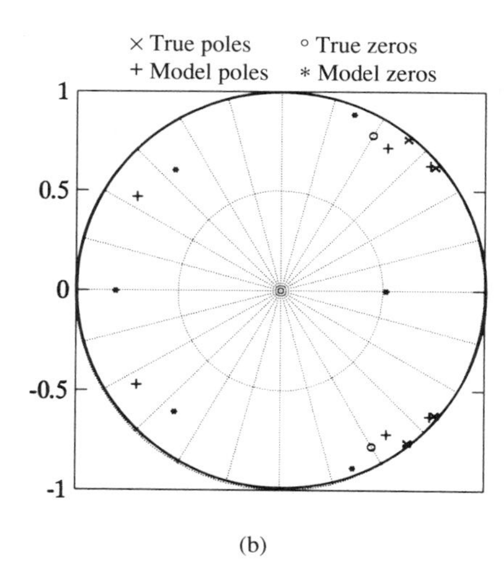

Figure 9.16 Results of basic Prony method applied to ARMA3 data: model order (6,6). (a) System impulse response. (b) Pole-zero plot.

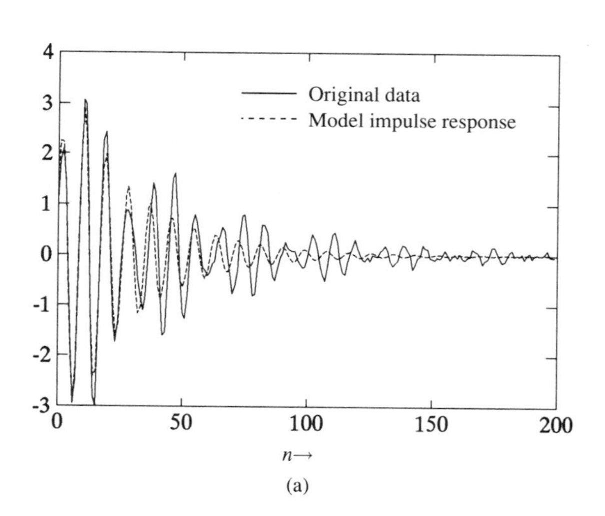

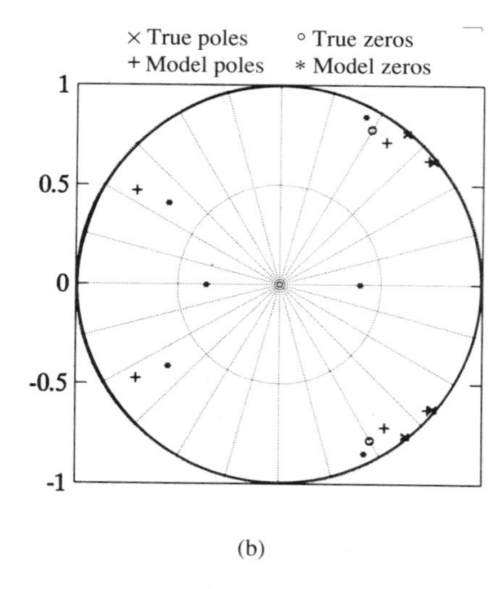

Figure 9.17 Results of Prony's method applied to ARMA3 data with Durbin's method used to find zeros: model order (6,6). (a) System impulse response. (b) Pole-zero plot.

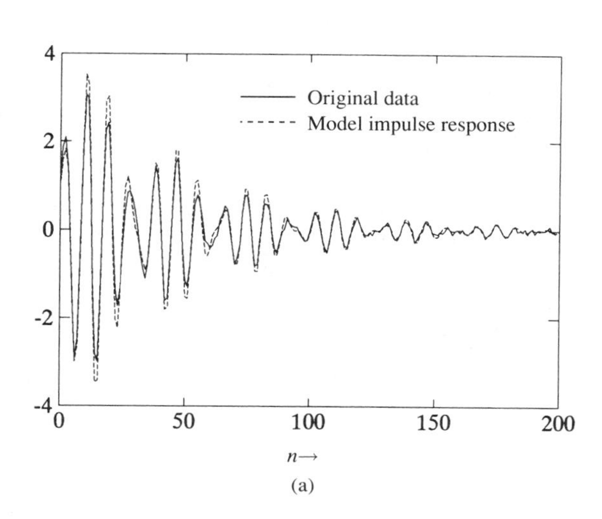

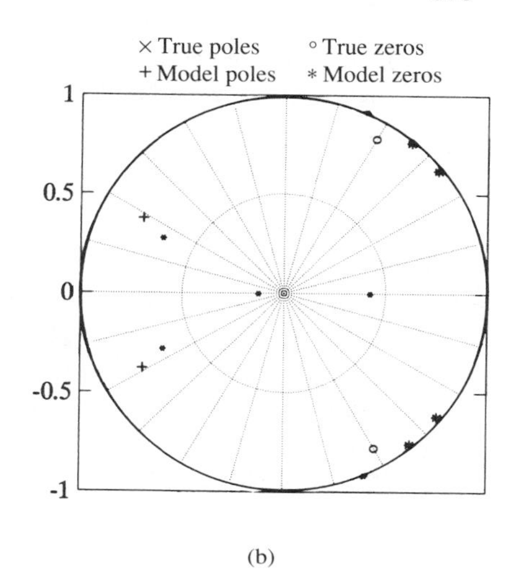

Figure 9.18 Results of least squares modified Yule–Walker method applied to ARMA3 data: model order (6,6). (a) System impulse response. (b) Pole-zero plot.

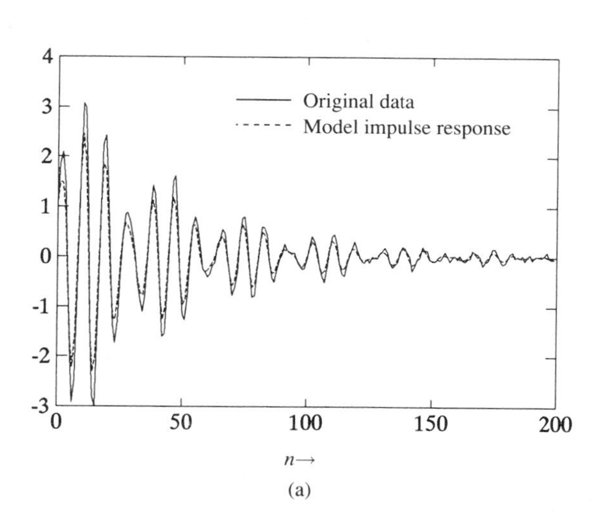

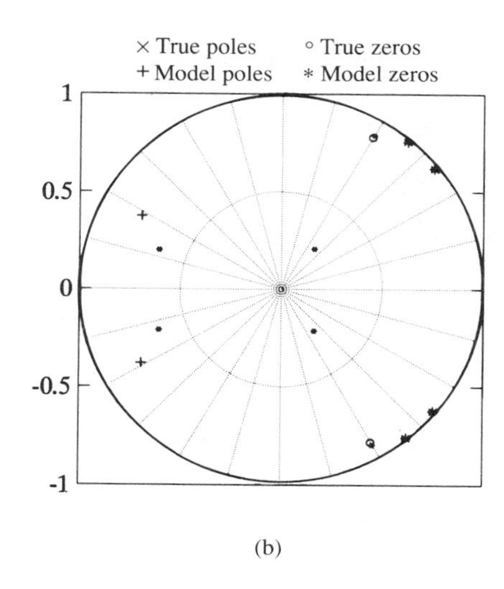

Figure 9.19 Results of least squares modified Yule–Walker method applied to ARMA3 data with Durbin's method used to find zeros: model order (6,6). (a) System impulse response. (b) Pole-zero plot.

Overall the least squares modified Yule–Walker method provided much better performance on this data than the Prony method. This is in spite of the fact that the former was developed as a procedure to model stochastic processes and the latter was developed as a method for matching a system impulse response! This underscores our earlier observation that the difference between stochastic and deterministic methods for modeling data is frequently not clear cut. Both methods applied here use correlation properties of the data but in entirely different ways. The distinction made in categorizing the methods is not nearly as important as their performance in a particular application.

For purposes of comparison the ARMA3 data was also modeled with a high-order AR process. In this example, the covariance method was used with a tenth-order model. Figure 9.20 shows the results of this AR model. The impulse response of the model matches the data quite well, perhaps even better than the Prony method results. The two critical poles are also quite accurately determined. The problem with using this method to represent ARMA data in general is that some of the extra poles could turn out to be located close to the unit circle, where they could have a significant effect on the impulse or frequency response. In addition, the effect of zeros on or near the unit circle is difficult to approximate with an all-pole model.

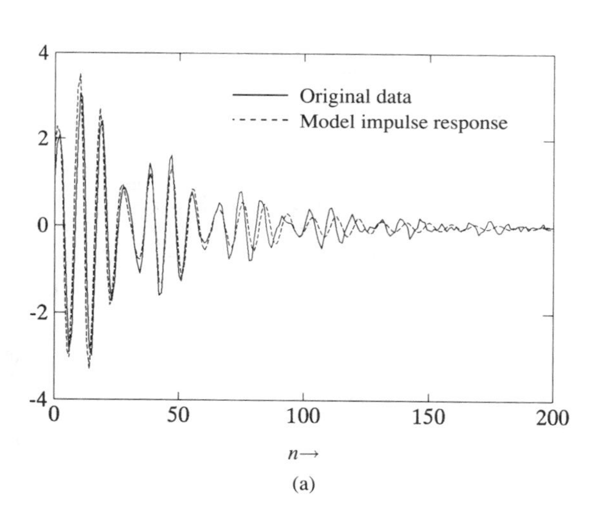

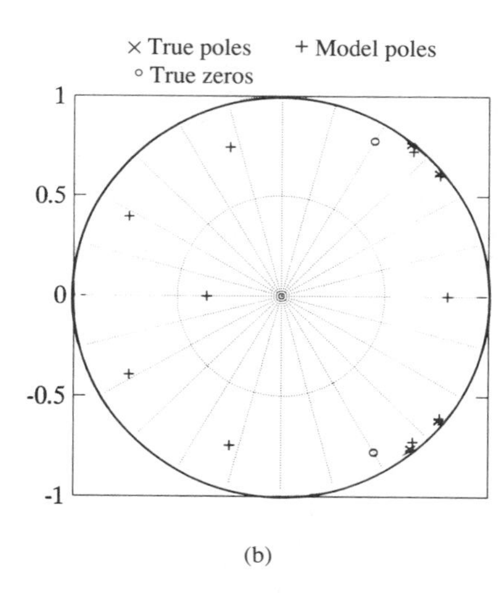

Figure 9.20 Results of the covariance method applied to ARMA3 data: 10th-order AR model. (a) System impulse response. (b) Pole-zero plot.

As a final example the least squares modified Yule–Walker method was applied to the early part of the ruler data shown in Fig. 9.13. This data is much more difficult to model because it is not derived from a mathematical equation and there is little guidance to be had concerning selection of the model order, the nature of the poles and zeros, and so on. If there is noise in the data, the true noise-free signal will never be known, so it is also impossible to compare the modeling results to the "true" results, as was done for the ARMA1 and ARMA3 cases. The ruler data is quite structured and required a relatively high order model (6,16) to obtain the results shown in Fig. 9.21. Shank's method was used to determine the MA parameters. This method allows zeros outside the unit circle which were important for determining the impulse response accurately. If we were dealing just with the stochastic properties of the data, those zeros could be placed in conjugate reciprocal locations inside the unit circle. As stated above, it is impossible to really determine the accuracy of the model since there is no "ground truth." The predominant effect in the impulse response is due to the two poles close to the unit circle near the positive real axis. The other poles may account for some high-frequency features, but overall have little effect since they are almost canceled by zeros.

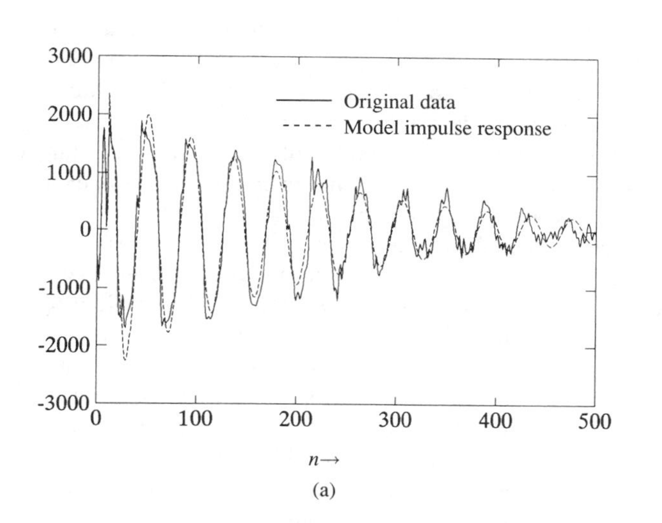

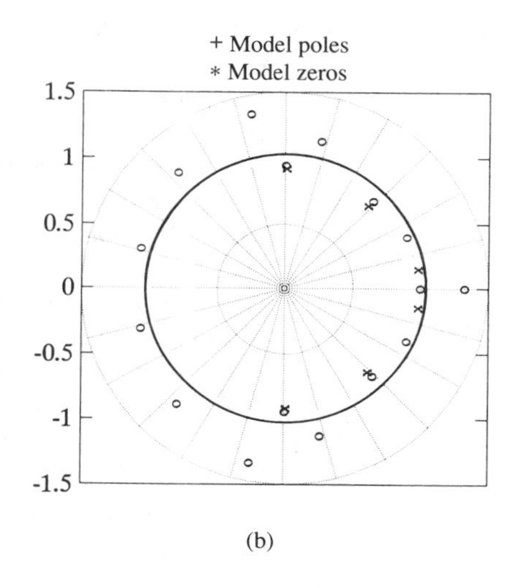

Figure 9.21 Results of least squares modified Yule–Walker method applied to ruler acoustic data with Shank's method to find zeros: model order (6,16). (a) System impulse response. (b) Pole-zero plot.

9.8 CHAPTER SUMMARY

The modeling of observed data sequences is a very important topic in statistical signal processing. This chapter develops a number of different methods to perform this modeling. Some of these methods are based on regarding the data as a realization of a random process while others treat the data as deterministic. In either case, statistical methods of one form or another are usually central to the modeling algorithms.

The most frequently used statistical signal models consist of a linear shift-invariant system driven by white noise. The system is assumed to have a rational transfer function that is usually chosen to be minimum-phase. Such linear models are characterized as either autoregressive ("all-pole"), moving average ("all-zero"), or autoregressive moving average ("pole-zero") processes. Autoregressive models were introduced in Chapter 8.

Their correlation function obeys a set of Yule–Walker equations that are equivalent to the Normal equations of linear prediction. Since the equations are linear, the model parameters can be computed easily if the correlation function for the process is known. The MA and ARMA models obey a larger set of Yule–Walker equations that are nonlinear in the model parameters. The Yule–Walker equations for an MA model can be solved by factoring the complex spectral density function. The Yule–Walker equations for an ARMA model require a combination of linear and nonlinear methods. The AR parameters are first found from the higher-order equations, which are linear. The MA parameters are then found by spectral factorization.

When the true correlation function for a process is not known, the model parameters have to be estimated from known realizations of the process (i.e., given data sequences). One approach is to estimate the correlation function from the data and use the Yule–Walker equations. Another approach is to estimate the model parameters directly from the data. If an unbiased estimate is sought, the best procedure is maximum likelihood. While maximum likelihood estimates are sometimes used or attempted, the difficulty in computing them, even in the Gaussian case, makes other methods seem quite attractive.

A number of practical methods are based on the mathematical methods of least squares. In least squares problems a sum of error terms computed from the data plays a role analogous to the mean square error for a stochastic process. The least squares optimal (Wiener) filtering problem serves as a prototype for a number of other similar problems that can be formulated using least squares principles. A least squares version of the orthogonality principle leads directly to a set of linear equations (the least squares Wiener–Hopf equations) that can be solved for the optimal FIR filter in this problem. The solution of these equations involves the pseudoinverse of the data matrix, and singular value decomposition plays an important role in computing the pseudoinverse. SVD methods also permit efficient computation of projection matrices. The projection matrix corresponding to the solution of the least squares optimal filtering problem projects the data to be estimated into a subspace spanned by the columns of the data matrix formed from the observations. A relatively new method known as total least squares holds considerable promise as a modeling technique. Its use in developing signal models for data has only recently begun to be explored, however.

Least squares forms of linear prediction are also developed in this chapter as a means for AR modeling. Two important techniques somewhat misleadingly known as the "autocorrelation method" and the "covariance method" are among those most frequently used in such problems. The autocorrelation method effectively windows the data and uses zeros in estimation of the correlation function to compute the model parameters. When the data length is short, this can lead to poor estimates of the true model parameters. However, since the method produces a positive definite Toeplitz correlation matrix, the model is always guaranteed to be minimum-phase. The covariance method, or the modified covariance method which involves both forward and backward prediction, can produce more accurate parameter estimates when the data is short. However, these methods do not result in a Toeplitz correlation matrix. This precludes use of the Levinson recursion to solve the equations and representation of the filter in lattice form. A clever and effective method due to Burg, however, combines the advantages of both procedures. This method avoids any windowing of the data but guarantees a minimum-phase model by estimating reflection coefficients in the lattice instead of the usual AR filter parameters. The form of the estimate is such that the resulting reflection coefficients are guaranteed to have magnitude less than 1.

Several approaches to ARMA modeling begin by regarding the data as the impulse response of a linear shift-invariant system. The Prony method, and the special case known as Padé approximation, involve a convenient separation of the AR and MA parts of the problem so that all coefficients can be found by solving linear equations. Prony's basic method involves a least squares solution for the AR parameters and a conventional non-least squares solution for the MA parameters. Variations that also result in a least squares solution for the MA parameters produce improvements to the basic method. Prony's method can be formulated in the signal domain as well, which leads to a representation of the sequence as a sum of damped exponentials.

Another method based on the impulse response and known as iterative prefiltering solves for both the AR and the MA parameters simultaneously. By introducing iteration this method is able to convert the inherently nonlinear problem for the filter parameters into a succession of problems involving linear equations.

Finally, a combination of ideas developed in this chapter leads to extending the Yule–Walker equations involving an estimated correlation function beyond their normal order and applying least squares methods to their solution. This method, known as the least squares modified Yule–Walker method, is quite easy to apply and seems to produce excellent results in practice.

REFERENCES

1. George E. P. Box and Gwilym M. Jenkins. *Time Series Analysis: Forecasting and Control*, rev. ed. Holden-Day, Oakland, California, 1976.
2. Charles L. Lawson and Richard J. Hanson. *Solving Least Squares Problems*. Prentice Hall, Inc., Englewood Cliffs, New Jersey, 1974.
3. Arthur A. Giordano and Frank M. Hsu. *Least Square Estimation with Applications to Digital Signal Processing*. John Wiley & Sons, New York, 1985.
4. S. T. Alexander. Fast adaptive filters: a geometrical approach. *IEEE ASSP Magazine*, 3(4), October 1986.
5. S. Thomas Alexander. *Adaptive Signal Processing Theory and Applications*. Springer-Verlag, New York, 1986.
6. Gene H. Golub and Charles F. Van Loan. *Matrix Computations*, 2nd ed. The Johns Hopkins University Press, Baltimore, Maryland, 1989.
7. Gilbert Strang. *Linear Algebra and Its Applications*, 3rd ed. Harcourt Brace Jovanovich, San Diego, California, 1976.
8. G. H. Golub and C. Reinsch. Singular value decomposition and least squares solutions. *Numerical Mathematics*, 14:403–420, 1970.
9. G. H. Golub and C. F. Van Loan. An analysis of the total least squares problem. *SIAM Journal of Numerical Analysis*, 17:883–893, 1980.

10. John Parker Burg. *Maximum Entropy Spectral Analysis.* Ph.D. thesis, Stanford University, Stanford, California, May 1975.

11. James H. McClellan. Parametric signal modeling. In Jae S. Lim and Alan V. Oppenheim, editors, *Advanced Topics in Signal Processing*, pages 1–57. Prentice Hall, Inc., Englewood Cliffs, New Jersey, 1988.

12. Albert H. Nuttall. *Spectral Analysis of a Univariate Process with Bad Data Points via Maximum Entropy and Linear Predictive Techniques.* Technical Report 5303, Naval Underwater Systems Center, New London, Connecticut, March 1976. (In volume entitled "NUSC Scientific and Engineering Studies, *Spectral Estimation.*")

13. S. Lawrence Marple, Jr. *Digital Spectral Analysis with Applications.* Prentice Hall, Inc., Englewood Cliffs, New Jersey, 1987.

14. Hirotugu Akaike. A new look at statistical model identification. *IEEE Transactions on Automatic Control*, AC-19:716–723, December 1974.

15. Emanual Parzen. Multiple time series modeling: determining the order of approximating autoregressive schemes. In P. R. Krishnaiah, editor, *Multivariate Analysis, IV*, pages 283–295. North Holland Publishing Co., New York, 1977. (Originally published by Academic Press, New York, 1969.)

16. Emanual Parzen. *An Approach to Time Series Modeling and Forecasting Illustrated by Hourly Electricity Demands.* Technical Report 37, State University of New York at Buffalo, Statistical Science Division, Amherst, New York, 1976.

17. H. Akaike. Statistical predictor identification. *Annals of the Institute of Statistical Mathematics*, 22:203–217, 1970.

18. Gideon Schwartz. Estimating the dimension of a model. *Annals of Statistics*, 6(2):461–464, 1978.

19. Jorma Rissanen. Modeling of shortest data description. *Automatica*, 14:465–471, 1978.

20. Jorma Rissanen. Stochastic complexity. *Journal of the Royal Statistical Society*, 49(3):223–239, 1987.

21. Steven M. Kay. *Modern Spectral Estimation: Theory and Application.* Prentice Hall, Inc., Englewood Cliffs, New Jersey, 1988.

22. Leland B. Jackson. *Digital Filters and Signal Processing*, 2nd ed. Kluwer, Boston, 1989.

23. H. E. Padé. Sur la représentation approchée d'une fonction par des fractions rationelles. *Annales Scientifique de l'Ecole Normale Supérieure*, 9(3):1–93; 1892 (supplement).

24. G. R. B. Prony. Essai expérimental et analytique sur les lois de la dilatabilité de fluides élastiques et sur celles de la force expansion de la vapeur de l'alcool, à différentes températures. *Journal de l'Ecole Polytechnique* (Paris), 1(2):24–76, 1795.

25. J. Durbin. Efficient estimation of parameters in moving-average models. *Biometrika*, 46:306–316, 1959.

26. John L. Shanks. Recursion filters for digital processing. *Geophysics*, 32(1):33–51, February 1967.

27. K. Steiglitz and L. E. McBride. A technique for the identification of linear systems. *IEEE Transactions on Automatic Control*, AC-10:461–464, October 1965.

28. K. Steiglitz. On the simultaneous estimation of poles and zeros in speech analysis. *IEEE Transactions on Acoustics, Speech, and Signal Processing*, ASSP-25:229–234, June 1977.

29. James A. Cadzow. Spectral estimation: an overdetermined rational model equation approach. *Proceedings of the IEEE*, 70:907–938, September 1982.

30. James A. Cadzow. Spectral analysis. In Douglas F. Elliott, editor, *Handbook of Digital Signal Processing*, pages 701–740. Academic Press, New York, 1987.

PROBLEMS [10]

9.1. A finite-length corelation function is the same as in Example 9.1 except that $R_x[\pm 1] = \frac{3}{2}$.
(a) Find the MA model for $Q = 2$.
(b) Does an MA model exist for $Q = 1$? If so, find it.

9.2. The correlation function for a real random process is known to be

$$R_x[l] = 0.1\,(0.8)^{|l|} + 0.1\,\delta[l]$$

(a) By formulating and solving the Yule–Walker equations, find an order (1,1) ARMA model for this process.
(b) What are the parameters produced by the Yule–Walker equations for an order (2,2) ARMA model for this process? Can you think of another way to verify your answer?

9.3. **(a)** Write out the specific form of (9.32) for an order (2,2) ARMA model with $N_s = 5$.
(b) For simplicity, assume that all parameters are real. By taking the derivative of (9.34) with respect to the parameters, show that the conditional maximum likelihood solution involves nonlinear equations.

9.4. Find the least squares optimal filter to estimate the data

$$\{\mathrm{d}[n]\,;0 \leq n \leq 4\} = \{1,\ 1,\ 1,\ 1,\ 1\}$$

from the observations

$$\{\mathrm{x}[n]\,;0 \leq n \leq 4\} = \{0.5,\ 1.5,\ 3.5,\ 3.5,\ 4.5\}$$

Take the order of the FIR filter to be $P = 2$, and the beginning and end points of the interval to be $n_I = 1$, $n_F = 4$. What is the minimum sum of squared errors $\mathcal{S}$?

*****9.5.** The sequence

$$\{\mathrm{d}[n]\,;0 \leq n \leq 4\} = \{1, -2,\ 3, -4,\ 5\}$$

is to be estimated from the sequence

$$\{\mathrm{x}[n]\,;0 \leq n \leq 4\} = \{1, -1,\ 1, -1,\ 1\}$$

Find an FIR optimal filter with $P = 2$ to perform this estimation. Take the end points of the interval to be $n_I = 1$, $n_F = 4$. What is the estimate for d and the corresponding sum of squared errors $\mathcal{S}$? Note that the solution to the Wiener–Hopf equations should be the *minimum norm least squares estimate* of the sequence.

9.6. Find the pseudoinverse $\mathbf{X}^{+}$, the projection matrices $\boldsymbol{P}_{\mathbf{X}}$ and $\boldsymbol{P}_{\mathbf{X}}^{\perp}$ corresponding to the data in Problem 9.4. Use these to verify the results of Problem 9.4. Also demonstrate that $\mathbf{X}^{+}\mathbf{X} = \mathbf{I}$.

[10]*** indicates that a computer program is necessary to perform the numerical computations.

9.7. Show that the following identities hold for the pseudoinverse:

$$\mathbf{X}^{+} = (\mathbf{X}^{*T}\mathbf{X})^{+}\mathbf{X}^{*T}$$

$$\mathbf{X}^{+} = \mathbf{X}^{*T}(\mathbf{X}\mathbf{X}^{*T})^{+}$$

Notice that in the second expression if $K < P$ and $\mathbf{X}$ has full row rank (K) that the identity implies that the pseudoinverse can be expressed as

$$\mathbf{X}^{+} = \mathbf{X}^{*T}(\mathbf{X}\mathbf{X}^{*T})^{-1}$$

*****9.8.** Use a computer program to find the SVD of the data matrix in Problem 9.4. What are the singular values? Compute $\mathbf{X}^{+}$ and $P_{\mathbf{X}}$ using the SVD to verify the results of Problem 9.6.

*****9.9.** **(a)** Compute the projection matrix corresponding to the data in Problem 9.5. Verify your estimate $\hat{\mathbf{d}}$ computed in that problem, using the projection matrix.
(b) Is it true that $\mathbf{X}^{+}\mathbf{X} = \mathbf{I}$ for this problem? State why or why not.

9.10. Show that the eigenvalues of any projection matrix P are either 1 or 0 and that the eigenvectors are any set of orthonormal vectors such that any vector is either completely contained within the subspace or orthogonal to it. Use this to show that any projection matrix has a canonical representation as

$$P = \sum_{i=1}^{P} \mathbf{e}_i \mathbf{e}_i^{*T}$$

where $\mathbf{e}_1, \ldots, \mathbf{e}_P$ are the eigenvectors spanning the subspace, assumed to be of dimension P.

9.11. Beginning with the relation

$$\|\mathbf{A}\|_F^2 = \text{tr}\ \mathbf{A}\mathbf{A}^{*T}$$

show that the squared Frobenius norm is equal to the sum of the squared singular values of the matrix $\mathbf{A}$.

9.12. Show that the matrix $\overline{\mathbf{\Delta}}$ given by (9.89) is the matrix of *minimum Frobenius norm* that makes the matrix $\overline{\mathbf{X}}$ of Section 9.3.5 rank deficient. You may first want to express any arbitrarily chosen $\overline{\mathbf{\Delta}}$ as

$$\overline{\mathbf{\Delta}} = \overline{\mathbf{U}}\mathbf{A}\overline{\mathbf{V}}^{*T}$$

where $\overline{\mathbf{U}}$ and $\overline{\mathbf{V}}$ are the matrices of left and right singular vectors of $\overline{\mathbf{X}}$, and show that

$$\|\overline{\mathbf{\Delta}}\|_F = \|\mathbf{A}\|_F$$

9.13. Beginning with the hypothesized data sequence

$$\{x[n]\} = \{-1,\ 2,\ 3,\ 4,\ 5,\ 6\}$$

demonstrate that the correlation matrix for the autocorrelation method is Toeplitz while the one for the covariance method is not. Solve the least squares Yule–Walker equations of the forward and backward models for the covariance method and show that the results are entirely different. You may use $P = 2$.

*****9.14.** In this problem you will need to use a calculator or computer with a high degree of precision (7 or 8 significant figures) to make a meaningful comparison of the results. Consider the data sequence

$$\{x[n]\} = \{1,\ \tfrac{1}{2},\ \tfrac{1}{4},\ \tfrac{1}{8},\ \tfrac{1}{16}\}$$

(a) Find a first-order AR model for this data using the autocorrelation method.
(b) Find a first-order AR model for this data using the covariance method.
(c) Find a second-order AR model for this data using the autocorrelation method.
(d) Find a second-order AR model for this data using the covariance method.
(e) Summarize your results.

9.15. Find first- and second-order AR models for the data in Problem 9.14 using the modified covariance method.

9.16. Find a lattice representation for the data in Problem 9.14 using Burg's method. Compute the parameters up to second order.

9.17. Show that if $\mathbf{u}$ and $\mathbf{v}$ are any complex vectors, then

$$\frac{2|\mathbf{u}^{*T}\mathbf{v}|}{\|\mathbf{u}\|^2 + \|\mathbf{v}\|^2} \leq 1$$

Thus show that the estimate for the reflection coefficient (9.130) computed in the Burg algorithm always has magnitude ≤ 1. (*Hint*: Consider the vectors defined by

$$\mathbf{w}_1 = \begin{bmatrix} \mathbf{u} & | & \mathbf{v} \end{bmatrix} \begin{bmatrix} c_1 \\ c_2 \end{bmatrix} \qquad \mathbf{w}_2 = \begin{bmatrix} \mathbf{u}^* & | & \mathbf{v}^* \end{bmatrix} \begin{bmatrix} c_1 \\ c_2 \end{bmatrix}$$

where c_1 and c_2 are any arbitrary complex numbers. Form the sum $\|\mathbf{w}_1\|^2 + \|\mathbf{w}_2\|^2$ and write it as a quadratic product involving the vector

$$\mathbf{c} = \begin{bmatrix} c_1 \\ c_2 \end{bmatrix}$$

Show that since the expression is nonnegative, the inequality of this problem is satisfied. Can you think of a simpler way to prove the result in the real case?)

9.18. Consider the data sequence

$$\mathrm{x}[n] = \begin{cases} \rho^n; & 0 \leq n \leq N_s - 1 \\ 0; & \text{otherwise} \end{cases}$$

where ρ is a real number.

(a) Find an expression for the parameter a_1 in the first-order AR model for this data using the covariance method. Under what conditions does this represent a stable model? What is the sum of squared errors over the interval over which the sum is minimized?
(b) Repeat part **(a)** for the autocorrelation method.

9.19. Repeat Problem 9.18(a) for the backward covariance method and the modified covariance method.

9.20. The prewindowed version of the covariance method minimizes the sum of squared errors over the interval $n = 0$ to $n = N_s - 1$. Find an expression for the parameter a_1 for the data of Problem 9.18 and the corresponding sum of squared errors.

9.21. Find an expression for the first-order reflection coefficient for the data of Problem 9.18 using Burg's method. Are the values of the higher-order reflection coefficients zero or nonzero?

9.22. Consider a data sequence from a Pth-order AR process with reflection coefficients given by

$$\gamma_1 = \gamma_2 = \cdots = \gamma_P = \frac{1}{\sqrt{2}}$$

Find the optimum model order P' given by each of the criteria in Table 9.1. Assume that $N_s >> P$.

9.23. Consider a signal with z-transform

$$\mathrm{X}(z) = \frac{1 + z^{-1}}{1 - \frac{1}{2}z^{-1}}$$

(a) Compute the first five values of the sequence $\mathrm{x}[0], \mathrm{x}[1], \mathrm{x}[2], \mathrm{x}[3], \mathrm{x}[4]$.

(b) Find a second-order AR model for this signal using Padé approximation. Compare the first five terms of the impulse response to those of the sequence $\mathrm{x}[n]$.

(c) Find a second-order MA model for this signal using Padé approximation. Again compare the first five terms of the impulse response to those of the sequence $\mathrm{x}[n]$.

(d) Write the equations for an order (2,2) ARMA model for this signal using Padé approximation. Note that the equations have multiple solutions.

(i) What is the minimum norm solution?

(ii) What is the solution of lowest order?

(iii) What is another possible solution?

Check to see that each of these solutions matches the five terms computed in part (a).

9.24. Consider the data sequence

$$\{\mathrm{x}[n]\,; 0 \le n \le 4\} = \{1,\ 2,\ 3,\ 2,\ 1\}$$

(a) Compute the parameters of a first-order (1,1) ARMA model by Padé approximation. Sketch the impulse response of the model and compare it to $\mathrm{x}[n]$ over the interval $0 \le n \le 4$.

(b) Compute the ARMA parameters using the basic Prony method. Again compare the model response to the data.

(c) Compute the ARMA parameters using the Prony method and Shanks' method to find the MA parameters. Compare the model response to the data and to the model of part (b).

9.25. Consider the sequence

$$\{\mathrm{x}[n]\,; 0 \le n \le 7\} = \{-2, -1,\ 1,\ 1,\ 2,\ 1, -1, -1\}$$

(a) Find a rational model with $P = Q = 2$ that will *exactly* match the first five values of the sequence. Compute the error $\mathrm{e}[n]$ produced by the model for $0 \le n \le 7$.

(b) Find an approximate model with $P = Q = 2$ by Prony's method. Compute the error sequence for $0 \le n \le 7$.

9.26. For the data in Problem 9.25, determine the MA parameters by Shanks' method. Use the AR parameters determined in part (b) of Problem 9.25 to form the all-pole part of the filter $1/A(z)$.

9.27. Consider the sequence

$$x[n] = n + 1; \qquad 0 \le n \le 7$$

Find the model coefficients for $P = Q = 2$ using Prony's method. What is the error sequence for $0 \le n \le 7$?

9.28. Find an order (1,1) ARMA model for the data sequence

$$x[n] = \begin{cases} 1; & 0 \le n \le N_s - 1 \\ 0; & \text{otherwise} \end{cases}$$

using Prony's method.

9.29. For the data in Problem 9.25, determine an approximation in the exponential form

$$x[n] \approx c_1 r_1^n + c_2 r_2^n$$

(a) Use only the data for $0 \le n \le 4$ to solve the problem.
(b) Use all of the data for $0 \le n \le 7$ to determine the roots r_1 and r_2 and use all of the data to find c_1 and c_2.

9.30. Compute an order (2,2) representation of the data in Problem 9.24 using the signal domain version of Prony's method.

COMPUTER ASSIGNMENTS

9.1. This computer assignment compares the autocorrelation method and the covariance method in developing AR models for data. The data sets to be used are called DATAA, DATAB, DATAC, DATAD, and DATAE and are on the enclosed diskette. You may want to make plots of these data sets and their estimated correlation functions.

(a) Apply both the autocorrelation method and the covariance method to each of the data sets to generate a fourth-order AR model (including the white noise variance). Do this for three cases:

(i) using only the first 20 samples of each data set.
(ii) using only the first 100 samples of each data set.

(b) Examine the correlation matrices produced by the two methods for the 20-sample case and note any differences.
(c) Make a table comparing the AR model parameters computed for each of the cases above.
(d) Compute the roots of the polynomials and show the poles in a polar plot containing the unit circle. Note any evidence of instability.
(e) (Optional) Repeat parts (a) through (d) for a sixth-order AR model for each of the data sets. Can you conclude anything about the possible order of the data from these experiments?

9.2. Repeat Computer Assignment 9.1 using the modified covariance method. Compare results to those of Computer Assignment 9.1.

9.3. Repeat Computer Assignment 9.1 using Burg's method. Compare results to those of Computer Assignments 9.1 and 9.2.

9.4. The sequences T01, T02, T03, and T04 on the diskette each represent the impulse response of a different linear shift-invariant system.

- **(a)** Plot each of the sequences.
- **(b)** Using the basic Prony method, compute an order (4,3) ARMA model for each sequence. Plot and compare the sequences.
- **(c)** Plot the poles and zeros of the models found in part (b).
- **(d)** Repeat parts (b) and (c) but use Shank's method (see Section 9.5.2) to find the MA parameters. How do the poles and zeros of these new models compare to those found by the basic Prony method?
- **(e)** Data sets T01_N, T02_N, T03_N, and T04_N have a small amount of added noise with a signal-to-noise ratio of 25 dB. Repeat parts (a) through (d) for these noisy data sets. How does the noise affect the results?

9.5. Find order (4,3) ARMA models for the data of Computer Assignment 9.4 by computing a correlation function from the data and applying the least squares modified Yule–Walker method to obtain the models. Compare your results to those of Computer Assignment 9.4.

9.6. Compute an ARMA model for the data sets DATAE and DATAG using the least squares modified Yule–Walker method to find the denominator terms and Durbin's method to find the numerator terms. Use order (2,2) for DATAE and order (4,2) for DATAG. Compare your results on DATAE to those obtained by AR modeling in Computer Assignment 9.1.

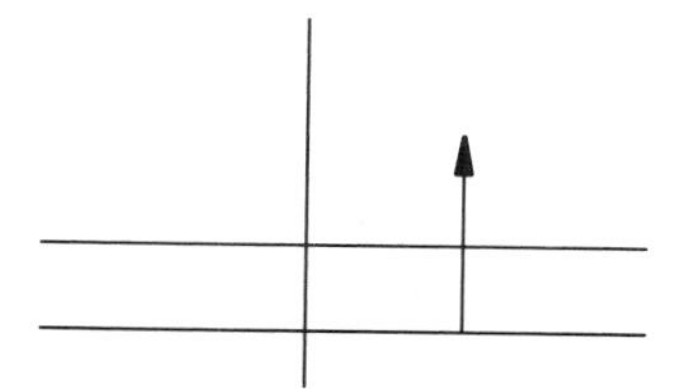

10

Spectrum Estimation

This chapter provides an introduction to a variety of methods of spectrum estimation. The original or "classical" methods are based directly on the Fourier transform. Some of these methods were used in the early 1950s and gained additional ground in the mid-1960s with the development of the fast Fourier transform.

The rest of the spectrum estimation methods presented in this chapter have all been developed since the late 1960s and some are as recent as the early 1980s. It is tempting to call them *modern* because the methods are generally newer than the Fourier-based methods; but "modern" is a relative term and these methods may soon be (if they are not already) regarded as classic. Other terms, such as *model-based, parametric, data adaptive,* and *high resolution*, have also been used to characterize these methods. The most recent methods are based on the linear algebraic concepts of subspaces associated with a data matrix or a correlation matrix and so have been called "subspace methods." All of the later methods are fundamentally different from the classical methods in that they are *not* based on Fourier transformation of the data sequence or its estimated correlation function.

All of the methods discussed in this chapter have strengths and weaknesses, so that sometimes one or another or a combination is best for a particular problem. Since entire volumes have been devoted to this subject it is impossible to be even nearly complete in a single chapter. However, after reading this you will at least be aware of the different types of techniques currently available and some of their relative advantages and disadvantages.

More information on the classical methods can be found in the books [1, 2]. An excellent source of information on most of the other methods is the review article by Kay and Marple [3] and these authors' two books [4, 5], which also cover the classical methods. A collection of some of the most important original papers since the late 1960s can be found in the two volumes [6, 7] published by the IEEE.

A word about notation is appropriate here before we begin. Throughout the chapter the variable N denotes the number of samples on which the spectral estimate is based while N_s denotes the total number of samples of the available data. In some cases these numbers will be the same, but in most cases an $N \times N$ correlation matrix needs to be estimated from the N_s data samples. The distinction is simple but important. When classical estimates are made by dividing the available data into segments, the segments are also of length N. In discussing the autoregressive or maximum entropy method in Section 10.2 the familiar variable P is used to refer to the order of the model. In this case $P = N - 1$. This variable is also used as the prediction order in Section 10.4.5. Keeping the size of the correlation matrix constant permits the various methods to be compared more easily.

10.1 CLASSICAL SPECTRUM ESTIMATION

Classical methods of spectrum estimation are based on the Fourier transform of the data sequence or its correlation function. In spite of all the developments in newer more "modern" techniques, classical methods are often the favorite when the data sequence is long and stationary. These methods are straightforward to apply and make no assumptions (other than stationarity) about the observed data sequence (i.e., the methods are *nonparametric*).

10.1.1 Correlogram and Periodogram Estimates

The power density spectrum is defined as the Fourier transform of the autocorrelation function. Since there are simple methods for estimating the correlation function, it seems that a resonable procedure for estimating the power density spectrum is to take the Fourier transform of the *estimated* correlation function. Recall that two forms of the sample correlation function are discussed in Chapter 6; one is biased and the other is not. Because the unbiased form is not guaranteed to be positive semidefinite, however, it can produce negative spectral estimates. Therefore, the *biased* form of the sample correlation function is usually preferred in spectrum estimation. This point is discussed in greater depth in Section 10.1.2.

Suppose that the total number of data samples is N_s. Then recall that the biased sample autocorrelation function is defined by

$$\hat{R}_x[l] = \frac{1}{N_s} \sum_{n=0}^{N_s-1-l} x[n+l]x^*[n]; \qquad 0 \leq l < N_s \tag{10.1}$$

with $\hat{R}_x[l] = \hat{R}_x^*[-l]$ for $l < 0$. This estimate is asymptotically unbiased and consistent. Specifically, it is shown in Chapter 6 that the expected value of the estimate is given by

$$\mathcal{E}\{\hat{R}_x[l]\} = \frac{N_s - |l|}{N_s} R_x[l] \tag{10.2}$$

and its variance for small values of lag decreases as $1/N_s$. It seems reasonable therefore to define a spectral estimate as

$$\hat{S}_x(e^{j\omega}) = \sum_{l=-L}^{L} \hat{R}_x[l]e^{-j\omega l} \; ; \qquad L < N_s \tag{10.3}$$

This estimate for the power density spectrum is known as the *correlogram.*[1] It is typically used with large N_s and relatively small values of L (say $L \leq 10\% N_s$), for reasons that will become clear shortly.

Now suppose that the maximum lag L is taken to be equal to $N_s - 1$. Then it can be shown (see Problem 10.1) that the spectral estimate has the form

$$\hat{S}_x(e^{j\omega}) = \sum_{l=-N_s+1}^{N_s-1} \hat{R}_x[l]e^{-j\omega l} = \frac{1}{N_s}|X(e^{j\omega})|^2 \tag{10.4}$$

where

$$X(e^{j\omega}) = \sum_{n=0}^{N_s-1} x[n]e^{-j\omega n} \tag{10.5}$$

is the Fourier transform of the data sequence. This estimate is called the *periodogram* and will be denoted by

$$\hat{P}_x(e^{j\omega}) = \frac{1}{N_s}|X(e^{j\omega})|^2 \tag{10.6}$$

The periodogram and variations of it are popular for spectrum estimation because the computation is especially convenient. In practice, an FFT program is used to compute the transform of the given data sequence and the formula (10.6) is applied to compute the spectral estimate. Note that the connection between the periodogram and the correlogram estimate exists only when the biased form (10.1) of the sample autocorrelation function is used.

10.1.2 Statistical Properties of the Periodogram

Although the sample correlation function is asymptotically unbiased and consistent, the spectral estimates derived from it do not necessarily inherit these properties. In particular, the lack of consistency is a fundamental problem in spectrum estimation and most classical spectrum estimation methods are concerned with ways to circumvent this difficulty. This issue is investigated briefly here.

[1]The term "correlogram" is also used by others to refer to a plot of the time history of the sample correlation function computed for nonstationary data. This type of correlogram is entirely different from what is defined by (10.3).

Mean of the Periodogram. The mean of the periodogram can be computed by taking the expectation of the first expression in (10.4) and using (10.2). This yields

$$\mathcal{E}\{\hat{P}_x(e^{\jmath\omega})\} = \sum_{l=-N_s+1}^{N_s-1} \mathcal{E}\{\hat{R}_x[l]\}\, e^{-\jmath\omega l} = \sum_{l=-N_s+1}^{N_s-1} \frac{N_s-|l|}{N_s} R_x[l] e^{-\jmath\omega l} \tag{10.7}$$

The latter result can be expressed more conveniently by using the Bartlett (triangular) window which is defined as

$$\mathrm{w}_B[l] = \begin{cases} \frac{N_s-|l|}{N_s}; & |l| < N_s \\ 0; & \text{otherwise} \end{cases} \tag{10.8}$$

This window is depicted in Fig. 10.1(a). With this definition (10.7) becomes

$$\mathcal{E}\{\hat{P}_x(e^{\jmath\omega})\} = \sum_{l=-\infty}^{\infty} \mathrm{w}_B[l] R_x[l] e^{-\jmath\omega l} \tag{10.9}$$

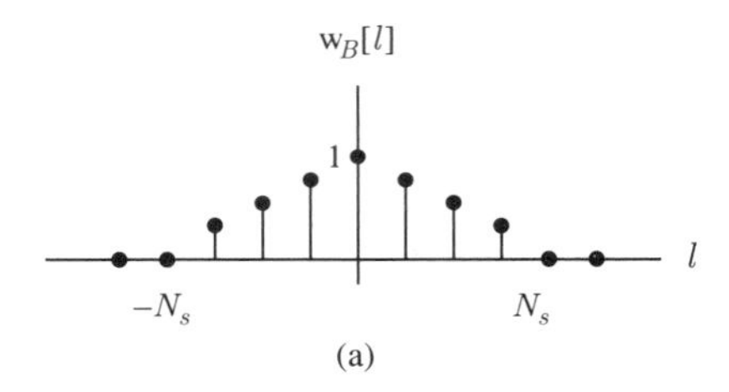

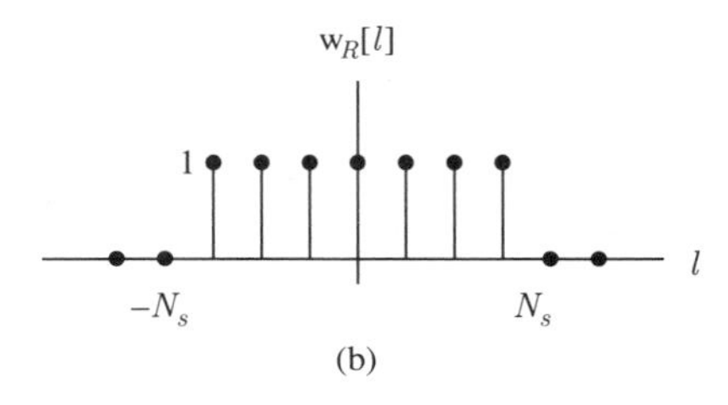

Figure 10.1 Types of windows. (a) Bartlett. (b) Rectangular.

or in the frequency domain,

$$\mathcal{E}\{\hat{P}_x(e^{\jmath\omega})\} = \frac{1}{2\pi} \mathrm{W}_B(e^{\jmath\omega}) \circledast S_x(e^{\jmath\omega}) \tag{10.10}$$

where W_B is the Fourier transform of the Bartlett window

$$\mathrm{W}_B(e^{\jmath\omega}) = \frac{1}{N_s} \frac{\sin^2(N_s\omega/2)}{\sin^2(\omega/2)} \tag{10.11}$$

and "$\circledast$" denotes the *periodic convolution*

$$\mathrm{W}_B(e^{\jmath\omega}) \circledast S_x(e^{\jmath\omega}) \stackrel{\text{def}}{=} \int_{-\pi}^{\pi} \mathrm{W}_B(e^{\jmath\theta}) S_x(e^{\jmath(\omega-\theta)}) d\theta \tag{10.12}$$

Equation 10.10 shows that the periodogram is a biased estimator, since the expected value of $\hat{P}_x(e^{\jmath\omega})$ is not equal to $S_x(e^{\jmath\omega})$. It is asymptotically unbiased, however, since as N_s becomes very large, $\mathrm{W}_B(e^{\jmath\omega})$ approaches an impulse in the frequency domain with area 2π. That is, in the limit of large N_s, the term on the right of (10.10) becomes

$$\lim_{N_s\to\infty} \frac{1}{2\pi}\mathrm{W}_B(e^{\jmath\omega})\circledast S_x(e^{\jmath\omega}) \;\rightarrow\; S_x(e^{\jmath\omega})$$

so the expected value of the periodogram is the true power spectral density function.

As a brief excursion, consider what happens if the unbiased sample correlation function instead of the biased sample correlation function is used to produce the spectral estimate. Call this estimate $\hat{S}'_x(e^{\jmath\omega})$. Following the procedure that began with (10.7) leads to

$$\begin{aligned}\mathcal{E}\{\hat{S}'_x(e^{\jmath\omega})\} &= \sum_{l=-N_s+1}^{N_s-1} \mathcal{E}\{\hat{R}'_x[l]\}\, e^{-\jmath\omega l} = \sum_{l=-N_s+1}^{N_s-1} R_x[l]e^{-\jmath\omega l} \\ &= \sum_{l=-\infty}^{\infty} \mathrm{w}_R[l]R_x[l]e^{-\jmath\omega l}\end{aligned} \tag{10.13}$$

where $\mathrm{w}_R[l]$ is the *rectangular* window

$$\mathrm{w}_R[l] = \begin{cases} 1; & |l| < N_s \\ 0; & \text{otherwise}\end{cases} \tag{10.14}$$

depicted in Fig. 10.1(b). Equation 10.13 can be written in the frequency domain as

$$\mathcal{E}\{\hat{S}'_x(e^{\jmath\omega})\} = \frac{1}{2\pi}\mathrm{W}_R(e^{\jmath\omega})\circledast S_x(e^{\jmath\omega}) \tag{10.15}$$

where $\mathrm{W}_R(e^{\jmath\omega})$ is the Fourier transform of the rectangular window

$$\mathrm{W}_R(e^{\jmath\omega}) = \frac{\sin((2N_s-1)\omega/2)}{\sin(\omega/2)} \tag{10.16}$$

This estimate is also asymptotically unbiased. The essential difference between (10.10) and (10.15), however, is that while the transform (10.11) of the Bartlett window is positive or zero for every value of ω, the transform (10.16) of the rectangular window is *negative* in certain regions. As a result, the expected value of the spectral estimate using the unbiased sample correlation function can also be negative! All of this is a manifestation of the fact that the unbiased sample correlation function is not guaranteed to be positive semidefinite. The unbiased correlation function is therefore rather undesirable for spectral estimation.

Variance and Covariance of the Periodogram. The computation of the variance and covariance of the periodogram is not too difficult for a Gaussian random process, although the procedure is a bit tedious. The result is very important, however, and leads to a clear understanding of how the available data should be used in spectral estimation. The result is first shown for a complex Gaussian white noise process $w[n]$ with variance σ_w^2, then extended to the case of an arbitrary Gaussian random process.

Denote the periodogram of $w[n]$ by $\hat{P}_w(e^{j\omega})$. Then to compute the covariance let us first evaluate the correlation between samples of the periodogram at two points ω_1 and ω_2. Begin with the definition

$$\mathcal{E}\{\hat{P}_w(e^{j\omega_1})\hat{P}_w(e^{j\omega_2})\} = \mathcal{E}\left\{\frac{1}{N_s^2}|W(e^{j\omega_1})|^2|W(e^{j\omega_2})|^2\right\} \tag{10.17}$$

where $W(e^{j\omega})$ is the Fourier transform of the data sequence $w[n]$. Then expanding this and rearranging yields

$$\begin{aligned}
&\mathcal{E}\{\hat{P}_w(e^{j\omega_1})\hat{P}_w(e^{j\omega_2})\} \qquad\qquad (10.18)\\
&= \frac{1}{N_s^2}\mathcal{E}\left\{\sum_{n_1=0}^{N_s-1} w[n_1]e^{-jn_1\omega_1}\sum_{k_1=0}^{N_s-1} w^*[k_1]e^{+jk_1\omega_1}\sum_{n_2=0}^{N_s-1} w[n_2]e^{-jn_2\omega_2}\sum_{k_2=0}^{N_s-1} w^*[k_2]e^{+jk_2\omega_2}\right\}\\
&= \frac{1}{N_s^2}\sum_{n_1=0}^{N_s-1}\sum_{k_1=0}^{N_s-1}\sum_{n_2=0}^{N_s-1}\sum_{k_2=0}^{N_s-1} \mathcal{E}\{w[n_1]w^*[k_1]w[n_2]w^*[k_2]\}\, e^{-j\omega_1(n_1-k_1)}e^{-j\omega_2(n_2-k_2)}
\end{aligned}$$

Now the following property for complex Gaussian random variables [8, p. 85]

$$\mathcal{E}\{v_1v_2^*v_3v_4^*\} = \mathcal{E}\{v_1v_2^*\}\,\mathcal{E}\{v_3v_4^*\} + \mathcal{E}\{v_1v_4^*\}\,\mathcal{E}\{v_3v_2^*\}$$

implies that

$$\mathcal{E}\{w[n_1]w^*[k_1]w[n_2]w^*[k_2]\} = \begin{cases} \sigma_w^4; & \text{if } n_1 = k_1 \text{ and } n_2 = k_2 \\ & \text{or } n_1 = k_2 \text{ and } n_2 = k_1 \\ 0; & \text{otherwise} \end{cases} \tag{10.19}$$

Substituting (10.19) into (10.18), evaluating the sums, and simplifying with the help of the Euler relation results in the expression

$$\mathcal{E}\{\hat{P}_w(e^{j\omega_1})\hat{P}_w(e^{j\omega_2})\} = \sigma_w^4\left[1 + \left(\frac{\sin(N_s(\omega_1-\omega_2)/2)}{N_s\sin((\omega_1-\omega_2)/2)}\right)^2\right] \tag{10.20}$$

This is the correlation between $\hat{P}_w(e^{j\omega_1})$ and $\hat{P}_w(e^{j\omega_2})$. The covariance is

$$\text{Cov}\left[\hat{P}_w(e^{j\omega_1}), \hat{P}_w(e^{j\omega_2})\right] = \mathcal{E}\{\hat{P}_w(e^{j\omega_1})\hat{P}_w(e^{j\omega_2})\} - \mathcal{E}\{\hat{P}_w(e^{j\omega_1})\}\,\mathcal{E}\{\hat{P}_w(e^{j\omega_2})\} \tag{10.21}$$

where from (10.7)

$$\mathcal{E}\{\hat{P}_w(e^{j\omega_1})\} = \sum_{l=-N_s+1}^{N_s-1}\frac{N_s-|l|}{N_s}R_w[l]e^{-j\omega l} = \sum_{l=-N_s+1}^{N_s-1}\frac{N_s-|l|}{N_s}\sigma_w^2\delta[l]e^{-j\omega l} = \sigma_w^2 \tag{10.22}$$

Finally, substituting (10.20) and (10.22) into (10.21) yields

$$\text{Cov}\left[\hat{P}_w(e^{\jmath\omega_1}), \hat{P}_w(e^{\jmath\omega_2})\right] = \sigma_w^4 \left(\frac{\sin(N_s(\omega_1 - \omega_2)/2)}{N_s \sin((\omega_1 - \omega_2)/2)}\right)^2 \tag{10.23}$$

A slightly different result obtains for the case of a real Gaussian white noise process; the covariance expression contains another term [2, 4, 5]. That case will not be discussed explicitly here since the results and conclusions are the same.

The variance of the periodogram is given by the limit of (10.23) as $\omega_1 \to \omega_2$. The variance is therefore

$$\text{Var}\left[\hat{P}_w(e^{\jmath\omega})\right] = \sigma_w^4 \tag{10.24}$$

Note that the standard deviation of $\hat{P}_w(e^{\jmath\omega})$ is σ_w^2, which is equal to the mean value of the estimate. In addition, the variance is a constant and *does not decrease with* N_s. Therefore, the estimate is not consistent. Further, the covariance (10.23) is seen to go to zero at points where $(\omega_1 - \omega_2)/2 = (k\pi)/N_s$. This happens whenever

$$\omega_1 = \frac{2\pi k_1}{N_s}\ ; \qquad \omega_2 = \frac{2\pi k_2}{N_s}\ ; \qquad k_1 \neq k_2$$

Therefore, as N_s gets larger, the points where the spectral estimate is uncorrelated get closer together. This gives a wilder-looking estimate for the spectrum for larger values of N_s (see Fig. 10.2).

For sequences other than Gaussian white noise, the properties of the variance are generally similar. For a regular Gaussian random process with arbitrary spectral density function the innovations representation of Fig. 10.3 applies; that is, the process can be thought of as the output of a causal linear filter $H_{ca}(z)$ driven by white noise. The periodogram of this more general process can then be written as

$$\hat{P}_x(e^{\jmath\omega}) = \frac{1}{N_s}|X(e^{\jmath\omega})|^2 \approx \frac{1}{N_s}|H_{ca}(e^{\jmath\omega})|^2|W(e^{\jmath\omega})|^2 = |H_{ca}(e^{\jmath\omega})|^2\hat{P}_w(e^{\jmath\omega}) \tag{10.25}$$

where the approximation occurs because N_s is *finite*. Therefore, the covariance is given (approximately) by

$$\text{Cov}\left[\hat{P}_x(e^{\jmath\omega_1}), \hat{P}_x(e^{\jmath\omega_2})\right] \approx |H_{ca}(e^{\jmath\omega_1})|^2|H_{ca}(e^{\jmath\omega_2})|^2\text{Cov}\left[\hat{P}_w(e^{\jmath\omega_1}), \hat{P}_w(e^{\jmath\omega_2})\right] \tag{10.26}$$

Substituting (10.23) with $\sigma_w^2 = \mathcal{K}_\text{o}$ now produces

$$\begin{aligned}\text{Cov}\left[\hat{P}_x(e^{\jmath\omega_1}), \hat{P}_x(e^{\jmath\omega_2})\right] &\approx |H_{ca}(e^{\jmath\omega_1})|^2|H_{ca}(e^{\jmath\omega_2})|^2\mathcal{K}_\text{o}^2\left(\frac{\sin(N_s(\omega_1 - \omega_2)/2)}{N_s \sin((\omega_1 - \omega_2)/2)}\right)^2 \\ &= S_x(e^{\jmath\omega_1})S_x(e^{\jmath\omega_2})\left(\frac{\sin(N_s(\omega_1 - \omega_2)/2)}{N_s \sin((\omega_1 - \omega_2)/2)}\right)^2\end{aligned} \tag{10.27}$$

The variance is therefore

$$\text{Var}\left[\hat{P}_x(e^{\jmath\omega})\right] \approx S_x^2(e^{\jmath\omega}) \tag{10.28}$$

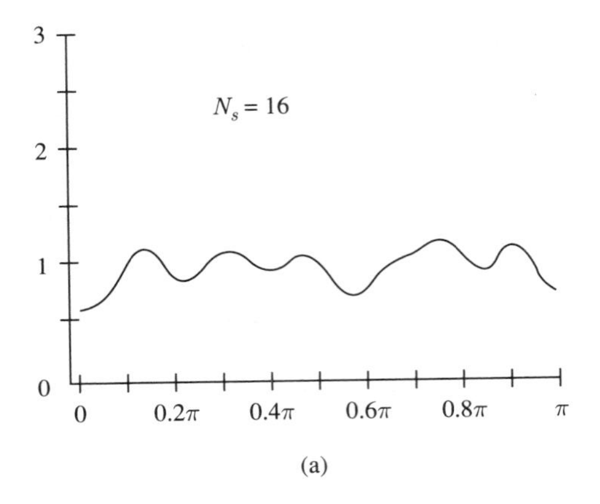

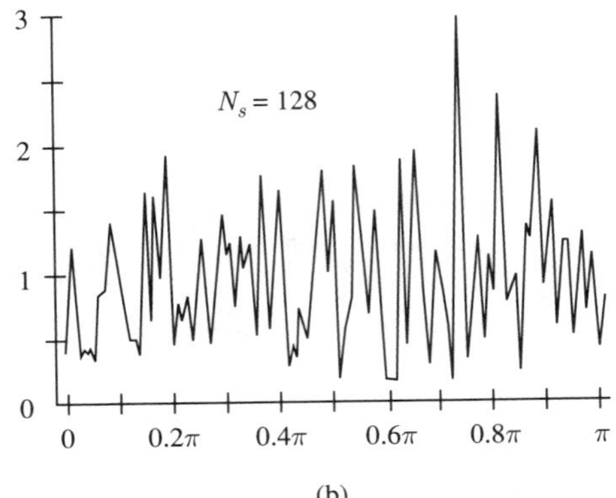

Figure 10.2 Periodogram spectral estimate for white noise for two values of N_s. (a) $N_s = 16$. (b) $N_s = 128$.

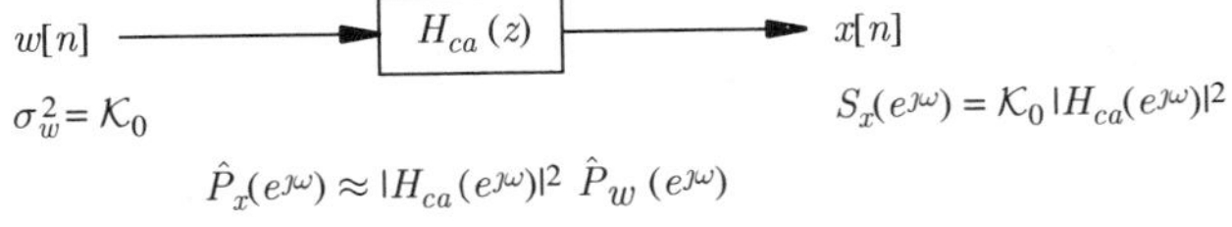

Figure 10.3 Innovations representation of a regular random process with power density spectrum $S_x(e^{j\omega})$.

It is seen that the variance of the periodogram estimate for the more general random process is also independent of N_s and the covariance has the same behavior with increasing N_s as that for the white noise.

10.1.3 Improving the Spectral Estimate

There are several ways to improve the basic periodogram or correlogram estimate. Three of the most common procedures are discussed here.

Averaging: the Bartlett Procedure. A simple way to reduce the variance of the periodogram spectral estimate is to average a number of different periodograms. Since the statistical properties of the periodogram do not improve with greater lengths of data, a more effective way to use a given long data segment is to break the data into smaller segments and average the resulting periodograms. This is known as the Bartlett procedure [9]. Figure 10.4 shows a data segment of length N_s that is broken into K segments each of length N. Let $\hat{P}_x^{(k)}(e^{\jmath\omega})$ be the periodogram of the kth data segment. Then the Bartlett estimate is computed as

$$\boxed{\hat{S}_B(e^{\jmath\omega}) = \frac{1}{K}\sum_{k=1}^{K} \hat{P}_x^{(k)}(e^{\jmath\omega})} \quad . \tag{10.29}$$

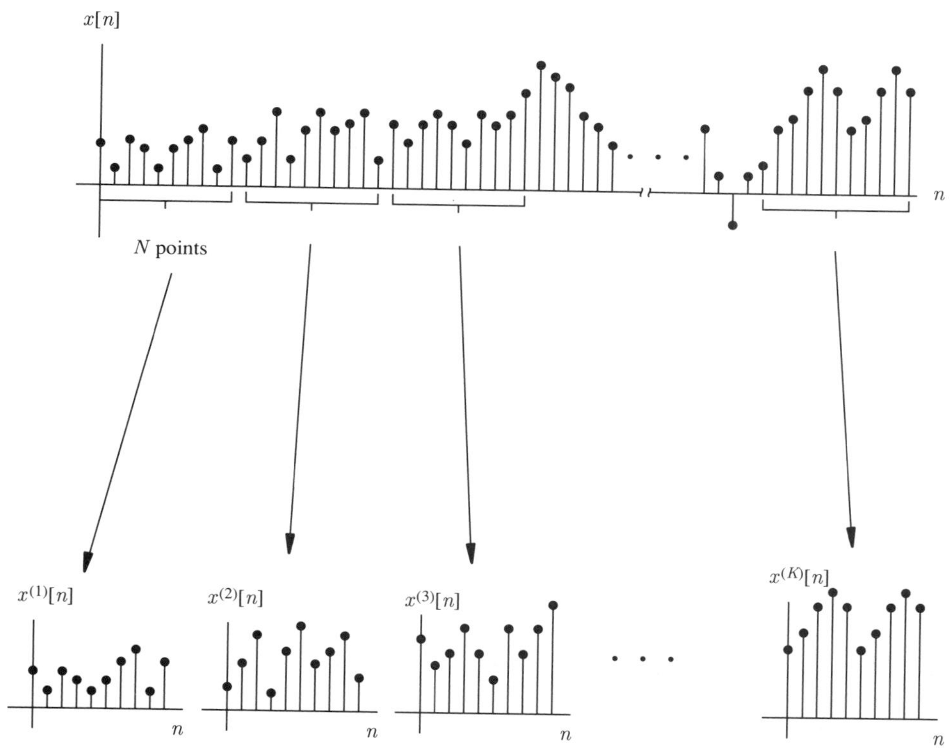

Figure 10.4 Breaking a long data segment into shorter segments.

If the data segments are uncorrelated, then the variance decreases by a factor of $1/K$; if they are not uncorrelated, the variance is still reduced but by less than a factor of $1/K$. Figure 10.5 shows the effect of averaging the periodograms from 32 adjacent data segments of a random process with a "low-pass" character. The variance appears to be reduced and the result is a much more reasonable estimate for the spectrum.

Windowing and Smoothing: the Blackman–Tukey Procedure. Another way to improve the spectral estimate is to apply a window to the estimated sample correlation

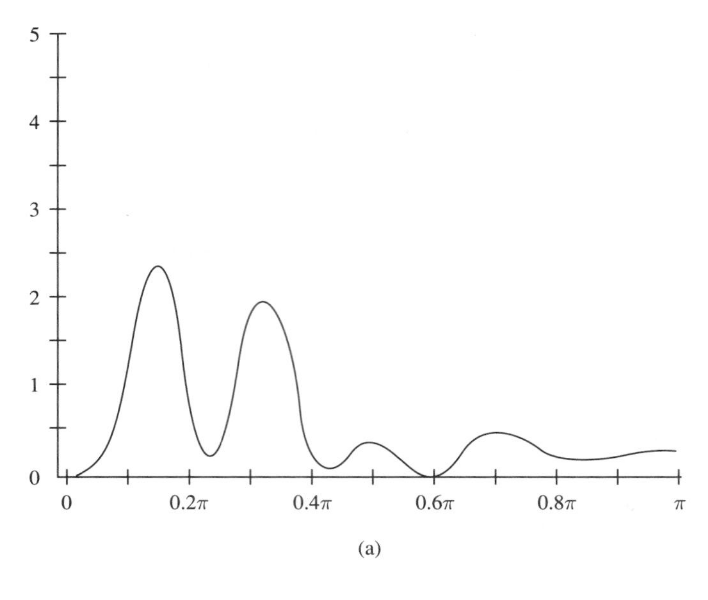

(a)

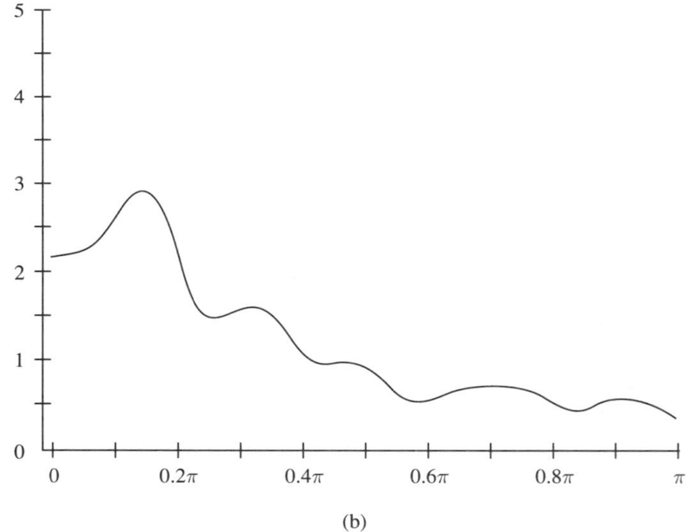

(b)

Figure 10.5 Reducing the variance of the spectral estimate by averaging. (a) One 16–point segment. (b) Thirty-two 16–point segments.

function before transforming. This method was promoted by Blackman and Tukey [10] and results in a smoothing of the spectral estimate by convolution with the transform of the window. The spectral estimate takes the form

$$\boxed{\hat{S}_{BT}(e^{\jmath\omega}) = \sum_{l=-L}^{L} \mathrm{w}[l]\hat{R}_x[l]e^{-\jmath\omega l}; \qquad L < N_s} \tag{10.30}$$

where $\mathrm{w}[l]$ is any suitable window (Bartlett, Hamming, Hann,[2] etc.) and L is the half-length of the window. The resulting estimate can also be computed directly in the frequency domain by convolving the transform of the window with the periodogram.

$$\hat{S}_{BT}(e^{\jmath\omega}) = \frac{1}{2\pi}\mathrm{W}(e^{\jmath\omega})\circledast\hat{P}_x(e^{\jmath\omega}) \tag{10.31}$$

When computed in this form the estimate is sometimes called the Danielle periodogram. In order to be effective in smoothing the periodogram, the length of the window in the signal domain needs to be considerably shorter than the length of the data N_s. The ordinary correlogram (10.3) may be thought of as computed by applying a rectangular window to the correlation function. This is the reason for restricting the size of the window in (10.30) or the maximum lag L in (10.3) to relatively small values.

The expected value of the smoothed spectral estimate is given by

$$\mathcal{E}\{\hat{S}_{BT}(e^{\jmath\omega})\} = \frac{1}{2\pi}\mathrm{W}(e^{\jmath\omega})\circledast\mathcal{E}\{\hat{P}_x(e^{\jmath\omega})\} = \frac{1}{4\pi^2}\mathrm{W}(e^{\jmath\omega})\circledast\mathrm{W}_B(e^{\jmath\omega})\circledast S_x(e^{\jmath\omega}) \tag{10.32}$$

The variance is much more difficult to compute. However, it can be argued [12] that when the transformed window W is narrow compared to the variations in the spectrum and the half length L of the window is much smaller than the data length N_s, the variance is reduced by approximately the factor

$$\frac{1}{N_s}\sum_{l=-L+1}^{L-1} \mathrm{w}^2[l] = \frac{1}{2\pi N_s}\int_{-\pi}^{\pi} |\mathrm{W}(e^{\jmath\omega})|^2 d\omega$$

Combined Windowing and Averaging: the Welch Procedure. A final strategy is to use a combination of windowing and averaging. This is the procedure studied by Welch [13]. In Welch's procedure the original data sequence is divided into a number K of possibly overlapping segments. A window is applied to these segments and the resulting modified periodograms are averaged. If $x^{(k)}[n]$ represents the kth data segment (of length N) and a normalized window $\mathrm{v}[n]$ is applied to this data, then the modified periodograms are defined as

$$\hat{P}_x^{\prime(k)}(e^{\jmath\omega}) = \frac{1}{N}\left|\sum_{n=0}^{N-1} \mathrm{v}[n]x^{(k)}[n]e^{-\jmath\omega n}\right|^2 \tag{10.33}$$

[2]This window, named after the Austrian meteorologist Julius von Hann, is frequently called the "Hanning" window. Tukey apparently introduced the term "Hanning" for von Hann's window as a parody on the name "Hamming" when describing work that Tukey and Hamming had performed together [11]. The term "Hanning" has since remained in use.

where v has the property

$$\frac{1}{N}\sum_{n=0}^{N-1} \mathrm{v}^2[n] = 1 \tag{10.34}$$

The spectral estimate is then taken as

$$\hat{S}_W(e^{\jmath\omega}) = \frac{1}{K}\sum_{k=1}^{K} \hat{P}_x'^{(k)}(e^{\jmath\omega}) \tag{10.35}$$

The expected value of this spectral estimate can be shown to be

$$\mathcal{E}\{\hat{S}_W(e^{\jmath\omega})\} = \frac{1}{2\pi}\mathrm{W}(e^{\jmath\omega})\circledast S_x(e^{\jmath\omega}) \tag{10.36}$$

(see Problem 10.5) where

$$\mathrm{W}(e^{\jmath\omega}) = \frac{1}{N}|\mathrm{V}(e^{\jmath\omega})|^2 \tag{10.37}$$

When the segments are nonoverlapping, the variance is approximately the same as that of the Bartlett estimate.

10.1.4 Computation Using the FFT

The FFT is the natural tool to use in computing the periodogram spectral estimate or any of its variations. However, if the data segment consists of N points, then an N-point FFT produces only N samples of the corresponding Fourier transform. The Fourier transform can be evaluated at a larger number of more closely spaced frequency values by "zero-padding" the data sequence and using a larger FFT. This is necessary when the goal is to produce a plot of the estimated spectrum.

As an example of this, suppose the data sequence consists of 16 points. For any of the periodogram methods, directly transforming the data using a 16-point FFT results in only 16 samples of the spectrum. If instead, 240 zeros are appended to the sequence and a 256-point FFT is used, then 256 samples of the Fourier transform are obtained, which is enough to make a good plot and show the shape of the spectral estimate. Zero-padding does not change the shape of the spectral estimate, which after all, is a *continuous* function of ω. It merely provides a way to evaluate the spectral estimate at a more closely spaced set of frequency values.

10.2 SPECTRUM ESTIMATION BASED ON LINEAR MODELS

The resolution, or ability to distinguish closely spaced features of the spectrum using classical methods is fundamentally limited by the length of data available. It was seen that the mean of the periodogram is equal to the true spectrum convolved with the Fourier transform of the window through which the data is observed. If this window is relatively narrow, owing to a short record of available data, the Fourier transform of the window is wide, so features of the spectrum that are closely spaced (approximately less than the reciprocal of the data length) are difficult or impossible to resolve. This effect is depicted in Fig. 10.6. Some more recently developed methods attempt to circumvent the fundamental limitation by developing parametric models for the random process and finding the associated spectrum. This is the fundamental approach of the methods discussed in this section. Throughout it is assumed that the random process is regular so that it has an innovations representation and the power spectral density function is continuous. Methods for dealing with discrete components in the spectrum are discussed in Section 10.4.

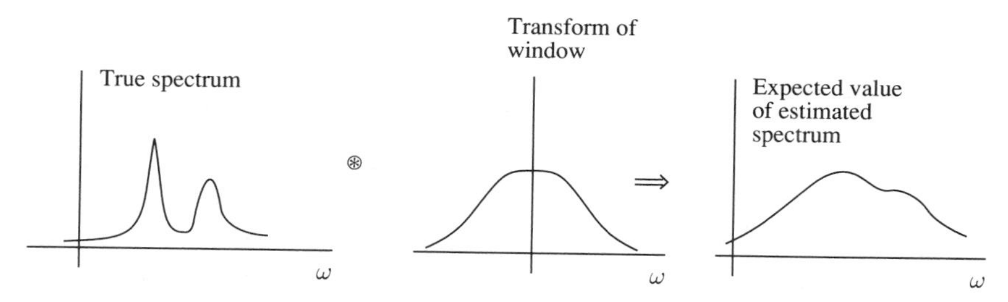

Figure 10.6 Illustration of resolution limitation of classical spectrum estimation methods due to data length.

10.2.1 ARMA, MA, and AR Spectrum Estimation

When a random process is represented by a linear filter $H(z)$ driven by white noise with variance σ_w^2, the power spectral density function of the model is given by

$$\sigma_w^2\,|H(e^{\jmath\omega})|^2$$

Since this last expression can be evaluated explicitly for the model, it provides an estimate of the spectrum of the random process. As stated in Chapter 9, it is conventional (at least for ARMA and MA models) to assume that $\sigma_w^2 = 1$ and provide the necessary gain in the numerator of the transfer function.

All of the types of models discussed in Chapter 9 can be used to generate spectral estimates. When an order (P, Q) ARMA model is used the corresponding spectral estimate for the random process has the form

$$\boxed{\hat{S}_{ARMA}(e^{\jmath\omega}) = \left|\frac{B(e^{\jmath\omega})}{A(e^{\jmath\omega})}\right|^2} \tag{10.38}$$

The ARMA spectral estimate is most effective when the data is known to have come from a rational model. Otherwise, the procedure can still be effective, but the problems of choosing an appropriate order for the model and selecting a method to estimate the model parameters are not always straightforward. All of these issues are discussed in Chapter 9 and are the same regardless of whether the goal of developing the model is spectrum estimation or something else.

If there is reason to believe that an MA model may be appropriate to the data, then a spectral estimate of the form

$$\boxed{\hat{S}_{MA}(e^{\jmath\omega}) = |B(e^{\jmath\omega})|^2} \tag{10.39}$$

can be used. MA or "all-zero" models may be the right choice when the predominant features of the spectrum are nulls at specific frequencies. These nulls are produced by zeros on or close to the unit circle.

If the predominant features of the spectrum are sharp peaks then an AR model may be most appropriate. This estimate has the form

$$\hat{S}_{AR}(e^{\jmath\omega}) = \frac{|b_0|^2}{|A(e^{\jmath\omega})|^2}$$

and the sharp peaks correspond to poles near the unit circle. For AR models the constant $|b_0|^2$ can be replaced by the error variance σ_P^2 in the corresponding linear prediction problem and written as

$$\boxed{\hat{S}_{AR}(e^{\jmath\omega}) = \frac{\sigma_P^2}{|A(e^{\jmath\omega})|^2}} \tag{10.40}$$

An ARMA spectral estimate can, of course, model spectra with both peaks and nulls. Since ARMA modeling is more difficult than either MA or AR modeling, however, frequently the simpler models (particularly the AR model) are used. If there are peaks in the spectrum and an MA model is chosen, it will need to be of high order to approximate those peaks. Correspondingly, if there are nulls in the spectrum and an AR model is used, its order will need to be high to approximate the nulls.

Although the techniques for developing all three types of models from given data sequences are covered in Chapter 9, the AR model is worthy of some further discussion in this chapter because of an additional interpretation that was discovered in the context of spectrum estimation. This interpretation is given in the next subsection.

10.2.2 The AR Model and Maximum Entropy

The AR model for a random process usually begins with a linear prediction problem (see Fig. 10.7). A linear predictive filter is constructed for the given random process $x[n]$ and the prediction order

$$P = N - 1$$

is chosen so that the output is approximately white noise. The filter is then (conceptually) inverted and driven by white noise with variance equal to the prediction error variance σ_P^2. The random process $x'[n]$ generated by the model satisfies the difference equation

$$x'[n] = -a_1 x'[n-1] - a_2 x'[n-2] - \cdots - a_P x'[n-P] + w[n] \tag{10.41}$$

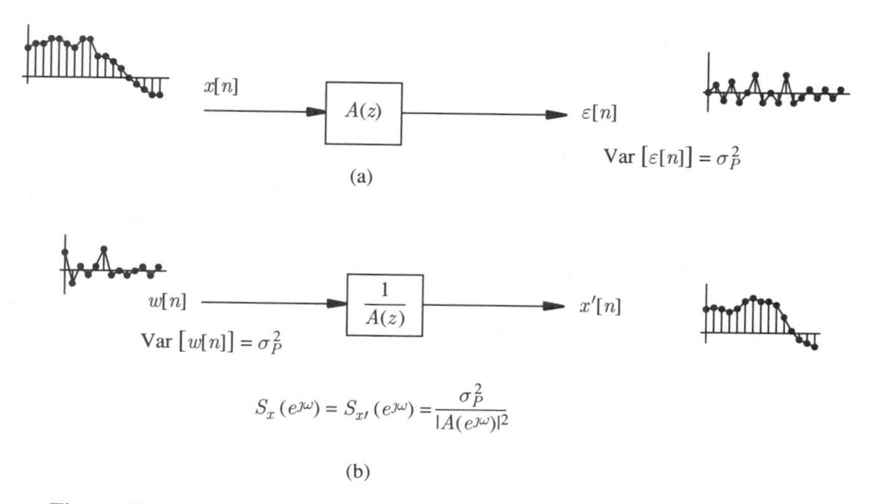

Figure 10.7 Spectrum estimation by AR modeling. (a) Prediction error filter. (b) Corresponding AR model and spectral estimate.

and the AR spectral estimate of $x[n]$ is taken to be the spectrum of the model, which is given by (10.40) with

$$A(z) = 1 + a_1 z^{-1} + \cdots + a_P z^{-P} \tag{10.42}$$

In practice, the AR model parameters are found by any of the methods discussed in previous chapters. This means that the correlation function is *estimated* (either explicitly or implicitly). For purposes of this section, however, the discussion is based on the theoretical correlation function and it is assumed that (at least a part of) it is known.

The AR model has two properties discussed in previous chapters which are central to its use in spectrum estimation. These are:

1. **Correlation Matching:** The correlation function of the AR process *matches* the correlation function of the original process for lags up to P, that is,

$$R_{x'}[l] = R_x[l]\ ; \qquad l = 0, \pm 1, \pm 2, \ldots, \pm P \tag{10.43}$$

   Although this property is implicit in the derivation of the Yule–Walker equations, it can be shown explicitly from other considerations that are taken up here.

2. **Correlation Extension:** The correlation function of the AR process *extends* the correlation function of the original process for lag values greater than P. Recall that the correlation function for the AR process satisfies the difference equation (see Section 8.2)

$$R_{x'}[l] + a_1 R_{x'}[l-1] + a_2 R_{x'}[l-2] + \cdots + a_P R_{x'}[l-P] = R_{wx'}[l]$$

Since $R_{wx'}[l] = 0$ for $l > 0$ it follows that

$$R_{x'}[l] = -a_1 R_{x'}[l-1] - a_2 R_{x'}[l-2] - \cdots - a_P R_{x'}[l-P] \qquad l > 0 \qquad (10.44)$$

This can be used to compute the extended values of the correlation function recursively for $l > P$. Values of $R_{x'}[l]$ for $l < -P$ can be obtained from the complex conjugate of this equation or by using the symmetry properties of the correlation function.

The second property shows that the AR method implicitly *extends* the correlation function before transforming it to obtain the spectral estimate (see Fig. 10.8). This is fundamentally different from the approach of classical methods, which window or *truncate* the correlation function before taking its Fourier transform. It is this property that gives rise to a new interpretation.

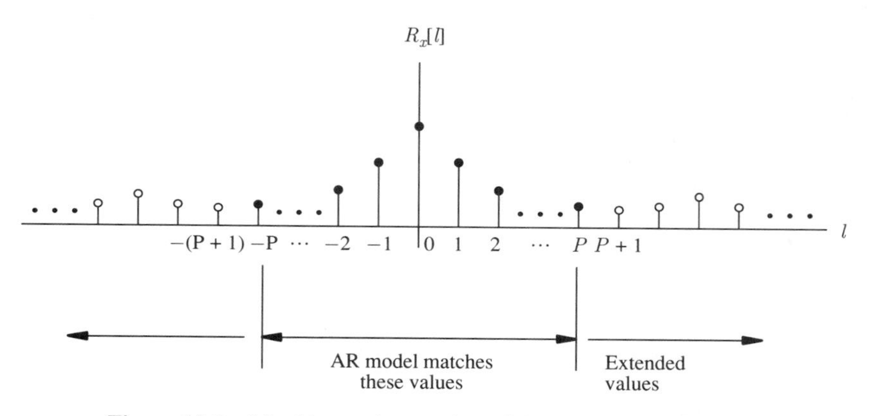

Figure 10.8 Matching and extension of the correlation function.

The new interpretation was discovered and aptly articulated by the geophysisist John Parker Burg who called his technique the "maximum entropy method" [14–16]. While AR spectrum estimation had also been explored by Parzen [17, 18] and others, it was Burg's approach that virtually caused an explosion of interest in new spectral estimation methods.[3] Burg showed that the AR process was unique in that among all possible random processes that could match the $2P+1$ given values of the correlation function and extend it, the AR process is the one that has maximum entropy. In other words, the AR process is the "most random" process that can still match the given correlation function values. Burg argued that this property was most appropriate since it is presumptuous to force the correlation function to assume specific values (such as zero) in the region where it is not known. Rather, an extension should be chosen which is "maximally noncommital."

Theoretically, the maximum entropy (ME) and autoregressive spectral estimates are exactly the same and are given by (10.40). In practice they usually differ in that what is generally called the ME spectral estimate is computed using Burg's method to estimate the

[3]It is also interesting that although Burg began promoting his method around 1967 and it became well known shortly after that, Burg himself published very little on the method until his Ph.D. thesis in 1975 [19].

parameters of the model, while the AR estimate normally uses another technique, such as the autocorrelation or covariance method. For purposes of the present theoretical discussion, however, the terms "ME spectral estimate" and "AR spectral estimate" are synonymous.

The maximum entropy property can be shown as follows [20]. It is known [21, 22] that the random process with maximum entropy is a Gaussian random process. Let us assume that the random process has zero mean. Then the entropy for a random vector $\boldsymbol{x}_p$ consisting of $p+1$ samples of a complex Gaussian random process (see Problem 2.17) is

$$\mathcal{H} = \mathcal{E}\{-\ln f_{\boldsymbol{x}_p}(\boldsymbol{x}_p)\} = (p+1)(1+\ln\pi) + \ln|\mathbf{R}_{\boldsymbol{x}}^{(p)}| \tag{10.45}$$

where $f_{\boldsymbol{x}_p}$ is the density function and $\mathbf{R}_{\boldsymbol{x}}^{(p)}$ is the correlation (equal to the covariance) matrix for the vector of $p+1$ samples.[4] This matrix is Toeplitz and has the specific form

$$\mathbf{R}_{\boldsymbol{x}}^{(p)} = \begin{bmatrix} R_x[0] & R_x[-1] & \cdots & R_x[-p] \\ R_x[1] & R_x[0] & \cdots & R_x[-p+1] \\ \vdots & \vdots & \vdots & \vdots \\ R_x[p] & R_x[p-1] & \cdots & R_x[0] \end{bmatrix} \tag{10.46}$$

Now the given values of the correlation function $R_x[0], R_x[\pm 1], \ldots, R_x[\pm P]$ define the entropy up to order P. The problem is to find an extension of the correlation function to maximize the entropy for any order $p > P$. Since the logarithm in (10.45) is monotonic, this is equivalent to maximizing $|\mathbf{R}_{\boldsymbol{x}}^{(p)}|$ for $p > P$. Let us thus proceed recursively and show that in order to maximize the entropy for $p = P+1, P+2, \ldots$ the new values of the correlation function must be defined by (10.44).

To begin, for $p = P+1$ the determinant is

$$|\mathbf{R}_{\boldsymbol{x}}^{(P+1)}| = \begin{vmatrix} R_x[0] & R_x[-1] & \cdots & R_x[-P] & R_x[-P-1] \\ R_x[1] & R_x[0] & \cdots & R_x[-P+1] & R_x[-P] \\ \vdots & \vdots & \vdots & \vdots & \vdots \\ R_x[P] & R_x[P-1] & \cdots & R_x[0] & R_x[-1] \\ R_x[P+1] & R_x[P] & \cdots & R_x[1] & R_x[0] \end{vmatrix} \tag{10.47}$$

The problem is to choose the new terms $R_x[P+1]$ and $R_x[-P-1] = R_x^*[P+1]$ to maximize this quantity. To accomplish this, first expand the determinant by cofactors of the top row. If $R_x[-P-1]$ is written as $R_x^*[P+1]$, this expansion produces

$$|\mathbf{R}_{\boldsymbol{x}}^{(P+1)}| = R_x^*[P+1](-1)^{P+1} \begin{vmatrix} R_x[1] & R_x[0] & \cdots & R_x[-P+1] \\ \vdots & \vdots & \vdots & \vdots \\ R_x[P] & R_x[P-1] & \cdots & R_x[0] \\ R_x[P+1] & R_x[P] & \cdots & R_x[1] \end{vmatrix}$$

$$+ \text{ other terms} \tag{10.48}$$

[4]The expression for a real Gaussian random process is slightly different. However, the dependence on $\ln|\mathbf{R}^{(p)}|$, which is our main concern, is the same.

Applying the scalar form of the complex gradient yields

$$\nabla_{R_x^*[P+1]}|\mathbf{R}_{\boldsymbol{x}}^{(P+1)}| = (-1)^{P+1} \begin{vmatrix} R_x[1] & R_x[0] & \cdots & R_x[-P+1] \\ \vdots & \vdots & \vdots & \vdots \\ R_x[P] & R_x[P-1] & \cdots & R_x[0] \\ R_x[P+1] & R_x[P] & \cdots & R_x[1] \end{vmatrix} \tag{10.49}$$

where all additional terms are zero because the "other terms" in (10.48) involve only the variables $R_x[-P]$ through $R_x[P]$ and $R_x[P+1]$ (unconjugated). Setting (10.49) equal to zero provides the necessary condition

$$\begin{vmatrix} R_x[1] & R_x[0] & \cdots & R_x[-P+1] \\ \vdots & \vdots & \vdots & \vdots \\ R_x[P] & R_x[P-1] & \cdots & R_x[0] \\ R_x[P+1] & R_x[P] & \cdots & R_x[1] \end{vmatrix} = 0 \tag{10.50}$$

If the solution to (10.50) is in fact a maximum, then the two conditions (see Appendix A, Section A.3)

$$\left(\nabla^2_{R_x R_x}|\mathbf{R}_{\boldsymbol{x}}^{(P+1)}|\right) \cdot \left(\nabla^2_{R_x^* R_x^*}|\mathbf{R}_{\boldsymbol{x}}^{(P+1)}|\right) - \left(\nabla^2_{R_x R_x^*}|\mathbf{R}_{\boldsymbol{x}}^{(P+1)}|\right)^2 < 0$$

and

$$\nabla^2_{R_x R_x^*}|\mathbf{R}_{\boldsymbol{x}}^{(P+1)}| < 0$$

must hold, where the argument $[P+1]$ on the subscripts of the ∇^2 operator have been dropped to avoid more cumbersome notation. Since (10.49) shows that $\nabla_{R_x^*[P+1]}|\mathbf{R}_{\boldsymbol{x}}^{(P+1)}|$ is not a function of $R_x^*[P+1]$, it follows that

$$\nabla^2_{R_x^* R_x^*}|\mathbf{R}_{\boldsymbol{x}}^{(P+1)}| = \nabla_{R_x^*[P+1]}\left(\nabla_{R_x^*[P+1]}|\mathbf{R}_{\boldsymbol{x}}^{(P+1)}|\right) = 0$$

Therefore, the first condition above is satisfied. To check the second condition, expand the determinant in (10.49) by cofactors of the first column. Since all terms except one are independent of $R_x[P+1]$, this yields

$$\begin{aligned} \nabla^2_{R_x R_x^*}|\mathbf{R}_{\boldsymbol{x}}^{(P+1)}| &= \nabla_{R_x[P+1]}\left(\nabla_{R_x^*[P+1]}|\mathbf{R}_{\boldsymbol{x}}^{(P+1)}|\right) \\ &= (-1)^{P+1}(-1)^P \begin{vmatrix} R_x[0] & \cdots & R_x[-P+1] \\ \vdots & \ddots & \vdots \\ R_x[P-1] & \cdots & R_x[0] \end{vmatrix} \\ &= -|\mathbf{R}_{\boldsymbol{x}}^{(P-1)}| < 0 \end{aligned}$$

The last inequality follows from the fact that since $\mathbf{R}_{\boldsymbol{x}}^{(P-1)}$ is strictly positive definite, its determinant is positive.[5] This shows that the stationary point defined by (10.50) is indeed a *maximum*.

[5]Recall that $x[n]$ is assumed to be a *regular* process and that the correlation function for a regular process is strictly positive definite.

Equation 10.50 tells how to choose $R_x[P+1]$ to maximize the entropy. However, it has further implications. Since the determinant in (10.50) is required to be zero, the columns must be linearly dependent. In other words, it must be true that

$$\begin{bmatrix} R_x[1] & R_x[0] & \cdots & R_x[-P+1] \\ \vdots & \vdots & \vdots & \vdots \\ R_x[P] & R_x[P-1] & \cdots & R_x[0] \\ R_x[P+1] & R_x[P] & \cdots & R_x[1] \end{bmatrix} \begin{bmatrix} 1 \\ c_1 \\ \vdots \\ c_P \end{bmatrix} = \begin{bmatrix} 0 \\ \vdots \\ 0 \\ 0 \end{bmatrix} \tag{10.51}$$

for some choice of the constants $c_1, \ldots, c_P$. Let us add the additional equation

$$R_x[0] + c_1 R_x[-1] + \ldots + c_P R_x[-P] = \sigma^2 \tag{10.52}$$

where σ^2 is another constant to be determined. If (10.52) is then combined with (10.51), these two equations become

$$\begin{bmatrix} R_x[0] & R_x[-1] & \cdots & R_x[-P] \\ R_x[1] & R_x[0] & \cdots & R_x[-P+1] \\ \vdots & \vdots & \vdots & \vdots \\ R_x[P] & R_x[P-1] & \cdots & R_x[0] \\ ---- & ---- & -- & ---- \\ R_x[P+1] & R_x[P] & \cdots & R_x[1] \end{bmatrix} \begin{bmatrix} 1 \\ c_1 \\ \vdots \\ c_P \end{bmatrix} = \begin{bmatrix} \sigma^2 \\ 0 \\ \vdots \\ 0 \\ -- \\ 0 \end{bmatrix} \tag{10.53}$$

The first $P+1$ rows (above the dashed lines) can be recognized as comprising the Yule–Walker equations for the AR model. These can be solved for the c_k and σ^2. The last row determines $R_x[P+1]$ and thus extends the correlation function:

$$R_x[P+1] = -c_1 R_x[P] - \cdots - c_P R_x[1]$$

($R_x[-P-1]$ is determined through the symmetry condition.)

To see that the maximum entropy requirement results in a further extension of the correlation function, assume that $R_x[0], \ldots, R_x[P+1]$ are known and let us now find $R_x[P+2]$ to maximize (10.45) for $p = P+2$. A repetition of the analysis above leads to an equation analogous to (10.53) for the new order:

$$\begin{bmatrix} R_x[0] & R_x[-1] & \cdots & R_x[-P-1] \\ R_x[1] & R_x[0] & \cdots & R_x[-P] \\ \vdots & \vdots & \vdots & \vdots \\ R_x[P+1] & R_x[P] & \cdots & R_x[0] \\ ---- & ---- & -- & ---- \\ R_x[P+2] & R_x[P+1] & \cdots & R_x[1] \end{bmatrix} \begin{bmatrix} 1 \\ c'_1 \\ \vdots \\ c'_P \\ c'_{P+1} \end{bmatrix} = \begin{bmatrix} \sigma'^2 \\ 0 \\ \vdots \\ 0 \\ -- \\ 0 \end{bmatrix}$$

where the rows above the dashed lines now represent the Yule–Walker equations for the model of order $P+1$. But observe that a solution that satisfies these equations is

$$\begin{aligned} c'_k &= c_k; \qquad 1 \le k \le P \\ c'_{P+1} &= 0 \\ \sigma'^2 &= \sigma^2 \end{aligned} \tag{10.54}$$

where the c_k's and σ^2 are the solution to (10.53). Since the solution to the Yule–Walker equations is unique, (10.54) is the *only* solution. Therefore, with this choice the bottom equation in the matrix becomes

$$R_x[P+2] = -c_1 R_x[P+1] - \cdots - c_P R_x[2]$$

which is again the extension provided by the Pth-order AR model.

Proceding in this way, we find that for each value $p > P$ the correlation function value $R_x[p]$ needed to maximize the entropy is given by

$$R_x[p] = -c_1 R_x[P-1] - \cdots - c_P R_x[p-P]$$

In other words, the values of the correlation function needed to maximize the entropy are provided by the AR model.

10.2.3 Burg's Proof of Maximum Entropy

Burg provided a proof of the maximum entropy property different from the one given in the preceding subsection [19]. As noted previously, it was Burg who pointed out the maximum entropy property and promoted the use of the maximum entropy or AR spectral estimate.[6] It is worthwhile to repeat Burg's proof here because it provides a view of the problem from the frequency domain.

It is possible to show [21, 22] that the entropy per sample for a random process with complex spectral density function $S_{x'}(z)$ is given by

$$\Delta\mathcal{H} = \frac{1}{2\pi}\int_{-\pi}^{\pi} \ln S_{x'}(e^{\jmath\omega})d\omega \;+\; \text{const.} \tag{10.55}$$

Now if $S_{x'}(e^{\jmath\omega})$ and $R_{x'}[l]$ represent the power spectral density and the correlation function of the maximum entropy random process, then the constraints on the correlation function can be written as

$$R_{x'}[l] = \frac{1}{2\pi}\int_{-\pi}^{\pi} S_{x'}(e^{\jmath\omega})e^{\jmath\omega l}d\omega = R_x[l]\;; \quad l = 0, \pm 1, \ldots, \pm P \tag{10.56}$$

where $R_x[l]$ for $l = 0, \pm 1, \ldots, \pm P$ represent the *given* values of the correlation function. The remaining values of $R_{x'}[l]$ (i.e., for $|l| > P$) are chosen to maximize (10.55).

A necessary condition for the maximum is that the (scalar) complex gradient with respect to all the free parameters is zero.[7]

$$\nabla_{R^*_{x'}[l]}\Delta\mathcal{H} = \frac{1}{2\pi}\int_{-\pi}^{\pi} \frac{1}{S_{x'}(e^{\jmath\omega})}\left(\nabla_{R^*_{x'}[l]} S_{x'}(e^{\jmath\omega})\right) d\omega = 0\;; \quad |l| > P \tag{10.57}$$

[6] It is interesting to observe however, that Burg's method of developing the AR model (see Section 9.4.4) is based directly on the data and completely bypasses any estimation of the correlation function or its extension.

[7] For the real case, the complex gradient can be replaced by the partial derivative with respect to the $R_{x'}[l]$. For the complex case, the formula for the gradient of the logarithm follows the formula for the derivative of the logarithm.

Since $S_{x'}(e^{\jmath\omega})$ can be written as

$$S_{x'}(e^{\jmath\omega}) = \sum_{k=-\infty}^{\infty} R_{x'}[k]e^{-\jmath\omega k} = \sum_{k'=-\infty}^{\infty} R_{x'}^*[k']e^{\jmath\omega k'}$$

it follows that

$$\nabla_{R_{x'}^*[l]} S_{x'}(e^{\jmath\omega}) = e^{\jmath\omega l}$$

and (10.57) therefore takes the form

$$\nabla_{R_{x'}^*[l]} \Delta\mathcal{H} = \frac{1}{2\pi}\int_{-\pi}^{\pi} \frac{1}{S_{x'}(e^{\jmath\omega})} e^{\jmath\omega l} d\omega = 0\,; \quad |l| > P \tag{10.58}$$

Now to make the following discussion a bit easier let $G(e^{\jmath\omega}) = 1/S_{x'}(e^{\jmath\omega})$. Then (10.58) is recognized as the inverse Fourier transform of $G(e^{\jmath\omega})$ and so corresponds to the sequence

$$g[l] = \frac{1}{2\pi}\int_{-\pi}^{\pi} G(e^{\jmath\omega})e^{\jmath\omega l} d\omega = \frac{1}{2\pi}\int_{-\pi}^{\pi} \frac{1}{S_{x'}(e^{\jmath\omega})} e^{\jmath\omega l} d\omega \tag{10.59}$$

Equation 10.58 states that this sequence has finite length.

Since $g[l]$ is nonzero only in the interval $[-P, P]$ its z-transform is of the form

$$G(z) = \frac{1}{S_{x'}(z)} = \sum_{l=-P}^{P} g[l]z^{-l} \tag{10.60}$$

or

$$S_{x'}(z) = \frac{1}{\sum_{l=-P}^{P} g[l]z^{-l}} \tag{10.61}$$

Finally, since it is possible to factor every complex spectral density function, (10.61) can be written as

$$S_{x'}(z) = \frac{\mathcal{K}_o}{C(z)C^*(1/z^*)} = \frac{\mathcal{K}_o}{\left(\sum_{n=0}^{P} c_n z^{-n}\right)\left(\sum_{k=0}^{P} c_k^* z^k\right)} \tag{10.62}$$

where $c_0 = 1$ and $\mathcal{K}_o$ is the appropriate constant. This implies that the maximum entropy random process is the output of a white noise-driven linear filter whose transfer function is

$$H_{ca}(z) = \frac{1}{\sum_{n=0}^{P} c_n z^{-n}} \tag{10.63}$$

Thus the maximum entropy random process is represented by an *all-pole model*.

To show that the parameters of this all-pole model satisfy Yule–Walker equations, write (10.62) as

$$\frac{\mathcal{K}_o}{\sum_{n=0}^{P} c_n z^{-n}} = \left(\sum_{k=0}^{P} c_k^* z^k\right) S_{x'}(z) = \sum_{k=0}^{P} c_k^* z^k \sum_{l=-\infty}^{\infty} R_{x'}[l]z^{-l} \tag{10.64}$$

where the definition of $S_{x'}(z)$ as the z-transform of $R_{x'}[l]$ has been substituted in the last equation. Then making the change of variables $l' = l - k$ and interchanging the order of summation yields

$$\frac{\mathcal{K}_o}{\sum_{n=0}^{P} c_n z^{-n}} = \sum_{l'=-\infty}^{\infty}\left(\sum_{k=0}^{P} c_k^* R_{x'}[l'+k]\right) z^{-l'} = \sum_{l'=-\infty}^{\infty}\left(\sum_{k=0}^{P} c_k R_{x'}[-l'-k]\right)^* z^{-l'} \tag{10.65}$$

Note now that since the term on the left is the z-transform of a causal sequence, the coefficient of $z^{-l'}$ on the right side of (10.65) must be zero for $l' < 0$. Therefore, the coefficients and the correlation function must satisfy the condition

$$\sum_{k=0}^{P} c_k R_{x'}[-l'-k] = 0; \qquad l' < 0 \tag{10.66}$$

Also, since $c_0 = 1$, the coefficient of the z^0 term in (10.65) can be found from the initial value theorem as

$$\lim_{z\to\infty} \frac{\mathcal{K}_o}{\sum_{n=0}^{P} c_n z^{-n}} = \mathcal{K}_o$$

Thus for $l' = 0$ the term on the right in (10.65) is

$$\sum_{k=0}^{P} c_k R_{x'}[-k] = \mathcal{K}_o \tag{10.67}$$

Finally, if (10.67) and (10.66) for $-1 \geq l' \geq -P$ are combined and written as a single matrix equation (using the original requirement that $R_{x'}[l] = R_x[l]$ for $|l| \leq P$), the result is the Yule–Walker equations.

$$\begin{bmatrix} R_x[0] & R_x[-1] & \cdots & R_x[-P] \\ R_x[1] & R_x[0] & \cdots & R_x[-P+1] \\ \vdots & \vdots & \vdots & \vdots \\ R_x[P] & R_x[P-1] & \cdots & R_x[0] \end{bmatrix} \begin{bmatrix} 1 \\ c_1 \\ \vdots \\ c_P \end{bmatrix} = \begin{bmatrix} \mathcal{K}_o \\ 0 \\ \vdots \\ 0 \end{bmatrix} \tag{10.68}$$

In summary, by beginning with the entropy formula (10.55) and the constraints (10.56) Burg's proof shows that the maximum entropy process must be defined by a white noise-driven all-pole model whose parameters satisfy the Yule–Walker equations. This is nothing other than the AR model.

It is interesting to give one more historical note at this point. Following the success of maximum entropy for one-dimensional signals, there was a considerable research effort to extend it to two-dimensional signals such as the spatial and spatial-temporal signals encountered in array and image processing. The extension turned out to be no simple matter and was thwarted by a number of factors. Not the least of these was the fact that a given finite two-dimensional correlation segment *may not have any* positive definite extension over the plane and therefore cannot be used to derive a maximum entropy model [23]. The matter was finally put to rest by Lang and McClellan [24], who devised a test for the extendibility and a procedure for computing the maximum entropy spectral estimate in the special cases when it exists.

10.2.4 Computation of Model-Based Spectral Estimates

The parameters of the models for the model-based spectral estimates are computed by any of the methods described in Chapter 9. Once the parameters of the model are determined, it is necessary to evaluate the formulas (10.38), (10.39), or (10.40) to compute or plot the spectral estimate. By far the most efficient way to do this is to use the FFT. For example, to compute the AR or ME spectral estimate the sequence of filter coefficients $\{a_k\}$ is zero-padded to a length equal to the desired number of frequency samples. The FFT of this extended sequence produces samples of the Fourier transform $A(e^{j\omega})$. The squared magnitude of the FFT samples are used in formula (10.40) to plot the spectrum.

In a similar manner, the estimates involving $B(e^{j\omega})$ are computed by zero-padding the $\{b_k\}$ sequence and using the FFT. Clearly, when an ARMA spectral estimate is being computed the size of the FFT for both numerator and denominator must be identical. Thus regardless of the orders P and Q, the lengths of the zero-padded sequences must be the same.

10.3 "MAXIMUM LIKELIHOOD" SPECTRUM ESTIMATION

Another of the "data adaptive" methods was proposed by J. Capon in the context of array processing [25] at about the same time that Burg devised the maximum entropy method. The procedure became known as the maximum likelihood method although the name is rather misleading. The term "maximum likelihood" refers to the fact that the filter involved produces a minimum variance unbiased estimate of an input signal in Gaussian noise. Since maximum likelihood is also used to refer to model-based methods, where the model *parameters* are found as maximum likelihood estimates (see Chapter 9), some people prefer to call Capon's method the "minimum variance" method or the "minimum variance distortionless response" method. We use the more traditional term "maximum likelihood" to refer to Capon's method and hope that it will not cause any confusion.[8]

10.3.1 The Maximum Likelihood Method

The maximum likelihood method begins with designing a very narrowband filter centered at some frequency ω_o. The impulse response of that filter is denoted by $h_{\omega_o}[n]$ (see Fig. 10.9). For purposes of spectrum estimation this filter is taken to be an FIR *causal* filter of length N, although other generalizations are possible.

The filter frequency response $H_{\omega_o}(e^{j\omega})$ is constrained to have a value of 1 (i.e., unit magnitude and zero phase) at the chosen frequency ω_o. With this constraint the filter is

[8]In his original description Capon referred to his method as simply the "high-resolution method" and used the term "maximum likelihood" only to describe the filter. However, the procedure soon became known as the maximum likelihood method and Capon himself called it that in later work (see, e.g., [26]).

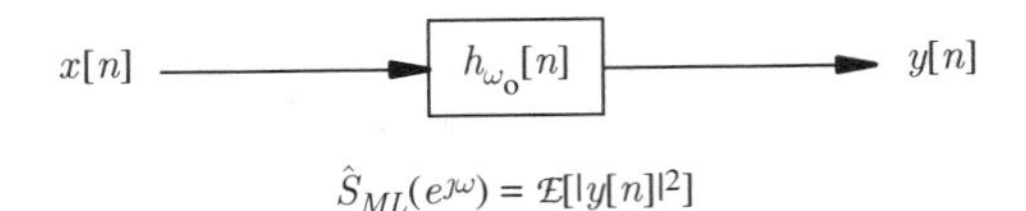

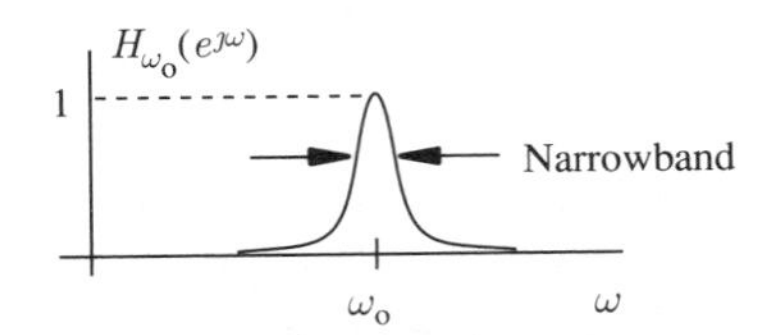

Figure 10.9 Narrowband filter used in maximum likelihood spectrum estimation.

designed to minimize the output power when the random process $x[n]$ is applied to it. This forces the filter to be as narrowband as possible. The power spectral estimate for $x[n]$ at the frequency $\omega = \omega_o$ is taken to be the average power output of the filter. That is,

$$\hat{S}_{ML}(e^{j\omega_o}) \stackrel{\text{def}}{=} \mathcal{P} = \mathcal{E}\left\{|y[n]|^2\right\} \tag{10.69}$$

Since the average output power of the filter is minimized while the gain at ω_o is kept equal to 1, the output power is due mostly to the power of the input at the frequency ω_o. Thus $\hat{S}_{ML}(e^{j\omega_o})$ as defined by (10.69) is a reasonable estimate of the power spectrum at ω_o.

To develop the details of the method let us define the vectors

$$\mathbf{h}_o = \begin{bmatrix} h_{\omega_o}[0] \\ h_{\omega_o}[1] \\ \vdots \\ h_{\omega_o}[N-1] \end{bmatrix} \tag{10.70}$$

$$\mathbf{w}_o = \begin{bmatrix} 1 \\ e^{j\omega_o} \\ \vdots \\ e^{j(N-1)\omega_o} \end{bmatrix} \tag{10.71}$$

$$\boldsymbol{x}[n] = \begin{bmatrix} x[n-N+1] \\ \vdots \\ x[n-1] \\ x[n] \end{bmatrix} \tag{10.72}$$

The output of the filter is then given by

$$y[n] = \sum_{k=0}^{N-1} h_{\omega_o}[k]x[n-k] = \mathbf{h}_o^T \tilde{\boldsymbol{x}}[n] \tag{10.73}$$

where $\tilde{\boldsymbol{x}}[n]$ is the reversal of $\boldsymbol{x}[n]$. It is desired to minimize the average power

$$\begin{aligned} \mathcal{P} = \mathcal{E}\left\{|y[n]|^2\right\} &= \mathbf{h}_o^T \mathcal{E}\{\tilde{\boldsymbol{x}}[n]\tilde{\boldsymbol{x}}^{*T}[n]\}\, \mathbf{h}_o^* \\ &= \mathbf{h}_o^T \tilde{\mathbf{R}}_x \mathbf{h}_o^* = \mathbf{h}_o^{*T} \mathbf{R}_x \mathbf{h}_o \end{aligned} \tag{10.74}$$

(where the last equality follows from conjugation because $\mathcal{P}$ is real) under the complex constraint

$$H_{\omega_o}(e^{\jmath\omega_o}) = \sum_{n=0}^{N-1} h_{\omega_o}[n]e^{-\jmath\omega_o n} = \mathbf{w}_o^{*T}\mathbf{h}_o = 1 \tag{10.75}$$

It is therefore required that $\mathbf{h}_o$ be a stationary point of

$$\mathcal{L} = \mathbf{h}_o^{*T}\mathbf{R}_x\mathbf{h}_o + \mu(1 - \mathbf{w}_o^{*T}\mathbf{h}_o) + \mu^*(1 - \mathbf{h}_o^{*T}\mathbf{w}_o) \tag{10.76}$$

where μ is a Lagrange multiplier, and the Lagrangian $\mathcal{L}$ is formed by appending both the complex constraint and its conjugate (see Appendix A). Taking the complex gradient by using the formulas in Table A.2 (Appendix A), and setting it to zero results in the neccesary condition

$$\nabla_{\mathbf{h}_o^*}\mathcal{L} = \mathbf{R}_x\mathbf{h}_o - \mu^*\mathbf{w}_o = \mathbf{0}$$

or

$$\mathbf{h}_o = \mu^*\mathbf{R}_x^{-1}\mathbf{w}_o \tag{10.77}$$

The Lagrange multiplier can be evaluated from the requirement (10.75)

$$\mathbf{w}_o^{*T}\mathbf{h}_o = \mu^*\mathbf{w}_o^{*T}\mathbf{R}_x^{-1}\mathbf{w}_o = 1$$

which yields

$$\mu^* = \mu = \frac{1}{\mathbf{w}_o^{*T}\mathbf{R}_x^{-1}\mathbf{w}_o} \tag{10.78}$$

Substituting this value in (10.77) then produces

$$\mathbf{h}_o = \frac{\mathbf{R}_x^{-1}\mathbf{w}_o}{\mathbf{w}_o^{*T}\mathbf{R}_x^{-1}\mathbf{w}_o} \tag{10.79}$$

Finally, the output power $\mathcal{P}$ obtained by substituting the value for $\mathbf{h}_o$ in (10.74) becomes

$$\mathcal{P} = \mathbf{h}_o^{*T}\mathbf{R}_x\mathbf{h}_o = \frac{\mathbf{w}_o^{*T}\mathbf{R}_x^{-1}\mathbf{R}_x\mathbf{R}_x^{-1}\mathbf{w}_o}{(\mathbf{w}_o^{*T}\mathbf{R}_x^{-1}\mathbf{w}_o)^2} = \frac{1}{\mathbf{w}_o^{*T}\mathbf{R}_x^{-1}\mathbf{w}_o} \tag{10.80}$$

Since the foregoing analysis holds for *any* choice of the frequency ω_o, the ML spectral estimate as defined by (10.69) is given by

$$\boxed{\hat{S}_{ML}(e^{\jmath\omega}) = \frac{1}{\mathbf{w}^{*T}\mathbf{R}_x^{-1}\mathbf{w}}} \tag{10.81}$$

where the generic vector $\mathbf{w}$ is defined to be

$$\mathbf{w} = \begin{bmatrix} 1 \\ e^{\jmath\omega} \\ \vdots \\ e^{\jmath(N-1)\omega} \end{bmatrix} \tag{10.82}$$

It is interesting to compare the form of the ML estimate to that of the conventional periodogram. For a data segment of length N the Fourier transform has the form

$$X(e^{\jmath\omega}) = \sum_{n=0}^{N-1} x[n]e^{-\jmath\omega n} = \mathbf{w}^{*T}\boldsymbol{x} \tag{10.83}$$

where the vector $\boldsymbol{x}$ is now taken to be

$$\boldsymbol{x} = \begin{bmatrix} x[0] \\ x[1] \\ \vdots \\ x[N-1] \end{bmatrix} \tag{10.84}$$

The periodogram is then

$$\hat{P}_x(e^{\jmath\omega}) = \frac{1}{N}|X(e^{\jmath\omega})|^2 = \frac{1}{N}X(e^{\jmath\omega})X^*(e^{\jmath\omega}) = \frac{1}{N}\mathbf{w}^{*T}\boldsymbol{x}\boldsymbol{x}^{*T}\mathbf{w} \tag{10.85}$$

and its expected value is

$$\mathcal{E}\{\hat{P}_x(e^{\jmath\omega})\} = \frac{1}{N}\mathbf{w}^{*T}\mathbf{R}_x\mathbf{w} \tag{10.86}$$

The correlation matrix $\mathbf{R}_x$ is the same one that appears in (10.81). Thus while the mean of the periodogram involves a quadratic product of the frequency vector $\mathbf{w}$ with the correlation matrix, the ML spectral estimate is the inverse of a quadratic product with the *inverse* correlation matrix. It will be seen later that when the true power spectrum has sharp closely spaced features, these features are better resolved by the ML method. In other words the ML method has better resolution.

Two facts are worth mentioning in passing. First, (10.86) provides an alternative expression to (10.10) which can sometimes be useful. Second, a set of relations similar to (10.81) and (10.86) but involving a correlation matrix *estimated from data* can be derived. This is explored in Problem 10.7.

Burg pointed out the relation between the ML and the ME spectral estimates [27]. The maximum entropy or autoregressive spectral estimate for some order p has the form

$$\hat{S}_{ME}^{(p)}(e^{\jmath\omega}) = \hat{S}_{AR}^{(p)}(e^{\jmath\omega}) = \frac{\sigma_p^2}{\left|\sum_{k=0}^{p} a_k^{(p)} e^{-\jmath\omega k}\right|^2} \tag{10.87}$$

where $\{a_k^{(p)}\}$ and σ_p^2 are the pth-order filter parameters and prediction error variance. Now recall from Section 8.11.1 that the upper–lower triangular decomposition of the correlation matrix results in a representation

$$\mathbf{R}_x = \mathbf{U}_1\mathbf{D}_U\mathbf{U}_1^{*T} \tag{10.88}$$

or

$$\mathbf{R}_x^{-1} = (\mathbf{U}_1^{-1})^{*T}\mathbf{D}_U^{-1}\mathbf{U}_1^{-1} \tag{10.89}$$

where the matrices $\mathbf{U}_1^{-1}$ and $\mathbf{D}_U^{-1}$ of order $P = N - 1$ are expressed in terms of the backward linear prediction parameters [see (8.247) and (8.248)]. These matrices can be written in terms of the AR model parameters as

$$\mathbf{U}_1^{-1} = \begin{bmatrix} 1 & a_1^{(N-1)*} & \cdots & a_{N-1}^{(N-1)*} \\ 0 & 1 & \cdots & a_{N-2}^{(N-2)*} \\ \vdots & \vdots & \ddots & \vdots \\ 0 & 0 & \cdots & 1 \end{bmatrix} \tag{10.90}$$

and

$$\mathbf{D}_U^{-1} = \begin{bmatrix} \frac{1}{\sigma_{N-1}^2} & 0 & \cdots & 0 \\ 0 & \frac{1}{\sigma_{N-2}^2} & \cdots & 0 \\ \vdots & \vdots & \ddots & \vdots \\ 0 & 0 & \cdots & \frac{1}{\sigma_0^2} \end{bmatrix} \tag{10.91}$$

Applying these results to (10.81) then yields

$$\frac{1}{\hat{S}_{ML}}(e^{\jmath\omega}) = \mathbf{w}^{*T}\mathbf{R}_x^{-1}\mathbf{w} = \mathbf{w}^{*T}(\mathbf{U}_1^{-1})^{*T}\mathbf{D}_U^{-1}\mathbf{U}_1^{-1}\mathbf{w} \tag{10.92}$$

$$= \mathbf{w}^{*T} \begin{bmatrix} 1 & 0 & \cdots & 0 \\ a_1^{(N-1)} & 1 & \cdots & 0 \\ \vdots & \vdots & \ddots & \vdots \\ a_{N-1}^{(N-1)} & a_{N-2}^{(N-2)} & \cdots & 1 \end{bmatrix} \begin{bmatrix} \frac{1}{\sigma_{N-1}^2} & 0 & \cdots & 0 \\ 0 & \frac{1}{\sigma_{N-2}^2} & \cdots & 0 \\ \vdots & \vdots & \ddots & \vdots \\ 0 & 0 & \cdots & \frac{1}{\sigma_0^2} \end{bmatrix} \begin{bmatrix} 1 & a_1^{(N-1)*} & \cdots & a_{N-1}^{(N-1)*} \\ 0 & 1 & \cdots & a_{N-2}^{(N-2)*} \\ \vdots & \vdots & \ddots & \vdots \\ 0 & 0 & \cdots & 1 \end{bmatrix} \mathbf{w}$$

Note that (10.92) can be written as a sum of terms of the form

$$\begin{bmatrix} 1 & e^{-\jmath\omega} & \cdots & e^{-\jmath(N-1)\omega} \end{bmatrix} \begin{bmatrix} 0 \\ \vdots \\ 1 \\ \vdots \\ a_p^{(p)} \end{bmatrix} \frac{1}{\sigma_p^2} \begin{bmatrix} 0 & \cdots & 1 & \cdots & a_p^{(p)*} \end{bmatrix} \begin{bmatrix} 1 \\ e^{\jmath\omega} \\ \vdots \\ e^{\jmath(N-1)\omega} \end{bmatrix}$$

$$= e^{-\jmath\omega(N-1-p)} \begin{bmatrix} 1 & \cdots & e^{-\jmath p\omega} \end{bmatrix} \begin{bmatrix} 1 \\ \vdots \\ a_p^{(p)} \end{bmatrix} \frac{1}{\sigma_p^2} \begin{bmatrix} 1 & \cdots & a_p^{(p)*} \end{bmatrix} \begin{bmatrix} 1 \\ \vdots \\ e^{\jmath p\omega} \end{bmatrix} e^{\jmath\omega(N-1-p)}$$

$$= \frac{\left|\sum_{k=0}^{p} a_k^{(p)} e^{-\jmath\omega k}\right|^2}{\sigma_p^2} = \frac{1}{\hat{S}_{ME}^{(p)}(e^{\jmath\omega})} \tag{10.93}$$

Finally, combining (10.93) with (10.92) produces

$$\boxed{\frac{1}{\hat{S}_{ML}(e^{\jmath\omega})} = \sum_{p=0}^{N-1} \frac{1}{\hat{S}_{ME}^{(p)}(e^{\jmath\omega})}} \tag{10.94}$$

This is the desired result.

The relation between the ME and the ML methods shows that the ML estimate is a kind of average of ME estimates of all orders up to $N-1$. This may lead you to suspect that the ML method has somewhat poorer resolution than the $(N-1)^{\text{st}}$ order ME or AR estimate. This suspicion is correct. It turns out that for short data sets the ML estimate typically provides more resolution than classical methods but less than the ME method. Figure 10.10 shows the spectral estimates produced by the classical, the ML, and the ME methods beginning with the known correlation function

$$R_x[l] = e^{-0.02l}\left(\cos 0.3\pi l + \frac{1}{15\pi}\sin 0.3\pi l\right) + 2e^{-0.04l}\left(\cos 0.6\pi l + \frac{1}{15\pi}\sin 0.6\pi l\right)$$

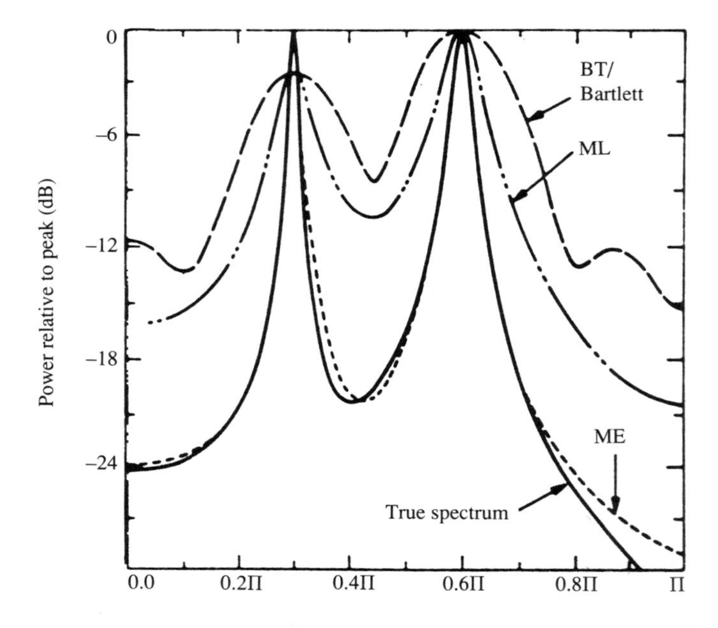

Figure 10.10 Comparison of spectral estimates corresponding to a known correlation function. (From R. T. Lacoss, Data adaptive spectral analysis methods, *Geophysics*, 36:661–675, August 1971. Reproduced by permission.)

and using samples at $l = 0, \pm 1, \pm 2, \ldots, \pm 10$. The classical method used is a Blackman–Tukey procedure with Bartlett window applied to the full length of the data. It is evident that the three spectral estimates exhibit resolution in the order stated above. In this case the all-pole ME model was close to reproducing the true spectrum of the random process; but the ordering of the methods with respect to resolution generally holds even when the underlying process is not close to an AR process. If the frequencies of the two damped sinusoids are moved closer together (say to 0.3π and 0.4π), it is usually not possible to tell from the classical estimate that two distinct sinusoids are present at all. The classical estimate appears as a single broad peak centered about midway between the two frequencies.

10.3.2 Computation of the ML Spectral Estimate

The maximum likelihood spectral estimate involves the reciprocal of the quadratic product $\mathbf{w}^{*T}\mathbf{R}_x^{-1}\mathbf{w}$, which must be computed over a range of values of frequency. If there are L frequency points this requires about N^2L multiplications. A more efficient way to perform the computation is as follows. Let the estimated inverse correlation matrix be represented in the form

$$\hat{\mathbf{R}}_x^{-1} = \sum_{i=1}^{N} \frac{1}{\lambda_i}\mathbf{e}_i\mathbf{e}_i^{*T} \tag{10.95}$$

where the λ_i are the eigenvalues and the $\mathbf{e}_i$ are the eigenvectors of the correlation matrix. If the correlation matrix is estimated in the form

$$\hat{\mathbf{R}}_x = \frac{1}{K}\mathbf{X}\mathbf{X}^{*T}$$

where K is the number of rows of $\mathbf{X}$, then the representation (10.95) can be obtained via the SVD of $\mathbf{X}$ (see Chapter 2). The quadratic product then has the form

$$\mathbf{w}^{*T}\hat{\mathbf{R}}_x^{-1}\mathbf{w} = \sum_{i=1}^{N} \frac{1}{\lambda_i}|\mathbf{w}^{*T}\mathbf{e}_i|^2 \tag{10.96}$$

Let the components of the vector $\mathbf{e}_i$ be regarded as a sequence $e_i[0], e_i[1], \ldots e_i[N-1]$. Then the Fourier transform of the sequence is

$$E_i(e^{\jmath\omega}) = \mathbf{w}^{*T}\mathbf{e}_i$$

so the term $|\mathbf{w}^{*T}\mathbf{e}_i|^2$ in (10.96) is just the squared magnitude of $E_i(e^{\jmath\omega})$. The L values of this term corresponding to the L desired frequency points can be found by zero-padding the sequence $e_i[n]$ to a length L and using an L-point FFT. A total of N such FFTs is needed to evaluate (10.96).

Since an L-point FFT can be computed with $\frac{L}{2}\log_2 L$ multiplications (if L is a power of 2), the total number of multiplications to find the ML spectral estimate is approximately $\frac{NL}{2}\log_2 L$. Comparing this to the number of multiplications (N^2L) required for direct evaluation of the formula shows that the FFT method is more efficient if

$$\log_2 L < 2N$$

or

$$L < 2^{2N}$$

The right-hand side of this inequality is a very large number for any typical values of N. (For $N = 8$ the number is 65,536, while for $N = 16$ it is more than 4 billion!) Thus there is usually a very significant computational saving by using the FFT.

An alternative method that saves an additional factor of N in the computations is to recognize that the quadratic product in the ML estimate can be written as

$$\mathbf{w}^{*T}\hat{\mathbf{R}}_x^{-1}\mathbf{w} = \sum_{k=-N}^{N} \varrho[k]e^{-\jmath\omega k}$$

where $\varrho[k]$ are coefficients resulting from combining terms in the inverse correlation matrix $\hat{\mathbf{R}}_x^{-1}$, and using the FFT to compute this expression directly. Even the terms $\varrho[k]$ can be computed efficiently using the Levinson recursion and the Gohberg–Semencul formula (see Problem 8.31). Details can be found in [28].

10.4 SUBSPACE METHODS: ESTIMATING THE DISCRETE COMPONENTS

When a spectral estimate is based on an underlying model for a random process, much can be said for the necessity of choosing the correct model. While this statement seems indisputable, applications of model-based methods to problems that do not fit the model are a common occurrence. A case in point is the estimation of random complex exponentials in additive noise. For example, consider the random process

$$x[n] = A_1 e^{\jmath\omega_1 n} + A_2 e^{\jmath\omega_2 n} + w[n]$$

where the amplitudes A_1 and A_2 are random variables and $w[n]$ is additive white noise. The most interesting components of the spectrum are discrete (the lines at ω_1 and ω_2). This type of process occurs in many practical problems, most notably in those dealing with arrays. The variables ω_1 and ω_2 may represent two actual frequency values or they may relate to spatial directions or bearings from the array of two sinusoidal sources. When ω_1 and ω_2 are closely spaced it is necessary to seek high-resolution methods that can resolve the two closely spaced spectral lines. Since methods such as the AR or ME methods provide much higher resolution than Fourier-based methods when the number of samples is limited, these methods have been applied to the estimation of spectra for the sinusoid in noise problems. In fact, many analyses have been made comparing the AR model performance to that of the maximum likelihood method and Fourier analysis on exactly this problem (see, e.g., [29, 30]). The use of the AR and ME methods to estimate the spectrum of sinusoids (or complex exponentials) in noise is fundamentally flawed, however, since these methods are based on the innovations representation, which applies to only the *continuous* part of the spectrum.

Model-based methods for estimating the discrete components of the spectrum did not appear for more than a decade after the introduction of the maximum entropy and maximum likelihood methods. Their final arrival filled a long-standing gap in the arsenal of techniques for spectrum analysis. Interestingly, while the AR and ME methods relate to the triangular decomposition of the correlation matrix, these new methods relate to the *eigenvector* decomposition of the correlation matrix (i.e., the DKLT).

A revealing comparison of the behavior of the correct model for a complex sinusoid in noise to that of an AR model was given by Richard Roy in his Ph.D. thesis [31]. The AR model for a process $x[n]$ can be described by the two equations

$$s[n] = as[n-1] + w[n]$$
$$x[n] = s[n]$$

where $w[n]$ is white noise, while the complex sinusoid in noise is characterized by

$$s[n] = as[n-1]$$
$$x[n] = s[n] + w[n]$$

Note that there is no white noise driving the difference equation in this last set of equations. In particular, if the pole location "a" is taken on the unit circle, the real and imaginary parts of $s[n]$ are undamped sinusoids. Figure 10.11 compares the trajectories of the output process for each of these models using the same white noise sequence and the initial condition $s[0] = 1$. The complex exponential in noise model stays close to its circular path while the AR model wanders widely over the entire plane. This wide deviation is not due just to the fact that the pole is on the unit circle (which is theoretically not allowed) but is truly a property of the model. A model with pole just slightly inside the unit circle has essentially the same behavior.

This section discusses methods specifically designed for the estimation of discrete components of the spectra: more particularly, for the estimation of complex exponentials in noise. The methods are based upon a decomposition of the observation vector space into a subspace associated with the signal and one associated only with the noise. Later it is shown that this same decomposition can also be used to improve the estimates of the continuous part of the spectrum.

10.4.1 Principles of Subspace Methods

In 1973 a Soviet scientist, V. F. Pisarenko, published a paper on a method that shed some new light on the estimation of complex exponentials in white noise [32]. The work was of interest not only for the estimation of spectra of time-series data but also in the estimation of direction of arrival of sinusoidal traveling waves impinging on an array of sensors. This spurred further work in the area. In the 1970s and early 1980s a number of methods inspired by Pisarenko's work were developed that approached the problem by separating the data into "signal and noise subspaces." Two of the most important of these techniques are the MUSIC method [33–35] and the method born of it, called *ESPRIT*

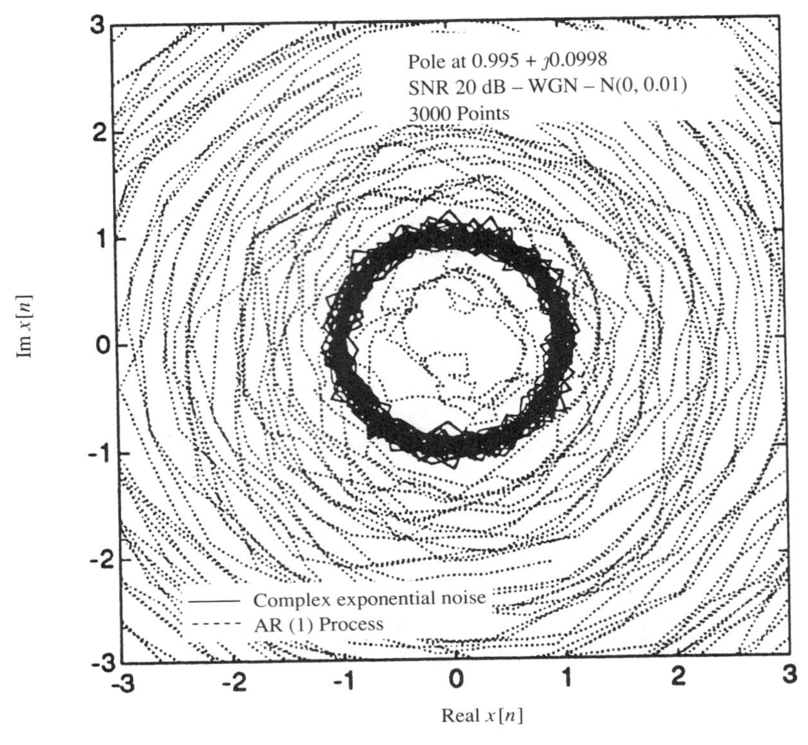

Figure 10.11 Trajectories for AR model and complex exponential in noise. [From Richard H. Roy. *ESPRIT —Estimation of Signal Parameters via Rotational Invariance Techniques*. Ph.D. thesis, Stanford University, Stanford, California, August 1987 (Information Systems Laboratory, Department of Electrical Engineering). Reproduced by permission.]

[31, 36]. These methods have elegant mathematical properties and important interpretations in the context of array processing beyond the scope of our discussion. The methods, however, provide an important new set of tools for the spectrum estimation problem that are not found in any of the previously discussed techniques. Two related methods, the minimum norm procedure [37, 38] and the method of principle components linear prediction [39, 40], are less general in the array context but are of significant interest in spectrum estimation. Finally, a general subspace methodology known as subspace fitting [41, 42] encompasses most of the other methods as special cases and with appropriately chosen weighting provides frequency estimates that asymptotically meet the Cramér-Rao bound. The specially weighted procedure, known as "weighted subspace fitting," is nonlinear but the corresponding estimate is not as difficult to compute as the (true) maximum likelihood estimate. An iterative procedure for the computation has been proposed.

Let us begin by considering a random sequence consisting of a single complex exponential in white noise. That is,

$$x[n] = As[n] + \eta[n] \tag{10.97}$$

where

$$s[n] = e^{j\omega_o n} \tag{10.98}$$

and the complex amplitude

$$A = |A|e^{j\phi} \tag{10.99}$$

is random (i.e., both $|A|$ and ϕ are random variables with ϕ uniformly distributed). The white noise is assumed to have zero mean and variance σ_o^2 and to be uncorrelated with the signal. The spectrum of the random process is shown in Fig. 10.12, where P_o is the variance of A. To see how the parameters defining this spectrum could be estimated, consider the vectors defined by N consecutive samples of the process, namely

$$\boldsymbol{x} = \begin{bmatrix} x[0] \\ x[1] \\ \vdots \\ x[N-1] \end{bmatrix} \tag{10.100}$$

$$\boldsymbol{\eta} = \begin{bmatrix} \eta[0] \\ \eta[1] \\ \vdots \\ \eta[N-1] \end{bmatrix} \tag{10.101}$$

and

$$\mathbf{s} = \begin{bmatrix} 1 \\ e^{j\omega_o} \\ \vdots \\ e^{j(N-1)\omega_o} \end{bmatrix} \tag{10.102}$$

Then

$$\boldsymbol{x} = A\mathbf{s} + \boldsymbol{\eta} \tag{10.103}$$

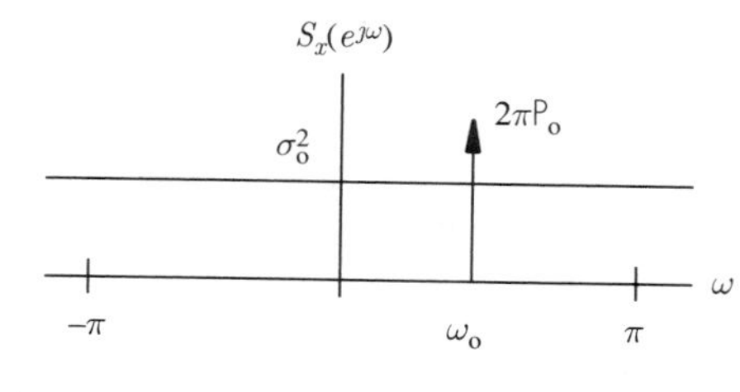

Figure 10.12 Power spectral density of a complex exponential in white noise.

Since the signal and noise are uncorrelated, the correlation matrix of the observation vector is

$$\mathbf{R}_x = \mathcal{E}\{A\mathbf{s}(A\mathbf{s})^{*T}\} + \mathcal{E}\{\boldsymbol{\eta}\boldsymbol{\eta}^{*T}\} = \mathsf{P}_o\mathbf{s}\mathbf{s}^{*T} + \sigma_o^2\mathbf{I} \tag{10.104}$$

where

$$\mathsf{P}_o \stackrel{\text{def}}{=} \mathcal{E}\{AA^*\} = \mathcal{E}\{|A|^2\} \tag{10.105}$$

Let us look at the eigenvectors of $\mathbf{R}_x$. One of these eigenvectors (although it is not normalized) is the signal vector $\mathbf{s}$. This follows because from (10.104),

$$\mathbf{R}_x\mathbf{s} = (\mathsf{P}_o\mathbf{s}\mathbf{s}^{*T} + \sigma_o^2\mathbf{I})\mathbf{s} = \mathsf{P}_o\mathbf{s}\mathbf{s}^{*T}\mathbf{s} + \sigma_o^2\mathbf{s} = (N\mathsf{P}_o + \sigma_o^2)\mathbf{s} \tag{10.106}$$

where the last step follows because $\mathbf{s}^{*T}\mathbf{s} = \|\mathbf{s}\|^2 = N$. The corresponding eigenvalue thus has a value of $N\mathsf{P}_o + \sigma_o^2$. Since the correlation matrix has a set of N orthonormal eigenvectors, all of the other eigenvectors are orthogonal to $\mathbf{s}$. If any one of these other eigenvectors is denoted by $\mathbf{e}_i$, then from (10.104)

$$\mathbf{R}_x\mathbf{e}_i = \mathsf{P}_o\mathbf{s}\mathbf{s}^{*T}\mathbf{e}_i + \sigma_o^2\mathbf{e}_i = \sigma_o^2\mathbf{e}_i \tag{10.107}$$

where the last step follows because $\mathbf{s}^{*T}\mathbf{e}_i = 0$. Thus all of the remaining eigenvalues of the correlation matrix are equal to σ_o^2.

The proceeding argument shows that the set of parameters (ω_o, P_o, and σ_o^2) that define the power density spectrum can be found by solving the eigenvalue problem for the correlation matrix. Specifically, as long as there is only one signal present, the following steps can be performed:

1. Form the correlation matrix and compute its eigenvalues and eigenvectors.
2. Identify the $N-1$ smallest eigenvalues. These all have the same value, namely σ_o^2.
3. Identify the remaining (largest) eigenvalue. It is equal to $N\mathsf{P}_o + \sigma_o^2$. Knowledge of its value and σ_o^2 determines P_o.
4. The eigenvector corresponding to the largest eigenvalue is a complex exponential with frequency ω_o. This in principle determines ω_o, although more specific methods to estimate ω_o will be derived shortly.

Now consider the problem of two signals in white noise. The observed sequence is of the form

$$x[n] = A_1 s_1[n] + A_2 s_2[n] + \eta[n] \tag{10.108}$$

where

$$s_i[n] = e^{\jmath\omega_i n} \tag{10.109}$$

and A_i is the complex amplitude of the ith signal

$$A_i = |A_i|e^{\jmath\phi_i} \tag{10.110}$$

satisfying the same assumptions as before. If the signal vectors $\mathbf{s}_i$ for $i = 1, 2$ are defined as

$$\mathbf{s}_i = \begin{bmatrix} 1 \\ e^{\jmath\omega_i} \\ \vdots \\ e^{\jmath(N-1)\omega_i} \end{bmatrix} \tag{10.111}$$

then the observation vector can be written as

$$\boldsymbol{x} = A_1\mathbf{s}_1 + A_2\mathbf{s}_2 + \boldsymbol{\eta} \tag{10.112}$$

Further, if the complex amplitudes of the signal are uncorrelated, then the correlation matrix of the observed sequence is

$$\mathbf{R}_x = \mathsf{P}_1\mathbf{s}_1\mathbf{s}_1^{*T} + \mathsf{P}_2\mathbf{s}_2\mathbf{s}_2^{*T} + \sigma_o^2\mathbf{I} \tag{10.113}$$

where

$$\mathsf{P}_i \stackrel{\text{def}}{=} \mathcal{E}\{A_iA_i^*\} = \mathcal{E}\{|A_i|^2\} \tag{10.114}$$

Now consider the eigenvectors of $\mathbf{R}_x$. If $\mathbf{s}_1$ and $\mathbf{s}_2$ are linearly independent (as they are when $\omega_1 \neq \omega_2$), then it is possible to find exactly $N-2$ eigenvectors in the N-dimensional vector space that are orthogonal to *both* $\mathbf{s}_1$ and $\mathbf{s}_2$. If $\mathbf{e}_i$ denotes one of these eigenvectors, then

$$\mathbf{R}_x\mathbf{e}_i = \mathsf{P}_1\mathbf{s}_1\mathbf{s}_1^{*T}\mathbf{e}_i + \mathsf{P}_2\mathbf{s}_2\mathbf{s}_2^{*T}\mathbf{e}_i + \sigma_o^2\mathbf{e}_i = \sigma_o^2\mathbf{e}_i \tag{10.115}$$

where the last step follows because $\mathbf{s}_1^{*T}\mathbf{e}_i = \mathbf{s}_2^{*T}\mathbf{e}_i = 0$. Thus these eigenvectors all correspond to eigenvalues $\lambda_i = \sigma_o^2$. The remaining two eigenvectors (let us call them $\mathbf{e}_1$ and $\mathbf{e}_2$) cannot be orthogonal to both $\mathbf{s}_1$ and $\mathbf{s}_2$. Therefore, they lie in the subspace defined by $\mathbf{s}_1$ and $\mathbf{s}_2$ and have eigenvalues greater than σ_o^2. This subspace is called the "signal subspace" (see Fig. 10.13). The complementary subspace, which is orthogonal to the signal subspace, is called the "noise subspace." The last term is slightly misleading because while the signal vectors lie totally within the signal subspace, the noise vector $\boldsymbol{\eta}$ has components in *both* subspaces.

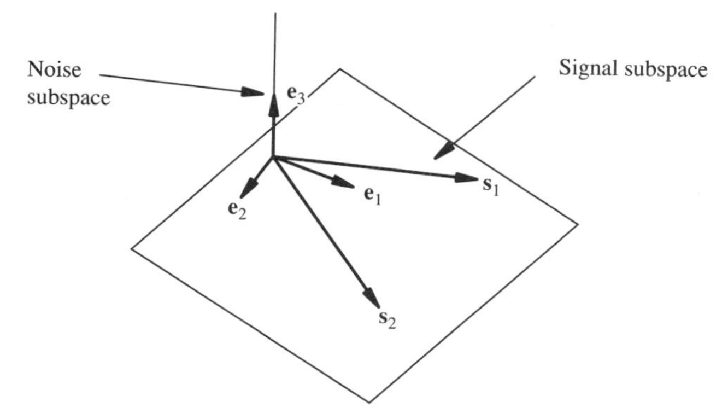

Figure 10.13 Illustration of signal and noise subspaces.

As a final generalization, consider the case of M independent signals in noise (where it is assumed that $M < N$). That is

$$x[n] = \sum_{i=1}^{M} A_i s_i[n] + \eta[n] \tag{10.116}$$

or

$$\boxed{\boldsymbol{x} = \sum_{i=1}^{M} A_i\mathbf{s}_i + \boldsymbol{\eta}} \tag{10.117}$$

where the quantities are defined by (10.109) through (10.111). This is the general signal model used in all of the subspace methods. If the noise is white, the correlation matrix is

$$\boxed{\mathbf{R}_x = \sum_{i=1}^{M} \mathsf{P}_i \mathbf{s}_i \mathbf{s}_i^{*T} + \sigma_o^2 \mathbf{I}} \tag{10.118}$$

where P_i is defined by (10.114). The last two equations can also be written with more compact matrix notation as

$$\boxed{\boldsymbol{x} = \mathbf{S}\begin{bmatrix} A_1 \\ A_2 \\ \vdots \\ A_M \end{bmatrix} + \boldsymbol{\eta}} \tag{10.119}$$

and

$$\boxed{\mathbf{R}_x = \mathbf{S}\mathbf{P}_o\mathbf{S}^{*T} + \sigma_o^2 \mathbf{I}} \tag{10.120}$$

where

$$\mathbf{S} = \begin{bmatrix} | & | & & | \\ \mathbf{s}_1 & \mathbf{s}_2 & \cdots & \mathbf{s}_M \\ | & | & & | \end{bmatrix} \tag{10.121}$$

and

$$\mathbf{P}_o = \begin{bmatrix} \mathsf{P}_1 & 0 & \cdots & 0 \\ 0 & \mathsf{P}_2 & \cdots & 0 \\ \vdots & \vdots & \ddots & \vdots \\ 0 & 0 & \cdots & \mathsf{P}_M \end{bmatrix} \tag{10.122}$$

By extension of the previous arguments, the correlation matrix $\mathbf{R}_x$ has M eigenvectors lying in the signal subspace, which is spanned by the $\mathbf{s}_i$, and $N - M$ eigenvectors lying in the orthogonal, complementary noise subspace. All $N - M$ noise subspace eigenvectors correspond to eigenvalues $\lambda_i = \sigma_o^2$ while all of the signal subspace eigenvectors correspond to eigenvalues $\lambda_i > \sigma_o^2$.

When the noise is not white, but is still uncorrelated with the signals, the foregoing development leads to the correlation matrix

$$\boxed{\mathbf{R}_x = \mathbf{S}\mathbf{P}_o\mathbf{S}^{*T} + \sigma_o^2 \boldsymbol{\Sigma}_\eta} \tag{10.123}$$

instead of (10.120). Here $\boldsymbol{\Sigma}_\eta$ is a normalized covariance matrix that represents the covariance structure of the noise vector. In this case the Mahalanobis or whitening transformation (see Chapter 2)

$$\boldsymbol{y} = \boldsymbol{\Sigma}_\eta^{-1/2} \boldsymbol{x} \tag{10.124}$$

leads to the correlation matrix

$$\mathbf{R}_y = \boldsymbol{\Sigma}_\eta^{-1/2}\mathbf{R}_x\boldsymbol{\Sigma}_\eta^{-1/2} = \mathbf{T}\mathbf{P}_\text{o}\mathbf{T}^{*T} + \sigma_\text{o}^2\mathbf{I} \tag{10.125}$$

where

$$\mathbf{T} = \boldsymbol{\Sigma}_\eta^{-1/2}\mathbf{S} \tag{10.126}$$

is the matrix with columns that are the transformed signal vectors

$$\mathbf{t}_k = \boldsymbol{\Sigma}_\eta^{-1/2}\mathbf{s}_k\,; \qquad k = 1, 2, \ldots, M \tag{10.127}$$

The relation between the original and the transformed vector space is depicted in Fig. 10.14. In the transformed vector space, the eigenvectors corresponding to the M largest eigenvalues span the signal subspace; those corresponding to the smallest eigenvalues (equal to σ_o^2) span the noise subspace. The eigenvectors and eigenvalues satisfy the equation

$$\mathbf{R}_y\mathbf{e}_k' = \left(\boldsymbol{\Sigma}_\eta^{-1/2}\mathbf{R}_x\boldsymbol{\Sigma}_\eta^{-1/2}\right)\mathbf{e}_k' = \lambda_k\mathbf{e}_k'$$

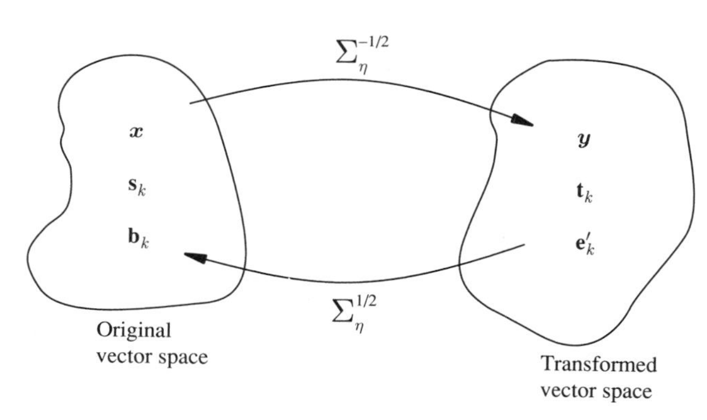

Figure 10.14 Whitening transformation for signals in colored noise.

which can be written as the generalized eigenvalue problem

$$\boxed{\mathbf{R}_x\mathbf{e}_k = \lambda_k\boldsymbol{\Sigma}_\eta\mathbf{e}_k} \tag{10.128}$$

where

$$\mathbf{e}_k = \boldsymbol{\Sigma}_\eta^{-1/2}\mathbf{e}_k' \tag{10.129}$$

The basis vectors that span the signal and noise subspaces in the original coordinate system are

$$\mathbf{b}_k = \boldsymbol{\Sigma}_\eta^{1/2}\mathbf{e}_k' = \boldsymbol{\Sigma}_\eta^{1/2}\left(\boldsymbol{\Sigma}_\eta^{1/2}\mathbf{e}_k\right)$$

or

$$\boxed{\mathbf{b}_k = \boldsymbol{\Sigma}_\eta\mathbf{e}_k} \tag{10.130}$$

Neither the eigenvectors $\mathbf{e}_k$ nor the basis vectors $\mathbf{b}_k$ are orthonormal in the usual sense.[9] The *eigenvalues* satisfy the same conditions, however. Those corresponding to the noise subspace are the $N-M$ smallest eigenvalues (equal to σ_o^2) and those corresponding to the signal subspace are the M largest eigenvalues (all greater than σ_o^2).

Let us now return to the case of white noise and the correlation matrix (10.120). It is convenient for our later discussion to define the matrices of eigenvectors

$$\mathbf{E}_{sig} = \begin{bmatrix} | & | & & | \\ \mathbf{e}_1 & \mathbf{e}_2 & \cdots & \mathbf{e}_M \\ | & | & & | \end{bmatrix} \tag{10.131}$$

and

$$\mathbf{E}_{noise} = \begin{bmatrix} | & | & & | \\ \mathbf{e}_{M+1} & \mathbf{e}_{M+2} & \cdots & \mathbf{e}_N \\ | & | & & | \end{bmatrix} \tag{10.132}$$

and also the two matrices of eigenvalues

$$\mathbf{\Lambda}_{sig} = \begin{bmatrix} \lambda_1 & 0 & \cdots & 0 \\ 0 & \lambda_2 & \cdots & 0 \\ \vdots & \vdots & \ddots & \vdots \\ 0 & 0 & \cdots & \lambda_M \end{bmatrix} \tag{10.133}$$

and

$$\mathbf{\Lambda}_{noise} = \begin{bmatrix} \lambda_{M+1} & 0 & \cdots & 0 \\ 0 & \lambda_{M+2} & \cdots & 0 \\ \vdots & \vdots & \ddots & \vdots \\ 0 & 0 & \cdots & \lambda_N \end{bmatrix} = \begin{bmatrix} \sigma_o^2 & 0 & \cdots & 0 \\ 0 & \sigma_o^2 & \cdots & 0 \\ \vdots & \vdots & \ddots & \vdots \\ 0 & 0 & \cdots & \sigma_o^2 \end{bmatrix} \tag{10.134}$$

The complete matrices of eigenvectors and eigenvalues are thus given by

$$\mathbf{E} = \begin{bmatrix} \mathbf{E}_{sig} & \mathbf{E}_{noise} \end{bmatrix} \tag{10.135}$$

and

$$\mathbf{\Lambda} = \begin{bmatrix} \mathbf{\Lambda}_{sig} & \mathbf{0} \\ \mathbf{0} & \mathbf{\Lambda}_{noise} \end{bmatrix} \tag{10.136}$$

Now observe that it is possible to write $\mathbf{R}_x$ as

$$\mathbf{R}_x = \mathbf{E}\mathbf{\Lambda}\mathbf{E}^{*T} = \mathbf{E}_{sig}\mathbf{\Lambda}_{sig}\mathbf{E}_{sig}^{*T} + \mathbf{E}_{noise}\mathbf{\Lambda}_{noise}\mathbf{E}_{noise}^{*T} \tag{10.137}$$

and to write $\mathbf{R}_x^{-1}$ [see (2.146)] as

$$\mathbf{R}_x^{-1} = \mathbf{E}\mathbf{\Lambda}^{-1}\mathbf{E}^{*T} = \mathbf{E}_{sig}\mathbf{\Lambda}_{sig}^{-1}\mathbf{E}_{sig}^{*T} + \mathbf{E}_{noise}\mathbf{\Lambda}_{noise}^{-1}\mathbf{E}_{noise}^{*T} \tag{10.138}$$

[9]These vectors satisfy the conditions $\mathbf{e}_k^{*T}\mathbf{\Sigma}_\eta\mathbf{e}_l = \mathbf{b}_k^{*T}\mathbf{\Sigma}_\eta^{-1}\mathbf{b}_l = 0$ if $k \neq l$ and 1 if $k = l$. That is, they are orthonormal *relative to* the matrices $\mathbf{\Sigma}_\eta$ and $\mathbf{\Sigma}_\eta^{-1}$.

Further, since the columns of $\mathbf{E}_{sig}$ are orthonormal and define the signal subspace, this matrix can be used to form a projection matrix for the signal subspace [see (9.62)]

$$\boldsymbol{P}_{sig} = \mathbf{E}_{sig}(\mathbf{E}_{sig}^{*T}\mathbf{E}_{sig})^{-1}\mathbf{E}_{sig}^{*T} = \mathbf{E}_{sig}\mathbf{E}_{sig}^{*T} \tag{10.139}$$

where the last step follows because $\mathbf{E}_{sig}^{*T}\mathbf{E}_{sig} = \mathbf{I}_{M\times M}$. Likewise, $\mathbf{E}_{noise}$ can be used to form a projection matrix for the noise subspace

$$\boldsymbol{P}_{noise} = \mathbf{E}_{noise}\mathbf{E}_{noise}^{*T} = \mathbf{I} - \boldsymbol{P}_{sig} \tag{10.140}$$

where the last equality follows because the subspaces are orthogonal complements of each other. The last definition is especially convenient for some of the discussions that follow.

10.4.2 Pisarenko Harmonic Decomposition

The subspace methods are concerned mainly with determining the *frequencies* $\omega_1, \omega_2, \ldots, \omega_M$ of the discrete components (lines) in the spectrum. Pisarenko's method was the original work in this area. It is based on the assumption that the number of discrete frequencies M is known and that the number of measurements N is given by

$$N = M + 1$$

This is not always a very practical assumption, but it is worth discussing because it motivates the more general subspace methods. Under the Pisarenko assumption there is only one eigenvector that defines the noise subspace. This is the eigenvector $\mathbf{e}_N$. Since each of the signal vectors is orthogonal to the noise subspace, then it follows that $\mathbf{s}_i^{*T}\mathbf{e}_N = 0$ for $i = 1, 2, \ldots, M$. Note also that if the frequency vector $\mathbf{w}$, defined by (10.82), is evaluated at $\omega = \omega_i$, then $\mathbf{w} = \mathbf{s}_i$ [compare (10.82) and (10.111)]; thus

$$\mathbf{w}^{*T}\mathbf{e}_N\big|_{\omega=\omega_i} = 0; \qquad i = 1, 2, \ldots, M \tag{10.141}$$

These considerations suggest the frequency estimation function

$$\boxed{\hat{P}_P(e^{j\omega}) = \frac{1}{|\mathbf{w}^{*T}\mathbf{e}_N|^2} = \frac{1}{\mathbf{w}^{*T}\mathbf{e}_N\mathbf{e}_N^{*T}\mathbf{w}}} \tag{10.142}$$

This function is called a "pseudospectrum." Since the denominator goes to zero at each of the frequencies ω_i, a plot of $\hat{P}_P(e^{j\omega})$ versus ω exhibits sharp (theoretically infinite) peaks at those frequencies (see Fig. 10.15). We emphasize that (10.142) is *not* a spectral estimate since it contains no information about the power P_i in the lines or the noise background σ_o^2. However, it does serve to locate the frequencies.

Another way to determine the frequencies is to define the function

$$E_N(z) = e_N[0] + e_N[1]z^{-1} + \cdots + e_N[N-1]z^{-(N-1)} \tag{10.143}$$

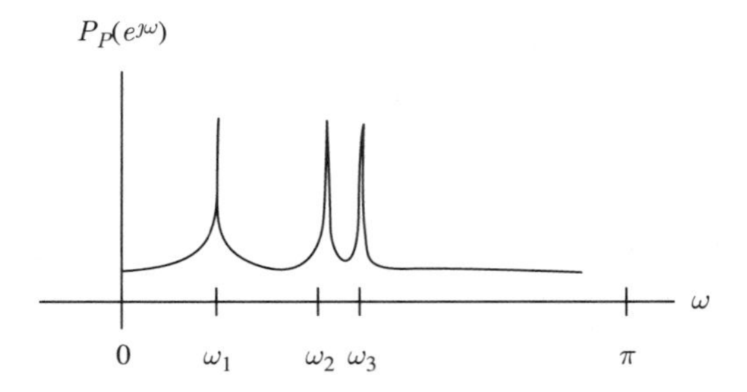

Figure 10.15 Illustration of Pisarenko pseudospectrum.

where the $e_N[n]$ are components of the eigenvector $\mathbf{e}_N$. This function is sometimes called an *eigenfilter* since it is formed from an eigenvector. Note that

$$E_N(e^{\jmath\omega}) = \mathbf{w}^{*T}\mathbf{e}_N$$

and so, from (10.142),

$$\hat{P}_P(e^{\jmath\omega}) = \frac{1}{|E_N(e^{\jmath\omega})|^2} = \frac{1}{E_N(e^{\jmath\omega})E_N^*(e^{\jmath\omega})} \tag{10.144}$$

Since the function $E_N(e^{\jmath\omega})$ goes to zero at $\omega_1, \omega_2, \ldots, \omega_M$, $E_N(z)$ has M roots at these positions on the unit circle. Thus an alternative way to determine the frequencies ω_i is to *find the roots of* $E_N(z)$.

Once the frequencies are known, either by plotting $\hat{P}_P(e^{\jmath\omega})$ or by finding the roots, then the signal powers can also be estimated from the eigenvalues if desired. For example, suppose that there are two signals and the two eigenvectors $\mathbf{e}_1$ and $\mathbf{e}_2$ defining the signal subspace have been found. Then the powers can be determined from the known relation

$$\mathbf{e}_i^{*T}\mathbf{R}_x\mathbf{e}_i = \lambda_i \tag{10.145}$$

Substituting (10.113) and evaluating (10.145) for $i = 1, 2$ produces the two equations

$$\begin{aligned} \mathsf{P}_1\mathbf{e}_1^{*T}\mathbf{s}_1\mathbf{s}_1^{*T}\mathbf{e}_1 + \mathsf{P}_2\mathbf{e}_1^{*T}\mathbf{s}_2\mathbf{s}_2^{*T}\mathbf{e}_1 + \sigma_o^2 &= \lambda_1 \\ \mathsf{P}_1\mathbf{e}_2^{*T}\mathbf{s}_1\mathbf{s}_1^{*T}\mathbf{e}_2 + \mathsf{P}_2\mathbf{e}_2^{*T}\mathbf{s}_2\mathbf{s}_2^{*T}\mathbf{e}_2 + \sigma_o^2 &= \lambda_2 \end{aligned} \tag{10.146}$$

These equations can be put in matrix form by defining the coefficients β_{ij} as

$$\begin{bmatrix} \beta_{11} & \beta_{12} \\ \beta_{21} & \beta_{22} \end{bmatrix} = \begin{bmatrix} - & \mathbf{e}_1^{*T} & - \\ - & \mathbf{e}_2^{*T} & - \end{bmatrix} \begin{bmatrix} | & | \\ \mathbf{s}_1 & \mathbf{s}_2 \\ | & | \end{bmatrix} \tag{10.147}$$

and writing (10.146) as

$$\begin{bmatrix} |\beta_{11}|^2 & |\beta_{12}|^2 \\ |\beta_{21}|^2 & |\beta_{22}|^2 \end{bmatrix} \begin{bmatrix} \mathsf{P}_1 \\ \mathsf{P}_2 \end{bmatrix} = \begin{bmatrix} \lambda_1 - \sigma_o^2 \\ \lambda_2 - \sigma_o^2 \end{bmatrix} \tag{10.148}$$

which can be solved for P_1 and P_2.

An example of the Pisarenko method follows. Although the example illustrates the steps in the method, you should be aware that since in most cases we have to *estimate* the correlation matrix, the results seldom turn out as neatly as they do here.

Example 10.1

It is desired to find the frequency and power of a single real sinusoidal signal in white noise. The correlation matrix for the data is

$$\mathbf{R}_x = \begin{bmatrix} 3 & 0 & -2 \\ 0 & 3 & 0 \\ -2 & 0 & 3 \end{bmatrix}$$

Note that since the real sinusoid consists of two complex exponentials at positive and negative frequencies, M is equal to 2 for this problem and the Pisarenko method thus requires a 3×3 correlation matrix.

The matrices of eigenvalues and eigenvectors of $\mathbf{R}_x$ are found to be

$$\mathbf{\Lambda} = \begin{bmatrix} 5 & 0 & 0 \\ 0 & 3 & 0 \\ 0 & 0 & 1 \end{bmatrix} \qquad \mathbf{E} = \begin{bmatrix} -\frac{1}{\sqrt{2}} & 0 & -\frac{1}{\sqrt{2}} \\ 0 & -1 & 0 \\ \frac{1}{\sqrt{2}} & 0 & -\frac{1}{\sqrt{2}} \end{bmatrix}$$

(You can check this by direct multiplication.) The noise variance is therefore

$$\sigma_o^2 = \lambda_3 = 1$$

and the noise eigenvector is

$$\mathbf{e}_3 = \begin{bmatrix} & -\frac{1}{\sqrt{2}} \\ 0 & -\frac{1}{\sqrt{2}} \end{bmatrix}$$

The Pisarenko pseudospectrum is

$$\hat{P}_P(e^{\jmath\omega}) = \frac{1}{|\mathbf{w}^{*T}\mathbf{e}_3|^2}$$

A plot of this function peaks at $\omega = \pm\pi/2$ because for these values

$$\mathbf{w} = \begin{bmatrix} 1 \\ e^{\pm\jmath\frac{\pi}{2}} \\ e^{\pm\jmath\frac{\pi}{2}2} \end{bmatrix} = \begin{bmatrix} 1 \\ \pm\jmath \\ -1 \end{bmatrix}$$

and

$$\mathbf{w}^{*T}\mathbf{e}_3 = \begin{bmatrix} 1 & \mp\jmath & -1 \end{bmatrix} \begin{bmatrix} -\frac{1}{\sqrt{2}} \\ 0 \\ -\frac{1}{\sqrt{2}} \end{bmatrix} = 0$$

The frequencies can also be located by finding the roots of the eigenfilter

$$E_3(z) = -\frac{1}{\sqrt{2}} - \frac{1}{\sqrt{2}}z^{-2}$$

The roots are

$$z = \pm \jmath = e^{\pm \jmath \frac{\pi}{2}}$$

which shows that $\omega = \pm\pi/2$. To find the power, use (10.147) and (10.148). The signal vectors are

$$\mathbf{s}_1 = \begin{bmatrix} 1 \\ e^{\jmath \frac{\pi}{2}} \\ e^{\jmath \pi} \end{bmatrix} = \begin{bmatrix} 1 \\ \jmath \\ -1 \end{bmatrix} \qquad \mathbf{s}_2 = \begin{bmatrix} 1 \\ e^{-\jmath \frac{\pi}{2}} \\ e^{-\jmath \pi} \end{bmatrix} = \begin{bmatrix} 1 \\ -\jmath \\ -1 \end{bmatrix}$$

Thus

$$\mathbf{S} = \begin{bmatrix} 1 & 1 \\ \jmath & -\jmath \\ -1 & -1 \end{bmatrix}$$

and

$$\mathbf{E}_{sig} = \begin{bmatrix} -\frac{1}{\sqrt{2}} & 0 \\ 0 & -1 \\ \frac{1}{\sqrt{2}} & 0 \end{bmatrix}$$

Therefore,

$$\mathbf{E}_{sig}^{*T}\mathbf{S} = \begin{bmatrix} -\frac{1}{\sqrt{2}} & 0 & \frac{1}{\sqrt{2}} \\ 0 & -1 & 0 \end{bmatrix} \begin{bmatrix} 1 & 1 \\ \jmath & -\jmath \\ -1 & -1 \end{bmatrix} = \begin{bmatrix} -\sqrt{2} & -\sqrt{2} \\ -\jmath & \jmath \end{bmatrix}$$

and (10.148) becomes

$$\begin{bmatrix} 2 & 2 \\ 1 & 1 \end{bmatrix} \begin{bmatrix} \mathsf{P}_1 \\ \mathsf{P}_2 \end{bmatrix} = \begin{bmatrix} 4 \\ 2 \end{bmatrix}$$

This has the solution

$$\mathsf{P}_1 = \mathsf{P}_2 = 1$$

10.4.3 MUSIC

A major improvement upon the Pisarenko procedure is the MUSIC (for MUltiple SIgnal Classification) method of Schmidt [33] and the similar method developed by Bienvenu and Kopp in France [43]. With the MUSIC method a correlation matrix of some size $N > M+1$ is formed and its eigenvalues and eigenvectors are found. As seen before, the eigenvalues are divided into two groups. Those corresponding to the signal subspace have values greater than σ_o^2, while those corresponding to the noise subspace have values just equal to σ_o^2 (see Fig. 10.16). In theory, if the number of signals is not known, it can be estimated by looking at the smallest eigenvalues and finding the set that are approximately equal. This number is equal to $N - M$. In practice, when the signal-to-noise ratio is low, statistical criteria such as the AIC and MDL have to be used (see Section 10.4.8). While the Pisarenko method involves projection of the signal vectors onto a single noise eigenvector, the MUSIC method involves projection of the signal onto the *entire noise subspace*.

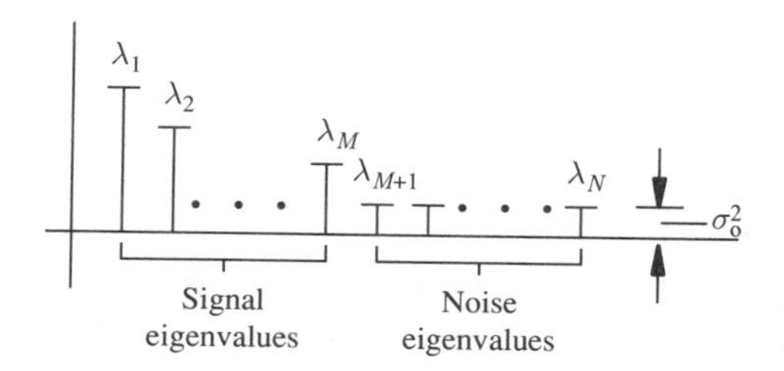

Figure 10.16 Signal and noise subspace eigenvalues for the MUSIC method.

The MUSIC procedure works like this. The squared magnitude of the projection of $\mathbf{w}$ onto the noise subspace is given using (10.140) by

$$\mathbf{w}^{*T}\boldsymbol{P}_{noise}\mathbf{w} = \mathbf{w}^{*T}\mathbf{E}_{noise}\mathbf{E}_{noise}^{*T}\mathbf{w}$$

Since each of the signals is orthogonal to the noise subspace, this quantity goes to zero for values of the frequency where $\mathbf{w} = \mathbf{s}_i$. The MUSIC pseudospectrum is defined as

$$\boxed{\hat{P}_{MU}(e^{\jmath\omega}) = \frac{1}{\mathbf{w}^{*T}\boldsymbol{P}_{noise}\mathbf{w}} = \frac{1}{\mathbf{w}^{*T}\mathbf{E}_{noise}\mathbf{E}_{noise}^{*T}\mathbf{w}}} \qquad (10.149)$$

and therefore exhibits sharp peaks at the signal frequencies where $\mathbf{w} = \mathbf{s}_i$. A similar expression holds when the noise is not white (see Problem 10.12). In this case the columns of the matrix $\mathbf{E}_{noise}$ correspond to the generalized eigenvectors [see (10.128)].

An alternative root-finding variation of the method called "*root MUSIC*" can be developed as follows. Define the eigenfilter $E_i(z)$ as

$$E_i(z) = e_i[0] + e_i[1]z^{-1} + \cdots + e_i[N-1]z^{-(N-1)} \qquad (10.150)$$

where the $e_i[n]$ are components of the eigenvector $\mathbf{e}_i$. The denominator of (10.149) can then be written using (10.132) as

$$\mathbf{w}^{*T}\mathbf{E}_{noise}\mathbf{E}_{noise}^{*T}\mathbf{w} = \sum_{i=M+1}^{N} \mathbf{w}^{*T}\mathbf{e}_i\mathbf{e}_i^{*T}\mathbf{w} = \sum_{i=M+1}^{N} E_i(e^{\jmath\omega})E_i^*(e^{\jmath\omega}) \qquad (10.151)$$

The MUSIC pseudospectrum can therefore be expressed as

$$\hat{P}_{MU}(e^{\jmath\omega}) = \left.\frac{1}{\sum_{i=M+1}^{N} E_i(z)E_i^*(1/z^*)}\right|_{z=e^{\jmath\omega}} \qquad (10.152)$$

Since the denominator goes to zero at $z = e^{\jmath\omega_i}$ $(i = 1, 2, \ldots, M)$, the denominator polynomial

$$\hat{P}_{MU}^{-1}(z) = \sum_{i=M+1}^{N} E_i(z)E_i^*(1/z^*) \qquad (10.153)$$

has M roots *lying on the unit circle.* These M roots (which are, in fact, double roots) correspond to the signal frequencies.

Note that since each eigenfilter $E_i(z)$ is an $(N-1)^{\text{th}}$-degree polynomial, it has a total of $N-1$ roots. M of these roots correspond to the $e^{\jmath\omega_i}$ and lie on the unit circle. The other $N-M-1$ roots not on the unit circle are called "spurious" roots and play no particular role in locating the spectral lines. In theory these roots are not a problem, but in practice some of them may lie close to the unit circle and could be mistakenly attributed to signals. The polynomial $\hat{P}_{MU}^{-1}(z)$ used in root MUSIC also has spurious roots. However, the effect of summing the eigenfilter terms in (10.153) is to move these spurious roots away from the unit circle. Only roots of the eigenfilters lying *on* the unit circle become roots of the polynomial $\hat{P}_{MU}^{-1}(z)$.

The following example illustrates the application of MUSIC.

Example 10.2

The correlation matrix corresponding to complex exponentials in white noise is given by

$$\mathbf{R}_x = \begin{bmatrix} 2 & -\jmath & -1 \\ \jmath & 2 & -\jmath \\ -1 & \jmath & 2 \end{bmatrix}$$

The matrices of eigenvalues and eigenvectors are found to be

$$\mathbf{\Lambda} = \begin{bmatrix} 4 & 0 & 0 \\ 0 & 1 & 0 \\ 0 & 0 & 1 \end{bmatrix} \qquad \mathbf{E} = \begin{bmatrix} -\frac{1}{\sqrt{3}}\jmath & \sqrt{\frac{2}{3}} & 0 \\ \frac{1}{\sqrt{3}} & -\frac{1}{\sqrt{6}}\jmath & \frac{1}{\sqrt{2}}\jmath \\ \frac{1}{\sqrt{3}}\jmath & \frac{1}{\sqrt{6}} & \frac{1}{\sqrt{2}} \end{bmatrix}$$

Since the two smallest eigenvalues are identical, the noise subspace has dimension 2, and there is only a single complex exponential present. The matrix of noise subspace eigenvectors is

$$\mathbf{E}_{noise} = \begin{bmatrix} \sqrt{\frac{2}{3}} & 0 \\ -\frac{1}{\sqrt{6}}\jmath & \frac{1}{\sqrt{2}}\jmath \\ \frac{1}{\sqrt{6}} & \frac{1}{\sqrt{2}} \end{bmatrix}$$

and the corresponding projection matrix for the noise subspace is

$$\boldsymbol{P}_{noise} = \mathbf{E}_{noise}\mathbf{E}_{noise}^{*T} = \begin{bmatrix} \frac{2}{3} & \frac{1}{3}\jmath & \frac{1}{3} \\ -\frac{1}{3}\jmath & \frac{2}{3} & \frac{1}{3}\jmath \\ \frac{1}{3} & -\frac{1}{3}\jmath & \frac{2}{3} \end{bmatrix}$$

The MUSIC pseudospectrum is given by

$$\hat{P}_{MU}(e^{\jmath\omega}) = \frac{1}{\mathbf{w}^{*T}\boldsymbol{P}_{noise}\mathbf{w}}$$

It can be verified that the denominator goes to zero for

$$\mathbf{w} = \begin{bmatrix} 1 \\ e^{\jmath\frac{\pi}{2}} \\ e^{\jmath\frac{\pi}{2}2} \end{bmatrix} = \begin{bmatrix} 1 \\ \jmath \\ -1 \end{bmatrix}$$

Therefore, the signal has frequency

$$\omega = \tfrac{\pi}{2}$$

To find the frequency by the root MUSIC procedure, it is necessary to form the denominator polynomial (10.153). Using the noise eigenvectors, the eigenfilters are

$$E_2(z) = \sqrt{\tfrac{2}{3}} - \tfrac{1}{\sqrt{6}}\jmath z^{-1} + \tfrac{1}{\sqrt{6}}z^{-2}$$

with

$$E_2(z)E_2^*(1/z^*) = \tfrac{1}{3}z^2 + \tfrac{1}{6}\jmath z + 1 - \tfrac{1}{6}\jmath z^{-1} + \tfrac{1}{3}z^{-2}$$

and

$$E_3(z) = \tfrac{1}{\sqrt{2}}\jmath z^{-1} + \tfrac{1}{\sqrt{2}}z^{-2}$$

with

$$E_3(z)E_3^*(1/z^*) = \tfrac{1}{2}\jmath z + 1 - \tfrac{1}{2}\jmath z^{-1}$$

The required polynomial is therefore

$$\begin{aligned} &E_2(z)E_2^*(1/z^*) + E_3(z)E_3^*(1/z^*) \\ &\qquad = \tfrac{1}{3}z^2 + \tfrac{2}{3}\jmath z + 2 - \tfrac{2}{3}\jmath z^{-1} + \tfrac{1}{3}z^{-2} \end{aligned}$$

This polynomial has a *double* root on the unit circle at $z = \jmath$ corresponding to the frequency $\omega = \pi/2$ and two other roots at $z = -0.2679\jmath$ and $z = -3.7321\jmath$. These last two are the spurious roots.

The MUSIC method can be related in a rather interesting way to some of the other methods of spectrum estimation encountered in this chapter. From (10.149) and (10.151) the MUSIC pseudospectrum can be written as

$$\hat{P}_{MU}(e^{\jmath\omega}) = \frac{1}{\mathbf{w}^{*T}(\sum_{i=M+1}^{N} \mathbf{e}_i\mathbf{e}_i^{*T})\mathbf{w}} \tag{10.154}$$

Since the eigenvalues for all the eigenvectors are equal to σ_o^2, a pseudospectrum for MUSIC that differs from this one by only a constant can be defined as

$$\hat{P}'_{MU}(e^{\jmath\omega}) = \frac{\sigma_o^2}{\mathbf{w}^{*T}(\sum_{i=M+1}^{N} \mathbf{e}_i\mathbf{e}_i^{*T})\mathbf{w}} = \frac{1}{\mathbf{w}^{*T}(\sum_{i=M+1}^{N} \frac{1}{\lambda_i}\mathbf{e}_i\mathbf{e}_i^{*T})\mathbf{w}} \tag{10.155}$$

where the last equality follows because $\lambda_i = \sigma_o^2,\ i = M+1, \ldots, N$. This representation, incidentally, is the same as the "eigenvector method" of Johnson and DeGraff [44]. The difference between the two algorithms is that in practice the estimated eigenvalues are not all exactly equal. Since it is possible to write the inverse correlation matrix in terms of its eigenvectors and eigenvalues [see (10.95)], the pseudospectrum (10.155) has a distinct relation to the ML spectral estimate. In particular, the MUSIC estimate is formed from that part of the inverse correlation matrix corresponding to the *noise subspace*. This fact leads to an interesting comparison between the ML, the ME, and the MUSIC methods. The

ML method uses the entire inverse covariance matrix to form the estimate. The ME and MUSIC methods relate to different decompositions of the inverse correlation matrix. The ME method uses a part of the *triangular* decomposition of the inverse correlation matrix; namely the part corresponding to the highest-order prediction error filter. The MUSIC method relates to the *eigenvector* decomposition of the correlation matrix and the part of the decomposition related to the noise subspace. The three methods are displayed for comparison in Table 10.1.

TABLE 10.1 COMPARISON OF MAXIMUM LIKELIHOOD, MAXIMUM ENTROPY, AND MUSIC SPECTRAL ESTIMATION METHODS

Maximum likelihood	$\hat{S}_{ML}(e^{\jmath\omega}) = \frac{1}{\mathbf{w}^{*T}\mathbf{R}_x^{-1}\mathbf{w}}$
Maximum entropy	$\hat{S}_{ME}(e^{\jmath\omega}) = \frac{\sigma_{N-1}^2}{\mathbf{w}^{*T}\mathbf{a}_{N-1}\mathbf{a}_{N-1}^{*T}\mathbf{w}}$
MUSIC	$\hat{P}'_{MU}(e^{\jmath\omega}) = \frac{1}{\mathbf{w}^{*T}\sum_{i=M+1}^{N}\frac{1}{\lambda_i}\mathbf{e}_i\mathbf{e}_i^{*T}\mathbf{w}}$

Usually, in the applications of MUSIC it is sufficient only to find the frequencies ω_i and not necessary to find the powers P_i. Theoretically, the powers can be found by a procedure similar to that developed in Section 10.4.2 (see Problem 10.11).

10.4.4 Minimum-Norm Procedure

For most cultural events in the United States of any significance, there is an east coast version and a west coast version. So it is with subspace methods. MUSIC and *ESPRIT* (see Section 10.4.6) were developed by researchers on the west coast. Others on the east coast, most notably Kumaresan and Tufts, also contributed considerably to the development of subspace methods. One important contribution of Kumaresan and Tufts was a rediscovery and interpretion of a method originally reported by Reddi [37] which has become known as the minimum-norm procedure [38]. Another of their contributions is described in the next subsection.

The idea behind the minimum-norm procedure is to find a single appropriately chosen vector **d** in the noise subspace and to define the pseudospectrum in terms of this vector as

$$\hat{P}_{MN}(e^{\jmath\omega}) \stackrel{\text{def}}{=} \frac{1}{|\mathbf{w}^{*T}\mathbf{d}|^2} = \frac{1}{\mathbf{w}^{*T}\mathbf{d}\mathbf{d}^{*T}\mathbf{w}} \tag{10.156}$$

The vector which lies in the subspace is chosen so that the squared magnitude $\|\mathbf{d}\|^2$ is minimized subject to the constraint that its first component is equal to 1. Why these characteristics should be desirable will be explained in a moment. First note, however, that if the components of $\mathbf{d}$ are denoted by $d[0], d[1], \ldots, d[N-1]$, with $d[0] = 1$, then the minimum-norm pseudospectrum can be equivalently expressed as

$$\hat{P}_{MN}(e^{j\omega}) = \frac{1}{|D(e^{j\omega})|^2} \tag{10.157}$$

where $D(z)$ is the polynomial

$$D(z) = \sum_{k=0}^{N-1} d[k] z^{-k} \tag{10.158}$$

Thus the frequencies can also be found as the roots of $D(z)$ lying on the unit circle.

Let us now explain the reason for the particular choice of $\mathbf{d}$. The polynomial $D(z)$ can be factored as

$$D(z) = D_1(z) \cdot D_2(z) \tag{10.159}$$

where $D_1(z)$ has only roots lying on the unit circle, and $D_2(z)$ has only roots *not* on the unit circle. [In other words $D_2(z)$ has just the *spurious* roots]. Notice that since the constant term of $D(z)$ is constrained to be equal to 1, the constant terms of $D_1(z)$ and $D_2(z)$ can be similarly constrained without any loss of generality. In other words, $D(z)$, $D_1(z)$, and $D_2(z)$ are *comonic* polynomials (in z^{-1}). By constraining $\mathbf{d}$ as above, it turns out that $D_2(z)$ is a minimum-phase polynomial, with all roots inside the unit circle. It can also be shown [45, 46] that these roots are approximately uniformly distributed around the unit circle in sectors where the M signal roots are not present. Thus the spurious roots tend to be well removed from the roots due to the signals.

The fact that all roots of $D_2(z)$ lie within the unit circle can be verified by noting that by Parseval's theorem, minimizing $||\mathbf{d}||^2$ is equivalent to minimizing

$$\frac{1}{2\pi}\int_{-\pi}^{\pi} |D(e^{j\omega})|^2 d\omega = \frac{1}{2\pi}\int_{-\pi}^{\pi} |D_1(e^{j\omega})|^2 |D_2(e^{j\omega})|^2 d\omega$$

where $D_1(z)$ is fixed and $D_2(z)$ is comonic. These conditions can be shown to be identical to the problem of linear prediction by the autocorrelation method (see Problem 10.13). $D_2(z)$ plays the role of a prediction error filter and therefore has all of its roots strictly inside the unit circle.

It remains now to find the optimum value of $\mathbf{d}$ that minimizes

$$||\mathbf{d}||^2 = \mathbf{d}^{*T}\mathbf{d}$$

subject to the constraints that $\mathbf{d}$ remains in the noise subspace and the first element of $\mathbf{d}$ is 1. These two constraints can be stated mathematically as

$$\mathbf{d} = \boldsymbol{P}_{noise}\mathbf{d} = \mathbf{E}_{noise}\mathbf{E}_{noise}^{*T}\mathbf{d}$$

and

$$\mathbf{d}^{*T}\boldsymbol{\iota} = 1$$

where $\boldsymbol{\iota}$ is the vector

$$\boldsymbol{\iota} = \begin{bmatrix} 1 & 0 & \cdots & 0 \end{bmatrix}^T \tag{10.160}$$

Further, the two constraints can be combined to form the single constraint

$$\mathbf{d}^{*T}\boldsymbol{\iota} = (\mathbf{E}_{noise}\mathbf{E}_{noise}^{*T}\mathbf{d})^{*T}\boldsymbol{\iota} = \mathbf{d}^{*T}\mathbf{E}_{noise}\mathbf{E}_{noise}^{*T}\boldsymbol{\iota} = 1$$

and we can now seek to minimize the Lagrangian (see Appendix A)

$$\mathcal{L} = \mathbf{d}^{*T}\mathbf{d} + \mu(1 - \mathbf{d}^{*T}\mathbf{E}_{noise}\mathbf{E}_{noise}^{*T}\boldsymbol{\iota}) + \mu^*(1 - \boldsymbol{\iota}^T\mathbf{E}_{noise}\mathbf{E}_{noise}^{*T}\mathbf{d})$$

where μ is a Lagrange multiplier. Taking the complex gradient with the help of the formulas in Table A.2 (Appendix A) produces

$$\nabla_{\mathbf{d}^*}\mathcal{L} = \mathbf{d} - \mu\mathbf{E}_{noise}\mathbf{E}_{noise}^{*T}\boldsymbol{\iota} = \mathbf{0}$$

or

$$\mathbf{d} = \mu\,\mathbf{E}_{noise}\mathbf{E}_{noise}^{*T}\boldsymbol{\iota} \tag{10.161}$$

where μ is selected so that the first component of $\mathbf{d}$ is 1. Now partition the matrix of noise eigenvectors as

$$\mathbf{E}_{noise} = \begin{bmatrix} \mathbf{c}^{*T} \\ \\ \mathbf{E}'_{noise} \end{bmatrix} \tag{10.162}$$

where $\mathbf{c}^{*T}$ is the top *row* (*not* an eigenvector). Thus from (10.162) the product $\mathbf{E}_{noise}^{*T}\boldsymbol{\iota}$ that appears in (10.161) is

$$\mathbf{E}_{noise}^{*T}\boldsymbol{\iota} = \mathbf{c}$$

With this observation it follows from (10.161) and (10.162) that the first component of $\mathbf{d}$ is equal to $\mu\mathbf{c}^{*T}\mathbf{c}$. Since this first component is constrained to be 1, the constant μ must be taken as $1/(\mathbf{c}^{*T}\mathbf{c})$ and (10.161) becomes

$$\boxed{\mathbf{d} = \frac{1}{\mathbf{c}^{*T}\mathbf{c}}\mathbf{E}_{noise}\mathbf{c} = \begin{bmatrix} 1 \\ \\ \mathbf{E}'_{noise}\mathbf{c}/(\mathbf{c}^{*T}\mathbf{c}) \end{bmatrix}} \tag{10.163}$$

An alternative expression can be written using the signal subspace eigenvectors (see Problem 10.14). This expression is

$$\boxed{\mathbf{d} = \frac{1}{1 - \mathbf{g}^{*T}\mathbf{g}}\left(\boldsymbol{\iota} - \mathbf{E}_{sig}\mathbf{g}\right) = \begin{bmatrix} 1 \\ \\ -\mathbf{E}'_{sig}\mathbf{g}/(1 - \mathbf{g}^{*T}\mathbf{g}) \end{bmatrix}} \tag{10.164}$$

where $\mathbf{g}$ and $\mathbf{E}'_{sig}$ are the partitions in

$$\mathbf{E}_{sig} = \begin{bmatrix} \mathbf{g}^{*T} \\ \\ \mathbf{E}'_{sig} \end{bmatrix} \tag{10.165}$$

Dowling and DeGroat [47] have also shown that when the problem involving the signals in noise is formulated as a linear prediction problem in the form (9.105), then the minimum-norm solution is equivalent to the total least squares solution of (9.105).

The following example illustrates the application of the minimum-norm procedure.

Example 10.3

The correlation matrix corresponding to the data in Example 10.2 is

$$\mathbf{R}_x = \begin{bmatrix} 2 & -\jmath & -1 \\ \jmath & 2 & -\jmath \\ -1 & \jmath & 2 \end{bmatrix}$$

Following the procedure in Example 10.2, the matrix of noise subspace eigenvectors is found to be

$$\mathbf{E}_{noise} = \begin{bmatrix} \sqrt{\frac{2}{3}} & 0 \\ -\frac{1}{\sqrt{6}}\jmath & \frac{1}{\sqrt{2}}\jmath \\ \frac{1}{\sqrt{6}} & \frac{1}{\sqrt{2}} \end{bmatrix}$$

The partitioning of (10.162) therefore defines the quantities

$$\mathbf{c} = \begin{bmatrix} \sqrt{\frac{2}{3}} \\ 0 \end{bmatrix}$$

and

$$\mathbf{E}'_{noise} = \begin{bmatrix} -\frac{1}{\sqrt{6}}\jmath & \frac{1}{\sqrt{2}}\jmath \\ \frac{1}{\sqrt{6}} & \frac{1}{\sqrt{2}} \end{bmatrix}$$

The terms needed in (10.163) are computed as

$$\begin{aligned} \mathbf{E}'_{noise}\mathbf{c}/(\mathbf{c}^{*T}\mathbf{c}) &= \begin{bmatrix} -\frac{1}{\sqrt{6}}\jmath & \frac{1}{\sqrt{2}}\jmath \\ \frac{1}{\sqrt{6}} & \frac{1}{\sqrt{2}} \end{bmatrix} \begin{bmatrix} \sqrt{\frac{2}{3}} \\ 0 \end{bmatrix} \Bigg/ \left(\sqrt{\frac{2}{3}}\right)^2 \\ &= \begin{bmatrix} -\frac{1}{3}\jmath \\ \frac{1}{3} \end{bmatrix} \Bigg/ \frac{2}{3} = \begin{bmatrix} -\frac{1}{2}\jmath \\ \frac{1}{2} \end{bmatrix} \end{aligned}$$

Therefore, the vector **d** [from (10.163)] is

$$\mathbf{d} = \begin{bmatrix} 1 \\ -\frac{1}{2}\jmath \\ \frac{1}{2} \end{bmatrix}$$

The pseudospectrum is then

$$\hat{P}_{MN}(e^{\jmath\omega}) \stackrel{\text{def}}{=} \frac{1}{|\mathbf{w}^{*T}\mathbf{d}|^2}$$

The product $\mathbf{w}^{*T}\mathbf{d}$ goes to zero for

$$\mathbf{w} = \begin{bmatrix} 1 \\ e^{\jmath\frac{\pi}{2}} \\ e^{\jmath\frac{\pi}{2}2} \end{bmatrix} = \begin{bmatrix} 1 \\ \jmath \\ -1 \end{bmatrix}$$

as in Example 10.2. Therefore, the frequency of the complex exponential is found to be

$$\omega = \frac{\pi}{2}$$

as before.

The frequency of the complex exponential can be found be the alternative method of forming the polynomial

$$D(z) = 1 - \tfrac{1}{2}\jmath z^{-1} + \tfrac{1}{2}z^{-2}$$

and finding its roots. These roots are located at $z = \jmath$ and $z = -\frac{1}{2}\jmath$. The former is on the unit circle and corresponds to the frequency $\omega = \pi/2$, while the latter, which is a spurious root, lies within the unit circle (as guaranteed for this method).

10.4.5 Principal Components Linear Prediction

Once the idea of signal and noise subspaces has been established, a technique involving simple modifications to some of the previously described methods suggests itself. The technique is the following. Whenever there is added noise, let the estimated correlation matrix be represented in terms of just its eigenvectors and eigenvalues pertaining to the *signal* subspace. This is sometimes referred to as the *principal components* approximation of the matrix:

$$\mathbf{R}_{\boldsymbol{x}}^{(M)} \stackrel{\text{def}}{=} \sum_{i=1}^{M} \lambda_i \mathbf{e}_i \mathbf{e}_i^{*T} \tag{10.166}$$

This procedure amounts to eliminating some of the noise and increasing the overall signal-to-noise ratio. For model-based methods a model derived from just the reduced rank correlation matrix can then be generated and used to estimate the spectrum. Related techniques are to use (10.166) to form a principal components version of the Bartlett estimate (see Problem 10.7) or to use the rank M pseudoinverse

$$\mathbf{R}_{\boldsymbol{x}}^{+(M)} = \sum_{i=1}^{M} \frac{1}{\lambda_i} \mathbf{e}_i \mathbf{e}_i^{*T} \tag{10.167}$$

in (10.81) to produce a principal components version of the maximum likelihood estimate.

Tufts and Kumaresan exploited this technique to the ultimate degree in the context of linear prediction and described a very effective method for the estimation of sinusoids or complex exponentials in noise [39, 40]. Since linear prediction can usually be interpreted as AR modeling, this procedure would seem contrary to our objections at the beginning of this section about using a white noise-driven linear model to estimate the *discrete* components of the spectrum. The method of Tufts and Kumaresan really *does not* attempt to develop an AR model for the data, however, and is in fact consistent with the theoretical description of the random process given in the Wold decomposition. Recall that linear prediction of the discrete components of a random process is not only possible, but also extremely effective; the resulting prediction error variance is identically zero!

To see how the method works, first consider the case of observing a random process consisting of M complex exponentials (with random complex amplitudes) and *no* added noise. The correlation matrix is

$$\mathbf{R}_x = \mathbf{R}_s = \sum_{i=1}^{M} \mathsf{P}_i \mathbf{s}_i \mathbf{s}_i^{*T} = \mathbf{S}^{*T}\mathbf{P}_{\text{o}}\mathbf{S} \tag{10.168}$$

where $\mathbf{S}$ and $\mathbf{P}_{\text{o}}$ are defined by (10.121) and (10.122), respectively. If $N > M$, a prediction error filter can be found such that the prediction error variance is zero. Therefore, the Normal equations can be written as

$$\mathbf{R}_x \mathbf{a} = \mathbf{0} \tag{10.169}$$

where $\mathbf{a}$ is the vector of prediction error filter coefficients

$$\mathbf{a} = \begin{bmatrix} 1 & a_1 & \cdots & a_{N-1} \end{bmatrix}^T \tag{10.170}$$

The following statements are then seen to be true:

1. $\mathbf{a}$ is an eigenvector corresponding to eigenvalue $\lambda = 0$.
2. $\mathbf{a}$ is orthogonal to the signal subspace and therefore lies in the noise subspace [follows from (10.168) and (10.169)].
3. The "noise subspace" is the null space of $\mathbf{R}_x$.

If a pseudospectrum of the form

$$\hat{P}_x(e^{\jmath\omega}) = \frac{1}{|\mathbf{w}^{*T}\mathbf{a}|^2} = \frac{1}{|A(e^{\jmath\omega})|^2}$$

is evaluated, or equivalently, the roots of $A(z)$ are located, then the exact frequencies $\omega_1, \omega_2, \ldots, \omega_M$ corresponding to the M complex exponentials will be found. The trouble is that for $N > M + 1$ the equations (10.169) have more than one solution. For reasons similar to those already discussed for the minimum norm method, it turns out that a most desirable solution is the minimum-norm solution to (10.169).

To review this last point briefly, observe that the solution that minimizes $\|\mathbf{a}^2\|$ is, by Parseval's theorem, the solution that minimizes

$$\frac{1}{2\pi}\int_{-\pi}^{\pi} |A(e^{\jmath\omega})|^2 d\omega$$

where $A(z)$ is of the form

$$A(z) = A_1(z) \cdot A_2(z)$$

and $A_1(z)$ is the fixed term that has M roots on the unit circle. Choosing the optimal value of $A_2(z)$ is analogous to solving for a prediction error filter by the autocorrelation method (see Problem 10.13). Therefore, $A_2(z)$ has all roots strictly *inside* the unit circle.

Let us now explore this minimum-norm solution. Let $\mathbf{a}'$ denote the vector of coefficients for the prediction error filter excluding the first component; that is,

$$\mathbf{a}' = \begin{bmatrix} a_1 & a_2 & \cdots & a_{N-1} \end{bmatrix}^T \tag{10.171}$$

The portion of the Normal equations that determines $\mathbf{a}'$ can be written in the form

$$\mathbf{R}'_x \mathbf{a}' = -\mathbf{r} \tag{10.172}$$

where $\mathbf{R}'_x$ and $\mathbf{r}$ are the appropriate partitions of $\mathbf{R}_x$. Since a_0 the first component of $\mathbf{a}$ is constrained to be equal to 1, minimizing $\|\mathbf{a}\|^2$ is equivalent to minimizing $\|\mathbf{a}'\|^2$. The minimum-norm solution is thus given in terms of the Moore–Penrose pseudoinverse as

$$\mathbf{a}' = -\mathbf{R}'^{+}_x \mathbf{r} \tag{10.173}$$

This solution can be written in terms of the eigenvalues and eigenvectors for the signal subspace (which are also the M nonzero singular values and corresponding singular vectors of $\mathbf{R}'_x$) as

$$\mathbf{a}' = -\sum_{i=1}^{M} \left(\frac{\mathbf{e}'^{*T}_i \mathbf{r}}{\lambda'_i} \right) \mathbf{e}'_i \tag{10.174}$$

Now consider the case where the noise *is* present. The correlation matrix is then [according to (10.118)]

$$\mathbf{R}_x = \sum_{i=1}^{M} \mathsf{P}_i \mathbf{s}_i \mathbf{s}^{*T}_i + \sigma_o^2 \mathbf{I} \tag{10.175}$$

and the Normal equations can be written as

$$\mathbf{R}_x \mathbf{a} = \sigma^2 \boldsymbol{\iota} \tag{10.176}$$

where σ^2 is the prediction error variance, now *not* equal to zero. The previous value of $\mathbf{a}$, which solves (10.169) when there is no noise, is still an eigenvector of the correlation matrix in (10.176), although the corresponding eigenvalue is now σ_o^2 instead of zero. If N is equal to $M+1$, then this is the only eigenvector in the noise subspace. If this eigenvector is used in (10.179), the previous sharp-peaked spectrum results (this is the Pisarenko case). This eigenvector does *not* satisfy the Normal equations (10.176), however, because from the form of $\mathbf{R}_x$ above

$$\mathbf{R}_x \mathbf{a} = \mathbf{0} + \sigma_o^2 \mathbf{a} \neq \sigma^2 \boldsymbol{\iota}$$

From this it is seen that even a small amount of noise disturbs the former relation.

To see more clearly what is happening in the general case ($N > M+1$), let us write the solution to (10.176) in terms of the eigenvectors and eigenvalues of $\mathbf{R}'_{\mathbf{x}}$.

$$\mathbf{a}' = -\sum_{i=1}^{P} \left(\frac{\mathbf{e}'^{*T}_i \mathbf{r}}{\lambda'_i} \right) \mathbf{e}'_i \tag{10.177}$$

where for convenience in the remainder of this discussion we use the prediction order variable

$$P = N - 1$$

instead of N. Note that the upper limit on the sum in (10.177) is taken to include *all* of the terms and that the eigenvectors and eigenvalues and $\mathbf{r}$ now pertain to the correlation matrix with added noise. It is known [48] that if the noise is of moderate level, then the eigenvectors in the signal subspace are not perturbed greatly by the noise. However, the eigenvectors in the noise subspace can be oriented in *any* set of mutually orthogonal directions within that subspace. From a computational point of view their directions are *arbitrary and unpredictable*. Further, if the noise eigenvalues are small, their reciprocals which appear in (10.177) tend to give large weight to the corresponding terms in the vector sum. Therefore, the value of $\mathbf{a}'$ computed even with a small amount of added noise can be far from the ideal solution for $\mathbf{a}'$ with no added noise.

The solution offered by Tufts and Kumaresan is to eliminate the terms in the noise subspace by use of the pseudoinverse. Because the noise is so dealt with, it is not only possible but also advantageous to use a high-order correlation matrix. When the value of P is taken to be close to the number of available data samples (N_s), the rank of the estimated correlation matrix is automatically reduced. The special situation where $P = N_s - M/2$ is called the *Kumaresan–Prony case* and results in a rank exactly equal to the number of signals.[10] This choice for P results in some computational advantages since it avoids the eigenvalue problem or the SVD of the data matrix. (See Section 10.4.8 for further discussion.) To maximally reduce the variance of the frequency estimates, however, a somewhat smaller value of $P \approx \frac{3}{4}N_s$ is recommended. The filter coefficient vector is thus computed from

$$\mathbf{a}' = -\sum_{i=1}^{M} \left(\frac{\mathbf{e}_i'^{*T}\mathbf{r}}{\lambda_i'} \right) \mathbf{e}_i' = -\mathbf{R}_{\boldsymbol{x}}'^{+(M)}\mathbf{r} \tag{10.178}$$

and the pseudospectrum is

$$\boxed{\hat{P}_{PCLP}(e^{\jmath\omega}) = \frac{1}{|\mathbf{w}^{*T}\mathbf{a}|^2} = \frac{1}{|A(e^{\jmath\omega})|^2}} \tag{10.179}$$

where

$$\mathbf{a} = \begin{bmatrix} 1 \\ \\ \mathbf{a}' \end{bmatrix} \tag{10.180}$$

If the number of signals M is not known, it has to be estimated from the eigenvalues (see Section 10.4.8). The signal frequencies are then determined by either finding the peaks of (10.179) or by finding the roots of $A(z)$.

A test case reported by Tufts and Kumaresan involves two closely spaced complex exponentials with frequencies of 1.00π and 1.04π radians in white noise. The signal is given by

$$x[n] = e^{\jmath 1.00\pi n + \pi/4} + e^{\jmath 1.04\pi n} + \eta[n]$$

where $\eta[n]$ is a realization of complex Gaussian white noise with variance σ_o^2. The signal-to-noise ratio defined as $-10\log_{10}\sigma_o^2$ was 10 dB. Twenty-five data samples were used

[10] This presumes that the modified covariance method is used. If the covariance method were used, the choice would be $P = N_s - M$.

(N_s = 25) and results of 50 independent trials were plotted. Figure 10.17 shows the locations of the zeros of $A(z)$ for eight different values of P using the modified covariance method of linear prediction with no attempt to reduce the rank of the correlation matrix. The true frequencies of the complex exponentials are shown by the two arrows. For the smallest-order case ($P = 4$) the two values tend to merge and the signals are not resolved. For larger values of P many spurious roots appear near the unit circle, making it generally impossible to find the two true frequencies. For values of P greater than 16 the rank of the correlation matrix becomes less than full. At the value $P = 24$, the Kumaresan–Prony case, something amazing happens [see Fig. 10.17(h)]. The roots form two distinct clusters close to the true location of the signals, and several other compact clusters evenly distributed inside the unit circle. This happens because when $P = 24$ the rank of the the correlation matrix is reduced to 2 (the number of signals). Thus only signal eigenvectors and no noise eigenvectors are used in the computation of the polynomial, and the solution for the prediction filter is close to what it would be if there were no added noise.

Figure 10.18 shows the results of the principal components linear prediction method for various values of P, where in each case the value of M in (10.178) was taken to be the known number of signals ($M = 2$). The roots tend to show more compact clusters at values of P somewhat below the largest (Kumaresan–Prony) value. A variance computation shows that the optimal value occurs at the value of $P = \frac{3}{4}N_s$ mentioned earlier and that this value is very close to the Cramér-Rao bound [4, 49].

Figure 10.19 shows pseudospectra for the same example using the value $P = 18$. In this case, however, the assumed number of signals [M in (10.178)] was allowed to vary. The figure shows the false peaks that occur due to spurious roots near the unit circle. The last case, corresponding to solution of the Normal equations without rank reduction, has several noise eigenvectors appearing in the solution and produces a particularly misleading pseudospectrum.

10.4.6 ESPRIT: Son of MUSIC

A spectrum estimation method inspired by MUSIC that added a new twist to the subspace techniques is known as *ESPRIT* (for Estimation of Signal Parameters via Rotational Invariance Techniques) [36, 50]. *ESPRIT* provides an elegant way to estimate the frequencies of complex exponentials in noise by exploiting an invariance principle that naturally exists for discrete sequences (e.g., time series). In the context of array processing, the method imposes a particular geometric constraint on the arrays, but has some tremendous computational advantages and eliminates the array calibration required in MUSIC. The original method was described by Paulraj, Roy, and Kailath [50, 51] and had some fundamental problems. Roy and Kailath later improved the technique significantly and developed the

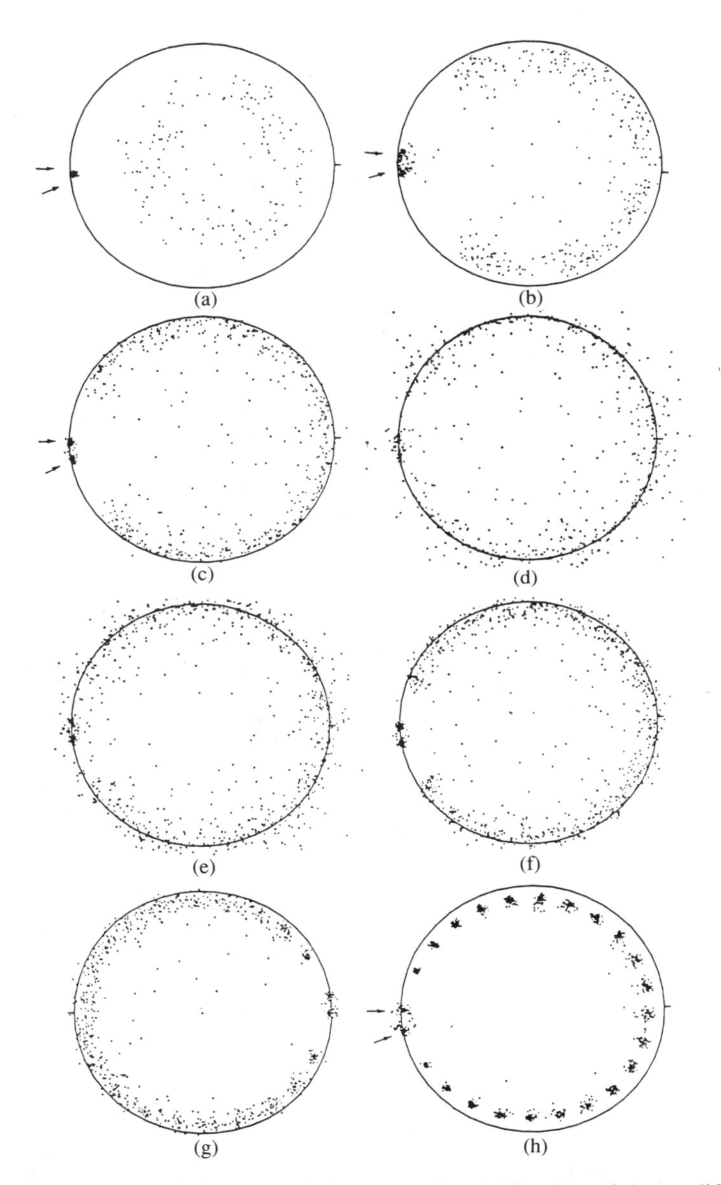

Figure 10.17 Zeros of the prediction error filter in 50 trials of the modified covariance method with $M = 2, N_s = 25$. (a) $P = 4$. (b) $P = 8$. (c) $P = 12$. (d) $P = 16$. (e) $P = 18$. (f) $P = 20$. (g) $P = 22$. (h) $P = 24$ (Kumaresan–Prony case). (From D. W. Tufts and R. Kumaresan, Estimation of frequencies of multiple sinusoids: making linear prediction behave like maximum likelihood, *Proceedings of the IEEE*, September 1982, ©IEEE 1982. Reproduced by permission.)

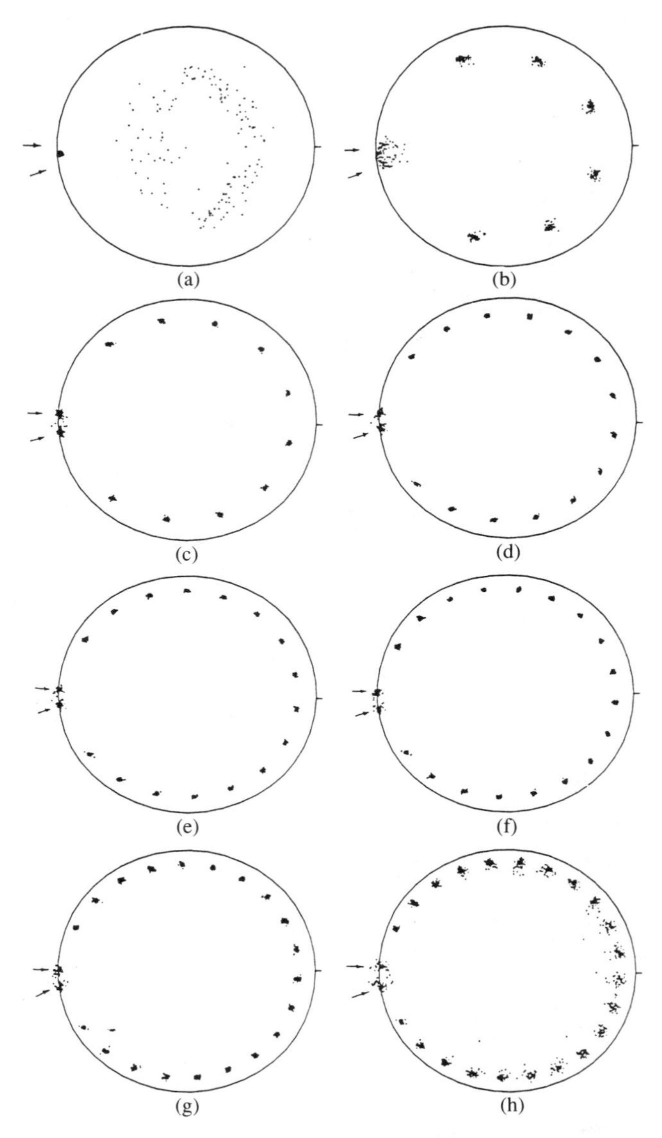

Figure 10.18 Zeros of the prediction error filter in 50 trials of the principal components linear prediction method of Tufts and Kumaresan. $M = 2$; $N_s = 25$. (a) $P = 4$. (b) $P = 8$. (c) $P = 12$. (d) $P = 16$. (e) $P = 18$. (f) $P = 20$. (g) $P = 22$. (h) $P = 24$ (Kumaresan–Prony case). (From D. W. Tufts and R. Kumaresan, Estimation of frequencies of multiple sinusoids: making linear prediction behave like maximum likelihood, *Proceedings of the IEEE*, September 1982, ©IEEE 1982. Reproduced by permission.)

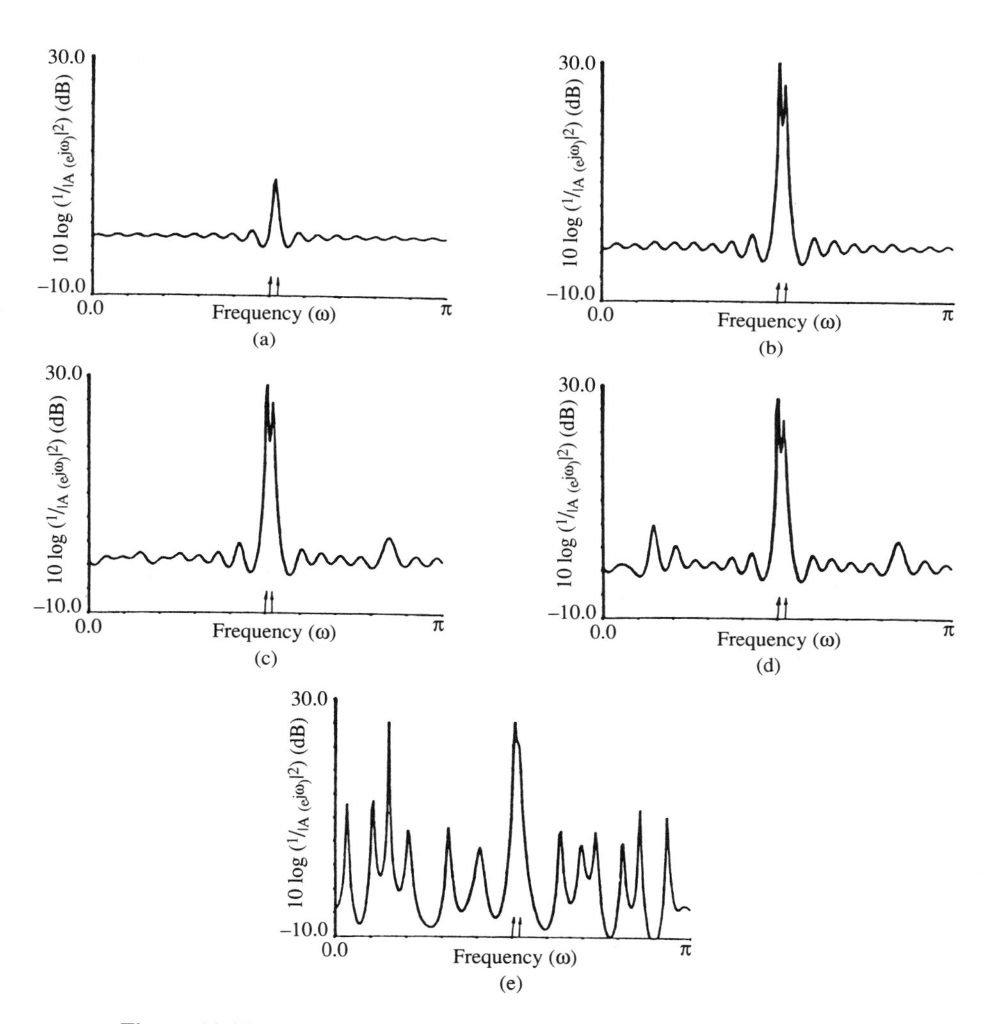

Figure 10.19 Pseudospectra for the principal components linear prediction method for $P = 18$ and various assumed values of M. (a) $M = 1$. (b) $M = 2$. (c) $M = 3$. (d) $M = 4$. (e) Minimum-norm solution of the Normal equtions without rank reduction. (From D. W. Tufts and R. Kumaresan, Singular value decomposition and improved frequency estimation using linear prediction, *IEEE Transactions on Acoustics, Speech, and Signal Processing*, August 1982, ©IEEE 1982. Reproduced by permission.)

current total least squares version of the method [31, 36, 52]. We discuss both versions of *ESPRIT* here; the original version because it is simplest to motivate and describe, and the later TLS version, which is the method of choice.

Original *ESPRIT*. The *ESPRIT* method starts with the signal model (10.116) and a set of $N+1$ data samples which are grouped as shown in Fig. 10.20.[11] The vector $\boldsymbol{x}$ consists of the samples $x[0]$ through $x[N-1]$ while the vector $\boldsymbol{x}'$ consists of the samples $x[1]$ through $x[N]$. The former vector is given by (10.117); the latter is given by

$$\boldsymbol{x}' = \sum_{i=1}^{M} A_i \mathbf{s}_i' + \boldsymbol{\eta}' \tag{10.181}$$

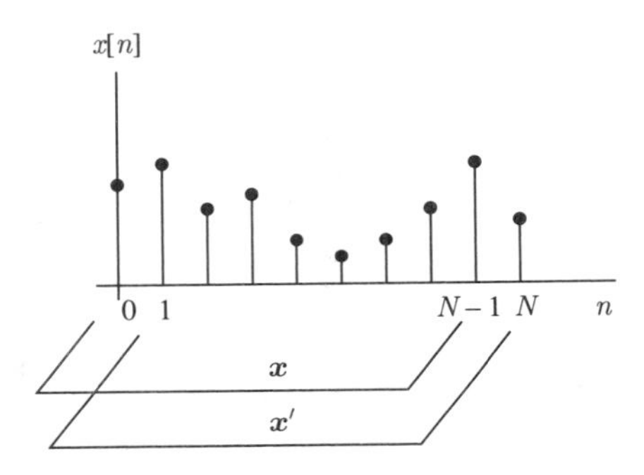

Figure 10.20 Grouping of data samples for *ESPRIT*.

where

$$\mathbf{s}_i' = \begin{bmatrix} e^{\jmath\omega_i} \\ e^{\jmath 2\omega_i} \\ \vdots \\ e^{\jmath N\omega_i} \end{bmatrix} = e^{\jmath\omega_i}\mathbf{s}_i \tag{10.182}$$

and $\boldsymbol{\eta}'$ is the vector of shifted noise samples.

The correlation matrix for $\boldsymbol{x}$ is given by (10.118) or (10.120). The *cross-correlation* matrix for the vectors $\boldsymbol{x}$ and $\boldsymbol{x}'$ is given by

$$\mathbf{R}_{\boldsymbol{x}\boldsymbol{x}'} = \sum_{i=1}^{M} \mathsf{P}_i e^{-\jmath\omega_i}\mathbf{s}_i\mathbf{s}_i^{*T} + \sigma_{\rm o}^2\mathbf{D}_{-1}$$

or

$$\mathbf{R}_{\boldsymbol{x}\boldsymbol{x}'} = \mathbf{S}\mathbf{P}_{\rm o}\boldsymbol{\Phi}^*\mathbf{S}^{*T} + \sigma_{\rm o}^2\mathbf{D}_{-1} \tag{10.183}$$

[11]The translation (shift) invariance between the two data sets leads to a rotational invariance between the two corresponding signal subspaces. Other methods to obtain this invariance, such as interleaving the data samples, could be used [53], but this method seems to be most efficient.

where $\mathbf{\Phi}$ is the diagonal matrix

$$\mathbf{\Phi} = \begin{bmatrix} e^{\jmath\omega_1} & 0 & \cdots & 0 \\ 0 & e^{\jmath\omega_2} & \cdots & 0 \\ \vdots & \vdots & \ddots & \vdots \\ 0 & 0 & \cdots & e^{\jmath\omega_M} \end{bmatrix} \tag{10.184}$$

and $\mathbf{D}_{-1}$ is the matrix with all ones immediately below the main diagonal and zeros elsewhere. Note that the matrix $\mathbf{\Phi}$ contains all of the information about the frequencies $\omega_1, \ldots, \omega_M$. The trick of *ESPRIT* is to extract the contents of this matrix neatly.

The original *ESPRIT* method begins by forming the matrices

$$\mathbf{R}_s \stackrel{\text{def}}{=} \mathbf{R}_x - \sigma_o^2 \mathbf{I} = \mathbf{S}\mathbf{P}_o\mathbf{S}^{*T} \tag{10.185}$$

and

$$\mathbf{R}_{ss'} \stackrel{\text{def}}{=} \mathbf{R}_{xx'} - \sigma_o^2 \mathbf{D}_{-1} = \mathbf{S}\mathbf{P}_o\mathbf{\Phi}^*\mathbf{S}^{*T} \tag{10.186}$$

These matrices can be computed by estimating the smallest eigenvalue of $\mathbf{R}_x$, namely σ_o^2, as in MUSIC and subtracting. It is easy to show then that the values $e^{\jmath\omega_k}$ are eigenvalues of the generalized eigenvalue problem

$$\mathbf{R}_s\dot{\mathbf{e}} = \dot{\lambda}\mathbf{R}_{ss'}\dot{\mathbf{e}} \tag{10.187}$$

This can be seen by writing (10.187) as

$$\begin{aligned} \mathbf{R}_s\dot{\mathbf{e}} - \dot{\lambda}\mathbf{R}_{ss'}\dot{\mathbf{e}} &= \mathbf{S}\mathbf{P}_o(\mathbf{I} - \dot{\lambda}\mathbf{\Phi}^*)\mathbf{S}^{*T}\dot{\mathbf{e}} \\ &= \mathbf{S}\mathbf{P}_o \begin{bmatrix} 1 - \dot{\lambda}e^{-\jmath\omega_1} & 0 & \cdots & 0 \\ 0 & 1 - \dot{\lambda}e^{-\jmath\omega_2} & \cdots & 0 \\ \vdots & \vdots & \ddots & \vdots \\ 0 & 0 & \cdots & 1 - \dot{\lambda}e^{-\jmath\omega_M} \end{bmatrix} \mathbf{S}^{*T}\dot{\mathbf{e}} = \mathbf{0} \end{aligned}$$

Since the values $\dot{\lambda}_k = e^{\jmath\omega_k}$ reduce the rank of this matrix, they are generalized eigenvalues. The steps in estimating the frequencies can then be summarized as follows.

1. Find the auto- and cross-correlation matrix estimates $\hat{\mathbf{R}}_x$ and $\hat{\mathbf{R}}_{xx'}$ from the data.
2. Estimate M and σ_o^2 by finding the eigenvalues of $\hat{\mathbf{R}}_x$ or by SVD of the data matrix.
3. Form estimates $\hat{\mathbf{R}}_s$ and $\hat{\mathbf{R}}_{ss'}$ according to (10.185) and (10.186).
4. Find the generalized eigenvalues of the problem (10.187). Those close to the unit circle are of the form $\dot{\lambda}_k = e^{\jmath\omega_k}$. The ω_k for $k = 1, \ldots, M$ are the angular positions of these eigenvalues.

Figure 10.21 shows results from [51] of estimating the frequencies of three complex exponentials at $\omega_1 = 0.10\pi, \omega_2 = 0.12\pi$, and $\omega_3 = 0.22\pi$ in additive white noise with an SNR of 20 dB. The size of the correlation matrices was taken to be $N = 7$, and 100

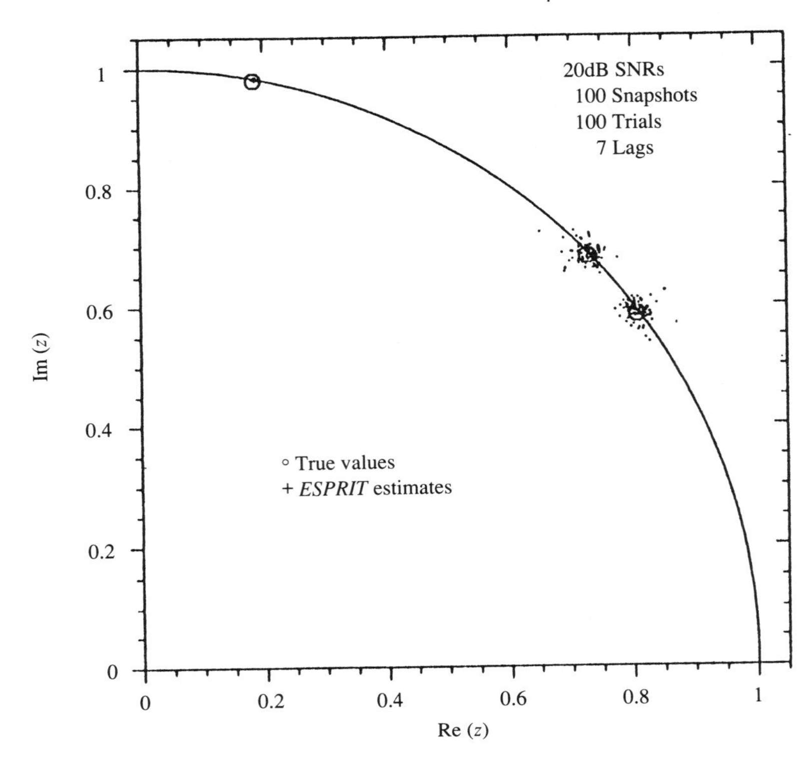

Figure 10.21 *ESPRIT* complex exponential frequency estimates. (From R. H. Roy, A. Paulraj, and T. Kailath, *ESPRIT*—a subspace rotational approach to estimation of parameters of cisoids in noise, *IEEE Transactions on Acoustics, Speech, and Signal Processing*, October 1986, ©IEEE 1982. Reproduced by permission.)

segments of data were used to estimate the correlation function. The results of 100 trials using independent data sets are shown in the figure. The estimates of the eigenvalues are clustered around the true values and the closely spaced signals are clearly resolved.

While the foregoing description serves to illustrate the basic principle of *ESPRIT*, the algorithm just described is not always practical. The trouble with this procedure is that the matrices $\mathbf{R}_s$ and $\mathbf{R}_{ss'}$ are not of full rank and so the generalized eigenvalue problem is ill-defined. Both matrices are of rank $M < N$ and have a common null space. This means that it is possible to find vectors $\dot{\mathbf{e}}$ so that the products $\mathbf{R}_s\dot{\mathbf{e}}$ and $\mathbf{R}_{ss'}\dot{\mathbf{e}}$ are both equal to $\mathbf{0}$. Consequently $\dot{\lambda}$ in (10.187) can take on *any value!* This produces a variety of problems in the numerical solution since it is difficult to identify the eigenvalues $\dot{\lambda} = e^{\jmath\omega_i}$ that lie on (or in practice, close to) the unit circle.

The method known as least squares *ESPRIT* provided an initial solution to this problem. The method based on total least squares supersedes this, however, and is about as easy to implement. Both methods are described here since they are closely related.

Least Squares and Total Least Squares *ESPRIT*. The later versions of *ESPRIT* exploit an invariance property of the signal subspaces corresponding to $\boldsymbol{x}$ and $\boldsymbol{x}'$ that is similar to the invariance property shared by the matrices $\mathbf{R}_s$ and $\mathbf{R}_{ss'}$. The original derivations of these methods involved the rather arcane concept of a matrix pencil and its rank-reducing numbers.[12] A somewhat more direct derivation was provided in later papers; the essence of it is as follows.

Define the vector $\bar{\boldsymbol{x}}$ comprised of the entire set of $N+1$ data samples:

$$\bar{\boldsymbol{x}} \stackrel{\text{def}}{=} \begin{bmatrix} \boldsymbol{x} \\ \\ x[N+1] \end{bmatrix} = \begin{bmatrix} x[0] \\ \\ \boldsymbol{x}' \end{bmatrix} \tag{10.188}$$

The signal subspace corresponding to this higher-dimensional set of data is spanned by the columns of the matrix

$$\bar{\mathbf{S}} \stackrel{\text{def}}{=} \begin{bmatrix} | & | & & | \\ \bar{\mathbf{s}}_1 & \bar{\mathbf{s}}_2 & \cdots & \bar{\mathbf{s}}_M \\ | & | & & | \end{bmatrix} \tag{10.189}$$

where

$$\bar{\mathbf{s}}_i = \begin{bmatrix} 1 \\ e^{\jmath\omega_i} \\ e^{\jmath 2\omega_i} \\ \vdots \\ e^{\jmath N\omega_i} \end{bmatrix} \tag{10.190}$$

It follows that $\bar{\mathbf{S}}$ can be written in partitioned forms as

$$\bar{\mathbf{S}} = \begin{bmatrix} \mathbf{S} \\ \\ - \ \mathbf{s}^{(N)T} \ - \end{bmatrix} = \begin{bmatrix} - \ \mathbf{s}^{(0)T} \ - \\ \\ \mathbf{S}\boldsymbol{\Phi} \end{bmatrix} \tag{10.191}$$

where $\mathbf{s}^{(0)T}$ and $\mathbf{s}^{(N)T}$ are the first and last *rows* of $\bar{\mathbf{S}}$ (not to be confused with the signal vectors $\mathbf{s}_i$). Now consider any orthonormal basis for the signal subspace and represent it as

$$\bar{\mathbf{B}} = \begin{bmatrix} | & | & & | \\ \bar{\mathbf{b}}_1 & \bar{\mathbf{b}}_2 & \cdots & \bar{\mathbf{b}}_M \\ | & | & & | \end{bmatrix} \tag{10.192}$$

In particular, these basis vectors can be found by forming the correlation matrix for the observations and solving the resulting eigenvalue problem as in MUSIC. Since the columns of $\bar{\mathbf{S}}$ and $\bar{\mathbf{B}}$ span the same subspace, they can be related by some nonsingular $M \times M$ transformation $\boldsymbol{\Upsilon}$, that is,

$$\bar{\mathbf{B}}\boldsymbol{\Upsilon} = \bar{\mathbf{S}} \tag{10.193}$$

[12] These concepts are closely related to generalized eigenvalue problems. If $\mathbf{M}$ and $\mathbf{N}$ are any two matrices of the same dimensions (not necessarily rectangular), then the set of matrices $\mathbf{M} - \mu\mathbf{N}$ defined by letting μ take on all possible values is called the matrix *pencil* corresponding to $\mathbf{M}$ and $\mathbf{N}$ [54]. The rank-reducing numbers are those values of μ that reduce the rank of the pencil from its maximum possible value to a smaller value.

Thus if we define partitions $\mathbf{B}$ and $\mathbf{B}'$ of $\bar{\mathbf{B}}$ corresponding to the partitions $\mathbf{S}$ and $\mathbf{S\Phi}$ in (10.191), the last equation can be written in the two forms

$$\bar{\mathbf{B}}\boldsymbol{\Upsilon} = \begin{bmatrix} & \mathbf{B} & \\ \times & \cdots & \times \end{bmatrix} \boldsymbol{\Upsilon} = \begin{bmatrix} & \mathbf{S} & \\ - & \mathbf{s}^{(N)T} & - \end{bmatrix}$$

and

$$\bar{\mathbf{B}}\boldsymbol{\Upsilon} = \begin{bmatrix} \times & \cdots & \times \\ & \mathbf{B}' & \end{bmatrix} \boldsymbol{\Upsilon} = \begin{bmatrix} - & \mathbf{s}^{(0)T} & - \\ & \mathbf{S\Phi} & \end{bmatrix}$$

where the symbols $\times$ represent a row of values that are not of direct concern. These equations provide the two conditions

$$\begin{aligned} \mathbf{B}\boldsymbol{\Upsilon} &= \mathbf{S} \; (a) \\ \mathbf{B}'\boldsymbol{\Upsilon} &= \mathbf{S\Phi} \; (b) \end{aligned} \tag{10.194}$$

Now notice that $\mathbf{B}$ and $\mathbf{B}'$ can be related by substituting (10.194)(a) into (10.194)(b) to obtain

$$\mathbf{B}'\boldsymbol{\Upsilon} = \mathbf{B}\boldsymbol{\Upsilon}\boldsymbol{\Phi}$$

This can be rewritten as

$$\boxed{\mathbf{B}\boldsymbol{\Psi} = \mathbf{B}'} \tag{10.195}$$

where

$$\boxed{\boldsymbol{\Psi} = \boldsymbol{\Upsilon}\boldsymbol{\Phi}\boldsymbol{\Upsilon}^{-1}} \tag{10.196}$$

These are the key relations that underlie ESPRIT. Equation 10.195 represents a set of linear equations that can be solved for $\boldsymbol{\Psi}$; equation 10.196 represents the *eigenvalue decomposition* of $\boldsymbol{\Psi}$. $\boldsymbol{\Phi}$ is the diagonal matrix of eigenvalues and $\boldsymbol{\Upsilon}$ is the matrix of *eigenvectors* of $\boldsymbol{\Psi}$. Solving (10.195) and (10.196) provides the desired frequencies through the relation (10.184).

In theory (10.195) is satisfied exactly. Thus although $\mathbf{B}$ and $\mathbf{B}'$ are rectangular matrices of size $N \times M$ with $N > M$, we can select any set of M rows of these matrices and solve the resulting linear equations. In practice, the matrices $\mathbf{B}$ and $\mathbf{B}'$ are derived from an *estimated* correlation matrix, so (10.195) does *not* hold exactly. That is, (10.195) represents an overdetermined set of linear equations. A first thought therefore is to solve the least squares problem

$$\mathbf{B}\boldsymbol{\Psi} \stackrel{\text{ls}}{=} \mathbf{B}' \tag{10.197}$$

If this seems like an unfamiliar formulation of a least squares problem, note that it is equivalent to M separate least squares problems of the form[13]

$$\mathbf{B}\boldsymbol{\psi}_k \stackrel{\text{ls}}{=} \mathbf{b}'_k; \qquad k = 1, 2, \ldots, M$$

[13]Problems of the form (10.197) are sometimes referred to as *multiple regression* problems.

where ψ_k and $\mathbf{b}'_k$ are the columns of $\mathbf{\Psi}$ and $\mathbf{B}'$, respectively. The solution to (10.197) is thus given by

$$\mathbf{\Psi}_{LS} = \mathbf{B}^{+}\mathbf{B}' = (\mathbf{B}^{*T}\mathbf{B})^{-1}\mathbf{B}^{*T}\mathbf{B}' \tag{10.198}$$

and (as was seen) the desired frequencies are related to the eigenvalues of $\mathbf{\Psi}_{LS}$. This method has been called *least squares ESPRIT*.

The least squares solution to this problem can be objected to for the following reasons, however. Recall from the discussion in Section 9.3.5 that the least squares problem implicitly assumes that there are errors on the right side of the equation but not on the left. Thus the least squares problem (10.197) can be written as

$$\mathbf{B\Psi} = \mathbf{B}' - \mathbf{\Delta}' \tag{10.199}$$

where $\mathbf{\Delta}'$ is a matrix of errors and the solution minimizes a norm involving the columns of $\mathbf{\Delta}'$. Since both $\mathbf{B}$ and $\mathbf{B}'$ are estimated from the same data, however, there seems to be a logical inconsistency in assuming that $\mathbf{B}'$ has errors while $\mathbf{B}$ has none. As was seen in Chapter 9, the *total least squares* approach rectifies this situation by allowing for errors in *both* $\mathbf{B}$ and $\mathbf{B}'$. Thus the total least squares problem here has the form

$$(\mathbf{B} - \mathbf{\Delta})\mathbf{\Psi} = \mathbf{B}' - \mathbf{\Delta}' \tag{10.200}$$

where $\mathbf{\Delta}$ represents the matrix of errors in $\mathbf{B}$. The solution $\mathbf{\Psi}_{TLS}$ is defined as that which minimizes the Frobenius norm of the error matrix

$$\left\| \begin{array}{cc} \mathbf{\Delta} & \mathbf{\Delta}' \end{array} \right\|_F$$

The solution is obtained via singular value decomposition [54]. Let $\mathbf{V}$ be the $2M \times 2M$ matrix of *right singular vectors* of the matrix $\begin{bmatrix} \mathbf{B} & \mathbf{B}' \end{bmatrix}$. If the matrix is divided into four $M \times M$ partitions as

$$\mathbf{V} = \begin{bmatrix} \mathbf{V}_{11} & \mathbf{V}_{12} \\ \mathbf{V}_{21} & \mathbf{V}_{22} \end{bmatrix}$$

then the solution to the total least squares problem is given by[14]

$$\mathbf{\Psi}_{TLS} = -\mathbf{V}_{12}\mathbf{V}_{22}^{-1} \tag{10.201}$$

A summary of the TLS *ESPRIT* algorithm is given in the box below. Note that the estimation of the basis for the signal subspace begins with the generalized eigenvalue problem for the *colored* noise problem in step 2 (see Section 10.4.1). It is assumed that the covariance matrix $\mathbf{\Sigma}_{\bar{\eta}}$ of the $N+1$ noise samples is known (at least to within a constant). When the noise is white, $\mathbf{\Sigma}_{\bar{\eta}}$ reduces to the identity matrix.

[14]A proof of this result is given in Theorem 12.2.1 in Golub and Van Loan [54].

ESPRIT (TLS version)

1. Define the $N+1$-dimensional random vector $\bar{x}$ pertaining to $N+1$ consecutive data samples $x[0], x[1], \ldots, x[N]$ and estimate the correlation matrix $\hat{\mathbf{R}}_{\bar{x}}$ from the data. [Usually, the covariance method or the modified covariance method should be used here, especially if the total length of the data record (N_s) is small.]
2. Compute the generalized eigenvectors and eigenvalues of $\hat{\mathbf{R}}_{\bar{x}}$:
$$\hat{\mathbf{R}}_{\bar{x}}\bar{\mathbf{e}}_k = \bar{\lambda}_k \boldsymbol{\Sigma}_{\bar{\eta}}\bar{\mathbf{e}}_k; \qquad k = 1, 2, \ldots, N+1$$
3. If necessary, estimate the number of signals M.
4. Generate a basis spanning the signal subspace and partition it as
$$\bar{\mathbf{B}} = \boldsymbol{\Sigma}_{\bar{\eta}} \begin{bmatrix} | & & | \\ \bar{\mathbf{e}}_1 & \cdots & \bar{\mathbf{e}}_M \\ | & & | \end{bmatrix} = \begin{bmatrix} \mathbf{B} \\ \\ \times \cdots \times \end{bmatrix} = \begin{bmatrix} \times \cdots \times \\ \\ \mathbf{B}' \end{bmatrix}$$
5. Compute the matrix $\mathbf{V}$ of right singular vectors of
$$\begin{bmatrix} \mathbf{B} & \mathbf{B}' \end{bmatrix}$$
and partition $\mathbf{V}$ into four $M \times M$ submatrices
$$\mathbf{V} = \begin{bmatrix} \mathbf{V}_{11} & \mathbf{V}_{12} \\ \mathbf{V}_{21} & \mathbf{V}_{22} \end{bmatrix}$$
6. Compute the eigenvalues $\lambda_1, \lambda_2, \ldots, \lambda_M$ of the matrix $\boldsymbol{\Psi}_{TLS} = -\mathbf{V}_{12}\mathbf{V}_{22}^{-1}$.
7. Find the desired frequencies from (10.192) as
$$\omega_k = \angle\lambda_k; \qquad k = 1, 2, \ldots, M$$

If it is more convenient, step 5 can be replaced by computation of the matrix of eigenvectors of

$$\begin{bmatrix} \mathbf{B}^{*T} \\ \mathbf{B}'^{*T} \end{bmatrix} \begin{bmatrix} \mathbf{B} & \mathbf{B}' \end{bmatrix}$$

although the SVD is frequently a more accurate computation. To obtain the least squares version of *ESPRIT*, you would simply replace steps 5 and 6 by the solution to the least squares problem (10.198). Other than reducing computation there is no reason for doing this, however, since the TLS *ESPRIT* solution has a lower bias than the least squares *ESPRIT* solution in general.

An interesting additional feature of *ESPRIT* is known as "signal copy," where the actual signals $s[n] = A_i e^{\jmath\omega_i n}$ can be estimated. Since the forms of the signals are determined once the ω_i are known, it is only necessary to estimate the complex amplitudes A_i. This can be accomplished by finding a set of weights which when applied to the observed data "nulls out" each of the signals except the ith. It is convenient to represent these weights in an $(N+1) \times M$ matrix whose i^{th} column is orthogonal to all signal vectors *except* $\bar{\mathbf{s}}_i$. The estimates $\hat{A}_i$ are then given by

$$\begin{bmatrix} \hat{A}_1 \\ \hat{A}_2 \\ \vdots \\ \hat{A}_M \end{bmatrix} = \mathbf{W}_{SC}^{*T} \bar{\boldsymbol{x}} \tag{10.202}$$

where the weighting matrix can be shown to be

$$\mathbf{W}_{SC} = \bar{\mathbf{S}}(\bar{\mathbf{S}}^{*T}\bar{\mathbf{S}})^{-1} \tag{10.203}$$

In the limit as the variance of the noise approaches zero, these estimates approach the true values of the random variables (see Problem 10.15).

10.4.7 Maximum Likelihood and Weighted Subspace Fitting

The subspace approach to estimation of complex exponentials in noise is fundamentally a nonlinear parameter estimation problem. The quantities to be estimated are the frequencies, and the powers of the signals and noise. An optimal procedure is therefore the maximum likelihood method.[15] This subsection first explores maximum likelihood in the contest of subspace methods and then shows how one form of maximum likelihood motivates the weighted subspace fitting approach.

Maximum Likelihood Methods. To apply the maximum likelihood procedure to the estimation of frequencies and other parameters it is assumed that the signals and the noise are zero mean Gaussian random processes. Then given a set of K independent N-dimensional sample vectors $\boldsymbol{x}^{(k)}$ of the form (10.119), the joint density is

$$f_{\boldsymbol{x}^{(1)},\ldots,\boldsymbol{x}^{(K)}}(\mathbf{x}^{(1)},\ldots,\mathbf{x}^{(K)}) = \prod_{k=1}^{K} \frac{1}{\pi^N |\mathbf{R}_{\boldsymbol{x}}|} e^{-\mathbf{x}^{(k)*T}\mathbf{R}_{\boldsymbol{x}}^{-1}\mathbf{x}^{(k)}} \tag{10.204}$$

Let $\boldsymbol{\omega}$ represent the vector of frequency parameters $\begin{bmatrix} \omega_1 & \omega_2 & \cdots & \omega_M \end{bmatrix}^T$. The log likelihood function is then

$$\ln f(\mathbf{x}^{(1)},\ldots,\mathbf{x}^{(K)};\boldsymbol{\omega},\mathbf{P}_{\text{o}},\sigma_{\text{o}}^2) = -NK\ln\pi - K\ln|\mathbf{R}_{\boldsymbol{x}}| - \sum_{k=1}^{K}\mathbf{x}^{(k)*T}\mathbf{R}_{\boldsymbol{x}}^{-1}\mathbf{x}^{(k)}$$

[15]We refer to the statistical maximum likelihood procedure for parameter estimation introduced in Chapter 6, not the maximum likelihood method of spectrum estimation discussed earlier in this chapter.

where from (10.120)

$$\mathbf{R}_x(\boldsymbol{\omega}, \mathbf{P}_o, \sigma_o^2) = \mathbf{S}(\boldsymbol{\omega})\mathbf{P}_o\mathbf{S}^{*T}(\boldsymbol{\omega}) + \sigma_o^2\mathbf{I}$$

and where in this last equation the explicit dependence of $\mathbf{R}_x$ and $\mathbf{S}$ on the parameters $\boldsymbol{\omega}$, $\mathbf{P}_o$, and σ_o^2 is shown. The log likelihood function can be rewritten as

$$\begin{aligned}\ln f(\mathbf{X}; \boldsymbol{\omega}, \mathbf{P}_o, \sigma_o^2)\\ = -NK \ln \pi - K \ln |\mathbf{R}_x(\boldsymbol{\omega}, \mathbf{P}_o, \sigma_o^2)| - \text{tr } \mathbf{R}_x^{-1}(\boldsymbol{\omega}, \mathbf{P}_o, \sigma_o^2)\mathbf{X}^{*T}\mathbf{X}\end{aligned} \tag{10.205}$$

where $\mathbf{X}$ is the data matrix

$$\mathbf{X} = \begin{bmatrix} - & \mathbf{x}^{(1)*T} & - \\ - & \mathbf{x}^{(2)*T} & - \\ & \vdots & \\ & \vdots & \\ - & \mathbf{x}^{(K)*T} & - \end{bmatrix} \tag{10.206}$$

Since the quantity $\mathbf{X}^{*T}\mathbf{X}$ is proportional to the correlation matrix *estimated* from the data, the log likelihood function has an interesting dependence upon the true and the estimated correlation matrices.

Maximizing (10.205) is a very difficult problem because of the highly nonlinear dependence on $\boldsymbol{\omega}$ and the other parameters. A easier problem results if the complex amplitudes A_i of the complex exponentials are treated as *parameters* instead of random variables. To distinguish between these two cases, the former problem (10.205) has been referred to in some of the literature as "stochastic maximum likelihood," while the problem now to be considered has been called "deterministic maximum likelihood."

For the deterministic maximum likelihood problem, the observations are still given by (10.119). Since $\boldsymbol{\eta}$ is the only *random* vector, however, the covariance matrix is

$$\mathbf{C}_x = \sigma_o^2\mathbf{I}$$

The remaining term in (10.119) represents a nonzero mean and is denoted here by

$$\mathbf{m}_x(\boldsymbol{\omega}) = \mathbf{S}(\boldsymbol{\omega}) \begin{bmatrix} A_1 \\ A_2 \\ \vdots \\ A_M \end{bmatrix} \tag{10.207}$$

The joint density for the sample vectors in this case becomes

$$f_{x^{(1)},\ldots,x^{(K)}}(\mathbf{x}^{(1)}, \ldots, \mathbf{x}^{(K)}) = \prod_{k=1}^{K} \frac{1}{\pi^N |\sigma_o^2\mathbf{I}|} e^{-\frac{1}{\sigma_o^2}(\mathbf{x}^{(k)} - \mathbf{m}_x(\boldsymbol{\omega}))^{*T}(\mathbf{x}^{(k)} - \mathbf{m}_x(\boldsymbol{\omega}))} \tag{10.208}$$

The maximization with respect to the parameter σ_o^2 is separable from the maximization with respect to the other parameters. If σ_o^2 is held fixed, the corresponding log likelihood function depends on only the term in the exponent. Therefore, the deterministic maximum likelihood estimate for the frequencies and their complex amplitudes can be found by simply *minimizing* the quantity

$$\sum_{k=1}^{K}(\mathbf{x}^{(k)} - \mathbf{m}_x(\boldsymbol{\omega}))^{*T}(\mathbf{x}^{(k)} - \mathbf{m}_x(\boldsymbol{\omega})) = \sum_{k=1}^{K}\left\|\mathbf{x}^{(k)} - \mathbf{m}_x(\boldsymbol{\omega})\right\|^2$$

with respect to the parameters.

The deterministic maximum likelihood problem can be written in a more convenient form as

$$[\hat{\boldsymbol{\omega}}, \hat{\mathbf{A}}] = \text{argmin } \|\mathbf{X}^{*T} - \mathbf{S}(\boldsymbol{\omega})\mathbf{A}\|_F^2 \tag{10.209}$$

where $\|\cdot\|_F$ is the Frobenius norm and $\mathbf{A}$ is the $M \times K$ matrix each of whose columns is the vector of complex amplitudes that appears in (10.207). This is still a nonlinear problem but an easier one than stochastic maximum likelihood since it is separable in the variables $\boldsymbol{\omega}$ and $\mathbf{A}$ and *linear* in $\mathbf{A}$. In particular, for any fixed value of $\boldsymbol{\omega}$, (10.209) represents a (multiple regression type) least squares problem for $\mathbf{A}$ with solution

$$\hat{\mathbf{A}} = \mathbf{S}^{+}(\boldsymbol{\omega})\mathbf{X}^{*T} \tag{10.210}$$

where $\mathbf{S}^{+}$ is the pseudoinverse of $\mathbf{S}$. Substitution of (10.210) into (10.209) eliminates the matrix $\mathbf{A}$ and permits writing the term on the right of (10.209) as

$$\left\|\mathbf{X}^{*T} - \mathbf{S}(\boldsymbol{\omega})\mathbf{S}^{+}(\boldsymbol{\omega})\mathbf{X}^{*T}\right\|_F^2 = \left\|(\mathbf{I} - \mathbf{S}(\boldsymbol{\omega})\mathbf{S}^{+}(\boldsymbol{\omega}))\mathbf{X}^{*T}\right\|_F^2 = \left\|\boldsymbol{P}_{\mathbf{S}}^{\perp}(\boldsymbol{\omega})\mathbf{X}^{*T}\right\|_F^2$$

When $\boldsymbol{\omega}$ is equal to the vector of *true* frequencies, the projection matrix $\boldsymbol{P}_{\mathbf{S}}^{\perp}(\boldsymbol{\omega})$ appearing in the last step is actually the projection matrix $\boldsymbol{P}_{noise}$ for the noise subspace. Then by using the identities $\|\mathbf{M}\|_F^2 = \text{tr } \mathbf{M}\mathbf{M}^{*T}$ and $\text{tr } \mathbf{M}\mathbf{N} = \text{tr } \mathbf{N}\mathbf{M}$ where $\mathbf{M}$ and $\mathbf{N}$ are any conformable matrices, the last expression becomes

$$\begin{aligned}\left\|\boldsymbol{P}_{\mathbf{S}}^{\perp}\mathbf{X}^{*T}\right\|_F^2 &= \text{tr } \boldsymbol{P}_{\mathbf{S}}^{\perp}\mathbf{X}^{*T}\mathbf{X}\boldsymbol{P}_{\mathbf{S}}^{\perp} \\ &= \text{tr } \boldsymbol{P}_{\mathbf{S}}^{\perp}\boldsymbol{P}_{\mathbf{S}}^{\perp}\mathbf{X}^{*T}\mathbf{X} = \text{tr } \boldsymbol{P}_{\mathbf{S}}^{\perp}\mathbf{X}^{*T}\mathbf{X}\end{aligned}$$

Therefore, an estimate for the frequencies which is equivalent to that provided by (10.209) is

$$\hat{\boldsymbol{\omega}} = \text{argmin tr } \boldsymbol{P}_{\mathbf{S}}^{\perp}(\boldsymbol{\omega})\mathbf{X}^{*T}\mathbf{X} \tag{10.211}$$

The estimate again in this case has a dependence on the *estimated* correlation matrix.

Weighted Subspace Fitting. Deterministic maximum likelihood, as expressed by (10.209), can be thought of as fitting the subspace spanned by the columns of $\mathbf{S}(\boldsymbol{\omega})$ to that spanned by the columns of $\mathbf{X}^{*T}$. This suggests consideration of a more general subspace fitting problem

$$[\hat{\boldsymbol{\omega}}, \hat{\mathbf{Z}}] = \text{argmin } \|\boldsymbol{\Xi} - \mathbf{S}(\boldsymbol{\omega})\mathbf{Z}\|_F^2 \tag{10.212}$$

where Ξ and $\mathbf{Z}$ are matrices analogous to $\mathbf{X}$ and $\mathbf{A}$ in (10.209). Then by proceding as in the preceding section, we can show that the estimate for the frequency vector has the form

$$\hat{\omega} = \text{argmin tr } \boldsymbol{P}_{\mathbf{S}}^{\perp}(\boldsymbol{\omega})\Xi\Xi^{*T} \tag{10.213}$$

It turns out that many subspace methods, including MUSIC, *ESPRIT*, and deterministic maximum likelihood, can be put in the form (10.212) by a suitable choice of the matrix Ξ and a suitable constraint on $\mathbf{S}$ [42, 55]. It can further be shown that a broad class of subspace methods can be defined by restricting the matrix Ξ to the form [42, 55]

$$\Xi = \hat{\mathbf{E}}_{sig}\boldsymbol{W}^{\frac{1}{2}} \tag{10.214}$$

where $\hat{\mathbf{E}}_{sig}$ is the signal subspace eigenvector matrix estimated from the data, and $\boldsymbol{W}$ is a suitable positive definite Hermitian weighting matrix. The particular choice

$$\boldsymbol{W} = \hat{\mathbf{\Lambda}}_{sig} - \hat{\sigma}_{\mathrm{o}}^2\mathbf{I}$$

makes the variance of the estimate asymptotically (i.e., as $K \to \infty$) equivalent to that of the deterministic maximum likelihood. Further it is shown [42, 56] that an *optimal* choice for $\boldsymbol{W}$ (that results in the asymptotic lowest possible variance) is

$$\boldsymbol{W}_{opt} = (\hat{\mathbf{\Lambda}}_{sig} - \hat{\sigma}_{\mathrm{o}}^2\mathbf{I})^2\hat{\mathbf{\Lambda}}_{sig}^{-1} \tag{10.215}$$

and further that the variance produced with this weight matrix achieves the Cramér-Rao lower bound [57, 58]. Thus this method is asymptotically equivalent to stochastic maximum likelihood.

The weighted subspace fitting method seeks to minimize the criterion function

$$V(\boldsymbol{\omega}) = \text{tr } \boldsymbol{P}_{\mathbf{S}}^{\perp}\Xi\Xi^{*T} \tag{10.216}$$

which appears in (10.213), where Ξ is given by (10.214) and (10.215). An iterative procedure (a form of Newton's method) can be set up to minimize $V(\boldsymbol{\omega})$ where the value of $\boldsymbol{\omega}$ at the $(k+1)^{\text{st}}$ iteration is computed according to

$$\boldsymbol{\omega}_{k+1} = \boldsymbol{\omega}_k - \mu_k\mathbf{H}^{-1}\boldsymbol{\nu}(\boldsymbol{\omega}_k) \tag{10.217}$$

In this last equation $\mathbf{H}$ is the Hessian matrix of the criterion function, $\boldsymbol{\nu}$ is the (real) gradient vector,

$$\boldsymbol{\nu}(\boldsymbol{\omega}_k) = \nabla_{\omega}V(\boldsymbol{\omega})\big|_{\boldsymbol{\omega}=\boldsymbol{\omega}_k}$$

and μ_k is a scalar step size. A form of the algorithm based on QR factorization [56] is given in the box below. In the third step of this algorithm the operator "$\odot$" represents element-by-element multiplication of the two matrices, and $\mathbf{D}$ is the matrix of column derivatives

$$\mathbf{D} = \begin{bmatrix} | & | & & | \\ \dfrac{d\mathbf{s}_1}{d\omega_1} & \dfrac{d\mathbf{s}_2}{d\omega_2} & \cdots & \dfrac{d\mathbf{s}_M}{d\omega_M} \\ | & | & & | \end{bmatrix} \tag{10.218}$$

The individual elements of this last matrix are

$$d_{l,k} = (l-1)e^{\jmath(l-1)\omega_k}$$

(with indices starting at 1).

Iteration for Weighted Subspace Fitting

For $k = 1, 2, 3, \ldots$

1. Factor $\mathbf{S}(\omega_k)$ as

$$\mathbf{S}(\omega_k) = \begin{bmatrix} \mathbf{Q}_1 & \mathbf{Q}_2 \end{bmatrix} \begin{bmatrix} \mathbf{R}_1 \\ \mathbf{0} \end{bmatrix}$$

where $\begin{bmatrix} \mathbf{Q}_1 & \mathbf{Q}_2 \end{bmatrix}$ is a unitary matrix and $\mathbf{R}_1$ is upper (right) triangular (*not* a correlation matrix).

2. Compute $\mathbf{Y} = \Xi^{*T}\mathbf{Q}_2$ and

$$V(\omega_k) = \text{tr } \mathbf{Y}\mathbf{Y}^{*T}$$

3. If $|V(\omega_k) - V(\omega_{k-1})| \approx 0$ then STOP
else compute

$$\mathbf{F} = \mathbf{Q}_2^{*T}\mathbf{D}$$

$$\mathbf{G} = \mathbf{R}_1^{-1}\mathbf{Q}_1^{*T}\Xi$$

$$\nu = \text{Re diag}\,(\mathbf{GYF})$$

$$\mathbf{H} = \text{Re }(\mathbf{F}^{*T}\mathbf{F}) \odot (\mathbf{G}\mathbf{G}^{*T})$$

$$\omega_{k+1} = \omega_k - \mu_k \mathbf{H}^{-1}\nu(\omega_k)$$

End loop.

The algorithm is initialized using results of one of the other subspace algorithms for a starting point. Since convergence could be to a local instead of a global minimum, it is important to begin with a good initial estimate for the parameter vector.

10.4.8 Computation of Subspace-Based Estimates

The foregoing subsections described subspace methods in terms of the theoretical or known correlation matrices. A number of specific issues are addressed here when (as is normally the case) the starting point is a given sequence of data.

Basic Computations. The correlation matrix estimate can be generated by any of the methods described in Chapters 6 and 9. It is sufficient to use an unnormalized estimate of the form

$$\hat{\mathbf{R}}_x = \mathbf{X}^{*T}\mathbf{X} \tag{10.219}$$

This saves the computation required by dividing by the number of rows K in the data matrix and may improve the final precision. For most of the methods an eigenvector decomposition of the correlation matrix is required. This can be accomplished by a singular value decomposition of the data matrix $\mathbf{X}$.

The principal components linear prediction method of Tufts and Kumaresan involves a reduced rank approximation to the data matrix and its pseudoinverse. The minimum norm least squares solution for the primed filter coefficient vector is

$$\mathbf{a}' = -\mathbf{X}_1^{+(M)}\mathbf{x}_0$$

where $\mathbf{x}_0$ and $\mathbf{X}_1$ are the partitions of the data matrix defined by

$$\mathbf{X} = \begin{bmatrix} | & \\ \mathbf{x}_0 & \mathbf{X}_1 \\ | & \end{bmatrix}$$

and the superscript, "$+(M)$" here denotes the rank M pseudoinverse of the data matrix obtained by setting the smallest $P - M$ singular values to zero. (Recall that the column dimension of $\mathbf{X}_1$ is equal to $P = N - 1$.) An alternative way to do the computation of $\mathbf{X}_1^{+(M)}$ is to use one of the formulas

$$\mathbf{X}_1^{+(M)} = (\mathbf{X}_1^{*T}\mathbf{X}_1)^{+(M)}\mathbf{X}_1^{*T} \tag{10.220}$$

or

$$\mathbf{X}_1^{+(M)} = \mathbf{X}_1^{*T}(\mathbf{X}_1\mathbf{X}_1^{*T})^{+(M)} \tag{10.221}$$

(see Problem 9.7). Depending on the values of the variables P and N_s, one or the other of the square matrices appearing in these equations may be small and yield an overall computational advantage. The SVD of the square matrix is the same as its eigenvalue decomposition.

For the special Kumaresan-Prony case no eigenvalue decomposition or SVD is required at all. In this case the data matrix has only M rows and is of rank M. Therefore, the pseudoinverse of the square matrix that appears in (10.221) becomes an ordinary inverse and $\mathbf{X}_1^{+(M)}$ can be computed from the simpler formula

$$\mathbf{X}_1^{+(M)} = \mathbf{X}_1^{+} = \mathbf{X}_1^{*T}(\mathbf{X}_1\mathbf{X}_1^{*T})^{-1} \tag{10.222}$$

Estimating the Number of Signals M. In the most common situation where the number of signals M is not known at the outset, this number must be estimated from the data. When the signal power is much larger than the noise power, estimating M is not a problem since the eigenvalues show a clear break with the signal eigenvalues larger than the noise eigenvalues. In other cases, it may be difficult to determine M just from inspection of the eigenvalues. Wax and Kailath developed a formulation of the Akaike information criterion (AIC) and the minimum description length (MDL) that is applicable to this problem [59].

The AIC is defined in terms of the likelihood function for the observations as

$$\text{AIC}(k) = -2\ln f_{x;\theta}(\mathbf{x};\boldsymbol{\theta}) + 2k \tag{10.223}$$

where k is the number of free parameters represented by $\boldsymbol{\theta}$. The MDL has a similar definition, namely

$$\text{MDL}(k) = -\ln f_{x;\theta}(\mathbf{x};\boldsymbol{\theta}) + \tfrac{1}{2}k\ln K \tag{10.224}$$

where K is the number of vectors used to estimate the parameters of the density (i.e., the number of rows of the data matrix). Wax and Kailath show in the context of the signals in noise problem that if the observation vector is Gaussian with zero mean, these criteria can be expressed in the explicit forms

$$\boxed{\text{AIC}(M) = -2K(N-M)\ln\varrho(M) + 2M(2N-M)} \tag{10.225}$$

and

$$\boxed{\text{MDL}(M) = -K(N-M)\ln\varrho(M) + \tfrac{1}{2}M(2N-M)\ln K} \tag{10.226}$$

where $\varrho(M)$ is the ratio of the geometric to the arithmetic mean of the eigenvalues

$$\varrho(M) = \frac{(\lambda_{M+1}\lambda_{M+2}\cdots\lambda_N)^{\frac{1}{N-M}}}{\frac{1}{N-M}\left(\lambda_{M+1}+\lambda_{M+2}+\cdots+\lambda_N\right)} \tag{10.227}$$

The value of M is chosen to minimize either of the quantities above.

Once a value of M has been chosen, the minimum eigenvalue (noise power σ_o^2) can be estimated as

$$\hat{\sigma}_o^2 = \frac{1}{N-M}\left(\lambda_{M+1}+\lambda_{M+2}+\cdots+\lambda_N\right) \tag{10.228}$$

Evaluating the Pseudospectrum. If the frequencies of the complex exponentials are to be found by plotting the pseudospectrum rather than finding the roots, then the FFT is the most computationally efficient method to compute the needed values (see Sections 10.2.4 and 10.3.2). In the case of MUSIC, the spectra can be written in the form

$$\hat{P}_{MU}(e^{\jmath\omega}) = \frac{1}{\sum_{i=M+1}^{N}|\mathbf{w}^{*T}\mathbf{e}_i|^2} \tag{10.229}$$

or

$$\hat{P}'_{MU}(e^{\jmath\omega}) = \frac{1}{\sum_{i=M+1}^{N}\frac{1}{\lambda_i}|\mathbf{w}^{*T}\mathbf{e}_i|^2} \tag{10.230}$$

[see (10.154) and (10.155)]. Then a procedure similar to that described in Section 10.3.2 for the maximum likelihood spectral estimate can be followed.

10.5 REFERENCES TO ADVANCED TOPICS

Several more advanced topics relating to spectrum estimation are cited in this section. Some of these topics are (or at least could be) the subjects of entire volumes in themselves. The goal is to provide a very brief description of each topic and give some references to the literature where more information can be found.

10.5.1 Cross-Spectrum Estimation: Computing the Spectral Matrix

The analysis of cross-spectra is important in applications such as system identification, time-delay estimation in signal reception, and in a number of other problems where it is needed to determine similarity or synchronism between frequency components of two random processes. The analysis of these effects is frequently found in papers dealing with the topic of coherence analysis. A brief outline of methods for estimating the cross-spectrum is given here. More description can be found in volumes dealing more specifically with spectrum estimation, such as [2, 4, 5].

Classical methods for estimation of the cross-spectrum are similar to methods for estimation of the autospectrum. Usually, the autospectra of each individual process is also estimated and the analysis is performed using the coherence function (see Chapter 4). One of the simplest methods for estimating the cross-spectrum is to transform the estimated cross-correlation function. If a window is applied before taking the transform, this results in the Blackman–Tukey type of estimate

$$\hat{S}_{xy}(e^{\jmath\omega}) = \sum_{l=-L}^{L} \mathrm{w}[l]\hat{R}_{xy}[l]e^{-\jmath\omega l} \; ; \qquad L < N_s \tag{10.231}$$

If the biased cross-correlation function estimate is used, the peak can have a fairly large bias if it does not occur near the origin. This can also induce a bias in the cross-spectrum estimate. Thus if the two random sequences tend to correlate at large lag values, it may be well to first align the data before estimating the cross-spectrum.

If the cross-periodogram is defined as

$$\hat{P}_{xy}(e^{\jmath\omega}) = \frac{1}{N}X(e^{\jmath\omega})Y^*(e^{\jmath\omega}) \tag{10.232}$$

then a Bartlett type of cross-spectrum estimate can be defined as

$$\hat{S}_{xy}(e^{\jmath\omega}) = \frac{1}{K}\sum_{k=1}^{K} \hat{P}_{xy}^{(k)}(e^{\jmath\omega}) \tag{10.233}$$

where $\hat{P}_{xy}^{(k)}(e^{\jmath\omega})$ is the cross-periodogram for corresponding data segments $x^{(k)}[n]$ and $y^{(k)}[n]$, and K is the number of those segments.

A cross-spectral estimate can also be generated by the Welch procedure in the form

$$\hat{S}_{xy}(e^{j\omega}) = \frac{1}{K}\sum_{k=1}^{K}\hat{P}'^{(k)}_{xy}(e^{j\omega}) \tag{10.234}$$

where

$$\hat{P}'^{(k)}_{xy}(e^{j\omega}) = \frac{1}{N}\left(\sum_{n=0}^{N-1} \mathrm{v}[n]x^{(k)}[n]e^{-j\omega n}\right)\left(\sum_{n=0}^{N-1} \mathrm{v}[n]y^{(k)}[n]e^{-j\omega n}\right)^{*} \tag{10.235}$$

and $\mathrm{v}[n]$ is a suitable window. The mean and variance of these cross-spectrum estimates have similar behavior to those statistics for the corresponding autospectra. Detailed analysis can be found in [2].

Model-based estimates of the cross-spectrum can be derived by multichannel modeling of the two random processes in question. A multichannel random process is defined to be the vector sequence

$$\boldsymbol{u}[n] = \begin{bmatrix} x[n] \\ y[n] \end{bmatrix} \tag{10.236}$$

and a spectral matrix of the form

$$\hat{\mathbf{S}}_{\boldsymbol{u}} = \begin{bmatrix} \hat{S}_x & \hat{S}_{xy}(e^{j\omega}) \\ \hat{S}_{yx}(e^{j\omega}) & \hat{S}_y(e^{j\omega}) \end{bmatrix} \tag{10.237}$$

is then estimated. This automatically provides all of the terms needed for a coherence estimate.[16] The procedure generalizes directly to the case when more than two random processes are involved. The spectral matrix then contains all of the autospectra and cross-spectra pertaining to the components of the multichannel random process.

For AR modeling $\boldsymbol{u}[n]$ is represented by a multichannel AR process which is described by the difference equation

$$\boldsymbol{u}[n] = -\mathbf{A}_1\boldsymbol{u}[n-1] - \mathbf{A}_2\boldsymbol{u}[n-2] - \cdots - \mathbf{A}_P\boldsymbol{u}[n-P] + \boldsymbol{w}[n] \tag{10.238}$$

In this equation $\mathbf{A}_i$ are *matrix* coefficients and $\boldsymbol{w}[n]$ is a multichannel white noise process with correlation function

$$\mathcal{E}\{\boldsymbol{w}[n]\boldsymbol{w}^{*T}[n-l]\} = \boldsymbol{\Sigma}_w\delta[l] \tag{10.239}$$

Notice that the correlation function for a multichannel random process is a sequence of *matrices*. Notice also that although the multichannel white noise process is uncorrelated with respect to lag l, it may be correlated *between channels*. In other words, the covariance matrix $\boldsymbol{\Sigma}_w$ *is not necessarily diagonal.*

The spectral estimate is taken to be the spectral matrix of the multichannel AR process, which has the form

$$\hat{\mathbf{S}}_{AR}(e^{j\omega}) = [\boldsymbol{A}(e^{j\omega})]^{-1}\boldsymbol{\Sigma}_w[\boldsymbol{A}^{*T}(e^{j\omega})]^{-1} \tag{10.240}$$

[16]See Section 4.4.1 for definition of coherence.

where the matrix $\boldsymbol{A}(e^{\jmath\omega})$ represents the multichannel filter

$$\boldsymbol{A}(e^{\jmath\omega}) = \mathbf{I} + \mathbf{A}_1 e^{-\jmath\omega} + \mathbf{A}_2 e^{-\jmath 2\omega} + \cdots + \mathbf{A}_P e^{-\jmath P\omega} \tag{10.241}$$

The model parameters are most conveniently found from the Normal equations of linear prediction, which have the form

$$\begin{bmatrix} \boldsymbol{R}_u[0] & \boldsymbol{R}_u[1] & \cdots & \boldsymbol{R}_u[P] \\ \boldsymbol{R}_u[-1] & \boldsymbol{R}_u[0] & \cdots & \boldsymbol{R}_u[P-1] \\ \vdots & \vdots & \cdots & \vdots \\ \boldsymbol{R}_u[-P] & \boldsymbol{R}_u[-P+1] & \cdots & \boldsymbol{R}_u[0] \end{bmatrix} \begin{bmatrix} 1 \\ \mathbf{A}_1^{*T} \\ \vdots \\ \mathbf{A}_P^{*T} \end{bmatrix} = \begin{bmatrix} \mathbf{\Sigma}_w \\ \mathbf{0} \\ \vdots \\ \mathbf{0} \end{bmatrix} \tag{10.242}$$

The correlation matrix that appears in this equation is block Toeplitz, where the blocks represent the multichannel correlation function

$$\boldsymbol{R}_u[l] = \mathcal{E}\{\boldsymbol{u}[n]\boldsymbol{u}^{*T}[n-l]\} \tag{10.243}$$

of the random process $\boldsymbol{u}[n]$. These terms can be estimated from

$$\hat{\boldsymbol{R}}_u[l] = \frac{1}{N} \sum_{n=0}^{N-1-l} \boldsymbol{u}[n+l]\boldsymbol{u}^{*T}[n]\,; \qquad 0 \le l < N \tag{10.244}$$

and the relation

$$\hat{\boldsymbol{R}}_u[-l] = \hat{\boldsymbol{R}}_u^{*T}[l] \tag{10.245}$$

A multichannel form of the Levinson recursion exists for solving these Normal equations (see [4, 5]); this is sometimes refered to as the Levinson–Wiggins–Robinson (LWR) algorithm. A multichannel form of the Burg algorithm known as the Nuttal–Strand algorithm also exists [60, 61]. Both of these involve a set of forward and backward linear prediction parameters corresponding to causal and anticausal AR models. However, the backward parameters do *not* bear any simple relation to the forward parameters as in the single-channel case. The spectral estimate can be defined in terms of either the causal or the anticausal AR model and it is curious that although the forward filter is not equal to the backward filter and the forward error covariance is not equal to the backward error covariance, the two spectral estimates *are exactly the same* [60].

Multichannel MA and ARMA models can also be used for simultaneous estimation of the autospectra and the cross-spectra (see [4, 5]). These methods however are generally less well developed.

A multichannel version of the maximum likelihood method is also possible. This spectral matrix is defined by

$$\hat{\mathbf{S}}_{ML}(e^{\jmath\omega}) = \left(\mathbf{W}^{*T}\mathbf{R}_u^{-1}\mathbf{W}\right)^{-1} \tag{10.246}$$

where $\mathbf{R}_u$ is the large correlation matrix that appears in (10.242) and $\mathbf{W}$ is defined by

$$\mathbf{W} = \begin{bmatrix} \mathbf{I} \\ \mathbf{I}e^{\jmath\omega} \\ \vdots \\ \mathbf{I}e^{\jmath P\omega} \end{bmatrix} \tag{10.247}$$

Not too much has been written on this method or its performance. A bit more discussion can be found in [4] and [5].

Relatively little has been reported in applying subspace methods to the cross-spectrum estimation problem. However, generalizations of the methods to the multichannel case to allow estimation of discrete terms in the cross-spectrum do not seem to be difficult.

10.5.2 Spectrum Estimation for Nonstationary Signals: Time-Frequency Methods

Many temporal signals that are inherently nonstationary can be regarded as stationary if viewed over a short period of time. Since the frequency content of stationary signals is a very powerful method of description, the time history of the power spectrum is an important tool for characterizing signals whose statistical properties are changing slowly with time. This method of characterization has been especially important for speech processing and for sonar signal processing. Some tutorial descriptions of various aspects of this type of processing and many further references can be found in [62–66].

The most direct approach to computing the time history of the power spectrum is to view the recorded data through a moving window whose length is commensurate to the time over which the data can be assumed to remain stationary. The Fourier transform of the windowed data is known as the short-time Fourier transform (STFT) [65]. The STFT of the given data is defined by[17]

$$\mathrm{X}[n, \omega) \stackrel{\text{def}}{=} \sum_{k=-\infty}^{\infty} \mathrm{x}[k]\mathrm{w}[n-k]e^{-\jmath\omega k} \tag{10.248}$$

where the window extends from $n - L$ to $n + L$ (see Fig. 10.22). The power spectral estimate is then given by the short-time periodogram

$$\hat{P}_x[n, \omega) = \frac{1}{2L+1}|\mathrm{X}[n, \omega)|^2 \tag{10.249}$$

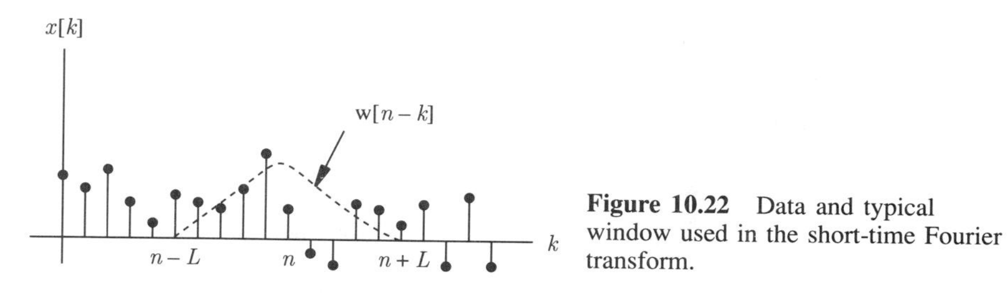

Figure 10.22 Data and typical window used in the short-time Fourier transform.

The short-time periodogram is usually displayed in one of the forms shown in Fig. 10.23. The representation in Fig. 10.23(a) shows the time history of the spectrum in three dimensions; the plot is usually discrete in time and continuous in frequency. The representation

[17]Mixed brackets are used for the arguments to the function to emphasize that n is *discrete* and ω is *continuous*.

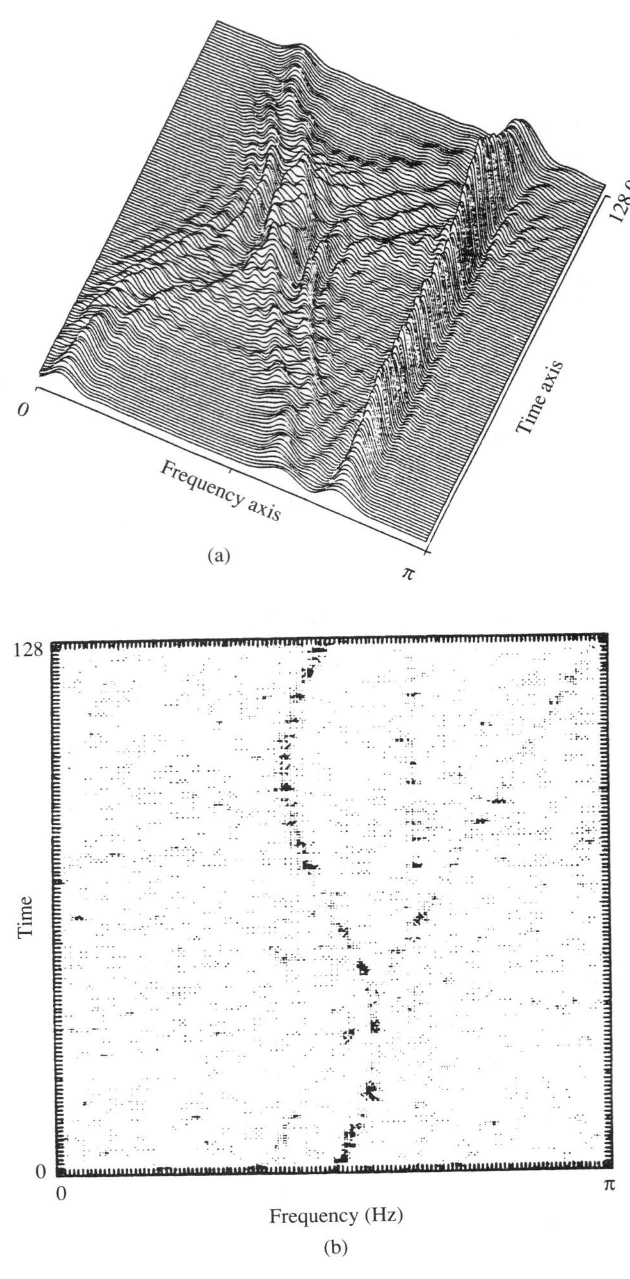

Figure 10.23 Representations for the short-time periodogram. (a) Three-dimensional plot of a constant-frequency sinusoid ("tone") and two frequency modulated "chirp" signals. (b) Spectrogram of two frequency-modulated signals in noise.

in Fig. 10.23(b) is called the *spectrogram.* This is a two-dimensional plot in the n, ω plane, where the gray level represents intensity (darker values represent regions of higher power). A version of this representation used in sonar signal processing (usually highly quantized in gray level) is called the "lofargram." Signals consisting of sinusoids of constant frequency appear as dark lines parallel to the time $[n]$ axis; narrowband signals with time-varying frequency produce sloped or curved lines in the plot.

Mathematically more sophisticated approaches capable of higher resolution for a given length of data employ what are known as (time-frequency) *distributions.* Considering for the moment a continuous-time signal, these distributions are functions $\mathcal{D}(t, f)$ that attempt to show how the energy of the signal is distributed in time and frequency. For a typical continuous signal $\mathrm{x}_c(t)$ it is desirable that the distribution $\mathcal{D}_{\mathrm{x}}(t, f)$ satisfy the "marginal" properties

$$\int_{-\infty}^{\infty} \mathcal{D}_{\mathrm{x}}(t, f) df = |\mathrm{x}_c(t)|^2 \; (a)$$

$$\int_{-\infty}^{\infty} \mathcal{D}_{\mathrm{x}}(t, f) dt = |\mathrm{X}_c(f)|^2 \; (b) \tag{10.250}$$

where $\mathrm{X}_c(f)$ is the ordinary Fourier transform of the signal. (The short-time periodogram does not satisfy either of these properties.)

The three most well-known distributions are the Wigner–Ville,

$$\mathcal{W}_{\mathrm{x}}(t, f) = \int_{-\infty}^{\infty} \mathrm{x}_c\left(t + \tfrac{\tau}{2}\right) \mathrm{x}_c^*\left(t - \tfrac{\tau}{2}\right) e^{-j2\pi f \tau} d\tau \tag{10.251}$$

the instantaneous power spectrum (IPS),

$$\mathcal{I}_{\mathrm{x}}(t, f) = \tfrac{1}{2} \int_{-\infty}^{\infty} \left[\mathrm{x}_c(t)\mathrm{x}_c^*(t - \tau) + \mathrm{x}_c^*(t)\mathrm{x}_c(t + \tau)\right] e^{-j2\pi f \tau} d\tau \tag{10.252}$$

and the Rihaczek distribution.

$$\mathcal{E}_{\mathrm{x}}(t, f) = \mathrm{x}_c(t)\mathrm{X}_c^*(f)e^{-j2\pi f t} \tag{10.253}$$

These three, and many others, fit into what is called "Cohen's class" of distributions [67]. Cohen's class is defined by the general relation

$$\mathcal{C}_{\mathrm{x}}(t, f) = \int_{-\infty}^{\infty} \int_{-\infty}^{\infty} \int_{-\infty}^{\infty} \phi_c(\nu, \tau)\mathrm{x}_c\left(\xi + \tfrac{\tau}{2}\right) \mathrm{x}_c^*\left(\xi - \tfrac{\tau}{2}\right) e^{j2\pi(\nu\xi - \nu t - f\tau)} d\nu d\xi d\tau \tag{10.254}$$

where $\phi_c(\nu, \tau)$ is called the *kernel.* The three previous specific distributions are obtained by choosing the kernel to be $\phi_c(\nu, \tau) = 1$, $\phi_c(\nu, \tau) = e^{j\pi\nu\tau}$, and $\phi_c(\nu, \tau) = \cos(\pi\nu\tau)$, respectively. Since $\cos(\pi\nu\tau) = \text{Re}[e^{j\pi\nu\tau}]$, it follows that IPS is the *real part* of the Rihaczek distribution.

Discrete forms of the Wigner–Ville and IPS can be formed by replacing the integrals by a discrete sum. In so doing the forms can be generalized by including a window function. For IPS the discrete form is

$$I_x[n,\omega] = \tfrac{1}{2}\sum_{k=-\infty}^{\infty}\{\mathrm{x}[n]\mathrm{x}^*[n-k] + \mathrm{x}^*[n]\mathrm{x}[n+k]\}\,\mathrm{w}[0]\mathrm{w}[k]e^{-j\omega k} \tag{10.255}$$

where ω is in the usual interval $-\pi \le \omega \le \pi$. For the Wigner–Ville the discrete form is [63]

$$W_x[n,\omega) = 2\sum_{k=-\infty}^{\infty}\mathrm{x}[n+k]\mathrm{x}^*[n-k]\mathrm{w}[k]\mathrm{w}[-k]e^{-j2\omega k} \tag{10.256}$$

The factor of 2 in front of the sum and in the exponential arises from the fact that in order to evaluate the arguments at $t \pm \tau/2$ the signal has to be sampled at twice the normal (Nyquist) rate. This expression, with the window included, is usually called the *pseudo Wigner–Ville* distribution (PWD). To keep the foldover frequency at the usual $\pm\pi$ instead of $\pm\pi/2$ the factor of 2 in the exponential can be dropped. The window is absolutely essential in the use of the IPS to prevent a "ringing" effect in the distribution that would otherwise make the distribution almost impossible to interpret. For the Wigner–Ville distribution it is frequently used to smooth the spectral estimate.

Although IPS and PWD are not based strictly on stochastic concepts, both represent (estimated) second moment properties of the data. Equation 10.255 shows that PWD is like a sum of instantaneous correlation estimates, while IPS is the sum of the average of two such estimates. Stochastic versions of PWD and IPS known as *evolutive spectra* can be formed by taking expectations of (10.256) and (10.255). Their practical computation, however, requires another windowing and averaging of the data. A brief discussion of this with respect to the Wigner–Ville distribution can be found in [62].

While the Wigner–Ville (or PWD) has been the distribution most thoroughly investigated in the literature (see, e.g., [68]), it suffers from the appearance of cross-terms between strong components that do not appear in other time-frequency representations. Thus in using the Wigner–Ville distribution it is almost imperative to work with only one side of the spectrum (the analytic signal), to avoid artifacts arising from the interaction of positive and negative frequency components. Hippenstiel and Oliveira [69] have compared IPS to PWD in a variety of test cases and concluded that while PWD has the advantage in resolution (i.e., spectral lines are narrower), IPS, which has so far received much less attention, has significant advantages in some problems. Figure 10.24 compares PWD to IPS for the combination of a sinusoid plus two chirp signals linearly and quadratically varying in frequency.

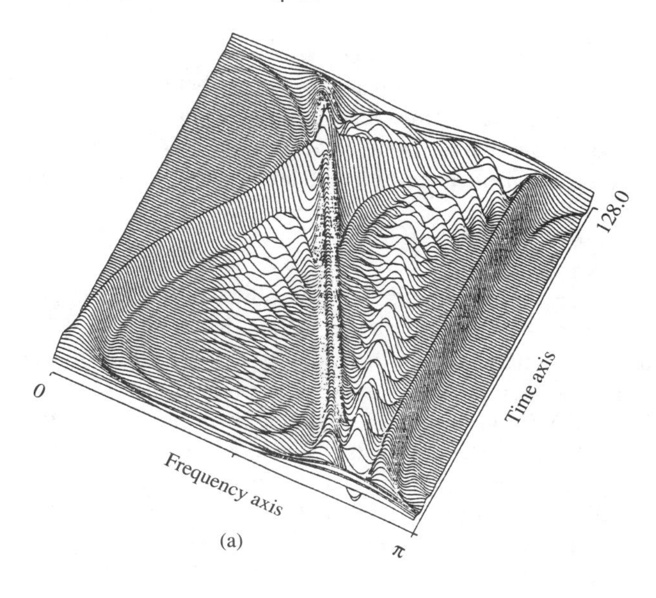

(a)

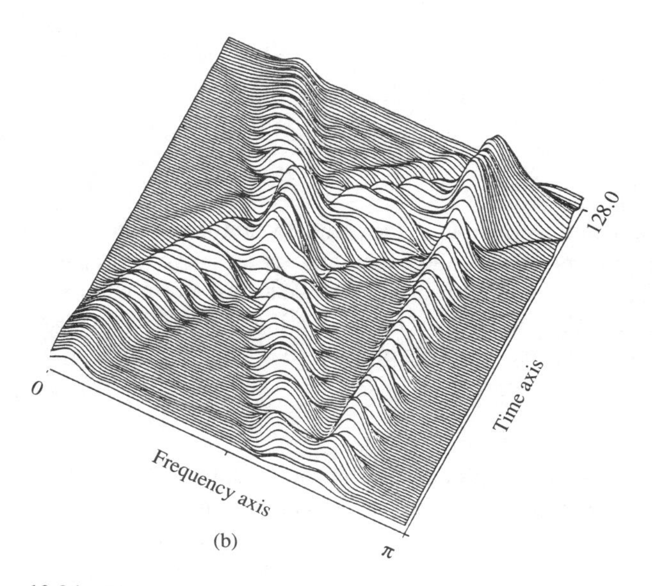

(b)

Figure 10.24 Time-frequency distributions for sinusoid plus linear and quadratic chirp. Time-smoothed Hamming window. (a) PWD. (b) IPS. (From R. D. Hippenstiel and P. M. Oliveira, Time-varying spectral estimation using the instantaneous power spectrum, *IEEE Transactions on Acoustics, Speech, and Signal Processing*, 38:10, October 1990, ©IEEE 1990. Reproduced by permission.)

[Fig. 10.23(a) shows the time-frequency spectrum computed via the STFT for the same data set.] Note that PWD shows extra peaks occurring between each of the legitimate frequency components which could be mistaken for true components of the spectrum. IPS, on the other hand, has ridges that are somewhat broader in frequency, but the cross-terms are not present. A comparison of several other features of IPS and PWD can be found in [69].

10.5.3 Array Processing[18]

While array processing may not seem like a branch of spectrum estimation, it bears such a close connection that it is appropriate to explore it briefly here. Mathematically, the problem is essentially the same as multidimensional spectrum estimation [70]. For purposes of the present brief discussion the focus is on array processing for narrowband signals.

To begin the discussion consider the uniform linear array discussed in Chapter 1 and depicted in Fig. 10.25. The continuous signals received at the N sensors are multiplied by complex weights $w_n^* = |w_n|e^{-\jmath n\varphi}$ and summed to produce the array output $s(t)$. Assume that the spatially propagating signal is in the far field of the radiating source so that the signal impinging on the array is in the form of a narrowband plane wave with center frequency f_o. The narrowband assumption means that the time for propagation across the array is very small compared to the time over which there is any significant change in the modulation. Therefore, for purposes of analysis, the signal received at the nth sensor can be represented by

$$s_n(t) = \mathrm{s}_n e^{\jmath 2\pi f_o t} \tag{10.257}$$

where s_n is the complex amplitude

$$\mathrm{s}_n = |\mathrm{s}_n| e^{\jmath \angle \mathrm{s}_n} \tag{10.258}$$

that *does not depend on* t.

Recall that the principle of the simple beamformer is to adjust the phase at each sensor so that the received signals add coherently. If the wave is propagating in a direction θ with respect to the centerline of the array, the planar wavefront will experience a delay τ between the time it arrives at adjacent sensors (see Fig. 10.25). The delay is given by the distance traveled divided by the speed of propagation. Thus

$$\tau = \frac{d \sin\theta}{c}$$

where c is the speed of propagation in the medium. The difference in phase of the plane wave between any two adjacent sensors is therefore

$$k = 2\pi f_o \tau = \frac{2\pi f_o d \sin\theta}{c} \tag{10.259}$$

This quantity k is known as the (normalized) wavenumber. If the phase of the array is adjusted to compensate for the propagation time of waves arriving from a particular

[18]Since the topic is specialized, notation in this subsection may not be totally consistent with that in the remainder of the chapter. Hopefully, this will not cause any confusion.

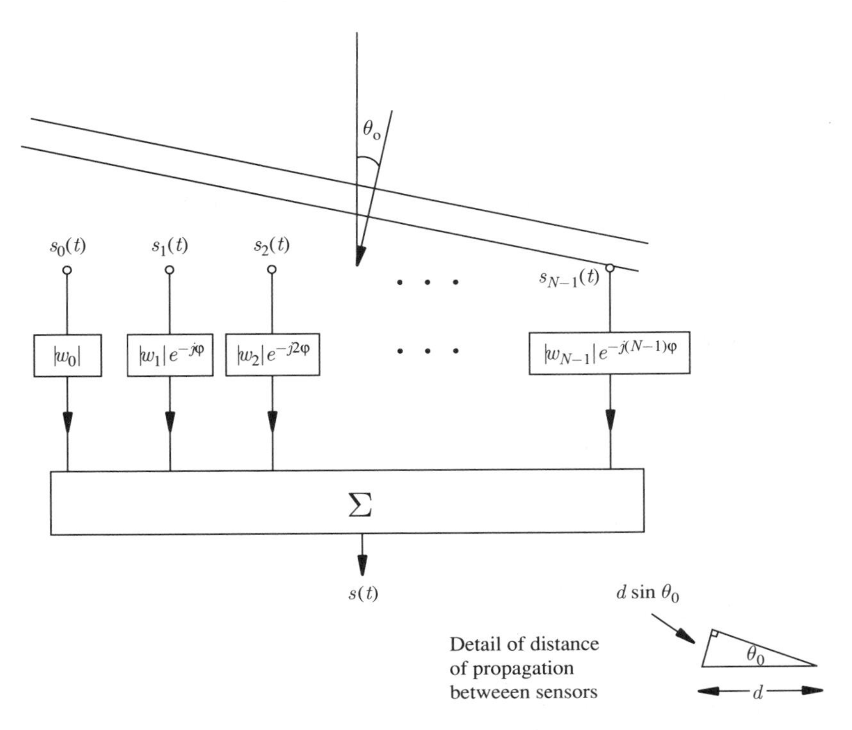

Figure 10.25 Uniform linear array with incident plane wave.

direction θ_o by setting $\varphi = k_o$, then plane waves arriving from direction θ_o will produce a large output from the array, while plane waves from other directions will generally result in smaller outputs. (Some may even result in no output at all.) Setting the phases of the array in this way is called "steering the beam."

Now assume that the array is set to some arbitrary phase φ and let us examine the effect on the received signals $s_n(t)$. Assume for the moment that the gains $|w_n|$ are all equal to 1. Since the inputs are narrowband, the output $s(t)$ will likewise be narrowband and can be represented as

$$s(t) = \mathrm{S}(\varphi)e^{\jmath 2\pi f_o t} \tag{10.260}$$

where the complex amplitude S is clearly a function of φ. The output of the beamformer is thus

$$s(t) = \mathrm{S}(\varphi)e^{\jmath 2\pi f_o t} = \sum_{n=0}^{N-1} \mathrm{s}_n e^{\jmath 2\pi f_o t} e^{-\jmath n\varphi}$$

Canceling the common term $e^{\jmath 2\pi f_o t}$ leaves an expression for the complex amplitude of the output, namely

$$\mathrm{S}(\varphi) = \sum_{n=0}^{N-1} \mathrm{s}_n e^{-\jmath n\varphi} \tag{10.261}$$

This is recognized as the Fourier transform of the spatially sampled signal s_n where the phase φ plays the role of digital frequency. Thus the effect of the beamformer as the phase φ is varied from $-\pi$ to π is to perform the rectangularly windowed *spatial Fourier transform of the signal.*

Now consider the effect of the array on the previous narrowband signal arriving from direction θ. Assume that the signal received at sensor 0 is

$$s_0(t) = Ae^{j2\pi f_o t} \tag{10.262}$$

where A is the complex amplitude. The signal at the nth sensor leads this signal in phase and is given by

$$s_n(t) = Ae^{j2\pi f_o t + nk} \tag{10.263}$$

where k is given by (10.259). Comparing (10.263) with (10.257) reveals that the complex amplitude of the signal is given by

$$\mathrm{s}_n = Ae^{jnk_o} \tag{10.264}$$

Now consider the complex amplitude of the output (10.261) in response to the narrowband signal above when the array is steered in a direction $\varphi = k_0$. From (10.261) and (10.264) this quantity is a function of both k and k_0 which we now denote by $\mathrm{S}(k; k_0)$. The squared magnitude is called the *beampattern.* If the array is very large, then the beampattern is very tall and narrow. In the limit as $N \to \infty$, $|\mathrm{S}(k; k_0)|$ becomes an impulse [the unwindowed Fourier transform of (10.264)] as depicted in Fig. 10.26(a). The array thus has perfect selectivity; it has infinite response to signals arriving from the chosen direction θ_o corresponding to k_0 and zero response to signals from any other direction, no matter how close. Of course this ideal situation can never be achieved because the number of sensors must be finite. For finite N the magnitude of the array response is given by

$$|\mathrm{S}(k; k_o)| = |A| \left| \frac{\sin N(k - k_o)/2}{\sin(k - k_o)/2} \right| \tag{10.265}$$

The corresponding beampattern is depicted in Fig. 10.26(b). Depending on the size N this array is much less selective. In addition, the subsidiary peaks or "sidelobes" that appear are of sufficient magnitude to cause a significant array response from signals arriving in directions corresponding to those values of k.

One way to suppress the sidelobes, at the expense of widening the main lobe, is to assign nonuniform weights to the signals. For an arbitrary set of weights $|w_n|$ assigned to the array (see Fig. 10.25) the response (10.261) is modified by multiplying the signals s_n by the weights $|w_n|$. This corresponds to a nonrectangular windowing of the signal and has the same effect as in spectrum analysis. Usually, standard windows such as Bartlett, Hamming, or others are chosen for the weights. The procedure is known as "shading" the array (like shading the lens on a camera) since the array becomes less sensitive to radiation from directions other than that to which it is steered. The effect is shown in Fig. 10.26(c).

The problem with conventional beamforming described here is the same as the problem with classical spectrum estimation. Namely, the resolution is determined fundamentally by

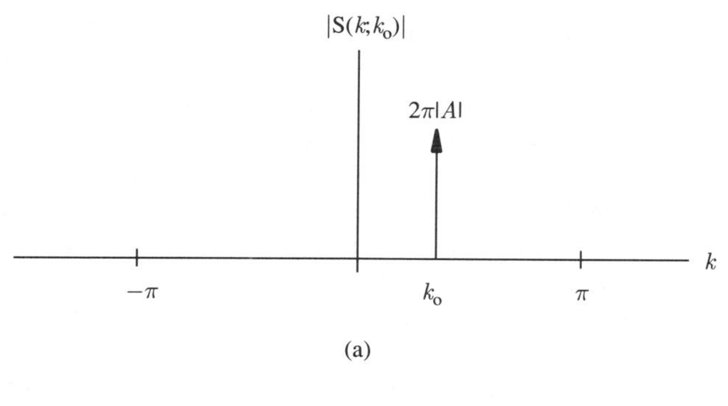

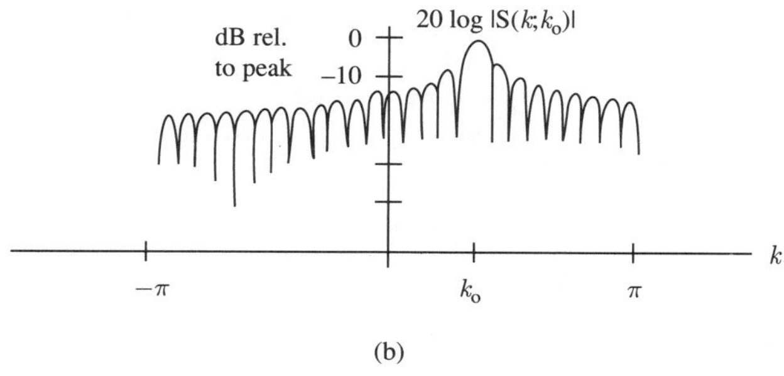

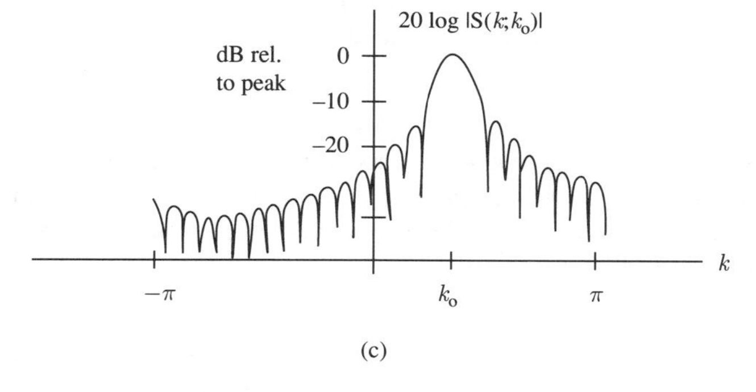

Figure 10.26 Array response. (a) $|S(k; k_0)|$ for $N \to \infty$. (b) Beampattern for finite N. (c) Beampattern for shaded array with finite N.

the number of points in the Fourier transform, in this case the size of the array. If there are two closely spaced sources (i.e., radiating from similar directions) they will not be resolved.

If the purpose of the array processing is to estimate the direction of arrival of multiple and closely spaced sources, then the problem is the same as the estimation of discrete sinusoidal components in a spectrum. In particular, since the samples s_n are available at the sensors, there is no reason to be restricted to conventional Fourier analysis (conventional beamforming). Rather, these "data samples" can be processed in any of the other possible ways described for spectrum estimation to obtain higher resolution. This is the principle behind all modern high-resolution methods of array processing.

The subject of array processing is vastly more complex than that of spectrum estimation, however. For one thing the sensors or array elements need not be arranged uniformly in a line. In fact, a linear array can locate sources in only one angular coordinate in space. To provide source location in more directions a planar or volumetric array is needed. Nonuniform or arbitrary placement of sensors provides more complexity. In the case of underwater towed arrays, the sensor locations cannot even be depended upon to remain fixed. Further complexity arises when it is desired to perform simultaneous frequency analysis and direction analysis. Finally, it is frequently necessary to structure the array response to produce a null in certain directions (say, corresponding to a jammer) while simultaneously maintaining high gain in a direction of choice. This kind of processing frequently requires adaptive signal processing techniques. Some further discussion of these and other issues as well as further references can be found in [71–73].

The recent subspace techniques fortunately adapt very well to the additional complexity inherent in array processing. MUSIC and its derivatives were fundamentally formulated for arrays with arbitrary geometry,[19] and apply to spectrum estimation as a special case. The subspace methods have elegant mathematical and geometric interpretations that provide a great deal of power in dealing with the more complex scenarios. Explanations can be found in several places, including [31, 33, 35, 36].

10.5.4 Estimation of Higher-Order Spectra

The topic of higher-order spectrum estimation has become a vast subject in itself and it is only possible to provide a glimpse of the area here. For purposes of brevity and clarity we restrict attention to estimation of the *bispectrum* and discuss so-called "conventional methods" and model-based methods. The approaches described can also be adapted to the trispectrum and higher-order spectra, however.

Conventional Methods. "Conventional" methods for estimating the bispectrum closely follow classical methods for estimating the power spectrum described in Section 10.1. Let $\{x[n], n = 0, 1, \ldots, N_s - 1\}$ represent a segment of a stationary ergodic random process. It is assumed that the mean, if nonzero, has been removed. An estimate for the third-order cumulant is then obtained from

$$\hat{C}_x^{(3)}[l_1, l_2] = \frac{1}{N_s} \sum_{n=n_I}^{n_F} x^*[n]x[n+l_1]x[n+l_2]; \tag{10.266}$$

where

$$n_I = \max(0, -l_1, -l_2), \qquad n_F = \max(N_s - 1, N_s - 1 - l_1, N_s - 1 - l_2)$$

The bispectrum estimate is then defined as

$$\hat{B}_x(\omega^{(1)}, \omega^{(2)}) = \sum_{l_1=-L}^{L} \sum_{l_2=-L}^{L} \hat{C}_x^{(3)}[l_1, l_2] \mathrm{w}_{2\mathrm{D}}[l_1, l_2] e^{-\jmath(\omega^{(1)} l_1 + \omega^{(2)} l_2)} \tag{10.267}$$

[19] *ESPRIT* requires that the array elements (sensors) occur in matched pairs.

where $\mathrm{w}_{2\mathrm{D}}[l_1, l_2]$ is an appropriate two-dimensional window. The computation of $\hat{C}_x^{(3)}[l_1, l_2]$ in (10.266) can be reduced by making use of its symmetry properties (see Section 4.10.2). With a single data segment, the variance of the spectral estimate is high. As with estimation of the regular power spectrum it is best to break a single long length of data into K smaller segments and average the results. Thus with K data segments, the bispectrum estimate is given by

$$\hat{B}_x(\omega^{(1)}, \omega^{(2)}) = \frac{1}{K} \sum_{l_1=-L}^{L} \sum_{l_2=-L}^{L} \sum_{k=1}^{K} \hat{C}_k^{(3)}[l_1, l_2] \mathrm{w}_{2\mathrm{D}}[l_1, l_2] e^{-\jmath(\omega^{(1)} l_{(1)} + \omega^{(2)} l_2)} \tag{10.268}$$

where $\hat{C}_k^{(3)}$ is the cumulant estimate for the kth segment. (The average over k is taken as the inner summation to avoid excess computation.)

The window choice is more difficult than for power spectrum estimation. In particular the window should satisfy [74, 75]

$$\begin{array}{ll} \mathrm{w}_{2\mathrm{D}}[l_1, l_2] = \mathrm{w}_{2\mathrm{D}}[l_2, l_1] & (a) \\ \quad \text{and, for real processes,} & \\ \quad \mathrm{w}_{2\mathrm{D}}[l_1, l_2] = \mathrm{w}_{2\mathrm{D}}[-l_1, l_2, -l_1] = \mathrm{w}_{2\mathrm{D}}[l_1 - l_2, -l_2] & \\ \mathrm{w}_{2\mathrm{D}}[l_1, l_2] = 0; \quad \max \{|l_1|, |l_2|, |l_1 - l_2|\} > L & (b) \\ \mathrm{w}_{2\mathrm{D}}[0, 0] = 1 & (c) \\ \mathrm{W}_{2\mathrm{D}}(e^{\jmath\omega^{(1)}}, e^{\jmath\omega^{(2)}}) \geq 0 \text{ for all values of } \omega^{(1)} and \omega^{(2)}. & (d) \end{array}$$

The first condition is simply the symmetry property for third-order cumulants. One such window satisfying these conditions is the uniform frequency domain window

$$\mathrm{W}_u(e^{\jmath\omega^{(1)}}, e^{\jmath\omega^{(2)}}) = \begin{cases} \frac{4}{3}\left(\frac{\pi}{\Omega_{\mathrm{o}}}\right)^2; & |\omega| \leq \Omega_{\mathrm{o}} \\ 0; & |\omega| > \Omega_{\mathrm{o}} \end{cases} \tag{10.269}$$

where $|\omega| \stackrel{\text{def}}{=} \max(|\omega^{(1)}|, |\omega^{(2)}|, |\omega^{(1)} + \omega^{(2)}|)$ and Ω_{o} is a constant parameter. Another is the class of *separable* windows defined by

$$\mathrm{w}_{2\mathrm{D}}[l_1, l_2] = \mathrm{w}[l_1]\mathrm{w}[l_2]\mathrm{w}[l_1 - l_2] \tag{10.270}$$

where $\mathrm{w}[l]$ is a one-dimensional window satisfying the properties

$$\begin{array}{lll} \mathrm{w}[l] = \mathrm{w}[-l] & & (a) \\ \mathrm{w}[l] = 0; & l > L & (b) \\ \mathrm{w}[0] = 1 & & (c) \\ \mathrm{W}[e^{\jmath\omega}] \geq 0; & \forall\omega & (d) \end{array} \tag{10.271}$$

Two suitable such windows are the Parzen window

$$\mathrm{w}_P[l] = \begin{cases} 1 - 6\left(\frac{|l|}{L}\right)^2 + 6\left(\frac{|l|}{L}\right)^3; & |l| \leq L/2 \\ 2\left(1 - \frac{|l|}{L}\right)^3; & L/2 < |l| \leq L \\ 0; & |l| > L \end{cases} \tag{10.272}$$

and the minimum bias supremum window due to Sasaki, Sato, and Yamashita [74]

$$\mathrm{w}_O[l] = \begin{cases} \frac{1}{\pi}\left|\sin\frac{\pi l}{L}\right| + \left(1 - \frac{|l|}{L}\right)\left(\cos\frac{\pi l}{L}\right); & |l| \leq L \\ 0; & |l| > L \end{cases} \tag{10.273}$$

Further discussion of these windows and the statistical properties of the corresponding bispectrum estimates can be found in [74].

Model-based Methods. Conventional methods of bispectrum estimation suffer from the same deficiencies as classical methods of power spectral estimation. Very long sequences of data are required to achieve good resolution and low variance simultaneously. The estimates can thus become computationally demanding. Model-based methods are therefore of interest when long data segments are not available or to reduce computation.

The basic idea in model-based estimation of the bispectrum is to represent a random process as a linear filter driven by *non-Gaussian* white noise (see Fig. 10.27). Third-order non-Gaussian white noise is defined by the conditions

$$\begin{aligned} \mathcal{E}\{w[n]\} &= 0 \ (a) \\ \mathcal{E}\{w^*[n]w[n+l]\} &= \sigma_o^2\delta[k] \ (b) \\ \mathcal{E}\{w^*[n]w[n+l_1]w[n+l_2]\} &= \beta_o\delta[l_1]\delta[l_2] \ (c) \end{aligned} \tag{10.274}$$

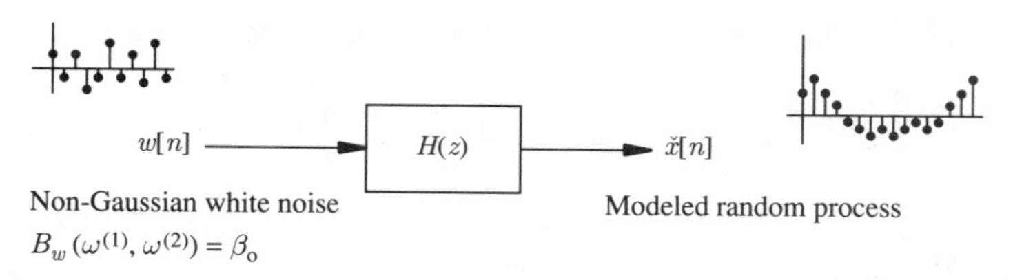

Figure 10.27 Linear model for estimation of the bispectrum.

The bispectrum of non-Gaussian white noise is therefore a constant

$$B_w(\omega^{(1)}, \omega^{(2)}) = \beta_o \tag{10.275}$$

The estimate of the bispectrum is taken to be the bispectrum of the model, that is,

$$\hat{B}_x(\omega^{(1)}, \omega^{(2)}) = \beta_o H(e^{\jmath\omega^{(1)}})H(e^{\jmath\omega^{(2)}})H^*(e^{\jmath(\omega^{(1)}+\omega^{(2)})}) \tag{10.276}$$

(see Section 5.6).

Although methods exist for AR, MA, and ARMA modeling, we discuss only the AR models here. A review of some procedures for the MA and ARMA cases can be found in [75]. Further discussion and related procedures can be found in [76] and [77].

For the AR case the linear system is described by the difference equation

$$x[n] + \sum_{k=1}^{P} a_k x[n-k] = w[n] \tag{10.277}$$

It is not difficult to show that the third-order cumulants satisfy the corresponding difference equation (see Problem 10.16)

$$C_x^{(3)*}[-l_1, -l_2] + \sum_{k=1}^{P} a_k C_x^{(3)*}[k-l_1, k-l_2] = \begin{cases} \beta_o^*; & l_1 = l_2 = 0 \\ 0; & l_1 > 0,\ l_2 > 0 \end{cases} \tag{10.278}$$

If the cumulants of the process are known, then by evaluating (10.278) for various values of l_1 and l_2 it is possible to obtain *linear* equations to solve for the a_k. For example, letting $l_1 = l_2 = 0, 1, \ldots, P$ yields

$$\begin{bmatrix} C_x^{(3)*}[0,0] & C_x^{(3)*}[1,1] & \cdots & C_x^{(3)*}[P,P] \\ C_x^{(3)*}[-1,-1] & C_x^{(3)*}[0,0] & \cdots & C_x^{(3)*}[P-1,P-1] \\ \vdots & \vdots & & \vdots \\ C_x^{(3)*}[-P,-P] & C_x^{(3)*}[-P+1,-P+1] & \cdots & C_x^{(3)*}[0,0] \end{bmatrix} \begin{bmatrix} 1 \\ a_1 \\ \vdots \\ a_P \end{bmatrix} = \begin{bmatrix} \beta_o^* \\ 0 \\ \vdots \\ 0 \end{bmatrix} \tag{10.279}$$

This represents a set of cumulant-based "Yule–Walker equations." The large matrix is Toeplitz, but *not symmetric nor Hermitian symmetric* since $C_x^{(3)}[k,k]$ is not equal to $C_x^{(3)}[-k,-k]$ or $C_x^{(3)*}[-k,-k]$ in general.

Another possibility is to use samples of the cumulants over the triangular region of support shown in Fig. 10.28. This leads to an entirely different set of Yule–Walker equations.

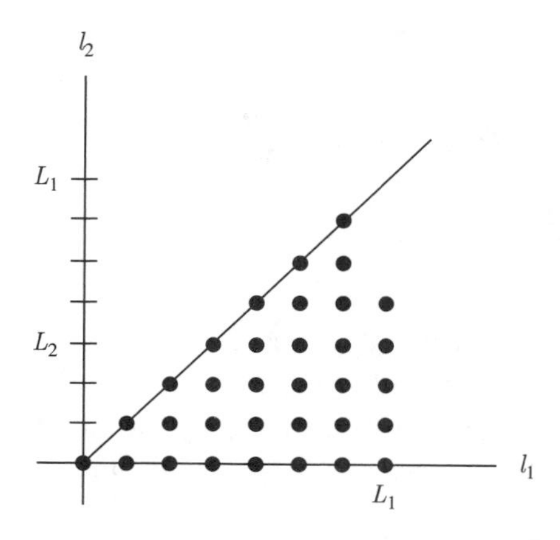

Figure 10.28 A possible region of support for the third-order cumulant in cumulant-based Yule–Walker equations.

In practice the values of the cumulants are not known and are replaced by the estimates (10.266). Once the model parameters have been solved for, the spectral estimate is given by (10.276) with $H(z)$ defined by

$$H(z) = \frac{1}{A(z)} = \frac{1}{1 + \mathrm{a}_1 z^{-1} + \cdots + \mathrm{a}_P^{-P}}$$

The AR bispectrum estimate is thus

$$\hat{B}_x(\omega^{(1)}, \omega^{(2)}) = \frac{\beta_o}{A(e^{\jmath\omega^{(1)}})A(e^{\jmath\omega^{(2)}})A^*(e^{\jmath(\omega^{(1)}+\omega^{(2)})})} \tag{10.280}$$

One problem with AR modeling of the bispectrum is similar to a problem that arises in AR modeling of the power spectrum for two-dimensional signals that was alluded to earlier. In particular, developing the Yule–Walker equations for different regions of support leads to different equations and in general *different* values for the model parameters. Moreover, there are not enough degrees of freedom in the parameters $\mathrm{a}_1, \mathrm{a}_2, \ldots, \mathrm{a}_P$ and β_o to match the possible number of different values of the cumulants that appear in the Yule–Walker equations. Therefore, unless the given cumulant values truly derive from a non-Gaussian AR process of order P, the given values are not exactly reproduced by the model. In the special case where the cumulant values *do* come from a Pth-order non-Gaussian AR process and are exactly known, Yule–Walker equations developed over *any* suitable region of support produce the same model parameters.

Another slightly different approach to finding the AR model is as follows. Consider the quantities

$$Q_n[l, k] \stackrel{\text{def}}{=} x^*[n-k]x^2[n-l] \tag{10.281}$$

and notice that

$$\mathcal{E}\{Q_n[l, k]\} = \mathcal{E}\{x^*[n-k]x[n-l]x[n-l]\} = C_x^{(3)}[k-l, k-l] \tag{10.282}$$

Now observe that for $l_1 = l_2 = l$ (10.278) and (10.282) imply that

$$\mathcal{E}\{Q_n[l, 0]\} + \sum_{k=1}^{P} \mathrm{a}_k^* \mathcal{E}\{Q_n[l, k]\} = \begin{cases} \beta_o; & l = 0 \\ 0; & l = 1, 2, \ldots, L \end{cases} \tag{10.283}$$

where L is some chosen number of lag values with $L \geq P$. The quantity

$$Q_n[l, 0] + \sum_{k=1}^{P} \mathrm{a}_k^* Q_n[l, k]$$

is called the *third-order error process*, and (10.283) states that for $l > 0$ its expected value is zero. Replacing the expectation by a signal average and defining the quantities

$$\overline{Q}[l, k] \stackrel{\text{def}}{=} \frac{1}{N_s - L} \sum_{n=L+1}^{N_s} Q_n[l, k] \tag{10.284}$$

converts (10.283) to the set of linear equations

$$\overline{Q}[l,0] + \sum_{k=1}^{P} a_k^* \overline{Q}[l,k] = 0; \qquad l = 1,2,\ldots,L \tag{10.285}$$

and

$$\beta_o = \overline{Q}[0,0] + \sum_{k=1}^{P} a_k^* \overline{Q}[0,k] \tag{10.286}$$

Equation 10.285 can be written as the least squares matrix equation

$$\begin{bmatrix} \overline{Q}[1,1] & \cdots & \overline{Q}[1,P] \\ \overline{Q}[2,1] & \cdots & \overline{Q}[2,P] \\ \vdots & & \vdots \\ \overline{Q}[L,1] & \cdots & \overline{Q}[L,P] \end{bmatrix} \begin{bmatrix} a_1^* \\ \vdots \\ a_P^* \end{bmatrix} \stackrel{\text{ls}}{=} \begin{bmatrix} \overline{Q}[1,0] \\ \overline{Q}[2,0] \\ \vdots \\ \overline{Q}[L,0] \end{bmatrix} \tag{10.287}$$

which can be solved for the filter coefficients. The constant β_o can then be found from (10.286). In practical implementation the factor $1/(N_s - L)$ in (10.284) can be canceled in (10.287) to simplify the computation. For the case of $L = P$ a solution can be found that satisfies (10.287) with no error, and the procedure is equivalent to solving (10.279) with appropriate estimates substituted for the cumulants.

10.6 CHAPTER SUMMARY

Spectrum estimation is a very broad and important part of statistical signal processing. Methods have been developed that approach the problem from many different points of view and the topic is still an area of intensive research.

Classical or nonparameteric methods of spectrum estimation are based on estimates of the correlation function and its Fourier transform. When the biased correlation function estimate is used for the largest possible number of lags, the corresponding spectral estimate is proportional to the squared magnitude of the transform of the data itself. This estimate is called the periodogram. The unmodified periodogram is a poor estimate of the spectrum because its variance remains constant as the data length is increased while the frequencies at which the estimates are uncorrelated move closer together. Practical methods of spectrum estimation based on estimates of the correlation function therefore require windowing, averaging (over multiple data segments), or some combination of these two procedures. The main drawback of classical methods is that resolution of closely spaced features in the spectrum is limited by the length of data available. When very long sequences of stationary data are available, however, the classical methods are hard to beat.

Model-based estimates of the spectrum are motivated by the innovations representation of a regular random process. By fitting a rational linear model to the data (i.e., a linear filter driven by white noise) the theoretical spectrum of the model becomes the *estimated* spectrum of the data. Model-based methods can provide very good resolution using only short lengths of data. The spectral estimates are only as good as the underlying model, however, and model order selection is a difficult problem in practice. Underfitting the data (choosing a model order too low) can result in poor resolution, while overfitting the data (choosing a model order too high) can result in artifacts such as spurious peaks in the spectral estimate. AR modeling in particular has received a lot of attention because of its simplicity and because of the maximum entropy property discoverd by Burg. The AR model generates a random process whose correlation function *matches* the given values of the correlation function (appearing in the Yule–Walker equations) and *extends* the correlation function in a way that maximizes the entropy. Thus the AR spectral estimate is also the maximum entropy spectral estimate.

The "maximum likelihood" method of spectrum estimation is conceptually based on a causal FIR linear filter that adapts to data at the frequency being estimated; in particular it passes the component of the data at the given frequency with no attenuation while minimizing the power from components at all other frequency values. In actual computation the estimate is a simple formula involving a quadratic product with the inverse correlation matrix. The reciprocal of this estimate can be shown to be the sum of the reciprocals of the maximum entropy spectral estimates of orders ranging from zero to the order of the correlation matrix.

An entire other class of spectral estimates is based on the concept of signal and noise subspaces associated with the correlation matrix for a random process. Subspace methods apply primarily to locating the discrete components (lines) of the spectrum. Pisarenko's harmonic decomposition was the first of these methods and motivated other improved methods such as MUSIC that followed. These methods are all based on the fact that when the data consists of complex exponentials in noise, the frequency vector $\mathbf{w}$ defined by (10.82) is an eigenvector of the correlation matrix. Its projection onto an orthogonal subspace complementary to the subspace defined by all of the signal vectors produces a null which can be exploited to estimate the frequency. Both Pisarenko's method and MUSIC can be formulated as a search for peaks in a function (called a pseudospectrum) or as a polynomial root-finding problem.

Subspace methods such as MUSIC, when viewed as a root-finding problem, are plagued by spurious roots that may cause problems in identifying the true signals. Tufts and Kumaresan did much to improve the methods by showing how to make the spurious roots move away from the roots corresponding to the signals. The minimum norm procedure projects the data onto a single vector in the noise subspace that optimizes this separation. Another method based on a principal components approximation to the correlation matrix shows once again that linear prediction when properly applied is a very important tool in statistical signal processing.

A method called *ESPRIT* provides an interesting twist on the subspace methods and leads to an elegant solution for the frequencies of interest. This method begins with a pair of data sets that bear a translation invariance relationship to each other in the signal domain and exploits the corresponding rotational invariance in the frequency domain to find the frequencies of the complex exponentials. The solution requires eigenanalysis or singular value decomposition but is more direct and computationally less expensive than other subspace methods, such as MUSIC. Both least squares and total least squares versions of *ESPRIT* have been developed. The total least squares version has lower bias in the frequency estimates and therefore is usually the method of choice.

A theory known as subspace fitting provides a common framework for most other subspace methods and has led to important results about their asymptotic performance. A particular procedure known as weighted subspace fitting has asymptotic variance that meets the Cramér-Rao bound and therefore performance equivalent to (stochastic) maximum likelihood. The criterion function to be optimized, however, is considerably simpler. This method, as well as MUSIC and *ESPRIT*, all have further interesting and elegant interpretations in the context of array processing for which they were originally developed.

Four further advanced topics may provide some answer to the question of "Where do we go from here?" Cross-spectrum estimation in general requires the consideration of multichannel random processes. Most of the methods for spectrum estimation discussed in this chapter extend to the multichannel case, although the generalization is not always straightforward. Short-time Fourier transforms and distributions such as the Wigner–Ville and instantaneous power spectrum provide some methods for representing the intuitively reasonable but mathematically elusive concept of a power spectrum that varies with time. Array processing is mathematically similar to spectrum estimation. However, the generality that arises in array geometry, sensor characteristics, multiple dimensions, and the necessarily sparse spatial sampling renders the topic one of much greater overall difficulty than that of ordinary spectral analysis. Finally, methods for estimation of higher-order spectra that parallel the classical and model-based methods are available, although the procedures are not as simple and the models are not unique as they are in the second moment case. In fact, the use of higher order moments and spectra may well be one of the new frontiers in not only signal processing, but also control, communications, statistical data analysis, and other related fields. While some theory has now been developed, the practical application of these methods in military and commercial products has yet make an appearance. Their introduction here may well be the prelude to a much more extensive study and use of these methods over the next couple of decades.

REFERENCES

1. L. H. Koopmans. *The Spectral Analysis of Time Series*. Academic Press, New York, 1974.
2. Gwilym M. Jenkins and Donald G. Watts. *Spectral Analysis and Its Applications*. Holden-Day, Oakland, California, 1968.
3. Steven M. Kay and S. Lawrence Marple, Jr. Spectrum analysis—a modern perspective. *Proceedings of the IEEE*, 69:1380–1419, November 1981.
4. Steven M. Kay. *Modern Spectral Estimation: Theory and Application*. Prentice Hall, Inc., Englewood Cliffs, New Jersey, 1988.
5. S. Lawrence Marple, Jr. *Digital Spectral Analysis with Applications*. Prentice Hall, Inc., Englewood Cliffs, New Jersey, 1987.

6. Donald G. Childers, editor. *Modern Spectrum Analysis.* IEEE Press, New York, 1978.
7. Stanislav B. Kesler, editor. *Modern Spectrum Analysis, II.* IEEE Press, New York, 1986.
8. Kenneth S. Miller. *Complex Stochastic Processes.* Addison-Wesley, Reading, Massachusetts, 1974.
9. M. S. Bartlett. Periodogram analysis and continuous spectra. *Biometrika*, 37:1–16, 1950.
10. R. B. Blackman and J. W. Tukey. *The Measurement of Power Spectra from the Point of View of Communications Engineering.* Dover Publications, New York, 1958.
11. R. W. Hamming. Private communication, August 1991.
12. Alan V. Oppenheim and Ronald W. Schafer. *Digital Signal Processing.* Prentice Hall, Inc., Englewood Cliffs, New Jersey, 1975.
13. P. D. Welch. The use of fast Fourier transform for the estimation of power spectra. *IEEE Transactions on Audio and Electroacoustics*, AU-15:70–73, June 1970.
14. John Parker Burg. Maximum entropy spectral estimation. In *Proceedings of the* 37th *Annual SEG Meeting*, 31 December 1967 (Oklahoma City; reprint appears in [6]).
15. John Parker Burg. A new analysis technique for time series data. In *NATO Advanced Study Institute on Signal Processing*, pp. 12–23, August 1968 (Enschede, The Netherlands; reprint appears in [6]).
16. John Parker Burg. New concepts in power spectra estimation. In *Proceedings of the* 40th *Annual SEG Meeting*, 11 November 1970 (New Orleans).
17. Emanual Parzen. *Multiple Time Series Modeling.* Technical Report 12, Stanford University, Stanford, California, 1968. On contract NONR-225-(80).
18. Emanual Parzen. Multiple time series modeling: determining the order of approximating autoregressive schemes. In P. R. Krishnaiah, editor, *Multivariate Analysis, IV*, pages 283–295. North Holland Publishing Co., New York, 1977. (Originally published by Academic Press, New York, 1969.)
19. John Parker Burg. *Maximum Entropy Spectral Analysis.* Ph.D. thesis, Stanford University, Stanford, California, May 1975.
20. A. van den Bos. Alternative interpretation of maximum entropy spectral analysis. *IEEE Transactions on Information Theory*, IT-17:92–93, July 1971.
21. Claude E. Shannon. A mathematical theory of communication (concluded). *Bell System Technical Journal*, 27(4):623–656, October 1948. (See also [22].)
22. Claude E. Shannon and Warren Weaver. *The Mathematical Theory of Communication.* University of Illinois Press, Urbana, Illinois, 1963.
23. B. W. Dickinson. Two-dimensional Markov spectrum estimates need not exist. *IEEE Transactions on Information Theory*, IT-26(1):120–121, January 1980.
24. Stephen W. Lang and James H. McClellan. Spectral estimation for sensor arrays. *IEEE Transactions on Acoustics, Speech, and Signal Processing*, ASSP-31:349–358, April 1983.
25. J. Capon. High-resolution frequency-wavenumber spectrum analysis. *Proceedings of the IEEE*, 57:119–129, August 1969.
26. J. Capon. Maximum-likelihood spectral estimation. In Simon Haykin, editor, *Nonlinear Methods of Spectral Analysis*, pages 155–179. Springer-Verlag, New York, 1983.
27. John Parker Burg. The relationship between maximum entropy and maximum likelihood spectra. *Geophysics*, 37:375–376, 1972.

28. Bruce R. Musicus. Fast MLM power spectrum estimation from uniformly spaced correlations. *IEEE Transactions on Acoustics, Speech, and Signal Processing*, ASSP-33:1333–1335, October 1985.

29. Richard T. Lacoss. Data adaptive spectral analysis methods. *Geophysics*, 36:661–675, August 1971.

30. Sophocles J. Orfanidis. *Optimum Signal Processing: An Introduction*, 2nd ed. Macmillan, New York, 1988.

31. Richard H. Roy. *ESPRIT—Estimation of Signal Parameters via Rotational Invariance Techniques*. Ph.D. thesis, Stanford University, Stanford, California, August 1987 (Information Systems Laboratory, Department of Electrical Engineering).

32. V. F. Pisarenko. The retrieval of harmonics from a covariance function. *Geophysical Journal of Royal Astronomical Society*, 33:347–366, 1973.

33. Ralph Schmidt. Multiple emitter location and signal parameter estimation. In *Proceedings of the RADC Spectrum Estimation Workshop*, pages 243–258, Rome Air Development Corp., 1979. See also [34].

34. Ralph Schmidt. Multiple emitter location and signal parameter estimation. *IEEE Transactions on Antennas and Propagation*, AP-34:276–290, March 1986. Reprint of [33].

35. Ralph Schmidt. *A Signal Subspace Approach to Multiple Emitter Location and Spectral Estimation*. Ph.D. thesis, Stanford University, Stanford, California, August 1981.

36. Richard H. Roy and Thomas Kailath. *ESPRIT*–estimation of signal parameters via rotational invariance techniques. *IEEE Transactions on Acoustics, Speech, and Signal Processing*, 37(7):984–995, July 1989.

37. S. S. Reddi. Multiple source location—a digital approach. *IEEE Transactions on Aerospace and Electronic Systems*, AES-15(1):134–139, January 1979.

38. Ramdas Kumaresan and Donald W. Tufts. Estimating the angles of arrival of multiple plane waves. *IEEE Transactions on Aerospace and Electronic Systems*, AES-19(1):134–139, January 1983.

39. Donald W. Tufts and Ramdas Kumaresan. Estimation of frequencies of multiple sinusoids: making linear prediction behave like maximum likelihood. *Proceedings of the IEEE*, 70:975–989, September 1982.

40. Donald W. Tufts and Ramdas Kumaresan. Singular value decomposition and improved frequency estimation using linear prediction. *IEEE Transactions on Acoustics, Speech, and Signal Processing*, ASSP-30:671–675, August 1982.

41. B. Ottersten and M. Viberg. Asymptotic results for multidimensional sensor array processing. In *Proceedings of the* 22nd *Asilomar Conference on Signals, Systems, and Computers*, pages 833–837, November 1988 (Pacific Grove, California).

42. M. Viberg and B. Ottersten. Sensor array processing based on subspace fitting. *IEEE Transactions on Signal Processing*, 39:1110–1121, May 1991.

43. Georges Bienvenu and Laurent Kopp. Principè de la goniomètre passive adaptive. In *Proc. 7'eme Colloque GRESIT*, pages 106/1–106/10, 1979. Nice, France.

44. Don H. Johnson and Stuart R. DeGraaf. Improving the resolution of bearing in passive sonar arrays by eigenvalue analysis. *IEEE Transactions on Acoustics, Speech, and Signal Processing*, ASSP-30(4):638–647, August 1982.

45. Ramdas Kumaresan. On the zeros of the linear prediction-error filter for deterministic signals. *IEEE Transactions on Acoustics, Speech, and Signal Processing*, ASSP-31(1):217–220, February 1983.

46. Ramdas Kumaresan. *Estimating the Parameters of Exponentially Damped/Undamped Sinusoidal Signals in Noise*. Ph.D. thesis, University of Rhode Island, Kingston, Rhode Island, August 1982.

47. Eric M. Dowling and Ronald D. DeGroat. The equivalence of total least squares and minimum norm methods. *IEEE Transactions on Signal Processing*, 39(8):1891–1892, August 1991.

48. J. H. Wilkinson. *The Algebraic Eigenvalue Problem*. Oxford University Press, New York, 1965.

49. D. C. Rife and R. R. Boorstyn. Multiple tone parameter estimation from discrete time observations. *Bell System Technical Journal*, 55:1389–1410, November 1976. (Computation of the Cramér-Rao bound.)

50. A. Paulraj, Richard H. Roy, and Thomas Kailath. Estimation of signal parameters via rotational invariance techniques – *ESPRIT*. In *Proceedings of the* 19th *Asilomar Conference on Circuits, Systems, and Computers*, pages 83–89, November 1985 (Pacific Grove, California).

51. Richard H. Roy, A. Paulraj, and Thomas Kailath. *ESPRIT* – a subspace rotational approach to estimation of parameters of cisoids in noise. *IEEE Transactions on Acoustics, Speech, and Signal Processing*, ASSP-34(5):1340–1342, October 1986.

52. Richard H. Roy and Thomas Kailath. Total least-squares *ESPRIT*. In *Proceedings of the* 21st *Asilomar Conference on Signals, Systems, and Computers*, pages 297–301, November 1987 (Pacific Grove, California).

53. Marc Goldburg and Richard Roy. Application of *ESPRIT* to parameter estimation from uniformly sampled data. In *Proceedings of the IEEE International Conference on Acoustics, Speech, and Signal Processing*, IEEE Signal Processing Society, April 1990 (Albuquerque). Paper presented but not published in proceedings; written paper received directly from authors.

54. Gene H. Golub and Charles F. Van Loan. *Matrix Computations*, 2nd ed. Johns Hopkins University Press, Baltimore, Maryland, 1989.

55. Björn E. Ottersten. *Parametric Subspace Fitting Methods for Array Signal Processing*. Ph.D. thesis, Stanford University, Stanford, California, December 1989 (Information Systems Laboratory, Deptartment of Electrical Engineering).

56. M. Viberg, B. Ottersten, and T. Kailath. Detection and estimation in sensor arrays using weighted subspace fitting. *IEEE Transactions on Signal Processing*, 39:2436–2449, November 1991.

57. B. Ottersten, B. Wahlberg, M. Viberg, and T. Kailath. Stochastic maximum likelihood in sensor arrays by weighted subspace fitting. In *Proceedings of the* 23rd *Asilomar Conference on Signals, Systems, and Computers*, pages 599–603, November 1989 (Pacific Grove, California).

58. B. Ottersten, M. Viberg, and T. Kailath. Analysis of subspace fitting and ML techniques for parameter estimation from sensor array data. *IEEE Transactions on Signal Processing*, 40(3), March 1992.

59. Mati Wax and Thomas Kailath. Detection of signals by information theoretic criteria. *IEEE Transactions on Acoustics, Speech, and Signal Processing*, ASSP-33:387–392, April 1985.

60. Albert H. Nuttall. *Multivariate Linear Predictive Spectral Analysis Employing Weighted Forward and Backward Averaging: A Generalization of Burg's Algorithm*. Technical Report 5501, Naval Underwater Systems Center, New London, CT, October 1976. (In volume entitled "NUSC Scientific and Engineering Studies, *Spectral Estimation*.")

61. O. N. Strand. Multichannel complex maximum entropy (autoregressive) spectral analysis. *IEEE Transactions on Automatic Control*, AC-22(4):634–640, August 1977.

62. Boualem Boashash. Time-frequency signal analysis. In Simon Haykin, editor, *Advances in Spectrum Analysis and Array Processing*, pages 418–517. Prentice Hall, Inc., Englewood Cliffs, New Jersey, 1991.

63. W. Mecklenbräuker. A tutorial on non-parametric bilinear time-frequency signal representations. In Jean-Louis Lacoume, Tariq Salim Durrani, and Raymond Stora, editors, *Traitement du Signal (Signal Processing), Les Houches, France 1985, Session XLV*, pages 277–336. North Holland (Elsevier Science Publishers), New York, 1987.

64. Leon Cohen. Time-frequency distributions—a review. *Proceedings of the IEEE*, 77(7):941–981, July 1989.

65. S. Hamid Nawab and Thomas F. Quatieri. Short-time Fourier transform. In Jae S. Lim and Alan V. Oppenheim, editors, *Advanced Topics in Signal Processing*, pages 289–337. Prentice Hall, Inc., Englewood Cliffs, New Jersey, 1988.

66. Ingrid Daubechies. The wavelet transform: a method for time-frequency localization. In Simon Haykin, editor, *Advances in Spectrum Analysis and Array Processing*, pages 366–417. Prentice Hall, Inc., Englewood Cliffs, New Jersey, 1991.

67. Leon Cohen. Generalized phase-space distribution functions. *Journal of Mathematical Physics*, 7:781–786, 1966.

68. G. F. Boudreaux-Bartels and Thomas W. Parks. Time-varying filtering and signal estimation using Wigner distribution synthesis techniques. *IEEE Transactions on Acoustics, Speech, and Signal Processing*, ASSP-34(3):442–451, June 1986.

69. Ralph D. Hippenstiel and Paulo M. Oliveira. Time-varying spectral estimation using the instantaneous power spectrum (IPS). *IEEE Transactions on Acoustics, Speech, and Signal Processing*, 38(10):1752–1759, October 1990.

70. Dan E. Dudgeon and Russell M. Mersereau. *Multidimensional Digital Signal Processing*. Prentice Hall, Inc., Englewood Cliffs, New Jersey, 1984.

71. Barry D. Van Veen and Kevin M. Buckley. Beamforming: a versatile approach to spatial filtering. *IEEE ASSP Magazine*, 5(2), April 1988.

72. Don H. Johnson and Dan E. Dudgeon. *Array Signal Processing: Concepts and Methods*. Prentice Hall, Inc., Englewood Cliffs, New Jersey, 1992.

73. Simon Haykin, editor. *Array Signal Processing*. Prentice Hall, Inc., Englewood Cliffs, New Jersey, 1985.

74. K. Sasaki, T. Sako, and Y. Yamashita. Minimum bias windows for bispectral estimation. *Journal of Sound and Vibration*, 40(1):139–148, 1975.

75. Chrysostomos L. Nikias and Mysore R. Raghuveer. Bispectrum estimation: a digital signal processing framework. *Proceedings of the IEEE*, 75(7):869–891, July 1987.

76. Georgios B. Giannakis. On the identifiablity of non-Gaussian ARMA models using cumulants. *IEEE Transactions on Automatic Control*, 35(1):18–26, January 1990.

77. Georgios B. Giannakis and Jerry M. Mendel. Identification of nonminimum phase systems using higher order statistics. *IEEE Transactions on Acoustics, Speech, and Signal Processing*, 37(3):360–377, March 1989.

PROBLEMS

10.1. Show that for $L = N_s - 1$ in (10.3) the correlogram estimate of the spectrum and the periodogram are the same thing.

10.2. In this problem you will explore the bias of the periodogram that arises when a window function is first applied to the data.

Given a finite segment of data $x[n]$ apply a real window to compute $x'[n] = \mathrm{v}[n] \cdot x[n]$. The periodogram is proportional to the magnitude squared of the Fourier transform of $x'[n]$, that is,

$$\hat{P}_{x'}(e^{\jmath\omega}) = \frac{1}{N_s}|X'(e^{\jmath\omega})|^2$$

The results of Problem 10.1 show that

$$\hat{P}_{x'}(e^{\jmath\omega}) = \sum_{l=-N_s+1}^{N_s-1} \hat{R}_{x'}[l]e^{-j\omega l}$$

where

$$\hat{R}_{x'}[l] = \frac{1}{N_s}\sum_{n=0}^{N_s-1-l} x'[n+l]x'^*[n]; \qquad 0 \leq l < N_s$$

$$\hat{R}_{x'}[l] = \frac{1}{N_s}\sum_{n=0}^{N_s-1-|l|} x'[n]x'^*[n+|l|]; \quad -N_s < l < 0$$

(a) By substituting $x'[n] = \mathrm{v}[n]x[n]$ and taking the expectations, show that the estimate $\hat{R}_{x'}[l]$ is biased, and in particular that

$$\mathcal{E}\left\{\hat{R}_{x'}[l]\right\} = \mathrm{w}[l] \cdot R_x[l]$$

where

$$\mathrm{w}[l] = \frac{1}{N_s}(\mathrm{v}[l] * \mathrm{v}[-l])$$

and $R_x[l]$ is the true correlation function of $x[n]$.

(b) Given that $\hat{P}_{x'}(e^{\jmath\omega})$ is the Fourier transform of $\hat{R}_{x'}[l]$, and that the expectation of $\hat{R}_{x'}[l]$ is given by the result of part (a), show that the spectral estimate is biased. That is, compute $\mathcal{E}\left\{\hat{P}_{x'}(e^{\jmath\omega})\right\}$. Express your answer in terms of the true spectrum and the Fourier transform $\mathrm{V}(e^{j\omega})$ of the window.

10.3. A real random process has the correlation function

$$R_x[l] = 2^{-|l|} + \delta[l]$$

(a) What is the power spectral density function of this random process?

(b) Assume that the correlation function was estimated perfectly. A correlogram spectral estimate is formed by using only the first three lag values, that is,

$$\hat{R}_x[l] = \begin{cases} R_x[l]; & l = 0, \pm 1, \pm 2 \\ 0; & \text{otherwise} \end{cases}$$

What is the expression for the correlogram spectral estimate? Write it in simplest form (without complex exponentials).

(c) What is the expression for the first-order AR spectral estimate ($P = 1$)?

A windowed correlogram spectral estimate is defined by the equation

$$\hat{S}_x(e^{\jmath\omega}) = \sum_{l=-N_s+1}^{N_s-1} \mathrm{w}[l]\hat{R}_x[l]e^{-\jmath\omega l}$$

where $\hat{R}_x[l]$ is given by (10.1) and the window $\mathrm{w}[l]$ is given by

$$\mathrm{w}[l] = \frac{\sin \omega_0 l}{\omega_0 l}; \qquad 0 \leq \omega_0 \leq \pi$$

Assume that N_s is very large so that we can let $N_s \to \infty$.

(a) What is the expected value of the spectral estimate? Express it as an integral involving the true spectrum.

(b) What is the bias?

(c) Can the expected value of this spectral estimate ever be negative?

10.5. Show that the expected value of the spectral estimate $\hat{S}_W(e^{\jmath\omega})$ of (10.35) is given by (10.36) and (10.37).

10.6. Show that the spectral estimate produced by the *anticausal* AR model for a random process is *identical* to that produced by the causal AR model.

10.7. Let a maximum likelihood spectral estimate be computed from a data sequence of length N_s by using the *covariance* method to estimate the $N \times N$ correlation matrix. The spectral estimate is therefore

$$\hat{S}_{ML}(e^{\jmath\omega}) = \frac{1}{\mathbf{w}^{*T}\hat{\mathbf{R}}_x^{-1}\mathbf{w}}$$

Now consider averaging a full set of periodogram estimates where each data segment consists of N points and overlaps with the previous segment in all but a single point. In other words, the first data segment consists of the samples $\mathrm{x}[0], \mathrm{x}[1], \ldots, \mathrm{x}[N-1]$; the second data segment consists of the points $\mathrm{x}[1], \mathrm{x}[2], \ldots, \mathrm{x}[N]$; and so on until all of the N_s samples are used. Show that the averaged periodogram (Bartlett) estimate then can be expressed as

$$\hat{S}_B(e^{\jmath\omega}) = \frac{1}{N}\mathbf{w}^{*T}\hat{\mathbf{R}}_x\mathbf{w}$$

where $\hat{\mathbf{R}}_x$ is the *same* correlation matrix that appears in the expression for the maximum likelihood estimate above. Use this formulation to show how a principal components version of the Bartlett estimate could be formed.

10.8. The correlation matrix for a single complex exponential in white noise is

$$\mathbf{R}_x = \begin{bmatrix} 2 & -\jmath \\ \jmath & 2 \end{bmatrix}$$

Find the signal power, the noise variance, and the frequency of the signal.

10.9. **(a)** Find the eigenvalues of the following correlation matrix:

$$\mathbf{R}_x = \begin{bmatrix} 2 & -1 & 1 \\ -1 & 2 & -1 \\ 1 & -1 & 2 \end{bmatrix}$$

(b) If the correlation matrix corresponds to a received sequence of the form

$$x[n] = \sum_{i=1}^{M} s_i[n] + \eta[n]$$

where $\eta[n]$ is white noise, how many signals are present and what are the powers and frequencies of these signals?

10.10. **(a)** What are the eigenfilters in Problem 10.9?
(b) What are the roots of the noise subspace eigenfilters? Are there any common roots?
(c) Use the root MUSIC procedure to find the signal frequency or frequencies. What are the spurious roots?

10.11. Following a procedure similar to that in Section 10.4.2, show that the powers of the signals for subspace methods can be found by solving the equation

$$\mathbf{B}_2 \begin{bmatrix} \mathsf{P}_1 \\ \mathsf{P}_2 \\ \vdots \\ \mathsf{P}_M \end{bmatrix} = \begin{bmatrix} \lambda_1 - \sigma_o^2 \\ \lambda_2 - \sigma_o^2 \\ \vdots \\ \lambda_M - \sigma_o^2 \end{bmatrix}$$

where $\mathbf{B}_2$ represents the matrix whose individual elements are the squared magnitudes of the elements of the matrix $\mathbf{E}_{sig}^{*T}\mathbf{S}$.

10.12. Show that when the noise is not white that the MUSIC pseudospectrum is given by (10.149) or (10.154), where the eigenvectors $\mathbf{e}_k$ are solutions to the generalized eigenvalue equation (10.128).
Hint: Consider what happens in the transformed coordinate system where the noise *is* white and develop the corresponding expression in the original coordinate system.

10.13. Let $\mathrm{x}[n]$ be a finite data segment and $\mathrm{X}(z)$ be its z-transform. In the autocorrelation method of linear prediction the error sequence is defined as

$$\epsilon[n] = \mathrm{x}[n] * a[n]$$

where $a[n]$ is the prediction error filter impulse response with $a[0] = 1$ and $\mathrm{x}[n]$ is the windowed data sequence. By applying Parseval's theorem, show that the problem of finding the prediction error filter can be stated as the problem of minimizing the integral

$$\frac{1}{2\pi}\int_{-\pi}^{\pi} |\mathrm{X}(e^{\jmath\omega})|^2 |A(e^{\jmath\omega})|^2 d\omega$$

subject to the constraint that $A(z)$ is a comonic polynomial (in z^{-1}).

10.14. By starting with (10.163) and using the identity

$$\begin{bmatrix} \mathbf{g}^{*T} & \mathbf{c}^{*T} \\ \\ \mathbf{E}'_{sig} & \mathbf{E}'_{noise} \end{bmatrix} \begin{bmatrix} \mathbf{g} & \mathbf{E}'^{*T}_{sig} \\ \mathbf{c} & \mathbf{E}'^{*T}_{noise} \end{bmatrix} = \mathbf{I}$$

derive the alternative form (10.164) for the parameter $\mathbf{d}$ in the minimum-norm procedure.

10.15. Show that when the variance of the noise approaches zero the signal copy formula (10.202) of *ESPRIT* becomes exact, that is, the complex amplitudes of the signals satisfy

$$\begin{bmatrix} A_1 \\ A_2 \\ \vdots \\ A_M \end{bmatrix} = \mathbf{W}_{SC}^{*T} \bar{\boldsymbol{x}}$$

where $\mathbf{W}_{SC}$ is given by (10.203).

10.16. By multiplying (10.277) by appropriate delayed versions of the random process x and taking expectations, show that the third-order cumulants satisfy (10.278).

COMPUTER ASSIGNMENTS

10.1. Compute power spectrum estimates for the real data sets S00, S01, S02, and S03 using the Bartlett procedure (i.e., use a rectangular window) with no overlap. In all cases zero-pad your data appropriately before giving it to the FFT routine so that you will have computed enough points of the spectrum to give a smooth plot. (Plotting say 256 points of the spectrum should be sufficient.)

(a) Use only one 16-point segment (the first 16 points) of each data set.
(b) Use 32 16-point segments.
(c) Use 16 32-point segments.
(d) Use 4 128-point segments.

Discuss the results of these experiments in light of what you learned in this chapter about the trade-off between resolution and variance in the spectral estimate. Which option—(a), (b), (c), or (d)—do you think provides the best estimate of the true power spectrum?

10.2. In Computer Assignment 7.1 you found the linear prediction parameters for the data sets S00, S01, S02, and S03.

(a) Compute and plot AR spectral estimates for these data sets using the first-order model parameters.
(b) Compute and plot AR spectral estimates for these data sets using the second-order model parameters. Are there any significant differences from the results of part (a)?

10.3. Using the AR model parameters determined for DATAC in Computer Assignment 9.1, compute and plot the AR spectral estimates for this particular data set. Compare the results for the autocorrelation method, the covariance method, and Burg's method. Do this for the parameters that were estimated using

(a) 20 samples of the sequence.
(b) 100 samples of the sequence.

10.4. The data sets R01, R10, R40 and I01, I10, I40 contain 32 samples of the real and imaginary parts respectively of the following complex signal in white noise:

$$x[n] = s[n] + Kw[n]$$

for $K = 0.01$, $K = 0.10$, and $K = 0.40$. The signal $s[n]$ is defined by

$$s[n] = e^{\frac{j3\pi n}{8}} + e^{\frac{j5\pi n}{8}} + e^{\frac{j\pi n}{2}}$$

and $w[n]$ is a zero-mean unit-variance white noise sequence. This kind of data (complex exponentials in noise) has been sometimes used to test the resolution of different types of spectrum estimation methods, regardless of the assumptions underlying the method. This problem shows the advantage of using the correct model.

Compute the ordinary periodogram for $K = 0.01$ and plot it in the range $0 \leq \omega \leq \pi$. This will be the reference plot. Observe that it is computed using all 32 samples and at the highest signal-to-noise ratio.

(a) Plot and compare the spectral estimates for *each* noise value using the following four methods:
  (i) An AR model using the autocorrelation method to estimate parameters.
  (ii) The maximum entropy method using Burg's method to estimate parameters.
  (iii) The maximum likelihood method.
  (iv) MUSIC.

  You will have to decide on what order models to use. Put the four plots for each noise value on a single page for ease of comparison.

(b) Repeat part (a) using only 16 time samples from each of the generated sequences.

Summarize the results of your experiments.

10.5. Repeat Computer Assignment 10.4 using the following additional methods:
  (i) An AR model using the covariance method to estimate the parameters.
  (ii) An AR model using the modified covariance method to estimate the parameters.
  (iii) The minimum-norm method.
  (iv) Principle components linear prediction. Find the roots of the denominator polynomial $A(z)$ in (10.179) using various order filters.

10.6. Apply the *ESPRIT* method (TLS *ESPRIT*) to the data of Computer Assignment 10.4 to estimate the frequencies of the complex exponentials. Use the signal copy feature of *ESPRIT* to estimate the signals themselves and compare them to the true signal sequences. Plot the real and imaginary parts of the error between the estimated and the true signal sequences.

Optimization of a Quantity with Respect to a Vector Parameter

The problem of minimizing or maximizing a scalar quantity with respect to a vector parameter occurs several times in this book. Frequently, the vector parameter is complex. Fundamentally, the optimization involves taking the partial derivative of the scalar quantity with respect to the real and imaginary parts of all the vector components and setting the results equal to zero. That leads to a set of equations whose solution determines the optimal value of the parameter. A formal definition of the gradient or set of derivatives with respect to the components of a vector, and some rules for differentiating commonly occurring matrix expressions, permit the operations to be carried out neatly in terms of matrix equations and makes the work easier.

Constrained optimization problems can be solved using the gradient concepts almost as easily as unconstrained optimization problems, but the formulation of these problems in terms of Lagrange multipliers requires some special care. The mathematical tools and techniques needed to solve both type of problems are developed in this appendix.

A.1 GRADIENT OF A SCALAR QUANTITY

This section develops the gradient concept for a scalar quantity which depends on a vector parameter. The first subsection discusses the gradient with respect to a real vector, which is easily defined. The next subsection develops the corresponding definitions with respect to a complex vector. Since the ordinary derivative with respect to the complex parameter is generally not defined for many scalar quantities of interest, this topic requires somewhat more care and discussion. Results for some common matrix (and scalar) expressions are listed in tables in both subsections for reference throughout the book. The final subsection deals with the conditions for checking for a maximum or minimum of a scalar quantity at a point where the gradient is zero.

A.1.1 Gradient with Respect to a Real Vector Parameter

Let Q be a scalar quantity depending on the real N-dimensional vector parameter $\mathbf{a}$. The *gradient* $\nabla_{\mathbf{a}} Q$ is defined as the vector of partial derivatives

$$\nabla_{\mathbf{a}} Q \stackrel{\text{def}}{=} \begin{bmatrix} \dfrac{\partial Q}{\partial \mathrm{a}_1} \\ \dfrac{\partial Q}{\partial \mathrm{a}_2} \\ \vdots \\ \dfrac{\partial Q}{\partial \mathrm{a}_N} \end{bmatrix} \qquad \text{(A.1)}$$

(real vector $\mathbf{a}$)

where $\mathrm{a}_1, \mathrm{a}_2, \ldots, \mathrm{a}_N$ are the vector components. This definition is convenient in optimization problems involving the parameter $\mathbf{a}$ since the single requirement $\nabla_{\mathbf{a}} Q = \mathbf{0}$ provides the necessary conditions for the minimum or maximum.

Expressions for the gradient in some commonly occurring cases are as follows. Suppose that $Q = \mathbf{b}^T\mathbf{a}$, where $\mathbf{b}$ is also a real vector. By writing out this inner product in terms of the vector components, and forming the partial derivatives in (A.1) it is easy to show that

$$\nabla_{\mathbf{a}}(\mathbf{b}^T\mathbf{a}) = \nabla_{\mathbf{a}}(\mathbf{a}^T\mathbf{b}) = \mathbf{b} \qquad \text{(A.2)}$$

(real vector $\mathbf{a}$)

By following a similar procedure it can also be shown that if Q is the quadratic product $\mathbf{a}^T\mathbf{B}\mathbf{a}$, then the gradient is

$$\nabla_{\mathbf{a}}(\mathbf{a}^T\mathbf{B}\mathbf{a}) = (\mathbf{B} + \mathbf{B}^T)\mathbf{a}$$

(real vector $\mathbf{a}$)

(You may want to prove this as an exercise.) When $\mathbf{B}$ is symmetric this reduces to

$$\nabla_{\mathbf{a}}(\mathbf{a}^T\mathbf{B}\mathbf{a}) = 2\mathbf{B}\mathbf{a} \qquad \text{(A.3)}$$

(real vector $\mathbf{a}$, symmetric matrix $\mathbf{B}$)

These results for real vectors are summarized in Table A.1.

TABLE A.1 GRADIENT RELATIONS FOR REAL VECTORS†

Quantity Q	$\mathbf{a}^T\mathbf{b}$	$\mathbf{b}^T\mathbf{a}$	$\mathbf{a}^T\mathbf{B}\mathbf{a}$
Gradient $\nabla_{\mathbf{a}}Q$	$\mathbf{b}$	$\mathbf{b}$	$2\mathbf{B}\mathbf{a}$

†Matrix **B** is symmetric.

A.1.2 Gradient with Respect to a Complex Vector Parameter

Defining the gradient for the case of complex vectors but can be somewhat troublesome mathematically. The basic difficulty arises because most of the quantities of interest depend on the complex parameter **a** in such a way that the resulting function is *not analytic*. In particular, many of the quantities depend on the complex conjugate of **a**, which is not an analytic function of **a**. Therefore, the derivative with respect to any of the components of the complex vector *does not exist.*[1]

The problem can be solved rather neatly and consistently however, thanks to a rigorous and lucid treatment of the subject by Brandwood [2]. First consider the case of a quantity Q which depends on a scalar parameter

$$a = a_{\mathrm{r}} + \jmath a_{\mathrm{i}}$$

and its conjugate

$$a^* = a_{\mathrm{r}} - \jmath a_{\mathrm{i}}$$

Let us assume that if a and a^* are treated as *independent parameters*, then the quantity Q is analytic in both a and a^*. (This is usually the case.) Therefore, the partial derivatives

$$\frac{\partial Q}{\partial a} \quad \text{and} \quad \frac{\partial Q}{\partial a^*}$$

are defined. Further assume that when the quantity Q is expressed in terms of the real and imaginary parts of a, the partial derivatives[2]

$$\frac{\partial Q}{\partial a_{\mathrm{r}}} \quad \text{and} \quad \frac{\partial Q}{\partial a_{\mathrm{i}}}$$

exist. In this case it is possible to show [2], by application of the chain rule, that these partial derivatives are related by

$$\frac{\partial Q}{\partial a} = \frac{1}{2}\left(\frac{\partial Q}{\partial a_{\mathrm{r}}} - \jmath\frac{\partial Q}{\partial a_{\mathrm{i}}}\right)$$

[1]See, e.g., [1] for a discussion of analytic functions of a complex variable and definition of the derivative.

[2]If Q is regarded as a *function* of a and a^*, then it should be written as a different function (say, Q') when we are speaking about its dependence on a_{r} and a_{i}. To keep the discussion simpler we avoid this distinction and refer to Q as a "quantity" that can be expressed in terms of either set of variables.

and

$$\frac{\partial Q}{\partial a^*} = \frac{1}{2}\left(\frac{\partial Q}{\partial a_\mathrm{r}} + \jmath \frac{\partial Q}{\partial a_\mathrm{i}}\right)$$

We *define* the scalar version of the complex gradient as these partial derivatives and write

$$\nabla_a Q \stackrel{\mathrm{def}}{=} \frac{1}{2}\left(\frac{\partial Q}{\partial a_\mathrm{r}} - \jmath \frac{\partial Q}{\partial a_\mathrm{i}}\right) \quad (a)$$

$$\nabla_{a^*} Q \stackrel{\mathrm{def}}{=} \frac{1}{2}\left(\frac{\partial Q}{\partial a_\mathrm{r}} + \jmath \frac{\partial Q}{\partial a_\mathrm{i}}\right) \quad (b) \qquad (A.4)$$

The definition (A.4) has the satisfying property that when Q does not depend on a^* and is an analytic function of a, the total derivative of the function exists and is equal to $\nabla_a Q$. It follows from (A.4) that if it is desired to find a maximum or minimum of Q with respect to the parameter a that is not at a boundary, then either

$$\nabla_a Q = 0 \qquad (A.5)$$

or

$$\nabla_{a^*} Q = 0 \qquad (A.6)$$

provides a necessary condition. This is clear because both of these last two equations require that

$$\frac{\partial Q}{\partial a_\mathrm{r}} = \frac{\partial Q}{\partial a_\mathrm{i}} = 0$$

The foregoing results can be generalized to the case of a scalar quantity $\mathcal{Q}$ that depends on a vector parameter

$$\mathbf{a} = \mathbf{a_r} + \jmath \mathbf{a_i}$$

To do so we define

$$\nabla_\mathbf{a} \mathcal{Q} \stackrel{\mathrm{def}}{=} \tfrac{1}{2}\left(\nabla_{\mathbf{a_r}} \mathcal{Q} - \jmath \nabla_{\mathbf{a_i}} \mathcal{Q}\right) \qquad (A.7)$$

(complex vector **a**)

and

$$\nabla_{\mathbf{a}^*} \mathcal{Q} \stackrel{\mathrm{def}}{=} \tfrac{1}{2}\left(\nabla_{\mathbf{a_r}} \mathcal{Q} + \jmath \nabla_{\mathbf{a_i}} \mathcal{Q}\right) \qquad (A.8)$$

(complex vector **a**)

where the gradients with respect to the real vectors $\mathbf{a_r}$ and $\mathbf{a_i}$ are as in (A.1). This definition clearly suits the purpose of minimizing or maximizing the quantity $\mathcal{Q}$, since the single statement

$$\nabla_\mathbf{a} \mathcal{Q} = \mathbf{0} \qquad (A.9)$$

or

$$\nabla_{\mathbf{a}^*} \mathcal{Q} = \mathbf{0} \qquad (A.10)$$

requires the partial derivatives with respect to all real and imaginary components of the vector to be zero.[3]

With the definition (A.7) of the gradient for complex vectors it follows that

$$\begin{aligned} \nabla_{\mathbf{a}}(\mathbf{b}^{*T}\mathbf{a}) &= \mathbf{b}^* \quad (a) \\ \nabla_{\mathbf{a}}(\mathbf{a}^{*T}\mathbf{b}) &= \mathbf{0} \quad (b) \\ & \quad \text{(complex vector } \mathbf{a}) \end{aligned} \tag{A.11}$$

and similarly from (A.8) it follows that

$$\begin{aligned} \nabla_{\mathbf{a}^*}(\mathbf{b}^{*T}\mathbf{a}) &= \mathbf{0} \quad (a) \\ \nabla_{\mathbf{a}^*}(\mathbf{a}^{*T}\mathbf{b}) &= \mathbf{b} \quad (b) \\ & \quad \text{(complex vector } \mathbf{a}) \end{aligned} \tag{A.12}$$

As before, these results can be shown by expressing the inner product in terms of its vector components (real and imaginary parts) and using the definitions (A.7), (A.8) and (A.1). Alternatively, if $\mathbf{b} = \mathbf{b_r} + \jmath\mathbf{b_i}$, then

$$\mathbf{b}^{*T}\mathbf{a} = (\mathbf{b_r}^T\mathbf{a_r} + \mathbf{b_i}^T\mathbf{a_i}) + \jmath(\mathbf{b_r}^T\mathbf{a_i} - \mathbf{b_i}^T\mathbf{a_r})$$

$$\mathbf{a}^{*T}\mathbf{b} = (\mathbf{b_r}^T\mathbf{a_r} + \mathbf{b_i}^T\mathbf{a_i}) - \jmath(\mathbf{b_r}^T\mathbf{a_i} - \mathbf{b_i}^T\mathbf{a_r})$$

and (A.11) and (A.12) follow by applying (A.2), (A.7), and (A.8). Since the quantities $\mathbf{b}^{*T}\mathbf{a}$ and $\mathbf{a}^{*T}\mathbf{b}$ are complex conjugates of one another, the differences in their gradients exhibited by (A.11) and (A.12) are really not surprising.

Let us now consider the complex gradient of a Hermitian quadratic form $\mathcal{Q} = \mathbf{a}^{*T}\mathbf{B}\mathbf{a}$. The results follow easily from (A.11) and (A.12) if you remember that in defining the complex gradient the two quantities $\mathbf{a}$ and $\mathbf{a}^*$ are treated as independent variables. Thus, if $\mathbf{a}^{*T}\mathbf{B}$ is treated as a constant vector, it follows from (A.11)(a) that

$$\nabla_{\mathbf{a}}\left(\mathbf{a}^{*T}\mathbf{B}\mathbf{a}\right) = \mathbf{B}^T\mathbf{a}^* = (\mathbf{B}\mathbf{a})^* \tag{A.13}$$

where the last equality holds because $\mathbf{B}$ is Hermitian. Similarly, by treating $\mathbf{Ba}$ as a constant, it follows from (A.12)(b) that

$$\nabla_{\mathbf{a}^*}\left(\mathbf{a}^{*T}\mathbf{B}\mathbf{a}\right) = \mathbf{B}\mathbf{a} \tag{A.14}$$

The results for complex vectors are summarized in Table A.2. Comparably useful results for scalar parameters are given in Table A.3. Although either of the operators $\nabla_{\mathbf{a}}$ or $\nabla_{\mathbf{a}^*}$ can be used to obtain necessary conditions for the minimization or maximization of a quantity, it turns out that the gradient operator $\nabla_{\mathbf{a}^*}$ is generally most convenient to use. Therefore, the latter operator has been used throughout the book.

[3]Other definitions for the complex gradient, such as

$$\nabla_{\mathbf{a}}\mathcal{Q} = \nabla_{\mathbf{a_r}}\mathcal{Q} + \jmath\nabla_{\mathbf{a_i}}\mathcal{Q}$$

have been used that serve the purpose of minimization and maximization as well as (A.7) (see [3], [4]). The definition (A.7) is the one given by Miller [5] and especially by Brandwood [2], who placed the topic on firm mathematical ground.

TABLE A.2 GRADIENT RELATIONS FOR COMPLEX VECTORS†

Quantity Q	$\mathbf{a}^{*T}\mathbf{b}$	$\mathbf{b}^{*T}\mathbf{a}$	$\mathbf{a}^{*T}\mathbf{B}\mathbf{a}$
Gradient $\nabla_{\mathbf{a}} Q$	$\mathbf{0}$	$\mathbf{b}^*$	$(\mathbf{B}\mathbf{a})^*$
Gradient $\nabla_{\mathbf{a}^*} Q$	$\mathbf{b}$	$\mathbf{0}$	$\mathbf{B}\mathbf{a}$

†Matrix **B** is Hermitian.

TABLE A.3 RELATIONS FOR SCALAR FORM OF THE COMPLEX GRADIENT

Quantity Q	a^*b	ab	$\lvert a\rvert^2 = aa^*$
Gradient $\nabla_a Q$	0	b	a^*
Gradient $\nabla_{a^*} Q$	b	0	a

A.1.3 Testing for a Minimum or Maximum

When a quantity Q depends on a complex scalar parameter a, the conditions (A.5) and (A.6) are *necessary* but not sufficient for a minimum or maximum of Q (i.e., they determine stationary points). This section develops conditions to check for an actual minimum or maximum.

Since Q actually depends on two real variables a_r and a_i, the following results obtain [6, p. 213]. First, at a minimum or maximum we must have[4]

$$\left(\frac{\partial^2 Q}{\partial a_\mathrm{r}^2}\right)\cdot\left(\frac{\partial^2 Q}{\partial a_\mathrm{i}^2}\right)-\left(\frac{\partial^2 Q}{\partial a_\mathrm{r}\partial a_\mathrm{i}}\right)^2>0 \tag{A.15}$$

If this holds, then the second derivatives $\frac{\partial^2 Q}{\partial a_\mathrm{r}^2}$ and $\frac{\partial^2 Q}{\partial a_\mathrm{i}^2}$ must have the same sign. The condition

$$\frac{\partial^2 Q}{\partial a_\mathrm{r}^2}>0 \quad \left(\text{and} \quad \frac{\partial^2 Q}{\partial a_\mathrm{i}^2}>0\right) \tag{A.16}$$

is then sufficient for a *minimum* while the condition

$$\frac{\partial^2 Q}{\partial a_\mathrm{r}^2}<0 \quad \left(\text{and} \quad \frac{\partial^2 Q}{\partial a_\mathrm{i}^2}<0\right) \tag{A.17}$$

determines that the point is a *maximum.*

Since Q is not usually written in terms of real and imaginary parts of the parameter, it is desirable to express these conditions in terms of a and a^*. To do this first observe that the two scalar equations (A.4) can be written as a single matrix equation

$$\begin{bmatrix}\nabla_a Q\\ \nabla_{a^*} Q\end{bmatrix}=\begin{bmatrix}\frac{1}{2} & -\jmath\frac{1}{2}\\ \frac{1}{2} & \jmath\frac{1}{2}\end{bmatrix}\begin{bmatrix}\frac{\partial Q}{\partial a_\mathrm{r}}\\ \frac{\partial Q}{\partial a_\mathrm{i}}\end{bmatrix} \tag{A.18}$$

[4]If this expression is < 0, then the point is neither a minimum nor a maximum; if it is equal to zero, then no conclusion can be drawn [6].

Now for any two complex parameters a and b, define the operator

$$\nabla^2_{ab} Q \overset{\text{def}}{=} \nabla_a (\nabla_b Q) \tag{A.19}$$

Then from these last two equations it is not hard to show that

$$\begin{bmatrix} \nabla^2_{aa} Q & \nabla^2_{aa^*} Q \\ \nabla^2_{a^*a} Q & \nabla^2_{a^*a^*} Q \end{bmatrix} = \begin{bmatrix} \frac{1}{2} & -\jmath\frac{1}{2} \\ \frac{1}{2} & \jmath\frac{1}{2} \end{bmatrix} \begin{bmatrix} \frac{\partial^2 Q}{\partial a_r^2} & \frac{\partial^2 Q}{\partial a_r \partial a_i} \\ \frac{\partial^2 Q}{\partial a_i \partial a_r} & \frac{\partial^2 Q}{\partial a_i^2} \end{bmatrix} \begin{bmatrix} \frac{1}{2} & \frac{1}{2} \\ -\jmath\frac{1}{2} & \jmath\frac{1}{2} \end{bmatrix} \tag{A.20}$$

from which it follows that

$$\nabla^2_{aa^*} Q = \frac{1}{4} \left[\frac{\partial^2 Q}{\partial a_r^2} + \frac{\partial^2 Q}{\partial a_i^2} \right] \tag{A.21}$$

To find the test for a minimum or maximum, first take determinants of (A.20) and apply (A.15) to obtain

$$\begin{vmatrix} \nabla^2_{aa} Q & \nabla^2_{aa^*} Q \\ \nabla^2_{a^*a} Q & \nabla^2_{a^*a^*} Q \end{vmatrix} = \begin{vmatrix} \frac{1}{2} & -\jmath\frac{1}{2} \\ \frac{1}{2} & \jmath\frac{1}{2} \end{vmatrix} \begin{vmatrix} \frac{\partial^2 Q}{\partial a_r^2} & \frac{\partial^2 Q}{\partial a_r \partial a_i} \\ \frac{\partial^2 Q}{\partial a_i \partial a_r} & \frac{\partial^2 Q}{\partial a_i^2} \end{vmatrix} \begin{vmatrix} \frac{1}{2} & \frac{1}{2} \\ -\jmath\frac{1}{2} & \jmath\frac{1}{2} \end{vmatrix}$$

$$= -\frac{1}{4} \begin{vmatrix} \frac{\partial^2 Q}{\partial a_r^2} & \frac{\partial^2 Q}{\partial a_r \partial a_i} \\ \frac{\partial^2 Q}{\partial a_i \partial a_r} & \frac{\partial^2 Q}{\partial a_i^2} \end{vmatrix} < 0$$

The inequality follows because since $\frac{\partial^2 Q}{\partial a_i \partial a_r} = \frac{\partial^2 Q}{\partial a_r \partial a_i}$, the determinant on the right is just equal to (A.15). Therefore, at a minimum or maximum it is required that

$$\left(\nabla^2_{aa} Q\right) \cdot \left(\nabla^2_{a^*a^*} Q\right) - \left(\nabla^2_{aa^*} Q\right)^2 < 0 \tag{A.22}$$

[note the change in the sense of the inequality compared to (A.15)]. Further, when this last condition holds, it follows from (A.16), (A.17), and (A.21) that

$$\nabla^2_{aa^*} Q \begin{cases} > 0 & \text{for a minimum} \\ < 0 & \text{for a maximum} \end{cases} \tag{A.23}$$

A.2 CONSTRAINED MINIMIZATION OR MAXIMIZATION

Frequently, minimization or maximization of a scalar quantity is subject to one or more constraints. The method of Lagrange multipliers [7, 8] applies here. Let us begin by reviewing this procedure for a simple scalar case involving functions of two real variables.

Suppose it is desired to optimize a certain function $Q(a_1, a_2)$ which depends on real scalar parameters a_1 and a_2, subject to a constraint of the form

$$C(a_1, a_2) = 0$$

A conceivable approach is to solve the last equation for one of the parameters, say

$$a_2 = g(a_1)$$

then substitute the result in Q and optimize the function $Q(a_1, g(a_1))$, which is now an unconstrained problem involving a single variable. While this procedure may be feasible in simple cases, in many cases this straightforward approach is either impossible or impractical. Optimization using Lagrange multipliers provides a much easier solution to the constrained optimization problem.

To solve the constrained optimization problem a Lagrange multiplier μ is introduced and the new function

$$L(a_1, a_2, \mu) = Q(a_1, a_2) + \mu C(a_1, a_2)$$

known as the *Lagrangian* is formed. The purpose of these manipulations is to convert a constrained optimization problem in two variables to an *unconstrained* optimization problem in three variables that is somewhat easier to solve. In particular, the necessary conditions for optimizing $L(a_1, a_2, \mu)$ with respect to the three parameters are

$$\frac{\partial L}{\partial a_1} = \frac{\partial Q}{\partial a_1} + \mu \frac{\partial C}{\partial a_1} = 0$$

$$\frac{\partial L}{\partial a_2} = \frac{\partial Q}{\partial a_2} + \mu \frac{\partial C}{\partial a_2} = 0$$

and

$$\frac{\partial L}{\partial \mu} = C(a_1, a_2) = 0$$

The last equation is just the original constraint. Since the solution of this set of equations minimizes L, which is equal to Q plus the constraint, and simultaneously forces C to be zero, it is also the solution to the original constrained optimization problem.

Let us now generalize this problem. Replace the scalar parameters a_1 and a_2 by the (real) vector parameter $\mathbf{a}$, and assume that there are multiple constraints. In this case a Lagrange multiplier is introduced for every constraint, all of these constraints are added to the original equation, and the Lagrangian is optimized with respect to all of the parameters and all of the Lagrange multipliers. For example, if a quantity $\mathcal{Q}$ depending on the real vector parameter $\mathbf{a}$ is to be optimized subject to two constraints

$$\mathcal{C}_1(\mathbf{a}) = 0 \tag{A.24}$$

and

$$\mathcal{C}_2(\mathbf{a}) = 0 \tag{A.25}$$

then the Lagrangian is

$$\mathcal{L} = \mathcal{Q}(\mathbf{a}) + \mu_1 \mathcal{C}_1(\mathbf{a}) + \mu_2 \mathcal{C}_2(\mathbf{a}) \tag{A.26}$$

and the equations to be solved are

$$\nabla_{\mathbf{a}} \mathcal{L} = \nabla_{\mathbf{a}} \mathcal{Q}(\mathbf{a}) + \mu_1 \nabla_{\mathbf{a}} \mathcal{C}_1(\mathbf{a}) + \mu_2 \nabla_{\mathbf{a}} \mathcal{C}_2(\mathbf{a}) = \mathbf{0} \tag{A.27}$$

$$\frac{\partial \mathcal{L}}{\partial \mu_1} = \mathcal{C}_1(\mathbf{a}) = 0 \tag{A.28}$$

$$\frac{\partial \mathcal{L}}{\partial \mu_2} = \mathcal{C}_2(\mathbf{a}) = 0 \tag{A.29}$$

where, in (A.27), the gradient of the real vector parameter replaces the partial derivative.

Now consider the case of a *complex* vector parameter **a**. As usual, we cannot proceed blindly to generalize the results for the real case. Instead, the conditions have to be developed by carefully examining the meaning of the problem in terms of *real* functions. To begin, assume that it is desired to minimize or maximize the scalar quantity $\mathcal{Q}$ with respect to the *complex* vector parameter **a** subject to the constraint

$$\mathcal{C}(\mathbf{a}) = 0 \tag{A.30}$$

where $\mathcal{C}(\mathbf{a})$ may in general be complex valued. The constraint (A.30) actually represents the pair of real constraints

$$\begin{aligned} \mathcal{C}_r(\mathbf{a}) &= 0 \\ \mathcal{C}_i(\mathbf{a}) &= 0 \end{aligned} \tag{A.31}$$

where $\mathcal{C}_r(\mathbf{a})$ and $\mathcal{C}_i(\mathbf{a})$ represent the real and imaginary parts of $\mathcal{C}(\mathbf{a})$, respectively. Therefore, to proceed using Lagrange multipliers, we form the complex Lagrangian

$$\mathcal{L} = \mathcal{Q}(\mathbf{a}) + \mu_1 \mathcal{C}_r(\mathbf{a}) + \mu_2 \mathcal{C}_i(\mathbf{a}) \tag{A.32}$$

and set the *complex gradient* $\nabla_{\mathbf{a}}\mathcal{L}$ or $\nabla_{\mathbf{a}^*}\mathcal{L}$ to zero.

The Lagrangian (A.32) can be written in a more compact form using a single complex parameter $\lambda = \lambda_r + \jmath\lambda_i$ as

$$\boxed{\mathcal{L} = \mathcal{Q}(\mathbf{a}) + \lambda \mathcal{C}(\mathbf{a}) + \lambda^* \mathcal{C}^*(\mathbf{a})} \tag{A.33}$$

The reason this can be done is that the two terms on the right of (A.33) can be written as

$$\lambda \mathcal{C}(\mathbf{a}) + \lambda^* \mathcal{C}^*(\mathbf{a}) = 2\text{Re}\, \lambda \mathcal{C}(\mathbf{a}) = 2\lambda_r \mathcal{C}_r(\mathbf{a}) - 2\lambda_i \mathcal{C}_i(\mathbf{a})$$

Therefore, (A.33) is equivalent to (A.32) with $\mu_1 = 2\lambda_r$ and $\mu_2 = -2\lambda_i$.

If the gradient operator $\nabla_{\mathbf{a}^*}$ is used, performing the optimization requires that

$$\nabla_{\mathbf{a}^*}\mathcal{L} = \nabla_{\mathbf{a}^*}\mathcal{Q} + \lambda \nabla_{\mathbf{a}^*}\mathcal{C} + \lambda^* \nabla_{\mathbf{a}^*}\mathcal{C}^* = \mathbf{0} \tag{A.34}$$

and

$$\nabla_\lambda \mathcal{L} = \mathcal{C}(\mathbf{a}) = 0 \tag{A.35}$$

The last equation simply regenerates the constraint.

In many cases of interest one of the last two terms in (A.34) is identically zero. Also, if $\mathcal{C}(\mathbf{a})$ is purely *real*, one of the last two terms in (A.33) and (A.34) is redundant and can be dropped. This is equivalent to applying the constraint with λ a single *real* Lagrange multiplier.

When multiple constraints are present, the additional constraints are simply added on to the Lagrangian, either as conjugate pairs as in (A.33) or singly if the constraint is real. Performing the optimization then requires taking the complex gradient with respect to the vector parameter and all the additional Lagrange multipliers to regenerate the constraints. If the quantity to be maximized depends on more than a single vector parameter $(\mathbf{a}_1, \mathbf{a}_2, \ldots)$, then the gradient with respect to each vector parameter must be taken.

To illustrate the ideas developed in this section, consider the following typical optimization problem. Suppose it is desired to maximize the Hermitian quadratic form

$$\mathcal{Q} = \mathbf{a}^{*T}\mathbf{B}\mathbf{a}$$

subject to the constraint

$$\mathbf{a}^{*T}\mathbf{a} = 1$$

which can be written as

$$\mathcal{C}(\mathbf{a}) = 1 - \mathbf{a}^{*T}\mathbf{a} = 0$$

Since the constraint is purely real, the Lagrangian can be written as

$$\mathcal{L} = \mathbf{a}^{*T}\mathbf{B}\mathbf{a} + \lambda(1 - \mathbf{a}^{*T}\mathbf{a})$$

where λ is a *real* Lagrange multiplier. Taking the complex gradient and using the results in Table A.2 then yields

$$\nabla_{\mathbf{a}^*}\mathcal{L} = \mathbf{B}\mathbf{a} - \lambda\mathbf{a} = \mathbf{0}$$

or

$$\mathbf{B}\mathbf{a} = \lambda\mathbf{a}$$

This shows that $\mathbf{a}$ must be an eigenvector of $\mathbf{B}$ and that λ is the corresponding eigenvalue. Substituting this last result in the expression for $\mathcal{Q}$ and recognizing the constraint $\mathbf{a}^{*T}\mathbf{a} = 1$ leads to

$$\mathcal{Q} = \mathbf{a}^{*T}(\mathbf{B}\mathbf{a}) = \mathbf{a}^{*T}(\lambda\mathbf{a}) = \lambda$$

To maximize $\mathcal{Q}$, λ should be the *largest* eigenvalue; therefore, the desired maximizing parameter $\mathbf{a}$ is the eigenvector corresponding to the largest eigenvalue of $\mathbf{B}$.

REFERENCES

1. Ruel V. Churchill and James Ward Brown. *Complex Variables and Applications*, 4th ed. McGraw-Hill Book Company, New York, 1984.
2. D. H. Brandwood. A complex gradient operator and its application in adaptive array theory. *IEE Proceedings*, 130(1):11–16, February 1983. Parts F and H.
3. Steven M. Kay. *Modern Spectral Estimation: Theory and Application*. Prentice Hall, Inc., Englewood Cliffs, New Jersey, 1988.

4. Simon Haykin. *Adaptive Filter Theory*, 2nd ed. Prentice Hall, Inc., Englewood Cliffs, New Jersey, 1991.
5. Kenneth S. Miller. *Complex Stochastic Processes*. Addison-Wesley, Reading, Massachusetts, 1974.
6. Angus E. Taylor and W. Robert Mann. *Advanced Calculus*, 3rd ed. McGraw-Hill Book Company, New York, 1983.
7. Gilbert Strang. *Introduction to Applied Mathematics*. Wellesley Cambridge Press, Wellesley, Massachusetts, 1986.
8. Francis B. Hildebrand. *Methods of Applied Mathematics*, 2nd ed. Prentice Hall, Inc., Englewood Cliffs, New Jersey, 1965.

Complex Representation of Bandpass Signals and Systems

Bandpass signals and systems represent an important class of entities in communications, and in radar and sonar. The description here attempts to develop an intuitive understanding for the processes involved rather than mathematical sophistication, and roughly follows the presentation in [1, Appendix] with some changes in notation.

B.1 COMPLEX REPRESENTATION OF BANDPASS SIGNALS

Bandpass signals are continuous real signals whose spectrum is nonzero only within a region of $\pm W$ hertz around some central (carrier) frequency $f_o > W$. The spectrum of a typical bandpass signal is shown in Fig. B.1(a). Although no signals encountered in the real world are strictly bandlimited, many signals have such low energy outside a band that the representation of Fig. B.1(a) is a suitable model. "Narrowband" signals are a special class of bandpass signals such that $W << f_o$.

Any bandpass signal x_b can be represented in the canonical form

$$x_b(t) = x_p(t) \cos 2\pi f_o t - x_q(t) \sin 2\pi f_o t \tag{B.1}$$

where $x_p(t)$ and $x_q(t)$ are low-pass signals bandlimited to $\pm W$. In common parlance x_p is referred to as the *in-phase* component and x_q is referred to as the *quadrature* component.

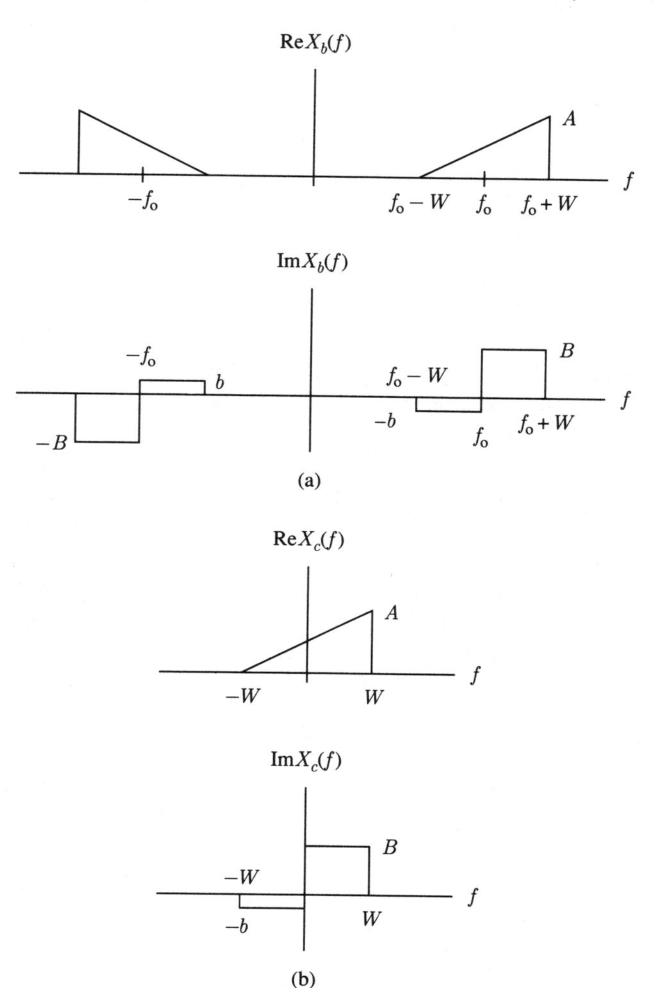

Figure B.1 Spectrum of a typical bandpass signal. (a) Bandpass signal. (b) Corresponding complex low-pass signal.

Together they are referred to as *quadrature components* or *quadrature representation* of the bandpass signal. In addition, the representation of the signal as low-pass components with frequencies limited to $\pm W$ is referred to as representation of the signal "at baseband."

It is clear that if x_p and x_q are low-pass signals bandlimited to $\pm W$, then x_b is bandlimited to $f_o \pm W$. It is less evident, however, that every bandlimited signal has a representation of the form (B.1) with unique quadrature components x_p and x_q. It will be shown that this is the case. Further, given knowledge of the center frequency f_o, the complex signal defined by

$$x_c(t) = \tfrac{1}{2}\left(x_p(t) + \jmath x_q(t)\right) \tag{B.2}$$

is an equivalent representation of the bandpass signal in that $x_b(t)$ can be obtained from $x_c(t)$, and vice versa. The signal $2x_c(t)$ is commonly known as the *complex envelope* of the original bandpass signal.[1] It will be shown that the spectrum $X_c(f)$ of the complex signal is just the positive part of the spectrum of the bandpass signal shifted down to the origin [see Fig. B.1(b)]. This can be written as

$$X_c(f) = X_b^+(f + f_o) \tag{B.3}$$

where $X_b^+(f)$ is defined as[2]

$$X_b^+(f) = \begin{cases} X_b(f) & \text{if } f > 0 \\ 0 & \text{if } f \leq 0 \end{cases} \tag{B.4}$$

Note that the real and imaginary parts of the spectrum of the complex signal are not necessarily even and odd functions of frequency because the corresponding time function $x_c(t)$ is complex [see Fig. B.1(b)].

To illustrate (but not prove) the representation of a bandpass signal in terms of its quadrature components, the signal of Fig. B.1(a) will be decomposed into its components x_p and x_q and then reconstructed according to (B.1). The quadrature components of the bandpass signal can be obtained (as will be shown later) by multiplying the signal by $2\cos 2\pi f_o t$ and $-2\sin 2\pi f_o t$ and low-pass filtering. This procedure, known as quadrature demodulation or "mixing", is shown in Fig. B.2 where the ideal low-pass filter has the frequency response shown in Fig. B.3. (You can verify the decomposition by representing the cosine and sine as a sum of two complex exponentials and carrying out the necessary convolutions with impulses in the frequency domain.) The bandpass signal can then be reconstructed as shown in Fig. B.4. The steps in the reconstruction are shown in Fig. B.5.

Given the spectra of the quadrature components, it follows from (B.2) that the spectrum of the complex signal can be found from

$$X_c(f) = \tfrac{1}{2}\left(X_p(f) + \jmath X_q(f)\right) \tag{B.5}$$

[Again you can verify that if the spectra of Fig. B.2 are combined according to (B.5), the result is the spectrum of Fig. B.1(b).] The result is the positive part of the bandpass signal spectrum shifted to the origin.

In order to *prove* the claims and observations made in the preceding discussion, begin by observing that the original bandpass signal (B.1) can be written as

$$\begin{aligned} x_b(t) &= \text{Re}\left[(x_p(t) + \jmath x_q(t))(\cos 2\pi f_o t + \jmath \sin 2\pi f_o t)\right] \\ &= 2\text{Re}\left[x_c(t)e^{\jmath 2\pi f_o t}\right] \end{aligned} \tag{B.6}$$

[1]The complex envelope is conventionally defined by (B.2) without the factor of $\frac{1}{2}$. The factor $\frac{1}{2}$ is included here because it leads to a simpler and more natural description of the processing of bandpass signals by bandpass linear filters. We therefore refer to this representation as the "complex signal" instead of the complex envelope.

[2]The term $2X_b^+(f)$ is the spectrum of what is known as the *analytic signal* or the *pre-envelope* of the bandpass signal [2]. The corresponding signal in the time domain can be expressed in terms of the original bandpass signal and its Hilbert transform [2–4].

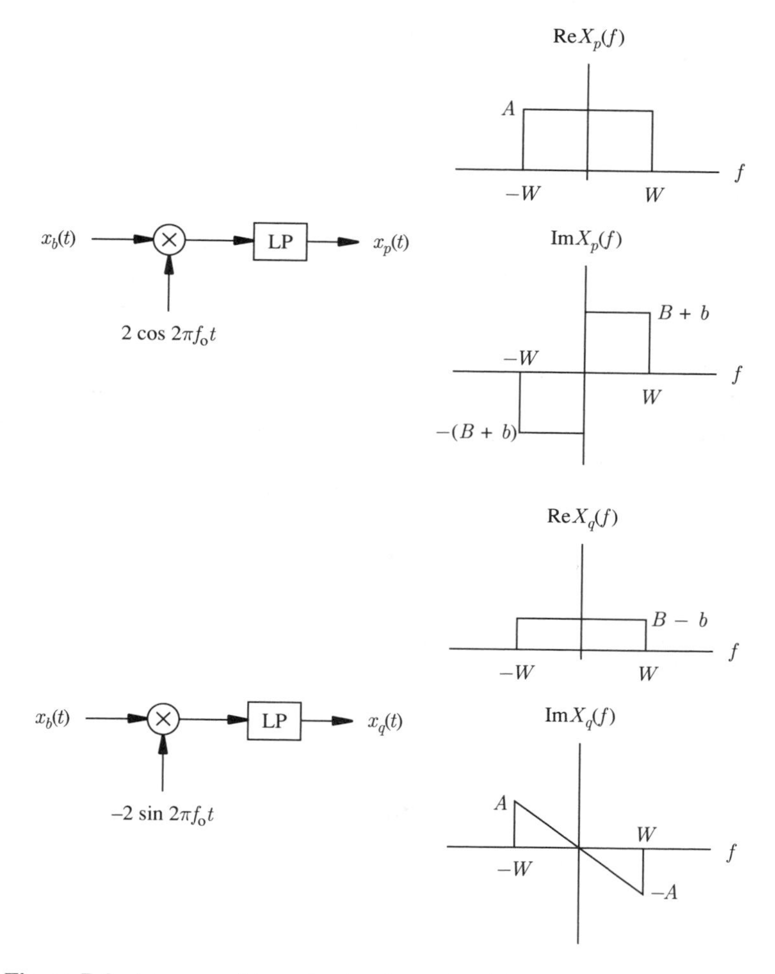

Figure B.2 Decomposition of a bandpass signal into its quadrature components.

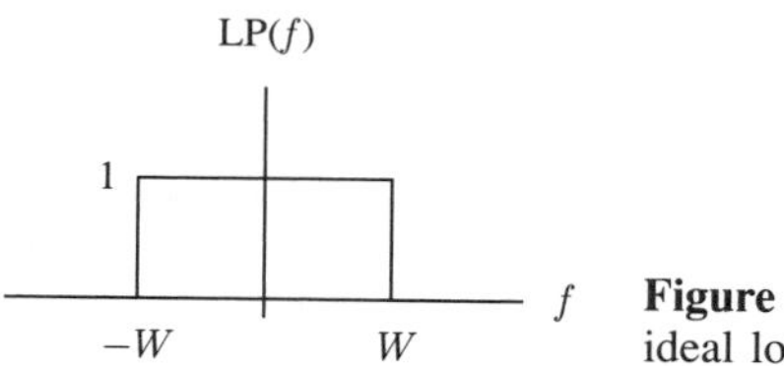

Figure B.3 Frequency response of ideal low-pass filter.

This shows that x_b can be written uniquely in terms of x_c. Notice from this form that x_b can also be expressed as

$$x_b(t) = 2|x_c(t)|\cos(2\pi f_o t + \angle x_c(t)) \tag{B.7}$$

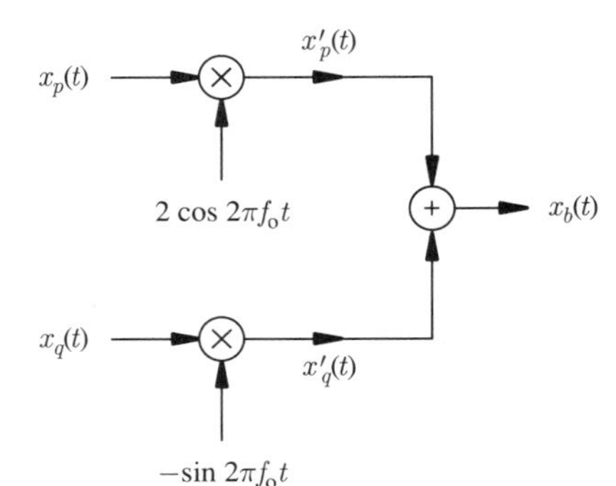

Figure B.4 Reconstruction of a bandpass signal from its quadrature components.

where the term $2|x_c(t)|$ is the (modulation) envelope of the bandpass signal and $\angle x_c(t)$ is the instantaneous phase. The time plot of a typical narrowband signal is shown in Fig. B.6.

It will now be shown that the complex signal can be derived by the system shown in Fig. B.7 and thus that the decomposition of x_b into x_p and x_q is unique. This system is equivalent to the two quadrature demodulators of Fig. B.2.

Begin by substituting the complex exponential expressions for the sine and cosine into (B.1) and writing it as

$$x_b(t) = \tfrac{1}{2}x_p(t)\left(e^{\jmath 2\pi f_o t} + e^{-\jmath 2\pi f_o t}\right) + \jmath\tfrac{1}{2}x_q(t)\left(e^{\jmath 2\pi f_o t} - e^{-\jmath 2\pi f_o t}\right) \tag{B.8}$$

Thus the spectrum of x_b has the form

$$X_b(f) = \tfrac{1}{2}\left(X_p(f+f_o) + X_p(f-f_o)\right) + \jmath\tfrac{1}{2}\left(X_q(f+f_o) + X_q(f-f_o)\right) \tag{B.9}$$

The output of the multiplier in Fig. B.7 is thus

$$X'_b(f) = \tfrac{1}{2}\left(X_p(f) + X_p(f-2f_o)\right) + \jmath\tfrac{1}{2}\left(X_q(f) + X_q(f-2f_o)\right) \tag{B.10}$$

which after low-pass filtering becomes

$$\tfrac{1}{2}X_p(f) + \jmath\tfrac{1}{2}X_q(f) = X_c(f) \tag{B.11}$$

Since the system of Fig. B.7 shifts $X_b(f)$ to the origin and low-pass filters it, it is evident that

$$X_c(f) = X_b^+(f+f_o)$$

and thus that $x_c(t)$ can be uniquely formed from $x_b(t)$, and vice versa.

B.2 LINEAR FILTERING OF BANDPASS SIGNALS

The main result of the preceding section is that any real bandpass signal can be represented by a complex signal whose real and imaginary parts are equal to one half of the quadrature components, and whose frequency spectrum is that of the original bandpass signal shifted

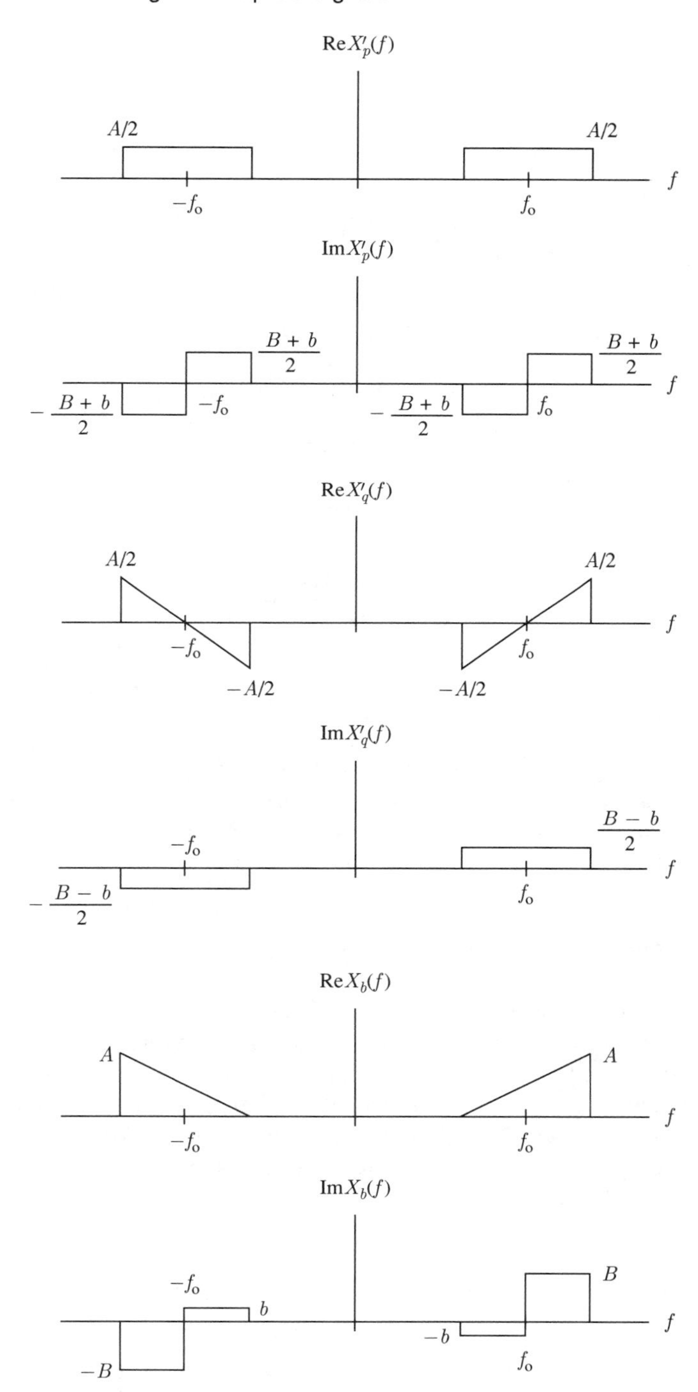

Figure B.5 Steps in the reconstruction of a bandpass signal.

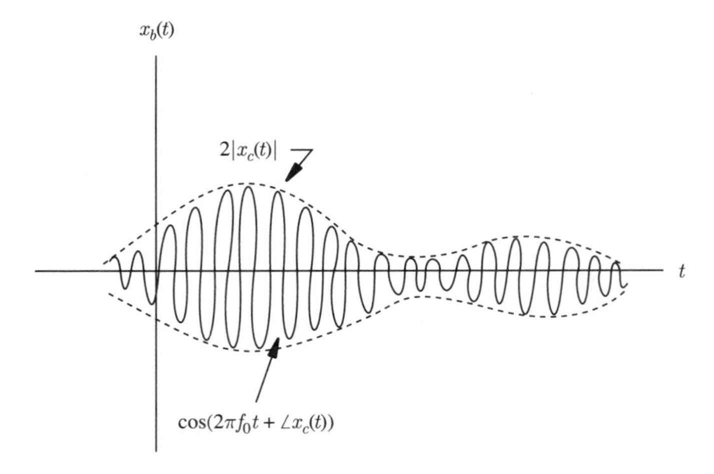

Figure B.6 Time plot of a bandpass signal.

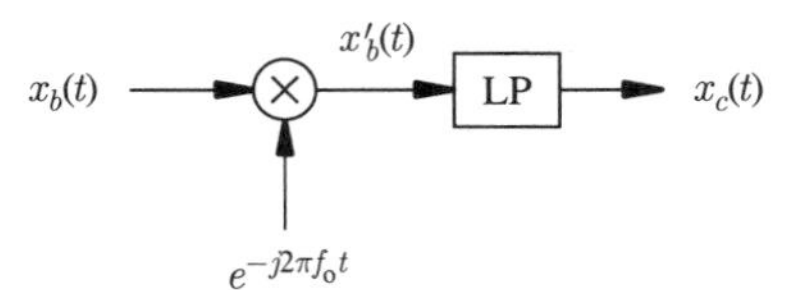

Figure B.7 System for deriving the low-pass complex signal corresponding to a bandpass signal.

down to the origin. Here we show that there is a similar representation for both time-invariant and time-varying linear systems. These facts permit the processing of bandpass signals by linear systems to be represented as equivalent operations on the complex signal at baseband.

B.2.1 Linear Time-Invariant Systems

It is not really surprising that linear time-invariant filtering operations on the bandpass signal can be represented as equivalent low-pass filtering operations applied to the complex signal. This is depicted in Fig. B.8. It is clear that if we shift both the spectrum of the input signal and the frequency response of the bandpass filter down to baseband, perform the linear filtering, and shift back up to f_o, the result is the same as if the filtering had been done at the carrier frequency. The frequency response of the complex low-pass filter is given by

$$H_c(f) = H_b^+(f + f_o) \tag{B.12}$$

and its impulse response is defined by the inverse Fourier transform

$$h_c(t) = \tfrac{1}{2}(h_p(t) + jh_q(t)) = \int_{-\infty}^{\infty} H_c(f)e^{j2\pi ft}df \tag{B.13}$$

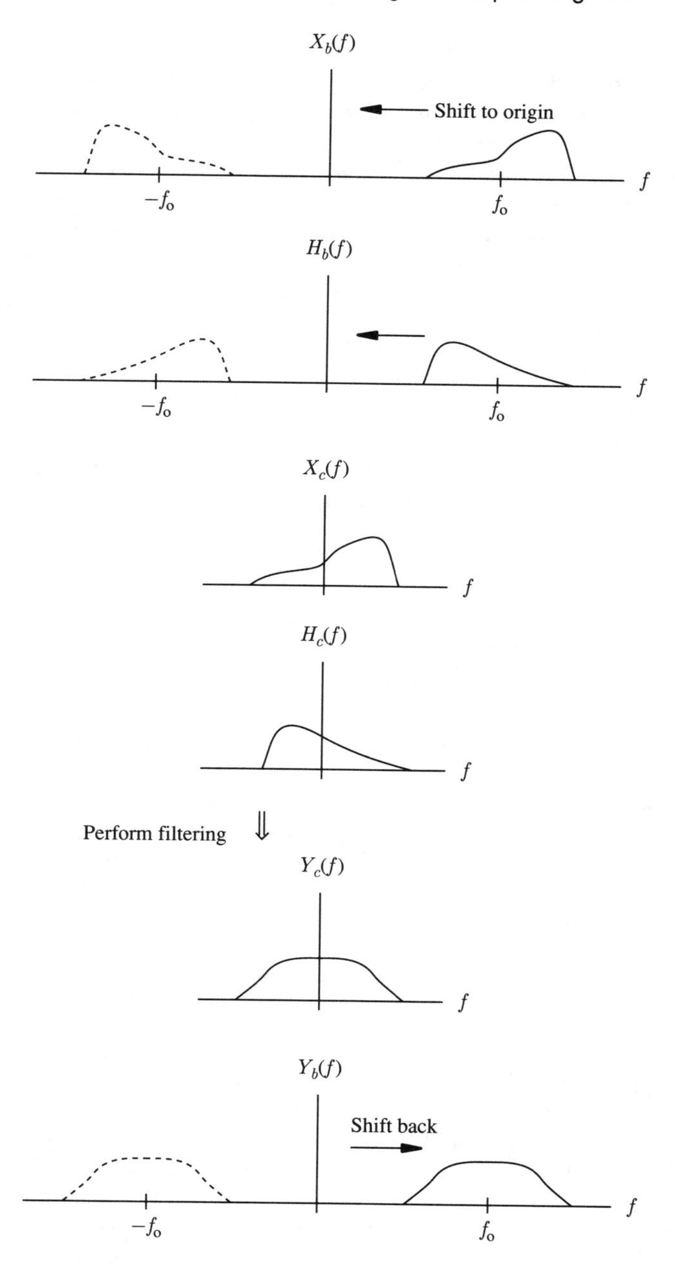

Figure B.8 Bandpass filtering as low-pass filtering of the complex signal.

This is just the complex signal representation of the impulse response of the bandpass filter. The operations on the complex signals can be written as the standard convolution

$$y_c(t) = \int_{-\infty}^{\infty} h_c(t - t')x_c(t')dt' \tag{B.14}$$

Since $x_c(t)$ is bandlimited, it can be sampled at a rate of $1/2W$ or higher and processed digitally. That is the reason for considering complex signals and systems in the study of digital signal processing. Figure B.9 depicts the overall processing scheme, although frequently either the demodulation or the reconstruction part of the processing is absent since the digital signal x may already be available, or the filtered signal y may be desired at baseband. (It is assumed that the A/D converters provide the necessary low-pass filtering, which is not shown explicitly.) Table B.1 shows the correspondences between the quadrature components, the complex signal at baseband, and the digital signal.

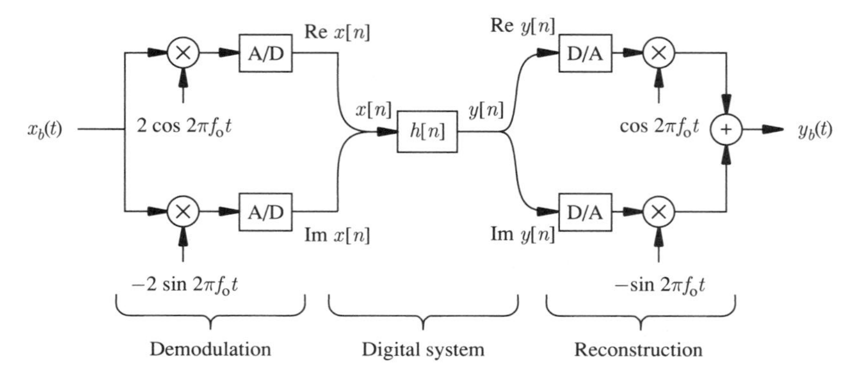

Figure B.9 Digital signal processing of a bandpass signal.

TABLE B.1 CORRESPONDING REPRESENTATIONS FOR BANDPASS SIGNALS AND SYSTEMS†

	Quadrature form	Complex signal	Digital signal
Input	$x_b(t) = \{x_p(t), x_q(t)\}$	$x_c(t) = \frac{1}{2}x_p(t) + \jmath\frac{1}{2}x_q(t)$	$x[n] = x_r[n] + \jmath x_i[n]$
Filter	$h_b(t) = \{h_p(t), h_q(t)\}$	$h_c(t) = \frac{1}{2}h_p(t) + \jmath\frac{1}{2}h_q(t)$	$h[n] = h_r[n] + \jmath h_i[n]$
Output	$y_b(t) = \{y_p(t), y_q(t)\}$	$y_c(t) = \frac{1}{2}y_p(t) + \jmath\frac{1}{2}y_q(t)$	$y[n] = y_r[n] + \jmath y_i[n]$

† In each representation y is the convolution of x and h.

B.2.2 Linear Time-Varying Systems

The preceding subsection provided a primarily graphical approach to show how bandpass filtering operations can be carried out by performing equivalent low-pass operations with the complex signals and systems at baseband. When time-varying filtering operations are involved the operations can likewise be carried out with complex signals and systems at baseband. However, the development of the results for time-varying systems needs to be carried out more carefully. (For example, it needs to be made clear what is meant by a time-varying bandpass system.) Since time-invariant systems are in fact a special case of time-varying systems, this development also serves to provide an analytic derivation of the results of the preceding subsection.

A time-varying linear transformation of a bandpass signal x_b can be represented by the integral

$$y_b(t) = \int_{-\infty}^{\infty} \mathrm{h}_b(t, t')x_b(t')dt' \tag{B.15}$$

where $\mathrm{h}_b(t, t')$ is the response at time t to an impulse at time t'. If the system is "slowly varying," then the response to an impulse at time t' is approximately dependent on the difference $t - t'$ for t in some interval around t': $[t' - \Upsilon, t' + \Upsilon]$. Let us refer the coordinate system to t' (see Fig. B.10) and define

$$\mathrm{h}'_b(u; t') = \mathrm{h}_b(u + t', t') \tag{B.16}$$

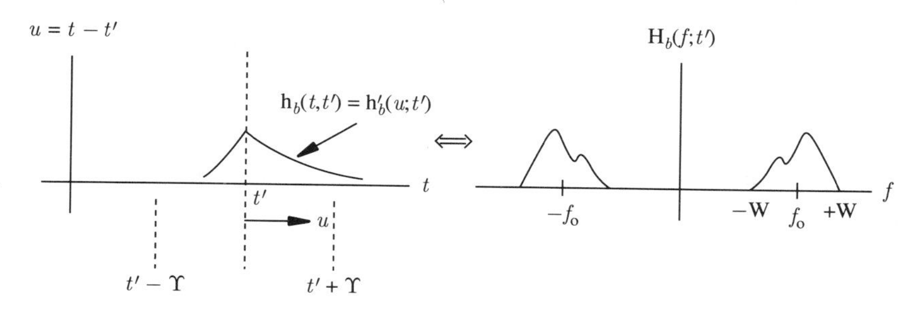

Figure B.10 Bandpass time-varying impulse response and corresponding short-time frequency response.

Then the "short-time frequency response" is defined as

$$\mathrm{H}_b(f; t') = \int_{-\Upsilon}^{\Upsilon} \mathrm{h}'_b(u; t')e^{-j2\pi f u}du \tag{B.17}$$

The system is defined to be a *time-varying bandpass linear system* if $\mathrm{H}_b(f; t')$ is bandlimited to $f_o \pm W$ for all values of t'. Figure B.10 is a depiction of such a system and its short-time frequency response.

Since H_b is bandpass, there exists a quadrature representation for the impulse response in the form

$$\mathrm{h}'_b(u; t') = \mathrm{h}'_p(u; t') \cos 2\pi f_o u - \mathrm{h}'_q(u; t') \sin 2\pi f_o u \tag{B.18}$$

Then if h_p and h_q are defined as

$$\begin{aligned} \mathrm{h}_p(t, t') &= \mathrm{h}'_p(t - t', t') \ (a) \\ \mathrm{h}_q(t, t') &= \mathrm{h}'_q(t - t', t') \ (b) \end{aligned} \tag{B.19}$$

and it is noted that $u = t - t'$, then (B.18) can be rewritten as

$$\mathrm{h}_b(t,t') = \mathrm{h}_p(t,t')\cos 2\pi f_o(t-t') - \mathrm{h}_q(t,t')\sin 2\pi f_o(t-t') \tag{B.20}$$

or

$$\mathrm{h}_b(t,t') = 2\mathrm{Re}\left[\mathrm{h}_c(t,t')e^{\jmath 2\pi f_o(t-t')}\right] \tag{B.21}$$

where

$$\mathrm{h}_c(t,t') = \tfrac{1}{2}\left(\mathrm{h}_p(t,t') + \jmath \mathrm{h}_q(t,t')\right) \tag{B.22}$$

To compute the output of the bandpass filter, substitute (B.6) and (B.21) into (B.15) to write

$$\begin{aligned} y_b(t) &= \int_{-\infty}^{\infty} \mathrm{h}_b(t,t')x_b(t')dt' \\ &= \int_{-\infty}^{\infty} \left(\mathrm{h}_c(t,t')e^{\jmath 2\pi f_o(t-t')} + \mathrm{h}_c^*(t,t')e^{-\jmath 2\pi f_o(t-t')}\right) \\ &\qquad \times \left(x_c(t')e^{\jmath 2\pi f_o t'} + x_c^*(t')e^{-\jmath 2\pi f_o t'}\right) dt' \end{aligned} \tag{B.23}$$

Now look at the "cross terms" in this expression, that is, the term

$$\int_{-\infty}^{\infty} \mathrm{h}_c(t,t')x_c^*(t')e^{\jmath 2\pi f_o(t-2t')}dt' = e^{\jmath 2\pi f_o t}\int_{-\infty}^{\infty} \mathrm{h}_c(t,t')x_c^*(t')e^{-\jmath 2\pi (2f_o)t'}dt'$$

and its complex conjugate. Since $\mathrm{h}_c(t,t')$ is slowly varying, let us assume that the Fourier transform of $\mathrm{h}_c(t,t')$ with respect to t' is zero for frequencies outside $\pm W$. Then notice that the final integral in the last equation is the Fourier transform of the product of time functions $\mathrm{h}_c(\cdot,t')$ and $x_c^*(t')$ evaluated at frequency $2f_o$. Since the spectrum of each of the time functions is limited to $\pm W$ the spectrum of their product is limited to $\pm 2W$. But since it was assumed that $2f_o > 2W$, the integral above is zero. The same is true, of course, for its complex conjugate.

In light of the discussion above, (B.23) reduces to

$$\begin{aligned} y_b(t) &= e^{\jmath 2\pi f_o t}\int_{-\infty}^{\infty} \mathrm{h}_c(t,t')x_c(t')dt' + e^{-\jmath 2\pi f_o t}\int_{-\infty}^{\infty} \mathrm{h}_c^*(t,t')x_c^*(t')dt' \\ &= 2\mathrm{Re}\left[e^{\jmath 2\pi f_o t}\int_{-\infty}^{\infty} \mathrm{h}_c(t,t')x_c(t')dt'\right] \end{aligned} \tag{B.24}$$

But since $y_b(t)$ has a representation

$$y_b(t) = 2\mathrm{Re}\left[y_c(t)e^{\jmath 2\pi f_o t}\right]$$

it follows from comparing these last two expressions that

$$y_c(t) = \int_{-\infty}^{\infty} \mathrm{h}_c(t,t')x_c(t')dt' \tag{B.25}$$

Thus the processing of a bandpass signal by a slowly time-varying bandpass linear filter can be represented by an equivalent operation on the complex signal at baseband.

Since the frequency response of the signals and the system involved are bandlimited to $\pm W$, the time functions can be sampled at a rate of $1/2W$ and processed digitally. This leads to the complex time-varying digital system representation

$$y[n] = \sum_{k=-\infty}^{\infty} h[n,k]x[k] \tag{B.26}$$

B.3 BANDPASS RANDOM PROCESSES

We now turn to the consideration of a bandpass random process and its representation and requirements. If both quadrature components $x_p(t)$ and $x_q(t)$ are low-pass random processes with power spectrum bandlimited to $\pm W$, then $x_b(t)$ is a bandpass random process of the form

$$x_b(t) = x_p(t)\cos 2\pi f_o t - x_q(t)\sin 2\pi f_o t \tag{B.27}$$

Although it is not neccesary to make any further assumptions for the analysis, it is frequently the case that $W << f_o$. The bandpass random process is in that case referred to as a *narrowband* random process.

Given prior knowledge of the carrier frequency, the real bandpass random process can conveniently be represented as the complex random process

$$x_c(t) = \tfrac{1}{2}x_p(t) + \jmath\tfrac{1}{2}x_q(t) \tag{B.28}$$

The real bandpass process $x_b(t)$ can then be written in terms of the complex random process as

$$x_b(t) = 2\text{Re}\left[x_c(t)e^{\jmath 2\pi f_o t}\right] = x_c(t)e^{\jmath 2\pi f_o t} + x_c^*(t)e^{-\jmath 2\pi f_o t} \tag{B.29}$$

Now consider the correlation function for the bandpass process

$$\begin{aligned} R_{x_b}^c(\tau) &= \mathcal{E}\{x_b(t)x_b(t-\tau)\} \qquad \text{(B.30)} \\ &= \mathcal{E}\left\{\left(x_c(t)e^{\jmath 2\pi f_o t} + x_c^*(t)e^{-\jmath 2\pi f_o t}\right)\left(x_c(t-\tau)e^{\jmath 2\pi f_o(t-\tau)} + x_c^*(t-\tau)e^{-\jmath 2\pi f_o(t-\tau)}\right)\right\} \end{aligned}$$

Expanding and collecting terms yields

$$\begin{aligned} R_{x_b}^c(\tau) &= \mathcal{E}\{x_c(t)x_c(t-\tau)\}\, e^{\jmath 2\pi f_o(2t-\tau)} + \mathcal{E}\{x_c^*(t)x_c^*(t-\tau)\}\, e^{-\jmath 2\pi f_o(2t-\tau)} \\ &\qquad + R_{x_c}^c(\tau)e^{\jmath 2\pi f_o\tau} + R_{x_c}^{c*}(\tau)e^{-\jmath 2\pi f_o\tau} \\ &= 2\text{Re}\left[R_{x_c}^c(\tau)e^{\jmath 2\pi f_o\tau}\right] + 2\text{Re}\left[\mathcal{E}\{x_c(t)x_c(t-\tau)\}\, e^{-\jmath 2\pi f_o(2t-\tau)}\right] \end{aligned} \tag{B.31}$$

In order for the bandpass random process to be stationary, the correlation function must be a function only of τ and not t.[3] Thus for stationarity it is required that the term

$$\mathcal{E}\{x_c(t)x_c(t-\tau)\} = 0 \tag{B.32}$$

Substituting (B.28) in (B.32) produces

$$\begin{aligned}\mathcal{E}\left\{\left(\tfrac{1}{2}x_p(t) + \jmath\tfrac{1}{2}x_q(t)\right)\left(\tfrac{1}{2}x_p(t-\tau) + \jmath\tfrac{1}{2}x_q(t-\tau)\right)\right\}\\ = \frac{1}{4}\left[R^c_{x_p}(\tau) - R^c_{x_q}(\tau) + \jmath\left(R^c_{x_px_q}(\tau) + R^c_{x_qx_p}(\tau)\right)\right] = 0\end{aligned}$$

which implies that

$$\begin{aligned} R^c_{x_p}(\tau) &= R^c_{x_q}(\tau) \ (a)\\ R^c_{x_px_q}(\tau) &= -R^c_{x_qx_p}(\tau) \ (b)\end{aligned} \tag{B.33}$$

When the complex signal is sampled, these are the same symmetry conditions that are required for the correlation functions of the real and imaginary parts of the discrete random process. Note that since $R^c_{x_px_q}(\tau)$ is always equal to $R^c_{x_qx_p}(-\tau)$, (B.33)(b) implies that both $R^c_{x_px_q}$ and $R^c_{x_qx_p}$ are *odd* functions of their arguments. On the other hand, $R^c_{x_p}$ and $R^c_{x_q}$ are both real-valued correlation functions and so are *even* functions of lag.

If (B.28) is substituted in the definition

$$R^c_{x_c}(\tau) = \mathcal{E}\{x_c(t)x_c^*(t-\tau)\} \tag{B.34}$$

for the correlation function the result is

$$\begin{aligned}R^c_{x_c}(\tau) &= \mathcal{E}\left\{\left(\tfrac{1}{2}x_p(t) + \jmath\tfrac{1}{2}x_q(t)\right)\left(\tfrac{1}{2}x_p(t-\tau) - \jmath\tfrac{1}{2}x_q(t-\tau)\right)\right\}\\ &= \frac{1}{4}\left[R^c_{x_p}(\tau) + R^c_{x_q}(\tau) + \jmath\left(R^c_{x_qx_p}(\tau) - R^c_{x_px_q}(\tau)\right)\right]\end{aligned}$$

Then because of (B.33) it follows that

$$R^c_{x_c}(\tau) = \tfrac{1}{2}R^c_{x_p}(\tau) - \jmath\tfrac{1}{2}R^c_{x_px_q}(\tau) \tag{B.35}$$

and

$$\begin{aligned} R^c_{x_p}(\tau) = R^c_{x_q}(\tau) &= 2\mathrm{Re}[R^c_{x_c}(\tau)] \ (a)\\ R^c_{x_px_q}(\tau) = -R^c_{x_qx_p}(\tau) &= -2\mathrm{Im}[R^c_{x_c}(\tau)] \ (b)\end{aligned} \tag{B.36}$$

[3]According to the definition of stationarity, the mean of the bandpass random process should also be independent of t. It can be seen from the definition (B.27) that this is possible only if $\mathcal{E}\{x_p(t)\} = \mathcal{E}\{x_q(t)\} = 0$. We can relax this condition and let x_p and x_q have nonzero mean if it is assumed that a phase coherence is maintained in the processing so that the nonzero mean values $m^c_{x_p}$ and $m^c_{x_q}$ can be removed prior to the processing. This permits the treatment of complex random processes that have nonzero mean but nevertheless satisfy the required conditions on their correlation or covariance functions.

Now from (B.31) it was seen that

$$R^c_{x_b}(\tau) = 2\text{Re}\left[R^c_{x_c}(\tau)e^{\jmath 2\pi f_o \tau}\right]$$
$$= 2\left[\text{Re}[R^c_{x_c}(\tau)]\cos 2\pi f_o\tau - \text{Im}[R^c_{x_c}(\tau)]\sin 2\pi f_o\tau\right]$$

Substituting (B.33) thus yields

$$R^c_{x_b}(\tau) = R^c_{x_p}(\tau)\cos 2\pi f_o\tau + R^c_{x_px_q}(\tau)\sin 2\pi f_o\tau \tag{B.37}$$

Equation B.37 shows that the autocorrelation function for the bandpass signal has a form analogous to the quadrature representation for the bandpass random signal with "quadrature components" equal to the autocorrelation and minus the cross-correlation of the signal components. Figure B.11 shows plots of the power spectral density for a typical bandpass process, the corresponding complex random process, and the complex discrete-time process derived from these. From conditions similar to those used in deriving (B.3) the power spectral density function of the complex process is just equal to the positive half of the power spectral density function for the bandpass random process shifted down to the origin. That is,

$$S^c_{x_c} = S^{c+}_{x_b}(f + f_o) \tag{B.38}$$

where

$$S^{c+}_{x_b}(f) = \begin{cases} S^c_{x_b}(f) & \text{if } f > 0 \\ 0 & \text{if } f \leq 0 \end{cases} \tag{B.39}$$

Because of this relation, bandpass linear filtering operations applied to the bandpass process can be thought of as equivalent low-pass filtering operations applied to the complex process, which in fact may be carried out in discrete time. The equivalence of these operations is depicted in Fig. B.12.

Before concluding this topic let us just mention how the power spectral density functions for the quadrature components $x_p(t)$ and $x_q(t)$ can be derived directly from the power spectral density function for the complex random process. Since $R^c_{x_p}(\tau)$ and $R^c_{x_px_q}(\tau)$ relate to the real and imaginary parts of $R^c_{x_c}(\tau)$, their Fourier transforms relate to the even and odd parts of the power spectrum. Specifically, it follows from (B.36) that

$$S^c_{x_p}(f) = 2\left[S^c_{x_c}(f)\right]_{\text{EVEN}} = S^c_{x_c}(f) + S^c_{x_c}(-f) \quad (a)$$
$$S^c_{x_px_q}(f) = 2\jmath\left[S^c_{x_c}(f)\right]_{\text{ODD}} = \jmath\left(S^c_{x_c}(f) - S^c_{x_c}(-f)\right) \quad (b) \tag{B.40}$$

For nonstationary random signals the condition $\mathcal{E}\{x_c(t)x_c(t-\tau)\} = 0$ that led to the development in the preceding paragraphs does not naturally arise. Nevertheless, it is convenient to assume the analogous condition

$$\mathcal{E}\{x_c(t_1)x_c(t_2)\} = 0; \qquad \forall t_1, t_2 \tag{B.41}$$

which frequently holds if the random process varies slowly with time. With this assumption, a development parallel to the one above shows that

$$\mathrm{R}^c_{x_c}(t_1, t_2) = \tfrac{1}{2}\mathrm{R}^c_{x_p}(t_1, t_2) - \jmath\tfrac{1}{2}\mathrm{R}^c_{x_px_q}(t_1, t_2) \tag{B.42}$$

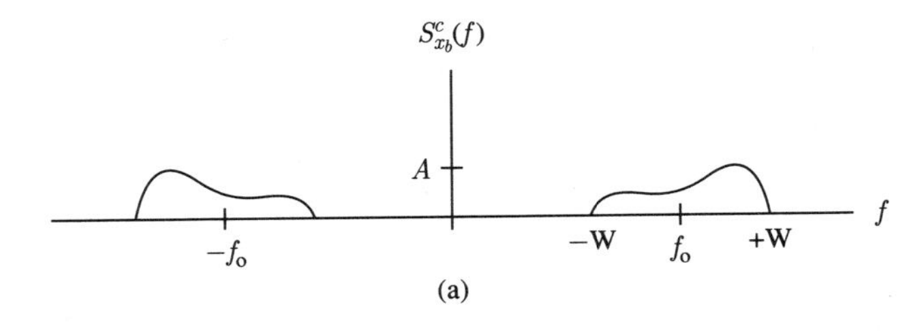

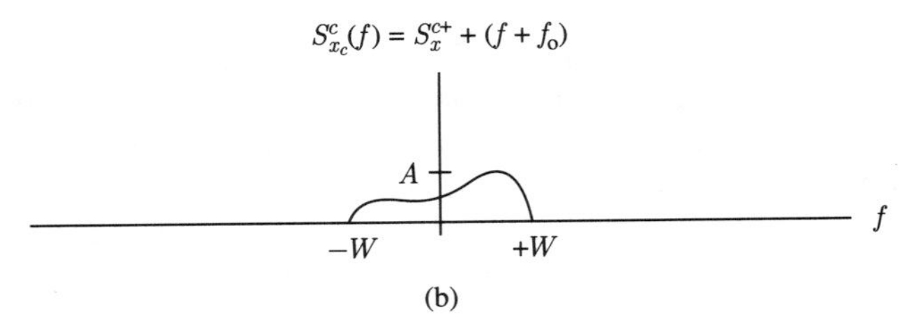

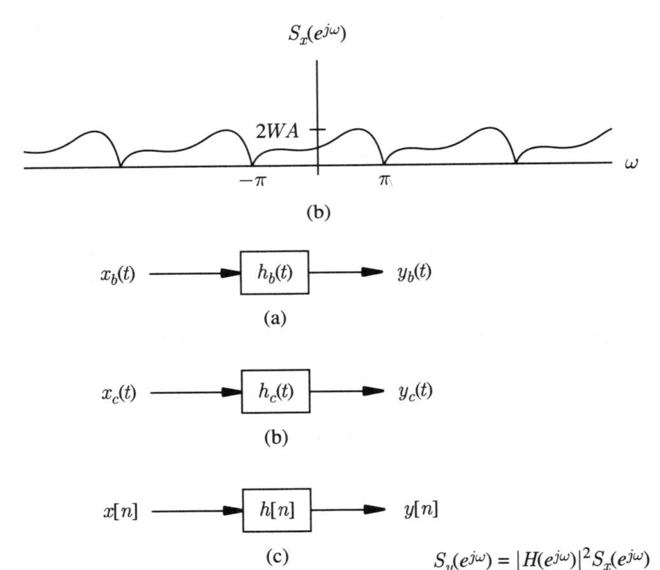

Figure B.11 Plots of power spectral density. (a) Bandpass random process. (b) Corresponding low-pass complex process. (c) Corresponding complex discrete process.

Figure B.12 Equivalent filtering operations. (a) Bandpass. (b) Complex low-pass. (c) Complex discrete.

and

$$\begin{aligned} \mathrm{R}^c_{x_p}(t_1, t_2) &= \mathrm{R}^c_{x_q}(t_1, t_2) = 2\mathrm{Re}[\mathrm{R}^c_{x_c}(t_1, t_2)] \ (a) \\ \mathrm{R}^c_{x_p x_q}(t_1, t_2) &= -\mathrm{R}^c_{x_q x_p}(t_1, t_2) = -2\mathrm{Im}[\mathrm{R}^c_{x_c}(t_1, t_2)] \ (b) \end{aligned} \tag{B.43}$$

where the complex correlation function for the process $x_c(t)$ is defined by

$$\mathrm{R}^c_{x_c}(t_1, t_2) = \mathcal{E}\{x_c(t_1) x^*_c(t_2)\} \tag{B.44}$$

REFERENCES

1. Harry L. Van Trees. *Detection, Estimation, and Modulation Theory*, Part III. John Wiley & Sons, New York, 1971.

2. J. Dugundji. Envelopes and pre-envelopes of real waveforms. *IRE Transactions on Information Theory*, IT-4:53–57, March 1958.

3. Simon Haykin. *Communication Systems*, 2nd ed. John Wiley & Sons, New York, 1983.

4. Athanasios Papoulis. *Probability, Random Variables, and Stochastic Processes*, 3rd ed. McGraw-Hill, New York, 1991.

Glossary

OPERATIONS

Pr[A]	probability of event A.
$\mathcal{E}\{x\}$	expectation of x.
$\langle x[n] \rangle$	signal average (see p. 90).
Var $[x]$	variance of x.
Cov $[x, y]$	covariance of x and y.
Re x	real part of x.
Im x	imaginary part of x.
$\|x\|$	magnitude of x.
$\angle x$	angle (phase) of x.
$\nabla_{\mathbf{a}}, \nabla_{\mathbf{a}^*}$	gradient with respect to vector parameter **a** or $\mathbf{a}^*$.
∇_a, ∇_{a^*}	complex gradient with respect to scalar parameter a or a^*.
∇^2_{ab}	defined as $\nabla_a \nabla_b$ (see Appendix A).

$\tilde{\mathbf{x}}, \tilde{\mathbf{A}}$ — reversal of $\mathbf{x}$ or $\mathbf{A}$.
$\mathbf{x}^T, \mathbf{A}^T$ — transpose of $\mathbf{x}$ or $\mathbf{A}$.
$\mathbf{x}^*, \mathbf{A}^*$ — conjugate of $\mathbf{x}$ or $\mathbf{A}$.
$\mathbf{x}^{*T}, \mathbf{A}^{*T}$ — conjugate (Hermitian) transpose of $\mathbf{x}$ or $\mathbf{A}$.
$\mathbf{A}^+$ — pseudoinverse of $\mathbf{A}$.
$\mathrm{tr}\ \mathbf{A}$ — trace of $\mathbf{A}$.
$|\mathbf{A}|$ or $\det \mathbf{A}$ — determinant of $\mathbf{A}$.
$\|\mathbf{x}\|, \|\mathbf{A}\|$ — Euclidean norm of $\mathbf{x}$ or $\mathbf{A}$.
$\|\mathbf{A}\|_F$ — Frobenius norm of $\mathbf{A}$.

$x[n] * h[n]$ — convolutio… wo sequences.
$X(e^{\jmath\omega}) \circledast H(e^{\jmath\omega})$ — cyclic … on of two Fourier transforms.
argmax $V(x)$ — value … maximizes $V(x)$.
argmin $V(x)$ — v… minimizes $V(x)$.

$\binom{k}{l}$ — … ient.

… b.
… at point $z = z_o$.
… g a new variable or function.
… east squares sense (see p. 522).

… ient.

… prediction coefficients of order p.
… s.
… xponential random signal.
… on coefficient.

… ar prediction coefficients of order p.
… s.

… n algorithms.
… a stationary random process $x[n]$.
… n for stationary processes $x[n]$ and $y[n]$.
$C^c_{x_c}(\tau)$ — covariance function of stationary continuous process $x_c(t)$.
$C^c_{x_c y_c}(\tau)$ — cross-covariance function for processes $x_c(t)$ and $y_c(t)$.
$C^{(k)}_x[l_1, \ldots]$ — k^{th} order cumulant of $x[n]$.
$\mathrm{C}_x[n_1, n_0]$ — covariance function for a general random process $x[n]$.

$\mathrm{C}_{xy}[n_1, n_0]$	cross-covariance function for general processes $x[n]$ and $y[n]$.
$\mathrm{C}^c_{x_c}(t_1, t_0)$	covariance function of general continuous process $x_c(t)$.
$\mathbf{C}_{\boldsymbol{x}}$	covariance matrix for random process $x[n]$ or random vector $\boldsymbol{x}$.
$\mathbf{C}_{\boldsymbol{xy}}$	cross-covariance matrix for $\boldsymbol{x}$ and $\boldsymbol{y}$.
$\mathbf{C}_{\boldsymbol{x}+\boldsymbol{y}}$	covariance matrix for sum of $\boldsymbol{x}$ and $\boldsymbol{y}$.
$\mathbf{C}_{\boldsymbol{y}\mid\boldsymbol{x}}$	conditional covariance matrix for $\boldsymbol{y}$ given $\boldsymbol{x}$.
$d[n]$ or $\mathrm{d}[n]$	"desired sequence" in optimal filtering.
$\mathbf{D}_L$	diagonal matrix in lower-upper triangular decomposition.
$\mathbf{D}_U$	diagonal matrix in upper-lower triangular decomposition.
$\mathbf{e}$ or $\mathbf{e}_k$	eigenvector or generalized eigenvector of correlation matrix.
$\check{\mathbf{e}}$ or $\check{\mathbf{e}}_k$	eigenvector or generalized eigenvector of covariance matrix.
$E_i(z)$	eigenfilter (filter formed from eigenvector).
$\mathbf{E}$	matrix of eigenvectors of correlation matrix.
$\check{\mathbf{E}}$	matrix of eigenvectors of covariance matrix.
f	frequency (in hertz).
$f_{\boldsymbol{x}}$	density function for random vector $\boldsymbol{x}$.
$f_{\boldsymbol{x};\boldsymbol{\theta}}$	density $\boldsymbol{x}$ depending on parameter $\boldsymbol{\theta}$ or likelihood function.
$f_{\boldsymbol{xy}}$	joint density function for $\boldsymbol{x}$ and $\boldsymbol{y}$.
$F_{\boldsymbol{x}}$	distribution function for random vector $\boldsymbol{x}$.
$F_{\boldsymbol{xy}}$	joint distribution function for $\boldsymbol{x}$ and $\boldsymbol{y}$.
$g_p[l], g_p^b[l]$	forward and backward functions in the Schur algorithm.
$\boldsymbol{g}_p, \boldsymbol{g}_b^\dagger, \boldsymbol{g}_p^b$	vector forms of functions in the Schur algorithm.
$h[n]$	impulse response of linear shift-invariant system.
$h_{ca}[n]$	impulse response corresponding to $H_{ca}(z)$ (see below).
$\mathrm{h}[n_1, n_0]$	impulse response of general linear system.
$\mathbf{h}$	vector of impulse response terms.
$H(e^{\jmath\omega})$	frequency response of linear shift-invariant system.
$H(z)$	system function of linear shift-invariant system.
$H_{ac}(z)$	anticausal stable system with anticausal stable inverse.
$H_{ap}(z)$	all-pass linear system.
$H_{ca}(z)$	causal stable system with causal stable inverse.
$H_{nc}(z)$	noncausal Wiener filter.
$\mathbf{I}$	identity matrix.
$\jmath$	square root of -1.
J	Jacobian.
$\mathbf{J}$	Fisher information matrix.
K_n	gain term in recursive (Kalman) filter.
K_p	constant in split Levinson algorithms.
$\mathcal{K}_0$	white noise variance in innovations representation.
$\mathcal{L}$	Lagrangian in constrained optimization problem.
$\mathbf{L}$	lower triangular matrix in (LU) triangular decomposition.
m_x	mean of stationary random process $x[n]$.
$m^c_{x_c}$	mean of stationary continuous random process $x_c(t)$.

$\mathrm{m}_x[n]$	mean of general random process $x[n]$.
$\mathrm{m}^c_{x_c}(t)$	mean of general continuous random process $x_c(t)$.
$\mathbf{m}_{\boldsymbol{x}}$	mean vector for $\boldsymbol{x}$.
$\mathbf{m}_{\boldsymbol{y}\mid\boldsymbol{x}}$	conditional mean vector for $\boldsymbol{y}$ given $\boldsymbol{x}$.
$M^{(k)}_x[\cdots]$	k^{th} order moment of $x[n]$.
N_s	number of samples of a random process or a given data segment.
$\mathbf{p}[n]$	vector of Markov state probabilities.
$\bar{\mathbf{p}}$	vector of limiting-state Markov state probabilities.
P	length of filter or period of periodic random process.
P_i	period of periodic random process.
$\hat{P}_x(e^{\jmath\omega})$	periodogram spectral estimate for random process $x[n]$.
P, P_o or P_i	discrete probability values.
$\mathrm{P}_{j\mid i}$	transition probability for a Markov process.
$\mathrm{P}^{(k)}_{j\mid i}$	k^{th} order transition probability for a Markov process.
$\mathsf{P}_o, \mathsf{P}_i$	Power of a complex exponential process.
$\boldsymbol{P}_{\boldsymbol{X}}$	projection matrix corresponding to matrix $\boldsymbol{X}$.
$\mathbf{P}$	matrix of transition probabilities $\mathbf{P} = \mathbf{\Pi}^T$.
$\mathbf{P}_o$	diagonal matrix of signal powers P_i.
$\mathbf{Q}_1$	first partition of unitary matrix in QR factorization.
$\mathbf{Q}_2$	second partition of unitary matrix in QR factorization.
$r[n]$	residual in recursive (Kalman) filter.
$R_x[l]$	correlation function of stationary random process $x[n]$.
$R_{xy}[l]$	cross-correlation function for stationary processes $x[n]$ and $y[n]$.
$\hat{R}_x[l]$	biased sample correlation function for random process $x[n]$.
$\hat{R}'_x[l]$	unbiased sample correlation function for random process $x[n]$.
$R^c_{x_c}(\tau)$	correlation function of stationary continuous process $x_c(t)$.
$R^c_{x_c y_c}(\tau)$	cross-correlation function for processes $x_c(t)$ and $y_c(t)$.
$R^{\mathrm{E}}_x[l]$	one-half real part of $R_x[l]$.
$R^{\mathrm{O}}_x[l]$	one-half imaginary part of $R_x[l]$.
$\mathrm{R}_x[n_1, n_0]$	correlation function of general random process $x[n]$.
$\mathrm{R}_{xy}[n_1, n_0]$	cross-correlation function for general processes $x[n]$ and $y[n]$.
$\mathrm{R}^c_{x_c}(t_1, t_0)$	correlation function of general continuous process $x_c(t)$.
$\mathrm{R}^{\mathrm{E}}_x[n_1, n_0]$	one-half real part of $\mathrm{R}_x[n_1, n_0]$.
$\mathrm{R}^{\mathrm{O}}_x[n_1, n_0]$	one-half imaginary part of $\mathrm{R}_x[n_1, n_0]$.
$\mathbf{R}_{\boldsymbol{x}}$	correlation matrix for $\boldsymbol{x}$ or random process $x[n]$.
$\hat{\mathbf{R}}_{\boldsymbol{x}}$	estimated correlation matrix for $\boldsymbol{x}$.
$\mathbf{R}_{\boldsymbol{xy}}$	cross-correlation matrix for $\boldsymbol{x}$ and $\boldsymbol{y}$ or $x[n]$ and $y[n]$.
$\mathbf{R}_{\boldsymbol{x}+\boldsymbol{y}}$	correlation matrix for sum of $\boldsymbol{x}$ and $\boldsymbol{y}$.
$\mathbf{R}^{\mathrm{E}}_{\boldsymbol{x}}$	one-half real part of $\mathbf{R}_{\boldsymbol{x}}$ (see Table 2.3).
$\mathbf{R}^{\mathrm{O}}_{\boldsymbol{x}}$	one-half imaginary part of $\mathbf{R}_{\boldsymbol{x}}$ (see Table 2.3).

$\mathbf{R}_1$	partition of right triangular matrix in QR factorization.
$\boldsymbol{s}_p$	vector variables in split Levinson algorithms.
$\mathbf{s}, \mathbf{s}_i$	signal vector (vector of signal samples).
$S_x(e^{j\omega})$	power spectral density function for $x[n]$.
$S'_x(e^{j\omega})$	continuous part of power density spectrum for $x[n]$.
$\hat{S}_x(e^{j\omega})$	an estimate for the power density spectrum.
$S_{xy}(e^{j\omega})$	cross-power spectral density function for $x[n]$ and $y[n]$.
$S_x(z)$	complex spectral density function for $x[n]$.
$S'_x(z)$	continuous part of the complex spectral density function.
$S_{xy}(z)$	complex cross-spectral density function for $x[n]$ and $y[n]$.
$S_x^{(k)}(\cdots)$	k^{th} order spectrum of $x[n]$.
$S^c_{x_c}(f)$	power spectral density function for process $x_c(t)$.
$S^c_{x_c y_c}(f)$	cross-power density function for processes $x_c(t)$ and $y_c(t)$.
S_k	possible values of a Markov chain.
$\mathrm{S}_x[k]$	sampled power spectral density function (DFT).
$\mathbf{S}$	matrix whose columns are the signal vectors $\mathbf{s}_i$.
$\mathbf{S}_1$	diagonal matrix of singular values (partition of $\boldsymbol{\Sigma}$).
T	interval for sampling a continuous random process.
$u[n]$	unit step function.
$\mathbf{u}_k$	left singular vector.
$\mathbf{U}$	matrix of left singular vectors.
$\mathbf{U}_1$	upper triangular matrix in (UL) triangular decomposition.
$\mathbf{v}_k$	right singular vector.
$\mathbf{V}$	matrix of right singular vectors.
$w[n]$	white noise process.
$w_c(t)$	continuous white noise process.
$\mathrm{w}[l]$	a window sequence.
$\mathbf{w}$	vector of complex exponential $e^{j\omega n}$.
W_P	"twiddle factor" in DFT $e^{j2\pi/P}$.
$\mathrm{W}(e^{j\omega})$	transform of a window sequence.
$x[n]$	a random process.
$x_c(t)$	a continuous random process.
x_k	k^{th} component of random vector $\boldsymbol{x}$.
$x_{\mathrm{i}}[n]$	imaginary part of random process $x[n]$.
$x_{\mathrm{r}}[n]$	real part of random process $x[n]$.
$\times$	unspecified element of vector or matrix.
$X[k]$	DFT coefficient for periodic random process.
$\boldsymbol{x}$	random vector or vector of samples of random process $x[n]$.
$\boldsymbol{x}_{\mathbf{i}}$	imaginary part of random vector $\boldsymbol{x}$.
$\boldsymbol{x}_{\mathbf{r}}$	real part of random vector $\boldsymbol{x}$.
$\boldsymbol{X}$	data matrix corresponding to random process $x[n]$.
$\mathbf{x}$	realization of a random vector $\boldsymbol{x}$.
$\mathbf{x}_{\mathbf{i}}$	imaginary part of realization $\mathbf{x}$.

$\mathbf{x}_\mathbf{r}$	real part of realization $\mathbf{x}$.
$\mathbf{x}_k$	columns of data matrix $\mathbf{X}$ (see below).
$\mathbf{x}^{(k)}$	sample realizations of $\boldsymbol{x}$.
$\mathbf{X}$	data matrix realization, with rows $\mathbf{x}^{(k)*T}$ and columns $\mathbf{x}_k$.
z	z-transform variable.
$\alpha_i[k]$	forward variable for a hidden Markov model.
α_p	variable in split Levinson algorithms.
$\beta_i[k]$	backward variable for a hidden Markov model.
β_p	variable in split Levinson algorithms.
γ_p, γ_p'	forward, backward reflection coefficients of order p.
$\Gamma_{xy}(e^{\jmath\omega})$	coherence function for x and y.
$\delta[n]$	discrete unit impulse (unit sample function).
$\delta_c(\cdot)$	continuous impulse function.
Δ_p, Δ_p'	forward, backward unnormalized partial correlations.
θ	parameter (in estimation), or direction of arrival (for array).
$\hat{\theta}_{ml}$	maximum likelihood estimate for θ.
$\boldsymbol{\theta}$	vector parameter (in estimation).
$\hat{\boldsymbol{\theta}}_N, \hat{\theta}_N$	estimate for $\boldsymbol{\theta}$ or θ based on N observations.
$\varepsilon[n]$	prediction error or error in estimation of a random process.
$\varepsilon_p[n]$	prediction error of order p. (In Chapter 7, error due to predictable part of random process.)
$\varepsilon_p^b[n]$	backward prediction error of order p ($\varepsilon_p^b[n] = \varepsilon_p'[n-p]$).
$\varepsilon_p'[n]$	error in backward linear prediction ($\varepsilon_p'[n] = \varepsilon_p^b[n+p]$).
$\boldsymbol{\varepsilon}$	random vector corresponding to $\varepsilon[n]$.
$\epsilon, \epsilon_p, \epsilon_p^b$	errors corresponding to $\varepsilon, \varepsilon_p, \varepsilon_p^b$ in least squares formulation.
$\boldsymbol{\epsilon}$	vector error in least squares problem, equation error.
$\eta[n]$	(white or colored) noise process.
$\boldsymbol{\eta}$	vector of noise samples.
$\boldsymbol{\iota}$	vector $[\,1\ 0\ \cdots\ 0\,]^T$.
κ_i	coefficient in orthonormal expansion of a random process.
$\boldsymbol{\kappa}$	vector of coefficients κ_i.
λ	Lagrange multiplier or eigenvalue of correlation matrix.
λ_k	eigenvalue or generalized eigenvalue of correlation matrix.
$\check{\lambda}$ or $\check{\lambda}_k$	eigenvalue or generalized eigenvalue of covariance matrix.
$\boldsymbol{\Lambda}$	matrix of eigenvalues of correlation matrix.
$\check{\boldsymbol{\Lambda}}$	matrix of eigenvalues of covariance matrix.
ν_o	noise spectral density of a continuous white noise process.
$\boldsymbol{\Pi}$	Markov transition matrix $\boldsymbol{\Pi} = \mathbf{P}^T$.
ρ	(normalized) correlation coefficient or, correlation parameter in exponential correlation function.
σ_k	singular values.

σ^2, σ_0^2	variance parameters (general).
σ_ε^2	prediction error variance, estimation error variance.
$\sigma_{\varepsilon_p}^2, \sigma_p^2$	prediction error variance (of order p).
$\sigma_p'^2$	backward prediction error variance (of order p).
σ_w^2	white noise variance.
$\mathbf{\Sigma}$	matrix of singular values.
$\mathbf{\Sigma}_\eta$	normalized noise covariance matrix.
$\boldsymbol{\upsilon}_p$	vector variables in split Schur algorithm.
ϕ or ϕ_i	phase of sinusoidal random process.
$\varphi_i[n]$	orthonormal basis function.
$\boldsymbol{\varphi}_i$	orthonormal basis vector.
$\boldsymbol{\Phi}$	matrix of orthonormal basis vectors.
$\mathbf{\Phi}$	rotation matrix in *ESPRIT* algorithm.
$\boldsymbol{\psi}$	scalar/vector/matrix quantity depending on random vector(s).
ω	radian frequency (digital frequency).
$\omega^{(k)}$	frequency variables in higher-order spectra.

Index

H

I

J

K

L

M